T0092777

MITOCHONDRIA IN LIVER DISEASE

EDITED BY
DERICK HAN
NEIL KAPLOWITZ

CRC Press
Taylor & Francis Group
Boca Raton London New York

CRC Press is an imprint of the
Taylor & Francis Group, an **informa** business

CRC Press
Taylor & Francis Group
6000 Broken Sound Parkway NW, Suite 300
Boca Raton, FL 33487-2742

First Issued in paperback 2019

© 2016 by Taylor & Francis Group, LLC
CRC Press is an imprint of Taylor & Francis Group, an Informa business

No claim to original U.S. Government works

ISBN-13: 978-1-4822-3697-2 (hbk)
ISBN-13: 978-0-367-37724-3 (pbk)

This book contains information obtained from authentic and highly regarded sources. Reasonable efforts have been made to publish reliable data and information, but the author and publisher cannot assume responsibility for the validity of all materials or the consequences of their use. The authors and publishers have attempted to trace the copyright holders of all material reproduced in this publication and apologize to copyright holders if permission to publish in this form has not been obtained. If any copyright material has not been acknowledged please write and let us know so we may rectify in any future reprint.

Except as permitted under U.S. Copyright Law, no part of this book may be reprinted, reproduced, transmitted, or utilized in any form by any electronic, mechanical, or other means, now known or hereafter invented, including photocopying, microfilming, and recording, or in any information storage or retrieval system, without written permission from the publishers.

For permission to photocopy or use material electronically from this work, please access www.copyright.com (http://www.copyright.com/) or contact the Copyright Clearance Center, Inc. (CCC), 222 Rosewood Drive, Danvers, MA 01923, 978-750-8400. CCC is a not-for-profit organization that provides licenses and registration for a variety of users. For organizations that have been granted a photocopy license by the CCC, a separate system of payment has been arranged.

Trademark Notice: Product or corporate names may be trademarks or registered trademarks, and are used only for identification and explanation without intent to infringe.

**Visit the Taylor & Francis Web site at
http://www.taylorandfrancis.com**

**and the CRC Press Web site at
http://www.crcpress.com**

MITOCHONDRIA IN LIVER DISEASE

OXIDATIVE STRESS AND DISEASE

Series Editors

LESTER PACKER, PhD
ENRIQUE CADENAS, MD, PhD

UNIVERSITY OF SOUTHERN CALIFORNIA SCHOOL OF PHARMACY
LOS ANGELES, CALIFORNIA

Contents

SECTION I   Overview of Mitochondria

SECTION II   The Role of Mitochondria in Liver Diseases

Series Preface

Oxidative stress is an underlying factor in health and disease. In this series, the significance of oxidative stress and disease associated with organ systems is highlighted by exploring the scientific evidence and the clinical applications of this knowledge. This series is intended for clinicians and researchers in the basic biomedical sciences. The potential of such knowledge on healthy aging and disease prevention warrants further knowledge about how oxidants and antioxidants modulate cell and tissue function, a theme explored in this book edited by Derick Han and Neil Kaplowitz.

The first section of the book, "Overview of Mitochondria," consisting of seven chapters, offers an excellent overview of mitochondria metabolism and their function in energy-dependent redox reactions, as well as the role of superoxide, hydrogen peroxide, and hydrogen sulfide in signal transduction and transcription and mitochondrial biogenesis. The second section of the book, "The Role of Mitochondria in Liver Diseases," focuses on, among other things, the role of mitophagy, epigenetic mitochondrial DNA modifications, mitochondrial dynamics, and sirtuins in several liver diseases, such as nonalcoholic fatty liver, viral hepatitis, diabetes, and ceramide and drug toxicity.

Derick Han and Neil Kaplowitz are congratulated for producing this excellent, well-organized, and timely book in the expanding field of mitochondrial metabolism and, especially, how liver energy metabolism—centered on diverse aspects of mitochondrial function—is affected by drug toxicity and disease states.

Lester Packer
Enrique Cadenas

Preface

The liver is a vital organ that is responsible for a wide range of functions, most of which are essential for survival. The multitude of functions the liver performs makes it vulnerable to a wide range of diseases. The liver is also responsible for metabolism and clearance of xenobiotic compounds and, thus, is susceptible to injury caused by drugs, alcohol, and herbal compounds. The liver is a major site for fatty acid synthesis, thus, is susceptible to developing fatty liver disease. In most of these diseases, mitochondria appear to play a central role. *Mitochondria in Liver Disease* gathers the most current knowledge regarding the role of mitochondria in liver diseases. This book is divided into two sections. The first section highlights the latest exciting developments in mitochondrial research. Cutting-edge topics such as the regulation of mitochondrial respiration by hydrogen sulfide and mitochondrial biogenesis are covered. In the second section, the most current research regarding the role of mitochondria on a wide range of liver diseases are reviewed. This book, written by experts in the field, provides a comprehensive overview of the latest research on mitochondria and its central role in mediating liver diseases.

Editors

Derick Han, PhD, is assistant professor in the Department of Biopharmaceutical Sciences at the School of Pharmacy, Keck Graduate Institute (Claremont, California), and a member of The Claremont Colleges. Dr. Han earned his PhD in molecular pharmacology and toxicology from the School of Pharmacy at the University of Southern California and a BA in sociology and biochemistry from the University of California, Berkeley. His research is focused on mitochondrial remodeling in the liver caused by metabolic stress, such as alcohol. He is also interested in mitochondrial reprogramming that occurs in cancer cells. He has published more than 70 research publications, reviews, and book chapters.

Neil Kaplowitz, MD, is director of the University of Southern California–National Institute of Diabetes and Digestive and Kidney Diseases (USC NIDDK)–sponsored Research Center for Liver Diseases (Los Angeles, California). He holds two endowed chairs, the Brem Professor of Medicine and the Budnick Chair in Liver Diseases, and is chief of the Division of Gastrointestinal and Liver Diseases. He is also professor of physiology and biophysics and pharmacology and pharmaceutical sciences at the Keck School of Medicine at the University of Southern California.

Dr. Kaplowitz has received a number of important honors and distinctions, including election to membership in the American Society for Clinical Investigation and the Association of American Physicians. He is the recipient of the Western Gastroenterology Research Prize, the William S. Middleton Award, the Solomon A. Berson Medical Alumni Achievement Award in Clinical Science from his alma mater, the Merit Award from the National Institutes of Health, the Mayo Soley Award from the Western Society for Clinical Investigation, the American Association for the Study of Liver Diseases (AASLD) Distinguished Achievement Award, and the American Liver Foundation (ALF) Distinguished Scientific Achievement Award. He has served as the president of the American Association for the Study of Liver Diseases and as vice chair for research at the American Liver Foundation. He has also served as associate editor of leading medical and scientific journals such as *Hepatology*, *Gastroenterology*, and the *American Journal of Physiology*.

In recent years, he has focused on the role of signal transduction, endoplasmic reticulum, and mitochondrial stress in the pathogenesis of liver injury. He has published more than 195 peer-reviewed scientific articles and 150 scholarly reviews and has edited 10 books related to liver diseases.

Contributors

Fernando Antunes
Laboratory of Biochemistry of Oxidants
  and Antioxidants
Centre and Department of Chemistry
  and Biochemistry
School of Sciences
University of Lisbon
Lisbon, Portugal

Bradley Blackshire
Department of Biology
University of La Verne
La Verne, California

Enrique Cadenas
Department of Pharmacology and
  Pharmaceutical Sciences
School of Pharmacy
University of Southern California
Los Angeles, California

Chien-Yu Chen
Department of Pharmacology and
  Pharmaceutical Sciences
School of Pharmacy
University of Southern California
Los Angeles, California

Jingyu Chen
Department of Pharmacology and
  Pharmaceutical Sciences
School of Pharmacy
University of Southern California
Los Angeles, California

Lily Dara
Division of Gastrointestinal and Liver
  Diseases
Department of Medicine
Research Center for Liver Diseases
Keck School of Medicine
University of Southern California
Los Angeles, California

Anketse Debebe
Department of Pharmacology and
  Pharmaceutical Sciences
School of Pharmacy
University of Southern California
Los Angeles, California

Carl Decker
Department of Biopharmaceutical
  Sciences
School of Pharmacy
Kcck Graduate Institute
Claremont, California

Wen-Xing Ding
Department of Pharmacology,
  Toxicology, and Therapeutics
School of Medicine
University of Kansas, Kansas City
Kansas City, Missouri

José C. Fernandez-Checa
Liver Unit, Hospital Clinic–IDIBAPS
Department of Cell Death and
  Proliferation
Institute of Biomedical Research of
  Barcelona
Barcelona, Spain

Andras Franko
Institute of Experimental Genetics
Helmholtz Zentrum München GmbH
and
German Center for Diabetes Research
Neuherberg, Germany

Bernard Fromenty
INSERM, U991
University of Rennes 1
Rennes, France

Carmen Garcia-Ruiz
Liver Unit, Hospital Clinic–IDIBAPS
Department of Cell Death and
    Proliferation
Institute of Biomedical Research of
    Barcelona
Barcelona, Spain

Jerome Garcia
Department of Biology
University of La Verne
La Verne, California

Eric S. Goetzman
Division of Medical Genetics
Department of Pediatrics
Children's Hospital of Pittsburgh
University of Pittsburgh Medical Center
Pittsburgh, Pennsylvania

Vivek Gupta
Department of Biopharmaceutical
    Sciences
School of Pharmacy
Keck Graduate Institute
Claremont, California

Kevork Hagopian
Department of Molecular Biosciences
School of Veterinary Medicine
University of California, Davis
Davis, California

María Isabel Hernández-Alvarez
Department of Biochemistry and
    Molecular Biology
University of Barcelona
and
Molecular Medicine Program of the
    Institute of Research in Biomedicine
IRB Barcelona
and
CIBER from Diabetes and Metabolic
    Associated Diseases
Health Care Institute Carlos III
Barcelona, Spain

Martin Hrabê de Angelis
Institute of Experimental Genetics
Helmholtz Zentrum München GmbH
and
German Center for Diabetes Research
Neuherberg, Germany

and

WZW–Center of Life and Food
    Sciences
Weihenstephan Technische Universität
    München
Weihenstephan, Germany

Hartmut Jaeschke
Department of Pharmacology,
    Toxicology, and Therapeutics
School of Medicine
University of Kansas, Kansas City
Kansas City, Missouri

Heather Johnson
Division of Gastrointestinal and Liver
    Diseases
Department of Medicine
Research Center for Liver Diseases
Keck School of Medicine
University of Southern California
Los Angeles, California

Zahra Karimi
Department of Biology
Lakehead University
Thunder Bay, Ontario, Canada

Bilon Khambu
Department of Pathology and
    Laboratory Medicine
School of Medicine
Indiana University
Indianapolis, Indiana

Mohsin Khan
Division of Infectious Diseases
Department of Medicine
University of California, San Diego
La Jolla, California

Seong-Jun Kim
Division of Infectious Diseases
Department of Medicine
University of California, San Diego
La Jolla, California

Christopher Kyaw
Department of Biopharmaceutical
    Sciences
School of Pharmacy
Keck Graduate Institute
Claremont, California

Ho Leung
Department of Biopharmaceutical
    Sciences
School of Pharmacy
Keck Graduate Institute
Claremont, California

Yang Li
Department of Pharmacology and
    Pharmaceutical Sciences
School of Pharmacy
University of Southern California
Los Angeles, California

Zhigang Liu
Department of Pharmacology and
    Pharmaceutical Sciences
School of Pharmacy
University of Southern California
Los Angeles, California

José Alberto López-Domínguez
Department of Molecular Biosciences
School of Veterinary Medicine
University of California, Davis
Davis, California

H. Susana Marinho
Laboratory of Biochemistry of Oxidants
    and Antioxidants
Centre and Department of Chemistry
    and Biochemistry
School of Sciences
University of Lisbon
Lisbon, Portugal

Mitchell R. McGill
Department of Pharmacology,
    Toxicology and Therapeutics
School of Medicine
University of Kansas, Kansas City
Kansas City, Missouri

Katalin Módis
Department of Biology
Lakehead University
Thunder Bay, Ontario, Canada

Sohail Mohammad
Division of Infectious Diseases
Department of Medicine
University of California, San Diego
La Jolla, California

Hong-Min Ni
Department of Pharmacology,
    Toxicology, and Therapeutics
School of Medicine
University of Kansas, Kansas City
Kansas City, Missouri

Claude A. Piantadosi
Departments of Anesthesiology,
    Medicine, and Pathology
Duke University Medical Center
Durham, North Carolina

Carlos J. Pirola
Department of Molecular Genetics and
    Biology of Complex Diseases
Institute of Medical Research A
    Lanari—IDIM
National Scientific and Technical
    Research Council
University of Buenos Aires
Buenos Aires, Argentina

Jon J. Ramsey
Department of Molecular Biosciences
School of Veterinary Medicine
University of California, Davis
Davis, California

Harsh Sancheti
Department of Pharmacology and
    Pharmaceutical Sciences
School of Pharmacy
University of Southern California
Los Angeles, California

Eric A. Schon
Departments of Neurology and Genetics
    and Development
Columbia University Medical Center
New York, New York

Samit Shah
Department of Biopharmaceutical
    Sciences
School of Pharmacy
Keck Graduate Institute
Claremont, California

Aleem Siddiqui
Division of Infectious Diseases
Department of Medicine
University of California, San Diego
La Jolla, California

Silvia Sookoian
Department of Clinical and Molecular
    Hepatology
Institute of Medical Research A
    Lanari—IDIM
National Scientific and Technical
    Research Council
University of Buenos Aires
Buenos Aires, Argentina

Bangyan Stiles
Department of Pharmacology and
    Pharmaceutical Sciences
School of Pharmacy
University of Southern California
Los Angeles, California

Hagir B. Suliman
Department of Anesthesiology and the
    Duke Cancer Institute
Duke University Medical Center
Durham, North Carolina

Gulam Syed
Division of Infectious Diseases
Department of Medicine
University of California, San Diego
La Jolla, California

Tin Aung Than
Division of Gastrointestinal and Liver
    Diseases
Department of Medicine
Research Center for Liver Diseases
Keck School of Medicine
University of Southern California
Los Angeles, California

Bhuvaneshwar Vaidya
Department of Biopharmaceutical
  Sciences
School of Pharmacy
Keck Graduate Institute
Claremont, California

Rui Wang
Department of Biology
Lakehead University
Thunder Bay, Ontario, Canada

Rudolf J. Wiesner
Center for Physiology and
  Pathophysiology
Institute of Vegetative Physiology
and
Center for Molecular Medicine Cologne
University of Köln
and
Cologne Excellence Cluster on
  Cellular Stress Responses in Aging-
  Associated Diseases
Köln, Germany

Jessica A. Williams
Department of Pharmacology,
  Toxicology, and Therapeutics
School of Medicine
University of Kansas, Kansas City
Kansas City, Missouri

Sanda Win
Division of Gastrointestinal and Liver
  Diseases
Department of Medicine
Research Center for Liver Diseases
Keck School of Medicine
University of Southern California
Los Angeles, California

Fei Yin
Department of Pharmacology and
  Pharmaceutical Sciences
School of Pharmacy
University of Southern California
Los Angeles, California

Xiao-Ming Yin
Department of Pathology and
  Laboratory Medicine
School of Medicine
Indiana University
Indianapolis, Indiana

Hao Zhang
Department of Pathology and
  Laboratory Medicine
School of Medicine
Indiana University
Indianapolis, Indiana

Antonio Zorzano
Department of Biochemistry and
  Molecular Biology
University of Barcelona
and
Molecular Medicine Program of the
  Institute of Research in Biomedicine
IRB Barcelona
and
CIBER from Diabetes and Metabolic
  Associated Diseases
Health Care Institute Carlos III
Barcelona, Spain

Abbreviations

ALT	Alanine aminotransferase
AMPK	AMP-activated protein kinase
AOAA	Aminooxyacetic acid
APAP	Acetaminophen
AST	Aspartate aminotransferase
AVG	Aminoethoxyvinylglycine
Bag4	Bcl2-associated athanogene 4
BCA	β-Cyano-L-alanine
Bcl-2	B-cell lymphoma 2
Bcl2L1	Bcl-2 like 1
BDL	Bile duct ligation
BH3	Bcl-2 homology 3 domain 3
BNIP3	Bcl-2/adenovirus E1B 19 kDa protein-interacting protein 3
CAT	Cysteine aminotransferase
CBS	Cystathionine β-synthase
CCCP	m-Chloro phenyl hydrazine
CI	Complex I
CII	Complex II
CIII	Complex III
CIV	Complex IV
CK2	Casein-kinase II
Clec16a	C-type lectin domain family 16, member A
CLP	Cecal ligation and puncture
CO	Carbon monoxide
CPTI	Carnitine palmitoyltransferase 1
CREB	cAMP-response element binding
CSE	Cystathionine γ-lyase
CV	Complex V
Cyp	Cytochrome p450
DAMP	Damage-associated molecular pattern
DAO	D-Amino acid oxidase
DNMTs	DNA methyltransferases
Dr4/5	Death receptors 4 and 5
Drp1	Dynamin-related protein 1
ETHE1	Persulfide dioxygenase
FAO	Fatty acid oxidation
FasL	Fas ligand
Fis1	Mitochondrial fission 1
FUNDC1	FUN14 domain-containing protein 1
GSH	Glutathione
GSK	Glycogen synthase kinase
GST	Glutathione S-transferase

HA	Hydroxylamine
H_2S	Hydrogen sulfide
HFD	High fat diet
Hif-1	Hypoxia-inducing factor 1
HMGB1	High mobility group box 1
HRE	Hif-1 responsive element
HSPA1L	Heat shock 70 kDa protein 1-like
HSPB1	Heat shock protein β-1
IR	Insulin resistance
IRS	Insulin receptor substrate
LC3	Microtubule-associated protein1 light-chain 3
LIR	LC3-interacting region
LPS	Bacterial lipopolysaccharide
MCD	Methionine-choline-deficient diet
Mff	Mitochondria fission factor
MFN1/2	Mitofusin 1 and 2
MFV	Mitochondria-derived vesicles
MPT	Membrane permeability transition
MPTP	Mitochondrial permeability transition pore
MST	3-Mercaptopyruvate sulfur transferase
mtDNA	Mitochondrial DNA
NAFLD	Nonalcoholic fatty liver disease
NAPQI	N-acetyl-p-benzoquinone imine
NASH	Nonalcoholic steatohepatitis
NMR	Nuclear magnetic resonance
NO	Nitric oxide
Nrdp1	Neuregulin receptor degradation protein 1
Nrf1/2	Nuclear respiratory factors 1 and 2
OPA1	Optic atrophy 1
OXPHOS	Oxidative phosphorylation
PAAT	Periaortic adipose tissue
PAG	DL-Propargylglycine
PARL	Presenilin-associated rhomboid-like
PARP	Poly (ADP-ribose) polymerase
PDEs	Phosphodiesterases
PGAM5	Phosphoglycerate mutase family member 5
PGC-1α	Peroxisome proliferator-activated receptor γ, coactivator 1 α
PINK1	PTEN-induced putative kinase 1α
PKA	cAMP-dependent protein kinase/protein kinase A
PPARGC1A	Peroxisome proliferator-activated γ coactivator-1α
RC	Respiratory chain
ROS	Reactive oxygen species
SIRT1	Sirtuin 1
Smurf1	Smad-specific E3 ubiquitin protein ligase 1
S-OPA1 and L-OPA1	Short and long form OPA1

SQR	Sulfide: quinone oxidoreductase
SQSTM1	Sequestosome 1
STS	Sodium thiosulfate
T1D	Type 1 diabetes
T2DM	Type 2 diabetes mellitus
tBid	Truncated bid
TCA	Tricarboxylic acid cycle
TFAM	Mitochondrial transcription factor A
TNF-R1	TNF-α receptor 1
TNF-α	Tumor necrosis factor-α
TOMM7	Translocase of outer mitochondrial membrane 7
TOMM20	Translocase of outer mitochondrial membrane 20
TPP$^+$	Triphenylphosphonium
TRAIL	TNF-related apoptosis-inducing ligand
UBA	Ubiquitin-associated
UCP	Uncoupling protein
ULK1	Unc-51 like autophagy activating kinase 1
Usp30	Ubiquitin-specific peptidase 30
VDAC	Voltage-dependent anion channel

Section I

Overview of Mitochondria

1 Metabolism of Superoxide Radicals and Hydrogen Peroxide in Mitochondria

H. Susana Marinho and Fernando Antunes

CONTENTS

ABSTRACT

We review the mechanisms of superoxide radical ($O_2^{·-}$) and hydrogen peroxide (H_2O_2) formation in and removal from mitochondria. The formation of $O_2^{·-}/H_2O_2$ in the mitochondrial inner membrane occurs through three respiratory chain complexes: I, II, and III—glycerol 3-phosphate dehydrogenase, the electron-transferring flavoprotein/electron-transferring flavoprotein ubiquinone oxidoreductase system, and dihydroorotate dehydrogenase. In the mitochondrial outer membrane, monoamine oxidase and cytochrome b_5 reductase also form $O_2^{·-}/H_2O_2$. Recently described is H_2O_2-forming NADPH oxidase isoform (NOX4) whose mitochondrial topology is uncertain. Finally, we briefly mention enzymatic systems present in the mitochondrial matrix that produce $O_2^{·-}/H_2O_2$. Concerning H_2O_2 removal, we compare the relative contribution of peroxiredoxins (Prx3 and Prx5) and glutathione peroxidases (Gpx1 and Gpx4) to the removal of H_2O_2. Calculations show that either glutathione peroxidases or peroxiredoxins may be the predominant enzymes depending on the rate of production of H_2O_2 and on the reduction state of thioredoxin, the electron donor for peroxiredoxins. At low H_2O_2 production rates, near 0.2 nmol/min/mg of protein, most H_2O_2 (estimated steady state near 1–2 nM) will be removed through Prx3 (nearly 90%). For 10-fold higher H_2O_2 production, both H_2O_2 removal and steady-state levels are strongly dependent on reduced thioredoxin concentrations: for 10 µM reduced thioredoxin, Prx3 reduces nearly 74% of the H_2O_2 formed, attaining a 40 nM steady state; for a 1 µM reduced thioredoxin, Prx3 reduces only 15% of the H_2O_2 formed, resulting in a 130 nM H_2O_2 steady state. Finally, the role of pyruvate as a metabolic sink for H_2O_2 is analyzed.

INTRODUCTION

The physiological production of H_2O_2 by mitochondrial membranes was first observed in 1966 by Jensen [1] who found that the antimycin-insensitive oxidation of reduced nicotinamide adenine dinucleotide (NADH) and succinate by bovine heart submitochondrial particles was coupled with the formation of H_2O_2. Soon after, Loschen et al. [2] demonstrated, for the first time, that isolated mitochondria produce H_2O_2 at rates mainly dependent on the metabolic state. Further studies made by Britton Chance's group [3] established that H_2O_2 production in intact mitochondria from rat liver or pigeon heart (about 0.3–0.6 nmol H_2O_2/min mg^{-1} protein) is maximal in state 4 [2–4], that is, when adenosine diphosphate (ADP) is depleted and the redox components of the respiratory chain are reduced, accounting for about 2% of the total oxygen utilization under these conditions [5]. Later, it was shown that this H_2O_2 arose from the dismutation of $O_2^{·-}$ generated within the mitochondria [6,7]. In this review, we describe the enzymatic systems that mediate the production of $O_2^{·-}$ and H_2O_2 as well as the antioxidant systems responsible for their removal.

GENERATION OF $O_2^{·-}$ AND H_2O_2 IN MITOCHONDRIA

Two of the respiratory electron transport complexes, complex I and complex III, have been frequently cited as being the most important sites of $O_2^{·-}$ formation in the mitochondria [8,9]. However, for mammalian cells, the use of inhibitors to both manipulate the redox states of particular $O_2{}^-/H_2O_2$-producing sites and prevent $O_2{}^-/$ H_2O_2 formation from others allowed the identification of at least 12 enzymes involved in $O_2{}^-/H_2O_2$ production in the mitochondria at measurable rates (reviewed in [8–13]) (Figure 1.1). All these enzymes are present ubiquitously in mammalian mitochondria, but their capacity in producing $O_2{}^-/H_2O_2$, as well as their expression, varies greatly among tissues and species [10,11,14] (Figure 1.2). Moreover, the production of $O_2{}^-/H_2O_2$ in isolated mitochondria is strongly dependent on the substrates being oxidized and on the metabolic state [14–19] (Figure 1.3). The overall rates of $O_2{}^-/H_2O_2$ production can differ by an order of magnitude between substrates and the relative contribution of each site can also be very different with different substrates [19].

GENERATION OF $O_2{}^{-}$ AND H_2O_2 BY MITOCHONDRIAL INNER MEMBRANE ENZYMES

The electron transport chain has been extensively studied regarding the formation of $O_2{}^-/H_2O_2$ (reviewed in [8,9,28,29]). With the exception of complex IV that is able to bind tightly partially reduced intermediates during oxygen reduction without significant release of reactive oxygen species [9,28], respiratory complexes form $O_2{}^-/H_2O_2$. Studies using isolated mitochondria from skeletal muscle showed that complex III (at site Q_o) has the greatest capacity to form $O_2{}^-/H_2O_2$ followed by complex I (at the ubiquinone-binding site) and complex II (at the flavin) [21–23,30] (Figure 1.1). In the mitochondrial inner membrane, $O_2{}^-/H_2O_2$ can also be formed by enzymes involved in electron transfer to the electron transport chain such as the electron-transferring flavoprotein (ETF) and electron-transferring flavoprotein ubiquinone oxidoreductase (ETFQOR) system or glycerol 3-phosphate dehydrogenase and also by dihydroorotate dehydrogenase (DHODH).

Complex I

Complex I (NADH–ubiquinone oxidoreductase, EC 1.6.5.3) catalyzes the first step in the respiratory electron transport chain in the mitochondria, the reduction of ubiquinone to ubiquinol by NADH. The free energy from this redox reaction is used to translocate four protons across the mitochondrial inner membrane, contributing to the protonmotive force (Δp) that is used to drive adenosine triphosphate (ATP) synthesis [31]:

$$NADH + H^+ + Q + 4H_{in}^+ \rightarrow NAD^+ + QH_2 + 4H_{out}^+ \tag{1.1}$$

In complex I from bovine heart mitochondria, which is a close model for the human enzyme, 45 different subunits have been identified [32]. Complex I contains at least 10 redox components, namely, flavin mononucleotide (FMN), 8 iron–sulfur clusters, and bound ubiquinone, all of which are present in the hydrophilic part of the complex exposed to the matrix [31].

Formation of $O_2{}^-$ by complex I has been shown to occur using the isolated complex [33,34], submitochondrial particles [35–37], and intact mitochondria isolated from different sources [14,15,38].

Complex I has a major role in $O_2{}^-/H_2O_2$ formation in the mitochondria. It has been estimated that complex I may account for about half of the total NADH-supported H_2O_2 formation in the mitochondria [39] and most of the $O_2{}^-$ produced by the respiratory chain complexes using succinate as a substrate [19].

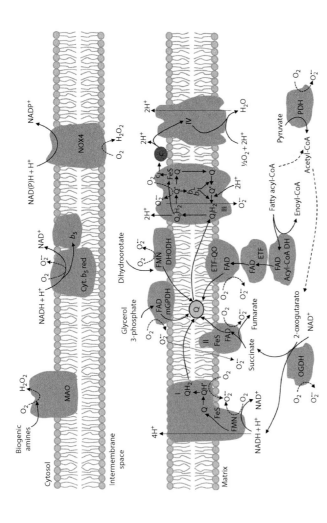

FIGURE 1.1 Enzymes involved in mitochondrial O_2^-/H_2O_2 formation in mammalian cells. Three respiratory chain complexes form O_2^-/H_2O_2. Complex I forms O_2^- at the flavin site and ubiquinone-binding site releasing it into the matrix. Complex II forms O_2^- into the matrix at its flavin site. Complex III forms O_2^- at the ubiquinone Qo site releasing it into both the matrix and intermembrane space. Mitochondrial glycerol 3-phosphate dehydrogenase (mGPDH) forms O_2^- to both sides of the mitochondrial inner membrane. The ETF and ETFQOR are involved in the final steps of β-oxidation and ETFQOR is the main candidate for O_2^-/H_2O_2 formation. DHODH also produces O_2^- during pyrimidine biosynthesis. NOX4 forms H_2O_2, but its mitochondrial topology and location is not known and it has been tentatively located in the mitochondrial outer membrane. In the mitochondrial outer membrane, MAO forms H_2O_2, while cytochrome b_5 reductase (Cyt. b_5 Red) forms O_2^-, and both are released into the cytosol. In the matrix, the enzyme complexes OGDH, PDH, and BCOADH (not shown) contain the DLDH subunit that produces O_2^-/H_2O_2. *Note:* FMN, flavin mononucleotide; FAD, flavin adenine dinucleotide.

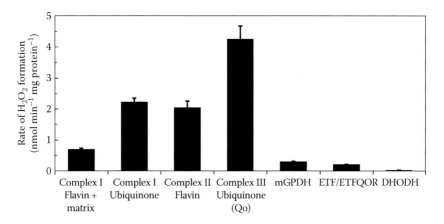

FIGURE 1.2 Maximum rates of O_2^-/H_2O_2 formation from different sites in isolated muscle mitochondria. Different combinations of substrates and inhibitors were used to obtain the maximum rates from each site and all rates were corrected for H_2O_2 consumption in the matrix through GSH peroxidases [20]. DHODH, dihydroorotate dehydrogenase; ETF/ETFQOR, electron-transferring flavoprotein and electron-transferring flavoprotein ubiquinone oxido-reductase; mGPDH, mitochondrial glycerol 3-phosphate dehydrogenase. Data are the mean ± SEM ($n \geq 3$) and rates were obtained for complex I from Treberg et al. [20,21], complex II from Quinlan et al. [22], complex III from Quinlan et al. [23], mGPDH from Orr et al. [24], DHODH from Hey-Mogensen et al. [25], and ETF/ETFQOR from Perevoshchikova et al. [26].

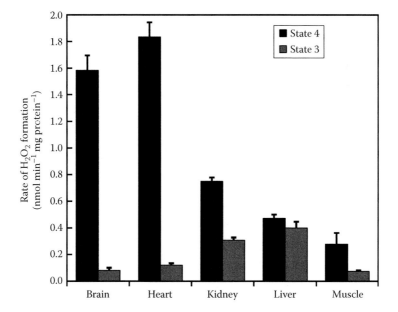

FIGURE 1.3 H_2O_2 formation in isolated mitochondria is tissue and state dependent. Data were obtained from Tahara et al. [27] and are the mean ± SEM ($n \geq 3$). Formation of H_2O_2 was measured in isolated mitochondria using 1 mM succinate in the presence of 1–6.3 µg of oligomycin (state 4) or 1 mM ADP (state 3).

Considering the complexity of complex I, it is not surprising that there is some debate regarding which of its redox components interacts with O_2 to generate O_2^-. Thus, the enzyme-bound NAD· radical [35], FMN [40,41], the iron–sulfur clusters N-1a and N-2 [38,42], and ubisemiquinone [17,43] have all been suggested as being involved in O_2^- production. Results supporting the presence of two separate sites of O_2^- production (ubisemiquinone and FMN) were obtained by Treberg et al. [21] using rat skeletal muscle intact mitochondria. Recently, Grivennikova and Vinogradov [11] proposed that both the purified bovine heart complex I and also the membrane-bound complex I produce not just O_2^- but also H_2O_2 at two different sites. According to Grivennikova and Vinogradov [11], FMNH$^-$ produces H_2O_2 through a two-electron reduction, while an iron–sulfur center, probably N-2, produces O_2^-. Moreover, they proposed that H_2O_2 would be the predominant species formed by complex I at physiologically relevant concentrations of NADH and/or NAD$^+$. In all possible mechanisms proposed thus far, formation of O_2^-/H_2O_2 by complex I is directed to the matrix.

Complex II

Succinate dehydrogenase (complex II, succinate–ubiquinone oxidoreductase, EC 1.3.5.1) is an enzyme involved both in the tricarboxylic acid cycle and in the mitochondrial electron transport chain that catalyzes the oxidation of succinate to fumarate and the reduction of ubiquinone to ubiquinol:

$$\text{Succinate} + Q \rightarrow \text{fumarate} + QH_2 \qquad (1.2)$$

Complex II has four subunits. Two of those subunits, the flavin adenine dinucleotide (FAD)-containing subunit SDHA and the SDHB subunit, which contains three iron–sulfur clusters, are hydrophilic and exposed to the mitochondrial matrix. The transmembrane proteins CybL (SDHC) and CybS (SDHD) are embedded in the inner mitochondrial membrane. They contain a heme b and the ubiquinone-binding site [44].

When compared to both complex I and complex III, complex II was shown to make little contribution to O_2^-/H_2O_2 formation by mammalian isolated mitochondria and submitochondrial particles under physiological conditions [14,15]. However, mutations in complex II can lead to higher rates of O_2^- formation and are associated with pathological states [45].

Recently, high rates of O_2^- production by complex II isolated mitochondria from rat skeletal muscle [22], under conditions that can occur in vivo in the resting muscle or during hypoxia, have been reported. Furthermore, for rat skeletal muscle isolated mitochondria using glycerol 3-phosphate as a substrate, a substantial portion of O_2^-/H_2O_2 formation, commonly attributed to mitochondrial glycerol 3-phosphate dehydrogenase (mGPDH), originates from electron flow into complex II [19,24]. Also, recent data indicate that complex II is partially responsible for the succinate-dependent O_2^-/H_2O_2 formation in the presence of rotenone, which had been previously attributed to a reverse electron flow from complex II to complex I [46]. O_2^- is formed at the enzyme flavin site [47].

Complex III

Complex III (ubiquinol–cytochrome c oxidoreductase, EC 1.10.2.2) catalyzes the reduction of cytochrome c by ubiquinol that is coupled to the transmembrane proton translocation and formation of Δp:

$$QH_2 + 2 \text{ ferricytochrome } c + 2H_{in}^+ \rightarrow Q + 2 \text{ ferrocytochrome } c + 4H_{out}^+ \qquad (1.3)$$

Complex III exists as a dimer with each monomer consisting of 11 polypeptides. The redox centers of complex I are cytochrome b, which contains two hemes, one with low potential (b_L) and another with a high potential (b_H), cytochrome c_1, the Rieske Fe–S protein, and the two ubiquinone centers. The ubiquinol reaction center (called Q_p or Q_o site) is located at the positive side of the membrane (intermembrane space), while the ubiquinone reduction center (Q_n or Q_i site) is located at the negative side of the membrane (matrix). Electron transfers within complex III occur according to the Q cycle proposed by Mitchell [48] starting with the oxidation of ubiquinol at site Q_o, through the donation of one electron to the Rieske Fe–S protein leading to ubisemiquinone formation, followed by the reduction of cytochrome b_L and formation of ubiquinone.

The rate of O_2^-/H_2O_2 formation by complex III is about 10% of that mediated by complex I for succinate-mediated respiration, but approximately equal to that of complex I for NADH-dependent respiration in isolated rat skeletal muscle mitochondria [19]. The use of inhibitors has led to the proposal that the site of O_2^- production by complex III is the ubisemiquinone formed at site Q_o [49]. Antimycin A is known to increase the rate of O_2^- formation by complex III [14,50,51]. Antimycin A blocks electron transfer from cytochrome b_L to the ubiquinone at Q_i, which leads to an increased concentration of ubisemiquinone at site Q_o [9], and this antimycin A–induced O_2^- formation is abolished by the Q_o site inhibitors stigmatellin and myxothiazol [52,53].

Unlike complex I that releases O_2^- exclusively to the matrix, studies using isolated mitochondria have shown that complex III releases O_2^- to both sides of the membrane [14,54]. Since the inner membrane is nonpermeable to O_2^- [55], to explain O_2^- release into the matrix, Muller et al. [54] proposed two possible mechanisms. In the first mechanism, a neutral ubisemiquinone would diffuse out of Q_o, along a hydrophobic tunnel, and at the lipid/aqueous phase interface, the ubisemiquinone would deprotonate and react with oxygen to form aqueously solvated O_2^-. In the second mechanism, O_2^- formed at Q_o can be protonated to form the hydroperoxyl radical that can diffuse along the membrane to be released in both the matrix and intermembrane space.

Mitochondrial Glycerol 3-Phosphate Dehydrogenase

mGPDH (EC 1.1.99.5) is an integral flavoprotein present in the outer leaflet of the mitochondrial inner membrane [56]. The enzyme is involved in lipid metabolism and in the glycerol phosphate shuttle that connects glycolysis with the mitochondrial respiratory chain. It catalyzes the oxidation of glycerol 3-phosphate to dihydroxyacetone phosphate and the reduction of ubiquinone to ubiquinol:

$$\alpha - \text{glycerophosphate} + Q \rightarrow \text{dihydroxyacetatone phosphate} + QH_2 \qquad (1.4)$$

The mitochondrial content of mGPDH has large variations among different mammalian tissues with the highest content found in brown adipose tissue, while almost negligible levels are present in tissues such as the heart, muscle, or liver [57].

Several studies have established that mGPDH, from both mammalian and insect mitochondria, forms O_2^-/H_2O_2 at levels comparable to those formed by complex III in the presence of antimycin A [24,58,59]. mGPDH produces mainly O_2^- that is released approximately equally toward each side of the mitochondrial inner membrane. This suggests the Q-binding pocket of mGPDH as the main site of O_2^- generation [24].

Electron-Transferring Flavoprotein and Electron-Transferring Flavoprotein Ubiquinone Oxidoreductase

During the 1970s, it was shown that the oxidation of palmitoyl carnitine by mitochondria leads to the generation of H_2O_2 [4]. Palmitoyl carnitine is metabolized by the β-oxidation pathway and electrons may enter the respiratory chain at two sites described as follows: complex I, from the NADH formed in the reaction catalyzed by 3-hydroxyacyl-CoA dehydrogenase, and the ubiquinone pool, from the ETF/ETFQOR system that acts as the electron acceptor from nine different mitochondrial FAD-containing acyl-CoA dehydrogenases of fatty acid β-oxidation [60]. Also, oxidation of the end product of β-oxidation, acetyl-CoA, in the tricarboxylic acid cycle leads to further electrons entering the respiratory chain through complex I and complex II [19,61]. Several studies, using isolated mitochondria, identified as sources of O_2^-/H_2O_2 during palmitoyl carnitine oxidation, complex I (site I_Q) [14,19,62], complex II [19], complex III [61,62], the ETF/ETFQOR (EC 1.5.5.1) system [14,27,61,62], and acyl-CoA dehydrogenases [27]. Although the ETF/ETF dehydrogenase system has been proposed as being a significant source of O_2^-/H_2O_2 into the matrix side of the membrane [14,27,61,62], it is possible that this may only occur either at high [O_2], due the high apparent K_M of ETFQOR for O_2 [62], or in the presence of respiratory chain inhibitors [26]. In fact, recent studies suggest that all O_2^-/H_2O_2 formed during oxidation of palmitoyl carnitine by mitochondria can be accounted for by complex I (site I_Q), complex II, and complex III [19,26].

Dihydroorotate Dehydrogenase

DHODH (EC 1.3.5.2) is ubiquitously distributed in mammalian tissues [63]. In upper eukaryotes, class 2 DHODH is an integral protein of the mitochondrial inner membrane with the dihydroorotate-binding site facing the intermembrane space and a hydrophobic tail inserted in the membrane [64]. The enzyme catalyzes the oxidation of dihydroorotate to orotate and the reduction of ubiquinone to ubiquinol during pyrimidine synthesis:

$$(S) - \text{dihydroorotate} + \text{ubiquinone} \rightleftarrows \text{orotate} + \text{ubiquinol} \qquad (1.5)$$

DHODHs have two redox-active sites: an FMN prosthetic group, which accepts two electrons from dihydroorotate, and a ubiquinone in the quinone-binding site, which accepts the electrons and subsequently joins the ubiquinone pool of the mitochondrial inner membrane [65]. During the DHODH catalytic cycle, the flavin semiquinone intermediate FMNH is likely formed [66]. Also, the midpoint potential of the flavin in DHODH is sufficiently negative (−310 mV for the class 2 enzyme purified from *Escherichia coli*) to enable reduction of oxygen to generate O_2^- and/or H_2O_2 by the enzyme [67] even though the activity of DHODH with quinone substrates is 14- to 58-fold higher than with O_2 [66]. In fact, in the absence of its physiological electron acceptor, reduced DHODH can produce H_2O_2 in vitro [63]. Studies made in the 1970s using isolated mitochondria found that DHODH also produced O_2^- [68,69], but later on this O_2^- formation was attributed to complex III [70]. Recently, it was shown that DHODH directly produces O_2^- and/or H_2O_2 at low rates but is also capable of indirect production at higher rates from other sites through its ability to reduce the ubiquinone pool [25]. In mitochondria isolated from rat skeletal muscle and in the presence of inhibitors of complex I, complex III, and complex II, DHODH generates O_2^-/H_2O_2 at a rate of about 20–40 pmol H_2O_2 min^{-1} mg protein^{-1}, from the ubiquinone-binding site [24,25].

GENERATION OF O_2^- AND H_2O_2 BY MITOCHONDRIAL OUTER MEMBRANE ENZYMES

There are two enzymes located in the mitochondrial outer membrane that may produce O_2^-/H_2O_2, cytochrome b_5 reductase, and monoamine oxidase (MAO).

Cytochrome b_5 Reductase

Cytochrome b_5 reductase (EC 1.6.2.2) is a flavoprotein widely distributed in mammalian tissues. It is an integral membrane protein present in the endoplasmic reticulum, plasma membrane, and also the mitochondrial outer membrane where it catalyzes the reduction of cytochrome b_5 by cytoplasmic NADH [71,72]:

$$NADH + 2 \text{ ferricytochrome } b_5 \rightleftarrows NAD^+ + H^+ + 2 \text{ ferrocytochrome } b_5 \quad (1.6)$$

However, the enzyme can also act as an NADH oxidase. In fact, rat brain mitochondrial cytochrome b_5 reductase was shown to produce O_2^- with a high rate of ~300 nmol O_2^- min^{-1} mg protein^{-1} [73]. Recently, cytochrome b_5 reductase was purified from pig liver microsomes and shown to catalyze the NADH-dependent production of superoxide anion with a $V_{max} = 3.0 \pm 0.5$ µmol O_2^- min^{-1}mg of purified cytochrome b_5 reductase and a K_M (NADH) $= 2.8 \pm 0.3$ µM NADH [74].

Monoamine Oxidases

MAO-A and MAO-B (EC 1.4.3.4) are flavoproteins ubiquitously expressed in various mammalian tissues that catalyze the oxidative deamination of primary aromatic amines along with long-chain diamines and tertiary cyclic amines. The oxidation of biogenic amines is accompanied by the release of H_2O_2 [75,76]:

$$RCH_2NHR' + H_2O + O_2 \rightleftarrows RCHO + R'NH_2 + H_2O_2 \quad (1.7)$$

Tyramine oxidation (0.2 mM) by rat brain mitochondria produces H_2O_2 at a rate of 2.71 nmol min^{-1} mg protein^{-1} [75], leading to a steady-state intramitochondrial H_2O_2 concentration of 0.8 µM [76], which is ~50-fold higher than that originating during succinate oxidation in the presence of antimycin A (0.016 µM) [77]. Therefore, in the brain, MAO-dependent H_2O_2 generation may far exceed that of other mitochondrial sources. However, it should be taken into account that dopamine, the physiological substrate, has a concentration of around 1 µM in the cytosol and leads to a 2.7 lower H_2O_2 formation rate by MAO than tyramine [78]. Under these physiological concentrations, MAO would be expected to form less H_2O_2 than the respiratory chain complexes. In other tissues, MAOs may also be a major source of H_2O_2 in the reperfusion following ischemia [79] and in aging [80].

NAD(P)H Oxidase

Reduced nicotinamide adenine dinucleotide phosphate (NADPH) oxidase 4 (NOX4) is a member of the NADPH oxidase family of enzymes NOX1–5 and DUOX1–2. NOX4, which has been reported to be present in the nucleus and endoplasmic reticulum [81], was recently found to have a mitochondrial localization sequence [82] and to be expressed in the mitochondria of cardiomyocytes [83], mesangial cells of the kidney [84], and neurons [85]. The exact location and topology of NOX in the mitochondria is not known and it may be possible that it is located in the inner mitochondrial membrane and not in the outer mitochondrial membrane [86]. NOX4, unlike most NADPH oxidases that catalyze the reduction of O_2 to $O_2^{\cdot-}$, mainly catalyzes the formation of H_2O_2 [87]:

$$NADPH + O_2 + H^+ \rightarrow NADP^+ + H_2O_2 \tag{1.8}$$

Recently, it was shown that a significant increase in both the protein level and activity of mitochondrial complex I–containing supercomplexes I_1III_2 and $I_1III_2IV_{0-1}$ occurred in Nox4-depleted endothelial cells [88]. This may indicate that Nox4-derived H_2O_2 can either damage the supercomplexes or impair complex I assembly into supercomplexes, which have much higher enzymatic activity than individual complexes.

GENERATION OF $O_2^{\cdot-}$ AND H_2O_2 BY MATRIX ENZYMES

Isolated mitochondrial 2-oxoglutarate dehydrogenase (OGDH) and pyruvate dehydrogenase (PDH) complexes produce $O_2^{\cdot-}/H_2O_2$ [89]. This formation of $O_2^{\cdot-}/H_2O_2$ has been associated with FAD-linked dihydrolipoamide dehydrogenase (DLDH, EC 1.8.1.4) [39], which catalyzes the reduction of NAD^+ by dihydrolipoamide to form NADH and lipoamide and is present in both these matrix enzyme complexes [89–92] and also in branched-chain 2-oxoacid dehydrogenase (BCOADH) [92]. In fact, DLDH-mediated $O_2^{\cdot-}/H_2O_2$ formation in rat brain mitochondria of heterozygous knockout mice deficient in DLDH (DLDH$^{+/-}$) was shown to be twofold lower than in control mice [89]. This importance of DLDH-mediated $O_2^{\cdot-}/H_2O_2$ formation was reinforced by recent results that showed that in skeletal muscle–isolated mitochondria oxidizing 2-oxoacids, these 2-oxoacid dehydrogenases can produce $O_2^{\cdot-}/H_2O_2$ at higher rates than complex I [92]. The OGDH complex has the greatest capacity followed by the PDH complex and the BCOADH complex.

REMOVAL OF $O_2^{\cdot-}$

The mitochondrial steady-state concentration of $O_2^{\cdot-}$ has been estimated to be in the range 0.08–0.2 nM while that of H_2O_2 is about 5 nM [76]. These low concentrations are the result of very efficient enzymatic antioxidant systems. In fact, most of the $O_2^{\cdot-}/H_2O_2$ produced in the mitochondria is eliminated through these enzymes, with only a minor fraction, estimated to be near 0.1% for $O_2^{\cdot-}$ and 0.001% for H_2O_2, being available to react with other molecules [93]. The superoxide radical main reaction is its dismutation catalyzed by superoxide dismutases (EC 1.15.1.1) present in the matrix (MnSOD) and in the intermembrane space (Cu,Zn-SOD) [94,95]:

$$O_2^{\cdot-} + O_2^{\cdot-} + 2H^+ \rightarrow H_2O_2 + O_2 \tag{1.8}$$

Other $O_2^{\cdot-}$ reactions may include reaction with iron–sulfur clusters in proteins [96] and with nitric oxide forming peroxynitrite (Equation 1.9) [97]. Also, in its protonated form, hydroperoxyl radical, it may initiate lipid peroxidation (Equation 1.10) through abstraction of an allylic hydrogen atom from an unsaturated fatty acid (RH) [93,98]:

$$O_2^{\cdot-} + {}^{\cdot}NO \rightarrow ONOO^- \tag{1.9}$$

$$HO_2^{\cdot} + RH \rightarrow H_2O_2 + R^{\cdot} \tag{1.10}$$

REMOVAL OF H_2O_2

In the case of H_2O_2, its main removal reaction is the reduction to H_2O catalyzed by glutathione peroxidases (EC 1.11.1.9) and peroxiredoxins (EC 1.11.1.15). Other reactions include the formation through Fenton chemistry of hydroxyl radical (Equation 1.11) [77], a very reactive species, or oxidation of thiols (Equation 1.12), a reaction that mediates a well-known regulatory role of H_2O_2 [99]:

$$Fe^{2+} + H_2O_2 + H^+ \rightarrow Fe^{3+} + HO^{\cdot} + H_2O \tag{1.11}$$

$$H_2O_2 + 2R-SH \rightarrow R-S-S-R + 2H_2O \tag{1.12}$$

Next, we describe the antioxidant systems responsible for H_2O_2 removal in the mitochondrial matrix. In liver mitochondria, initially H_2O_2 was assumed to be reduced to water via glutathione peroxidases, as the nonenzymatic reaction of H_2O_2 with thiols is much slower compared with the enzymatic reactions and catalase is absent in this organelle. After the discovery of peroxiredoxins, it became clear that a new player involved in H_2O_2 removal had to be taken into account. In addition to these two enzymatic systems, we will also discuss the role of the nonenzymatic reaction of H_2O_2 with pyruvate, a metabolite that is highly abundant in mitochondria.

GLUTATHIONE PEROXIDASES

Ultimately, both glutathione peroxidase and peroxiredoxin cycles convert H_2O_2 to water at the expense of NADPH, as described by

$$NADPH + H_2O_2 + H^+ \rightarrow 2H_2O + NADP^+ \tag{1.13}$$

The reducing power stored in NADPH, usually used for anabolic pathways, is diverted to detoxify the oxidant H_2O_2 to water. Next, we will describe in detail these two antioxidant systems.

There are two glutathione peroxidase isomers in the mitochondria, Gpx1, a tetramer, and Gpx4, a monomer; both are selenoproteins coded by two different nuclear genes [100]. Their reaction mechanism is similar, with H_2O_2 oxidizing the selenocysteine (–SeH) residue in the active center to a seleninic acid (–SeOH) (Equation 1.14). Then, seleninic acid intermediate is reduced back upon the sequential reaction with two molecules of reduced glutathione (GSH) forming oxidized glutathione (GSSG) (Equations 1.15 and 1.16):

$$H_2O_2 + Gpx - Cys - SeH \rightarrow Gpx - Cys - SeOH + H_2O \tag{1.14}$$

$$Gpx - Cys - SeOH + GSH \rightarrow Gpx - Cys - SeSG + H_2O \tag{1.15}$$

$$Gpx - Cys - SeSG + GSH \rightarrow Gpx - Cys - SeH + GSSG \tag{1.16}$$

GSSG is reduced back to GSH at the expense of NADPH (Equation 1.17), in a reaction catalyzed by glutathione reductase (EC 1.8.1.7), an enzyme present in the mitochondrial matrix [101]:

$$GSSG + NADPH + H^+ \rightarrow 2GSH + NADP^+ \tag{1.17}$$

In rat liver, Gpx1 activity is about 500-fold higher than that of Gpx4 when measured with H_2O_2 as a substrate [102]. The function of Gpx4 is probably the reduction of diacylated phospholipid hydroperoxides, which are not reduced by Gpx1 [103–106], while GPx1 besides removing H_2O_2 may also be responsible for the removal of lysophospholipid hydroperoxides [107,108].

PEROXIREDOXINS

Two peroxiredoxin isoforms are found in the mitochondria, Prx3 and Prx5; both are coded by two nuclear genes and their catalytic reaction mechanism involves two cysteine residues [109]. In the first step, H_2O_2 oxidizes one cysteine residue (–SH), the so-called peroxidatic cysteine (CysP), to a sulfenic acid (–SOH) (Equation 1.18), and in the second step, this intermediate reacts with a second cysteine residue, the resolving cysteine residue (CysR), to form a disulfide (Equation 1.19). These two cysteine residues are present in a single peroxiredoxin subunit, but while in typical two-cysteine peroxiredoxins like Prx3, the disulfide is formed between the two identical subunits forming a dimer, in Prx5, an atypical two-cysteine peroxiredoxin, an intramolecular disulfide is formed in the same unit.

$$H_2O_2 + Prx - CysP - SH \rightarrow Prx - CysP - SOH + H_2O \tag{1.18}$$

$$Prx - CysP - SOH + Prx - CysP - SH \rightarrow Prx - CysP - SS$$
$$- CysR - Prx + H_2O \tag{1.19}$$

Both mitochondrial peroxiredoxins are regenerated following reduction of the disulfide by thioredoxin 2 (Trx2) in a thiol exchange reaction forming a Trx2 intramolecular disulfide (Equation 1.20) [110,111]. The Trx disulfide is reduced back in a reaction catalyzed by thioredoxin 2 reductase (Trx2R, EC 1.8.1.9), a selenoprotein, using NADPH as the cosubstrate (Equation 1.21) [112]. Both Trx2 and its reductase are enzymes present in the mitochondrial matrix [113,114]. In addition, Prx3 is also reduced by the mitochondrial glutaredoxin 2, which may be reduced by glutathione or thioredoxin reductase [115].

$$Prx - CysP - SS - CysR - Prx + Trx2 - (Cys - SH)_2 \rightarrow Prx - CysP - SH$$
$$+ Prx - CysR - SH + Trx2 - (Cys - SS - Cys) \tag{1.20}$$

$$Trx2 - (Cys - SS - Cys) + NADPH + H^+ \rightarrow Trx2$$
$$- (Cys - SH)_2 + NADP^+ \tag{1.21}$$

Out of the two peroxiredoxins, Prx3 is mainly responsible for H_2O_2 elimination in the mitochondria because (1) the rate constant for the reaction of Prx3 with H_2O_2, 2×10^7 $M^{-1} s^{-1}$ [116] is two orders of magnitude higher than that of Prx5, $3 \times 10^5 M^{-1} s^{-1}$ [117] and (2) Prx3 is slightly more abundant in the mitochondria than Prx5 [109].

A distinctive characteristic of peroxiredoxin removal of H_2O_2 is that at large concentrations of H_2O_2, H_2O_2 may react with the sulfenic intermediate (Equation 1.22) before this intermediate is reduced back, inactivating the enzyme. Prx3 is more resistant than the cytosolic Prx2 to this inactivation [118,119] and Prx5 seems resistant to this inactivation [109]. This hyperoxidation facilitates the formation of decamers with chaperone activity. Peroxiredoxins play an important role as chaperones, but this alternative function has not been observed yet with mitochondrial peroxiredoxins [109]. Hyperoxidized Prx3 may be repaired by sulfiredoxin (EC 1.8.98.2) translocated from the cytosol at the expense of one ATP and GSH (Equation 1.23) [120]:

$$Prx - CysP - SOH + H_2O_2 \rightarrow Prx - CysP - SO_2H + H_2O \tag{1.22}$$

$$Prx - CysP - SO_2H + ATP + 2\ GSH \rightarrow Prx - CysP - SOH$$
$$+ ADP + PO_4^{3-} + GSSG \tag{1.23}$$

Relative Contribution of Glutathione Peroxidases and Peroxiredoxins to H_2O_2 Removal

In an excellent review, it was estimated that nearly 90% of mitochondrial H_2O_2 is removed by Prx3, with Gpx1 being responsible for approximately 9%, and the remaining 1% being accounted for other proteins [109]. This estimate of the order of magnitude was based on the reactivity of the reduced enzymes toward H_2O_2, taking also into account their abundance in the mitochondria. So, this estimate is a good approximation if most of the Gpx1 and Prx3 are in the reduced form, that is, if the antioxidant systems that regenerate the reduced enzymes are not rate limiting. Recently, the relative contributions of GPx/GSH and Prx/Trx were addressed experimentally in the heart mitochondria of the mouse, rat, and guinea pig, by measuring the increase in H_2O_2 emission from isolated mitochondria upon addition of auranofin, which inhibits Trx1/2, and dinitrochlorobenzene (DNCB), a GSH-depleting agent [121]. While in the mouse and rat heart mitochondria, addition of DNCB had a larger impact on H_2O_2 emission in either state 3 or state 4 respiration, thus suggesting that GPx1 is more important than Prx3 in eliminating H_2O_2; in guinea pig heart mitochondria, the opposite situation occurred. A mathematical model suggested that the two antioxidant systems act in concert, in which each system can partially replace the other and cooperate to eliminate H_2O_2 [121].

Next, we will estimate the relative contribution of GPx1 and Prx3, taking into account the whole kinetic cycle of both enzymes and focusing our analysis on liver mitochondria. For this, we will use the kinetics of GPx1 and Prx3. Kinetics for GPx1 and Prx3, as judged by *Plasmodium falciparum* peroxiredoxin [122], follows an ordered ping-pong mechanism as described by the following equation:

$$\frac{[E]}{v} = \phi_0 + \frac{\phi_1}{[H_2O_2]} + \frac{\phi_2}{[SH]} \tag{1.24}$$

where

[E] stands for the total enzyme concentration

v stands for the rate of reaction

[SH] is the concentration of the thiol responsible for enzyme regeneration

ϕ_0, ϕ_1, and ϕ_2 are the kinetic parameters characteristic of a ping-pong enzymatic mechanism (Dalziel coefficients)

Equation 1.24 can be rearranged into the more familiar Michaelis–Menten equation:

$$v = \frac{[SH]/(\phi_2 + \phi_0[SH])[E][H_2O_2]}{\phi_1[SH]/(\phi_2 + \phi_0[SH]) + [H_2O_2]} \tag{1.25}$$

For Prx3, $\phi_0 \neq 0$ and, consequently, a finite k_{cat} and K_M (for H_2O_2) are observed for a high thiol concentration (Equation 1.25). On the other hand, for GPx1, $\phi_0 = 0$ [123], this implies infinite k_{cat} and K_M (for H_2O_2), as can be seen from Equation 1.26, which was obtained from Equation 1.25 by letting $\phi_0 = 0$:

$$v = \frac{\left(\left[SH\right]/\phi_2\right)\left[E\right]\left[H_2O_2\right]}{\phi_1\left(\left[SH\right]/\phi_2\right)+\left[H_2O_2\right]} \tag{1.26}$$

Equations 1.25 and 1.26 are used to estimate the relative contribution of Gpx1 and Prx3. To achieve that, kinetic parameter values as well as enzyme and thiol concentrations are needed, while H_2O_2 will be let as an unknown variable. For rat liver mitochondria Gpx1, ϕ_1 and ϕ_2 have been estimated as 4.7×10^{-8} M s and 2.5×10^{-5} M s, respectively [124]. In rat liver mitochondria, the concentration of Gpx1 has been estimated at 10 μM (monomer concentration) [93] and GSH concentration is 10 mM [125]. This GSH concentration is in excess as a partial GSH depletion of 30%–40% is needed before an increase in mitochondrial H_2O_2 production can be observed [16].

Concerning Prx3, the only kinetic data available for the liver were obtained with saturating H_2O_2 concentrations. Under these conditions, Equation 1.25 simplifies to the following equation:

$$v = \frac{\left(\left[E\right]/\phi_0\right)\left[SH\right]}{\left(\phi_2/\phi_0\right)+\left[SH\right]} \tag{1.27}$$

Thus, ϕ_0 and ϕ_2 may be estimated from K_M (for Trx) and k_{cat} measured under H_2O_2 saturating conditions according to the following equations:

$$\phi_0 = \frac{1}{k_{cat}} \tag{1.28}$$

and

$$\phi_2 = \frac{K_M}{k_{cat}} \tag{1.29}$$

Kinetic parameters for Prx3 obtained by applying these equations under saturating H_2O_2 conditions are shown in Table 1.1. These values are close to those obtained experimentally for *P. falciparum* peroxiredoxin, for which ϕ_0 and ϕ_2 are 1.8 s and 1.85×10^{-5} M s, respectively [122]. Concerning ϕ_1, this parameter is the inverse of the rate constant measured between reduced Prx3 and H_2O_2, which is 2×10^7 M^{-1} s^{-1} [116]. One important information that can be taken from Table 1.1 is the estimation of the apparent K_M for H_2O_2. This value is lower than 10 μM because several investigators have observed that H_2O_2 concentrations of this magnitude saturate the enzyme. However, the value for the apparent K_M for H_2O_2 was never measured, because measuring the low reaction rate at low H_2O_2 concentrations has been challenging

TABLE 1.1

Kinetic Parameters for Purified Prx3 Obtained with Saturating H_2O_2 Concentrations in the Presence of Various Electron Donors

Electron Donor	Mitochondrial Trx (Trx2)		Cytosolic Trx (Trx1)		Glutaredoxin 2 (Grx2)
V_{max} (µmol min^{-1}mg protein^{-1})	1.1	—	2.4	13.3	1.2
k_{cat} (s^{-1})	0.39[a]	0.8–1.2	0.86[a]	4.8[a]	0.43[a]
K_M for Trx/Grx (µM)	11.2	—	4.0	4.3	23.8
ϕ_0 (s)	2.5	0.8–1.2	1.2	0.2	2.3
ϕ_1 (M s)	5.0×10^{-8}	5.0×10^{-8}	5.0×10^{-8}	5.0×10^{-8}	5.0×10^{-8}
ϕ_2 (M s)	2.8×10^{-5}	—	4.6×10^{-6}	9.0×10^{-7}	5.5×10^{-5}
K_M for H_2O_2 (µM)[b]	0.02	0.04–0.06	0.04	0.24	0.02
References	[115]	[126]	[115]	[111]	[115]

[a] Calculated from V_{max} assuming a molecular mass for Prx3 of 21.5 kDa.

[b] Estimated assuming a rate constant for the reaction between reduced Prx3 and H_2O_2 of 2×10^7 M^{-1} s^{-1}.

[109,111,126]. If the enzyme is assumed to be saturated with Trx2, then the K_M for H_2O_2 may be estimated as ϕ_1/ϕ_0 (Equation 1.25). Values obtained are in the range 0.02–0.24 µM (Table 1.1), which is between one and two orders of magnitude lower than the value observed for bacterial peroxiredoxin AhpC ($K_M = 1.4$ µM) [127].

The last parameters we need to estimate are Prx3 and Trx2 concentrations in liver mitochondria. Prx3 concentration in rat liver mitochondria can be estimated as 160 µM based on a level of 0.7 µg of Prx3/mg of rat liver protein [111]. Trx2 concentration has been estimated to be 10 µM in the mitochondria from bovine adrenal cortex [109,128]. The adrenal gland is extremely rich in Prx3 [111], raising the question whether the levels of its partner Trx2 are also increased compared with liver mitochondria. Taking into account the doubts concerning the concentration of reduced Trx2 in liver mitochondria, calculation with two values, 1 and 10 µM, was performed.

Using these estimations, the relative contribution of GPx1 and PRx3 was calculated (Figure 1.4). For low H_2O_2 concentrations, Prx3 removes most of the H_2O_2, while for larger concentrations, Gpx1 predominates. Table 1.2 summarizes the main data that may be read from Figure 1.4 in the range of endogenous H_2O_2 production. In the lower range of H_2O_2 production (0.2 nmol min^{-1} mg^{-1} of protein), H_2O_2 concentration is expected to be around 1–2 nM with Prx3 being the main contributor for its removal, around 90%, as predicted in [109]. At this condition, both Gpx1 and Prx3 are almost fully reduced, and removal of H_2O_2 may be predicted based simply on the reaction rate constants between the reduced enzymes and H_2O_2. In the high H_2O_2 production range, here assumed to be 2 nmol min^{-1} mg^{-1} of protein [121], the mitochondrial H_2O_2 concentration and the removal pathways of H_2O_2 are strongly dependent on the levels of reduced Trx2. For low Trx2 (1 µM), a high steady-state H_2O_2 is predicted (130 nM), Prx3 saturates, and the contribution of GPx1 becomes fundamental, with 85% of H_2O_2 being removed via Gpx1. This estimate of 130 nM

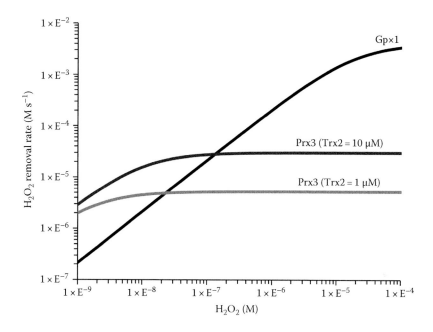

FIGURE 1.4 Removal of mitochondrial H_2O_2 by Prx3 and Gpx1. Equations 1.25 and 1.26 are used, respectively, for Prx3 and Gpx1. Parameters used for Gpx1 were $\phi_1 = 4.7 \times 10^{-8}$ M s, $\phi_2 = 2.5 \times 10^{-5}$ M s, [Gpx1] = 10 μM and [GSH] = 10 mM (black curve). For Prx3, $\phi_0 = 2.5$ s, $\phi_1 = 5.0 \times 10^{-8}$ M s, $\phi_2 = 2.8 \times 10^{-5}$ M s, [Prx3] = 160 μM, [Trx1] = 10 μM (dark gray curve), or [Trx1] = 1 μM (light gray curve).

TABLE 1.2
Estimates for H_2O_2 Concentrations and Relative Contributions for H_2O_2 Removal by Prx3 and Gpx1

H_2O_2 Production (nmol min⁻¹ mg⁻¹ of protein)	0.2		2	
Reduced Trx (μM)	1	10	1	10
[H_2O_2] (nM)	2	1	130	40
H_2O_2 *removal via* Prx3	87%	93%	15%	74%
Gpx1	13%	7%	85%	26%

Note: Data are taken from Figure 1.4. H_2O_2 production is assumed to be in the range 0.2–2 nmol min⁻¹ mg⁻¹ of protein, which converts to 3.3–33 μM s⁻¹.

is strongly dependent on the parameters used, with small changes having a significant impact on the predicted H_2O_2 steady state. If reduced Trx2 is assumed to be 10 μM, then H_2O_2 is predicted to be 40 nM, with around 75% of H_2O_2 being removed by Prx3. This dependency on the levels of Trx2 has been observed experimentally, as overexpression of Trx2 improves protection against mitochondrial generation of

H_2O_2 [129]. In other words, the levels of Trx2 are not in the saturating range but near its K_M or lower. This analysis is consistent with experimental observations that indicate that both GPx1 and Prx3 are important for H_2O_2 removal [121,130]. Tissue and species origin of mitochondria are certainly important factors when considering H_2O_2 catabolism.

PYRUVATE

In addition to enzymatic antioxidant systems, pyruvate has also been pointed out as having a possible antioxidant role by reacting with H_2O_2. 2-Oxoacids, like pyruvate, undergo a decarboxylation in the presence of H_2O_2, with H_2O_2 being reduced to water, a reaction first described more than a century ago:

$$CH_3COCOO^- + H_2O_2 \rightarrow CH_3COO^- + CO_2 + H_2O \qquad (1.30)$$

This reaction is relevant and several observations support a role for pyruvate as an H_2O_2 sink [131–134]. Interestingly, cultured cells export pyruvate to reach levels similar to those observed in human serum (60–150 µM), protecting themselves from added H_2O_2 [133]. Altogether these observations show that pyruvate can have a protective role against H_2O_2. In mitochondria, pyruvate is present in the millimolar range (1.5 mM), and so does pyruvate work as a sink for H_2O_2, competing with enzymatic antioxidant systems. The rate constant between pyruvate and H_2O_2 depends on the pH and ionic force, and a value of 2.2 M^{-1} s^{-1} was determined at pH 7.4 in 0.1 M phosphate buffer containing 0.1 mM DTPA at 37°C [135]. A similar value of 2.4 M^{-1} s^{-1} may be estimated from Figure 1.3 in Desagher et al. [134] at pH 7.4 in Krebs' bicarbonate buffer at 37°C. Thus, a pseudo-order reaction rate constant of 3.6×10^{-3} s^{-1} may be estimated, indicating that pyruvate does not compete with the enzymatic antioxidant systems in the mitochondria. This estimate and the experimental observations showing that pyruvate is effective in decreasing H_2O_2 concentrations when added in cell culture media or perfusion fluids [136], protecting the mitochondria from added H_2O_2 to cell media [137], are not incompatible. When pyruvate is present in the external media, the overall capacity of pyruvate to react with H_2O_2 is large because the ratio between the external media and the cellular volume is very large, and so the amount of pyruvate able to react with H_2O_2 is large, even if the rate constant with H_2O_2 is slow. For example, Desagher et al. [134] observed that only half of the 200 µM H_2O_2 initially present in the cellular growth medium remained after 2 min in the presence of 2 mM pyruvate. However, pyruvate cannot compete with antioxidant systems when present in the same compartment such as the mitochondria.

CONCLUSION

Nearly half a century has passed since the initial discovery of H_2O_2 production in mitochondria. However, a number of uncertainties still remain concerning both the mechanisms of the formation and removal of O_2^-/H_2O_2 in the mitochondria.

Some open questions are as follows: (1) To what extent is $O_2^{\cdot-}/H_2O_2$ formation an unavoidable consequence of O_2 reduction to H_2O in the respiratory chain or, alternatively, is this formation under the regulation with the formation rates of $O_2^{\cdot-}/H_2O_2$ having a relevant physiological role in cellular processes? (2) To what extent is the knowledge obtained from isolated mitochondria, respiratory complexes, or other isolated H_2O_2-forming enzymes a good picture of the in vivo situation, where the mitochondria are part of a dynamic network, at both functional and morphological levels, interacting with other cellular components? (3) To what extent do peroxiredoxins and glutathione peroxidases cooperate in the removal of H_2O_2?

ACKNOWLEDGMENTS

Supported by Fundação para a Ciência e a Tecnologia (FCT), Portugal (UID/MULTI/00612/2013).

REFERENCES

1. P. K. Jensen (1966) Antimycin-insensitive oxidation of succinate and reduced nicotinamide-adenine dinucleotide in electron-transport particles. I. pH dependency and hydrogen peroxide formation. *Biochim. Biophys. Acta*, 122, 157–166.
2. G. Loschen, L. Flohe, and B. Chance (1971) Respiratory chain linked H_2O_2 production in pigeon heart mitochondria. *FEBS Lett.*, 18, 261–264.
3. A. Boveris and A. Chance (1973) The mitochondrial generation of hydrogen peroxide. *Biochem. J.*, 134, 707–716.
4. A. Boveris, N. Oshino, and B. Chance (1972) The cellular production of hydrogen peroxide. *Biochem. J.*, 128, 617–630.
5. B. Chance, H. Sies, and A. Boveris (1979) Hydroperoxide metabolism in mammalian organs. *Physiol. Rev.*, 59, 527–605.
6. G. Loschen, A. Azzi, C. Richter, and L. Flohe (1974) Superoxide radicals as precursors of mitochondrial hydrogen peroxide. *FEBS Lett.*, 42, 68–72.
7. H. J. Forman and J. A. Kennedy (1974) Role of superoxide radical in mitochondrial dehydrogenase reactions. *Biochem. Biophys. Res. Commun.*, 60, 1044–1050.
8. M. P. Murphy (2009) How mitochondria produce reactive oxygen species. *Biochem. J.*, 417, 1–13.
9. M. D. Brand (2010) The sites and topology of mitochondrial superoxide production. *Exp. Gerontol.*, 45, 466–472.
10. A. Y. Andreyev, Y. E. Kushnareva, and A. A. Starkov (2005) Mitochondrial metabolism of reactive oxygen species. *Biochemistry (Mosc.)*, 70, 200–214.
11. V. G. Grivennikova and A. D. Vinogradov (2013) Partitioning of superoxide and hydrogen peroxide production by mitochondrial respiratory complex I. *Biochim. Biophys. Acta*, 1827, 446–454.
12. V. G. Grivennikova and A. D. Vinogradov (2013) Mitochondrial production of reactive oxygen species. *Biochemistry (Mosc.)*, 78, 1490–1511.
13. M. D. Brand, A. L. Orr, I. V. Perevoshchikova, and C. L. Quinlan (2013) The role of mitochondrial function and cellular bioenergetics in ageing and disease. *Br. J. Dermatol.*, 169 (Suppl 2), 1–8.
14. J. St Pierre, J. A. Buckingham, S. J. Roebuck, and M. D. Brand (2002) Topology of superoxide production from different sites in the mitochondrial electron transport chain. *J. Biol. Chem.*, 277, 44784–44790.

15. R. G. Hansford, B. A. Hogue, and V. Mildaziene (1997) Dependence of H_2O_2 formation by rat heart mitochondria on substrate availability and donor age. *J. Bioenerg. Biomembr.*, 29, 89–95.

16. D. Han, R. Canali, D. Rettori, and N. Kaplowitz (2003) Effect of glutathione depletion on sites and topology of superoxide and hydrogen peroxide production in mitochondria. *Mol. Pharmacol.*, 64, 1136–1144.

17. A. J. Lambert and M. D. Brand (2004) Inhibitors of the quinone-binding site allow rapid superoxide production from mitochondrial NADH:ubiquinone oxidoreductase (complex I). *J. Biol. Chem.*, 279, 39414–39420.

18. J. Hirst, M. S. King, and K. R. Pryde (2008) The production of reactive oxygen species by complex I. *Biochem. Soc. Trans.*, 36, 976–980.

19. C. L. Quinlan, I. V. Perevoshchikova, M. Hey-Mogensen, A. L. Orr, and M. D. Brand (2013) Sites of reactive oxygen species generation by mitochondria oxidizing different substrates. *Redox Biol.*, 1, 304–312.

20. J. R. Treberg, C. L. Quinlan, and M. D. Brand (2010) Hydrogen peroxide efflux from muscle mitochondria underestimates matrix superoxide production--a correction using glutathione depletion. *FEBS J.*, 277, 2766–2778.

21. J. R. Treberg, C. L. Quinlan, and M. D. Brand (2011) Evidence for two sites of superoxide production by mitochondrial NADH-ubiquinone oxidoreductase (complex I). *J. Biol. Chem.*, 286, 27103–27110.

22. C. L. Quinlan, A. L. Orr, I. V. Perevoshchikova, J. R. Treberg, B. A. Ackrell, and M. D. Brand (2012) Mitochondrial complex II can generate reactive oxygen species at high rates in both the forward and reverse reactions. *J. Biol. Chem.*, 287, 27255–27264.

23. C. L. Quinlan, A. A. Gerencser, J. R. Treberg, and M. D. Brand (2011) The mechanism of superoxide production by the antimycin-inhibited mitochondrial Q-cycle. *J. Biol. Chem.*, 286, 31361–31372.

24. A. L. Orr, C. L. Quinlan, I. V. Perevoshchikova, and M. D. Brand (2012) A refined analysis of superoxide production by mitochondrial sn-glycerol 3-phosphate dehydrogenase. *J. Biol. Chem.*, 287, 42921–42935.

25. M. Hey-Mogensen, R. L. Goncalves, A. L. Orr, and M. D. Brand (2014) Production of superoxide/H_2O_2 by dihydroorotate dehydrogenase in rat skeletal muscle mitochondria. *Free Radic. Biol. Med.*, 72, 149–155.

26. I. V. Perevoshchikova, C. L. Quinlan, A. L. Orr, A. A. Gerencser, and M. D. Brand (2013) Sites of superoxide and hydrogen peroxide production during fatty acid oxidation in rat skeletal muscle mitochondria. *Free Radic. Biol. Med.*, 61C, 298–309.

27. E. B. Tahara, F. D. Navarete, and A. J. Kowaltowski (2009) Tissue-, substrate-, and site-specific characteristics of mitochondrial reactive oxygen species generation. *Free Radic. Biol. Med.*, 46, 1283–1297.

28. J. F. Turrens (2003) Mitochondrial formation of reactive oxygen species. *J. Physiol.*, 552, 335–344.

29. T. R. Figueira, M. H. Barros, A. A. Camargo, R. F. Castilho, J. C. Ferreira, A. J. Kowaltowski, F. E. Sluse, N. C. Souza-Pinto, and A. E. Vercesi (2013) Mitochondria as a source of reactive oxygen and nitrogen species: From molecular mechanisms to human health. *Antioxid. Redox Signal*, 18, 2029–2074.

30. C. L. Quinlan, I. V. Perevoschikova, R. L. Goncalves, M. Hey-Mogensen, and M. D. Brand (2013) The determination and analysis of site-specific rates of mitochondrial reactive oxygen species production. *Methods Enzymol.*, 526, 189–217.

31. J. Hirst (2005) Energy transduction by respiratory complex I—An evaluation of current knowledge. *Biochem. Soc. Trans.*, 33, 525–529.

32. J. Hirst (2013) Mitochondrial complex I. *Annu. Rev. Biochem.*, 82, 551–575.

33. D. Esterhazy, M. S. King, G. Yakovlev, and J. Hirst (2008) Production of reactive oxygen species by complex I (NADH:ubiquinone oxidoreductase) from *Escherichia coli* and comparison to the enzyme from mitochondria. *Biochemistry*, 47, 3964–3971.

34. L. Kussmaul and J. Hirst (2006) The mechanism of superoxide production by NADH:ubiquinone oxidoreductase (complex I) from bovine heart mitochondria. *Proc. Natl. Acad. Sci. USA*, 103, 7607–7612.

35. G. Krishnamoorthy and P. C. Hinkle (1988) Studies on the electron transfer pathway, topography of iron–sulfur centers, and site of coupling in NADH-Q oxidoreductase. *J. Biol. Chem.*, 263, 17566–17575.

36. R. R. Ramsay and T. P. Singer (1992) Relation of superoxide generation and lipid peroxidation to the inhibition of NADH-Q oxidoreductase by rotenone, piericidin A, and MPP+. *Biochem. Biophys. Res. Commun.*, 189, 47–52.

37. A. Herrero and G. Barja (2000) Localization of the site of oxygen radical generation inside the complex I of heart and nonsynaptic brain mammalian mitochondria. *J. Bioenerg. Biomembr.*, 32, 609–615.

38. Y. Kushnareva, A. N. Murphy, and A. Andreyev (2002) Complex I-mediated reactive oxygen species generation: Modulation by cytochrome c and NAD(P)+ oxidation-reduction state. *Biochem. J.*, 368, 545–553.

39. V. G. Grivennikova, A. V. Kareyeva, and A. D. Vinogradov (2010) What are the sources of hydrogen peroxide production by heart mitochondria? *Biochim. Biophys. Acta*, 1797, 939–944.

40. A. P. Kudin, N. Y. Bimpong-Buta, S. Vielhaber, C. E. Elger, and W. S. Kunz (2004) Characterization of superoxide-producing sites in isolated brain mitochondria. *J. Biol. Chem.*, 279, 4127–4135.

41. K. R. Pryde and J. Hirst (2011) Superoxide is produced by the reduced flavin in mitochondrial complex I: A single, unified mechanism that applies during both forward and reverse electron transfer. *J. Biol. Chem.*, 286, 18056–18065.

42. M. L. Genova, B. Ventura, G. Giuliano, C. Bovina, G. Formiggini, C. G. Parenti, and G. Lenaz (2001) The site of production of superoxide radical in mitochondrial complex I is not a bound ubisemiquinone but presumably iron-sulfur cluster N2. *FEBS Lett.*, 505, 364–368.

43. S. T. Ohnishi, T. Ohnishi, S. Muranaka, H. Fujita, H. Kimura, K. Uemura, K. Yoshida, and K. Utsumi (2005) A possible site of superoxide generation in the complex I segment of rat heart mitochondria. *J. Bioenerg. Biomembr.*, 37, 1–15.

44. F. Sun, X. Huo, Y. Zhai, A. Wang, J. Xu, D. Su, M. Bartlam, and Z. Rao (2005) Crystal structure of mitochondrial respiratory membrane protein complex II. *Cell*, 121, 1043–1057.

45. P. Rustin, A. Munnich, and A. Rotig (2002) Succinate dehydrogenase and human diseases: New insights into a well-known enzyme. *Eur. J. Hum. Genet.*, 10, 289–291.

46. R. Moreno-Sanchez, L. Hernandez-Esquivel, N. A. Rivero-Segura, A. Marin-Hernandez, J. Neuzil, S. J. Ralph, and S. Rodriguez-Enriquez (2013) Reactive oxygen species are generated by the respiratory complex II--evidence for lack of contribution of the reverse electron flow in complex I. *FEBS J.*, 280, 927–938.

47. I. Siebels and S. Drose (2013) Q-site inhibitor induced ROS production of mitochondrial complex II is attenuated by TCA cycle dicarboxylates. *Biochim. Biophys. Acta*, 1827, 1156–1164.

48. P. Mitchell (1975) The protonmotive Q cycle: A general formulation. *FEBS Lett.*, 59, 137–139.

49. L. Bleier and S. Drose (2013) Superoxide generation by complex III: From mechanistic rationales to functional consequences. *Biochim. Biophys. Acta*, 1827, 1320–1331.

50. A. Herrero and G. Barja (1997) Sites and mechanisms responsible for the low rate of free radical production of heart mitochondria in the long-lived pigeon. *Mech. Ageing Dev.*, 98, 95–111.

51. Q. Chen, E. J. Vazquez, S. Moghaddas, C. L. Hoppel, and E. J. Lesnefsky (2003) Production of reactive oxygen species by mitochondria: Central role of complex III. *J. Biol. Chem.*, 278, 36027–36031.

52. M. Ksenzenko, A. A. Konstantinov, G. B. Khomutov, A. N. Tikhonov, and E. K. Ruuge (1983) Effect of electron transfer inhibitors on superoxide generation in the cytochrome bc1 site of the mitochondrial respiratory chain. *FEBS Lett.*, 155, 19–24.

53. J. F. Turrens, A. Alexandre, and A. L. Lehninger (1985) Ubisemiquinone is the electron donor for superoxide formation by complex III of heart mitochondria. *Arch. Biochem. Biophys.*, 237, 408–414.

54. F. L. Muller, Y. Liu, and H. Van Remmen (2004) Complex III releases superoxide to both sides of the inner mitochondrial membrane. *J. Biol. Chem.*, 279, 49064–49073.

55. M.-A. Takahashi and K. Asada (1983) Superoxide anion permeability of phospholipid membranes and chloroplast thylakoids. *Arch. Biochem. Biophys.*, 226, 558–566.

56. M. Klingenberg (1970) Localization of the glycerol-phosphate dehydrogenase in the outer phase of the mitochondrial inner membrane. *Eur. J. Biochem.*, 13, 247–252.

57. T. Mracek, E. Holzerova, Z. Drahota, N. Kovarova, M. Vrbacky, P. Jesina, and J. Houstek (2014) ROS generation and multiple forms of mammalian mitochondrial glycerol-3-phosphate dehydrogenase. *Biochim. Biophys. Acta*, 1837, 98–111.

58. Z. Drahota, S. K. Chowdhury, D. Floryk, T. Mracek, J. Wilhelm, H. Rauchova, G. Lenaz, and J. Houstek (2002) Glycerophosphate-dependent hydrogen peroxide production by brown adipose tissue mitochondria and its activation by ferricyanide. *J. Bioenerg. Biomembr.*, 34, 105–113.

59. S. Miwa and M. D. Brand (2005) The topology of superoxide production by complex III and glycerol 3-phosphate dehydrogenase in Drosophila mitochondria. *Biochim. Biophys. Acta*, 1709, 214–219.

60. N. J. Watmough and F. E. Frerman (2010) The electron transfer flavoprotein: Ubiquinone oxidoreductases. *Biochim. Biophys. Acta*, 1797, 1910–1916.

61. E. L. Seifert, C. Estey, J. Y. Xuan, and M. E. Harper (2010) Electron transport chain-dependent and -independent mechanisms of mitochondrial H_2O_2 emission during long-chain fatty acid oxidation. *J. Biol. Chem.*, 285, 5748–5758.

62. D. L. Hoffman and P. S. Brookes (2009) Oxygen sensitivity of mitochondrial reactive oxygen species generation depends on metabolic conditions. *J. Biol. Chem.*, 284, 16236–16245.

63. M. Loffler, C. Becker, E. Wegerle, and G. Schuster (1996) Catalytic enzyme histochemistry and biochemical analysis of dihydroorotate dehydrogenase/oxidase and succinate dehydrogenase in mammalian tissues, cells and mitochondria. *Histochem. Cell Biol.*, 105, 119–128.

64. J. Rawls, W. Knecht, K. Diekert, R. Lill, and M. Loffler (2000) Requirements for the mitochondrial import and localization of dihydroorotate dehydrogenase. *Eur. J. Biochem.*, 267, 2079–2087.

65. S. Liu, E. A. Neidhardt, T. H. Grossman, T. Ocain, and J. Clardy (2000) Structures of human dihydroorotate dehydrogenase in complex with antiproliferative agents. *Structure*, 8, 25–33.

66. O. Bjornberg, A. C. Gruner, P. Roepstorff, and K. F. Jensen (1999) The activity of *Escherichia coli* dihydroorotate dehydrogenase is dependent on a conserved loop identified by sequence homology, mutagenesis, and limited proteolysis. *Biochemistry*, 38, 2899–2908.

67. R. L. Fagan and B. A. Palfey (2009) Roles in binding and chemistry for conserved active site residues in the class 2 dihydroorotate dehydrogenase from *Escherichia coli*. *Biochemistry*, 48, 7169–7178.

68. H. J. Forman and J. Kennedy (1975) Superoxide production and electron transport in mitochondrial oxidation of dihydroorotic acid. *J. Biol. Chem.*, 250, 4322–4326.

69. H. J. Forman and J. Kennedy (1976) Dihydroorotate-dependent superoxide production in rat brain and liver. A function of the primary dehydrogenase. *Arch. Biochem. Biophys.*, 173, 219–224.

70. K. N. Dileepan and J. Kennedy (1985) Complete inhibition of dihydro-orotate oxidation and superoxide production by 1,1,1-trifluoro-3-thenoylacetone in rat liver mitochondria. *Biochem. J.*, 225, 189–194.

71. N. Borgese and J. Meldolesi (1980) Localization and biosynthesis of NADH-cytochrome b5 reductase, an integral membrane protein, in rat liver cells. I. Distribution of the enzyme activity in microsomes, mitochondria, and golgi complex. *J. Cell Biol.*, 85, 501–515.

72. G. L. Sottocasa, B. Kuylenstierna, L. Ernster, and A. Bergstrand (1967) An electron-transport system associated with the outer membrane of liver mitochondria. A biochemical and morphological study. *J. Cell Biol.*, 32, 415–438.

73. S. A. Whatley, D. Curti, G. F. Das, I. N. Ferrier, S. Jones, C. Taylor, and R. M. Marchbanks (1998) Superoxide, neuroleptics and the ubiquinone and cytochrome b5 reductases in brain and lymphocytes from normals and schizophrenic patients. *Mol. Psychiatry*, 3, 227–237.

74. A. K. Samhan-Arias and C. Gutierrez-Merino (2014) Purified NADH-cytochrome b5 reductase is a novel superoxide anion source inhibited by apocynin: Sensitivity to nitric oxide and peroxynitrite. *Free Radic. Biol. Med.*, 73, 174–189.

75. N. Hauptmann, J. Grimsby, J. C. Shih, and E. Cadenas (1996) The metabolism of tyramine by monoamine oxidase A/B causes oxidative damage to mitochondrial DNA. *Arch. Biochem. Biophys.*, 335, 295–304.

76. E. Cadenas and K. J. Davies (2000) Mitochondrial free radical generation, oxidative stress, and aging. *Free Radic. Biol. Med.*, 29, 222–230.

77. C. Giulivi, A. Boveris, and E. Cadenas (1995) Hydroxyl radical generation during mitochondrial electron transfer and the formation of 8-hydroxydesoxyguanosine in mitochondrial DNA. *Arch. Biochem. Biophys.*, 316, 909–916.

78. F. Antunes, D. Han, D. Rettori, and E. Cadenas (2002) Mitochondrial damage by nitric oxide is potentiated by dopamine in PC12 cells. *Biochim. Biophys. Acta*, 1556, 233 238.

79. O. R. Kunduzova, P. Bianchi, A. Parini, and C. Cambon (2002) Hydrogen peroxide production by monoamine oxidase during ischemia/reperfusion. *Eur. J. Pharmacol.*, 448, 225–230.

80. A. Maurel, C. Hernandez, O. Kunduzova, G. Bompart, C. Cambon, A. Parini, and B. Frances (2003) Age-dependent increase in hydrogen peroxide production by cardiac monoamine oxidase A in rats. *Am. J. Physiol Heart Circ. Physiol.*, 284, H1460–H1467.

81. K. Bedard and K. H. Krause (2007) The NOX family of ROS-generating NADPH oxidases: Physiology and pathophysiology. *Physiol. Rev.*, 87, 245–313.

82. K. A. Graham, M. Kulawiec, K. M. Owens, X. Li, M. M. Desouki, D. Chandra, and K. K. Singh (2010) NADPH oxidase 4 is an oncoprotein localized to mitochondria. *Cancer Biol. Ther.*, 10, 223–231.

83. J. Kuroda, T. Ago, S. Matsushima, P. Zhai, M. D. Schneider, and J. Sadoshima (2010) NADPH oxidase 4 (NOX4) is a major source of oxidative stress in the failing heart. *Proc. Natl. Acad. Sci. USA*, 107, 15565–15570.

84. K. Block, Y. Gorin, and H. E. Abboud (2009) Subcellular localization of NOX4 and regulation in diabetes. *Proc. Natl. Acad. Sci. USA*, 106, 14385–14390.

85. A. J. Case, S. Li, U. Basu, J. Tian, and M. C. Zimmerman (2013) Mitochondrial-localized NADPH oxidase 4 is a source of superoxide in angiotensin II-stimulated neurons. *Am. J. Physiol Heart Circ. Physiol.*, 305, H19–H28.

86. Y. Maejima, J. Kuroda, S. Matsushima, T. Ago, and J. Sadoshima (2011) Regulation of myocardial growth and death by NADPH oxidase. *J. Mol. Cell. Cardiol.*, 50, 408–416.

87. I. Takac, K. Schroder, L. Zhang, B. Lardy, N. Anilkumar, J. D. Lambeth, A. M. Shah, F. Morel, and R. P. Brandes (2011) The E-loop is involved in hydrogen peroxide formation by the NADPH oxidase NOX4. *J. Biol. Chem.*, 286, 13304–13313.

88. R. Koziel, H. Pircher, M. Kratochwil, B. Lener, M. Hermann, N. A. Dencher, and P. Jansen-Durr (2013) Mitochondrial respiratory chain complex I is inactivated by NADPH oxidase NOX4. *Biochem. J.*, 452, 231–239.

89. A. A. Starkov, G. Fiskum, C. Chinopoulos, B. J. Lorenzo, S. E. Browne, M. S. Patel, and M. F. Beal (2004) Mitochondrial α-ketoglutarate dehydrogenase complex generates reactive oxygen species. *J. Neurosci.*, 24, 7779–7788.

90. V. I. Bunik and C. Sievers (2002) Inactivation of the 2-oxo acid dehydrogenase complexes upon generation of intrinsic radical species. *Eur. J. Biochem.*, 269, 5004–5015.

91. L. Tretter and V. Adam-Vizi (2004) Generation of reactive oxygen species in the reaction catalyzed by α-ketoglutarate dehydrogenase. *J. Neurosci.*, 24, 7771–7778.

92. C. L. Quinlan, R. L. Goncalves, M. Hey-Mogensen, N. Yadava, V. I. Bunik, and M. D. Brand (2014) The 2-oxoacid dehydrogenase complexes in mitochondria can produce superoxide/hydrogen peroxide at much higher rates than complex I. *J. Biol. Chem.*, 289, 8312–8325.

93. F. Antunes, A. Salvador, H. S. Marinho, R. Alves, and R. E. Pinto (1996) Lipid peroxidation in mitochondrial inner membranes. I. An integrative kinetic model. *Free Radic. Biol. Med.*, 21, 917–943.

94. R. A. Weisiger and I. Fridovich (1973) Superoxide dismutase—organelle specificity. *J. Biol. Chem.*, 218, 3582–3592.

95. A. Okado-Matsumoto and I. Fridovich (2001) Subcellular distribution of superoxide dismutases (SOD) in rat liver: Cu,Zn-SOD in mitochondria. *J. Biol. Chem.*, 276, 38388–38393.

96. P. R. Gardner (1997) Superoxide-driven aconitase Fe-S center cycling. *Biosci. Rep.*, 17, 33–42.

97. R. Radi, M. Rodriguez, L. Castro, and R. Telleri (1994) Inhibition of mitochondrial electron transport by peroxynitrite. *Arch. Biochem. Biophys.*, 308, 89–95.

98. J. Aikens and T. A. Dix (1991) Perhydroxyl radical (HOO·) initiated lipid peroxidation. *J. Biol. Chem.*, 266, 15091–15098.

99. C. C. Winterbourn and M. B. Hampton (2008) Thiol chemistry and specificity in redox signaling. *Free Radic. Biol. Med.*, 45, 549–561.

100. R. Brigelius-Flohé (1999) Tissue-specific functions of individual glutathione peroxidases. *Free Radic. Biol. Med.*, 27, 951–965.

101. K. Kurosawa, H. Shibata, N. Hayashi, N. Sato, T. Kamada, and K. Tagawa (1990) Kinetics of hydroperoxide degradation by NADP-glutathione system in mitochondria. *J. Biochem.*, 108, 9–16.

102. L. Zhang, M. Maiorino, A. Roveri, and F. Ursini (1989) Phospholipid hydroperoxide glutathione peroxidase: Specific activity in tissues of rats of different age and comparison with other glutathione peroxidases. *Biochim. Biophys. Acta*, 1006, 140–143.

103. F. Antunes, A. Salvador, and R. E. Pinto (1995) PHGPx and phospholipase A2/GPx: Comparative importance on the reduction of hydroperoxides in rat liver mitochondria. *Free Radic. Biol. Med.*, 19, 669–677.

104. L. Zhao, H. P. Wang, H. J. Zhang, C. J. Weydert, F. E. Domann, L. W. Oberley, and G. R. Buettner (2003) L-PhGPx expression can be suppressed by antisense oligodeoxynucleotides. *Arch. Biochem. Biophys.*, 417, 212–218.

105. H. Liang, Q. Ran, Y. C. Jang, D. Holstein, J. Lechleiter, T. McDonald-Marsh, A. Musatov, W. Song, R. H. Van, and A. Richardson (2009) Glutathione peroxidase 4 differentially regulates the release of apoptogenic proteins from mitochondria. *Free Radic. Biol. Med.*, 47, 312–320.

106. K. Koulajian, A. Ivovic, K. Ye, T. Desai, A. Shah, I. G. Fantus, Q. Ran, and A. Giacca (2013) Overexpression of glutathione peroxidase 4 prevents β-cell dysfunction induced by prolonged elevation of lipids in vivo. *Am. J. Physiol. Endocrinol. Metab.*, 305, E254–E262.

107. H. S. Marinho, F. Antunes, and R. E. Pinto (1997) Role of glutathione peroxidase and phospholipid hydroperoxide glutathione peroxidase in the reduction of lysophospholipid hydroperoxides. *Free Radic. Biol. Med.*, 22, 871–883.

108. L. S. Huang, M. R. Kim, and D.-E. Sok (2009) Enzymatic reduction of polyunsaturated lysophosphati-dylcholine hydroperoxides by glutathione peroxidase-1. *Eur. J. Lipid Sci. Technol.*, 111, 584–592.

109. A. G. Cox, C. C. Winterbourn, and M. B. Hampton (2010) Mitochondrial peroxiredoxin involvement in antioxidant defence and redox signalling. *Biochem. J.*, 425, 313–325.

110. M. S. Seo, S. W. Kang, K. Kim, I. C. Baines, T. H. Lee, and S. G. Rhee (2000) Identification of a new type of mammalian peroxiredoxin that forms an intramolecular disulfide as a reaction intermediate. *J. Biol. Chem.*, 275, 20346–20354.

111. H. Z. Chae, H. J. Kim, S. W. Kang, and S. G. Rhee (1999) Characterization of three isoforms of mammalian peroxiredoxin that reduce peroxides in the presence of thioredoxin. *Diabetes Res Clin. Pract.*, 45, 101–112.

112. M. Luthman and A. Holmgren (1982) Rat liver thioredoxin and thioredoxin reductase: Purification and characterization. *Biochemistry*, 21, 6628–6633.

113. M. He, J. Cai, Y. M. Go, J. M. Johnson, W. D. Martin, J. M. Hansen, and D. P. Jones (2008) Identification of thioredoxin-2 as a regulator of the mitochondrial permeability transition. *Toxicol. Sci.*, 105, 44–50.

114. M. P. Rigobello, A. Donella-Deana, L. Cesaro, and A. Bindoli (2001) Distribution of protein disulphide isomerase in rat liver mitochondria. *Biochem. J.*, 356, 567–570.

115. E. M. Hanschmann, M. E. Lonn, L. D. Schutte, M. Funke, J. R. Godoy, S. Eitner, C. Hudemann, and C. H. Lillig (2010) Both thioredoxin 2 and glutaredoxin 2 contribute to the reduction of the mitochondrial 2-Cys peroxiredoxin Prx3. *J. Biol. Chem.*, 285, 40699–40705.

116. A. G. Cox, A. V. Peskin, L. N. Paton, C. C. Winterbourn, and M. B. Hampton (2009) Redox potential and peroxide reactivity of human peroxiredoxin 3. *Biochemistry*, 48, 6495–6501.

117. M. Trujillo, A. Clippe, B. Manta, G. Ferrer-Sueta, A. Smeets, J. P. Declercq, B. Knoops, and R. Radi (2007) Pre-steady state kinetic characterization of human peroxiredoxin 5: Taking advantage of Trp84 fluorescence increase upon oxidation. *Arch. Biochem. Biophys.*, 467, 95–106.

118. A. V. Peskin, N. Dickerhof, R. A. Poynton, L. N. Paton, P. E. Pace, M. B. Hampton, and C. C. Winterbourn (2013) Hyperoxidation of peroxiredoxins 2 and 3: Rate constants for the reactions of the sulfenic acid of the peroxidatic cysteine. *J. Biol. Chem.*, 288, 14170–14177.

119. A. C. Haynes, J. Qian, J. A. Reisz, C. M. Furdui, and W. T. Lowther (2013) Molecular basis for the resistance of human mitochondrial 2-Cys peroxiredoxin 3 to hyperoxidation. *J. Biol. Chem.*, 288, 29714–29723.

120. Y. H. Noh, J. Y. Baek, W. Jeong, S. G. Rhee, and T. S. Chang (2009) Sulfiredoxin translocation into mitochondria plays a crucial role in reducing hyperoxidized peroxiredoxin III. *J. Biol. Chem.*, 284, 8470–8477.

121. M. A. Aon, B. A. Stanley, V. Sivakumaran, J. M. Kembro, B. O'Rourke, N. Paolocci, and S. Cortassa (2012) Glutathione/thioredoxin systems modulate mitochondrial H_2O_2 emission: An experimental-computational study. *J. Gen. Physiol.*, 139, 479–491.

122. H. Sztajer, B. Gamain, K. D. Aumann, C. Slomianny, K. Becker, R. Brigelius-Flohe, and L. Flohe (2001) The putative glutathione peroxidase gene of *Plasmodium falciparum* codes for a thioredoxin peroxidase. *J. Biol. Chem.*, 276, 7397–7403.

123. Flohé, L. (1979) Glutathione peroxidase: Fact and fiction. In: Fitzsimmons, D. W. ed. *OxygenFreeRadicalsandTissueDamage,*ExcerptaMedica,Amsterdam,theNetherlands, pp. 95–122.

124. J. W. Forstrom, F. H. Stults, and A. L. Tappel (1979) Rat liver cytosolic glutathione peroxidase: Reactivity with linoleic acid hydroperoxide and cumene hydroperoxide. *Arch. Biochem. Biophys.,* 193, 51–55.

125. A. Wahllander, S. Soboll, and H. Sies (1979) Hepatic mitochondrial and cytosolic glutathione content and the subcellular distribution of GSH S-transferases. *FEBS Lett.,* 97, 138–140.

126. Z. Cao, D. Bhella, and J. G. Lindsay (2007) Reconstitution of the mitochondrial PrxIII antioxidant defence pathway: General properties and factors affecting PrxIII activity and oligomeric state. *J. Mol. Biol.,* 372, 1022–1033.

127. D. Parsonage, D. S. Youngblood, G. N. Sarma, Z. A. Wood, P. A. Karplus, and L. B. Poole (2005) Analysis of the link between enzymatic activity and oligomeric state in AhpC, a bacterial peroxiredoxin. *Biochemistry,* 44, 10583–10592.

128. S. Watabe, T. Hiroi, Y. Yamamoto, Y. Fujioka, H. Hasegawa, N. Yago, and S. Y. Takahashi (1997) SP-22 is a thioredoxin-dependent peroxide reductase in mitochondria. *Eur. J. Biochem.,* 249, 52–60.

129. J. M. Hansen, H. Zhang, and D. P. Jones (2006) Mitochondrial thioredoxin-2 has a key role in determining tumor necrosis factor-α-induced reactive oxygen species generation, NF-κB activation, and apoptosis. *Toxicol. Sci.,* 91, 643–650.

130. H. Zhang, Y. M. Go, and D. P. Jones (2007) Mitochondrial thioredoxin-2/peroxiredoxin-3 system functions in parallel with mitochondrial GSH system in protection against oxidative stress. *Arch. Biochem. Biophys.,* 465, 119–126.

131. U. Andrae, J. Singh, and K. Ziegler-Skylakakis (1985) Pyruvate and related α-ketoacids protect mammalian cells in culture against hydrogen peroxide-induced cytotoxicity. *Toxicol. Lett.,* 28, 93–98.

132. H. Herz, D. R. Blake, and M. Grootveld (1997) Multicomponent investigations of the hydrogen peroxide- and hydroxyl radical-scavenging antioxidant capacities of biofluids: The roles of endogenous pyruvate and lactate. Relevance to inflammatory joint diseases. *Free Radic. Res.,* 26, 19–35.

133. J. O'Donnell-Tormey, C. F. Nathan, K. Lanks, C. J. DeBoer, and J. de la Harpe (1987) Secretion of pyruvate. An antioxidant defense of mammalian cells. *J. Exp. Med.,* 165, 500–514.

134. S. Desagher, J. Glowinski, and J. Prémont (1997) Pyruvate protects neurons against hydrogen peroxide-induced toxicity. *J. Neurosci.,* 17, 9060–9067.

135. J. Vasquez-Vivar, A. Denicola, R. Radi, and O. Augusto (1997) Peroxynitrite-mediated decarboxylation of pyruvate to both carbon dioxide and carbon dioxide radical anion. *Chem. Res. Toxicol.,* 10, 786–794.

136. E. Bassenge, O. Sommer, M. Schwemmer, and R. Bünger (2000) Antioxidant pyruvate inhibits cardiac formation of reactive oxygen species through changes in redox state. *Am. J. Physiol. Heart Circ. Physiol.,* 279, H2431–H2438.

137. X. Wang, E. Perez, R. Liu, L.-J. Yan, R. T. Mallet, and S.-H. Yang (2007) Pyruvate protects mitochondria from oxidative stress in human neuroblastoma SK-N-SH cells. *Brain Res.,* 1132, 1–9.

2 Energy-Redox Axis in Mitochondria
Interconnection of Energy-Transducing Capacity and Redox Status

Zhigang Liu, Harsh Sancheti,
Enrique Cadenas, and Fei Yin

CONTENTS

ABSTRACT

Mitochondrial dysfunction is implicated in multiple pathological conditions, particularly those involving metabolically active tissues such as the brain, liver, and heart. Mitochondrial dysfunction is manifested by not only deficits in their energy transducing capacity, but also through perturbations to their redox homeostasis in terms of increased generation of O_2^- and H_2O_2, altered NADPH and GSH status, and increased macromolecule oxidative modification. The interconnectivity between these energy- and redox-related events is established by nicotinamide nucleotide transhydrogenase (NNT) and the inhibitory oxidation of mitochondrial bioenergetic enzymes. The energy-redox axis is, therefore, proposed as a unique perspective that describes concerted alterations in mitochondrial bioenergetic and redox-modulating properties under various circumstances. Moreover, the energy-redox axis is actively involved in the communications between mitochondria and the rest of the cell by an array of energy- and redox-sensitive kinase signaling and transcriptional pathways. Through these pathways, the energy-redox axis coordinates with the dynamic remodeling and quality control mechanisms of mitochondria for optimized metabolic and redox functions.

INTRODUCTION

Mitochondria became endosymbionts living inside the eukaryote over a billion years ago. Since then, their unique ability to implement oxidative mechanisms has made them the powerhouses of eukaryotic cell that meet the majority of the cellular energy demands. The most prominent metabolic process carried out by mitochondria is oxidative phosphorylation (OXPHOS) to generate ATP, the universal energy currency. Aside from their role as intracellular power plants, mitochondria play fundamental roles in multiple cellular processes such as apoptosis, calcium homeostasis, intermediary metabolism, and cell signaling (Cadenas, 2004). Mitochondria are also major cellular sources of oxidants such as superoxide ($O_2^{\cdot-}$) and hydrogen peroxide (H_2O_2), which are generated by the electron transfer chain (ETC) and can be removed by the mitochondrial redox component, which includes the glutathione- and thioredoxin-supported systems.

However, the role of mitochondria in energy transducing and redox regulation should not be viewed independently, considering the following strong interconnections between the energy and redox components of mitochondria: (a) The generation of $O_2^{\cdot-}$ and H_2O_2 by mitochondria depends on the respiratory state (higher H_2O_2 generation in state 4 respiration and lower generation in state 3 respiration) and shows an exponential dependence on the mitochondrial membrane potential (Boveris et al., 2006); (b) the reducing power (in the form of NADPH) of the mitochondrial redox systems to remove oxidants is ultimately from the tricarboxylic acid (TCA) cycle; and (c) altered mitochondrial redox status impacts on the energy-transducing capacity of the organelles through the post-translational modification of key elements of the bioenergetic machinery. Therefore, we proposed a mitochondrial *energy-redox axis* centered on mitochondrial energy metabolism and redox pathways as a dual pronged approach to describe the changes in mitochondrial function that occurs in aging and multiple pathologies (Yap et al., 2009).

Although high levels of H_2O_2 are involved in oxidative damage to macromolecules such as protein, lipids, and DNA, low to intermediate levels of H_2O_2 are required for the regulation of redox-sensitive signaling and transcription. Mitochondria is thus implicated in the regulation of the cellular redox status and cell function through the regulated production and removal of H_2O_2, thus transducing redox signals into a wide variety of responses, such as proliferation, adaptation, differentiation, and cell death (Cadenas, 2004; Yin et al., 2012a). In addition, mitochondria-originated signals to the cytosol and nucleus through messengers such as ADP/AMP and NAD^+, coordinate the biogenesis of mitochondria.

The purpose of this chapter is to review the pivotal roles of the energy-redox axis in (a) determining mitochondrial bioenergetic and redox functions, (b) modulating mitochondrial dynamic remodeling and quality control, and (c) communicating with the rest of the cell through energy- or redox-sensitive kinase signaling and transcriptional pathways.

MITOCHONDRIAL ENERGY-TRANSDUCING MACHINERY

The energy-transducing capacity of mitochondria meets the cellular energy demands, including supporting metabolic osmotic and mechanical functions, in the forms of ATP. The generation of ATP entails the oxidation of acetyl-CoA in the TCA cycle with the concomitant generation of reducing equivalents (NADH, $FADH_2$) that flow through the respiratory chain, generating a proton motive force (Mathews et al., 2000). In the liver, acetyl-CoA can be generated from glucose oxidation, ketone body metabolism, and fatty acid oxidation (Figure 2.1).

GLUCOSE OXIDATION

Glucose is the primary fuel for the human body, particularly for the brain. After being transported into the cytosol, glucose is metabolized by glycolysis in a multistep set of reactions resulting in the generation of pyruvate. Pyruvate will further undergo oxidative decarboxylation in mitochondria by the pyruvate dehydrogenase (PDH) complex to acetyl-CoA that feeds into the TCA cycle (Mathews et al., 2000).

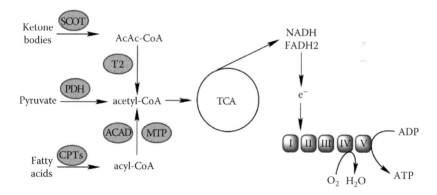

FIGURE 2.1 Mitochondrial metabolism of pyruvate, ketone bodies, and fatty acids. Pyruvate, generated from glucose by glycolysis in cytosol, undergoes oxidative decarboxylation by pyruvate dehydrogenase (PDH) to form acetyl-CoA. Fatty acid oxidation in mitochondria is catalyzed by carnitine palmitoyl transferase-1 (CPT1) and mitochondrial trifunctional protein (MTP) for acetyl-CoA generation. Ketone body metabolism is regulated primarily by succinyl-CoA: 3-oxoacid CoA transferase (SCOT) to form acetoacetyl-CoA (AcAc-CoA), then transferred to an acetyl group by mitochondrial acetoacetyl-CoA thiolase (T2), producing acetyl-CoA. Acetyl-CoA , which is generated by these three pathways enters the TCA cycle to produce NADH and $FADH_2$, which provide electrons to the electron transfer chain to build up the proton motive force for ATP synthesis.

In mammalian cells, PDH is subject to a continuous phosphorylation (inactivation) and dephosphorylation (activation) cycle catalyzed by the pyruvate dehydrogenase kinases (PDKs), dedicated regulatory enzymes that phosphorylate and inactivate PDH, and pyruvate dehydrogenase phosphatases (PDPs), which dephosphorylate and activate PDH (Holness and Sugden, 2003). PDK activation by the high mitochondrial acetyl-CoA/CoA and NADH/NAD$^+$ ratios that reflect high rates of long-chain fatty acid oxidation causes blockade of glucose oxidation (Holness and Sugden, 2003).

FATTY ACID OXIDATION

Fatty acids are energy-rich molecules, which are pivotal for energy storage and other cellular processes, such as the synthesis of membrane lipids. Fatty acids are one of the major energy sources for skeletal muscle and heart. The liver plays a critical role in the metabolism of fatty acids and provides alternative fuel for extrahepatic tissues (Reddy and Hashimoto, 2001). Fatty acids are oxidized by β-oxidation confined to mitochondria and peroxisomes, and the CYP4A-catalyzed ω-oxidation occurring in the endoplasmic reticulum (Reddy and Hashimoto, 2001). Fatty acids are completely oxidized to acetyl-CoA by mitochondrial β-oxidation, and the acetyl-CoA then either enters the TCA cycle for further oxidation or condenses to ketone bodies (acetoacetate, acetone, and β-hydroxybutyrate) in the liver to serve as oxidizable fuels for extrahepatic tissues, such as the brain (Hashimoto et al., 2000; Reddy and Hashimoto, 2001). The first step of mitochondrial β-oxidation is the α-β-dehydrogenation of the acyl-CoA ester by a family of four chain length-specific straight-chain acyl-CoA dehydrogenases (ACADs). The second, third, and fourth steps in the mitochondrial β-oxidation pathway are performed by 2-enoyl-CoA hydratase, 3-hydroxyacyl-CoA dehydrogenase, and 3-ketoacyl-CoA thiolase, respectively. All three of these enzyme activities are encompassed in a single mitochondrial trifunctional protein (MTP), which is a heterotrimeric protein that consists of four α-subunits and four β-subunits and catalyzes long-chain fatty acid oxidation (Reddy and Hashimoto, 2001). Mitochondrial β-oxidation is regulated by carnitine palmitoyl transferase-1 (CPT1), the carnitine concentration, and malonyl-CoA, which inhibits CPT1 (Reddy and Rao, 2006). Fatty acids, fatty acyl-CoA, and several structurally different synthetic compounds known as peroxisome proliferators, which activate PPAR-γ, regulate CPT1 levels in the liver (Reddy and Rao, 2006).

KETONE BODY METABOLISM

Ketone bodies are important vectors of energy from the liver to the brain when glucose supply is low (Yao et al., 2010; Yin et al., 2014). Acetoacetate and β-hydroxybutyrate are the two major types of ketone bodies (Fukao et al., 2014). In hepatocytes, excessive acetyl-CoA and acetoacetyl-CoA produced by β-oxidation are condensed to 3-hydroxy-3-methylglutaryl-CoA (HMG-CoA) by mitochondrial HMG-CoA synthase. Acetoacetate is produced from HMG-CoA by HMG-CoA lyase, and is in part reduced to form β-hydroxybutyrate. Both acetoacetate and β-hydroxybutyrate diffuse to the bloodstream. In extrahepatic tissues, β-hydroxybutyrate is converted back into acetoacetate, which is then activated to acetoacetyl-CoA by succinyl-CoA:

3-oxoacid CoA transferase (SCOT) (Fukao et al., 2014). Next, mitochondrial acetoacetyl-CoA thiolase (T2) transfers an acetyl group to free CoA, producing two molecules of acetyl-CoA. These steps are necessary for energy transducing from ketones in extrahepatic tissues, such as the brain, which has no other fatty acid–derived source of energy and ketone bodies are an alternative energy source for the fasted or developing brain.

Acetyl-CoA derived from glucose, fatty acids, and ketone bodies is then oxidized to CO_2 in the TCA cycle. Regulation of citrate synthase, isocitrate dehydrogenase, and α-ketoglutarate dehydrogenase controls the flow of metabolites in and out of the TCA cycle. The electron donors including NADH and $FADH_2$ generated from TCA cycle will further undergo OXPHOS (Mathews et al., 2000).

Mitochondrial OXPHOS is a process that encompasses electron transfer through the complexes I (NADH dehydrogenase), II (succinate dehydrogenase), III (cytochrome reductase), and IV (cytochrome oxidase) of the respiratory chain; this exergonic electron transfer is the driving force for the vectorial H^+ release into the intermembrane space and for the H^+ re-entry to the matrix through F_0 of complex V with ATP synthesis by F_1-ATP synthase (Mathews et al., 2000). It has been demonstrated that the components of the ETC (except complex II) exist as large macromolecular assemblies called supercomplexes (Schagger and Pfeiffer, 2000). The most common supercomplexes observed are complex I/III, complex III/IV, and complex I/III/IV (respirasome). The ultrastructure of these supercomplexes determines the activity of mitochondrial OXPHOS and thus plays a critical role in mitochondrial phenotype in aging and neurodegeneration (Gomez et al., 2009; Seelert et al., 2009). Interestingly, the plasticity of ETC organization was demonstrated by a recent study that the electron flux is modulated by the dynamic supercomplex assembly to adaptively optimize the use of available substrate to specific cell-type requirements (Lapuente-Brun et al., 2013).

MITOCHONDRIAL REDOX HOMEOSTASIS

Since the early 1970s, mitochondria have been found as an active source of H_2O_2 in the heart and liver (Boveris and Chance, 1973; Boveris et al., 1972), with $O_2^{\cdot-}$ as the stoichiometric precursor of mitochondrial H_2O_2. $O_2^{\cdot-}$ was primarily generated during ubisemiquinone auto-oxidation (Boveris and Cadenas, 1975; Boveris et al., 1976; Cadenas et al., 1977; Turrens and Boveris, 1980) by complex III and secondarily, by reverse electron transfer at the NADH dehydrogenase segment (Turrens, 2003). H_2O_2 and other mitochondrion-generated oxidants have a dual function: on the one hand, H_2O_2 is involved in the fine-tuning of signaling and transcription through the modulation of redox-sensitive pathways; and on the other hand, high levels of H_2O_2 that overwhelm the mitochondrial or cellular removal capacity cause oxidative damage to cell constituents, a well-documented phenomenon termed *oxidative stress.*

The mitochondrial redox component is the domain of H_2O_2 removal systems—mainly glutathione peroxidases and peroxiredoxin-3 (Prx3)—that use GSH and thioredoxin-2 (Trx2) as electron donors, respectively (Figure 2.2).

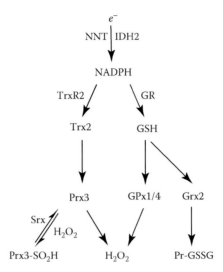

FIGURE 2.2 Mitochondrial redox component. NADPH generated by nicotinamide nucleo-tide transhydrogenase (NNT) or isocitrate dehydrogenase 2 (IDH2) is utilized by the Trx and GSH systems in the mitochondria to remove H_2O_2, through reactions catalyzed by Prx3 and GPx1/4. GSH can also be used by Grx2 to reduce protein-glutathione mixed disulfide (Pr-GSSG). Hyperoxidized Prx3 (sulfinic acid form) by high levels of H_2O_2 can be reduced to its sulfenic acid form by sulfiredoxin (Srx) translocated to mitochondria.

GSH-Supported System

GSH, synthesized by two-step reactions in cytosol catalyzed by γ glutamylcysteine synthetase and GSH synthase (Noctor et al., 1998), is transported into the mitochon-dria by dicarboxylate- and 2-oxoglutarate carriers (Lash, 2006). Depletion of mito-chondrial GSH decreases the cellular viability (Shan et al., 1993), whereas increasing mitochondrial GSH protects against oxidative and nitrosative stress (Muyderman et al., 2004). GSH supports the oxidant's reducing capacity of glutathione peroxi-dases (GPxs) in mitochondria (Rhee and Woo, 2011), primarily GPx1 localizes in the matrix, and GPx-4 existed in the inner mitochondrial membrane. GSH is regener-ated from oxidized glutathione (GSSG) by glutathione reductase (GR) with reducing power from NADPH.

Mitochondrial generation of oxidants and free radicals is associated with revers-ible and irreversible modifications of target proteins, including S-nitrosylation and S-glutathionylation of the cysteinyl residues and nitration of tyrosyl residues. Protein S-glutathionylation is specifically reduced by glutaredoxins (Grxs; Grx2 is the mitochondrial isoform) through a monothiol mechanism (Holmgren and Aslund, 1995). Oxidized Grx2 is reduced by GSH. Grx2 plays a critical role in mitochondrial redox homeostasis and cellular health as Grx2 knockdown leads to an increased sensitivity to cell death (Lillig et al., 2004), whereas Grx2 overexpres-sion decreases the susceptibility of cells to oxidants and inhibits intrinsic apoptosis activation (Enoksson et al., 2005).

TRX-SUPPORTED SYSTEM

In mitochondria, the reducing power to reduce H_2O_2 and other oxidants can also be transmitted through thiols of the mitochondrial Trx system: NADPH → thioredoxin reductase 2 (TrxR2) → Trx2 → peroxiredoxins (Prxs) (Zhang et al., 2007). Prx3 and Prx5 are the mitochondrial Prxs capable to remove H_2O_2, organic hydroperoxides, and $ONOO^-$ (Peng et al., 2004). Prx3 catalyzes up to 90% of H_2O_2 generated in the mitochondrial matrix with a high reaction rate especially at low levels of H_2O_2 (Cox et al., 2009b). At high concentrations of H_2O_2, hyperoxidation of Prx3 to sulfinic acid form Prx3-Cys-SO_2H, which results in the loss of its peroxidase activity (Cox et al., 2009a). Under these conditions, sulfiredoxin (Srx) translocates to the mitochondria and reduces the sulfinic acid form of Prx3 back to Prx3-Cys-SOH in an ATP-driven reaction (Noh et al., 2009; Rhee et al., 2007) (Figure 2.2).

As a typical 2-Cys Prx, the intermolecular disulfide formed on Prx3 upon the reduction of H_2O_2 to H_2O is reduced by Trx2, which is highly efficient at reducing disulfides in proteins via thiol-disulfide exchange (Miranda-Vizuete et al., 2000). Mitochondrial Trx2 is abundant in metabolically active tissues (Spyrou et al., 1997) and its oxidation is an early event in oxidative stress. Trx2 may be involved in the regulation of apoptosis through its interaction with apoptosis signal-regulating kinase-1 (ASK1): upon inflammatory (TNFα) or oxidative (H_2O_2) stimulation, mitochondrial ASK1 disassociates from Trx2 and mediates a JNK-independent caspase-mediated apoptotic pathway (Zhang et al., 2004).

Interestingly, initially known as paralleled redox systems, the mitochondrial GSH and Trx systems were recently found to cross talk and serve as a backup system to each other: it was documented that mitochondrial Grx2 can be a substrate of TrxR, and mitochondrial Prx3 can be reduced by not only Trx2 but also GSH-supported Grx2 (Lu and Holmgren, 2014).

INTERDEPENDENCE OF ENERGY AND REDOX COMPONENTS

The ultimate reductant for the two mitochondrial redox systems is NADPH, which is formed by three enzymes in mitochondria: $NADP^+$-dependent isocitrate dehydrogenase (IDH_2), malic enzyme, and nicotinamide nucleotide transhydrogenase (NNT). Since the majority of the mitochondrial NADPH pool is uncoupler sensitive, NNT-catalyzed reduction of $NADP^+$ accounts for more than 50% of the mitochondrial NADPH pool (Rydstrom, 2006). The inner mitochondrial membrane-located NNT catalyzes the reversible conversion between $NADP^+$ and NADPH coupled with the NADH-NAD^+ conversion and proton translocation across the inner membrane (Hoek and Rydström, 1988). Under physiological conditions, the proton gradient across the inner membrane strongly stimulates the forward reaction for NADPH generation (Ying, 2008). NNT activity thus links the mitochondrial energy-transducing activity to redox homeostasis by coupling NADPH generation to the TCA cycle and active respiration. NNT plays important roles in regulating cellular redox homeostasis, energy metabolism, and apoptotic pathways (Maack and Bohm, 2011), and its expression is upregulated in response to oxidative challenges such as cigarette smoke, a rich source of oxidants and electrophiles (Agarwal et al., 2012). The knockdown of NNT in PC12 cells results in decreased NADPH levels and GSH/GSSG ratios, increased H_2O_2 generation, as well as an impaired mitochondrial energy-transducing capacity. The activation of redox-sensitive JNK by H_2O_2 after NNT suppression induces intrinsic apoptosis (Yin et al., 2012b). Hence, the disruption of electron flux from fuel substrates to redox component due to NNT suppression not only compromises mitochondrial function, but also affects cellular functions (Yin et al., 2012b). This study potentially explains how NNT dysfunction leads to multiple pathologies such as familial glucocorticoid deficiency (Meimaridou et al., 2012).

On the other hand, posttranslational modifications to protein thiols due to changes in redox status can modulate the components of the energy metabolism. In mitochondria, aconitase (Han et al., 2005), α-ketoglutarate dehydrogenase (Nulton-Persson et al., 2003), isocitrate dehydrogenase (Kil and Park, 2005), SCOT (Garcia et al., 2010), aldehyde dehydrogenase (Wenzel et al., 2007), and complexes I (Taylor et al., 2003), II (Chen et al., 2007), and V (Garcia et al., 2010; West et al., 2006) can be inhibited upon S-glutathionylation. S-glutathionylation of SCOT and ATP synthase (F_1 complex, α-subunit) in brain mitochondria inhibits their activity and leads to a oxidized redox potential; supplementation of mitochondria with respiratory substrates to complex I or complex II increased NADH and NADPH levels, restored GSH levels through reduction of GSSG and deglutathionylation of mitochondrial proteins, and resulted in a more reducing mitochondrial environment (Garcia et al., 2010). Reversible glutathionylation of mitochondrial proteins could also serve as a protective mechanism that masks critical thiols from further irreversible oxidations (Queiroga et al., 2010) and enable these proteins to respond to the redox environment by reversible activation/inactivation (Thomas et al., 1995). In addition, bioenergetic enzymes such as SCOT and ATP synthase are inactivated as a function of age due to the nitration of their tyrosine residues by peroxynitrite ($ONOO^-$), which is formed by $O_2^{\cdot-}$ and NO (Lam et al., 2009).

Redox homeostasis also controls the utilization of energy fuels for ATP production, involving not only mitochondrial, but also cytosolic energy- and redox-related pathways, that is, glycolysis and the pentose phosphate pathway (Agarwal et al., 2013, 2014). Oxidative challenges by cigarette smoke or acrolein impair glycolysis and glucose-dependent respiration in lung cells by oxidatively inhibiting glyceraldehyde-3-phosphate dehydrogenase (GAPDH), thus enabling glucose to be rerouted to the pentose phosphate pathway for the production of NADPH, which is critical for alleviating oxidative stress (Agarwal et al., 2013, 2014). As a consequence, declined pyruvate supply to mitochondria leads to an increased utilization of fatty acids as energy fuel, as shown by increased expression of fatty acid transporters (CD36 and CPT1) and enhanced palmitate-supported mitochondrial respiration. Increased fatty acid consumption for energy transducing further affects the surfactant biosynthesis pathway, as evidenced by the decline in phosphatidylcholine levels and the increase in phospholipase A2 activity (Agarwal et al., 2013, 2014). These studies represent the complex interregulations within the mitochondrial energy-redox axis as well as the close connection of the axis with their cytosolic counterparts.

MITOCHONDRIAL ENERGY-REDOX AXIS
AND DYNAMIC REMODELING

Mitochondria are highly dynamic organelles and undergo fusion and fission throughout their life cycle. These processes regulate not only mitochondrial morphology, but also their biogenesis, transportation and localization, quality control, and degradation (Chan, 2012; Twig and Shirihai, 2011). Mitochondrial fission is mediated by dynamin-related protein-1 (Drp1), mitochondrial fission factor (Mff), and fission protein-1 homologue (Fis1), with Drp1 as scissors of mitochondrial membrane and the other two recruiting Drp1 to mitochondria (Ishihara et al., 2009). Mitochondrial fusion requires GTPases mitofusin-1/2 (Mfn1/2) and optic atrophy-1 (OPA1), which are responsible for the fusion of outer and inner mitochondrial membranes, respectively (Song et al., 2009). A coordinated balance between fission and fusion serves to maintain the quality of these organelles, while perturbations to the balance are associated with pathologies (Chen and Chan, 2010; Liesa et al., 2009).

Mitochondrial dynamics and the energy-redox axis mutually affect each other. OPA1 dysfunction leads to decreased mitochondrial ATP production and increased oxidant production (Zhang et al., 2011), whereas Mfn2 deficiency results in the loss of mitochondrial membrane potential and respiration (Pich et al., 2005). It was recently reported that OPA1 determines ETC supercomplexes assembly and respiration efficiency by controlling mitochondrial cristae shape (Cogliati et al., 2013). Deficiency of the fission machinery also impairs mitochondrial metabolism: downregulation of Drp1 leads to a loss of mtDNA content and mitochondrial membrane potential, a decrease of respiration and ATP production, as well as an oxidized cellular redox status (Frank et al., 2001; Parone et al., 2008); repression of Fis1 decreases mitochondrial respiration with accumulation of oxidized proteins (Twig et al., 2008).

In contrast, mitochondrial dynamics are affected by the energy and redox status. It was recently discovered that OXPHOS tunes mitochondrial inner membrane fusion by controlling the cleavage (activation) of OPA1 through YmeIL, an ATP-dependent protease that cleaves OPA1 more efficiently under high OXPHOS conditions but not glycolytic conditions (Mishra et al., 2014). These findings demonstrate a tight molecular link between cellular metabolism and mitochondrial fusion. In addition, Mfn2 is regulated transcriptionally by the mitochondrial biogenesis regulators PGC-1α/β and estrogen-related receptor α (ERRα), as a coordinated response to increased energy demands (Scarpulla et al., 2012). It has also been shown that increased generation of mitochondrial $O_2^{\cdot-}$ or H_2O_2 induces mitochondrial fragmentation (fission) and cell death, which is proposed to be due to an oxidative stress–induced transcriptional regulation of fusion and fission proteins (Jendrach et al., 2008; Pletjushkina et al., 2006). The coordination between mitochondrial dynamics machinery and the energy-redox axis therefore allows the mitochondria to meet various specialized and localized metabolic needs in a timely manner.

COMMUNICATIONS OF THE ENERGY-REDOX AXIS
WITH KINASE SIGNALING AND TRANSCRIPTION

Mitochondrion-generated second messengers, such as H_2O_2, regulate multiple cell signaling pathways, and a range of cell functions (Ghafourifar and Cadenas, 2005; Yin et al., 2014). H_2O_2 acts as an efficient redox molecule, since it can easily pass through the mitochondrial membranes. Additionally, H_2O_2 produced by one mitochondrion can diffuse to other mitochondria, thus relaying signals among mitochondria (Murphy, 2009). The cytosolic redox-sensitive signaling pathways that are regulated by mitochondrial originated H_2O_2 include the insulin/IGF1 signaling (IIS), the JNK signaling, and the AMPK signaling, as reviewed previously (Yin et al., 2013, 2014).

On the other hand, mitochondria are also recipients of cytosolic signaling that regulate mitochondrial metabolic and redox functions. The cytosolic modulation of mitochondrial bioenergetic functions is primarily carried out by components of the IIS. It is well known that the mitochondrial function is impaired during insulin resistance, an indicator of compromised insulin signaling (Lowell and Shulman, 2005). In addition to its role in regulating glucose metabolism, in the central nervous system, IIS has also been shown to influence neuronal survival and synaptic plasticity (van der Heide et al., 2006). Recent studies in our laboratory have shown that α-lipoic acid, an insulin-mimetic nutraceutical, is capable of rescuing mitochondrial dysfunction in brain aging (Jiang et al., 2013). α-Lipoic acid also restored brain metabolic deficits and synaptic impairments in a mouse model of Alzheimer's disease (Sancheti et al., 2013, 2014a). Moreover, α-lipoic acid also normalized a hypermetabolic state in the young triple transgenic mouse model of AD (Sancheti et al., 2014b).

In contrast to the IIS, which generally promotes mitochondrial function, JNK is a negative regulator of mitochondrial metabolic function. Anisomycin- or H_2O_2-activated JNK translocates to outer mitochondrial membrane from cytosol in primary neuron and initiates a cascade that leads to the inhibitory phosphorylation of the E1α subunit of the PDH complex, which results in a decrease in cellular ATP levels and a metabolic shift toward anaerobic glycolysis (Zhou et al., 2008).

The energy-redox axis is also linked to the nuclear transcription pathways, since mitochondrial and cellular energy charge modulate mitochondrial biogenesis (Leary et al., 1998). The vast majority of the 1500 mitochondrial proteins are encoded by nuclear genome (Calvo et al., 2006). The expression of these genes is directly or indirectly regulated by transcription factors, such as nuclear respiratory factor-1 and nuclear respiratory factor-2 (NRF-1 and NRF-2) and ERRα (Scarpulla, 2002). These transcriptional pathways are coordinated by members of the peroxisome-proliferator-activated receptor γ coactivator-1 (PGC-1) family, and PGC-1α is the best characterized member (Handschin and Spiegelman, 2006). As the master regulator of mitochondrial biogenesis, PGC-1α is regulated at multiple levels including transcription, posttranslational modification, localization, and degradation (Puigserver and Spiegelman, 2003). PGC-1α activity is highly regulated by energy-charge-related signals from mitochondria. These regulations involve energy messengers such as NAD^+ and AMP/ADP, which modulate mitochondrial biogenesis by regulating energy sensors including sirtuin 1 (Sirt1) and AMPK, respectively (Fernandez-Marcos and Auwerx, 2011; Yin et al., 2014).

CONCLUSION

In multiple pathological conditions involving tissue with high metabolic rates such as the liver, brain, and heart, an early and initiating event is mitochondrial dysfunction, entailing a decline in energy production accompanied by an altered redox status toward a prooxidant environment, which may be partly due to increased mitochondrial generation of $O_2{}^-$ and H_2O_2. The energy-redox axis represents a unique perspective that unifies alterations in mitochondrial bioenergetic capacity and redox status associated with these pathologies. The energy-redox axis not only determines the metabolic rate and redox tone of the mitochondria, but also integrates with the quality control mechanisms of these organelles, including biogenesis, fission and fusion, protein homeostasis, and mitophagy. Moreover, an intricate network encompassed by the energy-redox axis, cytosolic kinase signaling, and nuclear transcriptional pathways is critical for the achievement of optimal metabolic and redox homeostasis in the cell.

ACKNOWLEDGMENTS

This work was supported by U.S. National Institutes of Health Grants RO1AG016718 to EC and PO1AG026572 (to Roberta D. Brinton; Project 1 to EC).

REFERENCES

Agarwal, A.R., Yin, F., and Cadenas, E. 2013. Metabolic shift in lung alveolar cell mitochondria following acrolein exposure. *Am J Physiol Lung Cell Mol Physiol* 305: L764–L773.

Agarwal, A.R., Yin, F., and Cadenas, E. 2014. Short-term cigarette smoke exposure leads to metabolic alterations in lung alveolar cells. *Am J Respir Cell Mol* 51: 284–293.

Agarwal, A.R., Zhao, L., Sancheti, H., Sundar, I.K., Rahman, I., and Cadenas, E. 2012. Short-term cigarette smoke exposure induces reversible changes in energy metabolism and cellular redox status independent of inflammatory responses in mouse lungs. *Am J Physiol-Lung C* 303: L889–L898.

Boveris, A. and Cadenas, E. 1975. Mitochondrial production of superoxide anions and its relationship to the antimycin-insensitive respiration. *FEBS Lett* 54: 311–314.

Boveris, A., Cadenas, E., and Stoppani, A.O.M. 1976. Role of ubiquinone in the mitochondrial generation of hydrogen peroxide. *Biochem J* 156: 435–444.

Boveris, A. and Chance, B. 1973. The mitochondrial generation of hydrogen peroxide. General properties and effect of hyperbaric oxygen. *Biochem J* 134: 707–716.

Boveris, A., Oshino, N., and Chance, B. 1972. The cellular production of hydrogen peroxide. *Biochem J* 128: 617–630.

Boveris, A., Valdez, L.B., Zaobornyj, T., and Bustamante, J. 2006. Mitochondrial metabolic states regulate nitric oxide and hydrogen peroxide diffusion to the cytosol. *Biochim Biophys Acta* 1757: 535–542.

Cadenas, E. 2004. Mitochondrial free radical production and cell signaling. *Mol Aspects Med* 25: 17–26.

Cadenas, E., Boveris, A., Ragan, C.I., and Stoppani, A.O. 1977. Production of superoxide radicals and hydrogen peroxide by NADH- ubiquinone reductase and ubiquinol-cytochrome c reductase from beef- heart mitochondria. *Arch Biochem Biophys* 180: 248–257.

Calvo, S., Jain, M., Xie, X., Sheth, S.A., Chang, B., Goldberger, O.A., Spinazzola, A., Zeviani, M., Carr, S.A., and Mootha, V.K. 2006. Systematic identification of human mitochondrial disease genes through integrative genomics. *Nat Genet* 38: 576–582.

Chan, D.C. 2012. Fusion and fission: Interlinked processes critical for mitochondrial health. *Annu Rev Genet* 46: 265–287.

Chen, H. and Chan, D.C. 2010. Physiological functions of mitochondrial fusion. *Ann NY Acad Sci* 1201: 21–25.

Chen, Y.R., Chen, C.L., Pfeiffer, D.R., and Zweier, J.L. 2007. Mitochondrial complex II in the post-ischemic heart: Oxidative injury and the role of protein S-glutathionylation. *J Biol Chem* 282: 32640–32654.

Cogliati, S., Frezza, C., Soriano, M.E., Varanita, T., Quintana-Cabrera, R., Corrado, M., Cipolat, S. et al. 2013. Mitochondrial cristae shape determines respiratory chain supercomplexes assembly and respiratory efficiency. *Cell* 155: 160–171.

Cox, A.G., Pearson, A.G., Pullar, J.M., Jonsson, T.J., Lowther, W.T., Winterbourn, C.C., and Hampton, M.B. 2009a. Mitochondrial peroxiredoxin 3 is more resilient to hyperoxidation than cytoplasmic peroxiredoxins. *Biochem J* 421: 51–58.

Cox, A.G., Peskin, A.V., Paton, L.N., Winterbourn, C.C., and Hampton, M.B. 2009b. Redox potential and peroxide reactivity of human peroxiredoxin 3. *Biochemistry* 48: 6495–6501.

Enoksson, M., Fernandes, A.P., Prast, S., Lillig, C.H., Holmgren, A., and Orrenius, S. 2005. Overexpression of glutaredoxin 2 attenuates apoptosis by preventing cytochrome c release. *Biochem Biophys Res Commun* 327: 774–779.

Fernandez-Marcos, P.J. and Auwerx, J. 2011. Regulation of PGC-1α, a nodal regulator of mitochondrial biogenesis. *Am J Clin Nutr* 93: 884S–890S.

Frank, S., Gaume, B., Bergmann-Leitner, E.S., Leitner, W.W., Robert, E.G., Catez, F., Smith, C.L., and Youle, R.J. 2001. The role of dynamin-related protein 1, a mediator of mitochondrial fission, in apoptosis. *Dev Cell* 1: 515–525.

Fukao, T., Mitchell, G., Sass, J.O., Hori, T., Orii, K., and Aoyama, Y. 2014. Ketone body metabolism and its defects. *J Inherit Metab Dis* 37: 541–551.

Garcia, J., Han, D., Sancheti, H., Yap, L.P., Kaplowitz, N., and Cadenas, E. 2010. Regulation of mitochondrial glutathione redox status and protein glutathionylation by respiratory substrates. *J Biol Chem* 285: 39646–39654.

Ghafourifar, P. and Cadenas, E. 2005. Mitochondrial nitric oxide synthase. *Trends Pharmacol Sci* 26: 190–195.

Gomez, L.A., Monette, J.S., Chavez, J.D., Maier, C.S., and Hagen, T.M. 2009. Supercomplexes of the mitochondrial electron transport chain decline in the aging rat heart. *Arch Biochem Biophys* 490: 30–35.

Han, D., Canali, R., Garcia, J., Aguilera, R., Gallaher, T.K., and Cadenas, E. 2005. Sites and mechanisms of aconitase inactivation by peroxynitrite: Modulation by citrate and glutathione. *Biochemistry* 44: 11986–11996.

Handschin, C. and Spiegelman, B.M. 2006. Peroxisome proliferator-activated receptor γ coactivator 1 coactivators, energy homeostasis, and metabolism. *Endocr Rev* 27: 728–735.

Hashimoto, T., Cook, W.S., Qi, C., Yeldandi, A.V., Reddy, J.K., and Rao, M.S. 2000. Defect in peroxisome proliferator-activated receptor α-inducible fatty acid oxidation determines the severity of hepatic steatosis in response to fasting. *J Biol Chem* 275: 28918–28928.

Hoek, J.B. and Rydström, J. 1988. Physiological roles of nicotinamide nucleotide transhydrogenase. *Biochem J* 254: 1–10.

Holmgren, A. and Aslund, F. 1995. Glutaredoxin. *Meth Enzymol* 252: 283–292.

Holness, M.J. and Sugden, M.C. 2003. Regulation of pyruvate dehydrogenase complex activity by reversible phosphorylation. *Biochem Soc Trans* 31: 1143–1151.

Ishihara, N., Nomura, M., Jofuku, A., Kato, H., Suzuki, S.O., Masuda, K., Otera, H. et al. 2009. Mitochondrial fission factor Drp1 is essential for embryonic development and synapse formation in mice. *Nat Cell Biol* 11: 958–966.

Jendrach, M., Mai, S., Pohl, S., Voth, M., and Bereiter-Hahn, J. 2008. Short- and long-term alterations of mitochondrial morphology, dynamics and mtDNA after transient oxidative stress. *Mitochondrion* 8: 293–304.

Jiang, T., Yin, F., Yao, J., Brinton, R.D., and Cadenas, E. 2013. Lipoic acid restores age-associated impairment of brain energy metabolism through the modulation of Akt/JNK signaling and PGC1α transcriptional pathway. *Aging Cell* 12: 1021–1031.

Kil, I.S. and Park, J.W. 2005. Regulation of mitochondrial NADP+-dependent isocitrate dehydrogenase activity by glutathionylation. *J Biol Chem* 280: 10846–10854.

Lam, P.Y., Yin, F., Hamilton, R.T., Boveris, A., and Cadenas, E. 2009. Elevated neuronal nitric oxide synthase expression during ageing and mitochondrial energy production. *Free Radic Res* 43: 431–439.

Lapuente-Brun, E., Moreno-Loshuertos, R., Acin-Perez, R., Latorre-Pellicer, A., Colas, C., Balsa, E., Perales-Clemente, E. et al. 2013. Supercomplex assembly determines electron flux in the mitochondrial electron transport chain. *Science* 340: 1567–1570.

Lash, L.H. 2006. Mitochondrial glutathione transport: Physiological, pathological and toxicological implications. *Chem Biol Interact* 163: 54–67.

Leary, S.C., Battersby, B.J., Hansford, R.G., and Moyes, C.D. 1998. Interactions between bioenergetics and mitochondrial biogenesis. *Biochim Biophys Acta* 1365: 522–530.

Liesa, M., Palacin, M., and Zorzano, A. 2009. Mitochondrial dynamics in mammalian health and disease. *Physiol Rev* 89: 799–845.

Lillig, C.H., Lonn, M.E., Enoksson, M., Fernandes, A.P., and Holmgren, A. 2004. Short interfering RNA-mediated silencing of glutaredoxin 2 increases the sensitivity of HeLa cells toward doxorubicin and phenylarsine oxide. *Proc Natl Acad Sci* 101: 13227–13232.

Lowell, B.B. and Shulman, G.I. 2005. Mitochondrial Dysfunction and Type 2 Diabetes. *Science* 307: 384–387.

Lu, J. and Holmgren, A. 2014. The thioredoxin antioxidant system. *Free Radic Biol Med* 66: 75–87.

Maack, C. and Bohm, M. 2011. Targeting mitochondrial oxidative stress in heart failure throttling the afterburner. *J Am Coll Cardiol* 58: 83–86.

Mathews, C.K., Van Holde, K.E., and Ahern, K.G. 2000. *Biochemistry*, 3rd edn. San Francisco: Addison Wesley Longman.

Meimaridou, E., Kowalczyk, J., Guasti, L., Hughes, C.R., Wagner, F., Frommolt, P., Nurnberg, P. et al. 2012. Mutations in NNT encoding nicotinamide nucleotide transhydrogenase cause familial glucocorticoid deficiency. *Nat Genet* 44: 740–742.

Miranda-Vizuete, A., Damdimopoulos, A.E., and Spyrou, G. 2000. The mitochondrial thioredoxin system. *Antioxid Redox Signal* 2: 801–810.

Mishra, P., Carelli, V., Manfredi, G., and Chan, D.C. 2014. Proteolytic cleavage of Opa1 stimulates mitochondrial inner membrane fusion and couples fusion to oxidative phosphorylation. *Cell Metab* 19: 630–641.

Murphy, M. 2009. How mitochondria produce reactive oxygen species. *Biochem J* 417: 1–13.

Muyderman, H., Nilsson, M., and Sims, N.R. 2004. Highly selective and prolonged depletion of mitochondrial glutathione in astrocytes markedly increases sensitivity to peroxynitrite. *J Neurosci* 24: 8019–8028.

Noctor, G., Arisi, A.-C.M., Jouanin, L., Kunert, K.J., Rennenberg, H., and Foyer, C.H. 1998. Glutathione: Biosynthesis, metabolism and relationship to stress tolerance explored in transformed plants. *J Exp Bot* 49: 623–647.

Noh, Y.H., Baek, J.Y., Jeong, W., Rhee, S.G., and Chang, T.S. 2009. Sulfiredoxin translocation into mitochondria plays a crucial role in reducing hyperoxidized peroxiredoxin III. *J Biol Chem* 284: 8470–8477.

Nulton-Persson, A.C., Starke, D.W., Mieyal, J.J., and Szweda, L.I. 2003. Reversible inactivation of α-ketoglutarate dehydrogenase in response to alterations in the mitochondrial glutathione status. *Biochemistry* 42: 4235–4242.

Parone, P.A., Da Cruz, S., Tondera, D., Mattenberger, Y., James, D.I., Maechler, P., Barja, F., and Martinou, J.C. 2008. Preventing mitochondrial fission impairs mitochondrial function and leads to loss of mitochondrial DNA. *Plos One* 3: e3257.

Peng, Y., Yang, P.H., Guo, Y., Ng, S.S., Liu, J., Fung, P.C., Tay, D. et al. 2004. Catalase and peroxiredoxin 5 protect Xenopus embryos against alcohol-induced ocular anomalies. *Invest Ophth Vis Sci* 45: 23–29.

Pich, S., Bach, D., Briones, P., Liesa, M., Camps, M., Testar, X., Palacin, M., and Zorzano, A. 2005. The Charcot-Marie-Tooth type 2A gene product, Mfn2, up-regulates fuel oxidation through expression of OXPHOS system. *Hum Mol Genet* 14: 1405–1415.

Pletjushkina, O.Y., Lyamzaev, K.G., Popova, E.N., Nepryakhina, O.K., Ivanova, O.Y., Domnina, L.V., Chernyak, B.V., and Skulachev, V.P. 2006. Effect of oxidative stress on dynamics of mitochondrial reticulum. *Biochim Biophys Acta* 1757: 518–524.

Puigserver, P. and Spiegelman, B.M. 2003. Peroxisome proliferator-activated receptor-γ coactivator 1α (PGC-1α): Transcriptional coactivator and metabolic regulator. *Endocr Rev* 24: 78–90.

Queiroga, C.S., Almeida, A.S., Martel, C., Brenner, C., Alves, P.M., and Vieira, H.L. 2010. Glutathionylation of adenine nucleotide translocase induced by carbon monoxide prevents mitochondrial membrane permeabilization and apoptosis. *J Biol Chem* 285: 17077–17088.

Reddy, J.K. and Hashimoto, T. 2001. Peroxisomal β-oxidation and peroxisome proliferator-activated receptor α: An adaptive metabolic system. *Annu Rev Nutr* 21: 193–230.

Reddy, J.K. and Rao, M.S. 2006. Lipid metabolism and liver inflammation. II. Fatty liver disease and fatty acid oxidation. *Am J Physiol-Gastr L* 290: G852–G858.

Rhee, S.G., Jeong, W., Chang, T.S., and Woo, H.A. 2007. Sulfiredoxin, the cysteine sulfinic acid reductase specific to 2-Cys peroxiredoxin: Its discovery, mechanism of action, and biological significance. *Kidney Int Suppl* 106: S3–S8.

Rhee, S.G. and Woo, H.A. 2011. Multiple functions of peroxiredoxins: Peroxidases, sensors and regulators of the intracellular messenger HO, and protein chaperones. *Antioxid Redox Signal* 15: 781–794.

Rydstrom, J. 2006. Mitochondrial NADPH, transhydrogenase, and disease. *Biochim Biophys Acta* 1757: 721–726.

Sancheti, H., Akopian, G., Yin, F., Brinton, R.D., Walsh, J.P., and Cadenas, E. 2013. Age-dependent modulation of synaptic plasticity and insulin mimetic effect of lipoic acid on a mouse model of Alzheimer's disease. *Plos One* 8: e69830.

Sancheti, H., Kanamori, K., Patil, I., Diaz Brinton, R., Ross, B.D., and Cadenas, E. 2014a. Reversal of metabolic deficits by lipoic acid in a triple transgenic mouse model of Alzheimer's disease: A 13C NMR study. *J Cereb Blood Flow Metab* 34: 288–296.

Sancheti, H., Patil, I., Kanamori, K., Diaz Brinton, R., Zhang, W., Lin, A.L., and Cadenas, E. 2014b. Hypermetabolic state in the 7-month-old triple transgenic mouse model of Alzheimer's disease and the effect of lipoic acid: A (13)C-NMR study. *J Cereb Blood Flow Metab* 34: 1749–1760.

Scarpulla, R.C. 2002. Nuclear activators and coactivators in mammalian mitochondrial biogenesis. *Biochim Biophys Acta* 1576: 1–14.

Scarpulla, R.C., Vega, R.B., and Kelly, D.P. 2012. Transcriptional integration of mitochondrial biogenesis. *Trends Endocrinol Metab* 23: 459–466.

Schagger, H. and Pfeiffer, K. 2000. Supercomplexes in the respiratory chains of yeast and mammalian mitochondria. *EMBO J* 19: 1777–1783.

Seelert, H., Dani, D.N., Dante, S., Hauss, T., Krause, F., Schafer, E., Frenzel, M. et al. 2009. From protons to OXPHOS supercomplexes and Alzheimer's disease: Structure-dynamics-function relationships of energy-transducing membranes. *Biochim Biophys Acta* 1787: 657–671.

Shan, X., Jones, D.P., Hashmi, M., and Anders, M.W. 1993. Selective depletion of mitochondrial glutathione concentrations by (R,S)-3-hydroxy-4-pentenoate potentiates oxidative cell death. *Chem Res Toxicol* 6: 75–81.

Song, Z., Ghochani, M., McCaffery, J.M., Frey, T.G., and Chan, D.C. 2009. Mitofusins and OPA1 mediate sequential steps in mitochondrial membrane fusion. *Mol Biol Cell* 20: 3525–3532.

Spyrou, G., Enmark, E., Miranda-Vizuete, A., and Gustafsson, J. 1997. Cloning and expression of a novel mammalian thioredoxin. *J Biol Chem* 272: 2936–2941.

Taylor, E.R., Hurrell, F., Shannon, R.J., Lin, T.K., Hirst, J., and Murphy, M.P. 2003. Reversible glutathionylation of complex I increases mitochondrial superoxide formation. *J Biol Chem* 278: 19603–19610.

Thomas, J.A., Poland, B., and Honzatko, R. 1995. Protein sulfhydryls and their role in the antioxidant function of protein S-thiolation. *Arch Biochem Biophys* 319: 1–9.

Turrens, J.F. 2003. Mitochondrial formation of reactive oxygen species. *J Physiol* 552: 335–344.

Turrens, J.F. and Boveris, A. 1980. Generation of superoxide anion by the NADH dehydrogenase of bovine heart mitochondria. *Biochem J* 191: 421–427.

Twig, G., Elorza, A., Molina, A.J., Mohamed, H., Wikstrom, J.D., Walzer, G., Stiles, L. et al. 2008. Fission and selective fusion govern mitochondrial segregation and elimination by autophagy. *EMBO J* 27: 433–446.

Twig, G. and Shirihai, O.S. 2011. The interplay between mitochondrial dynamics and mitophagy. *Antioxid Redox Signal* 14: 1939–1951.

van der Heide, L.P., Ramakers, G.M., and Smidt, M.P. 2006. Insulin signaling in the central nervous system: Learning to survive. *Prog Neurobiol* 79: 205–221.

Wenzel, P., Hink, U., Oelze, M., Schuppan, S., Schaeuble, K., Schildknecht, S., Ho, K.K. et al. 2007. Role of reduced lipoic acid in the redox regulation of mitochondrial aldehyde dehydrogenase (ALDH-2) activity. Implications for mitochondrial oxidative stress and nitrate tolerance. *J Biol Chem* 282: 792–799.

West, M.B., Hill, B.G., Xuan, Y.T., and Bhatnagar, A. 2006. Protein glutathiolation by nitric oxide: An intracellular mechanism regulating redox protein modification. *FASEB J* 20: 1715–1717.

Yao, J., Hamilton, R.T., Cadenas, E., and Brinton, R.D. 2010. Decline in mitochondrial bioenergetics and shift to ketogenic profile in brain during reproductive senescence. *Biochim Biophys Acta* 1800: 1121–1126.

Yap, L.P., Garcia, J.V., Han, D., and Cadenas, E. 2009. The energy-redox axis in aging and age-related neurodegeneration. *Adv Drug Deliv Rev* 61: 1283–1298.

Yin, F., Boveris, A., and Cadenas, E. 2014. Mitochondrial energy metabolism and redox signaling in brain aging and neurodegeneration. *Antioxid Redox Signal* 20: 353–371.

Yin, F., Jiang, T., and Cadenas, E. 2013. Metabolic triad in brain aging: Mitochondria, insulin/IGF-1 signalling and JNK signalling. *Biochem Soc Trans* 41: 101–105.

Yin, F., Sancheti, H., and Cadenas, E. 2012a. Mitochondrial thiols in the regulation of cell death pathways. *Antioxid Redox Signal* 17: 1714–1727.

Yin, F., Sancheti, H., and Cadenas, E. 2012b. Silencing of nicotinamide nucleotide transhydrogenase impairs cellular redox homeostasis and energy metabolism in PC12 cells. *Biochim Biophys Acta* 1817: 401–409.

Ying, W. 2008. NAD+/NADH and NADP+/NADPH in cellular functions and cell death: Regulation and biological consequences. *Antioxid Redox Signal* 10: 179–206.

Zhang, H., Go, Y.M., and Jones, D.P. 2007. Mitochondrial thioredoxin-2/peroxiredoxin-3 system functions in parallel with mitochondrial GSH system in protection against oxidative stress. *Arch Biochem Biophys* 465: 119–126.

Zhang, R., Al-Lamki, R., Bai, L., Streb, J.W., Miano, J.M., Bradley, J., and Min, W. 2004. Thioredoxin-2 inhibits mitochondria-located ASK1-mediated apoptosis in a JNK-independent manner. *Circ Res* 94: 1483–1491.

Zhang, Z., Wakabayashi, N., Wakabayashi, J., Tamura, Y., Song, W.J., Sereda, S., Clerc, P. et al. 2011. The dynamin-related GTPase Opa1 is required for glucose-stimulated ATP production in pancreatic β cells. *Mol Biol Cell* 22: 2235–2245.

Zhou, Q., Lam, P.Y., Han, D., and Cadenas, E. 2008. c-Jun N-terminal kinase regulates mitochondrial bioenergetics by modulating pyruvate dehydrogenase activity in primary cortical neurons. *J Neurochem* 104: 325–335.

3 Signal Transduction Pathways That Regulate Mitochondrial Gene Expression

Chien-Yu Chen, Jingyu Chen, Anketse Debebe, Yang Li, and Bangyan Stiles

CONTENTS

ABSTRACT

The limited encoding capacity of mtDNA necessitates knowledge of the nuclear origin of regulatory factors governing mitochondrial gene expressions. Here, we review the nuclear factors that control the transcription of genes involved in mitochondrial and nuclear regulation of mitochondrial functions. We focus on key nuclear factors and cofactors and their regulation by signal transduction pathways. These include nuclear respiratory factors (NRF) and estrogen receptor–related receptors (ERR) and their regulation by PI3K, MAPK, and AMPK signaling pathways. We further review the involvement of these signals in the production of reactive oxygen species and mitochondrial gene dysfunction in the development of hepatocarcinoma.

Keywords: NRF, ERR, PI3K, MAPK, AMPK, HCC, hepatocarcinoma, ROS

Mitochondria contain their own genetic system, which undergoes cytoplasmic inheritance. The mitochondrial DNA (mtDNA) is a 16-kilobase circular double-stranded DNA and encodes 13 genes for essential subunits of mitochondrial respiratory chain, as well as 22 tRNAs and 2 rRNAs required for mtDNA transcription and translation (Clayton 2000). The proteins specified by mitochondrial genome only account for a small fraction of total proteins necessary for proper mitochondrial structure and function. For example, 39 out of 46 subunits of complex I, NADH dehydrogenase, and the entire complex II are nucleus encoded. The nuclear genes also specify 10 out of 11 complex III subunits and 10 out of 13 complex IV subunits (Bonawitz et al. 2006, Clayton 2003, Scarpulla 2011). Therefore, the majority of proteins required for maintaining the organelle architecture and respiratory function are specified by nuclear genome. In this chapter, we will summarize the mitochondrial gene transcription hierarchy and discuss the cell signaling networks that control this mitochondrial gene transcription network as well as the pathological implications of such signaling pathways in liver disease.

TRANSCRIPTIONAL NETWORKS THAT CONTROL MITOCHONDRIAL GENE TRANSCRIPTION

The limited encoding capacity of mtDNA necessitates the nuclear origin of regulatory factors governing mitochondrial gene expressions. The key transcription factors for mitochondrial biogenesis can be divided into two classes (Figure 3.1). The first class contains nucleus-encoded auxiliary factors that translocate into mitochondria and act upon mitochondrial genome to facilitate mtDNA transcription and translation. The second class includes nuclear transcription factors that act upon the nuclear genome and directly control respiratory gene expression. In addition to these two classes of transcription factors, another family of nuclear proteins required to maintain mitochondrial biogenesis is coactivators, which interact with transcription factors and induces their transcriptional activity.

TRANSCRIPTIONAL FACTORS ACTING ON THE MITOCHONDRIAL GENOME

The transcription of mtDNA is a bidirectional process that is tightly regulated by nucleus-encoded but mitochondria-localized factors such as mitochondrial DNA-directed RNA polymerase (POLRMT), mitochondrial transcription factor A (Tfam), mitochondrial transcription factor B1 (TFB1M), mitochondrial transcription factor B2 (TFB2M), and mitochondrial transcription terminator factor (mTERF) (Scarpulla 2011). The mitochondrial RNA polymerase, POLRMT, and the stimulatory factor that unwinds DNA, Tfam, are required to activate mtDNA transcription

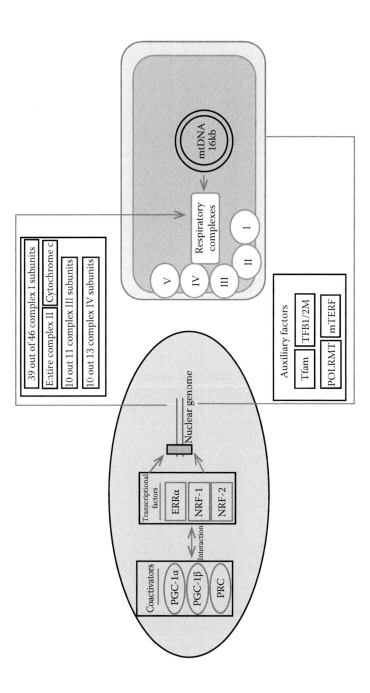

FIGURE 3.1 Transcriptional networks that control mitochondrial gene transcription. The expression of respiratory complexes is controlled by the processes in both the nucleus and mitochondria. In mitochondria, nucleus-encoded but mitochondria-localized factors such as POLRMT, Tfam, TFB1M, TFB2M, and mTERF can activate the transcription of mtDNA. In the nucleus, nuclear transcription factors such as estrogen-related receptor α, nuclear respiratory factor 1, and nuclear respiration factor 2 along with their coactivators such as peroxisomal proliferating activating factor γ coactivator-1α, PPARγ coactivator-1β, PPARγ, and PGC-1 related coactivator can synergistically induce the expressions of genes that directly constitute the mitochondrial respiratory apparatus.

(Tiranti et al. 1997, Wang et al. 2007). Two isoforms of dissociable specificity factors, TFB1M and TFB2M, directly contact with both POLRMT and Tfam to facilitate the transcription initiation on mtDNA promoters (McCulloch and Shadel 2003). Thus, the mtDNA transcription is directed by these three essential components. A high rate of mitochondrial rRNA synthesis relative to mitochondrial mRNA is an indication of more frequent mtDNA transcription. The mitochondrial transcription termination factor, mTERF, binds to the specific termination site at the end of the 16 rRNA and controls the ratio of mRNA to rRNA (Daga et al. 1993). Therefore, the steady levels of mitochondrial transcripts could be properly governed by the mTERF-regulated termination events. The homozygous knockout mice for these nuclear genes whose products reside in mitochondria have been generated. Deletion of either *Tfam*, *Polrmt*, *Tfb*, or *mTerf* results in embryonic lethality accompanied by mtDNA depletion, respiratory chain deficiency, and abnormal mitochondrial morphology (Larsson et al. 1998, Metodiev et al. 2009, Park et al. 2007), further highlighting the critical function of these factors to maintaining mitochondrial and cellular integrity.

Transcriptional Factors Acting on the Nuclear Genome

Besides nucleus-encoded auxiliary transcription factors that act on mtDNA to control transcription, nuclear transcription factors and coactivators synergistically induce expressions of genes that directly constitute the mitochondrial respiratory apparatus. In particular, transcription factors such as nuclear respiration factors 1 and 2 (NRF-1, NRF-2) and estrogen-related receptor α (ERRα) bind to target promoters and activate respiratory gene expression. The transcriptional coactivators, including peroxisomal proliferating activating factor γ [PPARγ] coactivator-1α (PGC-1α), PPARγ coactivator-1β (PGC-1β), and PGC-1 related coactivator (PRC), do not directly interact with DNA but rather bind to transcription factors to modulate their activity and expression. The interplay between these two nuclear protein families plays major roles in coordinating mitochondrial gene transcription.

NRF-1

The initial transcription factors implicated in the regulation of mitochondrial genes are the nuclear respiratory factors, NRF-1 and NRF-2. NRF-1 was identified through a search for transcription factors that control expressions of cytochrome c and cytochrome oxidase genes (Evans and Scarpulla 1990). NRF-1 was found to bind as a homodimer to a palindromic sequence within the cytochrome c promoter and function as a positive regulator of cytochrome c transcription. Subsequent studies have linked NRF-1 to expressions of a majority of genes that encode essential subunits of respiratory complexes I–IV. NRF-1 is also necessary for the transcription of genes encoding mitochondrial outer membrane components such as TOMM20 and COX17, gate keepers that control importation of proteins required for mitochondrial function (Kelly and Scarpulla 2004, Takahashi et al. 2002, Truscott et al. 2003). In addition, NRF-1 acts upon promoters of Tfam, TFB1/2M, and POLRMT, whose products, as mentioned earlier, play important roles in mitochondrial gene transcription (Gleyzer et al. 2005, Virbasius and Scarpulla 1994). Homozygous

NRF-1 knockout in mice resulted in embryonic lethality between E3.5 and 6.5, accompanied by the inability to maintain mitochondrial membrane potential and mtDNA copy number, indicating that NRF-1 is important for mtDNA maintenance. However, the early embryonic lethality by NRF-1 knockout may also result from altered expression of NRF-1 target genes involved in cell cycle regulation. A recent study utilized chromatin immunoprecipitation (ChIP) followed by microarray analysis and identified approximately 700 genes whose promoters exhibit NRF-1 occupancy (Cam et al. 2004). In addition to the mitochondrial components and mtDNA transcription regulators, a subset of NRF-1 target genes represents E2F-responsive genes required for DNA replication, mitosis, and so forth. Knockdown of NRF-1 led to reduced expression of E2F target genes along with cytochrome c and Tfam, indicating that NRF-1 positively regulates E2F signaling (Cam et al. 2004). NRF-1 exists in both phosphorylated and dephosphorylated states. Serum addition promotes phosphorylation of NRF-1 at multiple serine sites within the amino-terminal domain and enhances its DNA-binding and transcriptional activities (Herzig et al. 2000).

NRF-2

The second nuclear respiratory factor, NRF-2, was identified as a transcriptional activator of cytochrome oxidase IV and the human homolog of mouse GABP (Carter et al. 1992, LaMarco and McKnight 1989). NRF-2 is a multi-subunit protein that shares similar DNA-binding domain (DBD) and *trans*activation domain with NRF-1 (Gugneja et al. 1995). Subsequent chromatin immunoprecipitation studies revealed that NRF-2 is associated with all 10 cytochrome oxidase subunits' promoters and acts as an important regulatory factor for cytochrome oxidase complex expression (Ongwijitwat and Wong-Riley 2005). Although the respiratory genes initially defined as NRF-2 targets do not overlap with NRF-1 targets, a number of mitochondrial genes are found to utilize both NRFs to activate their expressions. The promoter occupancies of both NRF-1 and NRF-2 are crucial for expressions of human mito-chondrial cytochrome oxidase subunit V and entire complex II subunits (Au and Scheffler 1998, Virbasius et al. 1993). Moreover, functional NRF-2 sites have been identified on promoters of mitochondria-related factors, including Tfam, TFB1M, and TFB2M (Gleyzer et al. 2005, Larsson et al. 1998). Therefore, like NRF-1, NRF-2 coordinates the expressions of key respiratory chain components, as well as mito-chondrial transcription machinery. Like *Nrf-1* knockout mice, homozygous deletion of *Nrf-2* in mice resulted in lethality before implantation, suggesting its essential role for early embryogenesis.

ERRα

Estrogen-related receptor α (ERRα) is a member of the orphan nuclear receptor family, is expressed abundantly in high oxidative organs and has recently been recognized as a key regulator of adaptive energy metabolism (Villena and Kralli 2008). ERRα was first identified using the DBD of estrogen receptor α (ERα) as a screening probe (Giguere et al. 1988). The subsequent sequence analysis showed that ERRα shares about 68% homology in DBD and 33% homology in ligand-binding domain with ERα, which might explain the divergence in ligand-binding nature between these two receptors (Stein and McDonnell 2006). ERRα, by itself, is a weak transcriptional factor. Both its activity and expression are significantly increased when physically bound by a nuclear coactivator, PGC-1α, which will be discussed later. Nucleotide sequence binding analysis revealed that ERRα preferentially binds to the consensus DNA sequence 5'-TCAAGGTCA-3', termed estrogen-related response element (Sladek et al. 1997).

ERRα plays a predominant role in orchestrating mitochondrial transcription by either directly activating genes of mitochondrial structure or indirectly activating major transcription factors governing mitochondrial biogenesis. ERRα has been shown to regulate nuclear genes that function in the TCA cycle, fatty acid oxidation, mitochondrial oxidative phosphorylation, and respiratory chain. Recent studies using *ERRα*-null embryonic fibroblasts or ERRα knockdown in tumor cells showed that expressions of genes encoding mitochondrial protein, mtDNA level, and TCA enzyme activity were significantly decreased in response to reduced ERRα level (Rangwala et al. 2007). A genome-wide analysis for genomic locations of transcription factor occupancy revealed that ERRα is bound to the promoters of 195 genes, a majority of which are involved in the OXPHOS and TCA cycle as reported by previous studies, including cytochrome c, ATP synthase β (Atp5b), fumarate hydratase (Fh1), and succinate dehydrogenase (Sdha). Another subset of promoters bound by ERRα drives expressions of genes functioning in fatty acid oxidation, including Acadm, Slc25a29 (palmitoylcarnitine transporter), and fatty acid–binding protein 3 (Dufour et al. 2007).

The ERRα-binding sequence was also highly enriched at the NRF-2 promoters, suggesting that NRF-2 expression is subjected to regulation by ERRα. Moreover, a recent study discovered that NRF-1 also acts downstream of the PGC-1/ERRα complex in cultured muscle and liver cells (Mootha et al. 2004). In skeletal muscles, overexpression of NRF-1 itself is not sufficient to orchestrate a full transcriptional response to increase respiratory capacity (Baar et al. 2003). ERRα functions upstream of NRF-1 and NRF-2 and may act as a master regulator of mitochondrial gene transcription and is necessary for the full transcriptional activity of the NRF factors.

Unlike the NRFs, homozygous *Errα*-null mice are fertile and viable with reduced fat mass in peripheral organs and resistance to high-fat diet-induced obesity (Luo et al. 2003). Under cold exposure, *Errα*-null mice failed to initiate the adaptive thermogenesis, a process dependent on the adaptive biogenesis of mitochondria in brown adipose tissue. The inability of *Errα*-null mice to induce mitochondrial mass and mitochondrial DNA copy number together with decreased expression of respiratory chain components is thought to hinder the ability of these mice to cope with cold environment (Villena et al. 2007). These phenotypes of *Errα*-null mice confirm that ERRα plays a crucial role in controlling various mitochondrial functions, particularly those involved in adaptation. Recently, several clinical oncology studies have shown that an elevated ERRα expression was significantly associated with an unfavorable clinical outcome, hormonal insensitivity and increased risk of recurrence in human breast cancer, and ERRα might serve as a potent predictive biomarker of cancer therapy (Ariazi et al. 2002, Suzuki et al. 2004).

TRANSCRIPTIONAL COACTIVATORS

Coactivators are multifunctional nuclear proteins that act upon the DNA-binding transcription factors to regulate transcription. The major breakthrough in ERRα regulation research came with the discovery of PGC-1α, which is a member of a transcriptional coactivator family that also includes PGC-1β and PRC. The interaction between ERRα and PGC-1α is mediated by the N-terminal activation domain near leucine-rich LXXLL motifs on PGC-1α (Scarpulla 2011). Forced overexpression of PGC-1α not only potentiates ERRα transcriptional activity but also induces its mRNA and protein levels (Mootha et al. 2004). Moreover, PGC-1α is capable of inducing mitochondrial biogenesis, respiratory capacity, and fatty acid oxidation, supported by the results using both overexpression in cultured cells and tissue-conditional PGC-1α transgenic mice (Puigserver et al. 1998, Russell et al. 2004). Subsequent studies revealed that PGC-1α mediates such effects via direct interaction with the three key nuclear transcription factors promoting mitochondrial biogenesis, that is, ERRα, NRF-1, and NRF-2 (Finck and Kelly 2006). Mice carrying the germline deletion of PGC-1α exhibit normal mitochondrial volume and morphology in liver and fat tissues. Similar to *Errα*-null mice, PGC-1α-null mice are incapable to initiate the adaptive thermogenesis and elevate mitochondrial mass and function in response to physiological stimuli such as cold exposure and exercise (Leone et al. 2005, Lin et al. 2004). Therefore, PGC-1α by itself is not strictly required for the development and maintenance of mitochondria. However, PGC-1α serves as a central coordinator to optimize mitochondria homeostasis under the stimulation of physiological stress.

PGC-1β is closely related to PGC-1α and has been shown to interact with the same nuclear factors as PGC-1α and therefore shares a significant amount of downstream target genes with PGC-1α (Lin et al. 2002). Similar with PGC-1α-null mice, PGC-1β deletion led to normal energy expenditure and respiratory capacity but exhibit mitochondrial dysfunction under stress condition (Sonoda et al. 2007, Vianna et al. 2006). Interestingly, expression of PGC-1α is induced in the brown adipose tissue of PGC-1β knockout mice. However, this compensatory induction of PGC-1α failed to rescue the mitochondrial dysfunction phenotype in PGC-1β knockout mice, suggesting these two coactivators have nonredundant functions. In sharp contrast to the deletion of individual PGC-1s, PGC-1α/β double knockouts survived through embryogenesis. However, the majority of the double knockouts animals died within 24 h after birth due to cardiac failure, arguing that these two coactivators play important roles in postnatal cardiac function (Lai et al. 2008).

The third PGC-1 family member, PRC, was identified in a search for molecules with homology to PGC-1α. The PRC molecule shares homologous regions with the other two PGC-1 members in several domains, including acid amino-terminal domain, the LXXLL nuclear receptor–binding motif, RNA recognition motif, and a carboxyl-terminal RS (Ser-Arg) domain (Andersson and Scarpulla 2001). Besides binding to, and activating, mitochondria-associated transcription factors, such as NRF-1 and ERRα that overlap with PGC-1α/β, PRC plays an essential role in early embryonic development upon growth factor stimulation as the PRC expression peaks during the first day of embryonic body formation and germline PRC deletion result in peri-implantation lethality (He et al. 2012, Vercauteren et al. 2006, 2008).

CELL SIGNALING NETWORKS THAT CONTROL MITOCHONDRIAL GENE TRANSCRIPTION

These transcription factors and cofactors are coordinately regulated in the cells to control the catabolic and anabolic processes in response to insulin signaling as well as cell growth and survival in response to growth factor signals (Tjokroprawiro 2006). Several signaling pathways are found to regulate the mitochondrial transcriptional networks (Figure 3.2).

THE PI3K/AKT SIGNALING PATHWAY

The phosphatidylinositol 3-kinase (PI3K)/protein kinase B (AKT) pathway is a conserved signaling pathway that regulates various cellular processes including cell growth and survival, metabolism, immune response, and cell size. PI3K transduces receptor-binding signals from stimulated G-protein-coupled receptors

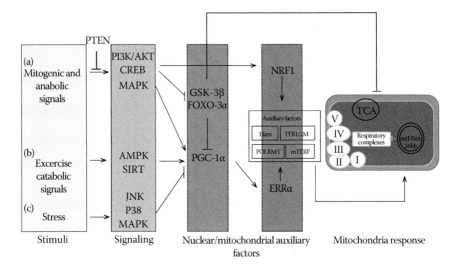

FIGURE 3.2 **(See color insert.)** Cell signaling networks that control mitochondrial gene transcription. (a) In cells exposed to mitogenic or anabolic signals, such as insulin or IGF-1, PI3K/AKT and mitogen-activated protein kinase signaling directly phosphorylates nuclear respiration factor 1 (NRF-1) and estrogen-related receptor α (ERRα) or increased levels of ERRα through non-PKA-dependent phosphorylation of the cAMP response element-binding protein. Both NRF and ERRα signals induce auxiliary factors that act on the mitochondrial genome. In addition, glycogen synthase kinase-3β and FOXO-3α, two substrates of AKT, may also function directly in the mitochondria. A phosphatase and tensin homolog, a tumor suppressor, inhibits this process via its inhibitory effect on the PI3K/AKT signaling pathway. (b) During an endurance exercise or under catabolic conditions, the need to supply ATP induces adenosine monophosphate–activated protein kinase (AMPK) and sirtuin 1 (SIRT1). Acting through deacetylation and activation of peroxisomal proliferating activating factor γ, coactivator-1α, and promoting glucose utilization, AMPK and SIRT1 induce mitochondrial functions to supply ATP demands. (c) Under stress conditions, c-Jun N-terminal kinase and p38MAPK inhibit mitochondrial respiration to slow down the production of reactive oxygen species.

and receptor tyrosine kinase by acting as a lipid kinase. Upon stimulation of the receptors, the activated PI3K catalyzes the reaction that converts membrane-bound phosphatidylinositol (3,4)-bisphosphate (PIP_2) into phosphatidylinositol (3,4,5)-trisphosphate (PIP_3). This action of PI3K results in increased cellular concentration of PIP_3. Acting as second messenger, PIP_3 recruits proteins containing the pleckstrin homology (PH) domains to the membrane. The best characterized PH domain containing protein and an important signaling molecule in PI3K-regulated signal is AKT. Subsequent to being recruited to the membrane by binding to PIP_3, AKT is activated by phosphorylation on two residues, Ser473 and Thr 308. The activated AKT mediates phosphorylation and inactivation of downstream targets including glycogen synthase kinase-3 (GSK-3), proapoptotic Bad and Bax (Romashkova and Makarov 1999, Tsuruta et al. 2002), and forkhead box (FOX) transcriptional factor such as FOX protein O1 (FOXO1) and FOX protein O3α (FOXO-3α) (Guertin et al. 2006). Through these actions, AKT regulates cell growth and metabolism as well as apoptosis. This action of PI3K is antagonized by a lipid phosphatase, phosphatase, and tensin homolog deleted on chromosome 10 (phosphatase and tensin homolog [PTEN]), which dephosphorylates PIP_3 and converts it back to PIP_2. PTEN is identified as a frequently mutated gene in many types of tumors in brain, breast, prostate as well as the liver (Certal et al. 2014, Li et al. 1997). Detailed signaling events of PI3K and AKT activation have been reviewed extensively elsewhere (Kok et al. 2009, Stiles 2009a, Stiles et al. 2004).

In recent years, studies that attempt to elucidate the molecular signals underlying "Warburg effects" have led to the discoveries of novel roles PI3K/AKT signaling plays in mitochondrial function (Stiles 2009b). In addition to regulating pro- and antiapoptotic factors located at the mitochondrial outer membrane (Stiles 2009b), AKT was found to promote binding of hexokinase II to the mitochondrial voltage-dependent anion channel (Gottlob et al. 2001), allowing rapid phosphorylation of available glucose molecules. Furthermore, AKT was also found to be localized in the inner membranes of the mitochondria (Antico Arciuch et al. 2009, Li et al. 2013a). In the mitochondria matrix, AKT is thought to phosphorylate mitochondrial pool of GSK-3β and regulate mitochondrial respiration through phosphorylation/dephosphorylation of pyruvate dehydrogenase (Li et al. 2013a). Interestingly, a FOXO-3 response element has been found on the promoter of a mitochondrial encoded gene, 3-hydroxy-3-methylglutaryl-CoA (HMG-CoA) (Nadal et al. 2002), suggesting that the mitochondria-localized AKT may also play a role in the transcription regulation of mitochondrial DNA.

In the nucleus, the PI3K/AKT signaling controls mitochondrial gene transcription network through multiple different mechanisms. Screening of colon cancer cells induced to express a constitutively active form of FOXO-3α showed that a large number of mitochondrial genes are downregulated when FOXO-3α is induced (Ferber et al. 2012). These genes include the nucleus-encoded auxiliary factors such as Tfam and TFB1M and TFB2M, as well as transcription cofactors PGC-1β and PRC. It is conceivable that FOXO-3α may directly suppress the expression of PGC-1β and PRC since FOXO-3α response element has been reported for the promoter of PGC-1α (Borniquel et al. 2010). Thus, through inhibiting this transcriptional repressor activity of FOXO-3α, AKT may regulate NRF-1/NRF-2 and ERRα transcriptional activity and subsequently the expression of Tfam and TFB1M and TFB2M.

In addition to FOXO-mediated regulation, AKT may also control mitochondrial gene transcription by phosphorylating and activating cAMP response element-binding protein (CREB) transcriptional factor (Du and Montminy 1998, Li et al. 2013b). We found that this effect is independent of the cAMP-mediated activation of protein kinase A (PKA), the common signal that induces CREB phosphorylation, and most likely due to direct phosphorylation of CREB by AKT kinase activity (Li et al. 2013b). When phosphorylated, CREB induces the transcription of PGC-1 in response to glucocorticoid signal to induce gluconeogenesis in fasted mammals (Herzig et al. 2001). In the case of growth factor–mediated AKT activation, induction of PGC-1 also increases the transcriptional activity of ERRα to promote transcription of genes encoding mitochondrial function, including Tfam and TFB1M and TFB2M (Li et al. 2013b). In addition, the AKT substrate consensus sequence has been found on NRF-1. In H4IIE hepatoma cells, phosphorylation of NRF-1 by AKT is reported to mediate the prooxidant t-BOOH-induced Tfam expression (Piantadosi and Suliman 2006). Thus, through directly phosphorylating FOXO and NRF-1 or indirectly induce PGC-1 and ERRα expression, AKT controls the gene transcriptional networks of mitochondria. Consistently, overexpression of NRF-1 and AKT has been shown to mimic the effect of TFAM to abrogate 1-methyl-4-phenyl-2,3-dihydropyridinium ion–induced mitochondrial damage (Piao et al. 2012), confirming a signaling relationship between the PI3K/AKT/FOXO signal and mitochondrial gene transcription regulation.

MAPK Signaling Pathways

ERRα, the transcriptional factor at the apex of the mitochondrial gene network hierarchy, has been studied in breast cancer samples and cell lines for its role on antagonizing the transcriptional activity of ER. Expression of ERRα is generally negatively correlated with breast cancer outcomes and is thought to be a biomarker for human breast cancer (Ariazi et al. 2002, Suzuki et al. 2004). In MCF-7 cells, ERRα was shown to actively modulate the estrogen response by either positively or negatively activating estrogen receptor response element (Kraus et al. 2002). Binding of ERRα to the promoter depends on the nucleotide sequence as well as ERRα phosphorylation (Barry and Giguere 2005, Barry et al. 2006). When bound as a monomer, ERRα is thought to act as a transcriptional repressor due to its inability to recruit a coactivator such as PGC-1 (Ariazi et al. 2007). Phosphorylation of ERRα induces its dimerization and allows its binding to coactivator PGC-1, converting it to a transcriptional activator (Barry et al. 2006). In breast cancer cell lines, phosphorylation of ERRα mediated by the mitogen-activated protein kinase (MAPK) signaling pathway induces its binding to a coactivator GRIP1 and allows gene transcription in an ER-independent manner.

MAPKs are ubiquitously expressed in all cell types and are involved in transducing growth factors, cytokines, and stress signals (Pearson et al. 2001). The activation of the MAPK signaling cascades normally involves a prototype phosphorylation relay with three components: MAPK kinase kinase (MAPKKK), MAPK kinase (MAPKK), and MAPK (Min et al. 2011). Once triggered by extracellular stimulations, the MAPKKK, MAPKK, and MAPK can be phosphorylated and activated in turn, leading to phosphorylation of the target substrates on their specific serine and threonine residues. Three major MAPKs, including extracellular-signal-regulated kinase 1 and 2, c-Jun N-terminal kinases (JNKs), and p38 mitogen-activated protein kinases (p38MAPKs), have been characterized. The Raf/MEK/ERK signaling pathway is sensitive to redox regulation and is regulated by reactive oxygen species (ROS) (Chu et al. 2004, Tamura et al. 2004). JNKs and p38MAPKs are extensively implicated in proapoptotic signaling pathway (Matsuzawa and Ichijo 2001), and the stimulus may include oxidative stress, irradiation, and proinflammatory cytokines such as tumor necrosis factor α (TNFα) (Horbinski and Chu 2005). In C2C12 myotubes, treatment with saturated fatty acids reduces the levels of PGC-1. Such effects of saturated fatty acids appear to be mediated by the stress kinases MEK1/2 and p38MAPK as pharmacological and molecular inhibition of MEK1/2 and p38MAPK reverses the saturated fatty acid–induced effects on PGC-1 (Coll et al. 2006, Crunkhorn et al. 2007). Thus, accumulation of saturated fatty acids may downregulate the mitochondrial functions through this MAPK-PGC-1-mediated signaling pathway to induce insulin resistance in the muscle.

AMPK AND SIRTUIN

Adenosine monophosphate–activated protein kinase (AMPK) is a master regulator of cellular energy homeostasis. This fuel-sensing enzyme responds to low energy levels by switching on catabolic pathways and turning off ATP-consuming pathways. It is mainly activated by stresses and stimuli that deplete ATP levels such as dietary restriction, hypoglycemia, ischemia, and heat shock. AMPK plays a vital role in lipid metabolism by stimulation of fatty acid oxidation and inhibition of lipogenesis. In skeletal muscle, APMK activation enhances fatty acid oxidation by inducing the expression of PPARα and mitochondrial PGC-1α (Jager et al. 2007, Lee et al. 2006). AMPK further phosphorylates PGC-1α at Thr177 and Ser538 to enhance the transcriptional activities of promoters that depends on PGC-1 coactivation signal (Jager et al. 2007). This induction appears to be important for the adaptation to endurance exercise training (Lee et al. 2006, Suwa et al. 2003, Winder et al. 2006). In epididymal fat, AMPK similarly controls the expression of PGC-1 and mitochondrial proteins (Wan et al. 2014). In general, energy deficiency leads to the activation of AMPK, which induces mitochondrial function through PGC-1α (Canto and Auwerx 2009).

AMPK can be activated by various stimuli such as exercise, drugs, caloric restriction, hormones, and cytokines. Caloric restriction is shown to induce the activation of sirtuin 1 (SIRT1), a NAD-dependent deacetylase of the sirtuin family, which further activates AMPK (Silvestre et al. 2014). It has been suggested that the sirtuins play a role in longevity, DNA repair, and the control of metabolic enzymes (Nassir and Ibdah 2014a). Three sirtuins SIRT3, SIRT4, and SIRT5 are localized within the mitochondria matrix (Hirschey et al. 2011), while others are distributed in other parts of the cells. SIRT1, originally characterized as a histone deacetylase, is thought to act as a cellular sensor for NAD that links metabolic alterations with transcriptional outcomes through controlling protein acetylation. SIRT1 can directly interact with and modify the activities of transcriptional factors that may play a role in mitochondrial gene transcription. These include the FOXO transcriptional factor and peroxisome proliferator-activated receptor γ (PPARγ). PGC-1α is also regulated by acetylation. Deacetylation of PGC-1α significantly enhances its transcriptional coactivator activity. While in vivo treatment that induces SIRT1 promotes PGC-1α deacetylation and increases its activity, the endogenous deacetylase for PGC-1α still needs to be confirmed.

SIGNIFICANCE AND CLINICAL IMPLICATIONS IN LIVER DISEASE

Mitochondrial abnormalities have been frequently reported to associate with the development and progression of liver diseases (Garcia-Ruiz et al. 2013, Hsu et al. 2013a, Nassir and Ibdah 2014b). These defects include altered expression of mitochondrial respiratory complexes, mtDNA mutations, abnormal production of ROS, and increased or decreased mitochondrial numbers.

MITOCHONDRIAL ROS PRODUCTION AND LIVER DISEASE

During aerobic respiration, oxidation of acetyl-CoA through TCA cycle generates reduced NADH and $FADH_2$, the electron donors that can transfer electrons to oxygen molecules via respiratory chain located in the inner membrane of mitochondria. Although most of the electrons end up in water, complexes I and III of the respiratory chain are not perfectly insulated from the oxygen (Fromenty and Pessayre 1995). ROS can be generated through the reduction of oxygen to superoxide at these two sites in mitochondria. About 1%–2% of mitochondrial oxygen consumption leads to ROS production under normal condition (Boveris and Chance 1973). This production of ROS can be upregulated when increased NADH and membrane potential are not coupled with an increase in ATP production, a condition in which excessive electrons accumulate at respiratory chain causes leakage of electrons in the mitochondria.

Since the mitochondrial electron transport chain (ETC) components, particularly complex I and complex III, are the major sites for ROS generation, decreased expression or reduced activity of ETC complexes often results in increased ROS production (Gasparre et al. 2008). Hepatocytes are rich in mitochondria given the fact that each hepatocyte contains about 800 mitochondria, occupying approximately 18% of the entire cell volume (Nassir and Ibdah 2014a). Mitochondrial dysfunction and excessive ROS production are highly associated with the development of insulin resistance and the pathogenesis of nonalcoholic steatohepatitis (NASH) when there is an increased oxidation of fatty acids (Begriche et al. 2006). Similarly, alcoholic steatohepatitis (ASH) where the levels of NADH increase due to an enhanced alcohol metabolism (Mansouri et al. 1999) is also highly associated with mitochondrial dysfunction and ROS production.

Mitochondrial dysfunction and ROS production is also proposed to induce oncogenic transformation and promote tumor progression and metastasis (Zager et al. 2006). In human SK-Hep1 hepatoma cell line, mtDNA depletion is able to induce resistance to oxidative stress and chemotherapeutic agents by increasing the expression of manganese superoxide dismutase (MnSOD) (Park et al. 2004). Consistently, inhibiting mitochondrial respiration resulted in cisplatin resistance in human hepatoma HepG2 cells and cell migration in other hepatoma cells due to upregulated amphiregulin, an epidermal growth factor, induced by mitochondrial dysfunction (Chang et al. 2009). Furthermore, it has been demonstrated that antioxidants and calcium chelators can block the expression of amphiregulin induced by mitochondrial dysfunction and therefore prevent cisplatin resistance and cell migration, which indicates that the increased ROS production and Ca^{2+} mobilization are involved in the malignant changes induced by mitochondrial dysfunction in hepatocellular carcinoma (HCC) (Chang et al. 2009).

MITOCHONDRIAL DNA MUTATION AND HEPATOCELLULAR CARCINOMA

The first identified mtDNA mutation which has a functional significance in cancer was the deletion of 294 nucleotides in NADH dehydrogenase-1 (ND1), a subunit of complex I, in renal adenocarcinoma patients (Horton et al. 1996). This mutation has led to decreased complex I activity, associated with high metastatic potential. Low expressions of complex III and complex IV were also reported in different cancer types afterward (Cavelier et al. 1995, Dasgupta et al. 2008). Among the identified point mutations in HCC, it is known that 76% are located in the D-loop region, 2% are located in rRNA genes, 3% are located in tRNA genes, and 19% are located in mRNA genes (Hsu et al. 2013a). As a hot spot for somatic mtDNA mutations in HCC, the D-loop region of mtDNA is the most susceptible site to the damage caused by ROS. Since the D-loop region regulates the replication and transcription of mtDNA (Shadel and Clayton 1997), a study has shown that the point mutations in the D-loop, especially the one close to the replication origin of the heavy-strand of mtDNA, can greatly affect the copy number of mtDNA, which is often altered in HCC (Lee et al. 2004).

Despite the strong association between mitochondrial defects and cancer, the current challenge is to determine whether mutated mitochondrial genome contribute to tumor progression. A study overexpressed a mutated mitochondrial cytochrome b with a 21-bp deletion in a mouse carcinoma cell line and showed that the overexpression led to increased ROS production, increased cell growth, and accelerated xenografted tumor formation (Dasgupta et al. 2008). Interestingly, the mutated cytochrome b overexpression simultaneously resulted in both elevated glycolysis and oxygen consumption, which together with recent clinical reports showing mitochondrial function remain intact or even enhanced in many cancer cells, might refute the hypothesis that the respiratory defects confer the glycolytic metabolism in tumor cells (Lynam-Lennon et al. 2014, Pasto et al. 2014). Furthermore, increased mtDNA copy numbers have been reported in various tumor types, including head and neck squamous cell carcinoma (Kim et al. 2004), thyroid carcinoma (Rogounovitch et al. 2002), lung cancer (Bonner et al. 2009), and liver cancer (Vivekanandan et al. 2010). However, decreased mtDNA copy numbers have also been reported in a number of cancer cases (Mambo et al. 2005). Whether the change in mtDNA copy numbers is an adaptive response to mitochondrial dysfunction or is due to dysregulated nuclear transcriptional machinery requires further investigation.

MITOCHONDRIAL DYSFUNCTION AND OTHER CELLULAR SIGNALING EFFECTS

In addition to ROS production and mtDNA damage, mitochondrial dysfunction can also reduce intracellular ATP content, which has been shown to subsequently repress the expression of HIF-1 via the activation of the AMPK-mTOR pathway in HepG2 cells (Hsu et al. 2013b). HIF-1 is a nuclear transcription factor that is critical for the process such as angiogenesis and metastasis during cancer progression. Hence, these findings suggest that the mitochondrial dysfunction induced activation of retrograde signaling from mitochondria to the nucleus may play a key role in the malignant progression of HCC (Hsu et al. 2013a).

Despite mtDNA dysfunctions that have been identified in HCC, it still remains unknown whether a specific mtDNA mutation is the driving force or just an indirect consequence of HCC progression. Therefore, further investigations are required to address this question in the future. Finally, mutations of the nuclear DNAs that encode genes involved in the mtDNA integrity and mitochondrial function could also play an important role in carcinogenesis. Studies have shown that defects in P53 (Achanta et al. 2005), mitochondrial DNA polymerase (Singh et al. 2009), and SIRT3 (Kim et al. 2010) can affect mtDNA integrity and promote tumorigenesis. Therefore, the interactions between mtDNA and nuclear DNA may play an important role in the development of liver disease, particularly HCC, and could underlie energy metabolism reprogramming in the tumor cells (Hsu et al. 2013a).

In summary, the mitochondrial transcription network is controlled at both the mitochondrial and nuclear DNA transcriptions. These signals are coordinately regulated by cell signaling pathways that control metabolism, cell growth, and migration, as well as other cellular processes. The dysfunction of mitochondria results in the production of ROS and affects mitochondrial signaling, membrane function, and mtDNA integrity. These effects ultimately affect the ability of the cells to function properly and contribute to the pathogenesis of various liver diseases including NASH, ASH, and HCC, among others.

ACKNOWLEDGMENTS

Dr. Stiles acknowledges the support from NCI (R01CA154986-01 and R01 DK084241-01), USC Liver Center (P30DK48522), and USC TREC (5U 54 CA11486). Anketse Debebe acknowledges funding from the California Institute of Regenerative Medicine.

REFERENCES

Achanta, G., R. Sasaki, L. Feng, J. S. Carew, W. Lu, H. Pelicano, M. J. Keating, and P. Huang. 2005. Novel role of p53 in maintaining mitochondrial genetic stability through interaction with DNA Pol γ. *EMBO J* 24 (19):3482–3492. doi: 10.1038/sj.emboj.7600819.

Andersson, U. and R. C. Scarpulla. 2001. Pgc-1-related coactivator, a novel, serum-inducible coactivator of nuclear respiratory factor 1-dependent transcription in mammalian cells. *Mol Cell Biol* 21 (11):3738–3749. doi: 10.1128/MCB.21.11.3738-3749.2001.

Antico Arciuch, V. G., S. Galli, M. C. Franco, P. Y. Lam, E. Cadenas, M. C. Carreras, and J. J. Poderoso. 2009. Akt1 intramitochondrial cycling is a crucial step in the redox modulation of cell cycle progression. *PLoS One* 4 (10):e7523. doi: 10.1371/journal.pone.0007523.

Ariazi, E. A., G. M. Clark, and J. E. Mertz. 2002. Estrogen-related receptor α and estrogen-related receptor γ associate with unfavorable and favorable biomarkers, respectively, in human breast cancer. *Cancer Res* 62 (22):6510–6518.

Ariazi, E. A., R. J. Kraus, M. L. Farrell, V. C. Jordan, and J. E. Mertz. 2007. Estrogen-related receptor α1 transcriptional activities are regulated in part via the ErbB2/HER2 signaling pathway. *Mol Cancer Res* 5 (1):71–85. doi: 10.1158/1541-7786.MCR-06-0227.

Au, H. C. and I. E. Scheffler. 1998. Promoter analysis of the human succinate dehydrogenase iron-protein gene—Both nuclear respiratory factors NRF-1 and NRF-2 are required. *Eur J Biochem* 251 (1–2):164–174.

Baar, K., Z. Song, C. F. Semenkovich, T. E. Jones, D. H. Han, L. A. Nolte, E. O. Ojuka, M. Chen, and J. O. Holloszy. 2003. Skeletal muscle overexpression of nuclear respiratory factor 1 increases glucose transport capacity. *FASEB J* 17 (12):1666–1673. doi: 10.1096/fj.03-0049com.

Barry, J. B. and V. Giguere. 2005. Epidermal growth factor-induced signaling in breast cancer cells results in selective target gene activation by orphan nuclear receptor estrogen-related receptor α. *Cancer Res* 65 (14):6120–6129. doi: 10.1158/0008-5472. CAN-05-0922.

Barry, J. B., J. Laganiere, and V. Giguere. 2006. A single nucleotide in an estrogen-related receptor α site can dictate mode of binding and peroxisome proliferator-activated receptor γ coactivator 1-α activation of target promoters. *Mol Endocrinol* 20 (2):302–310. doi: 10.1210/me.2005-0313.

Begriche, K., A. Igoudjil, D. Pessayre, and B. Fromenty. 2006. Mitochondrial dysfunction in NASH: Causes, consequences and possible means to prevent it. *Mitochondrion* 6 (1):1–28. doi: 10.1016/j.mito.2005.10.004.

Bonawitz, N. D., D. A. Clayton, and G. S. Shadel. 2006. Initiation and beyond: Multiple functions of the human mitochondrial transcription machinery. *Mol Cell* 24 (6):813–825. doi: 10.1016/j.molcel.2006.11.024.

Bonner, M. R., M. Shen, C. S. Liu, M. Divita, X. He, and Q. Lan. 2009. Mitochondrial DNA content and lung cancer risk in Xuan Wei, China. *Lung Cancer* 63 (3):331–334. doi: 10.1016/j.lungcan.2008.06.012.

Borniquel, S., N. Garcia-Quintans, I. Valle, Y. Olmos, B. Wild, F. Martinez-Granero, E. Soria, S. Lamas, and M. Monsalve. 2010. Inactivation of FOXO-3α and subsequent downregulation of PGC-1α mediate nitric oxide-induced endothelial cell migration. *Mol Cell Biol* 30 (16):4035–44. doi: 10.1128/MCB.00175-10.

Boveris, A. and B. Chance. 1973. The mitochondrial generation of hydrogen peroxide. General properties and effect of hyperbaric oxygen. *Biochem J* 134 (3):707–716.

Cam, H., E. Balciunaite, A. Blais, A. Spektor, R. C. Scarpulla, R. Young, Y. Kluger, and B. D. Dynlacht. 2004. A common set of gene regulatory networks links metabolism and growth inhibition. *Mol Cell* 16 (3):399–411. doi: 10.1016/j.molcel.2004.09.037.

Canto, C. and J. Auwerx. 2009. PGC-1α, SIRT1 and AMPK, an energy sensing network that controls energy expenditure. *Curr Opin Lipidol* 20 (2):98–105. doi: 10.1097/MOL.0b013e328328d0a4.

Carter, R. S., N. K. Bhat, A. Basu, and N. G. Avadhani. 1992. The basal promoter elements of murine cytochrome c oxidase subunit IV gene consist of tandemly duplicated ets motifs that bind to GABP-related transcription factors. *J Biol Chem* 267 (32):23418–23426.

Cavelier, L., E. E. Jazin, I. Eriksson, J. Prince, U. Bave, L. Oreland, and U. Gyllensten. 1995. Decreased cytochrome-c oxidase activity and lack of age-related accumulation of mitochondrial DNA deletions in the brains of schizophrenics. *Genomics* 29 (1):217–224. doi: 10.1006/geno.1995.1234.

Certal, V., J. C. Carry, F. Halley, A. Virone-Oddos, F. Thompson, B. Filoche-Romme, Y. El-Ahmad et al. 2014. Discovery and optimization of pyrimidone indoline amide PI3Kβ inhibitors for the treatment of phosphatase and tensin homologue (PTEN)-deficient cancers. *J Med Chem* 57 (3):903–920. doi: 10.1021/jm401642q.

Chang, C. J., P. H. Yin, D. M. Yang, C. H. Wang, W. Y. Hung, C. W. Chi, Y. H. Wei, and H. C. Lee. 2009. Mitochondrial dysfunction-induced amphiregulin upregulation mediates chemo-resistance and cell migration in HepG2 cells. *Cell Mol Life Sci* 66 (10):1755–1765. doi: 10.1007/s00018-009-8767-5.

Chu, C. T., D. J. Levinthal, S. M. Kulich, E. M. Chalovich, and D. B. DeFranco. 2004. Oxidative neuronal injury. The dark side of ERK1/2. *Eur J Biochem* 271 (11):2060–2066. doi: 10.1111/j.1432-1033.2004.04132.x.

Clayton, D. A. 2000. Vertebrate mitochondrial DNA-a circle of surprises. *Exp Cell Res* 255 (1):4–9. doi: 10.1006/excr.1999.4763.

Clayton, D. A. 2003. Mitochondrial DNA replication: What we know. *IUBMB Life* 55 (4–5):213–217. doi: 10.1080/1521654031000134824.

Coll, T., M. Jove, R. Rodriguez-Calvo, E. Eyre, X. Palomer, R. M. Sanchez, M. Merlos, J. C. Laguna, and M. Vazquez-Carrera. 2006. Palmitate-mediated downregulation of peroxisome proliferator-activated receptor-γ coactivator 1α in skeletal muscle cells involves MEK1/2 and nuclear factor-κB activation. *Diabetes* 55 (10):2779–2787. doi: 10.2337/db05-1494.

Crunkhorn, S., F. Dearie, C. Mantzoros, H. Gami, W. S. da Silva, D. Espinoza, R. Faucette, K. Barry, A. C. Bianco, and M. E. Patti. 2007. Peroxisome proliferator activator receptor γ coactivator-1 expression is reduced in obesity: Potential pathogenic role of saturated fatty acids and p38 mitogen-activated protein kinase activation. *J Biol Chem* 282 (21):15439–15450. doi: 10.1074/jbc.M611214200.

Daga, A., V. Micol, D. Hess, R. Aebersold, and G. Attardi. 1993. Molecular characterization of the transcription termination factor from human mitochondria. *J Biol Chem* 268 (11):8123–8130.

Dasgupta, S., M. O. Hoque, S. Upadhyay, and D. Sidransky. 2008. Mitochondrial cytochrome B gene mutation promotes tumor growth in bladder cancer. *Cancer Res* 68 (3):700–706. doi: 10.1158/0008-5472.CAN-07-5532.

Du, K. and M. Montminy. 1998. CREB is a regulatory target for the protein kinase Akt/PKB. *J Biol Chem* 273 (49):32377–32379.

Dufour, C. R., B. J. Wilson, J. M. Huss, D. P. Kelly, W. A. Alaynick, M. Downes, R. M. Evans, M. Blanchette, and V. Giguere. 2007. Genome-wide orchestration of cardiac functions by the orphan nuclear receptors ERRα and γ. *Cell Metab* 5 (5):345–356. doi: 10.1016/j.cmet.2007.03.007.

Evans, M. J. and R. C. Scarpulla. 1990. NRF-1: A trans-activator of nuclear-encoded respiratory genes in animal cells. *Genes Dev* 4 (6):1023–1034.

Ferber, E. C., B. Peck, O. Delpuech, G. P. Bell, P. East, and A. Schulze. 2012. FOXO3a regulates reactive oxygen metabolism by inhibiting mitochondrial gene expression. *Cell Death Differ* 19 (6):968–979. doi: 10.1038/cdd.2011.179.

Finck, B. N. and D. P. Kelly. 2006. PGC-1 coactivators: Inducible regulators of energy metabolism in health and disease. *J Clin Invest* 116 (3):615–622. doi: 10.1172/JCI27794.

Fromenty, B. and D. Pessayre. 1995. Inhibition of mitochondrial beta-oxidation as a mechanism of hepatotoxicity. *Pharmacol Ther* 67 (1):101–154.

Garcia-Ruiz, C., N. Kaplowitz, and J. C. Fernandez-Checa. 2013. Role of mitochondria in alcoholic liver disease. *Curr Pathobiol Rep* 1 (3):159–168. doi: 10.1007/s40139-013-0021-z.

Gasparre, G., E. Hervouet, E. de Laplanche, J. Demont, L. F. Pennisi, M. Colombel, F. Mege-Lechevallier et al. 2008. Clonal expansion of mutated mitochondrial DNA is associated with tumor formation and complex I deficiency in the benign renal oncocytoma. *Hum Mol Genet* 17 (7):986–995. doi: 10.1093/hmg/ddm371.

Giguere, V., N. Yang, P. Segui, and R. M. Evans. 1988. Identification of a new class of steroid hormone receptors. *Nature* 331 (6151):91–94. doi: 10.1038/331091a0.

Gleyzer, N., K. Vercauteren, and R. C. Scarpulla. 2005. Control of mitochondrial transcription specificity factors (TFB1M and TFB2M) by nuclear respiratory factors (NRF-1 and NRF-2) and PGC-1 family coactivators. *Mol Cell Biol* 25 (4):1354–1366. doi: 10.1128/MCB.25.4.1354-1366.2005.

Gottlob, K., N. Majewski, S. Kennedy, E. Kandel, R. B. Robey, and N. Hay. 2001. Inhibition of early apoptotic events by Akt/PKB is dependent on the first committed step of glycolysis and mitochondrial hexokinase. *Genes Dev* 15 (11):1406–1418. doi: 10.1101/gad.889901.

Guertin, D. A., D. M. Stevens, C. C. Thoreen, A. A. Burds, N. Y. Kalaany, J. Moffat, M. Brown, K. J. Fitzgerald, and D. M. Sabatini. 2006. Ablation in mice of the mTORC components raptor, rictor, or mLST8 reveals that mTORC2 is required for signaling to Akt-FOXO and PKCα, but not S6K1. *Dev Cell* 11 (6):859–871. doi: 10.1016/j.devcel.2006.10.007.

Gugneja, S., J. V. Virbasius, and R. C. Scarpulla. 1995. Four structurally distinct, non-DNA-binding subunits of human nuclear respiratory factor 2 share a conserved transcriptional activation domain. *Mol Cell Biol* 15 (1):102–111.

He, X., C. Sun, F. Wang, A. Shan, T. Guo, W. Gu, B. Cui, and G. Ning. 2012. Peri-implantation lethality in mice lacking the PGC-1-related coactivator protein. *Dev Dyn* 241 (5):975–983. doi: 10.1002/dvdy.23769.

Herzig, R. P., S. Scacco, and R. C. Scarpulla. 2000. Sequential serum-dependent activation of CREB and NRF-1 leads to enhanced mitochondrial respiration through the induction of cytochrome c. *J Biol Chem* 275 (17):13134–13141.

Herzig, S., F. Long, U. S. Jhala, S. Hedrick, R. Quinn, A. Bauer, D. Rudolph et al. 2001. CREB regulates hepatic gluconeogenesis through the coactivator PGC-1. *Nature* 413 (6852):179–183. doi: 10.1038/35093131.

Hirschey, M. D., T. Shimazu, E. Jing, C. A. Grueter, A. M. Collins, B. Aouizerat, A. Stancakova et al. 2011. SIRT3 deficiency and mitochondrial protein hyperacetylation accelerate the development of the metabolic syndrome. *Mol Cell* 44 (2):177–190. doi: 10.1016/j.molcel.2011.07.019.

Horbinski, C. and C. T. Chu. 2005. Kinase signaling cascades in the mitochondrion: A matter of life or death. *Free Radic Biol Med* 38 (1):2–11. doi: 10.1016/j.freeradbiomed.2004.09.030.

Horton, T. M., J. A. Petros, A. Heddi, J. Shoffner, A. E. Kaufman, S. D. Graham, Jr., T. Gramlich, and D. C. Wallace. 1996. Novel mitochondrial DNA deletion found in a renal cell carcinoma. *Genes Chromosomes Cancer* 15 (2):95–101. doi: 10.1002/(SICI)1098-2264(199602)15:2<95::AID-GCC3>3.0.CO;2-Z.

Hsu, C. C., H. C. Lee, and Y. H. Wei. 2013a. Mitochondrial DNA alterations and mitochondrial dysfunction in the progression of hepatocellular carcinoma. *World J Gastroenterol* 19 (47):8880–8886. doi: 10.3748/wjg.v19.i47.8880.

Hsu, C. C., C. H. Wang, L. C. Wu, C. Y. Hsia, C. W. Chi, P. H. Yin, C. J. Chang, M. T. Sung, Y. H. Wei, S. H. Lu, and H. C. Lee. 2013b. Mitochondrial dysfunction represses HIF-1α protein synthesis through AMPK activation in human hepatoma HepG2 cells. *Biochim Biophys Acta* 1830 (10):4743–4751. doi: 10.1016/j.bbagen.2013.06.004.

Jager, S., C. Handschin, J. St-Pierre, and B. M. Spiegelman. 2007. AMP-activated protein kinase (AMPK) action in skeletal muscle via direct phosphorylation of PGC-1α. *Proc Natl Acad Sci USA* 104 (29):12017–12022. doi: 10.1073/pnas.0705070104.

Kelly, D. P. and R. C. Scarpulla. 2004. Transcriptional regulatory circuits controlling mitochondrial biogenesis and function. *Genes Dev* 18 (4):357–368. doi: 10.1101/gad.1177604.

Kim, H. S., K. Patel, K. Muldoon-Jacobs, K. S. Bisht, N. Aykin-Burns, J. D. Pennington, R. van der Meer et al. 2010. SIRT3 is a mitochondria-localized tumor suppressor required for maintenance of mitochondrial integrity and metabolism during stress. *Cancer Cell* 17 (1):41–52. doi: 10.1016/j.ccr.2009.11.023.

Kim, M. M., J. D. Clinger, B. G. Masayesva, P. K. Ha, M. L. Zahurak, W. H. Westra, and J. A. Califano. 2004. Mitochondrial DNA quantity increases with histopathologic grade in premalignant and malignant head and neck lesions. *Clin Cancer Res* 10 (24):8512–8515. doi: 10.1158/1078-0432.CCR-04-0734.

Kok, K., B. Geering, and B. Vanhaesebroeck. 2009. Regulation of phosphoinositide 3-kinase expression in health and disease. *Trends Biochem Sci* 34 (3):115–127.

Kraus, R. J., E. A. Ariazi, M. L. Farrell, and J. E. Mertz. 2002. Estrogen-related receptor α 1 actively antagonizes estrogen receptor-regulated transcription in MCF-7 mammary cells. *J Biol Chem* 277 (27):24826–24834. doi: 10.1074/jbc.M202952200.

Lai, L., T. C. Leone, C. Zechner, P. J. Schaeffer, S. M. Kelly, D. P. Flanagan, D. M. Medeiros, A. Kovacs, and D. P. Kelly. 2008. Transcriptional coactivators PGC-1α and PGC-1β control overlapping programs required for perinatal maturation of the heart. *Genes Dev* 22 (14):1948–1961. doi: 10.1101/gad.1661708.

LaMarco, K. L. and S. L. McKnight. 1989. Purification of a set of cellular polypeptides that bind to the purine-rich *cis*-regulatory element of herpes simplex virus immediate early genes. *Genes Dev* 3 (9):1372–1383.

Larsson, N. G., J. Wang, H. Wilhelmsson, A. Oldfors, P. Rustin, M. Lewandoski, G. S. Barsh, and D. A. Clayton. 1998. Mitochondrial transcription factor A is necessary for mtDNA maintenance and embryogenesis in mice. *Nat Genet* 18 (3):231–236. doi: 10.1038/ng0398-231.

Lee, H. C., S. H. Li, J. C. Lin, C. C. Wu, D. C. Yeh, and Y. H. Wei. 2004. Somatic mutations in the D-loop and decrease in the copy number of mitochondrial DNA in human hepatocellular carcinoma. *Mutat Res* 547 (1–2):71–78. doi: 10.1016/j.mrfmmm.2003.12.011.

Lee, W. J., M. Kim, H. S. Park, H. S. Kim, M. J. Jeon, K. S. Oh, E. H. Koh et al. 2006. AMPK activation increases fatty acid oxidation in skeletal muscle by activating PPARα and PGC-1. *Biochem Biophys Res Commun* 340 (1):291–295. doi: 10.1016/j.bbrc.2005.12.011.

Leone, T. C., J. J. Lehman, B. N. Finck, P. J. Schaeffer, A. R. Wende, S. Boudina, M. Courtois et al. 2005. PGC-1α deficiency causes multi-system energy metabolic derangements: Muscle dysfunction, abnormal weight control and hepatic steatosis. *PLoS Biol* 3 (4):e101. doi: 10.1371/journal.pbio.0030101.

Li, C., Y. Li, L. He, A. R. Agarwal, N. Zeng, E. Cadenas, and B. L. Stiles. 2013a. PI3K/AKT signaling regulates bioenergetics in immortalized hepatocytes. *Free Radic Biol Med* 60:29–40. doi: 10.1016/j.freeradbiomed.2013.01.013.

Li, J., C. Yen, D. Liaw, K. Podsypanina, S. Bose, S. I. Wang, J. Puc et al. 1997. PTEN, a putative protein tyrosine phosphatase gene mutated in human brain, breast, and prostate cancer. *Science* 275 (5308):1943–1947.

Li, Y., L. He, N. Zeng, D. Sahu, E. Cadenas, C. Shearn, W. Li, and B. L. Stiles. 2013b. Phosphatase and tensin homolog deleted on chromosome 10 (PTEN) signaling regulates mitochondrial biogenesis and respiration via estrogen-related receptor α (ERRα). *J Biol Chem* 288 (35):25007–25024. doi: 10.1074/jbc.M113.450353.

Lin, J., P. Puigserver, J. Donovan, P. Tarr, and B. M. Spiegelman. 2002. Peroxisome proliferator-activated receptor γ coactivator 1β (PGC-1β), a novel PGC-1-related transcription coactivator associated with host cell factor. *J Biol Chem* 277 (3):1645–1648. doi: 10.1074/jbc.C100631200.

Lin, J., P. H. Wu, P. T. Tarr, K. S. Lindenberg, J. St-Pierre, C. Y. Zhang, V. K. Mootha et al. 2004. Defects in adaptive energy metabolism with CNS-linked hyperactivity in PGC-1α null mice. *Cell* 119 (1):121–135. doi: 10.1016/j.cell.2004.09.013.

Luo, J., R. Sladek, J. Carrier, J. A. Bader, D. Richard, and V. Giguere. 2003. Reduced fat mass in mice lacking orphan nuclear receptor estrogen-related receptor α. *Mol Cell Biol* 23 (22):7947–7956.

Lynam-Lennon, N., S. G. Maher, A. Maguire, J. Phelan, C. Muldoon, J. V. Reynolds, and J. O'Sullivan. 2014. Altered mitochondrial function and energy metabolism is associated with a radioresistant phenotype in oesophageal adenocarcinoma. *PLoS One* 9 (6):e100738. doi: 10.1371/journal.pone.0100738.

Mambo, E., A. Chatterjee, M. Xing, G. Tallini, B. R. Haugen, S. C. Yeung, S. Sukumar, and D. Sidransky. 2005. Tumor-specific changes in mtDNA content in human cancer. *Int J Cancer* 116 (6):920–924. doi: 10.1002/ijc.21110.

Mansouri, A., I. Gaou, C. De Kerguenec, S. Amsellem, D. Haouzi, A. Berson, A. Moreau, G. Feldmann, P. Letteron, D. Pessayre, and B. Fromenty. 1999. An alcoholic binge causes massive degradation of hepatic mitochondrial DNA in mice. *Gastroenterology* 117 (1):181–190.

Matsuzawa, A. and H. Ichijo. 2001. Molecular mechanisms of the decision between life and death: Regulation of apoptosis by apoptosis signal-regulating kinase 1. *J Biochem* 130 (1):1–8.

McCulloch, V. and G. S. Shadel. 2003. Human mitochondrial transcription factor B1 interacts with the C-terminal activation region of h-mtTFA and stimulates transcription independently of its RNA methyltransferase activity. *Mol Cell Biol* 23 (16):5816–5824.

Metodiev, M. D., N. Lesko, C. B. Park, Y. Camara, Y. Shi, R. Wibom, K. Hultenby, C. M. Gustafsson, and N. G. Larsson. 2009. Methylation of 12S rRNA is necessary for in vivo stability of the small subunit of the mammalian mitochondrial ribosome. *Cell Metab* 9 (4):386–397. doi: 10.1016/j.cmet.2009.03.001.

Min, L., B. He, and L. Hui. 2011. Mitogen-activated protein kinases in hepatocellular carcinoma development. *Semin Cancer Biol* 21 (1):10–20. doi: 10.1016/j.semcancer.2010.10.011.

Mootha, V. K., C. Handschin, D. Arlow, X. Xie, J. St Pierre, S. Sihag, W. Yang et al. 2004. Errα and Gabpa/b specify PGC-1α-dependent oxidative phosphorylation gene expression that is altered in diabetic muscle. *Proc Natl Acad Sci USA* 101 (17):6570–6575. doi: 10.1073/pnas.0401401101.

Nadal, A., P. F. Marrero, and D. Haro. 2002. Down-regulation of the mitochondrial 3-hydroxy-3-methylglutaryl-CoA synthase gene by insulin: The role of the forkhead transcription factor FKHRL1. *Biochem J* 366 (Pt 1):289–297. doi: 10.1042/BJ20020598.

Nassir, F. and J. A. Ibdah. 2014a. Role of mitochondria in alcoholic liver disease. *World J Gastroenterol* 20 (9):2136–2142. doi: 10.3748/wjg.v20.i9.2136.

Nassir, F. and J. A. Ibdah. 2014b. Role of mitochondria in nonalcoholic fatty liver disease. *Int J Mol Sci* 15 (5):8713–8742. doi: 10.3390/ijms15058713.

Ongwijitwat, S. and M. T. Wong-Riley. 2005. Is nuclear respiratory factor 2 a master transcriptional coordinator for all ten nuclear-encoded cytochrome c oxidase subunits in neurons? *Gene* 360 (1):65–77. doi: 10.1016/j.gene.2005.06.015.

Park, C. B., J. Asin-Cayuela, Y. Camara, Y. Shi, M. Pellegrini, M. Gaspari, R. Wibom et al. 2007. MTERF3 is a negative regulator of mammalian mtDNA transcription. *Cell* 130 (2):273–285. doi: 10.1016/j.cell.2007.05.046.

Park, S. Y., I. Chang, J. Y. Kim, S. W. Kang, S. H. Park, K. Singh, and M. S. Lee. 2004. Resistance of mitochondrial DNA-depleted cells against cell death: Role of mitochondrial superoxide dismutase. *J Biol Chem* 279 (9):7512–7520. doi: 10.1074/jbc.M307677200.

Pasto, A., C. Bellio, G. Pilotto, V. Ciminale, M. Silic-Benussi, G. Guzzo, A. Rasola et al. 2014. Cancer stem cells from epithelial ovarian cancer patients privilege oxidative phosphorylation, and resist glucose deprivation. *Oncotarget* 5 (12):4305–4319.

Pearson, G., F. Robinson, T. Beers Gibson, B. E. Xu, M. Karandikar, K. Berman, and M. H. Cobb. 2001. Mitogen-activated protein (MAP) kinase pathways: Regulation and physiological functions. *Endocr Rev* 22 (2):153–183. doi: 10.1210/edrv.22.2.0428.

Piantadosi, C. A. and H. B. Suliman. 2006. Mitochondrial transcription factor A induction by redox activation of nuclear respiratory factor 1. *J Biol Chem* 281 (1):324–333. doi: 10.1074/jbc.M508805200.

Piao, Y., H. G. Kim, M. S. Oh, and Y. K. Pak. 2012. Overexpression of TFAM, NRF-1 and myr-AKT protects the MPP(+)-induced mitochondrial dysfunctions in neuronal cells. *Biochim Biophys Acta* 1820 (5):577–585. doi: 10.1016/j.bbagen.2011.08.007.

Puigserver, P., Z. Wu, C. W. Park, R. Graves, M. Wright, and B. M. Spiegelman. 1998. A cold-inducible coactivator of nuclear receptors linked to adaptive thermogenesis. *Cell* 92 (6):829–839.

Rangwala, S. M., X. Li, L. Lindsley, X. Wang, S. Shaughnessy, T. G. Daniels, J. Szustakowski, N. R. Nirmala, Z. Wu, and S. C. Stevenson. 2007. Estrogen-related receptor α is essential for the expression of antioxidant protection genes and mitochondrial function. *Biochem Biophys Res Commun* 357 (1):231–236. doi: 10.1016/j.bbrc.2007.03.126.

Rogounovitch, T. I., V. A. Saenko, Y. Shimizu-Yoshida, A. Y. Abrosimov, E. F. Lushnikov, P. O. Roumiantsev, A. Ohtsuru, H. Namba, A. F. Tsyb, and S. Yamashita. 2002. Large deletions in mitochondrial DNA in radiation-associated human thyroid tumors. *Cancer Res* 62 (23):7031–7041.

Romashkova, J. A. and S. S. Makarov. 1999. NF-κB is a target of AKT in anti-apoptotic PDGF signalling. *Nature* 401 (6748):86–90. doi: 10.1038/43474.

Russell, L. K., C. M. Mansfield, J. J. Lehman, A. Kovacs, M. Courtois, J. E. Saffitz, D. M. Medeiros, M. L. Valencik, J. A. McDonald, and D. P. Kelly. 2004. Cardiac-specific induction of the transcriptional coactivator peroxisome proliferator-activated receptor γ coactivator-1α promotes mitochondrial biogenesis and reversible cardiomyopathy in a developmental stage-dependent manner. *Circ Res* 94 (4):525–533. doi: 10.1161/01. RES.0000117088.36577.EB.

Scarpulla, R. C. 2011. Metabolic control of mitochondrial biogenesis through the PGC-1 family regulatory network. *Biochim Biophys Acta* 1813 (7):1269–1278. doi: 10.1016/ j.bbamcr.2010.09.019.

Shadel, G. S. and D. A. Clayton. 1997. Mitochondrial DNA maintenance in vertebrates. *Annu Rev Biochem* 66:409–435. doi: 10.1146/annurev.biochem.66.1.409.

Silvestre, M. F., B. Viollet, P. W. Caton, J. Leclerc, I. Sakakibara, M. Foretz, M. C. Holness, and M. C. Sugden. 2014. The AMPK-SIRT signaling network regulates glucose tolerance under calorie restriction conditions. *Life Sci* 100 (1):55–60. doi: 10.1016/ j.lfs.2014.01.080.

Singh, K. K., V. Ayyasamy, K. M. Owens, M. S. Koul, and M. Vujcic. 2009. Mutations in mitochondrial DNA polymerase-γ promote breast tumorigenesis. *J Hum Genet* 54 (9):516–524. doi: 10.1038/jhg.2009.71.

Sladek, R., J. A. Bader, and V. Giguere. 1997. The orphan nuclear receptor estrogen-related receptor α is a transcriptional regulator of the human medium-chain acyl coenzyme A dehydrogenase gene. *Mol Cell Biol* 17 (9):5400–5409.

Sonoda, J., I. R. Mehl, L. W. Chong, R. R. Nofsinger, and R. M. Evans. 2007. PGC-1β controls mitochondrial metabolism to modulate circadian activity, adaptive thermogenesis, and hepatic steatosis. *Proc Natl Acad Sci USA* 104 (12):5223–5228. doi: 10.1073/ pnas.0611623104.

Stein, R. A. and D. P. McDonnell. 2006. Estrogen-related receptor α as a therapeutic target in cancer. *Endocr Relat Cancer* 13 (Suppl 1):S25–S32. doi: 10.1677/erc.1.01292.

Stiles, B., M. Groszer, S. Wang, J. Jiao, and H. Wu. 2004. PTENless means more. *Dev Biol* 273 (2):175–184. doi: 10.1016/j.ydbio.2004.06.008.

Stiles, B. L. 2009a. Phosphatase and tensin homologue deleted on chromosome 10: Extending its PTENtacles. *Int J Biochem Cell Biol* 41 (4):757–761.

Stiles, B. L. 2009b. PI-3-K and AKT: Onto the mitochondria. *Adv Drug Deliv Rev* 61 (14):1276–1282. doi: 10.1016/j.addr.2009.07.017.

Suwa, M., H. Nakano, and S. Kumagai. 2003. Effects of chronic AICAR treatment on fiber composition, enzyme activity, UCP3, and PGC-1 in rat muscles. *J Appl Physiol (1985)* 95 (3):960–968. doi: 10.1152/japplphysiol.00349.2003.

Suzuki, T., Y. Miki, T. Moriya, N. Shimada, T. Ishida, H. Hirakawa, N. Ohuchi, and H. Sasano. 2004. Estrogen-related receptor α in human breast carcinoma as a potent prognostic factor. *Cancer Res* 64 (13):4670–4676. doi: 10.1158/0008-5472.CAN-04-0250.

Takahashi, Y., K. Kako, H. Arai, T. Ohishi, Y. Inada, A. Takehara, A. Fukamizu, and E. Munekata. 2002. Characterization and identification of promoter elements in the mouse COX17 gene. *Biochim Biophys Acta* 1574 (3):359–364.

Tamura, Y., S. Simizu, and H. Osada. 2004. The phosphorylation status and anti-apoptotic activity of Bcl-2 are regulated by ERK and protein phosphatase 2A on the mitochondria. *FEBS Lett* 569 (1–3):249–255. doi: 10.1016/j.febslet.2004.06.003.

Tiranti, V., A. Savoia, F. Forti, M. F. D'Apolito, M. Centra, M. Rocchi, and M. Zeviani. 1997. Identification of the gene encoding the human mitochondrial RNA polymerase (h-mtRPOL) by cyberscreening of the Expressed Sequence Tags database. *Hum Mol Genet* 6 (4):615–625.

Tjokroprawiro, A. 2006. New approach in the treatment of T2DM and metabolic syndrome (focus on a novel insulin sensitizer). *Acta Med Indones* 38 (3):160–166.

Truscott, K. N., K. Brandner, and N. Pfanner. 2003. Mechanisms of protein import into mitochondria. *Curr Biol* 13 (8):R326–R337.

Tsuruta, F., N. Masuyama, and Y. Gotoh. 2002. The phosphatidylinositol 3-kinase (PI3K)-Akt pathway suppresses Bax translocation to mitochondria. *J Biol Chem* 277 (16):14040–14047. doi: 10.1074/jbc.M108975200.

Vercauteren, K., N. Gleyzer, and R. C. Scarpulla. 2008. PGC-1-related coactivator complexes with HCF-1 and NRF-2β in mediating NRF-2(GABP)-dependent respiratory gene expression. *J Biol Chem* 283 (18):12102–12111. doi: 10.1074/jbc.M710150200.

Vercauteren, K., R. A. Pasko, N. Gleyzer, V. M. Marino, and R. C. Scarpulla. 2006. PGC-1-related coactivator: Immediate early expression and characterization of a CREB/NRF-1 binding domain associated with cytochrome c promoter occupancy and respiratory growth. *Mol Cell Biol* 26 (20):7409–7419. doi: 10.1128/MCB.00585-06.

Vianna, C. R., M. Huntgeburth, R. Coppari, C. S. Choi, J. Lin, S. Krauss, G. Barbatelli et al. 2006. Hypomorphic mutation of PGC-1β causes mitochondrial dysfunction and liver insulin resistance. *Cell Metab* 4 (6):453–464. doi: 10.1016/j.cmet.2006.11.003.

Villena, J. A., M. B. Hock, W. Y. Chang, J. E. Barcas, V. Giguere, and A. Kralli. 2007. Orphan nuclear receptor estrogen-related receptor α is essential for adaptive thermogenesis. *Proc Natl Acad Sci USA* 104 (4):1418–1423. doi: 10.1073/pnas.0607696104.

Villena, J. A. and A. Kralli. 2008. ERRα: A metabolic function for the oldest orphan. *Trends Endocrinol Metab* 19 (8):269–276. doi: 10.1016/j.tem.2008.07.005.

Virbasius, J. V. and R. C. Scarpulla. 1994. Activation of the human mitochondrial transcription factor A gene by nuclear respiratory factors: A potential regulatory link between nuclear and mitochondrial gene expression in organelle biogenesis. *Proc Natl Acad Sci USA* 91 (4):1309–1313.

Virbasius, J. V., C. A. Virbasius, and R. C. Scarpulla. 1993. Identity of GABP with NRF-2, a multisubunit activator of cytochrome oxidase expression, reveals a cellular role for an ETS domain activator of viral promoters. *Genes Dev* 7 (3):380–392.

Vivekanandan, P., H. Daniel, M. M. Yeh, and M. Torbenson. 2010. Mitochondrial mutations in hepatocellular carcinomas and fibrolamellar carcinomas. *Mod Pathol* 23 (6):790–798. doi: 10.1038/modpathol.2010.51.

Wan, Z., J. Root-McCaig, L. Castellani, B. E. Kemp, G. R. Steinberg, and D. C. Wright. 2014. Evidence for the role of AMPK in regulating PGC-1 α expression and mitochondrial proteins in mouse epididymal adipose tissue. *Obesity* (*Silver Spring*) 22 (3):730–738. doi: 10.1002/oby.20605.

Wang, Z., J. Cotney, and G. S. Shadel. 2007. Human mitochondrial ribosomal protein MRPL12 interacts directly with mitochondrial RNA polymerase to modulate mitochondrial gene expression. *J Biol Chem* 282 (17):12610–12618. doi: 10.1074/jbc.M700461200.

Winder, W. W., E. B. Taylor, and D. M. Thomson. 2006. Role of AMP-activated protein kinase in the molecular adaptation to endurance exercise. *Med Sci Sports Exerc* 38 (11):1945–1949. doi: 10.1249/01.mss.0000233798.62153.50.

Zager, R. A., A. C. Johnson, S. Y. Hanson, and S. Lund. 2006. Acute nephrotoxic and obstructive injury primes the kidney to endotoxin-driven cytokine/chemokine production. *Kidney Int* 69 (7):1181–1188. doi: 10.1038/sj.ki.5000022.

4 Transcriptional Regulation of Mitochondrial Biogenesis and Quality Control

Hagir B. Suliman and Claude A. Piantadosi

CONTENTS

ABSTRACT

The capacity of a cell to renew or expand its mitochondrial population is important for cell survival during periods of intensified energy demand or after episodes of cell damage, particularly in organs with high energy utilization rates, such as the liver. This response is regulated by a bigenomic transcriptional program of mitochondrial biogenesis modulated by energy- and redox-dependent signals that coordinate mitochondrial mass with cellular energy demand. Energy demand can change rapidly in response to varying physiological or pathological conditions, such as work, growth, proliferation, and differentiation, or by the need to repair cell damage, including mitochondria, for example, from inflammation. Such events stimulate the coordinated activities of several multifunctional transcription factors and coactivators also involved in the elimination of defective mitochondria, as part of a unified mitochondrial quality- and damage-control network. Here, we present current information on the modes of action of known transcription factors that comprise the transcriptional machinery for nuclear-encoded mitochondrial genes and the actions of mitochondrial transcription factors and nuclear receptors on regulatory regions of the mitochondrial DNA. The importance of mitochondrial quality control is emphasized in supporting a cell's metabolic needs and improving its resistance to metabolic failure in the setting of inflammation and mitochondrial damage.

Keywords: liver, energy provision, mitochondria, mitochondrial biogenesis, transcription, mitochondrial DNA, mitophagy, mitochondrial quality control

A MODERN VIEW OF THE MITOCHONDRION

The traditional depiction of mitochondria as a static bean-shaped intracellular organelle has been superseded by a highly branched and dynamic network that moves throughout the cell and undergoes structural transitions, changing in length, shape, and size to meet the specific energy demands of the cell type (Figure 4.1). In addition, mitochondrial mass varies in relation to the extracellular milieu, increasing, for instance, through the process of mitochondrial biogenesis, which is activated physiologically and pathologically by stimuli that include but are not limited to development, division, exercise, thermogenesis, postnatal breathing, thyroid hormones, erythropoietin, calorie restriction, oxidative stress, and inflammation [1,2]. Effective regulation of mitochondrial biogenesis is brought about by several cell-signaling pathways, which collectively serve to support the basic homeostatic functions of the cell.

A unique feature of mitochondria among cellular organelles is the presence of their own genetic material. The mitochondrial genome consists of multiple copies of a ~16.6 kb circular double-stranded DNA (mitochondrial DNA, mtDNA) molecule, which encode the mRNA for 13 electron transport chain proteins as well as 2 rRNAs of the mitochondrial ribosome, and 22 tRNAs required for translation by mitochondrial ribosomes in the matrix [3]. The remaining ~80 oxidative phosphorylation (OXPHOS) subunits and ~1200+ other proteins present in mitochondria are encoded by the nuclear genome, translated by cytosolic ribosomes, and imported into

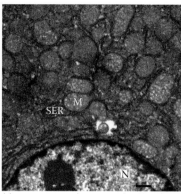 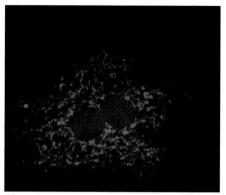

(a) In situ mouse liver mitochondria (EM)   (b)   Hep G cell mitochondria (mitosis)

FIGURE 4.1   (See color insert.) Mitochondrial structure in the liver. (a) Electron micrograph illustrating mitochondria in a normal mouse hepatocyte. Notice that the hepatocyte exhibits electron-dense mitochondria (M) surrounded by an abundant smooth endoplasmic reticulum. N = cell nucleus. Scale bars = 0.2 μm. (b) Fluorescence micrograph of human HepG cells in a culture stained with Mito Tracker Green to target the intracellular mitochondrial network. DAPI stains the cell nucleus, which is in mitosis. 300x.

mitochondria via specialized targeting and translocation mechanisms [4]. Because of the bigenomic origin of mitochondrial genes, mutations in either mitochondrial or nuclear genes can lead to mitochondrial dysfunction and can contribute to heritable metabolic disorders, the development of cancer and neurodegenerative diseases, and in aging and age-related diseases [5,6].

The genetic organization of mammalian mtDNA is highly conserved [7]. Genes are present on both strands, which are designated as a heavy (H)-strand a the light (L)-strand, based on their relative weight per nucleotide. The H-strand encodes 2 rRNAs, 12 mRNAs, and 14 tRNAs, while the L-strand encodes 1 mRNA and 8 tRNAs (Figure 4.2). The one major noncoding region in mtDNA, called the *D-loop regulatory region* [7], encompasses the *cis*-acting regulatory elements crucial for mtDNA transcription and replication. Because mitochondria are adaptive, most nuclear-encoded mitochondrial genes are expressed in a tissue- or cell-specific manner, implying that the mitochondrial proteome can be altered for specialized functions [1,2].

The regulation of mitochondrial biogenesis presents a unique transcriptional challenge in the cell because the control mechanisms require parallel coordination of a large set of inducible mitochondrial genes while also enabling the signal-specific induction of nuclear gene subsets. To allow for this process, mitochondrial biogenesis and energy production are regulated by genes controlled by a network of nuclear DNA–binding transcription factors and coregulators that are activated vigorously in response to diverse physiological conditions and by tissue- or cell-specific events involving mitochondrial function and/or mass. Hence, the integrity of both the nuclear and mitochondrial genomes and the cross talk between them are critical to implement and maintain the normal functions of mitochondria as well as to adjust mitochondrial phenotype in response to environmental stimuli.

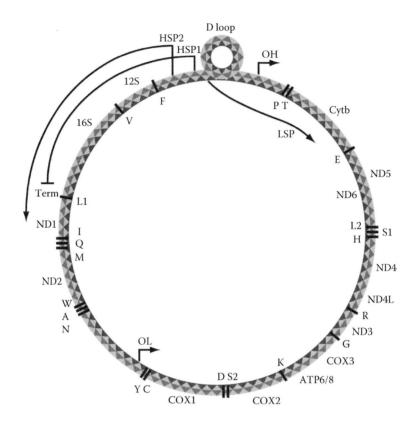

FIGURE 4.2   The human mitochondrial genome. Human mtDNA is depicted with the heavy (H)-strand in light gray and the light (L)-strand in dark gray. The individual rRNAs and mRNAs (black) and tRNAs labeled with letters (dark gray bars; letters represent cognate amino acids). H-strand transcription is initiated from two promoter sites, H-strand promoter (HSP) 1 and HSP2. HSP1 transcripts are terminated at the 22-bp termination sequence within the tRNA-Leu(UUR) gene where mitochondrial transcription termination factor 1 (MTERF1) binds. HSP2 transcripts generate near full-length polycistronic transcripts that are processed into the individual RNAs. L-strand transcription is initiated from a single promoter site LSP, which generates near full-length polycistronic messages that are processed. The primary replication origins of both the H- and L-strands are indicated (OH and OL, respectively). HSP, heavy-strand promoter; LSP, light-strand promoter; MTERF1, mitochondrial termination factor 1.

Apart from ATP production by OXPHOS, mitochondria perform other functions such as heme synthesis, β-oxidation, and metabolism of certain amino acids, and they are involved in Fe/S cluster formation, iron metabolism, and calcium homeostasis [8]. Additionally, mitochondria serve as a key checkpoint for intrinsic programmed cell death as well as a central platform for many other cell-signaling pathways, including portions of the innate immune response [9] and the genesis of autophagosomal membranes, for instance, during starvation-induced autophagy [10]. Moreover, the mitochondrion interacts and communicates with other cell constituents such as the cytoskeleton and the endoplasmic reticulum at specific domains called *mitochondria-associated membranes* [11], and it transports cargo to peroxisomes through mitochondria-derived vesicles [12]. Thus, the mechanisms controlling mitochondrial biogenesis are relevant not only to energy production but also to the other mitochondrial functions that are important for cellular homeostasis and the cellular response to environmental changes and pathological conditions.

The disruption of mitochondrial function or alterations in morphology or movement of the organelle can disturb a variety of cell functions and impair cell survival. In understanding these states, we must define the process of mitochondrial quality control (QC) as the regulation of mitochondrial mass and distribution through the production of functional (biogenesis) or degradation of nonfunctional (mitophagy) organelles. This QC process is represented in a set of crucial programs for the preservation of cell function and viability. Here, we have focused on the transcriptional regulation of mitochondrial biogenesis, although the entire QC process is becoming increasingly important to modern medical science because of its many links to human diseases that arise from defective mitochondrial function, redox homeostasis, and cellular energy provision.

MITOCHONDRIAL DNA REPLICATION AND TRANSCRIPTION

OVERVIEW OF THE MOLECULAR MACHINERY

MtDNA is replicated semiconservatively on the H- and L-strand from two primary origins of replication (OH and OL, respectively, Figure 4.2). Mitochondrial transcription factor A (TFAM) initiates a process that enables the catalytic subunit of the mitochondrial-specific polymerase, polymerase γ (POLG) A, to copy mtDNA [13]. This process is supported by at least three other proteins: POLGB (the accessory subunit of POLGA) [14]; the mitochondrial helicase, Twinkle [15]; and the mitochondrial single-stranded-binding protein (mtSSB) [16]. These factors contribute to the core mtDNA nucleoid, responsible for the transcription, replication, and packaging of the mitochondrial genome [17]. Other members of the core nucleoid include the transcription factors, TFB1M, TFB2M, mtRNA polymerase (POLRMT), and mitochondrial transcription termination factor 1 (MTERF1) [17]. Highly aerobic cells, such as heart, muscle, liver, and brain cells, contain a high number of mtDNA copies per mitochondrion [18]. This ensures sufficient ATP generation through OXPHOS to undertake complex and energy-demanding cellular functions [19]. For these cells, OXPHOS is the optimal process for ATP generation since it generates 30 or more ATP molecules from the oxidation of glucose compare to two ATP molecules by glycolysis alone [20].

Transcription of mtDNA is initiated at three promoters and generates polycistronic transcripts [3]. There is one L-strand promoter (LSP) and two H-strand promoters (HSP1 and HSP2) as indicated in Figure 4.2. Transcripts from HSP1 are terminated in a tRNA gene immediately after the rRNA genes, while the LSP and HSP2 transcripts represent nearly the complete transcript of their respective mtDNA strand templates [21]. These primary transcripts are processed into individual mRNA, rRNA, and tRNA by mitochondrial RNases through a tRNA punctuation model mechanism using excision of tRNAs flanking the mRNAs and rRNAs to liberate the 37 mature RNA species [5,22]. As with the process of mtDNA replication, the process of mtDNA transcription is entirely dependent on factors encoded for by the nuclear genes. In human and mouse mitochondria, transcription initiation involves two factors, TFAM and TFB2M, in addition to POLRMT that comprises the transcriptional initiation complex, which is organized around the LSP and HSP [23,24].

MITOCHONDRIAL TRANSCRIPTION FACTOR A

TFAM was the first mitochondrial transcription factor to be identified [25] and belongs to the high-mobility-group (HMG) family of DNA-binding proteins with the ability to wrap and bend DNA via its two HMG box domains [26]. TFAM is a sequence-specific transcription factor that is essential for the organization, replication, and transcriptional activation of mtDNA [3,27,28]. It has a central role in the production of transcripts from the LSP and the HSP1. The transcriptional activity of TFAM has been linked to its high affinity to bind and bend DNA at specific sites upstream of the mtDNA promoters [29,30], and its role in packaging is associated with its ability to bind nonspecifically throughout the mtDNA [29–32].

The interaction of TFAM with mtDNA is regulated both by posttranslational modification and by targeted degradation. TFAM is degraded by the Lon protease [33] after its phosphorylation at multiple serine residues in the HMG1 and HMG2 domains by protein kinase A (PKA) [34]. This defines when and where TFAM is bound to mtDNA, thus permitting precise regulation of mtDNA transcription, mtDNA packaging, and nucleoid structure via dynamic remodeling under various physiological conditions. TFAM activity may also be decreased by acetylation [34], similar to histones in the nucleus. TFAM is phosphorylated at serine 177 by extracellular signal-regulated protein kinases 1/2 resulting in decreased binding of TFAM to the LSP, thereby suppressing mitochondrial transcription [35]. Although TFAM is mtDNA specific, there is also some evidence for its regulation of transcription of certain nuclear genes such as sarcoplasmic/endoplasmic reticulum Ca^{2+}-ATPase 2 (SERCA2) [36]. This implies the coordinated regulation of the transcription of mitochondrial genes needed to produce ATP with the gene for a protein SERCA2a that consumes ATP.

MITOCHONDRIAL TRANSCRIPTION FACTOR B2: TFB1M AND TFB2M

Multiple lines of evidence suggest that TFB1M is the primary mitochondrial 12S rRNA methyltransferase important for small subunit ribosome biogenesis and that TFB2M is the primary mitochondrial transcription initiation factor [37]. TFB2M facilitates promoter conformation from a closed to an open DNA, which is needed for transcription initiation [38–40]. In this process, TFB2M interacts with POLRMT to enable an open complex formation by inducing promoter melting [39–41]; however, it does not significantly contribute to promoter recognition, a function that is inherent to POLRMT [42,43,44].

MITOCHONDRIAL RNA POLYMERASE

POLRMT are single-subunit, DNA-dependent, RNA polymerases with high sequence homology to the RNA polymerases of bacteriophages T3 and T7 [42]. Although evolutionarily related, POLRMT is different than bacteriophage enzymes in that by itself it cannot initiate promoter-specific transcription on double-stranded

DNA [39–41]. Like bacteriophage T7 [43], POLRMT generates RNA primers for initiation of DNA replication [44,46], thus coupling transcription and mtDNA replication. It should be noted that other schemes of priming and replication have been proposed for mammalian mtDNA as well [45,46].

MITOCHONDRIAL TRANSCRIPTION TERMINATION FACTORS FAMILY

The MTERF proteins are localized in mitochondria and regulate mtDNA expression by transcription termination. The family members share the MTERF motif, which contains leucine zipper-like heptads. Recent studies showed that MTERF's role is not restricted to transcription termination, but they can also promote transcription initiation and control mtDNA replication. MTERF3 is a key repressor of mtDNA transcription in the mouse but also enhances assembly of large mitochondrial ribosomal subunit. Thus, MTERF3 coordinates the cross talk between transcription and translation for the regulation of mammalian mtDNA gene expression. Functionally, this allows the fine-tuning of mtDNA replication and transcription in response to cellular demands by coordination of transcription-activating factors TFAM, TFB1/2M, and the repressor and termination factor MTERF3.

MITOCHONDRIAL TRANSCRIPTION ELONGATION FACTOR

The polycistronic RNA messages generated by the transcription of mtDNA are subsequently processed into individual rRNAs, tRNAs, and mRNAs. Studies have suggested that POLRMT alone is insufficient to generate long transcripts [47], and that mitochondria transcription elongation factors known as TEFM are also required. TEFM is imported into the mitochondrion [48] and interacts with the C-terminal domain of POLRMT in an RNA-independent manner to enhance its activity consistent with a role in elongation. Cells lacking TEFM have reduced levels of promoter-distal H- and L-strand mtDNA transcripts with resultant OXPHOS defects [48].

NUCLEAR-ENCODED FACTORS ACTING WITHIN MITOCHONDRIA

Loss-of-function studies in eukaryotic cells have established the vital role of mtDNA transcription and mitochondrial function early in embryonic development. Genetic ablation of TFAM results in severe OXPHOS defects coupled with abundant damaged and abnormal mitochondria and embryonic lethality at E10.5 associated with severe reduction in mtDNA content, demonstrating the necessity of TFAM in mtDNA maintenance in vivo [49]. Similarly, embryonic lethality was reported for mice with a germline deficiency in MTERF3 [50], TFB1M, encoding a dimethyltransferase [51], and MTERF4, encoding a regulator of ribosome biogenesis and translation through its actions with the 5-methylcytosine rRNA methyltransferase (NSUN4) [52]. Loss of these functions is accompanied by disruption of respiratory complexes I, III, IV, and V that depend on mitochondrial-encoded subunits. In total, genetic ablation studies evaluating the impact of the loss of function of nuclear-encoded factors involved in the regulation of the mitochondrial genome have revealed that these factors are largely essential for normal mitochondrial function and cell viability.

TRANSCRIPTIONAL REGULATION OF
MITOCHONDRIAL BIOGENESIS

NUCLEAR RESPIRATORY ACTIVATORS

The first nuclear transcription factors found to be involved in the expression of mito-chondrial genes were nuclear respiratory factors 1 and 2 (NRF-1 and NRF-2). NRF-1 was first identified as a transcriptional regulator of cytochrome c gene expression [53]. It also has a regulatory role for genes encoding respiratory subunits [54], TFAM [53], and for both TFB isoform genes [55] whose products are the major regula-tors of mtDNA transcription, ribosome assembly, and replication [38,51]. In addi-tion, NRF-1 promotes the expression of the chemokine receptors CXCR4 and 7 that bind stromal-derived factor 1, a proinflammatory chemokine involved in metastatic behavior of several cancers [56].

The NRF-2 protein complex identified by Scarpulla and colleagues is the human homologue of guanine and adenine-binding protein (GABP) [57], which is the only obligate multimer among the mammalian E26 transformation-specific (ETS) fac-tors [58]. The GABP complex is a tetramer of two distinct proteins; GABPα binds to DNA through its ETS domain and recruits GABPβ, which activates transcrip-tion through a glutamine-rich region in its carboxy terminus [59]. GABP regulates more than a dozen nuclear-encoded mitochondrial genes [60,61]. It has been dem-onstrated to directly bind and activate the transcription of several genes encoding key effectors of mtDNA transcription [61,62] including TFAM and MTERF, as well as mtDNA replication, including Twinkle, mtSSB, and the B subunit of POLG. Furthermore, GABP has been linked to the lineage-restricted myeloid and lym-phoid genes that are required for innate immunity [59] and for cell cycle control in fibroblasts [63].

GABP activity is regulated by the Hippo pathway and has been found to be critical for Yes-associated protein (YAP) gene expression. GABP-mediated YAP expression mitigates acetaminophen-induced liver injury, while both GABP and YAP expression are enhanced in human liver cancers. Therefore, GABP is a poten-tial therapeutic target for certain liver diseases [64]. Genetic disruption [65,66] and knockdown [66] of mouse *Gabpa* causes early embryonic lethality, most likely explained by mitochondrial dysfunction [65]. There is also evidence pointing to a direct role for both NRF-1 and NRF-2 in the expression of all 10 nuclear-encoded cytochrome oxidase subunits [67,68]. Both factors also participate in the expression of the mitochondrial import receptor complex and of putative cytochrome oxidase assembly factor COX17 [54]. Control of the mitochondrial transcription and impor-tation machinery by the NRFs are part of a mechanism that coordinates respiratory chain protein expression with biogenesis of the organelle itself (Figure 4.3). Genetic deletion of the NRFs and other NRFs, such as the transcriptional repressor protein Yin Yang 1 (YY1), results in peri-implantation lethality [65,69,70]. NRF-1 null blas-tocysts, for instance, fail to progress beyond E6.5 and display depleted mtDNA lev-els and diminished mitochondrial membrane potential, consistent with an electron transport chain deficiency [69].

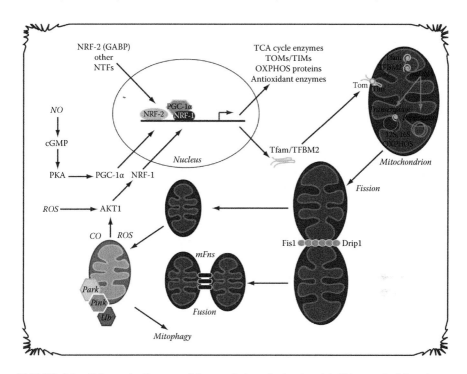

FIGURE 4.3 Schematic diagram of the regulation of mitochondrial biogenesis. Many intracellular and extracellular signals can stimulate mitochondrial biogenesis (see text). Illustrated are NO, CO, and reactive oxygen species (ROS) pathways including NO (or CO) activation of cyclic guanosine monophosphate, and resultant phosphorylation of peroxisome proliferator-activated receptor γ coactivator (PGC-1α) via protein kinase A. ROS, mainly H_2O_2, can activate Akt, leading to nuclear respiratory factors 1 (NRF-1) phosphorylation and nuclear translocation. NRF-1 and NRF-2 and other nuclear transcription factors are coactivated by PGC-1α leading to transcription of nuclear-encoded proteins such as mitochondrial transcription factor A, which activates transcription and replication of the mitochondrial genome. Nuclear-encoded proteins are imported into mitochondria through the outer or inner membrane transport machinery. Nuclear- and mitochondria-encoded subunits of the respiratory chain are then assembled. Mitochondrial fission through the dynamin-related protein 1 for the outer membrane and optic atrophy 1 for the inner membrane of mitochondria allows mitochondrial division, while mitofusins control mitochondrial fusion. Processes of fusion/fission lead to proper organization of the mitochondrial network. OXPHOS: oxidative phosphorylation. Damaged mitochondria are tagged by park and pink for elimination by mitophagy.

Mitochondrial Transcriptional Control by Myc

Earlier, we mentioned that mitochondrial genome transcription must adjust dynamically to changes in cellular requirements for energy and mitochondrial mass, as ATP and biosynthetics demand change with alterations in the cell cycle and with other events [71,72]. As the cell cycle rate increases, the cell must increase its rate of mitochondrial biogenesis so that progeny cells receive an appropriate complement of mitochondria. Consistent with the linkage of mtDNA transcription and biogenesis to proproliferative signaling pathways, the Myc oncoprotein is a direct regulator of these processes [73,74]. Myc is also implicated in the reprogramming of cellular metabolism that occurs in cancer [75], a process that may involve altered mitochondrial transcription.

Myc regulates mitochondrial genome replication and function, and 281 mitochondrial ontology genes were identified among 2679 Myc-responsive, nuclear-encoded genes [73]. Subsequent binding analysis of Myc to the nuclear genome revealed a link between Myc and nuclear-encoded mitochondrial genes by data showing that Myc directly occupies 107 nuclear genes encoding mitochondrial proteins. In the initial expression profiling study, the best characterized of these was TFAM. Experiments in human cells have demonstrated direct Myc binding to the TFAM locus and a tight correlation between Myc levels and TFAM transcription. While TFAM activation is central factor in the ability of Myc to regulate mitochondrial transcription, the authors suggested that TFAM induction may not explain the entirety of Myc's effects, and a number of genes among the 281 mitochondrial-ontology targets of Myc are certainly involved. For example, the RNA polymerase POLRMT/mtRNAP was among the Myc-regulated genes, and Myc's ability to control the expression of this enzyme may allow further modulation of mitochondrial transcription [76].

Myc also regulates nuclear-encoded mitochondrial programming largely through an interaction with NRF-1 [77] and by increasing the expression of Peroxisome proliferator-activated receptor γ coactivator 1-β (PGC-1β) [78]. NRF-1 and PGC-1β are among the most potent regulators of mitochondrial ontology. Through these indirect mechanisms coupled to the direct regulation of TFAM and POLRMT/mtRNAP, Myc is able to link its cell cycle and metabolic activities to critical components of mitochondrial transcription and mitochondrial function.

GLUCOCORTICOIDS, THYROID HORMONE, AND OTHER STEROID HORMONES

Many hormones regulate key physiological processes involving metabolism by exerting effects via receptors belonging to the nuclear receptor (NR) superfamily, which modulates nuclear gene expression. Upon ligand binding, these receptors are chaperoned into the nucleus where they bind to hormone-responsive elements (HREs) in the promoter regions of the target genes. Since hormones influence so many cellular metabolic processes, it is not surprising that many have been found to modulate mitochondrial function. Both steroid (type I) and nonsteroid (type II) NRs influence mitochondrial biogenesis in a variety of tissues. In addition, components of the OXPHOS complex have been shown to be under the control of these hormones [79]. Indeed, these hormones have also been shown to modulate the expression of key nuclear-encoded mitochondrial transcription factors, including TFAM, TFB1M, and TFB2M [80]. Moreover, certain NRs can be found in mitochondria, which together with the identification of HREs in the mitochondrial genome indicate that direct transcriptional effects also occur [81].

The first NR identified in mitochondria was the glucocorticoid receptor (GR), followed by the estrogen receptor and the thyroid receptor (TR) [82–84]. GR binds to the regulatory D-loop of mtDNA as shown by chromatin immunoprecipitation in HepG2 cells, which serves to increase the levels of COX I. These observations have been strengthened by experiments using α-amanitin to inhibit nuclear RNA polymerase-II function [80]. A similar role for thyroid hormone receptor has been found in mitochondria. A TR specifically targeted to mitochondria has been found to increase transcription in a thyroid hormone–dependent manner [85]. Additionally, thyroid hormone therapy was shown to stimulate significant increase in mitochondrial transcriptional activity in hypothyroid rats [86,87].

Proteomic studies have revealed that about half of the proteins affected by thyroid status are involved in substrate and energy metabolism [88]. Nongenomic and genomic signaling occurring via 3,5-diiodothyronine (3,5-T2) and T3 lead to increases in the rates of O_2 consumption of target tissues such as the liver [89]. T3 also contributes to induction of uncoupling proteins [90], energy expenditure [91], loss of energy due to higher active cation transport [92,94], and O_2 used to generate mitochondrial reactive oxygen species (ROS) [93]. T3-induced metabolism in target tissues occurs via gene expression orchestrated by binding to the NRs (TRα and TRβ), which are ligand-inducible transcription factors. In the presence of T3, thyroid hormone (TH) receptors bind to TH response elements (TRE) in the promoters of target genes and form coactivator complexes containing histone acetyltransferase activity to activate transcription [98]. For instance, T3 operating via TREs induces the expression of carnitine palmitoyltransferase 1a, phosphoenolpyruvate carboxykinase, and pyruvate dehydrogenase kinase 4 [95,96]. Additional factors including the CCAAT/enhancer-binding protein β, cAMP-responsive element-binding protein (CREB)-binding protein (CBP), and peroxisome proliferator-activated receptor (PPAR) γ coactivator (PGC-1α) are also involved in the T3 responsiveness of these genes [96–98]. CPT1a is a rate-limiting enzyme for mitochondrial transport and β-oxidation of fatty acids [99]. Thus, T3 stimulates free fatty acid shuttling into the mitochondria and therefore plays a regulatory role in free fatty acid metabolism [100]. While this process is well described, T3-regulated pathways that lead to the generation of free fatty acids from stored lipid droplets are not well understood. In hepatocytes, autophagy of lipid droplets, termed *lipophagy*, is a major pathway of lipid mobilization [101,102], and its inhibition is linked to the development of fatty liver and insulin resistance [103,104].

NUCLEAR HORMONE RECEPTOR ACTIVATORS OF MITOCHONDRIAL BIOGENESIS

The NR superfamily plays a key role in the transcriptional regulation of certain nuclear genes encoding mitochondrial enzymes and proteins. The PPARα was the first to be implicated in the transcriptional control of metabolism. Originally, discovered as a regulator of genes encoding peroxisomal fatty acid oxidation (FAO) enzymes, PPARα is now known to coordinately regulate nuclear genes encoding mitochondrial FAO enzymes (reviewed in [105]). PPARα is a member of a family of NRs including the ubiquitous PPARβ (or PPARδ) and PPARγ, an adipose-enriched transcription factor involved in adipocyte differentiation and the target of the insulin-sensitizing thiazolidinediones. A distinctive feature of the PPARs is their activation by a variety of fatty acid–containing lipid ligands. In fact, oxidized and nitrated fatty acids are also known to bind and activate PPARγ [106]. PPARs bind specific DNA response elements as heterodimers with the retinoid X receptor. The activation of PPARα and PPARβ by lipid ligands provides a mechanism for transducing changes in lipid metabolism to the genetic control of mitochondrial FAO, a key source of ATP production in heart and muscle [105].

Another family of NRs, the estrogen-related receptors or ERRs (ERRα, ERRβ, and ERRγ), also regulate nuclear genes encoding mitochondrial proteins involved in FAO, the tricarboxylic acid cycle, respiratory chain, and OXPHOS. Functional genome-wide analysis of transcription factor occupancy has revealed that ERRs are associated with several hundred genes regulating mitochondrial function [107]. ERR targets include genes involved in FAO that overlap with PPAR targets [108] as well as PPARα gene transcription itself [109], creating a cross-regulatory loop that amplifies the subset of ERR targets involved in mitochondrial FAO. All three ERRs are expressed in mitochondrial-rich tissues such as heart and skeletal muscle, with ERRγ being selectively expressed in oxidative muscles enriched in type I and IIa fibers [110]. ERRα null mice display a number of cardiac deficiencies when stressed by pressure overload but lack serious defects in organelle biogenesis [111], while ERRγ knockouts exhibit early postnatal lethality associated with defective OXPHOS in the heart [112]. Consistent with these findings, transgenic overexpression of ERRγ in skeletal muscle also promotes an oxidative phenotype with enhanced expression of mitochondrial FAO and respiratory chain genes [110,113].

OTHER NUCLEAR TRANSCRIPTION FACTORS

Several other nuclear transcription factors are linked to the expression of the respiratory apparatus including YY1 and myocyte enhancer factor-2 (MEF2). YY1 was first associated with the expression of cytochrome oxidase subunits Vb and VIIc [114,115], and subsequent analysis of 723 human core promoters revealed a preponderance of YY1 sites in nuclear genes encoding ribosomal subunits and mitochondrial proteins [116]. Functional studies have demonstrated YY1 to be an important component of nutrient sensing by the mammalian target of rapamycin (mTOR). Inhibition of mTOR decreases the expression of ERRα and the NRFs and blocks the interaction between YY1 and PGC-1α. In addition, YY1 silencing diminished both mitochondrial gene expression and respiration [117]. In cardiac and skeletal muscle, some cytochrome oxidase subunits also depend upon MEF2 [118]. NRF-1 regulates MEF2A in muscle possibly accounting for NRF-1 control over the expression of muscle-specific MEF2 target genes [119].

The nuclear factor erythroid 2-related factor 2 (Nrf2), a major transcription factor for antioxidant and cytoprotective responses, is normally sequestered in the cytosol by Kelch-like ECH-associated protein 1 (Keap1), an adaptor for a ubiquitin ligase complex, and is constitutively degraded through the ubiquitin–proteasome system [120]. Oxidative stress dissociates Nrf2 from Keap1; it then translocates to the nucleus to bind to antioxidant response elements (AREs) and activate antioxidant genes such as heme oxygenase-1 (HO-1) and NAD(P)H/quinone oxidoreductase 1 (Nqo1) [121]. Nrf2 occupies activating ARE motifs in the NRF-1 promoter [122], and under the influence of the HO-1/CO system, Nrf2, and NRF-1 along with GABPA (NRF-2) and PGC-1α stimulate mitochondrial biogenesis in response to oxidative stress [123,124]. Furthermore, Nrf2 signaling regulates mitochondrial biogenesis in rat brain and mouse cardiomyocytes. Nrf2 has also been found to potentiate recovery from sepsis in mice [123–126] and influence the innate immune response, survival, [127] and leukocyte function in response to sepsis [128].

Additional nuclear transcription factors linked to the control of mitochondrial respiratory function include CREB, which is involved along with NRF-1 in the growth-regulated expression of cytochrome c [129,130]. For instance, both the factors participate in cytochrome c induction in response to serum stimulation of quiescent cells [129]. CREB serves as a direct activator of the cytochrome c promoter and is also required for promoter activation by ERRα [108]. Aside from cytochrome c, CREB sites are not generally found in the proximal promoters of nuclear respiratory genes [129], and there is no selective enrichment among genomic fragments containing ERR binding sites [108]. In addition, a link between NRF-1 and the lipopolysaccharide-induced inflammatory response was demonstrated by the identification of NF-κB-responsive elements in the NRF-1 promoter region [131]. This response is facilitated by CREB binding to the promoter, and the functional interaction between NF-κB and CREB in NRF-1-dependent mitochondrial biogenesis may mitigate inflammatory mitochondrial damage. This association may be indicative of a general CREB-dependent mechanism for the control of respiratory chain protein expression during the inflammatory response.

PGC-1 COACTIVATOR FAMILY

The discovery of PGC-1α represented a breakthrough in our understanding of how transcription factors controlling mitochondrial biogenesis are coregulated. PGC-1α was first identified in brown adipocytes as a coactivator of PPARγ, [132]. PGC-1α is a member of a family of transcriptional coactivators that includes the closely related PGC-1β [133] and more distant relative, PGC-1-related coactivator (PRC) [54]. The three members are characterized by an N-terminal activation domain near leucine-rich LXXLL motifs required for NR interactions, an RNA recognition site (RRM), and a host cell factor-1 binding domain. As shown in Figure 4.4, PGC-1α exerts these effects through direct interaction with and coactivation of a range of transcription factors including PPARs, ERRs, NRF-1, NRF-2, MEF2, Forkhead box O1 (FOXO1), and YY1. The ability of PGC-1α to stimulate transcription factor activity provides the coactivator with the capacity to coordinate the large number of genes required for mitochondrial biogenesis [134]. PGC-1β interacts with and coactivates many of the same transcription factors as PGC-1α [133].

During starvation, PGC-1α expression is induced in the adult liver through glucagon and GR signaling [135]. In response, a complex program of metabolic changes during the transition from a fed to a fasted liver is orchestrated by PGC-1α including gluconeogenesis, fatty acid β-oxidation, ketogenesis, heme biosynthesis, and bile acid homeostasis. These effects of PGC-1α on fasting adaption are achieved by coactivating key hepatic transcription factors, such as HNF4α, PPARα, GR, FOXO1, FXR, and liver X receptor (LXR) [136]. Accordingly, PGC-1α KO mice and RNAi-mediated liver-specific PGC-1α knockdown mice display an impairment of gluconeogenic gene expression and hepatic glucose production [137,138], and upon

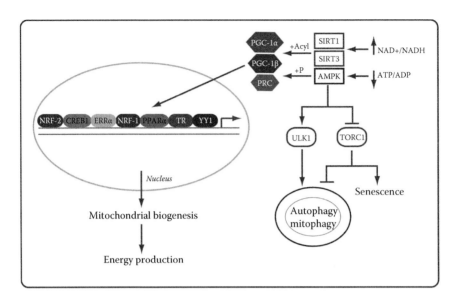

FIGURE 4.4   Proposed integration of energy-generating and antioxidant pathways by per-oxisome proliferator-activated receptor γ coactivator (PGC-1α). PGC-1α is activated via posttranscriptional phosphorylation by AMP-activated protein kinase (AMPK) or by deacet-ylation via sirtuin 1 in response to nutrient deprivation. Induction or activation of the coacti-vator can enhance mitochondrial biogenesis and oxidative function through the coactivation of multiple transcription factors involved in respiratory gene expression. PGC-1α activation also promotes an antioxidant environment, for instance, by coactivating estrogen-related receptor α to induce SIRT3, a mitochondrial sirtuin implicated in ROS detoxification. In addition, AMPK promotes autophagy through direct phosphorylation of unc-51-like kinases 1 or suppression of the target of rapamycin complex 1 (TORC1) kinase complex. Inhibition of the TORC1 pathway also has a negative effect on senescence. Increased autophagy, enhanced reactive oxygen species detoxification, and TORC1 inhibition are all associated with health and longevity. Under conditions of caloric excess, steroid receptor coactivator-3 induces GCN5 acetyltransferase, which inactivates PGC-1α through acetylation.

fasting, these mice develop hypoglycemia and hepatic steatosis [139]. Also, PGC-1α regulates the genes encoding homocysteine synthesis enzymes in the liver and mod-ulates plasma homocysteine levels [140].

   In mammals, many aspects of energy metabolism are regulated by the circadian clock including respiration and glucose and lipid homeostasis. Indeed, PGC-1α is a key component of the circadian pathway that integrates clock gene expression and energy metabolism [141]. PGC-1α coactivates the expression of Bmal1, a core clock gene in hepatocytes and muscle cells through coactivation of the tyrosine-protein kinase receptor (ROR) family of orphan NRs. Mice lacking PGC-1α have abnormal diurnal rhythms, body temperature, and metabolic rate. As PGC-1α expression is regulated by nutritional and hormonal cues, it is likely that it helps synchronize tis-sue metabolism with the circadian pacemaker.

   PGC-1α undergoes extensive posttranslational modifications, including acety-lation, phosphorylation, methylation, and SUMOylation in response to nutritional and hormonal signals. These modifications allow fine-tuning of PGC-1α activities in a context-dependent manner. The acetyl transferase general control of amino acid synthesis 5 (GCN5), acetylates PGC-1α at several lysine residues, alters its localization within the nucleus, and inhibits its cotranscriptional activity [142]. In contrast, PGC-1α deacetylation through sirtuin 1 (SIRT1) increases PGC-1α activity on gluconeogenic gene transcription in the liver [143]. Thus, both GCN5 and SIRT1 may be pharmacological targets to regulate the activity of PGC-1α, suggesting a potential metabolic approach for disorders in which hepatic glucose output is altered. In the liver, SIRT1 promotes lipid catabolism through PGC-1α and PPARα binding and activation [144]. In skeletal muscle, PGC-1α is phosphorylated by both p38 mitogen-activated protein kinase and AMP-activated protein kinase (AMPK) [145,146], leading to a more stable and active protein. In contrast, phosphorylation of PGC-1α by Akt/protein kinase B (PKB) downstream of insulin signaling in the liver may decrease its stability and transcriptional activity [147]. In addition, PGC-1α undergoes methylation at several arginine residues in the C-terminal region by protein arginine methyltransferase 1 [148]. Finally, PGC-1α transcriptional activity is attenuated by SUMOylation of conserved lysine residue 183 [149].

In experiments using C2C12 cells, it has been reported that AMPK-mediated phosphorylation primes PGC-1α for deacetylation by SIRT1 [150], suggesting that PGC-1α modifications likely communicate to coordinately regulate its activity. PGC-1β is also acetylated at multiple sites [151]; however, the biological significance of these events is less well defined. Hepatic PGC-1α overexpression also protects whole-body substrate switching, as well as maintains respiration after a high fat diet. Three mammalian sirtuins (SIRT3, 4, and 5) are located in the mitochondria [152–154], and SIRT4 can regulate hepatic lipid metabolism during changes in nutrient availability [155]. The decreased levels of SIRT4 in the liver during fasting and in SIRT4 KO mice are related to the expression of hepatic PPARα target genes linked to fatty acid catabolism. The augmented FAO observed in SIRT4-deficient hepatocytes requires SIRT1, revealing cross talk between mitochondrial and nuclear sirtuins.

PGC-1β, the second family member, is induced by dietary intake of fats leading to hyperlipidemia through activation of hepatic lipogenesis and very-low-density lipoprotein secretion [156]. Several factors mediate the effects of PGC-1β on plasma triglyceride metabolism, including sterol response element-binding protein, LXR, and FOXa2 [156,157]. It has been demonstrated that PGC-1β and its target gene apolipoprotein C3 (ApoC3) are downstream of nicotinic acid, a widely used hypotriglyceridemic drug [158]. Both acute injection and chronic feeding of mice with nicotinic acid suppress PGC-1β and ApoC3 expression in the liver [158]. This role for PGC-1β in modulating lipoprotein catabolism shows the relevance of this pathway in therapeutic action of nicotinic acid. Remarkably, systemic delivery of antisense oligonucleotide targeting PGC-1β improves metabolic homeostasis in fructose-induced insulin resistance [159]. There is also a novel MyD88-independent pathway that links toll-like receptor (TLR) 2 and TLR4 signaling in innate immunity in sepsis to PGC-1α, β gene regulation in a critical metabolic organ, the liver, by means of translocating chain-associated membrane protein (TRAM), toll-interleukin-1 receptor domain-containing adapter protein inducing IFN-β (TRIF), and interferon response factor (IRF)-7 [160].

The third PGC-1 family member, PRC, was discovered by a database search for sequence similarities with the carboxy-terminal arginine and serine rich (RS) domain and RNA recognition motif of PGC-1α [109]. Like PGC-1α and β, PRC binds and coactivates nuclear transcription factors implicated in the expression of mitochondrial function, including NRF-1, CREB, and ERRα [55,130,161]. PRC silencing reduces respiratory chain expression and ATP production and leads to a plethora of abnormal mitochondria in cultured cells [162]. These phenotypes resemble those observed upon tissue-specific disruption of TFAM, TFB1M, MTERF3, and MTERF4 whose products are localized to mitochondria and are required for expression and replication of the mitochondrial genome [49–52]. PRC can also direct mitochondrial biogenesis through its interaction with and activation of ERRα in cultured thyroid cells [163] and may function as a sensor of metabolic stress (Figure 4.4).

A germline knockout of the PRC gene in mice results in peri-implantation lethality [164], a phenotype similar to that observed in mouse knockouts of NRF-1, and several other essential transcription factors associated with mitochondrial biogenesis [53,54]. Thus, PRC differs from the other PGC-1 family members in that it is essential for early embryonic development. Among other growth-related functions, PRC may support the high rate of mitochondrial transcription that occurs during early embryogenesis.

The robust upregulation of PRC in response to multiple metabolic insults leads to the induction of a battery of genes involved in inflammation, cell stress, and proliferation [165]. The PRC-dependent inflammatory/stress genes include proinflammatory molecules such as interleukin (IL)-1α, IL-8, cyclooxygenase 2, and members of the proline-rich protein family (SPRR2D and SPRR2F). The latter are associated with the response to DNA damage and exit from the cell cycle and provide a protective antioxidant barrier to cellular damage, thereby promoting tissue remodeling [166]. Several of these genes are also associated with the inflammatory microenvironment in certain human cancers [167,168] and with cellular senescence [169]. These facts are consistent with the coinduction of PRC and c-Myc by chemical uncoupling [163] and with the elevated PRC levels observed in several human tumors [53,170]. Notably, PRC induction by respiratory chain uncoupling and the associated inflammatory/stress response is sensitive to antioxidants [163].

TRANSCRIPTIONAL REPRESSION OF MITOCHONDRIAL BIOGENESIS

The repression of nuclear-encoded mitochondrial genes can occur during hypoxia, but this is complex and involves FOXO3-induced repression. FOXO3 also activates the proapoptotic Bim, which constitutes an early mitochondrial damage signal that triggers ROS production. FOXO3 appears to regulate adaptation to low oxygen, slowly shutting down mitochondrial activity by antagonizing c-Myc function [171]. Conversely, FOXO3 can be phosphorylated by AMPK and subsequently imported into mitochondria where it forms a complex with SIRT3 and POLRMT. This complex activates mitochondria-encoded genes, increases mitochondrial respiration, and contributes to muscle adaptation during nutrient restriction [172].

The PPARs and ERRα are ligand-activated NRs that coordinately regulate gene expression. There is evidence that nuclear corepressors, NCoR, RIP140, and SMRT, repress NRs-mediated transcriptional activity on specific promoters, and thus regulate insulin sensitivity, adipogenesis, and mitochondrial number and activity in vivo. RIP140 interacts directly with PGC-1α to suppress its activity. This direct antagonism of PGC-1α by RIP140 provides a mechanism for regulating target gene transcription via NR-dependent and NR-independent pathways [173]. Furthermore, SUMOylation attenuates the transcriptional activity of PGC-1α by enhancing the PGC-1α interaction with RIP140. Prevention of SUMOylation also augments PGC-1α activity in the context of PPARγ-dependent transcription [149].

RIP140 overexpression can abrogate PGC-1α-mediated elevations of mitochondrial membrane potential and mitochondrial biogenesis and activate both autophagy and apoptosis. The zinc finger protein Parkin interacting substrate (PARIS) represses the expression of PGC-1α and the PGC-1α /NRF-1 target gene set, by binding to insulin response sequences in the PGC-1α promoter [174]. PARIS provides a mechanism by which inhibited mitophagy impairs mitochondrial biogenesis via repression of PGC-1α due to Parkin inactivation. PGC-1α may thus be an important connection between mitochondrial biogenesis and autophagy.

MITOPHAGY

The cellular abundance of mitochondria reflects not only organelle biogenesis but also the rates of mitophagy, a complex form of selective macroautophagy involved in degradation of mitochondrial components as well as the autophagic digestion of entire organelles [175].

The basal replacement of mitochondrial populations in liver, heart, and brain is roughly 9.3, 17.5, and 24.4 days, respectively [176,177], but basal mitochondrial turnover can be enhanced in response to stresses of various intensities such as the accumulation of unfolded or aggregated proteins that induce a specific, adaptive mitochondrial unfolded protein response [178]. Severe mitochondrial dysfunction can lead to parallel increases in mitophagy and biogenesis that are crucial for the recovery of mitochondrial homeostasis and maintenance of cell viability [179,180].

Mitophagy is a poorly understood process controlled by a program that recognizes damaged mitochondria by network morphology (e.g., fission) and bioenergetic parameters (e.g., decreased membrane potential). Because organelle degradation is a potentially dangerous undertaking for the cell, autophagy is highly regulated by more than 30 proteins that are important for synthesis and maturation of double membrane vesicles that fully encircle damaged mitochondria and result in the formation of autophagosomes [181]. These autophagosome fuses to lysosomes in order to allow their contents to be degraded (Figure 4.3).

Mammalian cells tag dysfunctional mitochondria and regulate selective mitophagy through certain outer membrane proteins such as Nix and FUN14 domain-containing protein 1 (FUNDC1) containing light chain 3 (LC3)-interacting region motifs and directly interacting with microtubule-associated protein 1 LC3 [182–184]. In addition to Nix and FUNDC1, perhaps the best studied signaling pathway for mammalian selective mitophagy is the PTEN-induced putative kinase 1 (PINK1)-Parkin-mediated mitophagy pathway. In this model, PINK1 is stabilized on depolarized mitochondria, which further recruits the ubiquitin E3 ligase Parkin to the mitochondria from the cytosol [184,185]. Parkin serves to promote mitophagy by the ubiquitination of a set of mitochondrial proteins that includes Miro, Mitofusin1/2, hFis1, voltage-dependent anion channel (VDAC) 1, and Tom20 [186,187]. Upon mitochondrial damage, mitochondrial localization of PINK1 and Parkin is accompanied by p62 translocation to mitochondria, which further facilitates the recognition of damaged mitochondria in preparation for mitophagy. This process is called mitochondrial priming. Specifically, p62 is thought to recruit LC3-positive autophagosomes to p62-ubiquitin decorated mitochondria for their degradation, although the exact role of p62 in mitophagy is controversial [184,188].

Also, Nix-tagged mitochondria directly connect to LC3 and γ-aminobutyric acid type A receptor-associated protein (GABARAP); LC3 and GABARAP are consistent components of the autophagy machinery. Nix also contributes to priming by controlling the mitochondrial translocation of Parkin [184]. Because mitochondria are quite elongated structures, their removal by autophagy is quite a great challenge. It is thus not surprising that work from both yeast and mammalian cells supports that mitochondrial fission can facilitate mitophagy [189,190]. Moreover, Parkin also promotes mitophagy by inducing mitochondrial fragmentation through proteasome-mediated mitofusin 1/2 degradation [191,192].

Removal of mitochondria by mitophagy also requires recruitment of the Unc-51-like kinases (ULK) 1/2 complex and an autophagy-related membrane protein Atg9A to the depolarized mitochondria. At a later stage, the recruitment of LC3 is associated with leading the dysfunctional mitochondria into the autophagosome [193]. During mitophagy, the capacity to identify badly damaged mitochondria is critical for selectivity. In mammals, loss of AMPK or ULK1 results in aberrant p62 accumulation and defective mitophagy. ULK1 phosphorylation by AMPK is required for mitochondrial homeostasis and cell survival during starvation [194]. In ischemia-reperfusion-induced cardiac damage, Parkin-mediated mitophagy was found to be protective against cardiomyocytes cell death. In contrast, TP53-induced glycolysis and apoptosis regulator, which was upregulated in ischemic myocardium, suppresses mitophagy and exacerbates heart injury [195]. Acetaminophen-induced liver injury is associated with increased mitochondrial damage with the resultant hepatic necrosis. Mitophagy was found to be critical for protecting against acetaminophen-induced liver damage by reducing ROS production [196]. Furthermore, impaired autophagic activity is featured in the livers of obese mice, and pharmacological inhibition of autophagy or ablation of Atg7 is associated with hepatic steatosis [101,103]. Induction of autophagy reduced alcohol-induced hepatic steatosis [192]. Although autophagy may selectively remove lipid droplets through lipophagy, the removal of damaged mitochondria may also contribute to the attenuation of steatosis. Therefore, activating mitophagy could serve as a novel approach for treating damaged mitochondria-mediated tissue injury in the future.

COREGULATION OF MITOPHAGY AND MITOCHONDRIAL BIOGENESIS

Progressive mitochondrial dysfunction may initiate a retrograde response, enabling cell adaptation through increased mitochondrial biogenesis as well as increased elimination of dysfunctional mitochondria by mitophagy [197]. The acute destruction of defective mitochondria through a ULK1-dependent stimulation of mitophagy, and the induction of mitochondrial biogenesis through PGC-1α-dependent transcription, can be triggered by AMP kinase [198]. The downstream mTOR complex 1 (mTORC1) promotes the expression of nuclear genes encoding mitochondrial proteins in resting muscle cells via the interaction of the mTORC1 components, mTOR and raptor, the transcription factor YY1, and PGC-1α [117]. On the other hand, p53 controls autophagy and contributes significantly to the regulation of mitochondrial content by interacting with TFAM. Overexpression of p53 in mouse myoblasts increases both TFAM and mtDNA levels [119,200]. It is noteworthy that certain types of regulators, such as ROS, p53, AMPK, and mTOR control autophagy, mitophagy, and mitochondrial biogenesis (Figure 4.5).

MITOCHONDRIAL BIOGENESIS IN DISEASE

INFLAMMATION AND MITOCHONDRIAL BIOGENESIS

The unique set of challenges of acute inflammation can lead to accelerated mitochondrial turnover. Mitochondria are susceptible to damage by several inflammatory mediators such as tumor necrosis factor α (TNF-α), by nitric oxide (NO), and by certain interleukins generated by innate immune cells like macrophages and Kupffer cells [201]. These damaged mitochondria generate higher levels of ROS and may release calcium and intrinsic apoptosis proteins [202,203]. In addition, inflammatory ROS and NO production can be sufficient to compromise mitochondrial structure and function by direct chemical oxidation of proteins and lipids or by the NO–superoxide reaction, which generates peroxynitrite ($ONOO^-$), a powerful oxidant [204].

Inflammation-induced oxidative or nitrosative stress results in the activation of NF-kB and other transcription factors that regulate inflammation [205,206]. Intense inflammation upregulates many antioxidant defenses, but the transcriptional activity of both PGC-1α and NRF-1 also increases [207,208]. PGC-1α promotes gene induction for certain ROS-detoxifying enzymes like mitochondrial superoxide dismutase (SOD2) and glutathione peroxidase-1 (GPx1). Additionally, data in neuronal cells support that CREB plays an important role in activating the Pargc1a (PGC-1α) gene promoter after oxidant exposure [207]. In macrophages, PGC-1β overexpression inhibits canonical NF-kB-dependent cytokine production, which perhaps would oppose acute inflammatory stress [209].

In the liver, a role for TLR-mediated gene expression of the PGC-1 family has been shown by the rapid upregulation of hepatocyte PGC-1α and PGC-1β in early *Staphylococcus aureus* peritonitis in wild-type mice, whereas both genes are concordantly deregulated in TLR2^{-mi} mice and increased in TLR4^{-mi} mice.

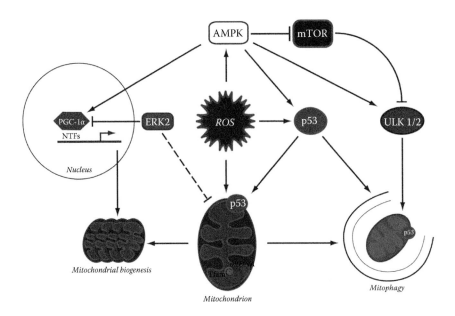

FIGURE 4.5 **(See color insert.)** Regulation of mitochondrial quality control via mitochondrial biogenesis and mitophagy. Mitophagy, in conjunction with mitochondrial biogenesis, regulates changes in mitochondrial number required to meet metabolic demand. AMP-activated protein kinase (AMPK) acutely triggers unc-51-like kinases 1 (ULK1)-dependent mitophagy and simultaneously triggers the biogenesis of new mitochondria via effects on peroxisome proliferator-activated receptor γ coactivator (PGC-1α)-dependent transcription. Conversely, a mammalian target of rapamycin (mTOR) represses mitochondrial biogenesis and ULK1-dependent mitophagy when nutrients are plentiful. The dual processes controlled by AMPK and mTOR determine the net effect of replacing defective mitochondria with new functional mitochondria. AMPK, AMP-activated protein kinase; mTOR, mammalian target of rapamycin; PGC-1α, PPARγ coactivator 1-α; ULK1, the mammalian Atg1 homologs, uncoordinated family member (unc)-51, like kinase 1; ERK2, the extracellular signal-regulated protein kinase 2.

Meanwhile, PRC is upregulated in all three strains [160]. PGC-1α and PGC-1β share micro-RNA binding sites for mmu-mir-202-3p, and mir-202-3p-mediated mRNA degradation has been implicated in PGC-1α and PGC-1β coregulation during inflammation.

Long periods of intense nonsterile inflammation lead to overlapping cycles of tissue damage and repair, accompanied by sterile inflammation, obscuring the timing and consistency of metabolic gene expression. In mice, sublethal but damaging lipopolysaccharide exposure transiently depletes hepatic mtDNA content and impairs mitochondrial transcription [210]. In response, redox-responsive mechanisms including Akt/PKB phosphorylation of NRF-1 and subsequent nuclear translocation increase gene expression for TFAM and other proteins of the mitochondrial transcriptome. After translation and mitochondrial importation of these proteins, mtDNA copy number is restored [211]. The initial loss of mtDNA content is related to TLR4 and NF-kB activation, and specifically TNF-α and NO production lead to mtDNA depletion. Genetic ablation of TLR4 reduces but does not eliminate this effect, and the recovery of mtDNA copy number is delayed in TLR4 null mice [212]. The cell responds to mtDNA depletion by the induction of the base excision repair glycosylase OGG1 by NRF-1 and NRF-2 binding to OGG1 promoter elements and mitochondrial importation of active enzyme [213]. Moreover, the loss of the mtDNA copy number can be abrogated by increasing SOD2 in mitochondria, by inhibiting NOS activity or by scavenging ONOO$^-$ [214]. In both authentic infections and after exposure to surrogate damage-associated molecular patterns (DAMPS), clear morphological evidence of accelerated mitochondrial turnover is seen involving autophagy and mitochondrial biogenesis, which restores mitochondrial mass in survivors within days [210,215,216].

Innate immune activation by TLR2 and TLR4 ligands is involved in the early upregulation of mitochondrial biogenesis [160,212] through several transcription factors including NF-kB [131,217], CREB [131], NRF-2 [218], and IRFs (IRF-3, IRF-7) [160]. TLR4-dependent activation of NF-kB and CREB coregulate the NRF-1 promoter with NF-kB intronic enhancement leading to NRF-1 synthesis and nuclear translocation, followed by target gene expression [130]. This also requires mitochondrial H_2O_2 production and identifies NRF-1 as an early-phase component of the host defense regulated by TLR signaling and by redox state.

Another key regulator of energy metabolism, the serine/threonine kinase AMPK [219], is activated by ATP depletion and stimulates glucose and lipid catabolism and blocks energy-utilizing pathways such as protein and fatty acid biosynthesis. AMPK promotes mitochondrial biogenesis [219–221], NO production [222], regulates autophagy [223], and opposes inflammation by interfering with NF-κB-dependent cytokine expression [224,225]. Conversely, loss of AMPK activity is associated with increased inflammation. The mechanism of NF-kB inhibition is not clear, but this may occur indirectly through processes involving PGC-1α, FOXO-type transcription factors, and/or SIRT1 [125,150,226,227].

The transcriptional activity of NRs, particularly ERRα, is repressed by RIP140 [228], which will suppress metabolic gene expression and mitochondrial biogenesis [229]. In macrophages, RIP140 coactivates certain cytokine genes, and RIP140 deficiency inhibits the inflammatory response [230]. RIP140 interacts with the RelA subunit of NF-kB and the histone acetylase CBP and cooperates with CBP coactivator complex on RelA-regulated promoters. RIP140 modulation of inflammatory gene expression is thus a good example of cell-specific integration of control pathways for metabolism and inflammation [230].

MITOCHONDRIA AND THE INFLAMMASOME

Many DAMPS, including those associated with infection and sterile tissue damage, lead to the assembly of inflammasomes that are required for full expression of host inflammation. The NLRP3 inflammasome and caspase-1 (and caspase-11 in mice) are required to generate IL-1β by the cleavage of its proform, and the assembly of the NLRP3 inflammasome influences metabolic pathways such as glycolysis and lipogenesis. Mitochondrial ROS production can activate the NOD-like receptor family, pyrin domain containing 3 (NLRP3) inflammasome, whereas removal of damaged or dysfunctional mitochondria by mitophagy negatively regulates the NLRP3 inflammasome. Two lines of evidence suggest that mitochondria are the main source of ROS for NLRP3 inflammasome activation and serve as a signal-integrating organelle for inflammasome activation [231–233]. In macrophages, inflammasome activation is impaired by inactivation of the outer membrane VDAC or by depletion of mtDNA [236]. Moreover, respiratory chain inhibition also activates the NLRP3 inflammasome [232].

To avoid autodestructive inflammation, the cell removes ROS-generating mitochondria by mitophagy [233]. Inhibition of mitophagy leads to the retention of ROS-generating mitochondria and activation of the NLRP3 inflammasome [232,233]. ROS-producing mitochondria removed by mitophagy must be replaced through mitochondrial biogenesis in order to avoid persistent inflammasome activation and inflammation, leading to energy failure.

METABOLIC DISORDERS

Mitochondria play critical roles in hepatocyte metabolism, and consequently, malfunction of liver mitochondria has a tremendous impact on body homeostasis [234]. Many studies on obese, diabetic, nonalcoholic steatohepatitis (NASH), or nonalcoholic fatty liver disease (NAFLD) patients have shown functional and structural abnormalities in hepatocyte mitochondria, such as OXPHOS impairment or megamitochondria [235]. Both increased and decreased oxidation has been reported as a feature of steatosis and insulin resistance in hepatocytes [236–238]. A decrease in oxidation activity induces diacylglycerol accumulation and steatosis in the hepatocyte with concurrent activation of protein kinase C and inhibition of insulin signaling [237]. In insulin-resistant patients, an increased activity of hepatic oxidation was observed, and this was correlated to an increase in ROS production [236,239]. High oxidation could be an adaptive mechanism to limit free fatty acid lipotoxicity and provide large amounts of reduced NADH regardless of energetic requirements, finally promoting ROS production due to impairment of respiratory chain [240]. On the other hand, mitochondrial ROS produced during OXPHOS promote insulin signaling through insulin receptor oxidation and inhibition of phosphatases, such as PTP1B and PTEN [241].

Insulin resistance, a hallmark of type 2 diabetes, is characterized by an impaired ability of insulin to inhibit glucose output from the liver and to promote glucose uptake in muscle [242,243]. PGC-1α is implicated in the onset of type 2 diabetes, and hepatic PGC-1α expression is elevated in mouse models of this disease where it promotes constitutive activation of gluconeogenesis and FAO through coactivation of NRs HNF-4 and PPAR-α, respectively [135,244,245]. Insulin suppresses gluconeogenesis stimulated by FOXO1/PGC-1α but coexpression of a mutant allele of FOXO1 that is insensitive to insulin completely reverses this suppression in hepatocytes or in transgenic mice. Recent findings support a role for PGC-1α in linking nutrient deprivation to mitochondrial oxidant production through its coactivation of ERRα to enhance SIRT3 expression [246,247]. SIRT3 localization to mitochondria optimizes key enzymatic activities in metabolic function [248]. In addition, SIRT3 can oppose oxidant stress by precipitating a series of reactions beginning with the activation of isocitrate dehydrogenase 2 and culminating in the detoxification of peroxides by glutathione peroxidase [249]. It also deacetylates manganese superoxide dismutase leading to increased activity and enhanced oxidant scavenging in the mitochondrial matrix [250,251]. Another key factor in metabolic orchestration is that PPARα is highly expressed in the liver, heart, and small intestine [252]. PPARα acts as a nutritional sensor, which promotes adaptation of the rates of fatty acid catabolism, lipogenesis, and ketone body synthesis in response to feeding and starvation to stimulate hepatic FAO and ketogenesis during fasting [253–255]. Indeed, the increase in PPARα would lead to the activation of its target genes involved in FAO (e.g., Cpt1a, Cpt2, Acadvl, Hadha) and ketogenesis (Hmgcs2, Hmgcl, Acat1). Moreover, PPARα negatively regulates proinflammatory and NF-κB and AP-1 signaling in systemic inflammation, atherosclerosis, and in NASH [256,257]. Therapeutic strategies to target mitochondria in the treatment of NAFLD and NASH are still in their very early stages.

CONCLUSION

The conditions discussed here should provide a broad sense of the wide-ranging pathways by which transcription of the compartmentalized nuclear and mitochondrial genomes are regulated and coordinated in health and disease. Real-time transcriptional regulation of the mitochondrial-encoded OXPHOS gene products is of significant benefit to organisms that must adapt rapidly to changes in nutrient availability and growth rates. While POLRMT, TFAM, and TFB2M comprise the basal transcriptosome for mitochondrial transcription, accessory factors that provide more intricate control of the system at multiple steps such as initiation, elongation, and termination have not been covered in detail in any part because how these factors interact with the primary mitochondrial transcription machinery to exert their specific effects is poorly understood. Nucleus-encoded regulatory factors are crucial regulators of mitochondrial biogenesis and function. In addition, the PGC-1 coactivators may balance the cellular response to oxidant stress by promoting an antioxidant environment or by orchestrating an inflammatory response to severe metabolic stress. All three PGC-1 family members activate gene expression through the NRFs (NRF-1, NRF-2) coherent with their biological activities involving changes in mitochondrial gene expression.

It is also clear that the PGC-1 family helps specify differentiation-induced mitochondrial content in tissues such as brown fat and skeletal muscle, but individually they do not seem to be the sole determinants of basal mitochondrial content in tissues. Perhaps, there is sufficient functional redundancy among the family members for the loss of one to be partially compensated for by the others. The transcriptional response to metabolic variations, changing energy demands, and environmental factors in vivo remains in its infancy, especially in terms of tissue-specific mitochondrial responses. This area of emerging interest may involve an increasing number of nuclear transcriptional regulators that will no doubt be important in deciphering how mitochondria are involved in normal physiology, disease states, and in aging. Moreover, as our understanding of the biochemical and molecular

events involved in mitochondrial transcription improves, we will no doubt identify additional mechanisms by which mitochondrial transcription can be controlled
in response to extramitochondrial signals. This understanding of the mechanisms
linking mitochondrial transcription to extramitochondrial events may eventually
promote the development of therapeutic modalities for treating patients with abnormal mitochondrial function.

REFERENCES

1. Piantadosi CA, Suliman HB. Transcriptional control of mitochondrial biogenesis and its interface with inflammatory processes. *Biochim Biophys Acta*. 2012; 1820:532–541.
2. Piantadosi CA, Suliman HB. Redox regulation of mitochondrial biogenesis. *Free Radic Biol Med*. 2012; 53:2043–2053.
3. Bonawitz ND, Clayton DA, Shadel GS. Initiation and beyond: Multiple functions of the human mitochondrial transcription machinery. *Mol Cell*. 2006; 24:813–825.
4. Becker T, Böttinger L, Pfanner N. Mitochondrial protein import: From transport path-ways to an integrated network. *Trends Biochem Sci*. 2012; 37:85–91.
5. Shutt TE, Shadel GS. A compendium of human mitochondrial gene expression machin-ery with links to disease. *Environ Mol Mutagen*. 2010; 51:360–379.
6. Wallace DC. A mitochondrial paradigm of metabolic and degenerative diseases, aging, and cancer: A dawn for evolutionary medicine. *Annu Rev Genet*. 2005; 39:359–407.
7. Shadel GS, Clayton DA. Mitochondrial DNA maintenance in vertebrates. *Annu Rev Biochem*. 1997; 66:409–435.
8. Duchen MR. Mitochondria in health and disease: Perspectives on a new mitochondrial biology. *Mol Aspects Med*. 2004; 25:365–451.
9. Hailey DW, Rambold AS, Satpute-Krishnan P, Mitra K, Sougrat R, Kim PK, Lippincott-Schwartz J. Mitochondria supply membranes for autophagosome biogenesis during starvation. *Cell*. 2010; 141:656–667.
10. West AP, Shadel GS, Ghosh S. Mitochondria in innate immune responses. *Nat Rev Immunol*. 2011; 11:389–402.
11. Vance JE. Phospholipid synthesis in a membrane fraction associated with mitochon-dria. *J Biol Chem*. 1990; 265:7248–7256.
12. Neuspiel M, Schauss AC, Braschi E, Zunino R, Rippstein P, Rachubinski RA, Andrade-Navarro MA, McBride HM. Cargo-selected transport from the mitochondria to peroxi-somes is mediated by vesicular carriers. *Curr Biol. CB*. 2008; 18:102–108.
13. Ekstrand MI, Falkenberg M, Rantanen A, Park CB, Gaspari M, Hultenby K, Rustin P, Gustafsson CM, Larsson NG. Mitochondrial transcription factor A regulates mtDNA copy number in mammals. *Hum Mol Genet*. 2004; 13:935–944.
14. Carrodeguas JA, Kobayashi R, Lim SE, Copeland WC, Bogenhagen DF. The accessory subunit of Xenopus laevis mitochondrial DNA polymerase γ increases processivity of the catalytic subunit of human DNA polymerase γ and is related to class II aminoacyl-tRNA synthetases. *Mol Cell Biol*. 1999; 19:4039–4046.
15. Spelbrink JN, Li FY, Tiranti V, Nikali K, Yuan QP, Tariq M, Wanrooij S et al. Human mitochondrial DNA deletions associated with mutations in the gene encoding Twinkle, a phage T7 gene 4-like protein localized in mitochondria. *Nat Genet*. 2001; 28:223–231.
16. Korhonen JA, Gaspari M, Falkenberg M. TWINKLE Has 5′ → 3′ DNA helicase activ-ity and is specifically stimulated by mitochondrial single-stranded DNA-binding pro-tein. *J Biol Chem*. 2003; 278:48627–48632.
17. Bogenhagen DF, Rousseau D, Burke S. The layered structure of human mitochondrial DNA nucleoids. *J Biol Chem*. 2008; 283:3665–3675.

18. Kelly RD, Mahmud A, McKenzie M, Trounce IA, St John JC. Mitochondrial DNA copy number is regulated in a tissue specific manner by DNA methylation of the nuclear-encoded DNA polymerase γ A. *Nucleic Acids Res.* 2012; 10124–10138.

19. Moyes CD, Battersby BJ, Leary SC. Regulation of muscle mitochondrial design. *J Exp Biol.* 1998; 201:299–307.

20. Pfeiffer T, Schuster S, Bonhoeffer S. Cooperation and competition in the evolution of ATP-producing pathways. *Science.* 2001; 292:504–507.

21. Montoya J, Gaines GL, Attardi G. The pattern of transcription of the human mitochondrial rRNA genes reveals two overlapping transcription units. *Cell.* 1983; 34:151–159.

22. Ojala D, oya J, Attardi G. tRNA punctuation model of RNA processing in human mitochondria. *Nature.* 1981; 290:470–474.

23. Correia RL, Oba-Shinjo SM, Uno M, Huang N, Marie SK. Mitochondrial DNA depletion and its correlation with TFAM, TFB1M, TFB2M and POLG in human diffusely infiltrating astrocytomas. *Mitochondrion.* 2011; 11:48–53.

24. Yakubovskaya E, Guja KE, Eng ET, Choi WS, Mejia E, Beglov D, Lukin M, Kozakov D, Garcia-Diaz M. Organization of the human mitochondrial transcription initiation complex. *Nucleic Acids Res.* 2014; 42:4100–4112.

25. Fisher RP, Clayton DA. A transcription factor required for promoter recognition by human mitochondrial RNA polymerase. Accurate initiation at the heavy- and light-strand promoters dissected and reconstituted in vitro. *J Biol Chem.* 1985; 260:11330–11338.

26. Parisi MA, Clayton DA. Similarity of human mitochondrial transcription factor 1 to high mobility group proteins. *Science.* 1991; 252:965–969.

27. Campbell CT, Kolesar JE, Kaufman BA. Mitochondrial transcription factor A regulates mitochondrial transcription initiation, DNA packaging, and genome copy number. *Biochim Biophys Acta.* 2012; 1819:921–929.

28. Rubio-Cosials A, Sola M. U-turn DNA bending by human mitochondrial transcription factor A. *Curr Opin Struct Biol.* 2013; 23:116–124.

29. Farge G, Laurens N, Broekmans OD, van den Wildenberg SM, Dekker LC, Gaspari M, Gustafsson CM, Peterman EJ, Falkenberg M, Wuite GJ. Protein sliding and DNA denaturation are essential for DNA organization by human mitochondrial transcription factor A. *Nat Commun.* 2012; 3:1013.

30. Malarkey CS, Bestwick M, Kuhlwilm JE, Shadel GS, Churchill ME. Transcriptional activation by mitochondrial transcription factor A involves preferential distortion of promoter DNA. *Nucleic Acids Res.* 2012; 40:614–624.

31. Ngo HB, Kaiser JT, Chan DC. The mitochondrial transcription and packaging factor Tfam imposes a U-turn on mitochondrial DNA. *Nat Struct Mol Biol.* 2011; 18:1290–1296.

32. Kaufman BA, Durisic N, Mativetsky JM, Costantino S, Hancock MA, Grutter P, Shoubridge EA. The mitochondrial transcription factor TFAM coordinates the assembly of multiple DNA molecules into nucleoid-like structures. *Mol Biol Cell.* 2007; 18:3225–3236.

33. Lu B, Lee J, Nie X, Li M, Morozov YI, Venkatesh S, Bogenhagen DF, Temiakov D, Suzuki CK. Phosphorylation of human TFAM in mitochondria impairs DNA binding and promotes degradation by the AAA(+) Lon protease. *Mol Cell.* 2013; 49:121–132.

34. Hebert AS, Dittenhafer-Reed KE, Yu W, Bailey DJ, Selen ES, Boersma MD, Carson JJ et al. Calorie restriction and SIRT3 trigger global reprogramming of the mitochondrial protein acetylome. *Mol Cell.* 2013; 49:186–199.

35. Wang KZ, Zhu J, Dagda RK, Uechi G, Cherra SJ 3rd, Gusdon AM, Balasubramani M, Chu CT. ERK-mediated phosphorylation of TFAM downregulates mitochondrial transcription: Implications for Parkinson's disease. *Mitochondrion.* 2014; 17:132–140.

36. Watanabe A, Arai M, Koitabashi N, Niwano K, Ohyama Y, Yamada Y, Kato N, Kurabayashi M. Mitochondrial transcription factors TFAM and TFB2M regulate Serca2 gene transcription. *Cardiovasc Res.* 2011; 90:57–67.

37. Cotney J, McKay SE, Shadel GS. Elucidation of separate, but collaborative functions of the rRNA methyltransferase-related human mitochondrial transcription factors B1 and B2 in mitochondrial biogenesis reveals new insight into maternally inherited deafness. *Hum Mol Genet.* 2009; 18:2670–2682.

38. Litonin D, Sologub M, Shi Y, Savkina M, Anikin M, Falkenberg M, Gustafsson CM, Temiakov D. Human mitochondrial transcription revisited: Only TFAM and TFB2M are required for transcription of the mitochondrial genes in vitro. *J Biol Chem.* 2010; 285:18129–18133.

39. Lodeiro MF, Uchida AU, Arnold JJ, Reynolds SL, Moustafa IM, Cameron CE. Identification of multiple rate-limiting steps during the human mitochondrial transcription cycle in vitro. *J Biol Chem.* 2010; 285:16387–16402.

40. Sologub M, Litonin D, Anikin M, Mustaev A, Temiakov D. TFB2 is a transient component of the catalytic site of the human mitochondrial RNA polymerase. *Cell.* 2009; 139:934–944.

41. Gaspari M, Falkenberg M, Larsson NG, Gustafsson CM. The mitochondrial RNA polymerase contributes critically to promoter specificity in mammalian cells. *EMBO J.* 2004; 23:4606–4614.

42. Tiranti V, Savoia A, Forti F, D'Apolito MF, Centra M, Rocchi M, Zeviani M. Identification of the gene encoding the human mitochondrial RNA polymerase (h-mtRPOL) by cyberscreening of the expressed sequence tags database. *Hum Mol Genet.* 1997; 6:615–625.

43. Kato M, Ito T, Wagner G, Richardson CC, Ellenberger T. Modular architecture of the bacteriophage T7 primase couples RNA primer synthesis to DNA synthesis. *Mol Cell.* 2003; 11:1349–1360.

44. Fuste JM, Wanrooij S, Jemt E, Granycome CE, Cluett TJ, Shi Y, Atanassova N, Holt IJ, Gustafsson CM, Falkenberg M. Mitochondrial RNA polymerase is needed for activation of the origin of light-strand DNA replication. *Mol Cell.* 2010; 37:67–78.

45. Holt IJ. Mitochondrial DNA replication and repair: All a flap. *Trends Biochem Sci.* 2009; 34:358–365.

46. Wong TW, Clayton DA. In vitro replication of human mitochondrial DNA: Accurate initiation at the origin of light-strand synthesis. *Cell.* 1985; 42:951–958.

47. Smidansky ED, Arnold JJ, Reynolds SL, Cameron CE. Human mitochondrial RNA polymerase: Evaluation of the single-nucleotide-addition cycle on synthetic RNA/DNA scaffolds. *Biochemistry.* 2011; 50:5016–5032.

48. Minczuk M, He J, Duch AM, Ettema TJ, Chlebowski A, Dzionek K, Nijtmans LG, Huynen MA, Holt IJ. TEFM (c17orf42) is necessary for transcription of human mtDNA. *Nucleic Acids Res.* 2011; 39:4284–4299.

49. Larsson NG, Wang J, Wilhelmsson H, Oldfors A, Rustin P, Lewandoski M, Barsh GS, Clayton DA. Mitochondrial transcription factor A is necessary for mtDNA maintenance and embryogenesis in mice. *Nat Genet.* 1998; 18:231–236.

50. Park CB, Asin-Cayuela J, Cámara Y, Shi Y, Pellegrini M, Gaspari M, Wibom R et al. MTERF3 is a negative regulator of mammalian mtDNA transcription. *Cell.* 2007; 130:273–285.

51. Metodiev MD, Lesko N, Park CB, Cámara Y, Shi Y, Wibom R, Hultenby K, Gustafsson CM, Larsson NG. Methylation of 12S rRNA is necessary for in vivo stability of the small subunit of the mammalian mitochondrial ribosome. *Cell Metab.* 2009; 9:386–397.

52. Camara Y, Asin-Cayuela J, Park CB, Metodiev MD, Shi Y, Ruzzenente B, Kukat C et al. MTERF4 regulates translation by targeting the methyltransferase NSUN4 to the mammalian mitochondrial ribosome. *Cell Metab.* 2011; 13:527–539.

53. Scarpulla RC. Nucleus-encoded regulators of mitochondrial function: Integration of respiratory chain expression, nutrient sensing and metabolic stress. *Biochim Biophys Acta.* 2012; 1819:1088–1097.

54. Kelly DP, Scarpulla RC. Transcriptional regulatory circuits controlling mitochondrial biogenesis and function. *Genes Dev.* 2004; 18:357–368.

55. Gleyzer N, Vercauteren K, Scarpulla RC. Control of mitochondrial transcription specificity factors (TFB1M and TFB2M) by nuclear respiratory factors (NRF-1 and NRF-2) and PGC-1 family coactivators. *Mole Cell Biol.* 2005; 25:1354–1366.

56. Tarnowski M, Grymula K, Reca R, Jankowski K, Maksym R, Tarnowska J, Przybylski G, Barr FG, Kucia M, Ratajczak MZ. Regulation of expression of stromal-derived factor-1 receptors: CXCR4 and CXCR7 in human rhabdomyosarcomas. *Mol Cancer Res.* 2010; 8:1–14.

57. Virbasius JV, Virbasius CA, Scarpulla RC. Identity of GABP with NRF-2, a multi-subunit activator of cytochrome oxidase expression, reveals a cellular role for an ETS domain activator of viral promoters. *Genes Dev.* 1993; 7:380–392.

58. Hsu T, Trojanowska M, Watson DK. Ets proteins in biological control and cancer. *J Cell Biochem.* 2004; 91:896–903.

59. Rosmarin AG, Resendes KK, Yang Z, McMillan JN, Fleming SL. GA-binding protein transcription factor: A review of GABP as an integrator of intracellular signaling and protein-protein interactions. *Blood Cells Mol Dis.* 2004; 32:143–154.

60. Ongwijitwat S, Wong-Riley MT. Is nuclear respiratory factor 2 a master transcriptional coordinator for all ten nuclear-encoded cytochrome c oxidase subunits in neurons? *Gene.* 2005; 360:65–77.

61. Yang Z-F, Drumea K, Mott S, Wang J, and Rosmarin AG. GABP transcription factor (nuclear respiratory factor 2) is required for mitochondrial biogenesis. *Mol Cell Biol.* September 2014; 34(17):3194–3201.

62. Bruni F, Polosa PL, Gadaleta MN, Cantatore P, Roberti M. Nuclear respiratory factor 2 induces the expression of many but not all human proteins acting in mitochondrial DNA transcription and replication. *J Biol Chem.* 2010; 285:3939–3948.

63. Yang Z-F, Mott S, Rosmarin AG. The Ets transcription factor GABP is required for cell-cycle progression. *Nat Cell Biol.* 2007; 9:339–346.

64. Wu H, Xiao Y, Zhang S, Ji S, Wei L, Fan F, Geng J et al. The Ets transcription factor GABP is a component of the hippo pathway essential for growth and antioxidant defense. *Cell Rep.* 2013; 3:1663–1677.

65. Ristevski S, O'Leary DA, Thornell AP, Owen MJ, Kola I, Hertzog PJ. The ETS transcription factor GABPα is essential for early embryogenesis. *Mol Cell Biol.* 2004; 24:5844–5849.

66. Xue HH, Bollenbacher J, Rovella V, Tripuraneni R, Du YB, Liu CY, Williams A, McCoy JP, Leonard WJ. GA binding protein regulates interleukin 7 receptor α-chain gene expression in T cells. *Nat Immunol.* 2004; 5:1036–1044.

67. Ongwijitwat S, Liang HL, Graboyes EM, Wong-Riley MT. Nuclear respiratory factor 2 senses changing cellular energy demands and its silencing down-regulates cytochrome oxidase and other target gene mRNAs. *Gene.* 2006; 374:39–49.

68. Dhar SS, Ongwijitwat S, Wong-Riley MT. Nuclear respiratory factor 1 regulates all ten nuclear-encoded subunits of cytochrome c oxidase in neurons. *J Biol Chem.* 2008; 283:3120–3129.

69. Huo L, Scarpulla RC. Mitochondrial DNA instability and peri-implantation lethality associated with targeted disruption of nuclear respiratory factor 1 in mice. *Mol Cell Biol.* 2001; 21:644–654.

70. Donohoe ME, Zhang X, McGinnis L, Biggers J, Li E, Shi Y. Targeted disruption of mouse yin yang 1 transcription factor results in peri-implantation lethality. *Mol Cell Biol.* 1999; 19:7237–7244.

71. Attardi G, Schatz G. Biogenesis of mitochondria. *Annu Rev Cell Biol*. 1988; 4:289–333.
72. Asin-Cayuela J, Gustafsson CM. Mitochondrial transcription and its regulation in mammalian cells. *Trends Biochem Sci*. 2007; 32:111–117.
73. Li F, Wang Y, Zeller KI, Potter JJ, Wonsey DR, O'Donnell KA, Kim JW, Yustein JT, Lee LA, Dang CV. Myc stimulates nuclearly encoded mitochondrial genes and mito- chondrial biogenesis. *Mol Cell Biol*. 2005; 25:6225–6234.
74. Kim J, Lee JH, Iyer VR. Global identification of Myc target genes reveals its direct role in mitochondrial biogenesis and its E-box usage in vivo. *PLoS One*. 2008; 3:e1798.
75. Wahlström T, Arsenian Henriksson M. Impact of MYC in regulation of tumor cell metabolism. *Biochim Biophys Acta*. July 17, 2014; pii:S1874-9399(14)00192-8.
76. Blomain ES, McMahon SB. Dynamic regulation of mitochondrial transcrip- tion as a mechanism of cellular adaptation. *Biochim Biophys Acta*. Sept–Oct 2012; 1819(9–10):1075–1079.
77. Morrish F, Giedt C, Hockenbery D. c-MYC apoptotic function is mediated by NRF-1 target genes. *Genes Dev*. 2003; 17:240–255.
78. Morrish F, Hockenbery D. MYC and mitochondrial biogenesis. *Cold Spring Harb Perspect Med*. May 1, 2014; 4(5):pii:a014225.
79. Weber K, Brück P, Mikes Z, Küpper JH, Klingenspor M, Wiesner RJ. Glucocorticoid hormone stimulates mitochondrial biogenesis specifically in skeletal muscle. *Endocrinology*. 2002; 143:177–184.
80. Psarra AM, Sekeris CE. Glucocorticoids induce mitochondrial gene transcription in HepG2 cells: Role of the mitochondrial glucocorticoid receptor. *Biochim Biophys Acta*. 2001; 1813:1814–1821.
81. Demonacos C, Djordjevic-Markovic R, Tsawdaroglou N, Sekeris CE. The mitochon- drion as a primary site of action of glucocorticoids: The interaction of the glucocorticoid receptor with mitochondrial DNA sequences showing partial similarity to the nuclear glucocorticoid responsive elements. *J Steroid Biochem Mol Biol*. 1995; 55:43–55.
82. Psarra AM, Solakidi S, Sekeris CE. The mitochondrion as a primary site of action of steroid and thyroid hormones: Presence and action of steroid and thyroid hormone receptors in mitochondria of animal cells. *Mol Cell Endocrinol*. 2006; 246:21–33.
83. Yang SH, Liu R, Perez EJ, Wen Y, Stevens SM Jr, Valencia T, Brun-Zinkernagel AM et al. Mitochondrial localization of estrogen receptor β. *Proc Natl Acad Sci USA*. 2004; 101:4130–4135.
84. Scheller K, Sekeris CE, Krohne G, Hock R, Hansen IA, Scheer U. Localization of glu- cocorticoid hormone receptors in mitochondria of human cells. *Eur J Cell Biol*. 2000; 79:299–307.
85. Casas F, Rochard P, Rodier A, Cassar-Malek I, Marchal-Victorion S, Wiesner RJ, Cabello G, Wrutniak C. A variant form of the nuclear triiodothyronine receptor c-ErbAα1 plays a direct role in regulation of mitochondrial RNA synthesis. *Mol Cell Biol*. 1999; 19:7913–7924.
86. Wiesner RJ, Kurowski TT, Zak R. Regulation by thyroid hormone of nuclear and mito- chondrial genes encoding subunits of cytochrome-c oxidase in rat liver and skeletal muscle. *Mol Endocrinol*. 1992; 6:1458–1467.
87. Mutvei A, Kuzela S, Nelson BD. Control of mitochondrial transcription by thyroid hormone. *Eur J Biochem*. 1989; 180:235–240.
88. Silvestri E, Burrone L, de Lange P, Lombarda A, Farina P, Chambery A, Parente A, Lanni A, Goglia F, Moreno M. Thyroid-state influence on protein-expression profile of rat skeletal muscle. *J Proteome Res*. 2007; 6:3187–3196.
89. Fernández V, Barrientos X, Kipreos K, Valenzuela A, Videla LA. Superoxide radical generation, NADPH oxidase activity, and cytochrome P450 content in an experimental hyperthyroid state: Relation to lipid peroxidation. *Endocrinology*. 1985; 117:496–501.

90. Lanni A, Moreno M, Lombardi A, Goglia F. Thyroid hormone and uncoupling proteins. *FEBS Lett.* 2003; 543:5–10.
91. Soboll S. Thyroid hormone action on mitochondrial energy transfer. *Biochim Biophys Acta.* 1993; 1144:1–16.
92. Davis PJ, Leonard JL, Davis FB. Mechanisms of nongenomic actions of thyroid hormone. *Front Neuroendocrinol.* 2008; 29:211–218.
93. Venditti P, De Rosa R, Di Meo S, De Leo T. Effect of thyroid state on H_2O_2 production by rat liver mitochondria. *Mol Cell Endocrinol.* 2003; 205:185–192.
94. Yen PM. Physiological and molecular basis of thyroid hormone action. *Physiol Rev.* 2001; 81:1097–1142.
95. Jansen MS, Cook GA, Song S, Park EA. Thyroid hormone regulates carnitine palmitoyltransferase Iα gene expression through elements in the promoter and first intron. *J Biol Chem.* 2000; 275:34989–34997.
96. Attia RR, Connnaughton S, Boone LR, Wang F, Elam MB, Ness GC, Cook GA, Park EA. Regulation of pyruvate dehydrogenase kinase 4 (PDK4) by thyroid hormone: Role of the peroxisome proliferator-activated receptor γ coactivator (PGC-1 α). *J Biol Chem.* 2010; 285:2375–2385.
97. Zhang Y, Ma K, Song S, Elam MB, Cook GA, Park EA. Peroxisomal proliferator-activated receptor-γ coactivator-1 α (PGC-1 α) enhances the thyroid hormone induction of carnitine palmitoyltransferase I (CPT-I α). *J Biol Chem.* 2004; 279:53963–53971.
98. Jurado LA, Song S, Roesler WJ, Park EA. Conserved amino acids within CCAAT enhancer-binding proteins (C/EBP(α) and β) regulate phosphoenolpyruvate carboxykinase (PEPCK) gene expression. *J Biol Chem.* 2002; 277:27606–27612.
99. Jackson-Hayes L, Song S, Lavrentyev EN, Jansen MS, Hillgartner FB, Tian L, Wood PA, Cook GA, Park EA. A thyroid hormone response unit formed between the promoter and first intron of the carnitine palmitoyltransferase-Iα gene mediates the liver-specific induction by thyroid hormone. *J Biol Chem.* 2003; 278:7964–7972.
100. Liu YY, Brent GA. Thyroid hormone crosstalk with nuclear receptor signaling in metabolic regulation. *Trends Endocrinol Metab.* 2010; 21:166–173.
101. Singh R, Kaushik S, Wang Y, Xiang Y, Novak I, Komatsu M, Tanaka K, Cuervo AM, Czaja MJ. Autophagy regulates lipid metabolism. *Nature.* 2009; 458:1131–1135.
102. Cahova M, Dankova H, Palenickova E, Papackova Z, Kazdova L. The autophagy-lysosomal pathway is involved in TAG degradation in the liver: The effect of high-sucrose and high-fat diet. *Folia Biol.* 2010; 56:173–182.
103. Yang L, Li P, Fu S, Calay ES, Hotamisligil GS. Defective hepatic autophagy in obesity promotes ER stress and causes insulin resistance. *Cell Metab.* 2010; 11:467–478.
104. Amir M, Czaja MJ. Autophagy in nonalcoholic steatohepatitis. *Expert Rev Gastroenterol Hepatol.* 2011; 5:159–166.
105. Madrazo JA, Kelly DP. The PPAR trio: Regulators of myocardial energy metabolism in health and disease. *J Mol Cell Cardiol.* 2008; 44:968–975.
106. Nettles KW. Insights into PPARγ from structures with endogenous and covalently bound ligands. *Nat Struct Mol Biol.* Sept 2008; 15:893–895.
107. Eichner LJ, Giguere V. Estrogen related receptors (ERRs): A new dawn in transcriptional control of mitochondrial gene networks. *Mitochondrion.* 2011; 11:544–552.
108. Dufour CR, Wilson BJ, Huss JM, Kelly DP, Alaynick WA, Downes M, Evans RM, Blanchette M, Giguère V. Genome-wide orchestration of cardiac functions by the orphan nuclear receptors ERRα and γ. *Cell Metab.* 2007; 5:345–356.
109. Andersson U, Scarpulla RC. PGC-1-related coactivator, a novel, serum-inducible coactivator of nuclear respiratory factor 1-dependent transcription in mammalian cells. *Mol Cell Biol.* 2001; 21:3738–3749.

110. Narkar VA, Fan W, Downes M, Yu RT, Jonker JW, Alaynick WA, Banayo E, Karunasiri MS, Lorca S, Evans RM. Exercise and PGC-1α-independent synchronization of type I muscle metabolism and vasculature by ERRγ. *Cell Metab*. 2011; 13:283–293.

111. Huss JM, Imahashi K, Dufour CR, Weinheimer CJ, Courtois M, Kovacs A, Giguère V, Murphy E, Kelly DP. The nuclear receptor ERRα is required for the bioenergetic and functional adaptation to cardiac pressure overload. *Cell Metab*. 2007; 6:25–37.

112. Alaynick WA, Kondo RP, Xie W, He W, Dufour CR, Downes M, Jonker JW, Giles W, Naviaux RK, Giguère V, Evans RM. ERRγ directs and maintains the transition to oxidative metabolism in the postnatal heart. *Cell Metab*. 2007; 6:13–24.

113. Rangwala SM, Wang X, Calvo JA, Lindsley L, Zhang Y, Deyneko G, Beaulieu V, Gao J, Turner G, Markovits J. Estrogen-related receptor γ is a key regulator of muscle mitochondrial activity and oxidative capacity. *J Biol Chem*. 2010; 285:22619–22629.

114. Basu A, Lenka N, Mullick J, Avadhani NG. Regulation of murine cytochrome oxidase Vb gene expression in different tissues and during myogenesis—Role of a YY-1 factor-binding negative enhancer. *J Biol Chem*. 1997; 272:5899–5908.

115. Seelan RS, Grossman LI. Structural organization and promoter analysis of the bovine cytochrome c oxidase subunit VIIc gene—A functional role for YY1. *J Biol Chem*. 1997; 272:10175–10181.

116. Xi H, Yu Y, Fu Y, Foley J, Halees A, Weng Z. Analysis of overrepresented motifs in human core promoters reveals dual regulatory roles of YY1. *Genome Res*. 2007; 17:798–806.

117. Cunningham JT, Rodgers JT, Arlow DH, Vazquez F, Mootha VK, Puigserver P. mTOR controls mitochondrial oxidative function through a YY1-PGC-1α transcriptional complex. *Nature*. 2007; 450:736–740.

118. Wan B, Moreadith RW. Structural characterization and regulatory element analysis of the heart isoform of cytochrome c oxidase VIa. *J Biol Chem*. 1995; 270:26433–26440.

119. Ramachandran B, Yu G, Gulick T. Nuclear respiratory factor 1 controls myocyte enhancer factor 2A transcription to provide a mechanism for coordinate expression of respiratory chain subunits. *J Biol Chem*. 2008; 283:11935–11946.

120. Itoh K, Wakabayashi N, Katoh Y, Ishii T, Igarashi K, Engel JD, Yamamoto M. Keap1 represses nuclear activation of antioxidant responsive elements by Nrf2 through binding to the amino-terminal Neh2 domain. *Genes Dev*. 1999; 13:76–86.

121. Dinkova-Kostova AT, Holtzclaw WD, Cole RN, Itoh K, Wakabayashi N, Katoh Y, Yamamoto M, Talalay P. Direct evidence that sulfhydryl groups of Keap1 are the sensors regulating induction of phase 2 enzymes that protect against carcinogens and oxidants. *Proc Natl Acad Sci USA*. 2002; 99:11908–11913.

122. Piantadosi CA, Carraway MS, Babiker A, Suliman HB. Heme oxygenase-1 regulates cardiac mitochondrial biogenesis via Nrf2-mediated transcriptional control of nuclear respiratory factor-1. *Circ Res*. 2008; 103:1232–1240.

123. MacGarvey NC, Suliman HB, Bartz RR, Fu P, Withers CM, Welty-Wolf KE, Piantadosi CA. Activation of mitochondrial biogenesis by heme oxygenase-1-mediated NF-E2-related factor-2 induction rescues mice from lethal *Staphylococcus aureus* sepsis. *Am J Respir Crit Care Med*. 2012; 185:851–861.

124. Scarpulla RC. Transcriptional paradigms in mammalian mitochondrial biogenesis and function. *Physiol Rev*. 2008; 88:611–638.

125. Scarpulla RC. Metabolic control of mitochondrial biogenesis through the PGC-1 family regulatory network. *Biochim Biophys Acta*. 2011; 1813:1269–1278.

126. Kwon J, Han E, Bui CB, Shin W, Lee J, Lee S, Choi YB et al. Assurance of mitochondrial integrity and mammalian longevity by the p62-Keap1-Nrf2-Nqo1 cascade. *EMBO Rep*. 2012; 13:150–156.

127. Thimmulappa RK, Lee H, Rangasamy T, Reddy SP, Yamamoto M, Kensler TW, Biswal S. Nrf2 is a critical regulator of the innate immune response and survival during experimental sepsis. *J Clin Invest*. 2006; 116:984–995.

128. Kong X, Thimmulappa R, Craciun F, Harvey C, Singh A, Kombairaju P, Reddy SP, Remick D, Biswal S. Enhancing Nrf2 pathway by disruption of Keap1 in myeloid leukocytes protects against sepsis. *Am J Respir Crit Care Med*. 2011; 184:928–938.

129. Herzig RP, Scacco S, Scarpulla RC. Sequential serum-dependent activation of CREB and NRF-1 leads to enhanced mitochondrial respiration through the induction of cytochrome c. *J Biol Chem*. 2000; 275:13134–13141.

130. Vercauteren K, Pasko RA, Gleyzer N, Marino VM, Scarpulla RC. PGC-1-related coactivator (PRC): Immediate early expression and characterization of a CREB/NRF-1 binding domain associated with cytochrome c promoter occupancy and respiratory growth. *Mol Cell Biol*. 2006; 26:7409–7419.

131. Suliman HB, Sweeney TE, Withers CM, Piantadosi CA. Co-regulation of nuclear respiratory factor-1 by NFκB and CREB links LPS-induced inflammation to mitochondrial biogenesis. *J Cell Sci*. 2010; 123:2565–2575.

132. Puigserver P, Wu Z, Park CW, Graves R, Wright M, Spiegelman BM. A cold-inducible coactivator of nuclear receptors linked to adaptive thermogenesis. *Cell*. 1998; 92:829–839.

133. Lin J, Puigserver P, Donovan J, Tarr P, Spiegelman BM. PGC-1α: A novel PGC-1-related transcription coactivator associated with host cell factor. *J Biol Chem*. 2002; 277:1645–1648.

134. Finck BN, Kelly DP. PGC-1 coactivators: Inducible regulators of energy metabolism in health and disease. *J Clin Invest*. 2006; 116:615–622.

135. Yoon JC, Puigserver P, Chen G, Donovan J, Wu Z, Rhee J, Adelmant G et al. Control of hepatic gluconeogenesis through the transcriptional coactivator PGC-1. *Nature*. 2001; 413:131–138.

136. Lin J, Handschin C, Spiegelman BM. Metabolic control through the PGC-1 family of transcription coactivators. *Cell Metab*. 2005; 1:361–370.

137. Lin J, Wu PH, Tarr PT, Lindenberg KS, St-Pierre J, Zhang CY, Mootha VK et al. Defects in adaptive energy metabolism with CNS-linked hyperactivity in PGC-1α null mice. *Cell*. 2004; 119:121–135.

138. Koo SH, Satoh H, Herzig S, Lee CH, Hedrick S, Kulkarni R, Evans RM. PGC-1 promotes insulin resistance in liver through PPAR-α-dependent induction of TRB-3. *Nat Med*. 2004; 10:530–534.

139. Leone TC, Lehman JJ, Finck BN, Schaeffer PJ, Wende AR, Boudina S, Courtois M et al. PGC-1α deficiency causes multisystem energy metabolic derangements: Muscle dysfunction, abnormal weight control and hepatic steatosis. *PLoS Biol*. 2005; 3:e101.

140. Li S, Arning E, Liu C, Vitvitsky V, Hernandez C, Banerjee R, Bottiglieri T et al. Regulation of homocysteine homeostasis through the transcriptional coactivator PGC-1α. *Am J Physiol Endocrinol Metab*. 2009; 296:E543–E548.

141. Liu C, Li S, Liu T, Borjigin J, Lin JD. Transcriptional coactivator PGC-1α integrates the mammalian clock and energy metabolism. *Nature*. 2007; 447:477–481.

142. Lerin C, Rodgers JT, Kalume DE, Kim SH, Pandey A, Puigserver P. GCN5 acetyltransferase complex controls glucose metabolism through transcriptional repression of PGC-1α. *Cell Metab*. 2006; 3:429–438.

143. Rodgers JT, Lerin C, Haas W, Gygi SP, Spiegelman BM, Puigserver P. Nutrient control of glucose homeostasis through a complex of PGC-1α and SIRT1. *Nature*. 2005; 434:113–118.

144. Purushotham A, Schug TT, Xu Q, Surapureddi S, Guo X, Li X. Hepatocyte-specific deletion of SIRT1 alters fatty acid metabolism and results in hepatic steatosis and inflammation. *Cell Metab*. 2009; 9:327–338.

145. Puigserver P, Rhee J, Lin J, Wu Z, Yoon JC, Zhang CY, Krauss S. Cytokine stimulation of energy expenditure through p38 MAP kinase activation of PPARγ coactivator-1. *Mol Cell*. 2001; 8:971–982.

146. Jager S, Handschin C, St-Pierre J, Spiegelman BM. AMP-activated protein kinase (AMPK) action in skeletal muscle via direct phosphorylation of PGC-1α. *Proc Natl Acad Sci USA*. 2007; 104:12017–12022.

147. Li X, Monks B, Ge Q, Birnbaum MJ. Akt/PKB regulates hepatic metabolism by directly inhibiting PGC-1α transcription coactivator. *Nature*. 2007; 447:1012–1016.

148. Teyssier C, Ma H, Emter R, Kralli A, Stallcup MR. Activation of nuclear receptor coactivator PGC-1α by arginine methylation. *Genes Dev*. 2005; 19:1466–1473.

149. Rytinki MM, Palvimo JJ. SUMOylation attenuates the function of PGC-1α. *J Biol Chem*. 2009; 284:26184–26193.

150. Cantó C, Gerhart-Hines Z, Feige JN, Lagouge M, Noriega L, Milne JC, Elliott PJ et al. AMPK regulates energy expenditure by modulating NAD+ metabolism and SIRT1 activity. *Nature*. 2009; 458:1056–1060.

151. Kelly TJ, Lerin C, Haas W, Gygi SP, Puigserver P. GCN5-mediated transcriptional control of the metabolic coactivator PGC-1β through lysine acetylation. *J Biol Chem*. 2009; 284:19945–19952.

152. Haigis MC, Mostoslavsky R, Haigis KM, Fahie K, Christodoulou DC, Murphy AJ, Valenzuela DM et al. SIRT4 inhibits glutamate dehydrogenase and opposes the effects of calorie restriction in pancreatic β cells. *Cell*. 2006; 126:941–954.

153. Michishita E, Park JY, Burneskis JM, Barrett JC, Horikawa I. Evolutionarily conserved and nonconserved cellular localizations and functions of human SIRT proteins. *Mol Biol Cell*. 2005; 16:4623–4635.

154. Schwer B, North BJ, Frye RA, Ott M, Verdin E. The human silent information regulator (Sir) 2 homologue hSIRT3 is a mitochondrial nicotinamide adenine dinucleotide-dependent deacetylase. *J Cell Biol*. 2002; 158:647–657.

155. Laurent G, de Boer VC, Finley LW, Sweeney M, Lu H, Schug TT, Cen Y, Jeong SM, Li X, Sauve AA, Haigis MC. SIRT4 represses peroxisome proliferator-activated receptor α activity to suppress hepatic fat oxidation. *Mol Cell Biol*. 2013; 33:4552–4561.

156. Lin J, Yang R, Tarr PT, Wu PH, Handschin C, Li S, Yang W et al. Hyperlipidemic effects of dietary saturated fats mediated through PGC-1β coactivation of SREBP. *Cell*. 2005; 120:261–273.

157. Wolfrum C, Stoffel M. Coactivation of Foxa2 through PGC-1β promotes liver fatty acid oxidation and triglyceride/VLDL secretion. *Cell Metab*. 2006; 3:99–110.

158. Hernandez C, Molusky M, Li Y, Li S, Lin JD. Regulation of hepatic ApoC3 expression by PGC-1β mediates hypolipidemic effect of nicotinic acid. *Cell Metab*. 2010; 12:411–419.

159. Nagai Y, Yonemitsu S, Erion DM, Iwasaki T, Stark R, Weismann D, Dong J et al. The role of peroxisome proliferator-activated receptor γ coactivator-1 β in the pathogenesis of fructose-induced insulin resistance. *Cell Metab*. 2009; 9:252–264.

160. Sweeney TE, Suliman HB, Hollingsworth JW, Piantadosi CA. Differential regulation of the PGC family of genes in a mouse model of *Staphylococcus aureus* sepsis. *PLoS One*. 2010; 5:e11606.

161. Vercauteren K, Gleyzer N, Scarpulla RC. PGC-1-related coactivator complexes with HCF-1 and NRF-2{β} in mediating NRF-2(GABP)-dependent respiratory gene expression. *J Biol Chem*. 2008; 283:12102–12111.

162. Vercauteren K, Gleyzer N, Scarpulla RC. Short hairpin RNA-mediated silencing of PRC (PGC-1-related coactivator) results in a severe respiratory chain deficiency associated with the proliferation of aberrant mitochondria. *J Biol Chem*. 2009; 284:2307–2319.

163. Mirebeau-Prunier D, Le Pennec S, Jacques C, Gueguen N, Poirier J, Malthiery Y, Savagner F. Estrogen-related receptor α and PGC-1-related coactivator constitute a novel complex mediating the biogenesis of functional mitochondria. *FEBS J*. 2010; 277:713–725.

164. He X, Sun C, Wang F, Shan A, Guo T, Gu W, Cui B, Ning G. Peri-implantation lethality in mice lacking the PGC-1-related coactivator protein. *Dev Dyn*. 2012; 241:975–983.

165. Gleyzer N, Scarpulla RC. PGC-1-related coactivator (PRC), a sensor of metabolic stress, orchestrates a redox-sensitive program of inflammatory gene expression. *J Biol Chem*. 2011; 286:39715–39725.

166. Vermeij WP, Backendorf C. Skin cornification proteins provide global link between ROS detoxification and cell migration during wound healing. *PLoS One*. 2010; 5:e11957.

167. Grivennikov SI, Greten FR, Karin M. Immunity, inflammation, and cancer. *Cell*. 2010; 140:883–899.

168. Mantovani A, Garlanda C, Allavena P. Molecular pathways and targets in cancer-related inflammation. *Ann Med*. 2010; 42:161–170.

169. Davalos AR, Coppe JP, Campisi J, Desprez PY. Senescent cells as a source of inflammatory factors for tumor progression. *Cancer Met Rev*. 2010; 29:273–283.

170. Savagner F, Mirebeau D, Jacques C, Guyetant S, Morgan C, Franc B, Reynier P, Malthièry Y. PGC-1-related coactivator and targets are upregulated in thyroid oncocytoma. *Biochem Biophys Res Commun*. 2003; 310:779–784.

171. Hagenbuchner J, Ausserlechner MJ. Mitochondria and FOXO3: Breath or die. *Front Physiol*. 2013; 4:147.

172. Tseng AH, Shieh SS, Wang DL. SIRT3 deacetylates FOXO3 to protect mitochondria against oxidative damage. *Free Radic Biol Med*. 2013; 63:222–234.

173. Hallberg M, Morganstein DL, Kiskinis E, Shah K, Kralli A, Dilworth SM, White R, Parker MG, Christian M. A functional interaction between RIP140 and PGC-1α regulates the expression of the lipid droplet protein CIDEA. *Mol Cell Biol*. 2008; 28:6785–6795.

174. Shin JH, Ko HS, Kang H, Lee Y, Lee YI, Pletinkova O, Troconso JC, Dawson VL, Dawson TM. PARIS (ZNF746) repression of PGC-1α contributes to neurodegeneration in Parkinson's disease. *Cell*. 2011; 144:689–702.

175. Kanki T, Klionsky DJ. The molecular mechanism of mitochondria autophagy in yeast. *Mol Microbiol*. 2010; 75:795–800.

176. Menzies RA, Gold PH. The turnover of mitochondria in a variety of tissues of young adult and aged rats. *J Biol Chem*. 1971; 246:2425–2429.

177. Miwa S, Lawless C, von Zglinicki T. Mitochondrial turnover in liver is fast in vivo and is accelerated by dietary restriction: Application of a simple dynamic model. *Aging Cell*. 2008; 7:920–923.

178. Haynes CM, Ron D. The mitochondrial UPR—Protecting organelle protein homeostasis. *J Cell Sci*. 2010; 123:3849–3855.

179. Luce K, Weil AC, Osiewacz HD. Mitochondrial protein quality control systems in aging and disease. *Adv Exp Med Biol*. 2010; 694:108–125.

180. Tatsuta T, Langer T. Quality control of mitochondria: Protection against neurodegeneration and ageing. *EMBO J*. 2008; 27:306–314.

181. Lee J, Giordano S, Zhang J. Autophagy, mitochondria and oxidative stress: Cross-talk and redox signalling. *Biochem J*. 2012; 441:523–540.

182. Novak I. Nix is a selective autophagy receptor for mitochondrial clearance. *EMBO Rep*. 2010; 11:45–51.

183. Liu L. Mitochondrial outer-membrane protein FUNDC1 mediates hypoxia-induced mitophagy in mammalian cells. *Nat Cell Biol*. 2012; 14:177–185.

184. Ding WX. Nix is critical to two distinct phases of mitophagy, reactive oxygen species-mediated autophagy induction and Parkin–ubiquitin–p62-mediated mitochondrial priming. *J Biol Chem*. 2010; 285:27879–27890.

185. Narendra D. Parkin is recruited selectively to impaired mitochondria and promotes their autophagy. *J Cell Biol*. 2008; 183:795–803.

186. Chan NC. Broad activation of the ubiquitin-proteasome system by Parkin is critical for mitophagy. *Hum Mol Genet.* 2011; 20:1726–1737.

187. Yoshii SR. Parkin mediates proteasome-dependent protein degradation and rupture of the outer mitochondrial membrane. *J Biol Chem.* 2011; 286:19630–19640.

188. Geisler S. PINK1/Parkin-mediated mitophagy is dependent on VDAC1 and p62/SQSTM1. *Nat Cell Biol.* 2010; 12:119–131.

189. Twig G, Hyde B, Shirihai OS. Mitochondrial fusion, fission and autophagy as a quality control axis: The bioenergetic view. *Biochimica et Biophysica Acta.* 2008; 1777:1092–1097.

190. Mao K. The scaffold protein atg11 recruits fission machinery to drive selective mitochondria degradation by autophagy. *Develop Cell.* 2013; 26:9–18.

191. Tanaka A. Proteasome and p97 mediate mitophagy and degradation of mitofusins induced by Parkin. *J Cell Biol.* 2010; 191:1367–1380.

192. Ding WX. Parkin and mitofusins reciprocally regulate mitophagy and mitochondrial spheroid formation. *J Biol Chem.* 2012; 287:42379–42388.

193. Itakura E, Kishi-Itakura C, Koyama-Honda I, Mizushima N. Structures containing Atg9A and the ULK1 complex independently target depolarized mitochondria at initial stages of Parkin-mediated mitophagy. *J Cell Sci.* 2012; 125:1488–1499.

194. Egan DF, Shackelford DB, Mihaylova MM, Gelino S, Kohnz RA, Mair W, Vasquez DS et al. Phosphorylation of ULK1 (hATG1) by AMP-activated protein kinase connects energy sensing to mitophagy. *Science.* 2011; 331:456–461.

195. Hoshino A, Matoba S, Iwai-Kanai E, Nakamura H, Kimata M, Nakaoka M, Katamura M et al. p53-TIGAR axis attenuates mitophagy to exacerbate cardiac damage after ischemia. *J Mol Cell Cardiol.* 2012; 52:175–184.

196. Ni HM, Bockus A, Boggess N, Jaeschke H, Ding WX. Activation of autophagy protects against acetaminophen-induced hepatotoxicity. *Hepatology.* 2012; 55:222–232.

197. Scheibye-Knudsen M, Fang EF, Croteau DL, Wilson DM 3rd, Bohr VA. Protecting the mitochondrial powerhouse. *Trends Cell Biol.* Dec 10, 2014; pii:S0962-8924(14)00195-0.

198. Mihaylova MM, Shaw RJ. The AMPK signalling pathway coordinates cell growth, autophagy and metabolism. *Nat Cell Biol.* 2011; 13:1016–1023.

199. Park JY, Wang PY, Matsumoto T, Sung HJ, Ma W, Choi JW, Anderson SA, Leary SC, Balaban RS, Kang JG, Hwang PM. P53 improves aerobic exercise capacity and augments skeletal muscle mitochondrial DNA content. *Circ Res.* 2009; 105:705–712.

200. Kulawiec M, Ayyasamy V, Singh KK. p53 regulates mtDNA copy number and mitocheckpoint pathway. *J Carcinog.* 2009; 8:8.

201. Schulze-Osthoff K, Bakker AC, Vanhaesebroeck B, Beyaert R, Jacob WA, Fiers W. Cytotoxic activity of tumor necrosis factor is mediated by early damage of mitochondrial functions. Evidence for the involvement of mitochondrial radical generation. *J Biol Chem.* 1992; 267:5317–5323.

202. Taylor DE, Ghio AJ, Piantadosi CA. Reactive oxygen species produced by liver mitochondria of rats in sepsis. *Arch Biochem Biophys.* 1995; 316:70–76.

203. Kantrow SP, Taylor DE, Carraway MS, Piantadosi CA. Oxidative metabolism in rat hepatocytes and mitochondria during sepsis. *Arch Biochem Biophys.* 1997; 345:278–288.

204. Kurose I, Miura S, Fukumura D, Yonei Y, Saito H, Tada S, Suematsu M, Tsuchiya M. Nitric oxide mediates Kupffer cell-induced reduction of mitochondrial energization in hepatoma cells: A comparison with oxidative burst. *Cancer Res.* 1993; 53:2676–2682.

205. Schreck R, Rieber P, Baeuerle PA. Reactive oxygen intermediates as apparently widely used messengers in the activation of the NF-κB transcription factor and HIV-1. *EMBO J* 1991; 10:2247–2258.

206. Koay MA, Christman JW, Segal BH, Venkatakrishnan A, Blackwell TR, Holland SM, Blackwell TS. Impaired pulmonary NF-κB activation in response to lipopolysaccharide in NADPH oxidase-deficient mice. *Infect Immun.* 2001; 69:5991–5996.

207. St-Pierre J, Drori S, Uldry M, Silvaggi JM, Rhee J, Jager S, Handschin C et al. Suppression of reactive oxygen species and neurodegeneration by the PGC-1 transcriptional coactivators. *Cell*. 2006; 127:397–408.

208. Piantadosi CA, Suliman HB. Mitochondrial transcription factor A induction by redox activation of nuclear respiratory factor 1. *J Biol Chem*. 2006; 281:324–333.

209. Vats D, Mukundan L, Odegaard JI, Zhang L, Smith KL, Morel CR, Wagner RA, Greaves DR, Murray PJ, Chawla A. Oxidative metabolism and PGC-1β attenuate macrophage-mediated inflammation. *Cell Metab*. 2006; 4:13–24.

210. Suliman HB, Carraway MS, Piantadosi CA. Postlipopolysaccharide oxidative damage of mitochondrial DNA. *Am J Respir Crit Care Med*. 2003; 167:570–579.

211. Suliman HB, Carraway MS, Welty-Wolf KE, Whorton AR, Piantadosi CA. Lipopolysaccharide stimulates mitochondrial biogenesis via activation of nuclear respiratory factor-1. *J Biol Chem*. 2003; 278:41510–41518.

212. Suliman HB, Welty-Wolf KE, Carraway MS, Schwartz DA, Hollingsworth JW, Piantadosi CA. Toll-like receptor 4 mediates mitochondrial DNA damage and biogenic responses after heat-inactivated *E. coli*. *FASEB J*. 2005; 19:1531–1533.

213. Bartz RR, Suliman HB, Fu P, Welty-Wolf K, Carraway MS, Macgarvey NC, Withers CM, Sweeney TE, Piantadosi CA. *Staphylococcus aureus* sepsis and mitochondrial accrual of OGG1 DNA repair enzyme in mice. *Am J Respir Crit Care Med*. 2011; 183:226–233.

214. Choumar A, Tarhuni A, Letteron P, Reyl-Desmars F, Dauhoo N, Damasse J, Vadrot N et al. Lipopolysaccharide-induced mitochondrial DNA depletion. *Antioxid Redox Signal*. 2011; 15:2837–2854.

215. Watanabe E, Muenzer JT, Hawkins WG, Davis CG, Dixon DJ, McDunn JE, Brackett DJ, Lerner MR, Swanson PE, Hotchkiss RS. Sepsis induces extensive autophagic vacuolization in hepatocytes: A clinical and laboratory-based study. *Lab Invest*. 2009; 89:549–561.

216. Haden DW, Suliman HB, Carraway MS, Welty-Wolf KE, Ali AS, Shitara H, Yonekawa H, Piantadosi CA. Mitochondrial biogenesis restores oxidative metabolism during *Staphylococcus aureus* sepsis. *Am J Respir Crit Care Med*. 2007; 176:768–777.

217. Bakkar N, Wang J, Ladner KJ, Wang H, Dahlman JM, Carathers M, Acharyya S, Rudnicki MA, Hollenbach AD, Guttridge DC. IKK/NF-κB regulates skeletal myogenesis via a signaling switch to inhibit differentiation and promote mitochondrial biogenesis. *J Cell Biol*. 2008; 180:787–802.

218. Piantadosi CA, Withers CM, Bartz RR, Macgarvey NC, Fu P, Sweeney TE, Welty-Wolf KE, Suliman HB. Heme oxygenase-1 couples activation of mitochondrial biogenesis to anti-inflammatory cytokine expression. *J Biol Chem*. 2011; 286:16374–16385.

219. Bergeron R, Ren JM, Cadman KS, Moore IK, Perret P, Pypaert M, Young LH, Semenkovich CF, Shulman GI. Chronic activation of AMP kinase results in NRF-1 activation and mitochondrial biogenesis. *Am J Physiol Endocrinol Metab*. 2001; 281:E1340–E1346.

220. Zong H, Ren JM, Young LH, Pypaert M, Mu J, Birnbaum MJ, Shulman GI. AMP kinase is required for mitochondrial biogenesis in skeletal muscle in response to chronic energy deprivation. *Proc Natl Acad Sci USA*. 2002; 99:15983–15987.

221. Reznick RM, Shulman GI. The role of AMP-activated protein kinase in mitochondrial biogenesis. *J Physiol*. 2006; 574:33–39.

222. Morrow VA, Foufelle F, Connell JM, Petrie JR, Gould GW, Salt IP. Direct activation of AMPactivated protein kinase stimulates nitric-oxide synthesis in human aortic endothelial cells. *J Biol Chem*. 2003; 278:31629–31639.

223. Kim J, Kundu M, Viollet B, Guan KL. AMPK and mTOR regulate autophagy through direct phosphorylation of Ulk1. *Nat Cell Biol*. 2011; 13:132–141.

224. Giri S, Nath N, Smith B, Viollet B, Singh AK, Singh I. 5-aminoimidazole-4-carboxamide-1-β-4-ribofuranoside inhibits proinflammatory response in glial cells: A possible role of AMP activated protein kinase. *J Neurosci.* 2004; 24:479–487.

225. Bai A, Ma AG, Yong M, Weiss CR, Ma Y, Guan Q, Bernstein CN, Peng Z. AMPK agonist downregulates innate and adaptive immune responses in TNBS-induced murine acute and relapsing colitis. *Biochem Pharmacol.* 2010; 80:1708–1717.

226. Salminen A, Hyttinen JM, Kaarniranta K. AMP-activated protein kinase inhibits NF-κB signaling and inflammation: Impact on healthspan and lifespan. *J Mol Med.* 2011; 89:667–676.

227. Barroso E, Eyre E, Palomer X, Vazquez-Carrera M. The peroxisome proliferator-activated receptor β/δ (PPARβ/δ) agonist GW501516 prevents TNF-α-induced NFκB activation in human HaCaT cells by reducing p65 acetylation through AMPK and SIRT1. *Biochem Pharmacol.* 2011; 81:534–543.

228. L'Horset F, Dauvois S, Heery DM, Cavailles V, Parker MG. RIP-140 interacts with multiple nuclear receptors by means of two distinct sites. *Mol Cell Biol.* 1996; 16:6029–6036.

229. Powelka AM, Seth A, Virbasius JV, Kiskinis E, Nicoloro SM, Guilherme A, Tang X, Straubhaar J, Cherniack AD, Parker MG, Czech MP. Suppression of oxidative metabolism and mitochondrial biogenesis by the transcriptional corepressor RIP140 in mouse adipocytes. *J Clin Invest.* 2006; 116:125–136.

230. Zschiedrich I, Hardeland U, Krones-Herzig A, Diaz MB, Vegiopoulos A, Muggenburg J, Sombroek D et al. Coactivator function of RIP140 for NFκB/RelA-dependent cytokine gene expression. *Blood.* 2008; 112:264–276.

231. Schroder K, Tschopp J. The inflammasomes. *Cell.* 2010; 140:821–832.

232. Zhou R, Yazdi AS, Menu P, Tschopp J. A role for mitochondria in NLRP3 inflammasome activation. *Nature.* 2011; 469:221–225.

233. Nakahira K, Haspel JA, Rathinam VA, Lee SJ, Dolinay T, Lam HC, Englert JA et al. Autophagy proteins regulate innate immune responses by inhibiting the release of mitochondrial DNA mediated by the NALP3 inflammasome. *Nat Immunol.* 2011; 12:222–230.

234. Grattagliano I, de Bari O, Bernardo TC, Oliveira PJ, Wang DQ, Portincasa P. Role of mitochondria in nonalcoholic fatty liver disease from origin to propagation. *Clin Biochem.* 2012; 45:610–618.

235. Begriche K, Igoudjil A, Pessayre D, Fromenty B. Mitochondrial dysfunction in NASH: Causes, consequences and possible means to prevent it. *Mitochondrion.* 2006; 6:1–38.

236. Fabbrini E, Sullivan S, Klein S. Obesity and nonalcoholic fatty liver disease: Biochemical, metabolic, and clinical implications. *Hepatology.* 2010; 51:679–689.

237. Zhang D, Liu ZX, Choi CS, Tian L, Kibbey R, Dong J, Cline GW, Wood PA, Shulman GI. Mitochondrial dysfunction due to long-chain Acyl-CoA dehydrogenase deficiency causes hepatic steatosis and hepatic insulin resistance. *Proc Natl Acad Sci USA.* 2007; 104:17075–17080.

238. Pessayre D. Role of mitochondria in non-alcoholic fatty liver disease. *J Gastroenterol Hepatol.* 2007; 22: S20–S27.

239. Pessayre D, Fromenty B, NASH: A mitochondrial disease. *J Hepatol.* 2005; 42:928–940.

240. Vial G, Dubouchaud H, Leverve XM. Liver mitochondria and insulin resistance. *Acta Biochim Pol.* 2010; 57:389–392.

241. Cheng Y, Zhou M, Tung CH, Ji M, Zhang F. Studies on two types of PTP1B inhibitors for the treatment of type 2 diabetes: Hologram QSAR for OBA and BBB analogues. *Bioorg Med Chem Lett.* 2010; 20:3329–3337.

242. Saltiel A, Kahn CR. Insulin signalling and the regulation of glucose and lipid metabolism. *Nature.* 2001; 414:799–806.

243. Hribal M, Oriente F, Accili D. Mouse models of insulin resistance. *Am J Physiol Endocrinol Metab.* 2002; 282:E977–E981.
244. Herzig S, Long F, Jhala US, Hedrick S, Quinn R, Bauer A, Rudolph D et al. CREB regulates hepatic gluconeogenesis through the coactivator PGC-1. *Nature.* 2001; 413:179–183.
245. Rhee J, Inoue Y, Yoon JC, Puigserver P, Fan M, Gonzalez FJ, Spiegelman BM. Regulation of hepatic fasting response by PPAR coactivator-1 (PGC-1): Requirement for hepatocyte nuclear factor 4 in gluconeogenesis. *Proc Natl Acad Sci USA.* 2003; 100:4012–4017.
246. Giralt A, Hondares E, Villena JA, Ribas F, az-Delfin J, Giralt M, Iglesias R, Villarroya F. Peroxisome proliferator-activated receptor-γ coactivator-1α controls transcription of the Sirt3 gene, an essential component of the thermogenic brown adipocyte phenotype. *J Biol Chem.* 2011; 286:16958–16966.
247. Kong X, Wang R, Xue Y, Liu X, Zhang H, Chen Y, Fang F, Chang Y. Sirtuin 3, a new target of PGC-1α, plays an important role in the suppression of ROS and mitochondrial biogenesis. *PLoS One.* 2010; 5:e11707.
248. Verdin E, Hirschey MD, Finley LW, Haigis MC. Sirtuin regulation of mitochondria: Energy production, apoptosis, and signaling. *Trends Biochem Sci.* 2010; 35:669–675.
249. Someya S, Yu W, Hallows WC, Xu J, Vann JM, Leeuwenburgh C, Tanokura M, Denu JM, Prolla TA. Sirt3 mediates reduction of oxidative damage and prevention of age-related hearing loss under caloric restriction. *Cell.* 2010; 143:802–812.
250. Qiu X, Brown K, Hirschey MD, Verdin E, Chen D. Calorie restriction reduces oxidative stress by SIRT3-mediated SOD2 activation. *Cell Metab.* 2010; 12:662–667.
251. Tao R, Coleman MC, Pennington JD, Ozden O, Park SH, Jiang H, Kim HS et al. Sirt3-mediated deacetylation of evolutionarily conserved lysine 122 regulates MnSOD activity in response to stress. *Mol Cell.* 2010; 40:893–904.
252. Kersten S. Integrated physiology and systems biology of PPARα. *Mol Metab.* 2014; 3:354–371.
253. Hashimoto T, Cook WS, Qi C, Yeldandi AV, Reddy JK, Rao MS. Defect in peroxisome proliferator-activated receptor α-inducible fatty acid oxidation determines the severity of hepatic steatosis in response to fasting. *J Biol Chem.* 2000; 275:28918–28928.
254. Patsouris D, Mandard S, Voshol PJ, Escher P, Tan NS, Havekes LM. PPARα governs glycerol metabolism. *J Clin Invest.* 2004; 114:94–103.
255. Sugden MC, Bulmer K, Gibbons GF, Knight BL, Holness MJ. Peroxisome-proliferator-activated receptor-α (PPARα) deficiency leads to dysregulation of hepatic lipid and carbohydrate metabolism by fatty acids and insulin. *Biochem J.* 2002; 364:361–368.
256. Gervois P, Kleemann R, Pilon A, Percevault F, Koenig W, Staels B, Kooistra T. Global suppression of IL-6-induced acute phase response gene expression after chronic in vivo treatment with the peroxisome proliferator-activated receptor-α activator fenofibrate. *J Biol Chem.* 2004; 279:16154–16160.
257. Ip E, Farrell GC, Robertson G, Hall P, Kirsch R, Leclercq I. Central role of PPARα-dependent hepatic lipid turnover in dietary steatohepatitis in mice. *Hepatology.* 2003; 38:123–132.

5 Regulation of Liver Mitochondrial Function by Hydrogen Sulfide

Katalin Módis, Zahra Karimi, and Rui Wang

CONTENTS

ABSTRACT

Hydrogen sulfide (H$_2$S) was conventionally viewed as a life-threatening environmental toxic gas. Studies in recent years have shown that with L-cysteine and/or homocysteine as the substrate, several enzymes catalyze H$_2$S production in mammalian cells. Once produced, H$_2$S participates in the regulation of numerous physiological and pathophysiological functions. The mitochondrion is where oxygen is used and ATP is produced. Also in this very same organelle, H$_2$S is consumed and produced, and both processes are oxygen-dependent. Thus, mitochondrial functions are closely related to and are regulated by oxygen-dependent mitochondrial H$_2$S metabolism. In liver mitochondria, hypoxia decreases H$_2$S consumption via suppressed H$_2$S oxidation and increases H$_2$S production via the accumulation of H$_2$S-generating enzymes. At physiologically relevant concentrations, H$_2$S serves as an electron donor to promote ATP production in the mitochondrion. H$_2$S also regulates mitochondrial cAMP homeostasis. Abnormal liver H$_2$S metabolism leads to or deteriorates the development of fatty liver disease, diabetic liver dysfunction, ischemia-reperfusion injury, acute liver failure, portal hypertension, and cirrhosis. It is perceived that clinical interventions based on mitochondrial H$_2$S metabolism and the cellular and molecular effects of H$_2$S donors will find profound and significant applications in preventing and treating various liver diseases.

INTRODUCTION

During the past decade, the infamous image of hydrogen sulfide (H$_2$S) as a noxious, malodorous, and toxic gas has been transformed into one of a novel and powerful signaling messenger and an indispensable life-supporting molecule in the mammalian system—a *gasotransmitter*. H$_2$S biology is definitely one of the most rapidly developing fields in biomedical research.

CHARACTERISTICS OF HYDROGEN SULFIDE

CHEMICAL PROPERTIES OF H$_2$S

H$_2$S (M$_W$ = 34.08) is a colorless, corrosive, and flammable gas. Its chemical structure is similar to that of water (H$_2$O) but with different chemical and physical properties. H$_2$S is much less polar than water due to lower electronegativity of sulfur compared to oxygen. It results in comparatively weaker intermolecular forces among H$_2$S molecules and much lower boiling point (−61°C), melting point (−82°C), and freezing point (−86°C) than those of water (boiling point, 99.98°C; melting/freezing point, 0°C), respectively.

H$_2$S has a characteristic rotten egg odor and sweetish taste. The human nose can smell H$_2$S at low concentrations (0.0005–0.3 ppm) in the air. High concentrations of H$_2$S (above 150 ppm) can paralyze the olfactory nerve and its ability to sense H$_2$S will be lost, which results in life-threatening toxicity (Beauchamp et al., 1984; Reiffenstein et al., 1992; Guidotti, 1994).

The aqueous solution of H_2S is a week acid, also known as hydrosulfuric acid. It dissociates to proton (H^+) and hydrosulfide anion (HS^-), which may further dissociate to a proton (H^+) and sulfide ion (S^{2-}) ($K_{a1} = 1.3 \times 10^{-7}$, $K_{a2} = 1 \times 10^{-19}$). At physiological pH and temperature, about 20%–30% of the total hydrogen sulfide remains in the form of H_2S and 70%–80% presents in dissociated HS^- form. S^{2-} is only present in a negligible quantity as dissociation of HS^- occurs at high pH values (Kabil and Banerjee, 2010; Nagy et al., 2014). However, the terms "sulfide" and "H_2S" are alternately used in the biomedical research of H_2S as well as in this chapter to refer to the H_2S and HS^- mixture.

TOXICOLOGICAL PROFILE OF H_2S

H_2S is present in sulfur springs, undersea vents, and swamps as well as in crude petroleum and natural gas. This poisonous gas is an environmental hazard when released as a by-product in industrial and petrochemical activities, and a metabolic hazard when individuals are exposed to high concentrations of environmental H_2S. H_2S enters body primarily through the breathing. Much smaller amounts can be absorbed through the skin. The inhaled H_2S enters the circulation directly across the alveolar-capillary barrier. Exposure to lower concentrations of hydrogen sulfide may cause irritation to the eyes, nose, or throat. High levels of hydrogen sulfide (above 500 ppm) cause loss of consciousness. In most cases, people after short-term intoxication appear to regain consciousness without any remaining effects. However, in some individuals, permanent or long-term effects such as headaches, poor attention span, memory, and motor function can persist (Turner and Fairhurst, 1980).

H_2S-RELEASING DONOR MOLECULES

To study the regulatory roles of H_2S in biological systems, investigators often use H_2S-releasing donor molecules. The application of various H_2S donor molecules in different disease models and their pharmacological effects have been reviewed by Modis et al. (2013b), Kashfi and Olson (2013), and Whiteman et al. (2011). Here, we will only provide a general concept of the H_2S-releasing pharmacological "tools." H_2S-releasing compounds can be classified into three main categories: (1) natural H_2S-releasing compounds, (2) inorganic sulfide salts, and (3) organic H_2S donors.

Natural H_2S-releasing compounds mainly come from plants and food. For instance, garlic (*Allium sativum*) contains many organic sulfur–containing compounds (e.g., allicin or diallyl thiosulfinate, diallyl disulfide, diallyl trisulfide, cysteine analogs such as *S*-allycysteine) and is proven to have beneficial pharmacological effects on various pathophysiological conditions (see details in the section on "Pathophysiological Roles of Hydrogen Sulfide in the Liver") (Benavides et al., 2007; Chuah et al., 2007; Wang et al., 2010; Kaschula et al., 2011; Lai et al., 2012). Isothiocyanate compounds are also found in broccoli (*Brassica oleracea*) (Shapiro et al., 2001; Ye et al., 2002; Mukherjee et al., 2010), rucola (*Eruca sativa*) (Azareko et al., 2014), wasabi (Li et al., 2010), mustard (Tian et al., 2013), and horseradish (Matsuda et al., 2007). However, it still needs to be confirmed whether isothiocyanate liberates H_2S in vivo. Nevertheless, their health benefits are quite similar to the sulfur-containing compounds in garlic.

Inorganic sulfide salts (NaHS, Na₂S, and CaS) are considered to be fast-releasing H₂S donors that release H₂S in aqueous solutions. Sodium hydrosulfide (NaHS) is the most frequently used hydrogen sulfide donor in H₂S biology. It must be noted that utilizing these inorganic sulfide salts has some disadvantages: (a) Their application in aqueous solution leads to the increase in pH, creating basic conditions in which some enzymes might not work properly. (b) Hydrolysis of ionic salts in aquatic conditions (in solutions, in vitro cell culture media, and in vivo circulation) is very rapid and it is difficult to obtain a sustained effect of H₂S.

Organic H₂S donors can be classified into two categories: (1) classical organic H₂S donors and (2) H₂S-releasing organic hybrid drugs. The *classical organic H₂S donors* are prodrugs. Some of them present endogenously (e.g., cysteine, methionine); others are synthesized (e.g., *N*-acetylcysteine or *N*-acetyl-penicillamine, *S*-allylcysteine, *S*-propargyl-cysteine, *S*-propyl-l-cysteine, thioglycine, and thioval). Their H₂S-releasing or H₂S-producing speed can be slow or fast. Classical fast-releasing/-producing organic H₂S donors are cysteine, methionine, and *N*-acetylcysteine. Slow-releasing/-producing organic H₂S donors include *S*-allylcysteine, *S*-propargyl-cysteine, *S*-propyl-l-cysteine, thioglycine, and thioval GYY4137. *H₂S-releasing organic hybrid drugs* combine the H₂S-releasing action with another mode of action provided by the parent compound independent of the H₂S-donating moiety. Most of them have excellent pharmacokinetic and pharmacodynamic properties associated with improved effectiveness and decreased side effects. Dithiolethiones are the commonly used chemical structures for H₂S delivery (Li et al., 2007; Wallace, 2012; Campolo et al., 2013; Chan and Wallace, 2013). In many cases, the parent compounds are nonsteroidal anti-inflammatory drugs (NSAIDs) coupled with H₂S-releasing moieties, such as H₂S-releasing aspirin (HS-ASA, Chattopadhyay et al., 2012a,b), *S*-naproxen (ATB-346, Wallace et al., 2010), *S*-diclofenac (ATB-337/ACS-5, Sidhapuriwala et al., 2007; Wallace et al., 2010), and *S*-mesalamine (ATB-429, Distrutti et al., 2006; Fiorucci et al., 2007). There are other pharmacologically active parent compounds linked to H₂S-releasing moieties, such as the L-DOPA hybrids ACS83, ACS84, ACS85, ACS86 (Lee et al., 2010), *S*-sildenafil (ACS-6) (Muzaffar et al., 2008), and *S*-valproate (Isenberg et al., 2007; Moody et al., 2010).

Over the years, compounds with active P–S bonds, some of which are also dithiolethione derivatives, such as Lawesson's reagent, GYY4137, and phosphorodithioate, have been developed. For instance, GYY 4137 is a widely used slow-releasing H₂S donor molecule (Li et al., 2008). GYY4137 is highly water soluble and its cellular permeability is limited. Its cytoprotective effects are usually noticed at concentrations higher than 100 μM. AP39 is a novel mitochondrion-targeted H₂S donor molecule (Le Trionnaire et al., 2014; Szczesny et al., 2014). It consists of a mitochondrion-targeted motif, triphenylphosphonium (TPP⁺), and a H₂S-donating moiety (dithiolethione). TPP⁺ is a molecular probe used for the determination of mitochondrial membrane potential (Smith et al., 2011). It can accumulate into the mitochondrion, depending on the mitochondrial membrane potential level. AP39 enters mitochondria and releases H₂S in a dose-dependent fashion (Szczesny et al., 2014). Furthermore, AP39 stimulates mitochondrial bioenergetics at low concentrations (30–100 nM) but inhibits it at concentrations above 300 nM. Oxidative stress increases mitochondrial ROS production and oxidative protein modifications in endothelial cells, which subsequently suppress mitochondrial bioenergetics. AP39 at low concentrations (100 nM) significantly reverses all these parameters and reduces oxidative mitochondrial DNA damage as well (Szczesny et al., 2014).

REGULATION OF HYDROGEN SULFIDE PRODUCTION AND METABOLISM IN THE LIVER

In mammalian cells, there are three different H_2S-producing enzymes. *Cystathionine β-synthase* (CBS, EC 4.2.1.22) was the first identified H_2S-producing enzyme, under the name of serine sulfhydrase in 1969 (Braunstein et al., 1969). It is generally considered a cytosolic enzyme in mammalian cells, although in certain cell types and in specialized conditions, it can also be found in the mitochondrion (Szabo et al., 2013; Teng et al., 2013). CBS catalyzes the production of cystathionine from homocysteine. A second enzyme, *cystathionine γ-lyase* (CSE, EC 4.4.1.1), also contributes to H_2S production. One of the functions of this enzyme is to convert cystathionine to L-cysteine. Both CBS and CSE can produce H_2S from L-cysteine. CSE is generally localized into the cytosol, but elevated intracellular calcium or lower oxygen partial pressure can trigger its translocation to the mitochondrion in certain types of cells (Fu et al., 2012). The third enzyme involved in cellular H_2S production is *3-mercaptopyruvate sulfur transferase* (MST, EC 2.8.1.2), a cytosolic and mitochondrial enzyme (Stipanuk and Beck, 1982; Shibuya et al., 2009). Its function is closely associated with the function of *cysteine aminotransferase* (CAT, EC 2.6.1.3), which produces 3-mercaptopyruvate (3-MP) from L-cysteine and α-ketoglutarate. Next, MST converts 3-MP to pyruvate and H_2S (Shibuya et al., 2009) (Figures 5.1 and 5.2). In liver tissues and in hepatocytes, all three enzymes (CBS, CSE, and MST) have been identified (Mani et al., 2014). In fact, the liver is one of the organs in the body that has the highest expression level of CBS (Ratnam et al., 2002). A putative fourth H_2S-producing enzyme may reside in mammalian cells, *mitochondrial thiosulfate sulfur transferase* (also called rhodanese, EC 2.8.1.1) (Mikami et al., 2011), which can convert thiosulfate to H_2S. However, the physiological or pathophysiological relevance of this alternative H_2S-producing mechanism remains to be elucidated.

H_2S can also be liberated nonenzymatically: It can be restored into or liberated from sulfane sulfur pools, different from *acid-labile* pools (Ogasawara et al., 1994; Kimura et al., 2013; Kimura, 2014). Acid-labile sulfur groups are iron–sulfur cluster proteins, which release sulfur under acidic, probably pathophysiological conditions. Iron–sulfur cluster proteins are concentrated in the mitochondrion, such as cytochrome components of the respiratory chain. The optimal pH for the sulfur release from acid-labile sulfur is less than pH 5.4 (Wang, 2012). However, the physiological pH in the mitochondrion is 7.2. Therefore, it may be difficult for sulfur to be released from acid-labile sulfur pool under physiological conditions. Acid-labile sulfur pools are mainly found in heart tissue with low levels in the liver (Ogasawara et al., 1994). Bound sulfane sulfur pools serve physiological roles, from which H_2S can be released under reducing conditions. L-cysteine and glutathione are the major cellular reducing agents. The H_2S-releasing capacity of various tissues can be increased after the addition of DTT (a reducing compound) (Ishigami et al., 2009). Bound sulfane sulfur pools have been identified in the cytosol of the kidneys, liver, and spleen (Ogasawara et al., 1994).

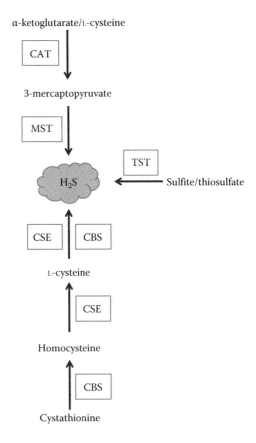

FIGURE 5.1 Schematic representation of endogenous H$_2$S production in mammalian cells. CBS, cystathionine-β-synthase (EC 4.2.1.22); CSE, cystathionine-γ-lyase (EC 4.4.1.1); MST, 3-mercaptopyruvate sulfur transferase (EC 2.8.1.2); CAT, cysteine aminotransferase (EC 2.6.1.3); TST, rhodanase/thiosulfate sulfur transferase (EC 2.8.1.1).

Certain drugs (e.g., ramipril, atorvastatin, paracetamol, carvedilol, metformin) have been shown to increase hepatic H$_2$S levels (Wiliński et al., 2010, 2011a,b,c, 2013). CBS levels in the liver have been extensively investigated in the context of hyperhomocysteinemia, a severe metabolic disease (Watanabe et al., 1995). These studies focused on the role of CBS as a homocysteine-degrading enzyme but without considering its role as an enzyme involved in H$_2$S production (Meier et al., 2003 Miles and Kraus, 2004; Singh and Banerjee, 2011). Mutations in CBS lead to the dysfunction of this enzyme, which, in turn, leads to the accumulation of homocysteine in the blood. Whether, in addition to the elevation of homocysteine, there is a concomitant lack of H$_2$S production in the liver remains to be investigated. Altered CBS protein expression has also been suggested to contribute to many pathophysiological conditions such as fatty liver, liver fibrosis, hepatic steatosis, hypercholesterolemia, and diabetes (Mani et al., 2014).

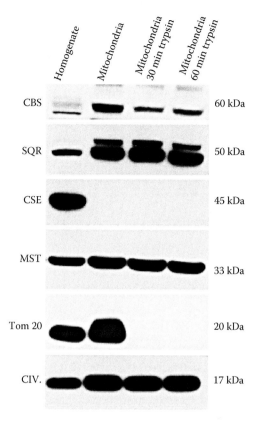

FIGURE 5.2 Mitochondrial localization of H_2S-producing enzymes. Representative Western blot analysis is performed on (1) rat liver homogenate, (2) rat liver–isolated mitochondria, (3) mitoplasts (mitochondrial preparation without the outer mitochondrial membrane). The tissue samples were obtained by in vitro partial trypsin digestion for 30 or 60 min–long period. Please note that CBS and MST are localized into the mitochondrial fraction. However, CSE was not mitochondrially associated. MST and SQR were recognized at the mitochondrial inner membrane. Other markers also show their respective expected localizations; Tom20 is a marker for mitochondrial inner membrane proteins and COX IV is the loading marker for mitochondrial fractions. (Reproduced with permission from Szabo, C. et al, *Br J Pharmacol*. 171(8):2099–2122, 2014.)

The cellular and tissue levels of H_2S are regulated both by its production and its elimination. The enzymatic and nonenzymatic degradation of H_2S is incompletely understood (Figure 5.3). Mitochondrial sulfide:quinone oxidoreductase protein, abbreviated SQR, plays a crucial role in H_2S oxidation. It forms an enzyme-bound persulfide that is then either converted to sulfite by persulfide dioxygenase (ETHE1, EC 1.13.11.18) or yield thiosulfate by rhodanese (EC 2.8.1.1) to (Hildebrandt and Grieshaber, 2008). Thiosulfate is subsequently cleaved by a glutathione-dependent thiosulfate reductase (EC 2.8.1.3), resulting in sulfite production. The sulfite is then oxidized to sulfate by sulfite oxidase (EC 1.8.3.1) (Uhteg and Westley, 1979).

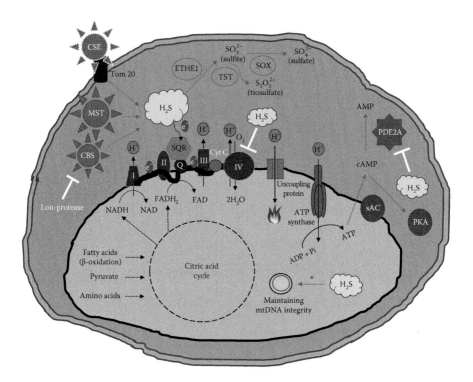

FIGURE 5.3 (See color insert.) Schematic representation of the mechanisms by which H$_2$S modulates mitochondrial functions. H$_2$S can be produced in the mitochondrion constitutively by MST and CBS. Moreover, CSE is capable of translocating to the outer mitochondrial membrane under certain stress conditions, resulting in an increment of the intramitochondrial H$_2$S level. The regulatory roles of H$_2$S in the mitochondria are diverse, based on its stimulatory or inhibitory actions. (1) H$_2$S donates electrons to mitochondrial electron transport chain (stimulatory effect). (2) H$_2$S oxidization results in sulfate and thiosulfate end products. (3) H$_2$S inhibits cytochrome c oxidase (complex IV), shutting down the respiration (inhibitory effect). (4) H$_2$S is responsible for the inhibition of mitochondrial PDE2A enzyme. This mode of action increases the intramitochondrial cAMP level that activates the PKA enzyme (stimulatory effect).

The major metabolites of H$_2$S are sulfite, sulfate, and thiosulfate. Thiosulfate is primarily excreted via the urine (Shibuya et al., 2009; Wang, 2012). A small, but clearly detectable amount of endogenous H$_2$S is excreted into the air via the lung (Insko et al., 2009; Toombs et al., 2010). Mitochondria play a crucial role in the metabolism of H$_2$S; part of this metabolic activity is coupled to the generation of cellular ATP. In an isolated perfused liver, it was shown that H$_2$S metabolism is associated with an increase in O$_2$ consumption, and the metabolism is suppressed during hypoxia (Norris et al., 2011). It has been suggested that the liver, in addition to the intestinal epithelial cells, plays a role in the neutralization of the H$_2$S that is produced from the intestinal microbiota and absorbed into the portal vein (Carbonero et al., 2012).

PHARMACOLOGICAL EFFECTS OF HYDROGEN SULFIDE

Prior to the last decade, the biological actions of H_2S were primarily discussed in the context of environmental toxicology (Beauchamp et al., 1984). H_2S is a reversible, noncompetitive inhibitor of cytochrome c oxidase (complex IV) (Cooper and Brown, 2008). It prevents the binding of oxygen to complex IV, which is the final electron acceptor of the mitochondrial electron transport chain. This can lead to reversible cellular metabolic suppression (sometimes called *cellular* or *tissue hibernation*) or culminate in cell death, when sustained.

Many of the biological or pharmacological effects of H_2S follow a biphasic/bell-shaped dose-dependent pattern. For example, at low (presumably physiological) concentrations, H_2S has a beneficial effect on mitochondrial respiration, vasorelaxation, cytoprotection, cellular viability, cellular antioxidant capacity, etc. In contrast, at higher (presumably pathophysiological/toxicological) concentrations, it acts as a mitochondrial poison, and a cytotoxic and genotoxic agent. It can induce excessive vasorelaxation, and respiratory or central nervous system suppression. The absolute concentrations of H_2S in these different pharmacological modes of action vary, which have been determined with a great uncertainty due to specific H_2S detection techniques used in each study. It needs to be emphasized that pathophysiological effects of H_2S reflect an alteration of endogenous H_2S level in certain tissues or plasma. Under certain disease conditions, decreased H_2S availability occurs, such as during hyperhomocysteinemia (Watanabe et al., 1995; Namekata et al., 2004; Robert et al., 2005; Yang et al., 2008), while under other conditions (e.g., in sepsis and burns), overproduction of H_2S is encountered (Hui et al., 2003; Zhang et al., 2006; Coletta and Szabo, 2013). In Table 5.1, we summarize only the most important pharmacological effects of H_2S in the liver.

HYDROGEN SULFIDE IN CELL SIGNALING

The signal transduction pathways of H_2S involve a wide array of molecular mechanisms. Figure 5.4 shows the most reported H_2S-mediated signaling pathways. This list is likely incomplete and additional mechanisms will be identified in the future. Among this list, three important H_2S-mediated signaling mechanisms are discussed here.

REACTIONS OF H_2S WITH METAL IONS

H_2S is known to interact with metalloproteins such as cytochrome c oxidase (complex IV) and carbonic anhydrase, which contain heme moiety and metal ions (Zn^{2+}, Fe^{3+}). Such interactions usually result in functional inhibition of the affected proteins, such as cytochrome c oxidase (Hill et al., 1984, also reviewed in Modis et al., 2013b; Szabo et al., 2014). Furthermore, sulfide can bind to hemoglobin or myoglobin, forming sulfhemoglobin or sulfmyoglobin. These modified proteins have decreased affinity to oxygen and diminished ability to transport oxygen in the body (Pietri et al., 2011). Nevertheless, the precise mechanisms in mammalian

TABLE 5.1

Summary of the Principal Pharmacological Effects of H$_2$S in the Liver

Biological System	Physiological Effects	Pathological Effects	References
Hepatic blood vessels	Vasorelaxation. Maintenance of the physiological intrahepatic circulation.	H$_2$S may contribute to pathological vasoconstriction during endotoxemia. Lowering of hepatic H$_2$S levels has been also proposed to contribute to the changes in hepatic microcirculation during liver cirrhosis.	Norris et al. (2011, 2013), Distrutti et al. (2008), Fiorucci et al. (2005), Wang et al. (2014), and Zhu et al. (2012)
Hepatocytes	Physiological H$_2$S production supports mitochondrial electron transport and protects against oxidative stress. H$_2$S donors also decrease glucose uptake and glycogen storage in hepatocytes. H$_2$S also maintains the circadian clock of hepatocytes. H$_2$S can exert antifibrotic effects.	Elevated concentration of H$_2$S or its prolonged exposure provide cytotoxic and proapoptotic effects.	Jurkowska et al. (2014), Shirozu et al. (2014), Módis et al. (2013a), Zhang et al. (2011, 2013), Wang et al. (2012), Shang et al. (2012), Caro et al. (2011), Morsy et al. (2010), Thompson et al. (2003), Zeng et al. (2013), Aslami et al. (2013a,b), Yan et al. (2012), Zhu et al. (2012)
Hepatic stellate cells	Inhibition of proliferation, protection against liver fibrosis.		Fan et al. (2013a,b)
Cellular bioenergetics	Stimulation of mitochondrial function via increasing the oxygen and ATP production.	Inhibition (mitochondrial toxin) of mitochondrial function at higher concentrations.	Fu et al. (2011), Teng et al. (2012), Helmy et al. (2014), Módis et al. (2013a,b,c, 2014), Caro et al. (2012), Lagoutte et al. (2010), Szabo et al. (2014)

cells that control heme–H$_2$S interactions are still unsettled. Intriguingly, interactions of H$_2$S with hemeproteins have been studied for many years in marine invertebrate organisms (Gail, 1993; Szabo et al., 2014).

PROTEIN S-SULFHYDRATION

Protein S-sulfhydration is a recently identified mechanism of posttranslational modification. H$_2$S covalently modifies the SH group of the cysteine amino acid, creating hydropersulfide (–SSH groups). It is similar to the process of protein S-nitrosylation

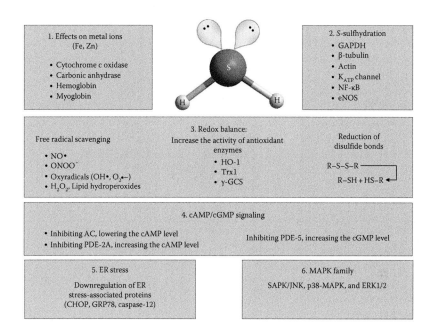

FIGURE 5.4 Signal transduction mechanisms of H_2S in mammalian cells. (1) H_2S can interact with metalloproteins. (2) H_2S induces S-sulfhydration of certain proteins. (3) H_2S has direct (free radical scavenger and reduction of disulfide bonds) and indirect (regulation of the expression of antioxidant genes) regulatory roles in redox balance. (4) H_2S regulates intracellular cAMP/cGMP levels. (5) H_2S also regulates ER stress and (6) MAPK superfamily enzymes. PDE, phosphodiesterases; CHOP, C/EBP homologous protein; GRP78, glucose-regulated protein 78; SAPK/JNK, stress-activated protein kinase/c-Jun NH2-terminal kinase; AC, adenylyl cyclase.

where NO covalently binds to the thiol group of the cysteine, forming S-nitrosothiols, that is, −SNO groups. Although both S-sulfhydration and S-nitrosylation are posttranslational modifications, which can be reversed by the addition of DTT, there are key differences between the two. First, more proteins in the liver can be S-sulfhydrated than those that can be S-nitrosylated. Generally, 10%–25% of liver proteins are S-sulfhydrated including actin, tubulin, and glyceraldehyde-3-phosphate dehydrogenase (GAPDH). In contrast, only 1%–2% of the total proteins are subject to S-nitrosylation (Mustafa et al., 2011). Second, S-sulfhydration is a more stable modification than S-nitrosylation mainly, because −S−NO bond is easier to be broken compared to −S−SH. Third, S-sulfhydration usually increases the activity of the target proteins, while S-nitrosylated proteins have, typically, decreased functionality either through direct effects or by triggering protein degradation. For example, Hara et al. demonstrated that activation of macrophages by endotoxin increases the intracellular NO production that triggers S-nitrosylation of GAPDH (Hara et al., 2005). The S-nitrosylated GAPDH (SNO-GAPDH) loses its glycolytic activity but binds to Siah1 (an E3 ubiquitin ligase). These proteins together translocate to the nucleus, triggering ubiquitin-mediated degradation of the targeted proteins and subsequent apoptosis.

It has been shown that S-sulfhydration of adenosine triphosphate (ATP)-sensitive potassium (K_{ATP}) channel leads to hyperpolarization of vascular smooth muscle cells and subsequent vasorelaxation (Jiang et al., 2010). Other ion channels that are S-sulfhydrated include intermediate (IK_{Ca}) and small conductance (SK_{Ca}) calcium-activated potassium channels in vascular endothelial cells (Mustafa et al., 2011). H_2S causes the S-sulfhydration of Keap1 that will release Nrf2, inducing its translocation to the nucleus. This mechanism is protective from oxidative stress and premature senescence, and is part of the process of ischemic preconditioning (Yang et al., 2013). Altaany et al. (2014) have recently demonstrated that eNOS (endothelial NO synthase) can be both S-sulfhydrated and S-nitrosylated on the same cysteine residue (Cys[443]), leading to the stimulation or inhibition of the same enzyme, respectively.

The effect of S-sulfhydration on protein function depends on the pH and the position of the target cysteine amino acids in the protein (e.g., its proximity to the active centrum of the enzyme). Moreover, the hydropersulfide product of S-sulfhydration (–SSH) might not be formed by direct interaction of H_2S and the –SH group of cysteine, because sulfur is in its lowest oxidation state, –2, in both H_2S and –SH group (Nagy and Winterbourn, 2010). Therefore, the presence of an oxidant species is mandatory to precede the S-sulfhydration reaction via one of these pathways: (a) direct oxidization of cysteine in the presence of oxidant species, (b) reduction of protein disulfide bonds by thiol derivates (Nagy and Winterbourn, 2010), and (c) production of polysulfides (through oxidation of sulfide) (Kimura et al., 2013; Nagy, 2013; Kimura, 2014). At physiological pH, most cysteine thiol groups are protonated (R-SH) and therefore display low reactivity toward oxidation. However, in some proteins where the cysteine residue is flanked by basic conditions, the deprotonated cysteine residue, as a thiolate anion (R-S⁻), appears. This thiolate anion is a strong nucleophile and undergoes oxidation to form sulfenic (SOH), sulfinic (SO_2H), or sulfonic acid (SO_3H) (Paul and Snyder, 2012). This mechanism suggests that mainly cysteine residues with low pK_a can undergo S-sulfhydration (Kimura, 2014). In conclusion, the mechanisms of S-sulfhydration are important and complex, requiring further explorations in the context of H_2S-mediated redox biology.

H_2S as an Antioxidant and Reducing Agent

H_2S is an antioxidant and reducing agent. It readily engages with 1 or 2 electrons in redox reactions. In other words, H_2S can drop 1 or 2 electrons for other molecules to pick them up. H_2S is a free-radical scavenger and can trap highly reactive free radical species. A large body of literature demonstrates that H_2S plays a key role in attenuating oxidative/nitrosative stress in different pathophysiological conditions (Whiteman et al., 2004; Yonezawa et al., 2007; Jha et al., 2008). The beneficial effect of H_2S in the reduction of mitochondrial ROS production has also been shown (Suzuki et al., 2011). H_2S can also react directly with various oxygen and nitrogen species under specific conditions (e.g., peroxynitrite, superoxide anion, hydroxyl radical, nitric oxide, lipid hydroperoxide).

MODULATION OF MITOCHONDRIAL FUNCTION BY HYDROGEN SULFIDE

The mitochondrion originated from a free-living bacterium, which was engulfed by a larger cell (prokaryote) around 1.5 billion years ago. In 1957, Dr. Siekevitz coined the concept of mitochondria as tiny, little powerhouses of cells (Sekevitz, 1957). Mitochondria are essential organelles in eukaryotes. They are responsible for the cellular energy production (ATP) and supply. There are about 300–400 mitochondria in each cell. They have their own genome, which codes for 13 key mitochondrial proteins. Since a single mitochondrion contains more than 13 proteins, obviously, the majority of the mitochondrial proteins are coded in the nucleus, synthesized in the cytosol, and transported into the mitochondrion.

Lack of proper mitochondrial function results in cellular dysfunction, the degree of which depends on the duration and severity of the energy restriction period. When mitochondrial ATP production is reduced, alternative (salvage) pathways come to the fore, such as glycolysis, pentose phosphate shunt, or production of ATP from purine nucleotides (Haun et al., 1996; Jurkowitz et al., 1998; Litsky et al., 1999; Módis et al., 2009). Although these non-mitochondrion-related ATP producing pathways are very limited for long-term energy supply, they serve as salvage routes and enable the cells to cope up with different stress-related insults. Improving the capability of these "rescue paths" and recuperating mitochondrial dysfunctions may be of therapeutic importance.

The widely accepted function of mitochondrial electron transport chain (Nicholls and Ferguson, 2013) is shown in Figure 5.3. Certain carbon-based substrates, such as pyruvate and succinate, produce $NADH+H^+$ and $FADH_2$ molecules through the Szent–Györgyi–Krebs cycle (also known as citric acid cycle) in the mitochondrial matrix. Then $NADH + H^+$ and $FADH_2$ donate electrons to complexes I and II, which are the initial components of mitochondrial electron transport chain. In the meantime, the protons derived from the degradation of $NADH + H^+$ and $FADH_2$ are being pumped out into the intermembrane space through complexes I, III, and IV. The proton concentration will increase in the intermembrane space, which creates an electrochemical gradient across the two sides of the inner mitochondrial membrane. This proton motive force is utilized by ATP synthase (complex V) to produce ATP via a process called chemiosmosis. It allows the return of protons to the matrix; the process generates significant amount of energy preserved in the form of ATP synthesis. In the meantime, the donated electrons are travelling on mitochondrial electron transport chain until they reach the final electron acceptor molecule, the oxygen, which is localized in complex IV. The oxygen, after accepting electrons, combines with protons, resulting in water production. When mitochondrial electron transport chain works optimally, the electrons mainly end up on the final electron acceptor (oxygen). A small amount of electrons can leave the electron transport chain through complexes II and III, resulting in free radical formation. Under normal conditions, thus generated free radicals are not dangerous, being neutralized by mitochondrial antioxidants (Murphy, 2009). However, during excessive oxidative/nitrosative stress, the production of mitochondrion-derived reactive oxygen species increases, leading to oxidatively modified and damaged mitochondrial proteins and mitochondrial

DNA, which can no longer fulfill their physiological functions. In extreme conditions, mitochondrial oxidants and free radicals can leave the mitochondrial compartment and travel to other cellular compartments as well.

To understand how H_2S regulates mitochondrial function, we need to first comprehend the function of complex II. Although only complexes I, III, and IV are so-called supercomplexes, complex II is composed of many highly organized and associated proteins (Dudkina et al., 2010). For example, succinate dehydrogenase (the single enzyme that is taking part in both the citric acid cycle and the electron transport chain) oxidizes succinate to fumarate via producing $FADH_2$, which will donate electrons immediately to complex II. Moreover, SQR, another complex II–related protein, has the capacity of oxidizing H_2S. Dr. Bouillaud coined the concept of the so-called sulfide oxidation unit (SOU), which consists of (1) SQR, (2) sulfur dioxygenase (ETHE1, also called dioxygenase ethylmalonic encephalopathy), (3) thiosulfate sulfur transferase (TST, also known as one isoenzyme of rhodanese), and (4) coenzyme Q that can be reduced by complex II and subsequently gives electrons to complex III (Bouillaud and Blachier, 2011) (Figure 5.3).

According to the current state-of-the-art knowledge, H_2S regulates mitochondrial function through three different pathways. First, H_2S can inhibit mitochondrial function at high, toxicological concentration. Second, H_2S acts as an inorganic mitochondrial substrate at low, physiological concentration. Third, H_2S regulates the mitochondrial cAMP-PKA pathways.

H_2S AS A MITOCHONDRIAL POISON

All known gasotransmitters (NO, H_2S, and CO) are considered as potent cytochrome c oxidase (complex IV) inhibitors (Cooper and Brown, 2008). They occupy the binding site for oxygen of this enzyme. The role of cytochrome c oxidase in the mitochondrial electron transport chain is crucial. It oxides complex III (also called coenzyme Q: cytochrome c—oxidoreductase) and transfers the electrons to oxygen. Blockage of cytochrome c oxidase by NO, H_2S, and CO uncouples mitochondria so that the oxidative phosphorylation is no longer linked to ATP production and cellular metabolism is suppressed.

The inhibitory effect of H_2S on cytochrome c oxidase is reversible by the application of nitrite and hyperbaric O_2 therapy as antidotes to H_2S toxicity (Reiffenstein et al., 1992). However, some in vivo studies show that H_2S can exert a nonreversible inhibitory effect mainly because H_2S can accumulate as labile sulfur in the mitochondrion that can subsequently induce long-term H_2S release, provoking a more prolonged inhibition (Szabo et al., 2014). The so-called detergent suicide involves mixing common household chemicals to produce H_2S gas at high concentration (Kamijo et al., 2013). However, at lower H_2S concentrations, a noncompetitive inhibition of cytochrome c oxidase also occurs (Petersen, 1977; Cooper and Brown, 2008). The toxicity of hydrogen sulfide is similar to cyanide. However, cyanide is an irreversible inhibitor of cytochrome c oxidase, five times more potent than CO.

The inhibitory effect of H_2S on cytochrome c oxidase occurs in a dose-dependent fashion and it is more pronounced in acidic conditions in rat liver mitochondria samples (Figure 5.5) (Szabo et al., 2014). Possibly, the acidosis shifts the proportion

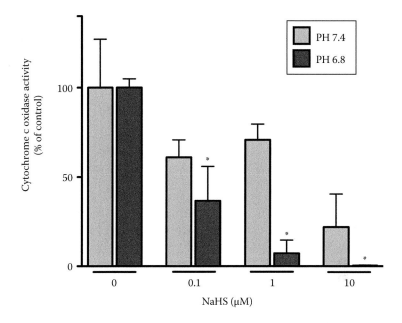

FIGURE 5.5   Inhibition of cytochrome c oxidase (complex IV) is more pronounced in acidic conditions. Enzyme activity measurement was conducted in isolated rat liver mitochondria utilizing the CYTOCOX1 cytochrome c oxidase kit (Sigma). NaHS (0.1 and 10 µM) was administered at two different pH levels, 7.4 (physiological) and 6.8 (pathological, moderate acidosis). The inhibitory effect of NaHS is potentiated by acidosis. Data are shown as means ± SEM (n = 3); *p < 0.05 represents the inhibitory effect of NaHS in acidosis. (Reproduced with permission from Szabo, C. et al, *Br J Pharmacol.* 171(8):2099–2122, 2014.)

of various forms of sulfide in solution (S^{2-}, SH^-, H_2S), and these various forms of sulfide may have different affinities to its enzymatic targets. Some diseases, such as shock and sepsis, are associated with an endogenous overproduction of H_2S, inducing the so-called cytopathic hypoxia. It means that even if O_2 is present in the blood and in the extracellular space among the cells, the mitochondrion cannot use the O_2. Diminishing endogenous production of H_2S by the application of specific pharmacological inhibitors of H_2S-producing enzymes might be a possible therapeutic approach in critical care to protect the liver (as well as other organs) (Hui et al., 2003; Coletta and Szabo, 2013, McCook et al., 2014).

H_2S AS AN INORGANIC SOURCE OF ENERGY

Some primitive creatures, certain bacteria (Kato et al., 2010; Sylvan et al., 2012), clams, and tubeworms can utilize H_2S as an inorganic, mitochondrial substrate to produce energy (ATP). These creatures are living in the environment lacking oxygen and light (there are some bacteria that can use light to generate a proton gradient, and therefore ATP), normally in near deep-sea hydrothermal vents (Gail, 1993). For example, tubeworms—living in an oxygen-poor but sulfide-rich environment in the deep sea—have worm hemoglobin proteins that carry H_2S to the symbionts, which are a group of bacteria possessing the capacity of oxidizing H_2S in order to provide energy to their host organism. In this case, the host organism is the tubeworm itself (Gail, 1993; Szabo et al., 2014).

In 2001, Searcy and his colleagues measured increased O_2 consumption and ATP production by the addition of H_2S in isolated chicken liver mitochondria. This was the first in vitro study that assigned a role to H_2S as a source of energy in mitochondria of higher species above worms, bacteria (Yong and Searcy, 2001). Furthermore, Searcy proposed that during evolution, a H_2S-releasing archaeon and a sulfide-oxidizing bacterium lived together in endosymbiosis in order to keep themselves alive in oxygen-poor circumstances (Searcy and Lee, 1998; Yong and Searcy, 2001).

The first suggestion that H_2S can act as an inorganic source of energy in mammalian cells came from the study in cultured colonic epithelial cells (Goubern et al., 2007). In permeabilized colon cells, the addition of H_2S significantly increased oxygen consumption rate and elevated mitochondrial membrane potentials. These two effects were blocked by antimycin and myxothiazol; both are widely used complex III inhibitors. A plausible explanation for the earlier findings is that an intramitochondrial enzyme, SQR, oxidizes hydrogen sulfide. From two H_2S molecules, two disulfides (–SSH) bounds are created on SQR that generates two electrons. Then the electrons enter mitochondrial electron transport chain through coenzyme Q, fostering mitochondrial ATP production. This positive bioenergetic effect of hydrogen sulfide occurs at lower H_2S concentration (nanomolar to micromolar concentration) (Lagoutte et al., 2010; Helmy et al., 2014) and proved to be physiologically relevant (Figure 5.3).

Fu et al. demonstrated that one of the H_2S-producing enzyme, cystathionine γ-lyase (CSE), can be translocated to the mitochondrion and produce H_2S intramitochondrially in vascular smooth muscle cells (SMCs) (Fu et al., 2012). The latter increases ATP production under hypoxic condition. The same study proved the pathological relevance of CSE translocation as it occurs when intracellular calcium level is increased or oxygen partial pressure drops.

Another study (Módis et al., 2013a) demonstrated that H_2S is a physiologically important mediator for keeping the basal, resting mitochondrial function in murine hepatocytes. In other words, partially suppressing one of the cellular source of H_2S (silencing MST) or interfering with the H_2S oxidizing capacity (silencing SQR) reduced mitochondrial function and ATP production in these cells.

CBS can also be constitutively present in mitochondria. Teng et al. (2013) showed that CBS has a low expression level in liver mitochondria under physiological, normoxic conditions. However, ischemia/hypoxia condition promotes the accumulation of CBS in mitochondria. According to their explanation, in normoxic conditions, oxygen binds to the heme prosthetic group of the CBS enzyme, inducing conformation changes of the protein structure. This oxygenated conformation can be recognizable by Lon protease, which causes the degradation of CBS proteins. During hypoxia, the lack of oxygen alters the oxygenation of CBS such that Lon protease cannot recognize the de-oxygenated CBS. CBS proteins are consequently accumulated in the mitochondrial matrix, increasing intramitochondrial H_2S level. The mitochondrial ATP production would be improved thereafter. The mitochondrial accumulation of heme oxygenase-1 (HO-1) is also regulated by Lon protease in the same mode. In another study, Szabo and his colleagues (Szabo et al., 2013) found that CBS enzyme is significantly overexpressed in colon cancer cells. CBS-derived H_2S production contributes to the production of mitochondrial ATP, thereby supporting the proliferation, migration, and invasion of colon cancer cells. When CBS expression level is reduced by shRNA-mediated silencing techniques or when CBS is pharmacologically inhibited, the intramitochondrial H_2S level is blunted and mitochondrial function (oxygen consumption rate, ATP level) reduced.

All aforementioned findings reach the same conclusion: H_2S, as an *inorganic source of energy* (Goubern et al., 2007), can provide electrons by its oxidation in mitochondria by SQR. Feeding mitochondrial electron transport chains with electrons results in increased ATP production, supporting cellular bioenergetics and maintaining cellular viability.

H_2S AS A REGULATOR OF MITOCHONDRIAL cAMP HOMEOSTASIS

A mitochondrial cAMP pool exists, which is not related to the cytosolic cAMP pool. These two distinct cAMP pools do not communicate with each other (Acin-Perez et al., 2009, 2011). Over the last decade, only a few studies have shown the importance of the cAMP signaling pathway for mitochondrial function. A mitochondrial-specific adenylyl cyclase isoform (Acin-Perez et al., 2009, 2011), a mitochondrial cAMP-dependent protein kinase/protein kinase A (PKA, Sardanelli et al., 1996; Acin-Perez et al., 2009), and phosphodiesterase 2A (PDE2A, Actin-Perez et al; 2011, Módis et al., 2013d) have been identified.

Several recent studies highlight the role of the mitochondrial cAMP-H_2S axis in the regulation of cellular bioenergetics. Módis et al. found that addition of NaHS to isolated rat liver mitochondria elevated cAMP level in the presence of ADP, which stimulates electron flow (Módis et al., 2013d). Similarly, a cAMP analog, Br-cAMP, stimulated mitochondrial electron transport chain. As cAMP activates PKA that can phosphorylate certain target proteins such as subunits of mitochondrial complexes, it was subsequently tested whether PKA inhibition reduces the cellular respiration. The PKA inhibitor, Rp-cAMP, reversed the positive bioenergetics effects of Br-cAMP and partly diminished the stimulating effect of NaHS on mitochondrial function (Figure 5.3).

PATHOPHYSIOLOGICAL ROLES OF HYDROGEN SULFIDE IN THE LIVER

In anabolic metabolism, the liver participates in protein synthesis, carbohydrate synthesis (gluconeogenesis, glycogenolysis, and glycogenesis), lipid synthesis (cholesterol synthesis, lipogenesis, and lipoprotein synthesis), and synthesis of coagulation factors, bile, and several hormones (insulin-like growth factor-1 [IGF-1], thrombopoietin, and angiotensinogen). In catabolic metabolism, the liver has a role in protein degradation, breakdown of insulin and other hormones, bilirubin elimination through bile excretion, xenobiotic metabolism (drug metabolism), and also removal of ammonia through the production of urea. There are other important functions of the liver such as maintaining glycogen storage;

vitamin A, D, B12, K pools, etc. The liver is also considered as a huge blood reservoir as well (Wang, 2012; Oren, 2014; Mani et al., 2014). In the following sections, various regulatory roles of H_2S in liver diseases are discussed. Whenever information is available, we are focusing on the effects of H_2S on mitochondrial function and dysfunction.

THERAPEUTIC EFFECTS OF H_2S IN DIABETIC LIVER DYSFUNCTION

Diabetes is considered one of the most critical socioeconomic problems that is associated with a major expense to the health-care system all over the world. According to the latest survey by World Health Organization in 2014, worldwide 347 million people have diabetes. Among this population, 90% is diagnosed with type 2 diabetes (also called non-insulin-dependent/adult-onset diabetes related to inefficient insulin utilization of cells). The other 10% people suffer from type 1 diabetes (previously known as insulin dependent/juvenile or childhood onset). The pathogenic roles of H_2S in both forms of diabetes have been studied.

For type 2 diabetes, H_2S has been studied in the following models:

- Diabetic Zucker rats with obesity and hyperinsulinemia (Wu et al., 2009)
- Obesity-induced diabetes murine model (db/db mice) (Liu et al., 2014)
- Insulin resistance–induced diabetic model in rats developed by administration of fructose (Padiya et al., 2011)

For type 1 diabetes, H_2S has been studied in the following models:

- Streptozotocin-induced diabetic model in rodents (Yusuf et al., 2005; Yang et al., 2007, 2011; Ou et al., 2010; Denizalti et al., 2011; El-Seweidy et al., 2011; Suzuki et al., 2011; Yuan et al., 2011)
- Nonobese diabetic murine model (NOD) (Brancaleone et al., 2008)
- Alloxan-induced diabetic rat model (Abdultawab and Ayuob, 2013)

However, most of the referenced studies investigated the role of H_2S in tissues other than the liver. For example, the potential pathogenic role of H_2S was studied in the context of pancreatic β cell damage, although the literature is somewhat contradictory (Kaneko et al., 2006, 2009; Ali et al., 2007; Yang et al., 2007; Rajpal et al., 2012; Okamoto et al., 2013). Moreover, the beneficial effects of H_2S have been demonstrated in rodent models of various diabetic complications such as endothelial dysfunction (Suzuki et al., 2011), diabetic nephropathy (Yuan et al., 2011), and cardiomyopathy (El-Seweidy et al., 2011). In the following paragraphs, we will focus on diabetic liver dysfunction in the context of H_2S biology.

In a streptozotocin-induced diabetic rat model, Yusuf et al. showed that H_2S levels are significantly increased in liver tissues and in the pancreas (Yusuf et al., 2005). Both CSE and CBS mRNA levels were upregulated in the liver. These alterations were reversed by insulin treatment. In contrast, plasma H_2S levels were slightly reduced in these diabetic animals. Another study on streptozotocin-diabetic rats found a significantly higher reactive oxygen species (ROS) production associated with reduced CSE protein expression and activity in liver tissues (Manna et al., 2014). These contradictory observations in liver tissues may be related to different experimental conditions, and the stage and severity of the disease. One important factor in the context of diabetic liver dysfunction relates to the modulation of H_2S bioavailability. Hyperglycemia leads to an increase in oxidative stress because of the overproduction of reactive oxygen and nitrogen species (ROS/RNS), coupled with a suppressed antioxidant capacity/defense (Giacco and Brownlee, 2010; Szabo, 2012; Gero et al., 2013). This enhanced oxidative/nitrative stress increases the consumption of H_2S, thereby decreasing its biological activity (Suzuki et al., 2011). This may, in theory, explain experimental conditions where, at the same time, H_2S-producing enzyme levels are elevated and circulating or tissue H_2S levels are reduced.

There are limited reports about how H_2S can modulate the oxidative/nitrosative stress in diabetic liver dysfunction. In this regard, some hints may be derived from garlic oil treatment of alloxan-induced type 1 diabetic rats (Abdultawab and Ayuob, 2013). This treatment lowered the elevated blood glucose, total lipids, triglycerides, and cholesterol. The elevated oxidative/nitrosative stress in plasma samples, reflected by the levels of lipid peroxide, NO, and ceruplasmin (a general indicator for plasma protein level that drops down in patients with hepatic disease), was also lowered by garlic oil treatment. Furthermore, the activities of glutathione *S*-transferase (GST) and superoxide dismutase (SOD) in liver homogenates were increased by garlic oil treatment. It should be clear that garlic oil and other dietary supplements cannot be considered "pure" H_2S donors as they have additional activities unrelated to H_2S donation (Baluchnejadmojarad and Roghani, 2003; Benavides et al., 2007; Chuah et al., 2007; Shaik et al., 2008; Ejaz et al., 2009; Ou et al., 2010; Chang et al., 2011; Abdultawab and Ayuob, 2013). Further studies, utilizing pure H_2S donors, are required to confirm and extend the potential hepatoprotective effect of H_2S in diabetes. In this context, we would like to reiterate an interesting experimental finding that the antidiabetic drug metformin increases hepatic H_2S levels (Wilinski et al., 2013).

H_2S IN LIVER ISCHEMIA/REPERFUSION INJURY

The role of H_2S in liver ischemia/reperfusion injury was initially reported in anesthetized mice subjected to 60 min of hepatic ischemia, followed by 5 h of reperfusion (Jha et al., 2008). The H_2S donor IK-1001 was administered 5 min before reperfusion. The beneficial effects of H_2S on hepatocellular integrity were evidenced by an attenuation of the elevated serum alanine aminotransferase (ALT) and aspartate aminotransferase (AST) levels. This cytoprotective effect was also associated with an improved balance between reduced and oxidized glutathione, decreased formation of lipid hydroperoxides, as well as increased expression of thioredoxin-1. H_2S administration increased the expression of heat shock protein-90 and Bcl-2 (Jha et al., 2008). Bcl-2 is an antiapoptotic protein that plays a role in maintaining mitochondrial membrane integrity, thereby preventing the release of proapoptotic proteins. A mitochondrial protective effect of H_2S is thus hinted.

Shaik et al. investigated the potential hepatoprotective effects of the H_2S donor, diallyl sulfide (DAS), in a rat model of 1 h long hepatic ischemia, followed by 3 h long reperfusion (Shaik et al., 2008). Similar to the effect of IK-1001, the protective effect of DAS was associated with a significant inhibition of lipid peroxidation and improvement of glutathione homeostasis, again suggesting an (direct and/or indirect) antioxidant effect of DAS or H_2S released. Kang et al. confirmed the cytoprotective, antiapoptotic, and antioxidant effects of H_2S in another rat model of hepatic ischemia/reperfusion. In addition, the application of CSE inhibitor, PAG, exacerbates the hepatic reperfusion response, indicating the protective role of *endogenously produced* H_2S in hepatic ischemia/reperfusion (Kang et al., 2009). Wang et al. further explored the mechanisms underlying the liver protective effect of H_2S, linking to the activation of the Akt-signaling pathway (Wang et al., 2012). The same study also revealed that H_2S markedly decreased the number of autophagosomes in hepatocytes subjected to anoxia/reoxygenation injury. Moreover, H_2S, in addition to being anti-inflammatory and cytoprotective, improves hepatic microcirculation and attenuates neutrophil chemotaxis (Zhu et al., 2012).

The direct evidence for a mitochondrial protective role of H_2S in hepatic reperfusion injury has been received (Zhang et al., 2013). In a rat model of hepatic ischemia/reperfusion, H_2S treatment markedly inhibited mitochondrial permeability transition pore (MPTP) opening. This action likely offers the protection against mitochondrial-related cell death and apoptosis by stimulating Akt and glycogen synthase kinase-3 β phosphorylation and by decreasing the release of mitochondrial cytochrome c and the levels of cleaved caspases 3 and 9.

In summary, H_2S exerts protective effects in hepatic ischemia/reperfusion, and some of the protective effects are related, directly or indirectly, to the stabilization of mitochondrial integrity and subsequent inhibition of cell death.

H_2S in Critical Illness

H_2S in Acute Liver Failure during Septic/Endotoxic Shock

Sepsis or septic shock is a systemic inflammatory response characterized by overwhelming systemic (typically bacterial) infection, progressive hypotension, and multiple organ failure (including liver failure), culminating in mortality. Although it is often debated about the most clinically relevant rodent model of sepsis, most laboratories use either the administration of bacterial lipopolysaccharide (LPS), a bacterial cell wall component, or cecal ligation and puncture (CLP model), which produces sepsis due to systemic bacteremia. Initial studies during 2005–2008 reported that in sepsis/systemic inflammation, upregulation of H_2S-producing enzymes occurred and the resulting H_2S overproduction exerted hepatotoxic effects and injured other parenchymal organs. For example, Collin and colleagues showed in a rat model of LPS-induced shock that hepatic expression of CBS and CSE increased and the application of the CSE inhibitor, PAG, attenuated the increases in ALT and aspartate aminotransferase (markers for hepatic injury) (Collin et al., 2005). Similar findings (including histopathological evidence of hepatoprotection by PAG) were reported in an LPS model of systemic inflammation in a mouse model (Li et al., 2005). The same authors reported that both mice and humans with septic shock exhibited elevated circulating levels of H_2S (Li et al., 2005). In contrast, other studies have shown that administration of the H_2S donor to mice subjected to CLP-induced sepsis did not prove to be protective. In fact, it increased plasma levels of ALT, and increased hepatic levels of myeloperoxidase, a marker of inflammatory mononuclear cell infiltration (Li et al., 2005; Zhang et al., 2008). In a CLP sepsis model, the pro-inflammatory effect of endogenously overproduced H_2S was shown as H_2S activated NF-kB and Erk1/2 (Zhang et al., 2006, 2007, 2008). H_2S also activated transient receptor potential vanilloid type 1–mediated neurogenic inflammation (Ang et al., 2010, 2011). These reports support the view that a toxic overproduction of H_2S appears in sepsis, and suggest that by pharmacologically inhibiting CSE and/or CBS, one may be able to decrease systemic inflammatory mediator production, suppress the oxidative stress and mononuclear cell infiltration into the liver, resulting in hepatic protection. The concept shares marked similarities with the previously demonstrated role of NO overproduction due to the upregulation of the inducible nitric oxide synthase, iNOS. In this case, the overproduced NO also exerts hepatotoxic effects in animal models of sepsis (Szabo et al., 1994; Kilbourn et al., 1997; Szabo and Módis, 2010). Yan et al. (2013) confirmed the elevation in systemic H_2S levels by LPS, the hepatoprotective effect of PAG, and the exacerbation of liver damage by coadministration of the H_2S donor NaHS. The hepatoprotective effect of PAG may be partially due to the normalization of hepatic microcirculation and hepatic tissue oxygen availability, as well as a reduction of sinusoidal leukocyte accumulation (Norris et al., 2013). Finally, Shirozu et al. (2014) demonstrated that CSE deficiency exerts hepatoprotective effects (as well as improves survival) in a mouse model of acute liver failure induced by the combination of galactosamine and LPS. The protection was associated with an inhibition of inflammatory cytokine production and inhibition of the activation of caspase 3 and poly (ADP-ribose) polymerase (PARP) in the liver. The beneficial effects of CSE deficiency were associated with an elevation of homocysteine levels, since CSE is involved in homocysteine metabolism. They were also associated with an elevation of thiosulfate levels. This is an unexpected finding since thiosulfate is the major metabolite of H_2S. It may be explained by a compensatory upregulation of another H_2S-producing enzyme, MST, in this model. The protection by CSE deficiency against TNF-induced cell death was confirmed in primarily cultured hepatocytes (Shirozu et al., 2014).

The slow-releasing H_2S donor, GYY4137, inhibited the proinflammatory signaling in the liver in a rat model of LPS-induced shock, and improved liver function and liver histopathology (Li et al., 2009a). One of the explanations of this finding was that the efficacy of the H_2S donor was markedly dependent on the time of its administration. GYY1437 did not exert protective effects before the administration of LPS, but it was protective when given 1 or 2 h after LPS. It was also postulated that the H_2S donor elevates local H_2S levels at particular locations in the body where endogenous H_2S levels may be low. The therapeutic usefulness of H_2S donation in sepsis was also shown by the protective effect of inhaled H_2S gas on LPS-induced hepatotoxicity (Tokuda et al., 2012). The mechanisms of the H_2S protective effect have been attributed to several factors, including anti-inflammatory effects (suppression of a wide range of proinflammatory factors in the plasma via inhibition of NF-kB and STAT3 activation) as well as to a reduction of systemic NO levels (Tokuda et al., 2012). Aslami et al. demonstrated the hepatoprotective effect of the H$_2$S donor NaHS in a model of sepsis induced by administration of live *Streptococcus pneumonia*. This protective effect was associated with a reduction in inflammation and oxidative stress in the liver (protein carbonyl levels were reduced) and maintenance of hepatic mitochondrial function in terms of ATP production (Aslami et al., 2013b). Interestingly, the protective effect of H$_2$S was evident in a short course of infusion, but extension of NaHS infusion time worsened some of the inflammatory responses and the degree of injury to some organs (Aslami et al., 2013a). This is another indication of the bell-shaped dose-dependent responses of H$_2$S, whereby elevation of H$_2$S levels beyond an optimal level creates deleterious responses. The mechanism for the hepatoprotective effect of H$_2$S donors may also involve the improved L-arginine handling (Bekpinar et al., 2013). To further increase the conflicting body of literature, Bekpinar et al. recently reported the aggravation of LPS-induced hepatic injury by PAG, in sharp contrast to the reported protective effect of PAG (Bekpinar et al., 2014). No clear explanation was offered for the discrepancy. In summary, even though dozens of studies have been conducted on the modulation of H$_2$S homeostasis in sepsis models, no clear conclusion has yet emerged on the pathophysiological role of this mediator, and on the potential future of therapeutic approaches.

H$_2$S in Acute Liver Failure during Hemorrhagic Shock

The role of H$_2$S in liver failure in hemorrhagic shock is also a topic of controversy. Mok et al. in a rat model of hemorrhagic shock detected a slight but statistically significant increase in circulating H$_2$S levels, and reported improved hemodynamic effects and reduced liver H$_2$S biosynthesis with the inhibitors PAG or BCA (Mok et al., 2004). Pretreatment of the animals with PAG elevated the hemodynamic function, attenuated the increase in plasma levels of TNF and IL-6, and reduced the expression of iNOS and the plasma levels of ALT. The liver myeloperoxidase levels were also lowered, indicating the suppressed inflammatory mononuclear cell infiltration (Mok and Moore, 2008). Another body of studies, in contrast, showed that H$_2$S levels are unchanged or are slightly decreased in hemorrhagic shock (Bracht et al., 2012; Chai et al., 2012), and that H$_2$S donation exerts hepatoprotective effects, both in a rodent model (Chai et al., 2012) and in a porcine model (Bracht et al., 2012).

H$_2$S in Acute Liver Failure during Burn Injury

Only three reports are available on the role of H$_2$S in the pathogenesis of burn-induced liver injury. Unfortunately, these reports are conflicting. In a murine model, 25% burn of the total body surface area caused a relatively slight increase in circulating H$_2$S levels due to the upregulation of CSE, associated with a severe systemic inflammatory reaction and functional impairment of the liver and other organs. PAG administration, both in a prophylactic and a therapeutic way, significantly reduced functional injury and improved histological profile of the liver. On the other hand, NaHS significantly aggravated the outcomes of the model (Zhang et al., 2010). In marked contrast, two subsequent studies on a 30% burn model found that circulating levels of H$_2$S were lower and CSE expression was lower in burned tissues. H_2S administration exerted hepatoprotective effects in terms of serum ALT and AST levels and reduction of systemic inflammation, while upregulating the circulating level of the anti-inflammatory cytokine IL-10 (Li et al., 2011; Zeng et al., 2013).

Critical Illness: H_2S Inhibition or H_2S Donation?

It is not easy to draw any simple conclusions from the aforementioned contradictory data on the roles of H_2S in rodent models of sepsis, hemorrhage, and burn. While part of the contradictories may be explained by experimental differences, such as different species and different models of critical illness used, it is likely that the explanations go well beyond methodological and species differences. At least three issues need to be considered when attempting to interpret the conflicting findings. (1) The bell-shaped dose-dependent effects of H_2S. Depending on the dose of H_2S used and the chemical nature of H_2S donors (e.g., fast- vs. slow-releasing H_2S donor), different biological responses may be achieved. (2) The specificity of pharmacological inhibitors of H_2S biosynthesis. Often these inhibitors inhibit additional enzymes unrelated to H_2S biosynthesis. They may also induce pharmacological effects on their own (Asimakopoulou et al., 2013). (3) The endogenous production of H_2S does not rely on a single enzyme. When one enzyme is knocked out, such as CSE, other H_2S-producing enzymes may undergo compensatory responses.

H_2S IN LIVER TRANSPLANTATION

Several preclinical studies have demonstrated that directly incubating the stored and transplanted organs (including the heart, kidney, skeletal muscle, and lung) with H_2S (H_2S gas or H_2S donors) can prolong storage time and/or improve the function of the transplanted organs. In 2011, Balaban et al. studied the effect of the H_2S donor diallyl disulfide (3.4 mM) on cold storage of rat livers in Wisconsin solution. They found that the hepatic clearance of bromosulphophtalein was enhanced by the diallyl disulfide treatment. The role of H_2S in human transplantation is also supported by a recent correlation analysis (van den Berg et al., 2014). It remains to be directly tested whether H_2S can be applied with benefit for the storage and transplantation of human organs. In addition, it must be kept in mind that thus far, no direct studies have measured the metabolic rate or mitochondrial activity of the stored organs, while it is assumed that some of the beneficial effects of H_2S are related to the inhibition of complex IV and a reversible shutdown of cellular metabolism. It is equally possible that additional pharmacological effects of H_2S (such as its effects as an antioxidant, antifibrotic, anti-inflammatory, and antiapoptotic mediator) may also play a role, in addition to or independently from any metabolic suppression action.

H_2S IN PORTAL HYPERTENSION AND LIVER CIRRHOSIS

One of the earliest recognized biological effects of H_2S is its regulatory effect on vascular tone. Although the magnitude of the effect is dependent on the vascular bed and the species, in general, H_2S exerts vasodilatory effects, and maintains a healthy vascular tone. Inhibition of this basal vasorelaxant effect results in vasoconstriction and hypertension (Hosoki et al., 1997; Zhao et al., 2001; Yang et al., 2008; Mustafa et al., 2011; Coletta et al., 2012; reviewed in Wang et al., 2012). In fact, the vaso-relaxant effect of H_2S on the portal vein was reported as early as 1997 (Hosoki et al., 1997) and similar findings were later reported for the hepatic artery as well (Siebert et al., 2008). The liver receives arterial blood through the hepatic artery and venous blood from the splanchnic circulation through the portal vein. Pathological increases in the tone of the portal vein lead to a pathophysiological state, portal hypertension, characterized by a combination of symptoms including ascites (fluid in the peritoneal cavity), splenomegaly, and compensatory portacaval anastomoses and varices. These are the consequences of the increased portal resistance forc-ing blood flow into alternative (and inferior) routes. In severe cases, the condi-tion can deteriorate into hepatorenal syndrome with increased risk of peritonitis. Ebrahimkhami et al. proposed that altered liver metabolism of H_2S may hold clues for the pathogenesis of portal hypertenions during the endotoxemia, a common feature of cirrhosis (Ebrahimkhami et al., 2005). Fiorucci et al. utilized a 4-week bile duct ligation (BDL) rat model of cirrhosis. This model produces an increase in portal pressure due to portal vein constriction (Fiorucci et al., 2005). Their studies demonstrated that in BDL rats, CSE was downregulated, and L-cysteine induced portal vein relaxation. These authors concluded that H_2S is involved in the maintenance of portal venous pressure in normal animals. The deficiency of this response in cirrhotic rats may contribute to the development of increased intrahe-patic resistance and portal hypertension (Fiorucci et al., 2005). In another study of carbon tetrachloride–induced hepatotoxicity, NaHS attenuated the hepatotoxicity and improved liver function and integrity, judged by the plasma levels of ALT and AST, the degree of liver fibrosis (histological evaluation, liver hydroxyproline con-tent), the production of inflammatory mediators (TNF, IL-1, IL-6, ICAM-1), and portal pressure (Tan et al., 2011). A recent clinical study showed a reduction (rather than the originally hypothesized increase) in circulating H_2S levels in patients with cirrhosis and portal hypertension (Wang et al., 2014). It appears that H_2S donation may be a potential therapeutic approach for the restoration of physiological portal pressure with liver cirrhosis.

CONCLUSION

H_2S biology is a recently emerging and rapidly developing research field. H_2S is one of the identified gasotransmitters in the body, in addition to NO, CO, and NH_3, which serves important roles in the regulation of various physiological and pathophysi-ological processes. In the liver, H_2S is generated from L-cysteine and homocysteine by CSE, CBS, and MST. CBS and MST are responsible for hepatic mitochondrial H_2S production. The hepatic mitochondrial functions are significantly regulated by H_2S. At high concentrations, H_2S inhibits the electron transport chain and the oxi-dative phosphorylation. On the other hand, as an inorganic mitochondrial energy substrate, H_2S sustains hepatocyte bioenergetics to cope up with hypoxic insult. Many of the cellular protective effects of H_2S can be explained by its interaction with metal ions, protein S-sulfhydration, and the antioxidant reaction. Altered liver metabolism of H_2S is involved in numerous liver diseases, including but not limited to diabetic liver complications, ischemia/reperfusion liver injury, acute liver failure with sepsis or shock, portal hypertension, and liver cirrhosis. In conclusion, H_2S is an essential gasotransmitter in the liver with multiple regulatory roles, many of which are intimately related to the modulation of mitochondrial function and cellular bioenergetics.

ACKNOWLEDGMENTS

This work has been supported by an Operating Grant from the Canadian Institutes of Health Research (CIHR) to R. Wang. K. Módis is a recipient of postdoctoral fellowship award from CIHR.

REFERENCES

Abdultawab, H. S. and N. N. Ayuob. 2013. Can garlic oil ameliorate diabetes-induced oxidative stress in a rat liver model? A correlated histological and biochemical study. *Food Chem Toxicol* 59:650–656.

Acin-Perez, R., Russwurm, M., Günnewig, K., Gertz, M., Zoidl, G., Ramos, L., Buck, J., Levin, L. R., Rassow, J., Manfredi, G., and C. Steegborn. 2011. A phosphodiesterase 2A isoform localized to mitochondria regulates respiration. *J Biol Chem* 286(35):30423–30432.

Acin-Perez, R., Salazar, E., Kamenetsky, M., Buck, J., Levin, L. R., and G. Manfredi. 2009. Cyclic AMP produced inside mitochondria regulates oxidative phosphorylation. *Cell Metab* 9(3):265–276.

Ali, M. Y., Whiteman, M., Low, C. M., and P. K. Moore. 2007. Hydrogen sulphide reduces insulin secretion from HIT-T15 cells by a KATP channel-dependent pathway. *J Endocrinol* 195(1):105–112.

Altaany, Z., Ju, Y., Yang, G., and R. Wang. 2014. The coordination of *S*-sulfhydration, *S*-nitrosylation, and phosphorylation of endothelial nitric oxide synthase by hydrogen sulfide. *Sci Signal* 7(342):ra87.

Ang, S. F., Moochhala, S. M, and M. Bhatia. 2010. Hydrogen sulfide promotes transient receptor potential vanilloid 1-mediated neurogenic inflammation in polymicrobial sepsis. *Crit Care Med* 38(2):619–628.

Ang, S. F., Moochhala, S. M., MacAry, P. A., and M. Bhatia. 2011. Hydrogen sulfide and neurogenic inflammation in polymicrobial sepsis: Involvement of substance P and ERK-NF-κB signaling. *PLoS One* 6(9):e24535.

Asimakopoulou, A., Panopoulos, P., Chasapis, C. T., Coletta, C., Zhou, Z., Cirino, G., Giannis, A., Szabo, C., Spyroulias, G. A., and A. Papapetropoulos. 2013. Selectivity of commonly used pharmacological inhibitors for cystathionine β synthase (CBS) and cystathionine γ lyase (CSE). *Br J Pharmacol* 169(4):922–932.

Aslami, H., Beurskens, C. J., de Beer, F. M., Kuipers, M. T., Roelofs, J. J., Hegeman, M. A., Van der Sluijs, K. F., Schultz, M. J., and N. P. Juffermans. 2013a. A short course of infusion of a hydrogen sulfide-donor attenuates endotoxemia induced organ injury via stimulation of anti-inflammatory pathways, with no additional protection from prolonged infusion. *Cytokine* 61(2):614–621.

Aslami, H., Pulskens, W. P., Kuipers, M. T., Bos, A. P., van Kuilenburg, A. B., Wanders, R. J., Roelofsen, J. et al. 2013b. Hydrogen sulfide donor NaHS reduces organ injury in a rat model of pneumococcal pneumosepsis, associated with improved bio-energetic status. *PLoS One* 8(5):e63497.

Azareko, O., Jordan, M. A., and L. Wilson. 2014. Erucin, the major isothiocyanate in arugula (*Eruca sativa*), inhibits proliferation of MCF7 tumor cells by suppressing microtubule dynamics. *PLoS One* 9:e100599

Balaban, C. L., Rodriguez, J. V., and E. E. Guibert. 2011. Delivery of the bioactive gas hydrogen sulfide during cold preservation of rat liver: Effects on hepatic function in an ex vivo model. *Artif Organs* 35(5):508–515.

Baluchnejadmojarad, T. and M. Roghani. 2003. Endothelium dependent and -independent effect of aqueous extract of garlic on vascular reactivity on diabetic rats. *Fitoterapia* 74(7–8):630–637.

Barnett, C. F. and R. F. Machado. 2006. Sildenafil in the treatment of pulmonary hypertension. *Vasc Health Risk Manag* 2(4):411–412.

Beauchamp, R. O. Jr., Bus, J. S., Popp, J. A., Boreiko, C. J., and D. A. Andjelkovich. 1984. A critical review of the literature on hydrogen sulfide toxicity. *Crit Rev Toxicol* 13(1):25–97.

Bekpinar, S., Develi-Is, S., Unlucerci, Y., Kusku-Kiraz, Z., Uysal, M., and F. Gurdol. 2013. Modulation of arginine and asymmetric dimethylarginine concentrations in liver and plasma by exogenous hydrogen sulfide in LPS-induced endotoxemia. *Can J Physiol Pharmacol* 91(12):1071–1075.

Bekpinar, S., Unlucerci, Y., Uysal, M., and F. Gurdol. 2014. Propargylglycine aggravates liver damage in LPS-treated rats: Possible relation of nitrosative stress with the inhibition of H_2S formation. *Pharmacol Rep* 66(5):897–901.

Benavides, G. A., Squadrito, G. L., Mills, R. W., Patel, H. D., Isbell, T. S., Patel, R. P., Darley-Usmar, V. M., Doeller, J. E., and D. W. Kraus. 2007. Hydrogen sulfide mediates the vasoactivity of garlic. *Proc Natl Acad Sci USA* 104(46):17977–17982.

Bhattacharyya, S., Saha, S., Giri, K., Lanza, I. R., Nair, K. S., Jennings, N. B., Rodriguez-Aguayo, C. et al. 2013. Cystathionine β-synthase (CBS) contributes to advanced ovarian cancer progression and drug resistance. *PLoS One* 8(11):e79167.

Bouillaud, F. and F. Blachier. 2011. Mitochondria and sulfide: A very old story of poisoning, feeding, and signaling? *Antioxid Redox Signal* 15(2):379–391.

Bracht, H., Scheuerle, A., Gröger, M., Hauser, B., Matallo, J., McCook, O., Seifritz, A. et al. 2012. Effects of intravenous sulfide during resuscitated porcine hemorrhagic shock. *Crit Care Med* 40(7):2157–21567.

Brancaleone, V., Roviezzo, F., Vellecco, V., De Gruttola, L., Bucci, M., and G. Cirino. 2008. Biosynthesis of H_2S is impaired in non-obese diabetic (NOD) mice. *Br J Pharmacol* 155(5):673–680.

Braunstein, A. E., Goryachenkova, E. V., and N. D. Lac. 1969. Reactions catalysed by serine sulfhydrase from chicken liver. *Biochim Biophys Acta* 171(2):366–368.

Campolo, M., Esposito, E., Ahmad, A., Di Paola, R., Wallace, J. L., and S. Cuzzocrea. 2013. A hydrogen sulfide-releasing cyclooxygenase inhibitor markedly accelerates recovery from experimental spinal cord injury. *FASEB J* 27(11):4489–4499.

Carbonero, F., Benefiel, A. C., Alizadeh-Ghamsari, A. H., and H. R. Gaskins. 2012. Microbial pathways in colonic sulfur metabolism and links with health and disease. *Front Physiol* 28;3:448.

Caro, A. A., Thompson, S., and J. Tackett. 2011. Increased oxidative stress and cytotoxicity by hydrogen sulfide in HepG2 cells overexpressing cytochrome P450 2E1. *Cell Biol Toxicol* 27(6):439–453.

Chai, W., Wang, Y., Lin, J. Y., Sun, X. D., Yao, L. N., Yang, Y. H., Zhao, H., Jiang, W., Gao, C. J., and Q. Ding. 2012. Exogenous hydrogen sulfide protects against traumatic hemorrhagic shock via attenuation of oxidative stress. *J Surg Res* 176(1):210–219.

Chan, M. V. and J. L. Wallace. 2013. Hydrogen sulfide-based therapeutics and gastrointestinal diseases: Translating physiology to treatments. *Am J Physiol Gastrointest Liver Physiol* 305(7):G467–G473.

Chang, S. H., Liu, C. J., Kuo, C. H., Chen, H., Lin, W. Y., and K. Y. Teng. 2011. Garlic oil alleviates MAPKs- and IL-6-mediated diabetes-related cardiac hypertrophy in STZ-induced DM rats. *Evid Based Complement Alternat Med* 2011:950150.

Chattopadhyay, M., Kodela, R., Nath, N., Barsegian, A., Boring, D., and K. Kashfi. 2012b. Hydrogen sulfide-releasing aspirin suppresses NF-κB signaling in estrogen receptor negative breast cancer cells in vitro and in vivo. *Biochem Pharmacol* 83(6):723–732.

Chattopadhyay, M., Kodela, R., Nath, N., Dastagirzada, Y. M., Velázquez-Martínez, C. A., Boring, D., and K. Kashfi. 2012a. Hydrogen sulfide-releasing NSAIDs inhibit the growth of human cancer cells: A general property and evidence of a tissue type-independent effect. *Biochem Pharmacol* 83(6):715–722.

Chuah, S. C., Moore, P. K., and Y. Z. Zhu. 2007. S-allylcysteine mediates cardioprotection in an acute myocardial infarction rat model via a hydrogen sulfide-mediated pathway. *Am J Physiol Heart Circ Physiol* 293(5):H2693–H2701.

Coletta, C., Papapetropoulos, A., Erdelyi, K., Olah, G., Módis, K., Panopoulos, P., Asimakopoulou, A., Gerö, D., Sharina, I., Martin, E., and C. Szabo. 2012. Hydrogen sulfide and nitric oxide are mutually dependent in the regulation of angiogenesis and endothelium-dependent vasorelaxation. *Proc Natl Acad Sci USA* 109(23):9161–9166.

Coletta, C. and C. Szabo. 2013. Potential role of hydrogen sulfide in the pathogenesis of vascular dysfunction in septic shock. *Curr Vasc Pharmacol* 11(2):208–221.

Collin, M., Anuar, F. B., Murch, O., Bhatia, M., Moore, P. K., and C. Thiemermann. 2005. Inhibition of endogenous hydrogen sulfide formation reduces the organ injury caused by endotoxemia. *Br J Pharmacol* 146(4):498–505.

Cooper, C. E. and G. C. Brown. 2008. The inhibition of mitochondrial cytochrome oxidase by the gases carbon monoxide, nitric oxide, hydrogen cyanide and hydrogen sulfide: Chemical mechanism and physiological significance. *J Bioenerg Biomembr* 40(5):533–539.

Denizalti, M., Bozkurt, T. E., Akpulat, U., Sahin-Erdemli, I., and N. Abacioglu. 2011. The vasorelaxant effect of hydrogen sulfide is enhanced in streptozotocin-induced diabetic rats. *Naunyn Schmiedebergs Arch Pharmacol* 383(5):509–517.

Distrutti, E., Sediari, L., Mencarelli, A., Renga, B., Orlandi, S., Russo, G., Caliendo, G., Santagada, V., Cirino, G., Wallace, J. L., and S. Fiorucci. 2006. 5-Amino-2-hydroxybenzoic acid 4-(5-thioxo-5H-[1,2]dithiol-3yl)-phenyl ester (ATB-429), a hydrogen sulfide-releasing derivative of mesalamine, exerts antinociceptive effects in a model of postinflammatory hypersensitivity. *J Pharmacol Exp Ther* 319(1):447–458.

Distrutti, E., Mencarelli, A., Santucci, L., Renga, B., Orlandi, S., Donini, A., Shah, V., and S. Fiorucci. 2008. The methionine connection: homocysteine and hydrogen sulfide exert opposite effects on hepatic microcirculation in rats. *Hepatology* 47(2): 659–667.

Dudkina, N. V., Kouril, R., Peters, K., Braun, H. P., and E. J. Boekema. 2010. Structure and function of mitochondrial supercomplexes. *Biochim Biophys Acta* 1797(6–7):664–670.

Ebrahimkhani, M. R., Mani, A. R., and K. Moore. 2005. Hydrogen sulphide and the hyperdynamic circulation in cirrhosis: A hypothesis. *Gut* 54(12):1668–1671.

Ejaz, S., Chekarova, I., Cho, J. W., Lee, S. Y., Ashraf, S., and C. W. Lim. 2009. Effect of aged garlic extract on wound healing: A new frontier in wound management. *Drug Chem Toxicol* 32(3):191–203.

El-Seweidy, M. M., Sadik, N. A., and O. G. Shaker. 2011. Role of sulfurous mineral water and sodium hydrosulfide as potent inhibitors of fibrosis in the heart of diabetic rats. *Arch Biochem Biophys* 506(1):48–57.

Fan, H. N., Wang, H. J., Ren, L., Ren, B., Dan, C. R., Li, Y. F., Hou, L. Z., and Y. Deng. 2013b. Decreased expression of p38 MAPK mediates protective effects of hydrogen sulfide on hepatic fibrosis. *Eur Rev Med Pharmacol Sci* 17(5):644–652.

Fan, H.N, Wang, H. J., Yang-Dan, C. R., Ren, L., Wang, C., Li, Y. F., and Y. Deng. 2013a. Protective effects of hydrogen sulfide on oxidative stress and fibrosis in hepatic stellate cells. *Mol Med Rep* 7(1):247–253.

Fiorucci, S., Antonelli, E., Mencarelli, A., Orlandi, S., Renga, B., Rizzo, G., Distrutti, E., Shah, V., and A. Morelli. 2005. The third gas: H$_2$S regulates perfusion pressure in both the isolated and perfused normal rat liver and in cirrhosis. *Hepatology* 42(3): 539–548.

Fiorucci, S., Orlandi, S., Mencarelli, A., Caliendo, G., Santagada, V., Distrutti, E., Santucci, L., Cirino, G., and J. L. Wallace. 2007. Enhanced activity of a hydrogen sulphide-releasing derivative of mesalamine (ATB-429) in a mouse model of colitis. *Br J Pharmacol* 150(8):996–1002.

Fu, M., Zhang, W., Wu, L., Yang, G., Li, H., and R. Wang. 2012. Hydrogen sulfide (H$_2$S) metabolism in mitochondria and its regulatory role in energy production. *Proc Natl Acad Sci USA* 109(8):2943–2948.

Gaill, F. 1993. Aspects of life development at deep sea hydrothermal vents. *FASEB J* 7(6):558–565.

George, T. J., Arnaoutakis, G. J., Beaty, C. A., Jandu, S. K., Santhanam, L., Berkowitz, D. E., and A. S. Shah. 2012. J Inhaled hydrogen sulfide improves graft function in an experimental model of lung transplantation. *Surg Res* 178(2):593–600.

Gerö, D., Szoleczky, P., Suzuki, K., Módis, K., Oláh, G., Coletta, C., and C. Szabo. 2013. Cell-based screening identifies paroxetine as an inhibitor of diabetic endothelial dysfunction. *Diabetes* 62(3):953–964.

Giacco, F. and M. Brownlee. 2010. Oxidative stress and diabetic complications. *Circ Res* 107(9):1058–1070.

Goubern, M., Andriamihaja, M., Nubel, T., Blachier, F., and F. Bouillaud. 2007. Sulfide, the first inorganic substrate for human cells. *FASEB J* 21(8):1699–1706.

Guidotti, T. L. 1994. Occupational exposure to hydrogen sulfide in the sour gas industry: Some unresolved issues. *Int Arch Occup Environ Health* 66(3):153–160.

Hara, M. R., Agrawal, N., Kim, S. F., Cascio, M. B., Fujimuro, M., Ozeki, Y., Takahashi, M.et al. 2005. *S*-nitrosylated GAPDH initiates apoptotic cell death by nuclear translocation following Siah1 binding. *Nat Cell Biol* 7(7):665–674.

Haun, S. E., Segeleon, J. E., Trapp, V. L., Clotz, M. A., and L. A. Horrocks. 1996. Inosine mediates the protective effect of adenosine in rat astrocyte cultures subjected to combined glucose-oxygen deprivation. *J Neurochem* 67(5):2051–2059.

Helmy, N., Prip-Buus, C., Vons, C., Lenoir, V., Abou-Hamdan, A., Guedouari-Bounihi, H., Lombès, A., and F. Bouillaud. 2014. Oxidation of hydrogen sulfide by human liver mitochondria. *Nitric Oxide* 41:105–112.

Hildebrandt, T. M. and M. K. Grieshaber. 2008. Three enzymatic activities catalyze the oxidation of sulfide to thiosulfate in mammalian and invertebrate mitochondria. *FEBS J* 275(13):3352–3361.

Hill, B. C., Woon, T. C., Nicholls, P., Peterson, J., Greenwood, C., and A. J. Thomson. 1984. Interactions of sulphide and other ligands with cytochrome c oxidase. An electron-paramagnetic-resonance study. *Biochem J* 224(2):591–600.

Hosoki, R., Matsuki, N., and H. Kimura. 1997. The possible role of hydrogen sulfide as an endogenous smooth muscle relaxant in synergy with nitric oxide. *Biochem Biophys Res Commun* 237(3):527–531.

Hui, Y., Du, J., Tang, C., Bin, G., and H. Jiang. 2003. Changes in arterial hydrogen sulfide (H(2)S) content during septic shock and endotoxin shock in rats. *J Infect* 47(2):155–160.

Insko, M. A., Deckwerth, T. L, Hill, P., Toombs, C. F., and C. Szabo. 2009. Detection of exhaled hydrogen sulphide gas in rats exposed to intravenous sodium sulphide. *Br J Pharmacol* 157(6):944–951.

Isenberg, J. S., Jia, Y., Field, L., Ridnour, L. A., Sparatore, A., Del Soldato, P., Sowers, A. L. et al. 2007. Modulation of angiogenesis by dithiolethione-modified NSAIDs and valproic acid. *Br J Pharmacol* 151(1):63–72.

Ishigami, M., Hiraki, K., Umemura, K., Ogasawara, Y., Ishii, K., and H. Kimura. 2009. A source of hydrogen sulfide and a mechanism of its release in the brain. *Antioxid Redox Signal* 11(2):205–214.

Jha, S., Calvert, J. W., Duranski, M. R., Ramachandran, A., and D. J. Lefer. 2008. Hydrogen sulfide attenuates hepatic ischemia-reperfusion injury: Role of antioxidant and antiapoptotic signaling. *Am J Physiol Heart Circ Physiol* 295(2):H801–H806.

Jiang, B., Tang, G., Cao, K., Wu, L., and R. Wang. 2010. Molecular mechanism for H(2) S-induced activation of K(ATP) channels. *Antioxid Redox Signal* 12(10):1167–1178.

Jurkowitz, M. S., Litsky, M. L., Browning, M. J., and C. M. Hohl. 1998. Adenosine, inosine, and guanosine protect glial cells during glucose deprivation and mitochondrial inhibition: Correlation between protection and ATP preservation. *J Neurochem* 71(2):535–548.

Jurkowska, H., Roman, H. B., Hirschberger, L. L., Sasakura, K., Nagano, T., Hanaoka, K., Krijt, J., and H. M. Stipanuk. 2014. Primary hepatocytes from mice lacking cysteine dioxygenase show increased cysteine concentrations and higher rates of metabolism of cysteine to hydrogen sulfide and thiosulfate. *Amino Acids* 46(5):1353–1365.

Kabil, O. and R. Banerjee. 2010. Redox biochemistry of hydrogen sulfide. *J Biol Chem* 285(29):21903–21907.

Kamijo, Y., Takai, M., Fujita, Y., Hirose, Y., Iwasaki, Y., and S. Ishihara. 2013. A multicenter retrospective survey on a suicide trend using hydrogen sulfide in Japan. *Clin Toxicol (Phila)* 51(5):425–258.

Kaneko, Y., Kimura, Y., Kimura, H., and I. Niki. 2006. L-cysteine inhibits insulin release from the pancreatic b-cell: Possible involvement of metabolic production of hydrogen sulfide, a novel gasotransmitter. *Diabetes* 55(5):1391–1397.

Kaneko, Y., Kimura, T., Taniguchi, S., Souma, M., Kojima, Y., Kimura, Y., Kimura, H., and I. Niki. 2009. Glucose-induced production of hydrogen sulfide may protect the pancreatic b-cells from apoptotic cell death by high glucose. *FEBS Lett* 583(2):377–382.

Kang, K., Zhao, M., Jiang, H., Tan, G., Pan, S., and X. Sun. 2009. Role of hydrogen sulfide in hepatic ischemia-reperfusion-induced injury in rats. *Liver Transpl* 15(10):1306–1314.

Kaschula, C. H., Hunter, R., Hassan, H. T., Stellenboom, N., Cotton, J., Zhai, X. Q., and M. I. Parker. 2011. Anti-proliferation activity of synthetic ajoene analogues on cancer celllines. *Anticancer Agents Med Chem* 11(3):260–266.

Kashfi, K. and K. R. Olson. 2013. Biology and therapeutic potential of hydrogen sulfide and hydrogen sulfide-releasing chimeras. *Biochem Pharmacol* 85(5):689–703.

Kato, S., Takano, Y., Kakegawa, T., Oba, H., Inoue, K., Kobayashi, C., Utsumi, M. et al. 2010. Biogeography and biodiversity in sulfide structures of active and inactive vents at deep-sea hydrothermal fields of the Southern Mariana Trough. *Appl Environ Microbiol* 76(9):2968–2979.

Kilbourn, R. G., Traber, D. L., and C. Szabó. 1997. Nitric oxide and shock. *Dis Mon* 43(5):277–348.

Kimura, H. 2014. Signaling molecules: Hydrogen sulfide and polysulfide. *Antioxid Redox Signal* 22(5):362–376 (Epub ahead of print).

Kimura, Y., Mikami, Y., Osumi, K., Tsugane, M., Oka, J., and H. Kimura. 2013. Polysulfides are possible H_2S-derived signaling molecules in rat brain. *FASEB J* 27(6):2451–2457.

Lagoutte, E., Mimoun, S., Andriamihaja, M., Chaumontet, C., Blachier, F., and F. Bouillaud. 2010. Oxidation of hydrogen sulfide remains a priority in mammalian cells and causes reverse electron transfer in colonocytes. *Biochim Biophys Acta* 1797(8):1500–1511.

Lai, K. C., Kuo, C. L., Ho, H. C., Yang, J. S., Ma, C. Y., Lu, H. F., Huang, H. Y., Chueh, F. S., Yu, C. C., and J. G. Chung. 2012. Diallyl sulfide, diallyl disulfide and diallyl trisulfide affect drug resistant gene expression in colo 205 human colon cancer cells in vitro and in vivo. *Phytomedicine* 19(7):625–630.

Le Trionnaire, S., Perry, A., Szczesny, B., Szabo, C., Winyard, G. P., Whatmore, L. J., Wood, E. M., and M. Whiteman. 2014. The synthesis and functional evaluation of a mitochondria targeted hydrogen sulfide donor, (10-oxo-10-(4-(3-thioxo-3H-1,2-dithiol-5-yl)phenoxy) decyl)triphenylphosphonium bromide (AP39). *MedChemComm* 5(6):728–736.

Lee, M., Tazzari, V., Giustarini, D., Rossi, R., Sparatore, A., Del Soldato, P., McGeer, E., and P. L. McGeer. 2010. Effects of hydrogen sulfide-releasing L-DOPA derivatives on glial activation: Potential for treating Parkinson disease. *J Biol Chem* 285(23):17318–17328.

Li, J., Zhang, G., Cai, S., and A. N. Redington. 2008. Effect of inhaled hydrogen sulfide on metabolic responses in anesthetized, paralyzed, and mechanically ventilated piglets. *Pediatr Crit Care Med* 9(1):110–112.

Li, L., Bhatia, M., Zhu, Y. Z., Zhu, Y. C., Ramnath, R. D., Wang, Z. J., Anuar, F. B., Whiteman, M., Salto-Tellez, M., and P. K. Moore. 2005. Hydrogen sulfide is a novel mediator of lipopolysaccharide-induced inflammation in the mouse. *FASEB J* 19(9):1196–1198.

Li, L., Lee, W., Lee, W. J., Auh, J. H., Kim, S. S., and J. Yoon. 2010. Extraction of allyl isothiocyanate from wasabi (Wasabia japonica Matsum) using supercritical carbon dioxide. *Food Sci Biotechnol* 19:405–410.

Li, L., Rossoni, G., Sparatore, A., Lee, L. C., Del Soldato, P., and P. K. Moore. 2007. Anti-inflammatory and gastrointestinal effects of a novel diclofenac derivative. *Free Radic Biol Med* 42(5):706–719.

Li, L., Salto-Tellez, M., Tan, C. H., Whiteman, M., and, P. K. Moore. 2009a. GYY4137, a novel hydrogen sulfide-releasing molecule, protects against endotoxic shock in the rat. *Free Radic Biol Med* 47(1):103–113.

Li, L., Whiteman, M., Guan, Y. Y., Neo, K. L., Cheng, Y., Lee, S. W., Zhao, Y., Baskar, R., Tan, C. H., and P. K. Moore. 2008. Characterization of a novel, water-soluble hydrogen sulfide-releasing molecule (GYY4137): New insights into the biology of hydrogen sulfide. *Circulation* 117(18):2351–2360.

Li, L., Whiteman, M., and P. K. Moore. 2009b. Dexamethasone inhibits lipopolysaccharide-induced hydrogen sulphide biosynthesis in intact cells and in an animal model of endotoxic shock. *J Cell Mol Med* 13(8B):2684–2692.

Li, Y., Wang, H. J., Song, X. F., and H. L. Yang. 2011. Influence of hydrogen sulfide on important organs in rats with severe burn. *Zhonghua Shao Shang Za Zhi* 27(1):54–58.

Litsky, M. L., Hohl, C. M., Lucas, J. H., and M. S. Jurkowitz. 1999. Inosine and guanosine preserve neuronal and glial cell viability in mouse spinal cord cultures during chemical hypoxia. *Brain Res* 821(2):426–432.

Liu, F., Chen, D. D., Sun, X., Xie, H. H., Yuan, H., Jia, W., and A. F. Chen. 2014. Hydrogen sulfide improves wound healing via restoration of endothelial progenitor cell functions and activation of angiopoietin-1 in type 2 diabetes. *Diabetes* 63(5):1763–1778.

Mani, S., Cao, W., Wu, L., and R. Wang. 2014. Hydrogen sulfide and the liver. *Nitric Oxide* 41C:62–71.

Manna, P., Gungor, N., McVie, R., and S. K. Jain. 2014. Decreased cystathionine-γ-lyase (CSE) activity in livers of type 1 diabetic rats and peripheral blood mononuclear cells (PBMC) of type 1 diabetic patients. *J Biol Chem* 2014 289(17):11767–11778.

Matsuda, H., Ochi, M., Nagatomo, A., and M. Yoshikawa. 2007. Effects of allyl isothiocyanate from horseradish on several experimental gastric lesions in rats. *Eur J Pharmacol* 561(1–3):172–181.

McCook, O., Radermacher, P., Volani, C., Asfar, P., Ignatius, A., Kemmler, J., Möller, P. et al. 2014. H₂S during circulatory shock: Some unresolved questions. *Nitric Oxide* pii: S1089-8603(14)00183-9.

Meier, M., Oliveriusova, J., Kraus, J. P., and P. Burkhard. 2003. Structural insights into mutations of cystathionine β-synthase. *Biochim Biophys Acta* 1647(1–2):206–213.

Mikami, Y., Shibuya, N., Kimura, Y., Nagahara, N., Ogasawara, Y., and H. Kimura. 2011. Thioredoxin and dihydrolipoic acid are required for 3-mercaptopyruvate sulfurtransferase to produce hydrogen sulfide. *Biochem J* 439(3):479–485.

Miles, E. W. and J. P. Kraus. 2004. Cystathionine β-synthase: Structure, function, regulation, and location of homocystinuria-causing mutations. *J Biol Chem* 279(29): 29871–29874.

Módis, K., Asimakopoulou, A., Coletta, C., Papapetropoulos, A., and C. Szabo. 2013c. Oxidative stress suppresses the cellular bioenergetic effect of the 3-mercaptopyruvate sulfurtransferase/hydrogen sulfide pathway. *Biochem Biophys Res Commun* 433(4):401–407.

Módis, K., Bos, E. M., Calzia, E., van Goor, H., Coletta, C., Papapetropoulos, A., Hellmich, M. R., Radermacher, P., Bouillaud, F., amd C. Szabo. 2014. Regulation of mitochondrial bioenergetic function by hydrogen sulfide. Part II. Pathophysiological and therapeutic aspects. *Br J Pharmacol* 171(8):2123–2146.

Módis, K., Coletta, C., Erdélyi, K., Papapetropoulos, A., and C. Szabo. 2013a. Intramitochondrial hydrogen sulfide production by 3-mercaptopyruvate sulfurtransferase maintains mitochondrial electron flow and supports cellular bioenergetics. *FASEB J* 27(2):601–611.

Módis, K., Gero, D., Nagy, N., Szoleczky, P., Tóth, Z. D., and C. Szabó. 2009. Cytoprotective effects of adenosine and inosine in an in vitro model of acute tubular necrosis. *Br J Pharmacol* 158(6):1565–1578.

Módis, K., Panopoulos, P., Coletta, C., Papapetropoulos, A., and C. Szabo. 2013d. Hydrogen sulfide-mediated stimulation of mitochondrial electron transport involves inhibition of the mitochondrial phosphodicstcrasc 2A, clevation of cAMP and activation of protcin kinase A. *Biochem Pharmacol* 86(9):1311–1319.

Módis, K., Wolanska, K., and R. Vozdek. 2013b. Hydrogen sulfide in cell signaling, signal transduction, cellular bioenergetics and physiology in C. elegans. *Gen Physiol Biophys* 32(1):1–22.

Mok, Y. Y. and P. K. Moore. 2008. Hydrogen sulphide is pro-inflammatory in haemorrhagic shock. *Inflamm Res* 57(11):512–518.

Mok, Y. Y., Atan, M. S., Yoke Ping, C., Zhong Jing, W., Bhatia, M., Moochhala, S., and P. K. Moore. 2004. Role of hydrogen sulphide in haemorrhagic shock in the rat: Protective effect of inhibitors of hydrogen sulphide biosynthesis. *Br J Pharmacol* 143(7):881–889.

Moody, T. W., Switzer, C., Santana-Flores, W., Ridnour, L. A., Berna, M., Thill, M., Jensen, R. T. et al. 2010. Dithiolethione modified valproate and diclofenac increase E-cadherin expression and decrease proliferation of non-small cell lung cancer cells. *Lung Cancer* 68(2):154–160.

Morsy, M. A., Ibrahim, S. A., Abdelwahab, S. A., Zedan, M. Z., and I. H. Elbitar. 2010. Curative effects of hydrogen sulfide against acetaminophen-induced hepatotoxicity in mice. *Life Sci* 87(23–26):692–698.

Mukherjee, S., Lekli, I., Ray, D., Gangopadhyay, H., Raychaudhuri, U., and D. K. Das. 2010. Comparison of the protective effects of steamed and cooked broccolis on ischaemia-reperfusion-induced cardiac injury. *Br J Nutr* 103(6):815–823.

Murphy, M. P. 2009. How mitochondria produce reactive oxygen species. *Biochem J* 417(1):1–13.

Mustafa, A. K., Sikka, G., Gazi, S. K., Steppan, J., Jung, S. M., Bhunia, A. K., Barodka, V. M. et al. 2011. Hydrogen sulfide as endothelium-derived hyperpolarizing factor sulfhydrates potassium channels. *Circ Res* 109(11):1259–1268.

Muzaffar, S., Jeremy, J. Y., Sparatore, A., Del Soldato, P., Angelini, G. D., and N. Shukla. 2008. H$_2$S-donating sildenafil (ACS6) inhibits superoxide formation and gp91phox expression in arterial endothelial cells: Role of protein kinases A and G. *Br J Pharmacol* 155(7):984–994.

Nagy, P. 2013. Kinetics and mechanisms of thiol-disulfide exchange covering direct substitution and thiol oxidation-mediated pathways. *Antioxid Redox Signal* 18(13):1623–1641.

Nagy, P. and C. C. Winterbourn. 2010. Rapid reaction of hydrogen sulfide with the neutrophil oxidant hypochlorous acid to generate polysulfides. *Chem Res Toxicol* 23(10):1541–1543.

Nagy, P., Pálinkás, Z., Nagy, A., Budai, B., Tóth, I., and A. Vasas. 2014. Chemical aspects of hydrogen sulfide measurements in physiological samples. *Biochim Biophys Acta* 1840(2):876–891.

Namekata, K., Enokido, Y., Ishii, I., Nagai, Y., Harada, T., and H. Kimura. 2004. Abnormal lipid metabolism in cystathionine β-synthase-deficient mice, an animal model for hyperhomocysteinemia. *J Biol Chem* 279(51):52961–52969.

Nicholls, D. and S. Ferguson. 2013. *Bioenergetics*, 4th Edition, Academic Press, London, UK.

Norris, E. J., Culberson, C. R., Narasimhan, S., and M. G. Clemens. 2011. The liver as a central regulator of hydrogen sulfide. *Shock* 36(3):242–250.

Norris, E. J., Feilen, N., Nguyen, N. H., Culberson, C. R., Shin, M. C., Fish, M., and M. G. Clemens. 2013. Hydrogen sulfide modulates sinusoidal constriction and contributes to hepatic microcirculatory dysfunction during endotoxemia. *Am J Physiol Gastrointest Liver Physiol* 304(12):G1070–G1078.

Ogasawara, Y., Isoda, S., and S. Tanabe. 1994. Tissue and subcellular distribution of bound and acid-labile sulfur, and the enzymic capacity for sulfide production in the rat. *Biol Pharm Bull* 17(12):1535–1542.

Okamoto, M., Yamaoka, M., Takei, M., Ando, T., Taniguchi, S., Ishii, I., Tohya, K., Ishizaki, T., Niki, I., and T. Kimura. 2013. Endogenous hydrogen sulfide protects pancreatic β-cells from a high-fat diet-induced glucotoxicity and prevents the development of type 2 diabetes. *Biochem Biophys Res Commun* 442(3–4):227–233.

Oren, T. 2014. *Liver Metabolism and Fatty Liver Disease*, Series: Oxidative Stress and Disease, CRC Press, Taylor & Francis, Hoboken, NJ.

Ou, H. C., Tzang, B. S., Chang, M. H., Liu, C. T., Liu, H. W., Lii, C. K., Bau, D. T., Chao, P. M., and W. W Kuo. 2010. Cardiac contractile dysfunction and apoptosis in streptozotocin-induced diabetic rats are ameliorated by garlic oil supplementation. *J Agric Food Chem* 58(19):10347–10355.

Padiya, R., Khatua, T. N., Bagul, P. K., Kuncha, M., and S. K. Banerjee. 2011. Garlic improves insulin sensitivity and associated metabolic syndromes in fructose fed rats. *Nutr Metab* (*Lond*) 8:53.

Paul, B. D. and S. H. Snyder. 2012. H$_2$S signalling through protein sulfhydration and beyond. *Nat Rev Mol Cell Biol* 13(8):499–507.

Petersen, L. C. 1977. The effect of inhibitors on the oxygen kinetics of cytochrome c oxidase. *Biochim Biophys Acta* 460(2):299–307.

Pietri, R., Román-Morales, E., and J. López-Garriga. 2011. Hydrogen sulfide and hemeproteins: Knowledge and mysteries. *Antioxid Redox Signal* 15(2):393–404.

Rajpal, G., Schuiki, I., Liu, M., Volchuk, A., and P. Arvan. 2012. Action of protein disulfide isomerase on proinsulin exit from endoplasmic reticulum of pancreatic β-cells. *J Biol Chem.* 287(1):43–47.

Ratnam, S., Maclean, K. N., Jacobs, R. L., Brosnan, M. E., Kraus, J. P., and J. T. Brosnan. 2002. Hormonal regulation of cystathionine β-synthase expression in liver. *J Biol Chem* 277(45):42912–42918.

Reiffenstein, R. J., Hulbert, W. C., and S. H. Roth. 1992. Toxicology of hydrogen sulfide. *Annu Rev Pharmacol Toxicol* 32:109–134.

Robert, K., Nehmé, J., Bourdon, E., Pivert, G., Friguet, B., Delcayre, C., Delabar, J. M., and N. Janel. 2005. Cystathionine β synthase deficiency promotes oxidative stress, fibrosis, and steatosis in mice liver. *Gastroenterology* 128(5):1405–1415.

Sardanelli, A. M., Technikova-Dobrova, Z., Speranza, F., Mazzocca, A. Scacco, S., and S. Papa. 1996. Topology of the mitochondrial cAMP-dependent protein kinase and its substrates. *FEBS Lett* 396(2–3):276–278.

Searcy, D. G. and S. H. Lee. 1998. Sulfur reduction by human erythrocytes. *J Exp Zool* 282(3):310–322.

Shaik, I. H., George, J. M., Thekkumkara, T. J., and R. Mehvar. 2008. Protective effects of diallyl sulfide, a garlic constituent, on the warm hepatic ischemia-reperfusion injury in a rat model. *Pharm Res* 25(10):2231–2242.

Shang, Z., Lu, C., Chen, S., Hua, L., and R. Qian. 2012. Effect of H_2S on the circadian rhythm of mouse hepatocytes. *Lipids Health Dis* 11:23.

Shapiro, T. A., Fahey, J. W., Wade, K. L., Stephenson, K. K., and P. Talalay. 2001. Chemoprotective glucosinolates and isothiocyanates of broccoli sprouts: Metabolism and excretion in humans. *Cancer Epidemiol Biomarkers Prev* 10(5):501–508.

Shibuya, N., Tanaka, M., Yoshida, M., Ogasawara, Y., Togawa, T., Ishii, K., and H. Kimura. 2009. 3-Mercaptopyruvate sulfurtransferase produces hydrogen sulfide and bound sulfane sulfur in the brain. *Antioxid Redox Signal* 11(4):703–714.

Shirozu, K., Tokuda, K., Marutani, E., Lefer, D., Wang, R., and F. Ichinose. 2014. Cystathionine γ-lyase deficiency protects mice from galactosamine/lipopolysaccharide-induced acute liver failure. *Antioxid Redox Signal* 20(2):204–216.

Sidhapuriwala, J., Li, L., Sparatore, A., Bhatia, M., and P. K. Moore. 2007. Effect of S-diclofenac, a novel hydrogen sulfide releasing derivative, on carrageenan-induced hindpaw oedema formation in the rat. *Eur J Pharmacol* 569(1–2):149–154.

Siebert, N., Cantré, D., Eipel, C., and B. Vollmar. 2008. H_2S contributes to the hepatic arterial buffer response and mediates vasorelaxation of the hepatic artery via activation of K(ATP) channels. *Am J Physiol Gastrointest Liver Physiol* 295(6): G1266–G1273.

Siekevitz, P. (1957). Powerhouse of the cell. *Sci Am* 197:131–140.

Singh, S. and R. Banerjee. 2011. PLP-dependent H_2S biogenesis. *Biochim Biophys Acta* 1814(11):1518–1527.

Smith, R. A., Hartley, R. C., and M. P. Murphy. 2011. Mitochondria-targeted small molecule therapeutics and probes. *Antioxid Redox Signal* 15(12):3021–3038.

Sparatore, A., Santus, G., Giustarini, D., Rossi, R., and P. Del Soldato. 2011. Therapeutic potential of new hydrogen sulfide-releasing hybrids. *Expert Rev Clin Pharmacol* 4(1):109–121.

Stipanuk, M. H. and P. W. Beck 1982. Characterization of the enzymic capacity for cysteine desulphhydration in liver and kidney of the rat. *Biochem J* 206(2):267–277.

Suzuki, K., Olah, G., Modis, K., Coletta, C., Kulp, G., Gero, D., Szoleczky, P. et al. 2011. Hydrogen sulfide replacement therapy protects the vascular endothelium in hyperglycemia by preservingmitochondrial function. *Proc Natl Acad Sci USA* 108(33):13829–13834.

Sylvan, J. B., Toner, B. M., and K. J. Edwards. 2012. Life and death of deep-sea vents: Bacterial diversity and ecosystem succession on inactive hydrothermal sulfides. *MBio* 3(1):e00279-11.

Szabo, C. 2012. Roles of hydrogen sulfide in the pathogenesis of diabetes mellitus and its complications. *Antioxid Redox Signal* 17(1):68–80.

Szabo, C. 2007. Hydrogen sulphide and its therapeutic potential. *Nat Rev Drug Discov* 6(11):917–935 (Review).

Szabó, C. and K. Módis. 2010. Pathophysiological roles of peroxynitrite in circulatory shock. *Shock* 34(Suppl 1):4–14.

Szabo, C., Coletta, C., Chao, C., Módis, K., Szczesny, B., Papapetropoulos, A., and M. R. Hellmich. 2014. Regulation of mitochondrial bioenergetic function by hydrogen sulfide. Part I. Biochemical and physiological mechanisms. *Br J Pharmacol* 171(8):2099–2122. (Review).

Szabo, C., Ransy, C., Módis, K., Andriamihaja, M., Murghes, B., Coletta, C., Olah, G., Yanagi, K., and F. Bouillaud. 2014. Regulation of mitochondrial bioenergetic function by hydrogen sulfide. Part I. Biochemical and physiological mechanisms. *Br J Pharmacol* 171(8):2099–2122.

Szabó, C., Southan, G. J., and C. Thiemermann. 1994. Beneficial effects and improved survival in rodent models of septic shock with S-methylisothiourea sulfate, a potent and selective inhibitor of inducible nitric oxide synthase. *Proc Natl Acad Sci USA* 91(26):12472–12476.

Szabó, G., Veres, G., Radovits, T., Gero, D., Módis, K., Miesel-Gröschel, C., Horkay, F., Karck, M., and C. Szabó. 2011. Cardioprotective effects of hydrogen sulfide. *Nitric Oxide* 25(2):201–210.

Szczesny, B., Módis, K., Yanagi, K., Coletta, C., Le Trionnaire, S., Perry, A., Wood, M. E., Whiteman, M., and C. Szabo. 2014. AP39, a novel mitochondria-targeted hydrogen sulfide donor, stimulates cellular bioenergetics, exerts cytoprotective effects and protects against the loss of mitochondrial DNA integrity in oxidatively stressed endothelial cells in vitro. *Nitric Oxide* 41:120–130

Tan, G., Pan, S., Li, J., Dong, X., Kang, K., Zhao, M., Jiang, X., Kanwar, J. R., Qiao, H., Jiang, H., and Sun X. Hydrogen sulfide attenuates carbon tetrachloride-induced hepatotoxicity, liver cirrhosis and portal hypertension in rats. *PLoS One* 2011;6(10):e25943.

Teng, H., Wu, B., Zhao, K., Yang, G., Wu, L., and R. Wang. 2013. Oxygen-sensitive mitochondrial accumulation of cystathionine β-synthase mediated by Lon protease. *Proc Natl Acad Sci USA* 110(31):12679–12684.

Thompson, R. W., Valentine, H. L., and M. W. Valentine. 2003. Cytotoxic mechanisms of hydrosulfide anion and cyanide anion in primary rat hepatocyte cultures. *Toxicology* 188(2–3):149–159.

Tian, M., Hanley, A. B., and M. W. Dodds. 2013. Allyl isothiocyanate from mustard seed is effective in reducing the levels of volatile sulfur compounds responsible for intrinsic oral malodor. *J Breath Res* 7(2):026001.

Tokuda, K., Kida, K., Marutani, E., Crimi, E., Bougaki, M., Khatri, A., Kimura, H., and F. Ichinose. 2012. Inhaled hydrogen sulfide prevents endotoxin-induced systemic inflammation and improves survival by altering sulfide metabolism in mice. *Antioxid Redox Signal* 17(1):11–21.

Toombs, C. F., Insko, M. A., Wintner, E. A., Deckwerth, T. L, Usansky, H., Jamil, K., Goldstein, B., Cooreman, M., and C. Szabo. 2010. Detection of exhaled hydrogen sulphide gas in healthy human volunteers during intravenous administration of sodium sulphide. *Br J Clin Pharmacol* 69(6):626–636.

Turner, R. and S. Fairhurst. 1980. *Toxicology of Substances in Relation to Major Hazards: Hydrogen Sulphide.* HSE Books, London, UK.

Uhteg, L. C. and J. Westley. 1979. Purification and steady-state kinetic analysis of yeast thiosulfate reductase. *Arch Biochem Biophys* 195(1):211–222.

van den Berg, E., Pasch, A., Westendorp, W. H, Navis, G., Brink, E. J., Gans, R. O., van Goor, H., and S. J. Bakker. 2014. Urinary sulfur metabolites associate with a favorable cardiovascular risk profile and survival benefit in renal transplant recipients. *J Am Soc Nephrol* 25(6):1303–1312.

Wallace, J. L. 2012. Hydrogen sulfide: A rescue molecule for mucosal defence and repair, *Dig Dis Sci* 57:1432–1434.

Wallace, J. L., Caliendo, G., Santagada, V., and G. Cirino. 2010. Markedly reduced toxicity of a hydrogen sulphide-releasing derivative of naproxen (ATB-346). *Br J Pharmacol* 159(6):1236–1246.

Wang, C., Han, J., Xiao, L., Jin, C. E., Li, D. J., and Z. Yang. 2014. Role of hydrogen sulfide in portal hypertension and esophagogastric junction vascular disease. *World J Gastroenterol* 20(4):1079–1087.

Wang, Q., Wang, X. L., Liu, H. R., Rose, P., and Y. Z. Zhu. 2010. Protective effects of cysteine analogues on acute myocardial ischemia: Novel modulators of endogenous H_2S production. *Antioxid Redox Signal* 12(10):1155–1165.

Wang, R. 2012. Physiological implications of hydrogen sulfide: A whiff exploration that blossomed. *Physiol Rev* 92(2):791–896.

Wang, R. 2014. Gasotransmitters: Growing pains and joys. *Trends Biochem Sci* 39(5):227–232.

Watanabe, M., Osada, J., Aratani, Y., Kluckman, K., Reddick, R., Malinow, M. R., and N. Maeda. 1995. Mice deficient in cystathionine β-synthase: Animal models for mild and severe homocyst(e)inemia. *Proc Natl Acad Sci USA* 92(5):1585–1589.

Whiteman, M., Armstrong, J. S., Chu, S. H., Jia-Ling, S., Wong, B. S., Cheung, N. S., Halliwell, B., and P. K. Moore. 2004. The novel neuromodulator hydrogen sulfide: An endogenous peroxynitrite 'scavenger'? *J Neurochem* 90(3):765–768.

Whiteman, M., Le Trionnaire, S., Chopra, M., Fox, B., and J. Whatmore. 2011. Emerging role of hydrogen sulfide in health and disease: Critical appraisal of biomarkers and pharmacological tools. *Clin Sci (Lond)* 121(11):459–488.

Wiliński, B., Wiliński, J., Somogyi, E., Góralska, M., and J. Piotrowska. 2010. Ramipril affects hydrogen sulfide generation in mouse liver and kidney. *Folia Biol (Krakow)* 58(3–4):177–180.

Wiliński, B., Wiliński, J., Somogyi, E., Góralska, M., and J. Piotrowska. 2011b. Paracetamol (acetaminophen) decreases hydrogen sulfide tissue concentration in brain but increases it in the heart, liver and kidney in mice. *Folia Biol (Krakow)* 59(1–2):41–44.

Wiliński, B., Wiliński, J., Somogyi, E., Piotrowska, J., and M. Góralska. 2011a. Atorvastatin affects the tissue concentration of hydrogen sulfide in mouse kidneys and other organs. *Pharmacol Rep* 63(1):184–188.

Wiliński, B., Wiliński, J., Somogyi, E., Piotrowska, J., and W. Opoka. 2013. Metformin raises hydrogen sulfide tissue concentrations in various mouse organs. *Pharmacol Rep* 65(3):737–742.

Wiliński, B., Wiliński, J., Somogyi, E., Piotrowska, J., Góralska, M., and B. Macura. 2011c. Carvedilol induces endogenous hydrogen sulfide tissue concentration changes in various mouse organs. *Folia Biol (Krakow)* 59(3–4):151–155.

Wu, L., Yang, W., Jia, X., Yang, G., Duridanova, D., Cao, K., and R. Wang. 2009. Pancreatic islet overproduction of H_2S and suppressed insulin release in Zucker diabetic rats. *Lab Invest* 89(1):59–67.

Yan, Y., Chen, C., Zhou, H., Gao, H., Chen, L., Chen, L., Gao, L., Zhao, R., and Y. Sun. 2013. Endogenous hydrogen sulfide formation mediates the liver damage in endotoxemic rats. *Res Vet Sci* 94(3):590–595

Yan, J., Teng, F., Chen, W., Ji, Y., and Z. Gu. 2012. Cystathionine β-synthase-derived hydrogen sulfide regulates lipopolysaccharide-induced apoptosis of the BRL rat hepatic cell line in vitro. *Exp Ther Med* 4(5):832–838.

Yang, G., Tang, G., Zhang, L., Wu, L., and R. Wang. 2011. The pathogenic role of cystathionine c-lyase/hydrogen sulfide in streptozotocin-induced diabetes in mice. *Am J Pathol* 179(2):869–879.

Yang, G., Wu, L., Jiang, B., Yang, W., Qi, J., Cao, K., Meng, Q. et al. 2008. H_2S as a physiologic vasorelaxant: Hypertension in mice with deletion of cystathionine γ-lyase. *Science* 322(5901):587–590.

Yang, G., Yang, W., Wu, L., and R. Wang. 2007. H_2S, endoplasmic reticulum stress, and apoptosis of insulin-secreting b cells. *J Biol Chem* 282(22):16567–16576.

Yang, G., Zhao, K., Ju, Y., Mani, S., Cao, Q., Puukila, S., Khaper, N., Wu, L., and R. Wang. 2013. Hydrogen sulfide protects against cellular senescence via *S*-sulfhydration of Keap1 and activation of Nrf2. *Antioxid Redox Signal* 18(15):1906–1919.

Ye, L., Dinkova-Kostova, A. T., Wade, K. L., Zhang, Y., Shapiro, T. A., and P. Talalay. 2002. Quantitative determination of dithiocarbamates in human plasma, serum, erythrocytes and urine: Pharmacokinetics of broccoli sprout isothiocyanates in humans. *Clin Chim Acta* 2002 316(1–2):43–53.

Yonezawa, D., Sekiguchi, F., Miyamoto, M., Taniguchi, E., Honjo, M., Masuko, T., Nishikawa, H., and A. Kawabata. 2007. A protective role of hydrogen sulfide against oxidative stress in rat gastric mucosal epithelium. *Toxicology* 241(1–2):11–18.

Yong, R. and D. G. Searcy. 2001. Sulfide oxidation coupled to ATP synthesis in chicken liver mitochondria. *Comp Biochem Physiol B Biochem Mol Biol* 129(1):129–137.

Yuan, P., Xue, H., Zhou, L., Qu, L., Li, C., Wang, Z., Ni, J., Yu, C., Yao, T., Huang, Y., Wang, R., and L. Lu. 2011. Rescue of mesangial cells from high glucose-induced overproliferation and extracellular matrix secretion by hydrogen sulfide. *Nephrol Dial Transplant* 26(7):2119–2126.

Yusuf, M., Kwong Huat, B. T., Hsu, A., Whiteman, M., Bhatia M. and P. K. Moore. 2005. Streptozotocin-induced diabetes in the rat is associated with enhanced tissue hydrogen sulfide biosynthesis. *Biochem Biophys Res Commun* 333(4):1146–1152.

Zeng, J., Lin, X., Fan, H., and C. Li. 2013. Hydrogen sulfide attenuates the inflammatory response in a mouse burn injury model. *Mol Med Rep* 8(4):1204–1208.

Zhang, H., Moochhala, S. M., and M. Bhatia. 2008. Endogenous hydrogen sulfide regulates inflammatory response by activating the ERK pathway in polymicrobial sepsis. *J Immunol* 181(6):4320–4331.

Zhang, H., Zhi, L., Moochhala, S., Moore, P. K., and M. Bhatia. 2007. Hydrogen sulfide acts as an inflammatory mediator in cecal ligation and puncture-induced sepsis in mice by upregulating the production of cytokines and chemokines via NF-κB. *Am J Physiol Lung Cell Mol Physiol* 292(4):L960–L971

Zhang, H., Zhi, L., Moore, P. K., and M. Bhatia. 2006. Role of hydrogen sulfide in cecal ligation and puncture-induced sepsis in the mouse. *Am J Physiol Lung Cell Mol Physiol* 290(6):L1193–L1201.

Zhang, J., Sio, S. W., Moochhala, S., and M. Bhatia. 2010. Role of hydrogen sulfide in severe burn injury-induced inflammation in mice. *Mol Med* 16(9–10):417–424.

Zhang, J., Xie, Y., Xu, Y., Pan, Y., and C. Shao. 2011. Hydrogen sulfide contributes to hypoxia-induced radioresistance on hepatoma cells. *J Radiat Res* 52(5):622–628.

Zhang, L., Yang, G., Untereiner, A., Ju, Y., Wu, L., and R. Wang. 2013. Hydrogen sulfide impairs glucose utilization and increases gluconeogenesis in hepatocytes. *Endocrinology* 154(1):114–126.

Zhang, Q., Fu, H., Zhang, H., Xu, F., Zou, Z., Liu, M., Wang, Q., Miao, M., and X. Shi. 2013. Hydrogen sulfide preconditioning protects rat liver against ischemia/reperfusion injury by activating Akt-GSK-3β signaling and inhibiting mitochondrial permeability transition. *PLoS One* 8(9):e74422.

Zhao, W., Zhang, J., Lu, Y., and R. Wang. 2001. The vasorelaxant effect of H_2S as a novel endogenous gaseous K(ATP) channel opener. *EMBO J* 20(21):6008–6016.

Zhu, J. X., Kalbfleisch, M., Yang, Y. X., Bihari, R., Lobb, I., Davison, M., Mok, A., Cepinskas, G., Lawendy, A. R., and A. Sener. 2012. Detrimental effects of prolonged warm renal ischaemia-reperfusion injury are abrogated by supplemental hydrogen sulphide: An analysis using real-time intravital microscopy and polymerase chain reaction. *BJU Int* 110(11 Pt C):E1218–E1227.

6 The Impact of Aging and Calorie Restriction on Liver Mitochondria

Jon J. Ramsey, José Alberto López-Domínguez, and Kevork Hagopian

CONTENTS

ABSTRACT

Impaired mitochondrial function is often considered to be a factor contributing to aging in post-mitotic tissues. However, there is considerable evidence that liver mitochondria also undergo age-related changes that may impact function. There is also evidence that mitochondrial and cristae numbers are decreased and mitochondrial volume is increased in hepatocytes with aging. Several studies have shown that mitochondrial respiration and membrane potential are decreased in hepatocytes from old versus young animals. Mitochondrial proton leak, reactive oxygen species production, and oxidative damage to mitochondrial macromolecules are increased with aging in liver. Furthermore, mitochondrial membrane fatty acid composition is altered with aging in a manner that increases susceptibility to peroxidation. Calorie restriction (CR) opposes many of these age-related changes in liver mitochondria. In particular, CR prevents age-related decreases in the number of mitochondria and cristae. There is also evidence that CR opposes age-related increases in mitochondrial proton leak and oxidative damage. Age-related changes in liver mitochondria are consistent with the idea that impaired mitochondrial function could contribute to liver aging.

INTRODUCTION

Aging is associated with many changes in energy metabolism, including a decrease in energy expenditure (Ramsey and Hagopian 2007). Mitochondria play a central role in energy metabolism, and liver mitochondria in particular may be important contributors to the age-related decline in energy expenditure at the whole animal level. This reflects the fact that liver is responsible for more than 20% of the total energy expenditure in humans (Elia 1992; Grande 1980; Ramsey et al. 2000), and thus, age-related changes in liver energy metabolism are expected to have a significant impact on whole animal energy metabolism. Mitochondria also play a prominent role in aging because they are a site of reactive oxygen species (ROS) production. The free radical theory (Harman 1956) is one of the most prominent theories of aging and oxidative damage to mitochondria is a major factor that contributes to age-related changes in liver, and whole animal energy metabolism. Over the past 50 years, several studies have investigated the influence of aging on liver mitochondria, have characterized typical mitochondrial changes with aging, and have also identified possible mechanisms which may contribute to loss of physiological function with aging. The purpose of this chapter is to provide an overview of this previous work and summarize available information regarding changes in liver mitochondrial structure and function with aging.

This chapter also provides a summary of the influence of calorie restriction (CR) on liver mitochondria. CR increases maximum life span and prevents or delays the onset of pathophysiological changes in a variety of animal species (Sohal and Weindruch 1996). The mechanism(s) responsible for life span extension with CR are not entirely known, although mitochondria play a central role in many of the proposed mechanisms. Several studies have investigated CR-induced changes in liver mitochondria as a step toward determining whether or not alterations in mitochondria may contribute to the beneficial effects of CR. This chapter provides a brief review of these studies and indicates which mitochondrial changes with aging are opposed by CR.

MITOCHONDRIAL NUMBER AND MORPHOLOGY

AGING

Age-related changes in mitochondrial number and structure (size, volume, surface area) provide potential indicators of impaired mitochondrial function with aging. In 1964, light and electron microscopy studies found that mitochondria from hepatocytes of old rats were fewer in number and larger in size than mitochondria from young adult animals (Tauchi et al. 1964). Studies using human liver tissue also found that mitochondrial number was decreased and mitochondrial size was increased in people more than 60 years of age when compared to individuals less than 50 years of age (Tauchi and Sato 1968). In this study, mitochondrial number per unit of tissue was decreased by approximately 36% and mitochondrial size was increased by approximately 45% in over 70-year-old compared under 50-year-old people. Electron microscopy studies using liver samples from mice (Herbener 1976; Tate and Herbener 1976) and humans (Sato and Tauchi 1975) and measures of mitochondrial DNA and protein as markers of mitochondrial content (Stocco and Hutson 1978) provide further support for a decrease in liver mitochondrial number with aging. A few studies, however, have reported no change in mitochondrial number with aging (Nagata and Ma 2005; Sastre et al. 1996). Such studies either used indirect measures of mitochondrial number (mitochondrial membrane mass) (Sastre et al. 1996) or used mice (2 years old for the oldest age group) that were likely not old enough to see major mitochondrial changes with aging (Nagata and Ma 2005). Thus, available data support a decrease in liver mitochondrial numbers per unit of tissue with aging.

Average mitochondrial circumferences and/or areas were reported to increase with aging in mice (Wilson and Franks 1975a,b) and humans (Sato and Tauchi 1975; Tauchi and Sato 1968). In particular, aging increased variation in mitochondrial size in mouse livers and altered mitochondrial distribution toward a greater number of large mitochondria (Wilson and Franks 1975a). However, a few studies have indicated that any changes in mitochondrial morphology are insufficient to counter decreases in mitochondrial number with aging, and thus, volume density (volume of mitochondria per volume of cytosol) of liver mitochondria is decreased in mice (Herbener 1976; Tate and Herbener 1976) and rats (Meihuizen and Blansjaar 1980). Furthermore, the percentage of mitochondrial volume occupied by cristae is decreased in old versus young rats (de la Cruz et al. 1990) and the surface density of mitochondrial cristae per unit of either cytosol or mitochondria is decreased with aging in mice (Tate and Herbener 1976). Thus, decreases in mitochondrial number and cristae area could be factors contributing to impaired energy metabolism in liver with aging. The age at which these mitochondrial changes occur is not well defined, although in humans it has been reported that decreases in mitochondrial number occur after 60 years of age (Sato and Tauchi 1975; Tauchi and Sato 1968) while changes in morphology may occur a few years later (Sato and Tauchi 1975). In mice, changes in mitochondrial number occur by 30 months of age (Herbener 1976); however, there is not sufficient data to currently determine the precise age at which this change first appears.

There are a number of questions remaining regarding the influence of aging on liver mitochondrial morphology, and in particular, the following three factors should be taken into consideration in future experiments. (1) It is likely that changes in number and morphology with aging are not uniform among all populations of mitochondria. It has been reported that mitochondrial volume density is decreased in the midzonal but not central or peripheral areas of the right liver lobe in 35 versus 3-month-old rats (Meihuizen and Blansjaar 1980), and thus, location within the liver appears to have an important influence on age-related changes in mitochondria. Additional work is needed to determine the influence of mitochondrial location within the cell and the location of the cell within the organ on age-related changes in both mitochondrial number and morphology. (2) Diet composition could have a significant influence on alterations in mitochondrial morphology with age. In mice-fed diets where the fat source was either olive oil or sunflower oil, only the sunflower oil group showed decreased mitochondrial cristae number with aging (Quiles et al. 2006). Few studies have considered diet as a potential factor that influences age-related changes in mitochondrial morphology and number. (3) Isolation of mitochondria may cause loss of large mitochondria and produce results that are not representative of tissue mitochondrial populations. It has been reported that large mitochondria are lost during isolation procedures and age-related changes in mitochondrial size are greatly diminished in isolated mitochondria versus liver tissue (Wilson and Franks 1975b). Isolated mitochondria are used in many studies investigating the influence of aging on mitochondrial composition and function, and it is important to at least consider the possible impact on the results of the study of greater mitochondrial loss during isolation in older age groups.

Calorie Restriction

The influence of CR on mitochondrial biogenesis is an area of great interest since a CR-induced change in mitochondrial mass could potentially provide a way to combat the age-related decline in mitochondrial number. Therefore, several studies have used a wide range of approaches to determine if CR alters mitochondrial content in liver. Transmission electron microscopy (TEM) is considered the gold standard for providing detailed information about both mitochondrial morphology (area, volume, cristae number) and number, and a few studies have used this approach to determine mitochondrial response to CR. The first of these studies reported that liver from 3- to 7-month-old CR mice had larger mitochondria than control groups (Weindruch et al. 1980). However, the influence of CR on mitochondrial size appears to be complex and may depend on diet composition, age, or duration of CR with one study finding no change in 6-month CR versus control mice (Khraiwesh et al. 2013) and another study by the same group finding an increase in average mitochondrial volume in 18-month CR versus control mice (Khraiwesh et al. 2014). Mitochondrial volume density (volume of mitochondria per volume of cell) was increased in liver of mice at 6 (Khraiwesh et al. 2013) and 18 (Khraiwesh et al. 2014) months of CR because CR induced a greater than 30% increase in mitochondrial numerical density. Consistent with this finding, another study reported that mitochondrial number was increased by greater than 50% in hepatocytes from 24-month-old CR compared to control rats (Lopez-Lluch et al. 2006). In addition to changes in mitochondrial number, CR also alters mitochondrial structure with liver mitochondria from CR animals showing an increase in mean number of cristae per mitochondrion (Khraiwesh et al. 2013, 2014; Lopez-Lluch et al. 2006) and a decrease in circularity coefficient when compared to control animals (Lopez-Lluch et al. 2006). Certain approaches, such as amount of mitochondrial DNA, activity of marker enzymes, expression of mitochondrial proteins, or expression of mRNA for regulators of mitochondrial biogenesis, have also been used to determine the influence of CR on mitochondrial content. Studies using one or more of these approaches, however, have produced mixed results with one concluding that CR increases mitochondrial content (Nisoli et al. 2005) and another showing no increase in liver mitochondria with CR (Hancock et al. 2011). A limitation of these studies is that they are not capable of determining if the observed CR-induced changes are due to alterations in number of mitochondria, mitochondrial surface area, mitochondrial DNA (mtDNA) copy number, or the composition of mitochondrial lipids and proteins. However, electron microscopy studies provide the best evidence that CR increases liver mitochondrial number, volume density, and mean number of cristae per mitochondrion. These changes are the opposite of the mitochondrial changes that occur with aging in liver and may contribute to the positive effects of CR on aging.

MITOCHONDRIAL RESPIRATION

AGING

A decrease in mitochondrial electron transfer chain activity is commonly considered to occur with aging, and this conclusion is at least partly derived from studies which report a decrease in State 3 (substrate plus ADP) respiration with aging (albeit often in tissues other than liver) (Navarro and Boveris 2007). The finding that liver mitochondrial respiration under either State 3 or 4 (all ADP converted to ATP) conditions shows a linear decrease from 31 to 76 years of age in humans (Yen et al. 1989) has also been used to support the conclusion that mitochondrial respiration decreases with aging (Lee and Wei 2012). However, there is no uniform agreement among studies regarding the influence of aging on mitochondrial respiration in liver (Hansford 1983). Since the 1960s many studies have used classic approaches to determine mitochondrial respiratory capacity under State 3 and State 4 conditions (Chance and Williams 1955), in addition to assessing mitochondrial efficiency using respiratory control ratio (State 3/State 4) and P/O ratio. These studies have produced mixed results with a nearly equal number of studies reporting either a decrease or no change in respiration, and respiratory control ratio with aging in liver (Table 6.1).

There are several possible reasons for the lack of agreement among studies regarding the influence of aging on mitochondrial respiration. *First*, it is possible that age-related changes in liver mitochondrial respiration depend on the substrate used in the study. This means that age-related changes in substrate transport or activities of key enzymes in substrate oxidation pathways are primarily responsible for any observed changes in mitochondrial respiration. However, there is no overall evidence to support the conclusion that the substrate being used to support respiration is a contributing factor to the differences in age-related changes in respiration between studies. The primary substrates used for liver mitochondrial respiration studies are succinate and glutamate/malate, and they either cause a decrease or no change in liver mitochondrial respiration with aging (Table 6.1). *Second*, it is possible that age-related changes in mitochondrial respiration only occur at very high rates of respiration (State 3 or uncoupled respiration). Indeed, it has also been suggested that concentrations of ADP have been too low to detect age-related changes in mitochondrial respiration which were reported to only occur near maximal respiration rate (Darnold et al. 1990). A greater number of studies have reported decreases in State 3 than State 4 respiration with aging (Table 6.1), providing some support for the notion that age-related impairments in mitochondrial respiration may be most apparent under

TABLE 6.1

Influence of Aging on Mitochondrial Respiration

Substrate	Respiratory State	Change with Aging	References
Succinate	State 3	↓	Alemany et al. (1988), Darnold et al. (1990), de la Cruz et al. (1990), Garcia-Fernandez et al. (2011), Goodell and Cortopassi (1998), Grattagliano et al. (2004), Hamelin et al. (2007), Horton and Spencer (1981), Lopez-Torres et al. (2002), Modi et al. (2008), and Yen et al. (1989)
Succinate	State 3	↔	Bakala et al. (2003), Capozza et al. (1994), Guerrieri et al. (1996), Lambert et al. (2004b), Paradies and Ruggiero (1991), Puche et al. (2008), Sanchez-Roman et al. (2012), Serviddio et al. (2007), Vorbeck et al. (1982b), Weindruch et al. (1980), and Wilson and Franks (1975)
Succinate	State 4	↓	Alemany et al. (1988), Hamelin et al. (2007), Sanchez-Roman et al. (2012), and Yen et al. (1989)
Succinate	State 4	↔	Bakala et al. (2003), Capozza et al. (1994), Garcia-Fernandez et al. (2011), Guerrieri et al. (1996), Horton and Spencer (1981), Lambert et al. (2004b), Lopez-Torres et al. (2002), Modi et al. (2008), Paradies and Ruggiero (1991), Puche et al. (2008), Serviddio et al. (2007), Vorbeck et al. (1982b), Weindruch et al. (1980), and Wilson and Franks (1975)
Glutamate/malate	State 3	↓	de Cavanagh et al. (2008), Delaval et al. (2004), Hamelin et al. (2007), Horton and Spencer (1981), and Yen et al. (1989)
Glutamate/malate	State 3	↔	Grattagliano et al. (2004), Puche et al. (2008), Sanchez-Roman et al. (2012), and Serviddio et al. (2007)
Glutamate/malate	State 4	↓	Sanchez-Roman et al. (2012) and Yen et al. (1989)
Glutamate/malate	State 4	↔	Delaval et al. (2004), Hamelin et al. (2007), Horton and Spencer (1981), Puche et al. (2008), and Serviddio et al. (2007)
Pyruvate/malate	States 3 and 4	↔	Lopez-Torres et al. (2002), Modi et al. (2008), Rocha-Rodrigues et al. (2013), and Weindruch et al. (1980)
β-HB[a]	State 3	↓	Weindruch et al. (1980)
β-HB[a]	State 3	↔	Gold et al. (1968)

[a] β-HB refers to β-hydroxybutyrate.

conditions of high oxygen consumption. However, any differences in the magnitude of changes in State 3 and State 4 respiration with aging must be very small since the majority of studies find no difference in liver mitochondrial respiratory control ratios with aging (Table 6.1). Also, it seems unlikely that failure to induce high respiration rates is responsible for the lack of age-related changes observed in some studies, since higher rates of respiration are not consistently observed in studies which reported an increase versus those that found no change in liver mitochondrial respiration with aging. *Third*, it is possible that studies which found no age-related decrease in mitochondrial respiration did not use old enough animals. This does not appear to be the case since studies using liver mitochondria from 30- to 32-month-old mice (Wilson and Franks 1975) or 27- (Bakala et al. 2003), 28- (Paradies and Ruggiero 1991), or 29- (Vorbeck et al. 1982b) month-old rats did not report an age-related decrease in mitochondrial respiration. The ages of the animals in these experiments meet or exceed the ages of the animals in other studies which have reported age-related decreases in mitochondrial respiration. Thus, there is no age at which all studies find a decrease in mitochondrial respiration and animal age does not explain differences in results between studies. *Fourth*, loss of mitochondrial populations impacted by aging during mitochondrial isolation procedures could dampen or eliminate age-related changes in respiration. If aging damages mitochondria or changes mitochondrial size, it becomes quite likely that these mitochondria could be lost during isolation steps. In support of this idea, it has been reported that large mitochondria, which are more common in older animals, are lost during mitochondrial isolation (Wilson and Franks 1975a,b). Additional studies are needed to determine the extent to which mitochondrial isolation procedures influence the results of studies investigating age-related changes in mitochondria.

One way to get around the potential problem of loss of some mitochondrial populations in older animals is to measure oxygen consumption in intact hepatocytes. A few studies have taken this approach. It has been reported that resting oxygen consumption is decreased in hepatocytes from 30- versus 3-month-old mice supplemented with glucose and pyruvate (Harper et al. 1998). Similarly, oxygen consumption and respiratory control ratio was decreased in hepatocytes from 20- to 28-month versus 3- to 5-month-old rats supplemented with glucose and glutamate (Hagen et al. 1998). Another study found that differences in mitochondrial respiration were most evident in hepatocytes from old- (36 months) versus middle-aged (12 months) rats at high respiration rates induced by the uncoupler dintrophenol, while no differences between age groups were observed under basal respiration or following complex V inhibition with oligomycin (Brouwer et al. 1977). However, it has also been reported that nicotinamide adenine dinucleotide (NADH) oxidation is substantially decreased in hepatocytes from 28- versus 5-month-old mice (Ramanujan and Herman 2007). The fact that all studies measuring oxygen consumption in intact hepatocytes found some level of age-related decrease in respiration strongly supports the conclusion that aging leads to a decrease in liver mitochondrial respiration. This decrease in respiration does not occur in all mitochondria (which would easily be picked up in isolated mitochondria studies), but rather appears to be limited to a subpopulation of mitochondria. Additional studies are needed to determine the morphology and cellular location of the mitochondrial populations that show an age-related decrease in respiration.

CALORIE RESTRICTION

The vast majority of studies have found no changes in liver mitochondrial respiration with CR (Caro et al. 2009; Chen et al. 2013; Gomez et al. 2007; Gredilla et al. 2001; Hagopian et al. 2011; Lopez-Torres et al. 2002; Ramsey et al. 2004). No studies have found a consistent increase (or decrease) in mitochondrial oxygen consumption with all substrates in CR animals, although a few studies have reported that the mito-chondrial respiration is selectively decreased with succinate (Crescenzo et al. 2012), pyruvate/malate (Sanz et al. 2005), or β-hydroxybutyrate (β-HB) (Weindruch et al. 1980) and increased with glutamate/malate (Grattagliano et al. 2004) or pyruvate/malate (Weindruch et al. 1980). However, these substrate-specific changes in respi-ration are not consistently observed among studies and there is insufficient data at this time to support the conclusion that CR influences mitochondrial respiration in a substrate-specific manner. One limitation of the existing CR studies is that they are primarily completed in animals less than 12 months of age (middle age), and it is not possible in these studies to determine if CR opposes age-related decreases in mitochondrial respiration. It has been reported that succinate-supported respiration is decreased in 17-month-old CR rats (Lambert et al. 2004b). However, other studies have found that succinate-supported mitochondrial respiration is not altered with CR in 24- (Lopez-Torres et al. 2002) or 28-month (Grattagliano et al. 2004) old rats. CR studies suffer from the same concerns regarding loss of some mitochondrial popula-tions in isolation procedures as those encountered in studies of aging using isolated mitochondria. Few studies have measured oxygen consumption using approaches that do not require isolation of mitochondria. Considering the fact that mitochondrial isolation may be influencing CR results, it was found that CR-induced increases in State 3 respiration in liver homogenates were lost following isolation of mitochon-dria (Crescenzo et al. 2012). In contrast to this finding, it has been reported that hepa-tocyte oxygen consumption (expressed per dry weight of cells) is not altered with CR (Lambert and Merry 2005). Overall, studies in young adult animals indicate that CR has no effect on liver mitochondrial oxygen consumption. However, there is an insufficient number of studies in older animals to make strong conclusions about the influence of CR on age-related changes in liver mitochondrial respiration.

MITOCHONDRIAL MEMBRANE POTENTIAL AND PROTON LEAK

AGING

A decrease in mitochondrial membrane potential is one possible outcome of age-related impairments in liver mitochondrial function. This decrease in membrane potential could be due to either decreased capacity for mitochondrial respiration or changes in the mitochondrial membrane composition which increase proton perme-ability. To determine if decreases in membrane potential occur with aging, a few studies have measured mitochondrial membrane potential in isolated mitochon-dria or hepatocytes from young and old rats. Studies in isolated mitochondria have reported a decrease in liver mitochondrial membrane potential in old (24–26 months of age) versus young adult (4–8 months old) animals (Garcia-Fernandez et al. 2011;

Meng et al. 2007; Puche et al. 2008; Serviddio et al. 2007). This change in membrane potential occurred primarily under conditions which promote maximal membrane potential (State 4 conditions or inhibition of complex V), and no age-related change was observed following the decrease in mitochondrial membrane potential induced by initiating State 3 respiration (Garcia-Fernandez et al. 2011). Similarly, age-related decreases in membrane potential were only observed with succinate but not with glutamate/malate as substrates and this appeared to be primarily due to the fact that higher membrane potentials were obtained in both age groups in this study with succinate compared to glutamate/malate (Puche et al. 2008). Only one study reported no change in mitochondrial membrane potential between young (19-week old) and old (106-week old) rats, although the conditions under which these measurements were completed are poorly defined making it difficult to compare these results with other studies (Rocha-Rodrigues et al. 2013). The same limitations of using isolated mitochondria for respiration experiments, namely loss of mitochondria in older animals during mitochondrial isolation steps, also apply to membrane potential measurements. Nonetheless, the available evidence indicates that membrane potential is decreased in the mitochondrial populations isolated from old versus young animals under State 4 conditions.

Similar to the results observed in isolated liver mitochondria, it has been reported that mitochondrial membrane potential is lower in intact hepatocytes from old (20–28 months old) versus young (3–5 months old) rats (Hagen et al. 1997, 1998; Sastre et al. 1996). This age-related decrease in mitochondrial membrane potential does not appear to be due to uniform changes in all hepatocyte populations since cells from older animals could be separated into three distinct subpopulations which showed either a major decrease (largest two subpopulations) or no change (smallest subpopulation) in membrane potential compared to young animals (Hagen et al. 1997, 1998). Only one study reported no changes in mitochondrial membrane potential between young (6 months old) and old (24 months old) rats (Cavazzoni et al. 1999). The reason for the different conclusions in this study compared to other studies is not clear, although it is important to consider the method used to measure membrane potential as a potential contributing factor. This study, and several other studies investigating aging, used RH-123 (a fluorescent cationic dye) to assess membrane potential. It can be difficult to accurately determine mitochondrial membrane potential using flow cytometry and RH-123 since many factors need to be considered when using this dye (Nicholls and Ward 2000), and it is possible that differences between studies are more of a technical than physiological nature. Tetraphenylphosphonium (TPP) electrodes provide an approach which avoids some of the potential problems with fluorescent dyes and studies using TPP electrodes in isolated mitochondria (Serviddio et al. 2007) and hepatocytes (Hagen et al. 1997) have reported an age-related decrease in mitochondrial membrane potential. Overall, the available data point toward a decrease in liver mitochondrial membrane potential with aging, although additional studies are needed to confirm this conclusion and determine under which conditions age-related changes in membrane potential occur.

Mitochondrial proton leak is one factor that could contribute to age-related decreases in membrane potential under resting conditions (approaching State 4 respiration). State 4 respiration provides an indication of proton leak-dependent respiration but does not actually give information about mitochondrial membrane proton permeability since this can only be accurately assessed by simultaneously measuring both mitochondrial membrane potential and oxygen consumption. Liver mitochondrial proton leak kinetics has been assessed in isolated mitochondria (Garcia-Fernandez et al. 2011; Puche et al. 2008; Serviddio et al. 2007) and intact hepatocytes (Harper et al. 1998) from young and old rats. The results of all these studies indicate that aging is associated with an increase in mitochondrial proton leak. This increase in proton leak could be an important factor contributing to impaired liver mitochondrial function with aging.

CALORIE RESTRICTION

Liver mitochondrial membrane potential measurements in CR animals have primarily been completed while determining mitochondrial proton leak kinetics. Mitochondrial proton leak consists of a basal leak and regulated leak which are controlled by the uncoupling proteins (Brookes 2005). Little is known about the influence of aging and CR on regulated proton leak and most of the existing work has focused on CR-related changes in basal proton leak. It has been proposed that CR may decrease basal proton leak as a way to conserve energy (Ramsey et al. 2000) or CR may oppose age-related increases in basal proton leak (Harper et al. 2004). In contrast, it has also been proposed that CR may increase proton leak (and decrease membrane potential) as a mechanism to lower ROS (Ash and Merry 2011). While regulated proton leak could perform this function, it is highly unlikely that the increase in energy expenditure which accompanies a chronic increase in basal proton leak would provide a viable option to combat ROS production in an animal with a severely constrained energy supply. Furthermore, the available information supports the notion that proton leak is either unaltered or decreased in liver mitochondria from CR animals. It has been reported in hepatocytes that nonphosphorylating oxygen consumption and membrane potential are not altered in rats following 4 months of CR, consistent with no CR-related change in proton leak (Lambert and Merry 2005). Consistent with this finding, it has been shown that liver mitochondrial proton leak is not changed following 3 days of 50% CR (Dumas et al. 2004). Age, however, may have an impact on CR-related changes in proton leak since proton leak was decreased in liver following 12 or 18 months of 40% CR (Hagopian et al. 2005) while no changes in mitochondrial proton leak were observed in rats maintained on CR for 1 or 6 months (Ramsey et al. 2004). These results are consistent with the idea that CR may oppose age-related increases in liver mitochondrial proton leak. In contrast to these results, one study has reported that liver mitochondrial proton leak is increased in rats following 4 or 16 months of CR induced by weight clamping the CR group to 55% of the body weight of control animals. The reason for the discrepancy between this study and other published studies is not entirely known, although the CR protocol should be considered as a potential factor. This study initiated CR in young animals (60 days of age) and targeted a weight differential between the CR and control groups that is larger than typical CR studies which often use 30%–40% CR. Weight clamping, unlike most CR protocols, may also induce fluctuation in food intake throughout adulthood. Thus, the available data indicates that CR does not alter liver mitochondrial proton leak or may oppose age-related increases in proton leak in animals maintained under the more typical CR protocols.

ELECTRON TRANSPORT CHAIN ENZYMES

AGING

Age-related increases in oxidative damage to mitochondrial DNA (Harman 2006; Lenaz 1998; Shigenaga et al. 1994) and proteins (Navarro and Boveris 2007; Shigenaga et al. 1994) have been proposed to play an important role in aging by decreasing the level and/or activity of mitochondrial electron transport chain (ETC) enzymes. However, it is uncertain if age-related impairments in ETC enzyme activities occur in mitotic tissues, such as the liver. To address this issue, several studies have determined the influence of aging on the activities of ETC enzymes in liver homogenates or isolated liver mitochondria (Table 6.2). There is no uniform agreement among these studies regarding age-related changes in ETC enzyme activities. This may at least partially reflect the fact that most studies do not measure the activities of all ETC enzymes and it is possible that some ETC enzyme complexes are more susceptible than others to alterations in activity with aging. The majority of studies have reported an age-related decrease in complex V (F_0F_1 ATP synthase) activity (Table 6.2), with only two studies finding no change in the activity of this enzyme when comparing young adult (5–6 months old) with old (24–25 month old) rats (Barogi et al. 1995; Zhao et al. 2014). However, in one of these studies, older animals had a decreased response to the ATPase activator, bicarbonate, suggesting at least some level of age-related impairment in this enzyme (Barogi et al. 1995). There is also considerable support for an age-related decrease in liver mitochondrial complex I activity (Braidy et al. 2011; Genova et al. 1995; Miquel et al. 1995; Torres-Mendoza et al. 1999). Evidence for an age-related decrease in the activities of the other ETC complexes is less clear, with an equal number of studies indicating either a decrease or no change in Complex IV activity with aging (Table 6.2). In contrast, the majority of studies find no age-related changes in the activities of complex II or III with aging in liver mitochondria (Table 6.2). Thus, the results of existing studies support the idea that in liver some ETC complexes are prone to age-related decreases in activity while others are relatively resistant to changes in activity with aging.

One factor that should be considered when comparing ETC enzyme activity data between studies is the time course for changes in enzyme activity with age. It is common in the literature for comparisons between only two age groups, with one group being young- or middle-aged animals and the other group being older animals. This two age groups comparison can be risky since it does not provide information about the trajectory of change in enzyme activity and it may give misleading information if the data do not adequately capture the ages at which changes in enzyme activity occur. It has been shown that liver complex V activity dramatically increases in rats from 1 to 3 months of age, and then remains stable through 12 months of

TABLE 6.2

Influence of Aging on Mitochondrial Electron Transport Chain (ETC) Enzyme Activities

ETC Enzyme	Change with Aging	Isolated Mitochondria or Tissue Homogenate	References
Complex I	↓	Mitochondria	Braidy et al. (2011), Genova et al. (1995), Miquel et al. (1995), Rafique et al. (2004), Torres-Mendoza et al. (1999)
Complex I	↔	Mitochondria	Davies et al. (2001) and Kwong and Sohal (2000)
Complex I	↑	Mitochondria	Zhao et al. (2014)
		Homogenate	Laurent et al. (2012)
Complex II	↓	Mitochondria	Braidy et al. (2011)
Complex II	↔	Mitochondria	Armeni et al. (2003), Davies et al. (2001), Kwong and Sohal (2000), and Zhao et al. (2014)
		Homogenate	Laurent et al. (2012)
Complex III	↓	Mitochondria	Armeni et al. (2003) and Braidy et al. (2011)
Complex III	↔	Mitochondria	Davies et al. (2001), Genova et al. (1995), and Kwong and Sohal (2000)
Complex IV	↓	Mitochondria	Braidy et al. (2011), Miquel et al. (1995), Navarro et al. (2004), Puche et al. (2008), Rafique et al. (2004), Tian et al. (1998), and Weindruch et al. (1980)
		Homogenate	Navarro et al. (2004) and Vorbeck et al. (1982b)
Complex IV	↔	Mitochondria	Davies et al. (2001), Genova et al. (1995), Gold et al. (1968), Kwong and Sohal (2000), Vorbeck et al. (1982b), Wilson and Franks (1975), and Zhao et al. (2014)
		Homogenate	Barazzoni et al. (2000) and Wilson and Franks (1975)
Complex IV	↑	Mitochondria	Armeni et al. (2003) and Dani et al. (2010)
Complex V	↓	Mitochondria	Capozza et al. (1994), Davies et al. (2001), Guerrieri et al. (1992), Haynes et al. (2010), Miquel et al. (1995), and Puche et al. (2008)
Complex V	↔	Mitochondria	Barogi et al. (1995) and Zhao et al. (2014)

age before decreasing by 24 months of age (Guerrieri et al. 1992). In this case, the impact of aging on complex V activity would depend on the age group with which the 24-month-old animals are compared. Similarly, it has been reported that liver mitochondrial complex IV activity is increased in rats from 1 to 6 months of age and then begins to decrease with a significant difference being observed between 6- and 24-month-old rats (Tian et al. 1998). The age at which significant decreases in complex IV activity occur is not well defined, with one study indicating that it occurs in rats by late middle age (17.5 months of age) (Vorbeck et al. 1982b) and another study finding by regression analysis significant decreases in both complex I and IV activities by 1 year of age (Rafique et al. 2004). There is evidence that the activities of some ETC enzymes in rat liver show age-related decreases in activity beginning in middle age. Thus, the magnitude of decreases in ETC enzyme activity in older animals may be heavily influenced by the age group to which the data from the older animals are being compared.

Similar to mitochondrial respiration and morphology studies, the mitochondria isolation procedures have the potential to significantly impact the results of studies investigating the influence of age on ETC enzyme activities. It has been reported that only 30%–40% of total liver complex IV activity is recovered in isolated mito-chondria when compared to tissue homogenates (Navarro et al. 2004; Vorbeck et al. 1982b). Therefore, an assumption must be made that measurements in isolated mito-chondria studies are representative of the total liver mitochondrial population. This does not appear to be the case in aging studies. It has been shown that complex IV activity significantly decreases in rats from 8.5 to 17.5 and 17.5 to 29 months of age when measured in liver homogenates. However, in this same study, no significant age-related changes in complex IV activity were observed in isolated mitochondria (Vorbeck et al. 1982b). Similarly, complex IV activity decreased in rats from 28 to 60 weeks and 60 to 92 weeks of age when measured in liver homogenates, while decreases in complex IV activity were only observed at 92 weeks of age in isolated mitochondria (Navarro et al. 2004). These studies indicate that liver complex IV activity decreases with age and suggest that age-related changes in ETC enzyme activities may be dampened or lost when relying only on measurements completed in isolated mitochondria.

It is often assumed that age-related increases in oxidative damage to mtDNA may lead to impaired mitochondrial function by decreasing the levels of ETC enzymes encoded by mtDNA. If this were true, it would be expected that enzymes contain-ing several mtDNA-encoded proteins (such as complexes I and IV) would show age-related losses in function prior to enzymes that are not encoded by mtDNA (such as complex II). The existing data do not refute this idea since few studies have reported age-related decreases in liver complex II activity while several studies have found age-related decreases in activities of Complexes I and IV (Table 6.2). However, there is insufficient data at this point to determine if age-related decreases in ETC enzyme activities are primarily due to decreased levels of mtDNA-encoded proteins. Nonetheless, a few studies have measured protein levels or the levels of specific cytochromes to determine if changes in ETC enzyme activities are accompanied by changes in the amount of enzymes. It has been reported that the levels of complex I proteins are decreased with aging (Bakala et al. 2013; Torres-Mendoza et al. 1999) and decreased levels of complex I proteins were consistent with lower complex I activity in liver mitochondria from old rats (Torres-Mendoza et al. 1999). Lack of change in cytochrome b and cytochrome $c + c_1$ levels (indica-tors of complex III amount) was consistent with no age-related changes in rat liver Complex III activity (Armeni et al. 2003; Genova et al. 1995), while one study found that age-related increases in complex III activity were accompanied by increased levels of complex III proteins (Torres-Mendoza et al. 1999). It has also been shown that age-related decreases in complex V activity are accompanied by decreased levels of complex V subunits (capozza et al. 1994; Guerrieri et al. 1992) and age-related decreases (Vorbeck et al. 1982b) or increases (Armeni et al. 2003; Dani et al. 2010) complex IV activity are parallel with changes in cytochrome a + a_3 levels (an indicator of amount of complex IV). These studies provide support for the idea that the level of ETC proteins is a factor contributing to age-related changes in the activities of ETC enzymes.

Posttranslational modification of proteins is another factor which could influence age-related changes in ETC enzyme activities. In particular, increased complex V β-subunit nitration appears to contribute to decreases in liver complex V activity in middle-aged compared to young rats (Haynes et al. 2010). It is possible that other forms of oxidative damage or protein posttranslational modifications could also contribute to changes in ETC enzyme activity across the life span.

Diet composition is rarely considered as a major factor contributing to possible differences between studies in age-related changes in ETC enzyme activities, although it has been shown that changes in complex IV activity with age are influenced by dietary lipid source (Quiles et al. 2006). Dietary lipids could exert influence on ETC enzymes through their effects on membrane composition or gene transcription. Currently, however, there is insufficient data to determine the extent to which diet composition can modulate age-related changes in ETC enzyme activities.

Overall, aging does not produce a uniform change in the activities of all ETC complexes. However, several studies have reported that the activities of ETC complexes I, IV and V decrease in liver with aging suggesting that some impairment in ETC enzyme activity is a common occurrence with aging. The substantial loss of ETC enzyme activity during isolation of mitochondria is a concern and it is possible that studies in isolated mitochondria do not adequately reflect the magnitude of age-related changes in mitochondria which occur in vivo. Few studies have used more than two age groups to make conclusions about the influence of aging on ETC enzyme activities and the trajectory of age-related changes in ETC enzyme activities remains to be determined.

CALORIE RESTRICTION

CR has been shown in rat skeletal muscle to induce an initial decrease in the activities of ETC complexes I, III, and IV and to prevent a further age-related decline in the activities of these ETC complexes (Desai et al. 1996). Similar studies have not been completed in liver from CR animals and information is only available on changes in ETC complexes induced by short-term (<3 months) CR. No significant differences in the activities of any of the ETC complexes were observed for mice maintained on 5% or 40% CR diets for 1 month (Chen et al. 2013). However, it was reported that the influence of CR on the activities of ETC complexes depends on the mitochondrial population being studied with large mitochondria showing a CR-related decrease in complex I activity while no CR-related changes in complex I activity were observed in small mitochondria following 2 months of CR. Furthermore, in this study, complex II activity was increased with CR regardless of mitochondrial size

(Hagopian et al. 2013). It has been shown that complex IV activity in liver homogenates is not altered by 14 days of 50% CR (Crescenzo et al. 2012) or 8 weeks of 40% CR (Rojas et al. 1993). To further determine the influence of CR on the liver mitochondrial ETC, several studies have measured the levels of ETC proteins. It was reported that complex IV levels are increased in liver from mice fed every other day and assumed to undergo 30%–40% CR (Nisoli et al. 2005). This result, however, has not been confirmed by other studies using either standard, controlled CR approaches or every other day feeding. However, 7 weeks of every other day feeding was reported to decrease the protein levels of all ETC complexes in mouse liver (Caro et al. 2008). Similarly, 14 weeks of 30% CR decreased the protein levels of complexes I and IV in rat liver (Hancock et al. 2011). In contrast to this finding, 2 months of 30% CR only decreased complex III protein level without altering the protein levels of complexes I and IV in mouse liver (Jove et al. 2014). The results of these studies suggest that short-term CR does not have a dramatic impact on liver ETC complexes.

OXIDATIVE STRESS

AGING

Studies in isolated mitochondria (Bejma et al. 2000; de Cavanagh et al. 2008; Garcia-Fernandez et al. 2011; Haak et al. 2009; Puche et al. 2008) and submitochondrial particles (Sohal et al. 1990) indicate that aging is associated with a dramatic increase in liver mitochondrial ROS production. This increase is also observed in isolated hepatocytes (Cavazzoni et al. 1999; Hagen et al. 1997, 1998) and liver homogenates (Bejma et al. 2000; Ko et al. 2008) from old versus young adult animals. In fact, mitochondrial isolation may dampen the age-related change in ROS production, since the magnitude of the increase in oxidant production is higher in liver homogenates than in isolated mitochondria (Bejma et al. 2000). The mitochondrial site(s) responsible for this age-related increase in liver ROS production have not been extensively studied and are not entirely known. In one study, it was found that liver mitochondrial superoxide anion production was increased with aging when mitochondria were respiring on NADH, NADH + rotenone (complex I inhibitor) or NADH + antimycin A (complex III inhibitor), but not when respiring on succinate with or without inhibitors (Haak et al. 2009). These results suggest that age-related increases in ROS production occur primarily in complex I during forward electron flow. However, another study found that age-related increases in mitochondrial ROS production occurred when mitochondria were respiring on succinate + ADP, succinate without rotenone, or succinate + TTFA (complex II inhibitor). These results suggest that complex II or possibly electron backflow into complex I could also be contributors to age-related increases in liver ROS production (Lopez-Torres et al. 2002). Further work in this area is needed to determine if a particular mitochondrial site is consistently responsible for age-related increases in ROS production. There is considerable evidence that mitochondrial ROS production is increased in livers from old (≥24 months of age) rats and mice, although several questions remain regarding the cause of these changes and the typical time course over which changes in mitochondrial ROS production occur.

TABLE 6.3

The Influence of Aging on Antioxidants in Isolated Liver Mitochondria

Antioxidant	Change with Aging[a]	References
Superoxide dismutase	↓	Ko et al. (2008), Navarro et al. (2004), and Tian et al. (1998)
	↔	De and Darad (1991) and Ji et al. (1990)
	↑	Quiles et al. (2006), Rikans et al. (1992), Rocha-Rodrigues et al. (2013), and Yen et al. (1994)
α-Tocopherol	↓	Kamzalov and Sohal (2004) and Ko et al. (2008)
	↑	Quiles et al. (2006)
Catalase	↓	Tian et al. (1998)
	↔	De and Darad (1991) and Ji et al. (1990)
	↑	Quiles et al. (2006)
Glutathione peroxidase	↓	Bejma et al. (2000) and Rikans et al. (1992)
	↔	Quiles et al. (2006) and Tian et al. (1998)
	↑	Grattagliano et al. (2004), Ji et al. (1990), Quiles et al. (2006), and Rikans et al. (1992)
Glutathione reductase	↓	Bejma et al. (2000) and Rikans et al. (1992)
	↔	Rikans et al. (1992)
GSH[b]	↓	Capozza et al. (1994), Grattagliano et al. (2004), and Ko et al. (2008)
	↔	Bejma et al. (2000), Dani et al. (2010), and Rebrin et al. (2007)

[a] A change in enzyme activity (superoxide dismutase, catalase, glutathione peroxidase, glutathione reductase) or change in amount of antioxidant (α-tocopherol, GSH).

[b] GSH = reduced glutathione.

Some studies have indicated that the increase in ROS levels in hepatocytes from older animals is at least partially due to decreased ROS removal (Lopez-Cruzan and Herman 2013; Ramanujan and Herman 2007). Age-related deficits in mitochondrial antioxidants could slow ROS removal and contribute to increased oxidative stress. However, there is no uniform change in antioxidant activities or levels with aging among studies (Table 6.3) indicating that factors, other than simply age, have a major influence on the mitochondrial antioxidant systems in older animals. Although there is no agreement as to whether levels of reduced glutathione (GSH) are decreased (Capozza et al. 1994; Grattagliano et al. 2004; Ko et al. 2008) or not changed (Bejma et al. 2000; Dani et al. 2010; Rebrin et al. 2007) with aging, available evidence suggests that aging is associated with increased levels of oxidized glutathione (GSSG) and a reduced ratio of GSH to GSSG in liver mitochondria (Capozza et al. 1994; Dani et al. 2010; Grattagliano et al. 2004; Rebrin et al. 2007). These changes are indicative of chronic oxidative stress and could influence cellular function by altering redox state and influencing redox-sensitive signaling pathways.

The oxidative stress theory (Harman 1972) continues to be a widely cited theory of aging. This theory postulates that oxidative stress influences aging through damage to proteins, lipids, and DNA, leading to impaired cellular function. Mitochondrial DNA (mtDNA) mutations resulting from oxidative damage are a potential contributor to aging. However, mtDNA mutations are generally considered to accumulate primarily in post-mitotic tissues (Cortopassi and Wong 1999; Lee and Wei 2012), and mtDNA deletions are often not considered to be major contributors to aging in liver. Nonetheless, studies have shown that levels of 8-oxo-2'-deoxyguanosine (8-oxo-DG), a common marker of DNA oxidative damage, are increased in liver mitochondria from old (≥23 months of age) rats and mice (de Souza-Pinto et al. 2001; Hamilton et al. 2001; Lopez-Torres et al. 2002). These increases in 8-oxo-DG appear to be due primarily to increased ROS levels with aging and not impaired mtDNA repair capacity since the activities of mtDNA repair enzymes are either unchanged or increased with aging (Souza-Pinto et al. 1999; Szczesny et al. 2010; Szczesny and Mitra 2005). Oxidative damage to mtDNA may also lead to mtDNA deletions, and there is evidence for increased 4834 bp deletions (Cassano et al. 2004; de Cavanagh et al. 2008; Yowe and Ames 1998) and ND4 gene deletions (Quiles et al. 2006) in liver mitochondria from old rats. Furthermore, 4977 and 6063 bp liver mtDNA deletions have been shown to increase in humans from 27 to 78 years of age (Yen et al. 1994), and mtDNA lesions (base modifications and single- or double-strand breaks) have been shown to increase in livers from 19- to 24-year-old rhesus monkeys compared to young adult animals (Castro Mdel et al. 2012). A common criticism, however, for mtDNA mutations playing a role in aging is that these mutations occur in only a small percentage of DNA. In some post-mitotic tissues, mtDNA mutations may occur in clusters and impair function of specific cells (Cao et al. 2001; Wanagat et al. 2001), but it is uncertain if this occurs in liver. It has also been reported that mtDNA fragments insert into nuclear DNA and mtDNA deletions may influence aging by disrupting nuclear DNA (Caro et al. 2010). This theory remains to be tested and at this time the contribution of liver mtDNA deletions to aging remains unknown.

Lipid oxidation may also contribute to aging by impairing mitochondrial membrane function. The vast majority of studies have reported an increase in markers of lipid peroxidation with aging in liver mitochondria from rats (Armeni et al. 2003; Garcia-Fernandez et al. 2011; Lambert et al. 2004a; Navarro et al. 2004; Puche et al. 2008; Quiles et al. 2006; Rocha-Rodrigues et al. 2013; Sanchez-Roman et al. 2012; Sawada and Carlson 1987; Yu et al. 1996) and mice (Li et al. 2012; Uysal et al. 1989), with a smaller number of studies finding that markers of lipid peroxidation are either unchanged (Haak et al. 2009; Ji et al. 1990) or decreased (Barrett and Horton 1975) with aging. It has also been reported that peroxidation induced by AAPH [2,2'-azobis(amidinopropane hydrochloride)] (Armeni et al. 2003) or ascorbic acid (Barrett and Horton 1975) is increased in liver mitochondria from old versus young adult rats. The existing data support the idea that mitochondrial lipid peroxidation could be an important contributor to aging in the liver.

Impaired protein function is often considered to be a factor through which oxidative, nitrative, or glycative stress influences aging. Several studies have shown that aging is associated with increased levels of markers of protein oxidative damage in liver mitochondria (Bakala et al. 2003; Grattagliano et al. 2004; Navarro et al. 2004; Sanchez-Roman et al. 2012), although there is increasing appreciation for the fact that specific proteins often bear the brunt of oxidative damage and global measures of protein damage may miss important alterations in key proteins. It has been shown that rat liver mitochondrial peroxiredoxin III undergoes extensive oxidation with aging (Musicco et al. 2009) and complex V F1 β-subunit was identified as undergoing nitration in mouse liver mitochondria with aging (Marshall et al. 2013). Recently, several liver mitochondrial proteins have also been shown to undergo extensive glycative damage with aging. These proteins include glutamate dehydrogenase, malate dehydrogenase, aconitase, aspartate aminotransferase, and six enzymes of fatty acid beta-oxidation (Bakala et al. 2013). Although there is evidence that liver mitochondrial proteins undergo oxidative or other forms of damage with aging, questions still remain as to the impact of this damage on protein function. A decrease in enzyme activity has been demonstrated for a few enzymes which undergo extensive glycative damage with aging (Bakala et al. 2013), but further work is needed to determine the extent to which age-related damage to specific proteins contributes to impairments in cellular function.

CALORIE RESTRICTION

The effect of CR on oxidative stress in liver, and other tissues, has recently been summarized in an extensive review (Walsh et al. 2014). While there is a substantial amount of information about the influence of CR on liver ROS production, antioxidants and oxidative damage, there is very little information specifically about the impact of CR on these parameters in liver mitochondria from late middle-aged or old CR animals.

The studies which have investigated long-term CR (≥12 months) in old (24 months) rats have reported that CR-related decreases in ROS production are limited to specific substrate and inhibitor combinations (Hagopian et al. 2005; Lopez-Torres et al. 2002). Furthermore, there is no agreement among the studies regarding the specific sites of ROS production altered with CR in older animals, suggesting that changes in mitochondrial ROS production are heavily influenced by factors such as diet, duration of CR, or rodent strain.

It has been reported that levels of the antioxidant GSH are not altered in liver mitochondria from 22-month-old CR mice compared to controls (Rebrin et al. 2003). There is a paucity of data for other mitochondrial antioxidants in older CR animals. In the case of GSH, there is evidence that levels of GSSG and protein–SSG are decreased and the GSH/GSSG ratio is increased in liver mitochondria from old CR versus control rats (Rebrin et al. 2003) indicating that CR protects the older mitochondria from oxidative stress.

CR has been shown to decrease oxidative damage to mtDNA (8-oxo-DG) in liver from 24-month-old rats (Chung et al. 1992; Hamilton et al. 2001) and mice (Hamilton et al. 2001). CR has also been reported to decrease lipid peroxidation (Laganiere and Yu 1987; Li et al. 2012) and resist the induction of peroxidation in liver mitochondria by iron and ascorbic acid (Laganiere and Yu 1987) in older animals. However, there is not complete agreement among the studies regarding the influence of CR on mitochondrial lipid peroxidation with other studies finding no change (Hagopian et al. 2005) or a slight increase (Lambert et al. 2004a) in old CR compared to control rats. In regard to proteins, it has been reported that oxidative damage is decreased in liver mitochondria from old CR versus control rats (Hagopian et al. 2005; Li et al. 2012) with one study indicating the CR-related changes depend on the marker measured with only N$^\varepsilon$-(carboxymethyl)lysine (a marker of lipid and protein oxidation) show-ing a decrease in old CR rats (Lambert et al. 2004a). The majority of the data support the notion that CR decreases age-related oxidative damage to mitochondria although the mechanisms responsible CR-related change are not clear.

FATTY ACID COMPOSITION

Aging

Age-related changes in mitochondrial phospholipid classes and fatty acid composi-tion could impact membrane function and susceptibility to peroxidation. There is evidence that aging increases the amount of liver mitochondrial cholesterol (Grinna 1977; Modi et al. 2008; Paradies and Ruggiero 1991; Vorbeck et al. 1982a). Aging also alters phospholipid composition in liver mitochondria with studies reporting a decrease in the amount of cardiolipin (Hagen et al. 1998; Vorbeck et al. 1982a) and either a decrease (Paradies and Ruggiero 1991) or no change (Grinna 1977) in cardiolipin as a proportion of total phospholipids. There is no agreement regarding the influence of aging on the relative amount of phosphatidylcholine (PC) and phos-phatidylethanolamine (PE), the major phospholipids in mitochondria, with studies finding an increase (Paradies and Ruggiero 1991) or no change (Grinna 1977) in PC and a decrease (Modi et al. 2008; Paradies and Ruggiero 1991) or no change (Grinna 1977) in PE with aging. The changes in cholesterol and cardiolipin observed in liver mitochondria from older animals may contribute to potential age-related changes in membrane fluidity and the activities of membrane proteins.

In addition to changes in phospholipid classes, aging is associated with alterations in membrane phospholipid fatty acid composition. In particular, it has been reported that aging increases the relative amount of double bonds (Barrett and Horton 1975), polyunsaturated fatty acids (PUFA) (Armeni et al. 2003), and n-3 PUFAs (Armeni et al. 2003; Lambert et al. 2004a) and decreases the relative amount of saturated fatty acids (Armeni et al. 2003; Lambert et al. 2004a) in liver mitochondria from old rats. The specific fatty acids altered by aging include a decrease in linoleic acid (C18:2 n-6) and an increase in 22 carbon n-3 PUFAs (Grinna 1977; Laganiere and Yu 1989, 1993; Paradies and Ruggiero 1991) in livers from older animals. These changes are consis-tent with an age-related increase in susceptibility to peroxidation in liver mitochondria.

Calorie Restriction

It has been proposed that CR increases life span by altering mitochondrial mem-brane composition in a manner which opposes lipid peroxidation (Yu et al. 2002). It was initially shown that the relative levels of linoleic acid were increased and long-chain n-3 fatty acids were decreased in liver mitochondria from 18- (Laganiere and Yu 1989) and 24-month-old (Laganiere and Yu 1993) CR rats compared to controls. The influence of CR on mitochondrial phospholipid fatty acid composi-tion, however, appears to depend on duration of CR, diet composition, and magni-tude of CR. A few studies have reported only minor changes in liver mitochondrial fatty acids with short-term CR (≤6 months) (Chen et al. 2013; Lambert et al. 2004a;

Ramsey et al. 2004) while one study found that CR-related changes in mitochondrial fatty acids increase with duration of CR and/or age (Lambert et al. 2004a). Another study showed that changes in mitochondrial fatty acid composition increased with greater levels of CR (Faulks et al. 2006). It has also been shown that dietary lipid composition has a major influence on liver mitochondrial fatty acid composition in CR animals (Chen et al. 2013). Overall, the existing data indicate that decreases in long-chain polyunsaturated fatty acids in liver mitochondria from old CR animals could play a role in preventing age-related oxidative damage to mitochondria.

CONCLUSION

There is considerable evidence that liver mitochondria undergo age-related changes which may have a negative impact on function. Liver mitochondrial and cristae numbers are decreased and mitochondrial volume is increased in hepatocytes with aging. Several studies have also shown that mitochondrial respiration and membrane potential are decreased in hepatocytes from old versus young animals. There is evidence that the activities of some ETC enzymes (complexes I, IV, and V) are decreased with aging and mitochondrial proton leak, ROS production, and oxidative damage to mitochondrial macromolecules are increased in liver mitochondria from older animals. Furthermore, mitochondrial membrane fatty acid composition is altered with aging in a manner which increases susceptibility to peroxidation. CR opposes many of these age-related changes in liver mitochondria. In particular, CR prevents age-related decreases in mitochondrial and cristae number. Several studies have shown that CR opposes age-related increases in mitochondrial proton leak and oxidative damage. There is also evidence that long-term CR decreases the levels of long-chain n-3 polyunsaturated fatty acid in mitochondrial membranes and may oppose age-related increases in mitochondrial susceptibility to peroxidation. Age-related changes in liver mitochondria are consistent with the idea that impaired mitochondrial function contributes to liver aging.

REFERENCES

Alemany, J., M. J. de la Cruz, I. Roncero, and J. Miquel. 1988. Effects of aging on respiration, ATP levels and calcium transport in rat liver mitochondria. Response to theophylline. *Exp Gerontol* 23 (1):25–34.

Armeni, T., G. Principato, J. L. Quiles, C. Pieri, S. Bompadre, and M. Battino. 2003. Mitochondrial dysfunctions during aging: Vitamin E deficiency or caloric restriction—Two different ways of modulating stress. *J Bioenerg Biomembr* 35 (2):181–191.

Ash, C. E. and B. J. Merry. 2011. The molecular basis by which dietary restricted feeding reduces mitochondrial reactive oxygen species generation. *Mech Ageing Dev* 132 (1–2):43–54.

Bakala, H., E. Delaval, M. Hamelin, J. Bismuth, C. Borot-Laloi, B. Corman, and B. Friguet. 2003. Changes in rat liver mitochondria with aging. Lon protease-like reactivity and N(epsilon)-carboxymethyllysine accumulation in the matrix. *Eur J Biochem* 270 (10):2295–2302.

Bakala, H., R. Ladouce, M. A. Baraibar, and B. Friguet. 2013. Differential expression and glycative damage affect specific mitochondrial proteins with aging in rat liver. *Biochim Biophys Acta* 1832 (12):2057–2067.

Barazzoni, R., K. R. Short, and K. S. Nair. 2000. Effects of aging on mitochondrial DNA copy number and cytochrome c oxidase gene expression in rat skeletal muscle, liver, and heart. *J Biol Chem* 275 (5):3343–3347.

Barogi, S., A. Baracca, G. Parenti Castelli, C. Bovina, G. Formiggini, M. Marchetti, G. Solaini, and G. Lenaz. 1995. Lack of major changes in ATPase activity in mitochondria from liver, heart, and skeletal muscle of rats upon ageing. *Mech Ageing Dev* 84 (2):139–150.

Barrett, M. C. and A. A. Horton. 1975. Age-related changes in lipids peroxidation in rat liver mitochondria. *Biochem Soc Trans* 3 (1):124–126.

Bejma, J., P. Ramires, and L. L. Ji. 2000. Free radical generation and oxidative stress with ageing and exercise: Differential effects in the myocardium and liver. *Acta Physiol Scand* 169 (4):343–351.

Braidy, N., G. J. Guillemin, H. Mansour, T. Chan-Ling, A. Poljak, and R. Grant. 2011. Age related changes in NAD+ metabolism oxidative stress and Sirt1 activity in wistar rats. *PLoS One* 6 (4):e19194.

Brookes, P. S. 2005. Mitochondrial H(+) leak and ROS generation: An odd couple. *Free Radic Biol Med* 38 (1):12–23.

Brouwer, A., C. F. Van Bezooijen, and D. L. Knook. 1977. Respiratory activities of hepatocytes isolated from rats of various ages. A brief note. *Mech Ageing Dev* 6 (4):265–269.

Cao, Z., J. Wanagat, S. H. McKiernan, and J. M. Aiken. 2001. Mitochondrial DNA deletion mutations are concomitant with ragged red regions of individual, aged muscle fibers: Analysis by laser-capture microdissection. *Nucleic Acids Res* 29 (21):4502–4508.

Capozza, G., F. Guerrieri, G. Vendemiale, E. Altomare, and S. Papa. 1994. Age related changes of the mitochondrial energy metabolism in rat liver and heart. *Arch Gerontol Geriatr* 19 (Suppl. 1):31–38.

Caro, P., J. Gomez, A. Arduini, M. Gonzalez-Sanchez, M. Gonzalez-Garcia, C. Borras, J. Vina, M. J. Puertas, J. Sastre, and G. Barja. 2010. Mitochondrial DNA sequences are present inside nuclear DNA in rat tissues and increase with age. *Mitochondrion* 10 (5):479–486.

Caro, P., J. Gomez, M. Lopez-Torres, I. Sanchez, A. Naudi, M. Portero-Otin, R. Pamplona, and G. Barja. 2008. Effect of every other day feeding on mitochondrial free radical production and oxidative stress in mouse liver. *Rejuvenation Res* 11 (3):621–629.

Caro, P., J. Gomez, I. Sanchez, A. Naudi, V. Ayala, M. Lopez-Torres, R. Pamplona, and G. Barja. 2009. Forty percent methionine restriction decreases mitochondrial oxygen radical production and leak at complex I during forward electron flow and lowers oxidative damage to proteins and mitochondrial DNA in rat kidney and brain mitochondria. *Rejuvenation Res* 12 (6):421–434.

Cassano, P., A. M. Lezza, C. Leeuwenburgh, P. Cantatore, and M. N. Gadaleta. 2004. Measurement of the 4,834-bp mitochondrial DNA deletion level in aging rat liver and brain subjected or not to caloric restriction diet. *Ann N Y Acad Sci* 1019:269–273.

Castro Mdel, R., E. Suarez, E. Kraiselburd, A. Isidro, J. Paz, L. Ferder, and S. Ayala-Torres. 2012. Aging increases mitochondrial DNA damage and oxidative stress in liver of rhesus monkeys. *Exp Gerontol* 47 (1):29–37.

Cavazzoni, M., S. Barogi, A. Baracca, G. Parenti Castelli, and G. Lenaz. 1999. The effect of aging and an oxidative stress on peroxide levels and the mitochondrial membrane potential in isolated rat hepatocytes. *FEBS Lett* 449 (1):53–56.

Chance, B. and G. R. Williams. 1955. Respiratory enzymes in oxidative phosphorylation. III. The steady state. *J Biol Chem* 217 (1):409–427.

Chen, Y., K. Hagopian, D. Bibus, J. M. Villalba, G. Lopez-Lluch, P. Navas, K. Kim, R. B. McDonald, and J. J. Ramsey. 2013. The influence of dietary lipid composition on liver mitochondria from mice following 1 month of calorie restriction. *Biosci Rep* 33 (1):83–95.

Chung, M. H., H. Kasai, S. Nishimura, and B. P. Yu. 1992. Protection of DNA damage by dietary restriction. *Free Radic Biol Med* 12 (6):523–525.

Cortopassi, G. A. and A. Wong. 1999. Mitochondria in organismal aging and degeneration. *Biochim Biophys Acta* 1410 (2):183–193.

Crescenzo, R., F. Bianco, I. Falcone, P. Coppola, A. G. Dulloo, G. Liverini, and S. Iossa. 2012. Mitochondrial energetics in liver and skeletal muscle after energy restriction in young rats. *Br J Nutr* 108 (4):655–665.

Dani, D., I. Shimokawa, T. Komatsu, Y. Higami, U. Warnken, E. Schokraie, M. Schnolzer, F. Krause, M. D. Sugawa, and N. A. Dencher. 2010. Modulation of oxidative phosphory-lation machinery signifies a prime mode of anti-ageing mechanism of calorie restriction in male rat liver mitochondria. *Biogerontology* 11 (3):321–334.

Darnold, J. R., M. L. Vorbeck, and A. P. Martin. 1990. Effect of aging on the oxidative phos-phorylation pathway. *Mech Ageing Dev* 53 (2):157–167.

Davies, S. M., A. Poljak, M. W. Duncan, G. A. Smythe, and M. P. Murphy. 2001. Measurements of protein carbonyls, ortho- and meta-tyrosine and oxidative phosphorylation complex activity in mitochondria from young and old rats. *Free Radic Biol Med* 31 (2):181–190.

De, A. K. and R. Darad. 1991. Age-associated changes in antioxidants and antioxidative enzymes in rats. *Mech Ageing Dev* 59 (1–2):123–128.

de Cavanagh, E. M., I. Flores, M. Ferder, F. Inserra, and L. Ferder. 2008. Renin-angiotensin system inhibitors protect against age-related changes in rat liver mitochondrial DNA content and gene expression. *Exp Gerontol* 43 (10):919–928.

de la Cruz, J., I. Buron, and I. Roncero. 1990. Morphological and functional studies during aging at mitochondrial level. Action of drugs. *Int J Biochem* 22 (7):729–735.

de Souza-Pinto, N. C., B. A. Hogue, and V. A. Bohr. 2001. DNA repair and aging in mouse liver: 8-oxodG glycosylase activity increase in mitochondrial but not in nuclear extracts. *Free Radic Biol Med* 30 (8):916–923.

Delaval, E., M. Perichon, and B. Friguet. 2004. Age-related impairment of mitochondrial matrix aconitase and ATP-stimulated protease in rat liver and heart. *Eur J Biochem* 271 (22):4559–4564.

Desai, V. G., R. Weindruch, R. W. Hart, and R. J. Feuers. 1996. Influences of age and dietary restriction on gastrocnemius electron transport system activities in mice. *Arch Biochem Biophys* 333 (1):145–151.

Dumas, J. F., D. Roussel, G. Simard, O. Douay, F. Foussard, Y. Malthiery, and P. Ritz. 2004. Food restriction affects energy metabolism in rat liver mitochondria. *Biochim Biophys Acta* 1670 (2):126–131.

Elia, M. 1992. Organ and tissue contribution to metabolic rate. In *Energy Metabolism: Tissue Determinants and Cellular Corollaries*, M. Elia, J. M. Kinney, and H. N. Tucker (eds.). New York: Raven. pp. 61–80.

Faulks, S. C., N. Turner, P. L. Else, and A. J. Hulbert. 2006. Calorie restriction in mice: Effects on body composition, daily activity, metabolic rate, mitochondrial reactive oxygen spe-cies production, and membrane fatty acid composition. *J Gerontol A Biol Sci Med Sci* 61 (8):781–794.

Garcia-Fernandez, M., I. Sierra, J. E. Puche, L. Guerra, and I. Castilla-Cortazar. 2011. Liver mitochondrial dysfunction is reverted by insulin-like growth factor II (IGF-II) in aging rats. *J Transl Med* 9:123.

Genova, M. L., C. Castelluccio, R. Fato, G. Parenti Castelli, M. Merlo Pich, G. Formiggini, C. Bovina, M. Marchetti, and G. Lenaz. 1995. Major changes in complex I activity in mitochondria from aged rats may not be detected by direct assay of NADH:coenzyme Q reductase. *Biochem J* 311 (Pt 1):105–109.

Gold, P. H., M. V. Gee, and B. L. Strehler. 1968. Effect of age on oxidative phosphorylation in the rat. *J Gerontol* 23 (4):509–512.

Gomez, J., P. Caro, A. Naudi, M. Portero-Otin, R. Pamplona, and G. Barja. 2007. Effect of 8.5% and 25% caloric restriction on mitochondrial free radical production and oxidative stress in rat liver. *Biogerontology* 8 (5):555–566.

Goodell, S. and G. Cortopassi. 1998. Analysis of oxygen consumption and mitochondrial permeability with age in mice. *Mech Ageing Dev* 101 (3):245–256.

Grande, F. 1980. Energy expenditure of organs and tissues. In *Assessment of Energy Metabolism in Health and Disease: Report of the First Ross Conference on Medical Research*, J. M. Kinney (ed.). Columbus, OH: Ross Laboratories. pp. 88–92.

Grattagliano, I., P. Portincasa, T. Cocco, A. Moschetta, M. Di Paola, V. O. Palmieri, and G. Palasciano. 2004. Effect of dietary restriction and *N*-acetylcysteine supplementation on intestinal mucosa and liver mitochondrial redox status and function in aged rats. *Exp Gerontol* 39 (9):1323–1332.

Gredilla, R., G. Barja, and M. Lopez-Torres. 2001. Effect of short-term caloric restriction on H_2O_2 production and oxidative DNA damage in rat liver mitochondria and location of the free radical source. *J Bioenerg Biomembr* 33 (4):279–287.

Grinna, L. S. 1977. Age related changes in the lipids of the microsomal and the mitochondrial membranes of rat liver and kidney. *Mech Ageing Dev* 6 (3):197–205.

Guerrieri, F., G. Capozza, M. Kalous, and S. Papa. 1992. Age-related changes of mitochondrial F_0F_1 ATP synthase. *Ann N Y Acad Sci* 671:395–402.

Guerrieri, F., G. Vendemiale, N. Turturro, A. Fratello, A. Furio, L. Muolo, I. Grattagliano, and S. Papa. 1996. Alteration of mitochondrial F_0F_1 ATP synthase during aging. Possible involvement of oxygen free radicals. *Ann N Y Acad Sci* 786:62–71.

Haak, J. L., G. R. Buettner, D. R. Spitz, and K. C. Kregel. 2009. Aging augments mitochondrial susceptibility to heat stress. *Am J Physiol Regul Integr Comp Physiol* 296 (3):R812–R820.

Hagen, T. M., C. M. Wehr, and B. N. Ames. 1998. Mitochondrial decay in aging. Reversal through supplementation of acetyl-L-carnitine and *N*-tert-butyl-α-phenyl-nitrone. *Ann N Y Acad Sci* 854:214–223.

Hagen, T. M., D. L. Yowe, J. C. Bartholomew, C. M. Wehr, K. L. Do, J. Y. Park, and B. N. Ames. 1997. Mitochondrial decay in hepatocytes from old rats: Membrane potential declines, heterogeneity and oxidants increase. *Proc Natl Acad Sci USA* 94 (7):3064–3069.

Hagopian, K., Y. Chen, K. Simmons Domer, R. Soo Hoo, T. Bentley, R. B. McDonald, and J. J. Ramsey. 2011. Caloric restriction influences hydrogen peroxide generation in mitochondrial sub-populations from mouse liver. *J Bioenerg Biomembr* 43 (3):227–236.

Hagopian, K., M. E. Harper, J. J. Ram, S. J. Humble, R. Weindruch, and J. J. Ramsey. 2005. Long-term calorie restriction reduces proton leak and hydrogen peroxide production in liver mitochondria. *Am J Physiol Endocrinol Metab* 288 (4):E674–E684.

Hagopian, K., R. Soo Hoo, J. A. Lopez-Dominguez, and J. J. Ramsey. 2013. Calorie restriction influences key metabolic enzyme activities and markers of oxidative damage in distinct mouse liver mitochondrial sub-populations. *Life Sci* 93 (24):941–948.

Hamelin, M., J. Mary, M. Vostry, B. Friguet, and H. Bakala. 2007. Glycation damage targets glutamate dehydrogenase in the rat liver mitochondrial matrix during aging. *FEBS J* 274 (22):5949–5961.

Hamilton, M. L., H. Van Remmen, J. A. Drake, H. Yang, Z. M. Guo, K. Kewitt, C. A. Walter, and A. Richardson. 2001. Does oxidative damage to DNA increase with age? *Proc Natl Acad Sci USA* 98 (18):10469–10474.

Hancock, C. R., D. H. Han, K. Higashida, S. H. Kim, and J. O. Holloszy. 2011. Does calorie restriction induce mitochondrial biogenesis? A reevaluation. *FASEB J* 25 (2):785–791.

Hansford, R. G. 1983. Bioenergetics in aging. *Biochim Biophys Acta* 726 (1):41–80.

Harman, D. 1956. Aging: A theory based on free radical and radiation chemistry. *J Gerontol* 11 (3):298–300.

Harman, D. 1972. The biologic clock: The mitochondria? *J Am Geriatr Soc* 20 (4):145–147.

Harman, D. 2006. Free radical theory of aging: An update: Increasing the functional life span. *Ann N Y Acad Sci* 1067:10–21.

Harper, M. E., L. Bevilacqua, K. Hagopian, R. Weindruch, and J. J. Ramsey. 2004. Ageing, oxidative stress, and mitochondrial uncoupling. *Acta Physiol Scand* 182 (4):321–331.

Harper, M. E., S. Monemdjou, J. J. Ramsey, and R. Weindruch. 1998. Age-related increase in mitochondrial proton leak and decrease in ATP turnover reactions in mouse hepatocytes. *Am J Physiol* 275 (2 Pt 1):E197–E206.

Haynes, V., N. J. Traaseth, S. Elfering, Y. Fujisawa, and C. Giulivi. 2010. Nitration of specific tyrosines in F_0F_1 ATP synthase and activity loss in aging. *Am J Physiol Endocrinol Metab* 298 (5):E978–E987.

Herbener, G. H. 1976. A morphometric study of age-dependent changes in mitochondrial population of mouse liver and heart. *J Gerontol* 31 (1):8–12.

Horton, A. A. and J. A. Spencer. 1981. Decline in respiratory control ratio of rat liver mitochondria in old age. *Mech Ageing Dev* 17 (3):253–259.

Ji, L. L., D. Dillon, and E. Wu. 1990. Alteration of antioxidant enzymes with aging in rat skeletal muscle and liver. *Am J Physiol* 258 (4 Pt 2):R918–R923.

Jove, M., A. Naudi, O. Ramirez-Nunez, M. Portero-Otin, C. Selman, D. J. Withers, and R. Pamplona. 2014. Caloric restriction reveals a metabolomic and lipidomic signature in liver of male mice. *Aging Cell* 13 (5):828–837.

Kamzalov, S. and R. S. Sohal. 2004. Effect of age and caloric restriction on coenzyme Q and α-tocopherol levels in the rat. *Exp Gerontol* 39 (8):1199–1205.

Khraiwesh, H., J. A. Lopez-Dominguez, L. Fernandez del Rio, E. Gutierrez-Casado, G. Lopez-Lluch, P. Navas, R. de Cabo, J. J. Ramsey, M. I. Buron, J. M. Villalba, and J. A. Gonzalez-Reyes. 2014. Mitochondrial ultrastructure and markers of dynamics in hepatocytes from aged, calorie restricted mice fed with different dietary fats. *Exp Gerontol* 56:77–88.

Khraiwesh, H., J. A. Lopez-Dominguez, G. Lopez-Lluch, P. Navas, R. de Cabo, J. J. Ramsey, J. M. Villalba, and J. A. Gonzalez-Reyes. 2013. Alterations of ultrastructural and fission/fusion markers in hepatocyte mitochondria from mice following calorie restriction with different dietary fats. *J Gerontol A Biol Sci Med Sci* 68 (9):1023–1034.

Ko, K. M., N. Chen, H. Y. Leung, E. P. Leong, M. K. Poon, and P. Y. Chiu. 2008. Long-term schisandrin B treatment mitigates age-related impairments in mitochondrial antioxidant status and functional ability in various tissues, and improves the survival of aging C57BL/6J mice. *Biofactors* 34 (4):331–342.

Kwong, L. K. and R. S. Sohal. 2000. Age-related changes in activities of mitochondrial electron transport complexes in various tissues of the mouse. *Arch Biochem Biophys* 373 (1):16–22.

Laganiere, S. and B. P. Yu. 1987. Anti-lipoperoxidation action of food restriction. *Biochem Biophys Res Commun* 145 (3):1185–1191.

Laganiere, S. and B. P. Yu. 1989. Effect of chronic food restriction in aging rats. II. Liver cytosolic antioxidants and related enzymes. *Mech Ageing Dev* 48 (3):221–230.

Laganiere, S. and B. P. Yu. 1993. Modulation of membrane phospholipid fatty acid composition by age and food restriction. *Gerontology* 39 (1):7–18.

Lambert, A. J. and B. J. Merry. 2005. Lack of effect of caloric restriction on bioenergetics and reactive oxygen species production in intact rat hepatocytes. *J Gerontol A Biol Sci Med Sci* 60 (2):175–180.

Lambert, A. J., M. Portero-Otin, R. Pamplona, and B. J. Merry. 2004a. Effect of ageing and caloric restriction on specific markers of protein oxidative damage and membrane peroxidizability in rat liver mitochondria. *Mech Ageing Dev* 125 (8):529–538.

Lambert, A. J., B. Wang, J. Yardley, J. Edwards, and B. J. Merry. 2004b. The effect of aging and caloric restriction on mitochondrial protein density and oxygen consumption. *Exp Gerontol* 39 (3):289–295.

Laurent, C., B. Chabi, G. Fouret, G. Py, B. Sairafi, C. Elong, S. Gaillet, J. P. Cristol, C. Coudray, and C. Feillet-Coudray. 2012. Polyphenols decreased liver NADPH oxidase activity, increased muscle mitochondrial biogenesis and decreased gastrocnemius age-dependent autophagy in aged rats. *Free Radic Res* 46 (9):1140–1149.

Lee, H. C. and Y. H. Wei. 2012. Mitochondria and aging. *Adv Exp Med Biol* 942:311–327.

Lenaz, G. 1998. Role of mitochondria in oxidative stress and ageing. *Biochim Biophys Acta* 1366 (1–2):53–67.

Li, X. D., I. Rebrin, M. J. Forster, and R. S. Sohal. 2012. Effects of age and caloric restriction on mitochondrial protein oxidative damage in mice. *Mech Ageing Dev* 133 (1):30–36.

Lopez-Cruzan, M. and B. Herman. 2013. Loss of caspase-2 accelerates age-dependent alterations in mitochondrial production of reactive oxygen species. *Biogerontology* 14 (2):121–130.

Lopez-Lluch, G., N. Hunt, B. Jones, M. Zhu, H. Jamieson, S. Hilmer, M. V. Cascajo, J. Allard, D. K. Ingram, P. Navas, and R. de Cabo. 2006. Calorie restriction induces mitochondrial biogenesis and bioenergetic efficiency. *Proc Natl Acad Sci USA* 103 (6):1768–1773.

Lopez-Torres, M., R. Gredilla, A. Sanz, and G. Barja. 2002. Influence of aging and long-term caloric restriction on oxygen radical generation and oxidative DNA damage in rat liver mitochondria. *Free Radic Biol Med* 32 (9):882–889.

Marshall, A., R. Lutfeali, A. Raval, D. N. Chakravarti, and B. Chakravarti. 2013. Differential hepatic protein tyrosine nitration of mouse due to aging—Effect on mitochondrial energy metabolism, quality control machinery of the endoplasmic reticulum and metabolism of drugs. *Biochem Biophys Res Commun* 430 (1):231–235.

Meihuizen, S. P. and N. Blansjaar. 1980. Stereological analysis of liver parenchymal cells from young and old rats. *Mech Ageing Dev* 13 (2):111–118.

Meng, Q., Y. T. Wong, J. Chen, and R. Ruan. 2007. Age-related changes in mitochondrial function and antioxidative enzyme activity in fischer 344 rats. *Mech Ageing Dev* 128 (3):286–292.

Miquel, J., M. L. Ferrandiz, E. De Juan, I. Sevila, and M. Martinez. 1995. *N*-acetylcysteine protects against age-related decline of oxidative phosphorylation in liver mitochondria. *Eur J Pharmacol* 292 (3–4):333–335.

Modi, H. R., S. S. Katyare, and M. A. Patel. 2008. Ageing-induced alterations in lipid/phospholipid profiles of rat brain and liver mitochondria: Implications for mitochondrial energy-linked functions. *J Membr Biol* 221 (1):51–60.

Musicco, C., V. Capelli, V. Pesce, A. M. Timperio, M. Calvani, L. Mosconi, L. Zolla, P. Cantatore, and M. N. Gadaleta. 2009. Accumulation of overoxidized peroxiredoxin III in aged rat liver mitochondria. *Biochim Biophys Acta* 1787 (7):890–896.

Nagata, T. and H. Ma. 2005. Electron microscopic radioautographic study of RNA synthesis in hepatocyte mitochondria of aging mouse. *Microsc Res Technol* 67 (2):55–64.

Navarro, A. and A. Boveris. 2007. The mitochondrial energy transduction system and the aging process. *Am J Physiol Cell Physiol* 292 (2):C670–C686.

Navarro, A., C. Gomez, J. M. Lopez-Cepero, and A. Boveris. 2004. Beneficial effects of moderate exercise on mice aging: Survival, behavior, oxidative stress, and mitochondrial electron transfer. *Am J Physiol Regul Integr Comp Physiol* 286 (3):R505–R511.

Nicholls, D. G. and M. W. Ward. 2000. Mitochondrial membrane potential and neuronal glutamate excitotoxicity: Mortality and millivolts. *Trends Neurosci* 23 (4):166–174.

Nisoli, E., C. Tonello, A. Cardile, V. Cozzi, R. Bracale, L. Tedesco, S. Falcone et al. 2005. Calorie restriction promotes mitochondrial biogenesis by inducing the expression of eNOS. *Science* 310 (5746):314–317.

Paradies, G. and F. M. Ruggiero. 1991. Effect of aging on the activity of the phosphate carrier and on the lipid composition in rat liver mitochondria. *Arch Biochem Biophys* 284 (2):332–337.

Puche, J. E., M. Garcia-Fernandez, J. Muntane, J. Rioja, S. Gonzalez-Baron, and I. Castilla Cortazar. 2008. Low doses of insulin-like growth factor-I induce mitochondrial protection in aging rats. *Endocrinology* 149 (5):2620–2627.

Quiles, J. L., J. J. Ochoa, M. C. Ramirez-Tortosa, J. R. Huertas, and J. Mataix. 2006. Age-related mitochondrial DNA deletion in rat liver depends on dietary fat unsaturation. *J Gerontol A Biol Sci Med Sci* 61 (2):107–114.

Rafique, R., A. H. Schapira, and J. M. Coper. 2004. Mitochondrial respiratory chain dysfunction in ageing; influence of vitamin E deficiency. *Free Radic Res* 38 (2):157–165.

Ramanujan, V. K. and B. A. Herman. 2007. Aging process modulates nonlinear dynamics in liver cell metabolism. *J Biol Chem* 282 (26):19217–19226.

Ramsey, J. J. and K. Hagopian. 2007. Bioenergetics. In *Encyclopedia of Gerontology*, J. E. Birren (ed.). Boston, MA: Elsevier/Academic Press. pp. 170–177.

Ramsey, J. J., K. Hagopian, T. M. Kenny, E. K. Koomson, L. Bevilacqua, R. Weindruch, and M. E. Harper. 2004. Proton leak and hydrogen peroxide production in liver mitochondria from energy-restricted rats. *Am J Physiol Endocrinol Metab* 286 (1):E31–E40.

Ramsey, J. J., M. E. Harper, and R. Weindruch. 2000. Restriction of energy intake, energy expenditure, and aging. *Free Radic Biol Med* 29 (10):946–968.

Rebrin, I., C. Bregere, S. Kamzalov, T. K. Gallaher, and R. S. Sohal. 2007. Nitration of tryptophan 372 in succinyl-CoA:3-ketoacid CoA transferase during aging in rat heart mitochondria. *Biochemistry* 46 (35):10130–10144.

Rebrin, I., S. Kamzalov, and R. S. Sohal. 2003. Effects of age and caloric restriction on glutathione redox state in mice. *Free Radic Biol Med* 35 (6):626–635.

Rikans, L. E., C. D. Snowden, and D. R. Moore. 1992. Effect of aging on enzymatic antioxidant defenses in rat liver mitochondria. *Gerontology* 38 (3):133–138.

Rocha-Rodrigues, S., E. Santos-Alves, P. M. Coxito, I. Marques-Aleixo, E. Passos, J. T. Guimaraes, M. J. Martins, P. J. Oliveira, J. Magalhaes, and A. Ascensao. 2013. Combined effects of aging and in vitro non-steroid anti-inflammatory drugs on kidney and liver mitochondrial physiology. *Life Sci* 93 (8):329–337.

Rojas, C., S. Cadenas, R. Perez-Campo, M. Lopez-Torres, R. Pamplona, J. Prat, and G. Barja. 1993. Relationship between lipid peroxidation, fatty acid composition, and ascorbic acid in the liver during carbohydrate and caloric restriction in mice. *Arch Biochem Biophys* 306 (1):59–64.

Sanchez-Roman, I., A. Gomez, I. Perez, C. Sanchez, H. Suarez, A. Naudi, M. Jove, M. Lopez-Torres, R. Pamplona, and G. Barja. 2012. Effects of aging and methionine restriction applied at old age on ROS generation and oxidative damage in rat liver mitochondria. *Biogerontology* 13 (4):399–411.

Sanz, A., R. Gredilla, R. Pamplona, M. Portero-Otin, E. Vara, J. A. Tresguerres, and G. Barja. 2005. Effect of insulin and growth hormone on rat heart and liver oxidative stress in control and caloric restricted animals. *Biogerontology* 6 (1):15–26.

Sastre, J., F. V. Pallardo, R. Pla, A. Pellin, G. Juan, J. E. O'Connor, J. M. Estrela, J. Miquel, and J. Vina. 1996. Aging of the liver: Age-associated mitochondrial damage in intact hepatocytes. *Hepatology* 24 (5):1199–11205.

Sato, T. and H. Tauchi. 1975. The formation of enlarged and giant mitochondria in the aging process of human hepatic cells. *Acta Pathol Jpn* 25 (4):403–412.

Sawada, M. and J. C. Carlson. 1987. Changes in superoxide radical and lipid peroxide formation in the brain, heart and liver during the lifetime of the rat. *Mech Ageing Dev* 41 (1–2):125–137.

Serviddio, G., F. Bellanti, A. D. Romano, R. Tamborra, T. Rollo, E. Altomare, and G. Vendemiale. 2007. Bioenergetics in aging: Mitochondrial proton leak in aging rat liver, kidney and heart. *Redox Rep* 12 (1):91–95.

Shigenaga, M. K., T. M. Hagen, and B. N. Ames. 1994. Oxidative damage and mitochondrial decay in aging. *Proc Natl Acad Sci USA* 91 (23):10771–10778.

Sohal, R. S., L. A. Arnold, and B. H. Sohal. 1990. Age-related changes in antioxidant enzymes and prooxidant generation in tissues of the rat with special reference to parameters in two insect species. *Free Radic Biol Med* 9 (6):495–500.

Sohal, R. S. and R. Weindruch. 1996. Oxidative stress, caloric restriction, and aging. *Science* 273 (5271):59–63.

Souza-Pinto, N. C., D. L. Croteau, E. K. Hudson, R. G. Hansford, and V. A. Bohr. 1999. Age-associated increase in 8-oxo-deoxyguanosine glycosylase/AP lyase activity in rat mitochondria. *Nucleic Acids Res* 27 (8):1935–1942.

Stocco, D. M. and J. C. Hutson. 1978. Quantitation of mitochondrial DNA and protein in the liver of Fischer 344 rats during aging. *J Gerontol* 33 (6):802–809.

Szczesny, B. and S. Mitra. 2005. Effect of aging on intracellular distribution of abasic (AP) endonuclease 1 in the mouse liver. *Mech Ageing Dev* 126 (10):1071–1078.

Szczesny, B., A. W. Tann, and S. Mitra. 2010. Age- and tissue-specific changes in mitochondrial and nuclear DNA base excision repair activity in mice: Susceptibility of skeletal muscles to oxidative injury. *Mech Ageing Dev* 131 (5):330–337.

Tate, E. L. and G. H. Herbener. 1976. A morphometric study of the density of mitochondrial cristae in heart and liver of aging mice. *J Gerontol* 31 (2):129–134.

Tauchi, H. and T. Sato. 1968. Age changes in size and number of mitochondria of human hepatic cells. *J Gerontol* 23 (4):454–461.

Tauchi, H., T. Sato, M. Hoshino, H. Kobayashi, F. Adachi, J. Aoki, and T. Masuko. 1964. Studies on correlation between ultrastructure and enzymatic activities of the parenchymal cells in senescence. In *Age with a Future*, P. F. Hansen (ed.). Copenhagen, Denmark: Munksgaard. pp. 203–211.

Tian, L., Q. Cai, and H. Wei. 1998. Alterations of antioxidant enzymes and oxidative damage to macromolecules in different organs of rats during aging. *Free Radic Biol Med* 24 (9):1477–1484.

Torres-Mendoza, C. E., A. Albert, and M. J. de la Cruz Arriaga. 1999. Molecular study of the rat liver NADH: Cytochrome c oxidoreductase complex during development and ageing. *Mol Cell Biochem* 195 (1–2):133–142.

Uysal, M., S. Seckin, N. Kocak-Toker, and H. Oz. 1989. Increased hepatic lipid peroxidation in aged mice. *Mech Ageing Dev* 48 (1):85–89.

Vorbeck, M. L., A. P. Martin, J. W. Long, Jr., J. M. Smith, and R. R. Orr, Jr. 1982a. Aging-dependent modification of lipid composition and lipid structural order parameter of hepatic mitochondria. *Arch Biochem Biophys* 217 (1):351–361.

Vorbeck, M. L., A. P. Martin, J. K. Park, and J. F. Townsend. 1982b. Aging-related decrease in hepatic cytochrome oxidase of the Fischer 344 rat. *Arch Biochem Biophys* 214 (1):67–79.

Walsh, M. E., Y. Shi, and H. Van Remmen. 2014. The effects of dietary restriction on oxidative stress in rodents. *Free Radic Biol Med* 66:88–99.

Wanagat, J., Z. Cao, P. Pathare, and J. M. Aiken. 2001. Mitochondrial DNA deletion mutations colocalize with segmental electron transport system abnormalities, muscle fiber atrophy, fiber splitting, and oxidative damage in sarcopenia. *FASEB J* 15 (2):322–332.

Weindruch, R. H., M. K. Cheung, M. A. Verity, and R. L. Walford. 1980. Modification of mitochondrial respiration by aging and dietary restriction. *Mech Ageing Dev* 12 (4):375–392.

Wilson, P. D. and L. M. Franks. 1975a. The effect of age on mitochondrial ultrastructure. *Gerontologia* 21 (2):81–94.

Wilson, P. D. and L. M. Franks. 1975b. The effect of age on mitochondrial ultrastructure and enzyme cytochemistry. *Biochem Soc Trans* 3 (1):126–128.

Yen, T. C., Y. S. Chen, K. L. King, S. H. Yeh, and Y. H. Wei. 1989. Liver mitochondrial respiratory functions decline with age. *Biochem Biophys Res Commun* 165 (3):944–1003.

Yen, T. C., K. L. King, H. C. Lee, S. H. Yeh, and Y. H. Wei. 1994. Age-dependent increase of mitochondrial DNA deletions together with lipid peroxides and superoxide dismutase in human liver mitochondria. *Free Radic Biol Med* 16 (2):207–214.

Yowe, D. L. and B. N. Ames. 1998. Quantitation of age-related mitochondrial DNA deletions in rat tissues shows that their pattern of accumulation differs from that of humans. *Gene* 209 (1–2):23–30.

Yu, B. P., J. J. Chen, C. M. Kang, M. Choe, Y. S. Maeng, and B. S. Kristal. 1996. Mitochondrial aging and lipoperoxidative products. *Ann N Y Acad Sci* 786:44–56.

Yu, B. P., B. O. Lim, and M. Sugano. 2002. Dietary restriction downregulates free radical and lipid peroxide production: Plausible mechanism for elongation of life span. *J Nutr Sci Vitaminol (Tokyo)* 48 (4):257–264.

Zhao, L., X. Zou, Z. Feng, C. Luo, J. Liu, H. Li, L. Chang, H. Wang, Y. Li, J. Long, and F. Gao. 2014. Evidence for association of mitochondrial metabolism alteration with lipid accumulation in aging rats. *Exp Gerontol* 56:3–12.

7 Drug Delivery to Mitochondria

A Promising Approach for the Treatment of Liver Disease

*Bhuvaneshwar Vaidya, Samit Shah,
and Vivek Gupta*

CONTENTS

ABSTRACT

In recent years, mitochondria have gained significant attention from researchers due to their specific role in various physiological and cellular processes. It has also been reported that mitochondria play a critical role in liver diseases and, thus, have also been considered a potential target to treat liver diseases. Mitochondria, being cellular organelles, are located in the cell cytoplasm, which makes drug entry into mitochondria a challenging task due to the presence of various biological barriers. Therefore, several strategies have been designed to combat these barriers and to specifically target mitochondria, which include direct drug modification and encapsulation of drugs in novel carriers. Further, to increase cell specificity, these carriers can also be modified with ligands specific to cell surface receptors. In this chapter, we discuss mitochondrial properties which are exploited to develop novel approaches and compile studies involving the use of these approaches for the delivery of drugs/DNA to the mitochondria. We also discuss cell-specific delivery systems to target mitochondria of liver cells, specifically for the treatment of liver diseases related to mitochondrial dysfunction.

INTRODUCTION

Recent advances in molecular biology, cell biology, and drug delivery have led to the development of drugs aimed at exploiting novel cellular and subcellular targets. Delivery of such drugs to their cellular target may result in enhanced therapeutic efficacy while reducing nontarget interactions and toxicities. These drugs should be designed in a way that they have high specificity to meet their target. Delivering therapeutics to the liver has always been an interesting area of research, mainly due to the inability of current therapies to provide adequate treatment for a variety of severe liver ailments. It has been investigated that cellular organelles like mitochondria play a central role in pathology of various liver diseases (Sastre et al., 2007). Several diseases are associated with altered mitochondrial structure and function, which leads to the production of low energy and highly reactive oxygen species (ROS). Mutations in mitochondrial DNA (mtDNA) may also cause changes in normal physiology (Serviddio et al., 2010). Therefore, drug delivery to liver mitochondria so as to act on specific target involved in pathophysiology of liver diseases can be beneficial for an effective therapeutic intervention. However, targeted delivery of a drug molecule to the mitochondria is quite difficult without a transporter molecule at the surface of the membrane. Furthermore, targeting drugs to mitochondria of specific cells such as hepatocytes requires drugs to be encapsulated in efficient carrier systems, decorated with precise targeting moieties, so as to target the mitochondria-related liver diseases.

Targeting of drugs to any cellular organelle can be approached by either active or passive targeting. Active targeting relies on exploiting specific properties of the target where drug needs to be delivered. These properties may include antigen–antibody interactions and ligand–receptor interactions. On the other hand, passive targeting is based on the physicochemical properties of the carrier like size, shape, mass, charge, and solubility. These properties can be modified according to the physiology of target organs, cells, or organelles. Although, small molecular weight drugs can also be targeted to the mitochondria by passive targeting, the use of novel carriers such as lipid and polymeric nanocarriers is the most effective approach (Heller et al., 2012).

DRUG DELIVERY TO MITOCHONDRIA: EXPLOITING THE PHYSIOLOGICAL CHARACTERISTICS OF MITOCHONDRIA

Structurally, mitochondria are thread-like elongated structures containing four major compartments, that is, an outer membrane, inner membrane, intermembrane space, and matrix. Like all membranes, mitochondrial membranes are composed of a phospholipid bilayer with embedded proteins. The outer membrane, like plasma membrane, is highly porous and permeable to small molecules (molecular weight <6 kDa). These molecules diffuse through pores in the membrane formed by a membrane spanning protein called *porin*. The inner membrane is highly impermeable and differs in composition from the outer membrane. The inner membrane contains a higher protein to lipid ratio (3:1) and an unusual phospholipid cardiolipin (Mukhopadhyay and Weiner, 2007). In addition, it also contains enzymes of oxidative phosphorylation (respiratory chain) and metabolite transport proteins. Cristae formation in mitochondria happens due to the nature of maximum surface area acquirement by the membrane, while impermeable nature of the membrane helps in balancing the distribution of protons between matrix and cytosol that develop a driven force for ATP synthesis (Paliwal et al., 2007). Intermembrane space contains large numbers of specialized proteins. The mitochondrial matrix is the innermost space enclosed by cristae membrane and contains enzymes for a number of metabolic process including the citric acid cycle, oxidation of fatty acid molecules, and the urea cycle. The matrix also contains several mtDNA copy numbers (Vaidya et al., 2009a).

Two specific properties of mitochondria, high inner membrane potential and protein import machinery, help in the delivery of any drug/carrier system into the mitochondria. These mitochondrial properties can also be utilized in the transport of any therapeutic molecule from cytoplasm of specific cells into mitochondria, or into isolated mitochondria in in vitro studies. High mitochondrial membrane potential helps in the accumulation of positively charged lipophilic cationic molecules or carriers selectively inside the mitochondria, whereas protein import machinery helps to deliver specific peptides to the mitochondria (Vaidya et al., 2009a).

MITOCHONDRIAL MEMBRANE POTENTIAL

As discussed earlier, mitochondria, the powerhouse of the cell, synthesizes ATP by oxidative phosphorylation through respiratory electron transport chain (ETC). Electron transfer through respiratory chain is coupled with the pumping of protons from the mitochondrial matrix to the intermembrane space through inner mitochondrial membrane against a concentration gradient. This process leads to generation of an electrochemical potential gradient, comprising a membrane potential (negative inside) and a pH gradient (basic inside) across the inner membrane (Mitchell, 1961). The potential gradient promotes ATP synthesis by F_0F_1 ATP synthase. When required, protons reenter the matrix through F_0 portion of ATP synthase whereas F_1 portion helps in the conversion of ADP into ATP.

The membrane potential in isolated mitochondria has been experimentally measured to be around 180–200 mV, which is the maximum potential any lipid membrane can bear without breaking (Murphy, 1989). However, it is relatively low in living cells (130–150 mV) (Azzone et al., 1984; Murphy and Smith, 2000). High negative mitochondrial membrane potential as compared to membrane potential of plasma membrane (30–60 mV) potentiates preferential accumulation of lipophilic cations toward the mitochondria (Burns and Murphy, 1997).

The distribution of lipophilic cations across the mitochondrial membrane is promoted by high negative potential of mitochondrial membrane as described by the following Nernst equation:

$$\text{Membrane potential (mV)} = 61.5 \times \log 10 \times \frac{[(\text{cation})_{\text{in}}]}{[(\text{cation})_{\text{out}}]}$$

The previous equation describes the equilibrium distribution of lipophilic cations at 37°C, when they do not undergo any metabolism or selective transport. This equation also illustrates that accumulation of lipophilic cations molecules increases 10 times for every increment of 61.5 mV in mitochondrial membrane potential, resulting in 100–1000 times higher accumulation of molecules in the mitochondria as compared to the cytoplasm of the cell (Figure 7.1) (Murphy and Smith, 2000). The ability of mitochondria to preferentially accumulate lipophilic cations has been exploited for mitochondrial delivery of biomolecules using mitochondriotropics (lipophilic cations) as carrier systems (Murphy, 2008), which will be discussed in detail in the "Mitochondriotropics" section of this chapter.

MITOCHONDRIAL PROTEIN IMPORT MACHINERY

Most of the polypeptides utilized by mitochondria during the metabolic process are synthesized by nuclear DNA, with only 13 polypeptides being synthesized by the mitochondria genome. These polypeptides are transported from the cytoplasm to the

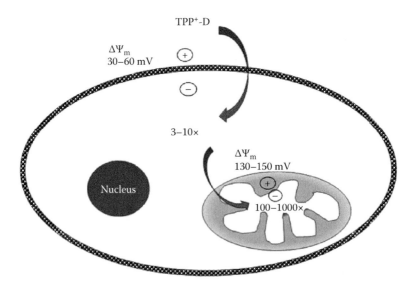

FIGURE 7.1 Schematic representation of drug uptake by mitochondria within cells due to mitochondrial membrane potential. Drug (D) conjugated with mitochondriotropics (triphenylphosphonium cations) can also be delivered. Δψ represents mitochondrial and plasma membrane potentials.

mitochondrial membrane and are then processed and assembled at the mitochondrial site of oxidative phosphorylation. The nuclear DNA generated polypeptides are transported from cytoplasm to the mitochondria utilizing a specific complex mitochondrial protein import machinery. This machinery also helps in the processing and assembling of these polypeptides to the correct protein and placed at right place (Neupert, 1997; Pfanner et al., 1994, 1997; Schatz, 1996). Proteins translated in the cytoplasm either by mtDNA or nuclear DNA are transported to the mitochondria through a specialized targeting sequence or signal peptide known as the *Mitochondria Leader Sequence* (MLS).

The polypeptide transport process is a three-step process (Figure 7.2). The first step involves binding of the target sequence, with the help of cytosolic chaperones, to the receptor on the mitochondrial outer membrane and passage of protein through a hydrophilic pore in the outer membrane by translocase of the outer mitochondrial membrane machinery (TOM). In the second step, target sequence enters into the inner membrane translocase of the inner mitochondrial membrane machinery (TIM) in a membrane potential-dependent process, and the protein is threaded through a hydrophilic pore in the inner membrane. The third step of the process is cleavage of signal sequence in the matrix by processing peptides and assembling of polypeptides into the final protein (Weissig et al., 2004; Wiedemann et al., 2004). The detailed process of mitochondrial protein import machinery is found in published review literature (Dudek et al., 2013; Harbauer et al., 2014). These signal peptides have been explored by researchers for the delivery of various therapeutic proteins or peptides to the mitochondria, which are discussed in detail later in the chapter.

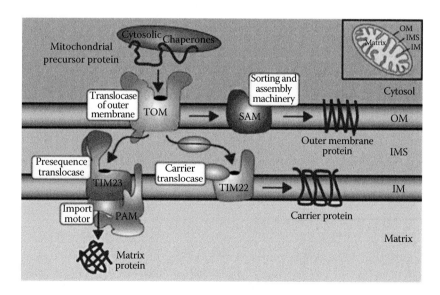

FIGURE 7.2 (**See color insert.**) Protein import machinery into mitochondria. (From Wiedemann, N. et al., *J. Biol. Chem.*, 279, 14473, 2004.)

STRATEGIES FOR DRUG DELIVERY TO MITOCHONDRIA

Many therapeutic agents have been shown to target mitochondria for the treatment of various diseases that are known to progress by altering the functioning of mitochondrial membrane, mtDNA, or matrix. These agents include antioxidants, plasmid DNA, and apoptosis-inducing agents. A number of studies dealing with the exploration of various routes and strategies for the delivery of such agents and studying their efficacy have been reported. Numerous delivery approaches have been designed in order to exploit physiological characteristics of mitochondria. As summarized in Figure 7.3, these may include conjugation of therapeutic molecules to the mitochondriotropic or MLS, and encapsulation of drug/DNA within carriers such as DQAsomes and liposomes that have been engineered to selectively target mitochondria (Figure 7.3). In addition, carrier systems can be custom designed to deliver molecules selectively to mitochondria. However, each carrier system also suffers from some drawbacks that should be considered before selection of suitable carriers for mitochondrial delivery of therapeutic molecules (Table 7.1).

Mitochondriotropics

As discussed earlier in the chapter, mitochondriotropics are molecules that have a specificity toward mitochondria. They accumulate into mitochondria by utilizing the high negative membrane potential of mitochondrial membrane. They are naturally amphiphilic; that is, they have a hydrophilic charged center along with a hydrophobic core. Further, the π-electron charge density for mitochondriotropics extends to three or more atoms and positive charge distributes to two or more atoms. Because of delocalization of positive charge, these compounds are also known as Delocalized Cations (DLCs) (Weissig and Torchilin, 2001). Delocalization of the positive charge helps to reduce the free energy change when moving from an aqueous (cytoplasm) to a hydrophobic environment (mitochondrial membrane). Structural features of these molecules have been considered to be critical for their localization into the mitochondrial matrix. Sufficient lipophilicity, combined with positive charge delocalization, helps in the accumulation of these compounds into the mitochondria. Mitochondriotropics accumulate in mitochondria in response to mitochondrial membrane potential and accumulation (greater or lesser) is dependent on the degree of charge localization. Several compounds in this category have been explored by different research groups for transportation of therapeutic molecules to the mitochondria and have been reviewed by various authors (Paliwal et al., 2007; Weissig and Torchilin, 2001). Most commonly explored compounds include triphenyl phosphonium ions (TPP$^+$), dequalinium chloride, rhodamine 123, and cyanine dyes such as N,N'-bis(2-ethyl-1,3-dioxolane)kryptocyanine and Victoria Blue BO (Figure 7.4). These molecules have the ability to carry small molecule weight drugs (molecular weight <500 Da) with them through lipid bilayer and to deliver drugs to the mitochondria within cells.

Various research groups have explored mitochondriotropics for the delivery of antioxidants selectively to the mitochondria of live cells (James et al., 2005; Kelso et al., 2001; Murphy and Smith, 2007; Smith et al., 1999, 2003). It has been reported that conjugation of vitamin E to the lipophilic cation, TPP$^+$, enhances the

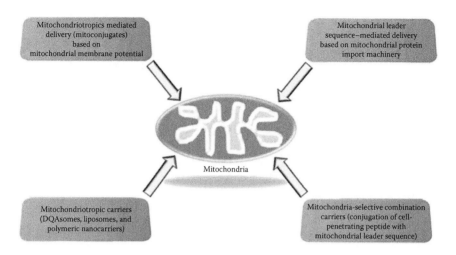

FIGURE 7.3   Strategies for the selective delivery of drug molecules to the mitochondria.

TABLE 7.1
Summary of Mitochondria Targeting Carrier Systems

Types of Carrier Systems	Advantages	Limitations
Mitochondriotropics	• Simple conjugation • Small targeting vector • Easy access to blood–brain barrier • Low toxicity and rapid uptake in vivo	• High molecular weight drugs cannot be delivered
Mitochondrial leader sequence (MLS) or mitochondrial-targeting signal (MTS)	• Simple solid phase synthesis • Easily tunable for different cargoes and applications • Simple conjugation • Relatively small targeting vector	• Limited in vivo data • High molecular weight drugs cannot be delivered
Liposomes	• Can deliver high molecular weight or charged drugs • No conjugation required • Tunable for different cargoes and applications	• Large size can lead to immune response or toxicity • Complex synthesis
Nanoparticles	• Can deliver high mol. wt. or charged drugs • No conjugation required • Tunable for different cargoes and applications	• Limited in vivo data • Complex synthesis

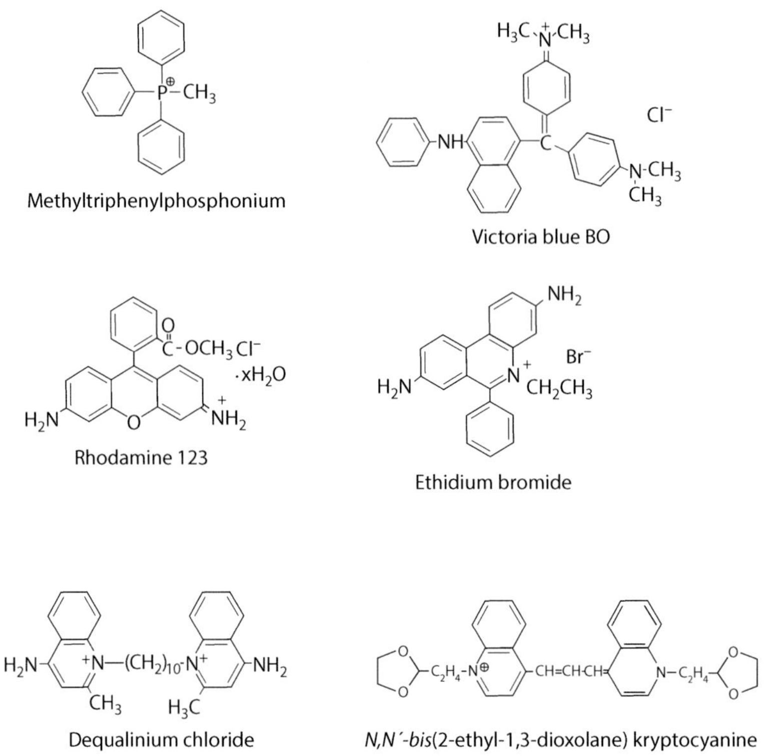

FIGURE 7.4 Chemical structures of representative mitochondriotropic compounds.

accumulation of vitamin E into the mitochondria up to 1000-fold. This might be because of the high negative transmembrane potential (200 mV) of isolated mitochondria. Later, Skulachev et al. (2009) reported that by conjugating antioxidant (plastoquine) to the mitochondria-penetrating cations through a linker, it can be targeted to the mitochondria for treatment of senescence and age-related diseases. In this study, they used TPP+ and rhodamine 19 as mitochondria-penetrating cations.

Mitochondriotropics have also been tested in various liver diseases using either mitochondria isolated from liver cells or intact liver cells. Filipovska et al. (2005) reported that peroxidase mimetic ebselen conjugated with TPP+, termed MitoPeroxidase, protects liver mitochondrial membranes from lipid peroxidation. Further, it has also been reported that mitochondria-targeted antioxidant MitoQ protects liver damage in a lipopolysaccharide–peptidoglycan model of sepsis in vivo in rat model, by preventing mitochondrial membrane potential depletion (Lowes et al., 2008) (Figure 7.5). In this model, damage to the organs (kidney and liver) was measured by release of creatinine and alanine amino transferase (ALT), respectively from organs in the plasma, whereas damage of mitochondria was assessed by their ability to sustain membrane potential across the inner membrane using JC-1 dye, which accumulates in intact mitochondria and fluoresces when excited.

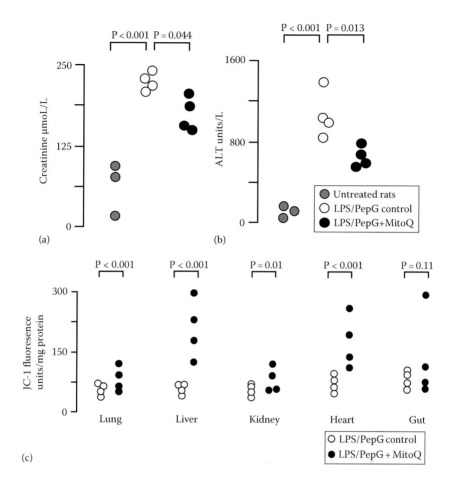

FIGURE 7.5 (a) Plasma creatinine concentrations and (b) plasma ALT activity in untreated rats and rats treated with LPS/PepG+ dTPP control or LPS/PepG+ MitoQ. (c) Mitochondrial membrane potential in mitochondria from organs from rats treated with LPS/PepG+ dTPP control and rats treated with LPS/PepG+ MitoQ. Results from individual rats are shown. (From Lowes, D.A. et al., *Free Radic. Biol. Med.*, 45, 1559, 2008. With permission.)

In another study, Whiteman et al. reported that MitoQ protects liver cells against hypochlorous acid–induced cell death and mitochondrial damage, which suggests the use of MitoQ in the preservation of hepatocytes during chronic inflammation (Whiteman et al., 2008). Resveratrol (3,4′,5-trihydroxy-*trans*-stilbene) has also been conjugated to TPP[+], and it was shown to preferentially accumulate in highly energized mitochondria of liver cells and demonstrated antiproliferative and proapoptotic effects against fast growing cells, suggesting its application as anticancer agent against xenografts and mutagen-induced cancers (Biasutto et al., 2008). Blaikie et al. (2006) also reported that conjugation of TPP[+] with a protonophore 2,4-dinitrophenol (DNP) and MitoDNP suppresses its protonophoric activity. Protonophores are molecules that increase proton leak from mitochondria thus leading to toxicity.

Very recently, a small molecule mitochondrial kinase inhibitor dichloroacetate (DCA) has also been targeted to mitochondria by using TPP$^+$ as a mitochondria-targeting moiety (Pathak et al., 2014). In this study, Mito-DCA was designed by incorporating multiple molecules of DCA via *tris*(hydroxymethyl)-aminomethane (*Tris*) using esterase-labile ester bonds whereas TPP cation was introduced via a stable amide linkage. The results indicated that targeting mitochondrial metabolic inhibitors (DCA) selectively to the mitochondria using TPP can improve the antitumor immunity of the DCA. Further, it has also been suggested that targeted molecular scaffold developed using multiple copies of DCA, Mito-DCA, can be a promising approach to deliver multiple copies of DCA using a single molecule of TPP$^+$ as targeting moiety. The results from this study could be extended for the delivery of several therapeutic molecules to the mitochondria by the use of mitochondriotropics.

CONJUGATION TO THE MITOCHONDRIAL LEADER SEQUENCE/ MITOCHONDRIAL-TARGETING SIGNAL

Another strategy for targeted delivery of bioactive molecules to the mitochondria is to explore mitochondrial protein import machinery for the transport of therapeutic proteins and peptides to the mitochondrial matrix. The signaling peptides also known as mitochondrial-targeting signal (MTS) or MLS are short peptides (3–60 amino acids long) consisting of hydrophobic amino acids and positively charged amino acids in an alternating pattern that can translocate into mitochondria (Vaidya et al., 2009a).

The MTS can be used for delivery of other nonmitochondrial proteins to the mitochondrial matrix via the protein import pathway. Green fluorescent protein and cytosolic enzymes such as dihydrofolate reductase and cytochrome C oxidase have been targeted to the mitochondria with the help of MTS (Murphy, 1997; Verner and Lemire, 1989). The MTS can also be utilized for gene therapy to repair/replace defects in the mitochondrial genome. Several studies have explored MTS to introduce ODN, dsDNA, or PNA to the isolated mitochondria by conjugating these biomolecules to the MTS. These conjugates utilize mitochondrial protein import machinery for the transport from cytoplasm to the mitochondrial membrane (Flierl et al., 2003; Vestweber and Schatz, 1989). It has been reported that the previous strategy can be utilized to transport DNA of 17–322 bp in length (Seibel et al., 1995). Thus, based on the earlier studies, it can be concluded that macromolecules such as DNA, protein therapeutics, and enzymes can be targeted to mitochondria after coupling with MTS. Utilization of protein import machinery and coupling therapeutic molecules with MTS thus seems to be a promising approach for the treatment of various liver diseases involving mtDNA like chronic cholestasis, viral hepatitis, chronic ethanol exposure, and non-alcohol-induced steatohepatitis (NASH).

Earlier reports have suggested that cell-penetrating peptides (CPP) or protein transduction domain (PTD) could be utilized to deliver biomolecules or carrier systems to the cytosol directly. These peptides have also been utilized in combination with MTS for the delivery of exonuclease III protein (Exo III) to the mitochondria matrix

(Shokolenko et al., 2005). A MTS–Exo III–TAT-fusion protein was constructed by the fusion of MTS and TAT with Exo III at the N- and C-terminus, respectively. In this system, PTD enhances cytoplasmic delivery while MTS helps to transport from cytoplasm to mitochondria. Later, other combination systems consisting of mitochondrial leader peptide (LP) and polyethyleneimine (PEI) have also been developed for the delivery of DNA to mitochondria (Choi et al., 2006). Low molecular weight PEI (<2000 Da) was used to synthesize mitochondrial LP–conjugated PEI (PEI-LP). Results of the study demonstrated that PEI-LP is more effective at delivering DNA to the mitochondria as compared to naked DNA or PEI alone. This study demonstrated the use of PEI-LP as a novel carrier system with low toxicity for mitochondrial gene therapy (Choi et al., 2006).

The previous discussed strategies are particularly important for macromolecular delivery to the mitochondria, but can very well be extended for delivery of small molecular weight therapeutics. In a recent study, Lin et al. (2015) developed a dual peptide conjugation strategy for delivery of therapeutic agents and diagnostic agents to the mitochondria. In this study, 5-carboxyfluorescein (5-FAM), non-membrane-penetrating dye, was used as a model drug. Peptides, MTS as MTS and R8 as CPP sequences, were chemically conjugated with the dye for enhancing mitochondrial delivery. Results of the study revealed improved uptake and effective mitochondrial targeting of 5-FAM. This strategy can effectively be applied for targeted delivery of therapeutics to mitochondria of intact cells.

NOVEL CARRIER SYSTEMS FOR MITOCHONDRIAL DELIVERY OF THERAPEUTICS

Designing of mitochondria-selective novel carriers systems, instead of conjugating bioactive moieties to a mitochondriotropic molecule, has recently evolved as another interesting approach in the field of targeted mitochondrial therapeutics (Pathak et al., 2015; Wongrakpanich et al., 2014). In this section, a variety of these strategies will be discussed.

DQAsome: A Mitochondria-Selective Vesicular System

DQAsomes are di-cationic amphiphilic vesicles which localize in mitochondria. DQAsomes were first investigated by Weissig et al. (1998), and have since been explored for the delivery of drugs and DNA to the mitochondria. DQAsomes are composed of dequalinium chloride, a bolaamphiphile which self-assembles and forms liposomes like vesicles system upon sonication. Initially, these vesicles were investigated for intracellular delivery of bioactive molecules. However, later it was observed that DQAsomes may also be useful for mitochondrial targeting (D'Souza et al., 2003; Weissig et al., 2000, 2001). DQAsomes have the ability to bind tightly to the DNA and protect it from deoxyribonuclease-I (DNAse-I) digestion (Lasch et al., 1999). It has been suggested that DQAsomes exhibit endosomolytic activity and release encapsulated content in the vicinity of mitochondrial membrane. Further, it has also been reported that DQAsomes release DNA only after contact with cardiolipin-rich mitochondrial membranes but not on contact with anionic lipid–rich inner cytosolic membrane (D'Souza et al., 2003). This could be attributed to the destabilization of DQAsomes–DNA complex after interaction with the mitochondrial membrane, thus resulting in the release of DNA. Based on the earlier studies, DQAsomes have been considered as a suitable candidate for the selective delivery of DNA to the mitochondria. However, DQAsomes delivered DNA in the proximity of mitochondria. Hence, to deliver DNA inside mitochondria or to facilitate transport of DNA from cytoplasm to the mitochondrial matrix, other approaches have been described based on the conjugation of DNA to the MLS. In a study by D'Souza et al. (2005), MLS-conjugated oligonucleotides and plasmid DNA were loaded in DQAsomes. Results of the study exhibited that MLS-conjugated DNA colocalize with mitochondria in mammalian cells when delivered using DQAsomes (Figure 7.6). This strategy verified the hypothesis that DQAsomes deliver DNA into the close proximity of mitochondria followed by the uptake of DNA into mitochondria matrix by MLS through protein import machinery.

DQAsomes have also been studied for delivery of small molecular weight drugs to the mitochondria (Cheng et al., 2005; D'Souza et al., 2008). In a very first study by Cheng et al. (2005), paclitaxel, a lipophilic anticancer agent, was encapsulated in the DQAsomes and was evaluated in an animal model for therapeutic efficacy. Results of the study demonstrated that paclitaxel induced apoptosis by releasing cytochrome

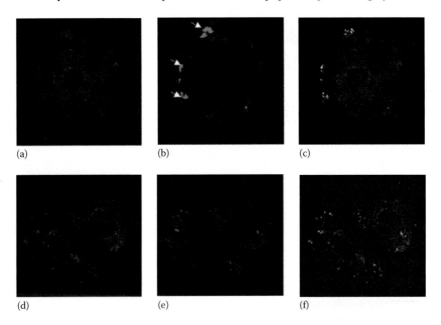

(a)   (b)   (c)

(d)   (e)   (f)

FIGURE 7.6 (**See color insert.**) Representative confocal fluorescence micrographs of BT20 cells stained with Mitotracker® Red CMXRos (red) after exposure for 10 h to fluorescein labeled MLS-pDNA conjugate (green) complexed with cyclohexyl-DQAsomes; (a–c) circular pDNA conjugate, (d–f) linearized pDNA conjugate. (a, d) Green channel, (b, e) red channel, (c, f) overlay of red and green channels with white indicating co-localization of red and green fluorescence. (From D'Souza, G.G. et al., *Mitochondrion*, 5, 352, 2005. With permission.)

c from mitochondrial membrane in a permeability transition pore–dependent manner. In vivo studies in a nude mice model showed that paclitaxel-loaded DQAsomes inhibited human colon carcinoma growth by 50% as compared to control animals, thus underlining the efficacy of DQAsomes as a carrier for small molecular weight drugs. Until now, no study has been reported using DQAsomes for the delivery of drug/DNA to liver disease. However, the ability of DQAsomes to selectively deliver DNA/drug to the mitochondria can encourage scientists to test their applications in liver disease in which mitochondrial DNA is depleted like chronic cholestasis, viral hepatitis, chronic ethanol exposure, and NASH (Serviddio et al., 2010).

Liposomes as Mitochondriotropic Carriers

Liposomes can undoubtedly be regarded as the most investigated and the most successful carrier systems for the delivery of therapeutics to the disease site. Conventional liposomes are bilayer vesicles mainly composed of phospholipids and cholesterol. In general, drug-loaded liposomes are taken up by cells through endocytic pathways, thus leading to the encapsulation of liposomes in lysosomes and other acidic vesicles like endosomes. The liposomes release encapsulated drug molecules when they are degraded in these acidic vesicles. Some of the intact liposomes, however, escape from acidic environment, and may interact with other cell organelles like mitochondria in a nonspecific manner. Several studies reported that a fraction of total liposomes delivered nonspecifically interact with mitochondria (Cudd and Nicolau, 1985, 1986). At the same time, liposomes having cell-specific ligands at their surface are not able to specifically interact with mitochondria. As discussed earlier, mitochondriotropic triphenylphosphonium (TPP) cation can be conjugated to biomolecules to increase selective localization of these molecules into the mitochondria. Hence, it can be hypothesized that conjugation of several mitochondriotropic molecules at the surface of carrier systems such as liposomes will increase their accumulation into the mitochondria. In a very first study, Boddapati et al. (2005) reported the conjugation of TPP$^+$ at the surface of liposomes to target mitochondria. TPP conjugated to a stearyl residue, to form stearyltriphenyl phosphonium (STPP) (an amphiphilic molecule), was incorporated into lipid bilayers. Results showed that STPP molecules get attached to a polar head layer of liposome surface while protecting their mitochondriotropic properties. Further, it has also been observed that at least partially intact liposomes accumulate in the vicinity of mitochondria when conjugated with mitochondriotropic molecules. In another study, Yamada et al. (2005) proposed a different approach for cell-selective mitochondrial targeting with liposomes. They reported the use of a cell-specific ligand transferrin in combination with a pH-sensitive fusogenic peptide (GALA) for targeted delivery of liposomes to the mitochondria of the tumor cells. Mastoparan (a potent facilitator of mitochondrial permeability transition) was used as a model antitumor drug. This study validated that mastoparan possessed potent activity in a homogenate of K-562 cells as well as in isolated mitochondria. Further, it has also been observed that mastoparan in free state releases cytochrome c in the cytosol as well as extracellularly while encapsulated mastoparan releases cytochrome c selectively only from mitochondria to cytosol. Result of this study revealed that cell-selective mitochondrial targeting might be a better approach than the use of mitochondriotropic delivery system alone. This approach can also be extended for the delivery of drug in in vivo conditions. Detailed discussion of liposome and liposome-like vesicles as suitable carriers for the mitochondrial delivery of drug/DNA has been published in several reviews (Pathak et al., 2015; Weissig et al., 2006; Wongrakpanich et al., 2014). Recently, liposomes have also been modified with other mitochondriotropic molecules including dequalinium and rhodamine 123 for the selective delivery of therapeutic molecules (Table 7.2).

DF-MITO-PORTER: A NANOCARRIER FOR MITOCHONDRIAL DELIVERY VIA A STEPWISE MEMBRANE FUSION PROCESS

Mitochondria being an intracellular organelle is not an easily accessible target for the delivery of therapeutic molecules. Before reaching the target, carriers have to pass through various barriers, that is, cellular membrane and endosomal membrane. Hence, in order to achieve mitochondrial delivery of therapeutics, a novel carrier having property of not being degraded by endolysosomal pathways needs to be developed. Further, the optimal carrier system should also have property of being internalized through macropinocytosis so as to avoid endolysosomal degradation.

A novel carrier, MITO-Porter, has been described by Yamada et al. (2008). It is a liposome-based carrier consisting of a high density of octaarginine (R8, a cell-penetrating peptide) over the surface. Liposomes are made of highly fusogenic lipid DOPE. High density of R8 helps in the internalization of liposomes by micropinocytosis and easy escape from macropinosomes with intact encapsulated compounds. After being released from macropinosomes, MITO-Porter interacts with the mitochondria via an electrostatic interaction and fuse with the mitochondrial membrane. GFP was encapsulated as a model drug and was tested in the

TABLE 7.2
Mitochondria-Targeted Liposomes for Selective Delivery of Therapeutics

Liposomes Modified With	Drug Delivered	Application	References
Triphenylphosphonium (TPP)	Ceramide	Anticancer	Boddapati et al. (2008)
Stearyl triphenyl phosphonium (STPP)	Sclareol	Anticancer	Patel et al. (2010)
D-α-Tocopheryl polyethylene glycol 1000 succinate-triphenylphosphine conjugate (TPGS1000-TPP)	Paclitaxel	Resistant lung cancer	Zhou et al. (2013)
Nanoemulsion modified with TPP	Cyclosporin and Vitamin E	Myocardial protection	Faulk et al. (2013)
Rhodamine-123-PEG-DOPE	Paclitaxel	Anticancer	Biswas et al. (2011)
Dequalinium	Daunorubicin and quinacrine	Treatment of relapsed breast cancer	Zhang et al. (2012)
Dequalinium	Resveratrol	Resistant lung cancer	Wang et al. (2011)

mitochondria of an isolated rat liver. Results of the study revealed that GFP was detected in both the outer membrane and intermembrane space fraction of the mitochondria when delivered using MITO-Porter, whereas it was only detected in the outer membrane fraction when delivered using low-fusion liposomes containing Egg PC instead of DOPE.

Later, Yamada and Harashima (2012) also reported a novel dual function (DF)-MITO-Porter to target mitochondria genome (mtDNA). DF-MITO-Porter is a novel core shell nanocarrier system utilizing complicated intracellular processes for selective mitochondrial delivery of bioactive molecules. This nanocarrier is composed of different layers/assemblies to escape different intracellular trafficking events. The innermost core consists of cargoes that are made up of mitochondria-fusogenic lipid envelopes. Inner cargoes are coated with an endosomal-fusogenic lipid envelope. The outer membrane of this carrier is modified with a high density of octaarginine (a type of cell-penetrating peptide), which permits the particle to be efficiently internalized by cells through macropinocytocis. The mechanism of mitochondrial targeting using developed carrier is summarized in Figure 7.7 (Kajimoto et al., 2014).

Yamada and Harashima (2012) have encapsulated DNase protein in DF-MITO-Porter and expected that mtDNA would be digested when delivered to the mitochondrial matrix, thus leading to the reduction in mitochondrial activity. Results demonstrated that the use of DF-MOTO-Porter significantly reduces the mitochondrial activity, measured by mitochondrial dehydrogenase activity, as compared to conventional MITO-Porter (Figure 7.8). This study clearly depicts that these novel carriers could be useful for the delivery of therapeutics to treat diseases.

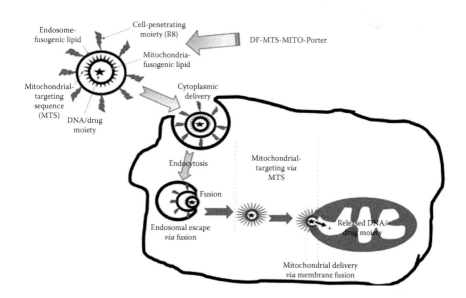

FIGURE 7.7 Schematic presentation of a mechanism of drug targeting to mitochondria using DF-MTS-MITO-Porter.

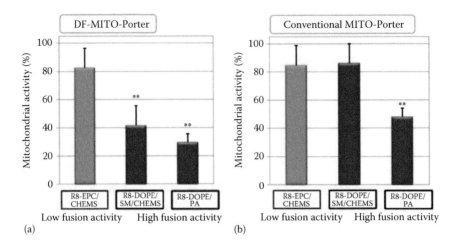

FIGURE 7.8 Mitochondrial delivery of DNase I using DF-MITO-Porter and conventional MITO-Porter. (a) DNase I (0.5 μg) encapsulated in DF-MITO-Porter (SM), DF-MITO-Porter (PA), or a control carrier with low mitochondrial fusion activity were incubated with HeLa cells. Mitochondrial activity was then evaluated by measuring mitochondrial dehydrogenase activity. (b) Mitochondrial activities were also evaluated, when DNase I (6.25 μg) encapsulated in conventional MITO-Porter and a control carrier with low mitochondrial fusion activity were used. Data are represented as the mean ± S.D. (n = 3–4). **Significant differences between the control carrier (EPC/CHEMS/STR-R8) and other carriers (p < 0.01 by one-way ANOVA, followed by Bonferroni correction). (From Yamada, Y. and Harashima, H., *Biomaterials*, 33, 1589, 2012. With permission.)

POLYMERIC NANOPARTICLES FOR MITOCHONDRIAL DELIVERY OF THERAPEUTICS

Polymeric nanocarriers are drug delivery systems that are formulated using biodegradable polymers to achieve controlled and site-specific delivery of therapeutics. In line with their application in targeted drug delivery, polymeric nanocarriers (micelles/nanoparticles) have also been tested for the selective delivery of molecules to the mitochondria. These particles have an additional advantage over other carriers as they can very easily be modified to allow for chemically conjugating targeting moiety at the surface of carriers. As discussed earlier, by attaching mitochondriotropics at the surface of vesicles, selective delivery of molecules to the mitochondria can be achieved. Hence, similar strategies have been tested with polymeric nanocarriers by conjugating mitochondriotropic and MTS peptides for mitochondrial delivery of therapeutics.

Sharma et al. (2012) developed mitochondria-targeted CoQ10-loaded polymeric micelles based on ABC miktoarm star-shaped polymers (A = poly(ethylene glycol) [PEG], B = polycaprolactone [PCL], and C = triphenylphosphonium bromide [TPPBr]). Polymeric micelles were characterized for their drug loading capacity (60 wt.%), micelle size (25–60 nm), and stability. The high CoQ10 loading efficiency allowed testing of micelles within a broad concentration range and provided evidence for CoQ10 effectiveness in two different experimental paradigms: oxidative stress and inflammation. Combined results from chemical, analytical, and biological experiments suggest that the new miktoarm-based carriers provide a suitable means of CoQ10 delivery to mitochondria without loss of drug effectiveness. Dhar and associates also designed a mitochondria-targeted nanoparticle system by blending a targeted poly(D,L-lactic-co-glycolic acid)-*block* (PLGA-*b*)–PEG–TPP polymer with nontargeted polymers, and attempted to deliver various mitochondria-active therapeutics including curcumin, the mitochondrial decoupler 2,4-dinitrophenol, and lonidamine (Marrache and Dhar, 2012; Marrache et al., 2013). These nanoparticles synthesized using cationic TPP-conjugated polymer can be accumulated in mitochondria, driven by the transmembrane electric potential of negatively charged mitochondria. From these results (Figure 7.9), it is clearly depicted that by increasing the charge on the nanoparticle surface, the accumulation of the encapsulated compound (a probe quantum dot, QD) in the mitochondria also increased. Further, it was also reported that the particles with PLGA-*b*–PEG–TPP polymer internalized into cells via endocytosis and escaped from the endosome via proton sponge effect.

MTS-conjugated polyethyleneimine (PEI) for delivering DNA to mitochondria has been developed by Lee et al. (2007). Localization of the MTS-conjugated PEI/DNA complex to the mitochondrial site was confirmed in the living cells. Results of the study also suggested that carriers were internalized into the cells via endocytosis whereas proton sponge effect was exploited to escape from the endosomes. Similar strategies have also been used by other researchers to deliver different molecules/DNA to the mitochondria (Kajimoto et al., 2014).

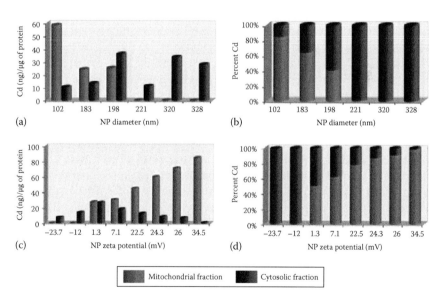

FIGURE 7.9 Mitochondrial and cytosolic distribution of targeted PLGA-*b*–PEG–TPP/PLGA-*b*–PEG–QD blended NPs in HeLa cells by ICP-MS analysis. (a) Effect of size on uptake of NPs. (b) Overall cellular uptake of size-varying NPs. (c) Effect of zeta potential on cellular trafficking of NPs. (d) Overall cellular uptake of zeta potential-varying NPs. (From Marrache, S. and Dhar, S., *Proc. Natl. Acad. Sci. USA*, 109, 16288, 2012. With permission.)

CELL-/ORGAN-SELECTIVE MITOCHONDRIAL TARGETING: LIVER TARGETING

Designing mitochondria-selective carriers for enhanced therapeutic efficacy depends on the selectivity of the developed carriers for specific cells, in addition to mitochondrial targeting. Hence, for the delivery of drugs to the mitochondria of specific cells, a carrier with affinity to that specific cell surface as well as to the mitochondria is required. There are various physiological and pathological constraints that need to be circumvented by the carriers so as to reach their intended target, that is, mitochondria of the specific cell type. Various approaches have been explored by researchers so as to modify the characteristics of carriers to target cells of specific organs. These characteristics include size, shape, surface charge, and surface properties, which can be modified by attaching specific ligands for receptors present at the surface of cells.

Liver is the major organ in the body equipped for uptake, detoxification, and metabolic transformation, thus resulting in high hepatic concentration of all the drugs. However, many drugs are rapidly cleared from the blood and display high first pass clearance by the liver. At the same time, total hepatic uptake of the small molecules predominantly depends on hepatocytes, whereas Kupffer cells largely contribute to the hepatic uptake of particulate material. Therefore, the drugs that enter the liver as such or in the form of covalent carrier conjugates will not necessarily reach the same and/or required cell type. Thus, the challenge is to obtain selective accumulation of drugs in one specific cell type and to sustain intracellular levels for longer period. Different hepatic diseases involve different cell types in liver (e.g., hepatocytes, Kupffer cell, hepatic stellate cell, sinusoidal endothelial cells, and so on). Thus, it is imperative to design or select proper materials to target these different diseased cell types.

Several studies report total liver uptake of a compound and use that as a surrogate for hepatocyte targeting. Total liver uptake of a compound is measured without proper identification of the cell type that actually takes up the drug. Although hepatocytes represent more than 80% of the total number of resident hepatic cells, uptake in other cell types like Kupffer cells may occur as well. High uptake of viruses, antibodies, or other biological compounds into Kupffer cells often leads to complete degradation of such compounds, which in some cases destroys their therapeutic efficacy (Poelstra et al., 2012). Therefore, for specific cell targeting, there should be a specific delivery system and ligand for targeting. Table 7.3 provides details of different cells involved in liver diseases and receptors present on that cells. Hence, by designing carrier systems with specific ligands for these receptors, mitochondria-specific carriers can be targeted to the specific cells for a more effective treatment of diseases.

At the time of this writing, no studies were found that described the conjugation of liver cell–specific ligands at the surface of mitochondria-selective carriers. However, it has already been shown by the researchers that by conjugating specific ligands (transferrin and folic acid) at the surface of DQAsomes or mitochondriotropic liposomes, selectivity to the specific cells (cancer cells) can be increased (Malhi et al., 2012; Vaidya et al., 2009b; Vaidya and Vyas, 2012). Hence, these strategies demonstrate that it might be possible to target mitochondria-selective carriers

TABLE 7.3
Cells Involved in the Diseases and Target Receptors Present on the Surface of Cells

Liver Cells	Diseases Associated with Cells	Receptors Present on Cells
Hepatocytes	Viral hepatitis (hepatitis A, B, or C), alcohol-induced steatohepatitis (ASH), non-alcohol-induced steatohepatitis (NASH), some genetic diseases like Wilson disease, hemochromatosis, $\alpha1$ antitrypsin deficiency, and several other metabolic disorders	Asialoglycoprotein (ASGP), HDL, LDL, Scavenger (Class BI), Transferrin, Insulin
Kupffer cells	Sepsis, acute inflammation in alcohol-induced hepatitis, and chronic inflammation leading to liver fibrosis or NASH	Mannose/N-acetyl glucose amine, galactose particle, Fc (immune complexes, opsonized material), scavenger (Class AI, BI, BII, MARCO CD36 and macrosialin), complement (C3b and C1q) LPS $\alpha2$ macroglobulin, LDL matrix compounds (fibronectin)
Endothelial cells	Viral hepatitis, steatohepatitis, alcoholic liver disease, intrahepatic cholestasis, and activation or rejection of the liver during liver transplantation and liver fibrosis	Mannose/N-acetyl glucose amine, scavenger (Class A1 and A11), Fc immune complexes, matrix compound (hyaluronan fibronectin, denatured collagen PIIINP)
Hepatic stellate cells (HSC)	Liver fibrosis, cirrhosis	M6P/IGF II R, $\alpha2$ macroglobulin, ferritin, uro-plasminogen, thrombin, RBP matrix compounds (integrin, collagen type VI, fibronectin CD44)

Sources: Serviddio, G. et al., *Curr. Med. Chem.*, 17, 2325, 2010; Poelstra, K. et al., *J. Control. Rel.*, 161, 188, 2012; Mishra, N. et al., *Biomed. Res. Int.*, 2013, 20 p., 2013, Article ID 382184, http://dx.doi.org/10.1155/2013/382184.

to hepatic cells by conjugating them to liver cell–specific ligands. Liver cell–specific mitochondria-targeted delivery can thus be considered as a promising approach for the treatment of liver diseases.

CONCLUSION

In addition to being a major source of energy, mitochondria have also been recognized for its role in apoptosis (cell death) and generation of ROS. Recent developments in mitochondrial research have suggested a strong role for mitochondria in various diseases including liver diseases. Hence, the development of molecules acting directly on mitochondria is an interesting field for researchers. Many molecules have been investigated and shown to act by targeting the mitochondrial membrane or the mitochondrial matrix. These molecules can be used for the treatment of various diseases involving mitochondria. The repurposing of drugs by targeting mitochondria cannot be possible without selectively delivering these agents to the mitochondria of specific cell type. Hence, designing of novel methods for the delivery of molecules to the mitochondria is a promising area of research. Until now, several studies have reported the use of different novel carriers for the delivery of drugs and shown considerably high therapeutic efficacy. One of these (MitoQ) has passed phase II clinical studies and demonstrated long-term safety. However, other approaches developed for mitochondrial delivery of therapeutics require further preclinical studies to reach a conclusion about their potential for therapeutic use. Furthermore, the development of surface-modified carriers to specifically target a diseased organ/cell could be beneficial to increase therapeutic efficacy and simultaneously to reduce the side effects of drugs. Further developments in this field could result in a variety of new approaches for specific delivery of drugs for the treatment of mitochondria-related liver diseases.

REFERENCES

Azzone, G.F., Pietrobon, D., and Zoratti, M. 1984. Determination of the proton electrochemical gradient across biological membranes. *Curr Top Bioenergy* 13:1–77.

Biasutto, L., Mattarei, A., Marotta, E., Bradaschia, A., Sassi, N., Garbisa, S., Zoratti, M., and Paradisi, C. 2008. Development of mitochondria targeted derivatives of resveratrol. *Bioorg Med Chem Lett* 18:5594–5597.

Biswas, S., Dodwadkar, N.S., Sawant, R.R., Koshkaryev, A., and Torchilin, V.P. 2011. Surface modification of liposomes with rhodamine-123-conjugated polymer results in enhanced mitochondrial targeting. *J Drug Target* 19:552–561.

Blaikie, F.H., Brown, S.E., Samuelsson, L.M., Brand, M.D., Smith, R.A., and Murphy, M.P. 2006. Targeting dinitrophenol to mitochondria: Limitations to the development of a self-limiting mitochondrial protonophore. *Biosci Rep* 26:231–243.

Boddapati, S.V., D'Souza, G.G., Erdogan, S., Torchilin, V.P., and Weissig, V. 2008. Organelle-targeted nanocarriers: Specific delivery of liposomal ceramide to mitochondria enhances its cytotoxicity in vitro and in vivo. *Nano Lett* 8:2559–2563.

Boddapati, S.V., Tongcharoensirikul, P., Hanson, R.N., D'Souza, G.G., Torchilin, V.P., and Weissig, V. 2005. Mitochondriotropic liposomes. *J Liposome Res* 15:49–58.

Burns, R.J. and Murphy, M.P. 1997. Labeling of mitochondrial proteins in living cells by the thiol probe thiobutyltriphenylphosphonium bromide. *Arch Biochem Biophys* 339:33–39.

Cheng, S.M., Pabba, S., Torchilin, V.P., Fowle, W., Kimpfler, A., Schubert, R., and Weissig, V. 2005. Towards mitochondria specific delivery of apoptosis inducing agents: DQAsomal incorporated paclitaxel. *J Drug Del Sci Technol* 15:81–86.

Choi, J.S., Choi, M.J., Ko, K.S. et al. 2006. Low molecular weight polyethylenimine-mitochondrial leader peptide conjugate for DNA delivery to mitochondria. *Bull Korean Chem Soc* 27:1335–1340.

Cudd, A. and Nicolau, C. 1985. Intracellular fate of liposome-encapsulated DNA in mouse liver. Analysis using electron microscope autoradiography and subcellular fractionation. *Biochim Biophys Acta* 845:477.

Cudd, A. and Nicolau, C. 1986. Interaction of intravenously injected liposomes with mouse liver mitochondria. A fluorescence and electron microscopy study. *Biochim Biophys Acta* 860:201.

D'Souza, G.G., Boddapati, S.V., and Weissig, V. 2005. Mitochondrial leader sequence-plasmid DNA conjugates delivered into mammalian cells by DQAsomes co-localize with mitochondria. *Mitochondrion* 5:352–358.

D'Souza, G.G., Cheng, S.M., Boddapati, S.V., Horobin, R.W., and Weissig, V. 2008. Nanocarrier assisted sub-cellular targeting to the site of mitochondria improves the proapoptotic activity of paclitaxel. *J Drug Target* 16:578–585.

D'Souza, G.G., Rammohan, R., Cheng, S.M., Torchilin, V.P., and Weissig, V. 2003. DQAsome-mediated delivery of plasmid DNA toward mitochondria in living cells. *J Control Rel* 92:189–197.

Dudek, J., Rehlinga, P., and van der Laan, M. 2013. Mitochondrial protein import: Common principles and physiological networks. *Biochim Biophys Acta: Mol Cell Res* 833:274–285.

Faulk, A., Weissig, V., and Elbayoumi, T. 2013. Mitochondria-specific nano-emulsified therapy for myocardial protection against doxorubicin-induced cardiotoxicity. In V. Weissig et al. (eds.), *Cellular and Subcellular Nanotechnology: Methods and Protocols. Methods in Molecular Biology*, Humana press, 991:99–102.

Filipovska, A., Kelso, G.F., Brown, S.E., Beer, S.M., Smith, R.A., and Murphy, M.P. 2005. Synthesis and characterization of a triphenylphosphoniumconjugated peroxidase mimetic: Insights into the interaction of ebselen with mitochondria. *J Biol Chem* 280:24113–24126.

Flierl, A., Jackson, C., Cottrell, B. et al. 2003. Targeted delivery of DNA to the mitochondrial compartment via import sequence conjugated peptide nucleic acid. *Mol Ther* 7:550–557.

Harbauer, A.B., Zahedi, R.P., Sickmann, A., Pfanner, N., and Meisinger, C. 2014. The protein import machinery of mitochondria—A regulatory hub in metabolism, stress, and disease. *Cell Metab* 19:357–372.

Heller, A., Brockhoff, G., and Goepferich, A. 2012. Targeting drugs to mitochondria. *Eur J Pharm Biopharm* 82:1–18.

James, A.M., Cochemè, H.M., Smith, R.A., and Murphy, M.P. 2005. Interactions of mitochondria-targeted and untargeted ubiquinones with the mitochondrial respiratory chain and reactive oxygen species. Implications for the use of exogenous ubiquinones as therapies and experimental tools. *J Biol Chem* 280:21295–21312.

Kajimoto, K., Sato, Y., Nakamura, T., Yamada, Y., and Harashima, H. 2014. Multifunctional envelope-type nano device for controlled intracellular trafficking and selective targeting in vivo. *J Control Rel* 190:593–606.

Kelso, G.F., Porteous, C.M., Coulter, C.V. et al. 2001. Selective targeting of a redox-active ubiquinone to mitochondria within cells. *J Biol Chem* 276:4588–4596.

Lasch, J., Meye, A., Taubert, H. et al. 1999. Dequalinium vesicles form stable complexes with plasmid DNA which are protected from DNase attack. *Biol Chem* 380:647–652.

Lee, M., Choi, J.S., Choi, M.J., Pak, Y.K., Rhee, B.D., and Ko, K.S. 2007. DNA delivery to the mitochondria sites using mitochondrial leader peptide conjugated polyethylenimine. *J Drug Target* 15:115–122.

Lin, R., Zhang, P., Cheetham, A.G., Walston, J., Abadir, P., and Cui, H. 2015. Dual peptide conjugation strategy for improved cellular uptake and mitochondria targeting. *Bioconj Chem* 26:71–77.

Lowes, D.A., Thottakam, B.M., Webster, N.R., Murphy, M.P., and Galley, H.F. 2008. The mitochondria-targeted antioxidant MitoQ protects against organ damage in a lipopolysaccharide peptidoglycan model of sepsis. *Free Radic Biol Med* 45:1559–1565.

Malhi, S.S., Budhiraja, A., Arora, S., Chaudhari, K.R., Nepali, K., Kumar, R., Sohi, H., and Murthy, R.S. 2012. Intracellular delivery of redox cycler-doxorubicin to the mitochondria of cancer cell by folate receptor targeted mitocancerotropic liposomes. *Int J Pharm* 432:63–74.

Marrache, S. and Dhar, S. 2012. Engineering of blended nanoparticle platform for delivery of mitochondria-acting therapeutics. *Proc Natl Acad Sci USA* 109:16288–16293.

Marrache, S., Tundup, S., Harn, D.A., and Dhar, S. 2013. Ex vivo programming of dendritic cells by mitochondria-targeted nanoparticles to produce interferon-γ for cancer immunotherapy. *ACS Nano* 7:7392–7402.

Mishra, N., Yadav, N.P., Rai, V.K., Sinha, P., Yadav, K.S., Jain, S., and Arora, S. 2013. Efficient hepatic delivery of drugs: Novel strategies and their significance. *BioMed Res Int* 2013:20 p. Article ID 382184, http://dx.doi.org/10.1155/2013/382184.

Mitchell, P. 1961. Coupling of phosphorylation to electron and hydrogen transfer by a chemiosmotic type of mechanism. *Nature* 191:144–148.

Mukhopadhyay, A. and Weiner, H. 2007. Delivery of drugs and macromolecules to mitochondria. *Adv Drug Deliv Rev* 59:729–738.

Murphy, M.P. 1989. Slip and leak in mitochondrial oxidative phosphorylation. *Biochim Biophys Acta* 977:123–141.

Murphy, M.P. 1997. Targeting bioactive compounds to mitochondria. *Trends Biotechnol* 15:326–330.

Murphy, M.P. 2008. Targeting lipophilic cations to mitochondria. *Biochim Biophys Acta* 1777:1028–1031.

Murphy, M.P. and Smith, R.A. 2007. Targeting antioxidants to mitochondria by conjugation to lipophilic cations. *Annu Rev Pharmacol Toxicol* 47:629–656.

Murphy, M.P. and Smith, R.A.J. 2000. Drug delivery to mitochondria: The key to mitochondrial medicine. *Adv Drug Deliv Rev* 41:235–250.

Neupert, W. 1997. Protein import into mitochondria. *Annu Rev Biochem* 66:863–917.

Paliwal, R., Rai, S., Vaidya, B., Mahor, S., Gupta, P.N., Rawat, A., and Vyas, S.P. 2007. Cell-selective mitochondrial targeting: Progress in mitochondrial medicine. *Curr Drug Deliv* 4:211–224.

Patel, N.R., Hatziantoniou, S., Georgopoulos, A., Demetzos, C., Torchilin, V.P., Weissig, V., and D'Souza, G.G. 2010. Mitochondria-targeted liposomes improve the apoptotic and cytotoxic action of sclareol. *J Liposome Res* 20:244–249.

Pathak, R.K., Kolishetti, N., and Dhar, S. 2015. Targeted nanoparticles in mitochondrial medicine. *Wiley Interdiscip Rev Nanomed Nanobiotechnol* 7:315–329.

Pathak, R.K., Marrache, S., Harn, D.A., and Dhar, S. 2014. Mito-DCA: A mitochondria targeted molecular scaffold for efficacious delivery of metabolic modulator dichloroacetate. *ACS Chem Biol* 9:1178–1187.

Pfanner, N., Craig, E.A., and Honlinger, A. 1997. Mitochondrial preprotein translocase. *Annu Rev Cell Dev Biol* 13:25–51.

Pfanner, N., Craig, E.A., and Meijer, M. 1994. The protein import machinery of the mitochondrial inner membrane. *Trends Biochem Sci* 19:368–372.

Poelstra, K., Prakash, J., and Beljaars, L. 2012. Drug targeting to the diseased liver. *J Control Rel* 161:188–197.

Sastre, J., Serviddio, G., Pereda, J., Minana, J.B., Arduini, A., Vendemiale, G., Poli, G., Pallardo, F.V., and Vina, J. 2007. Mitochondrial function in liver disease. *Front Biosci* 12:1200–1209.

Schatz, G.J. 1996. The protein import system of mitochondria. *J Biol Chem* 271:31763–31766.

Seibel, P., Trappe, J., Villani, G. et al. 1995. Transfection of mitochondria: Strategy towards a gene therapy of mitochondrial DNA diseases. *Nucleic Acids Res* 23:10–17.

Serviddio, G., Bellanti, F., Sastre, J., Vendemiale, G., and Altomare, E. 2010. Targeting mitochondria: A new promising approach for the treatment of liver diseases. *Curr Med Chem* 17:2325–2337.

Sharma, A., Soliman, G.M., Al-Hajaj, N., Sharma, R., Maysinger, D., and Kakkar, A. 2012. Design and evaluation of multifunctional nanocarriers for selective delivery of coenzyme Q10 to mitochondria. *Biomacromolecules* 13:239–252.

Shokolenko, I.N., Alexeyev, M.F., LeDoux, S.P., and Wilson, G.L. 2005. TAT mediated protein transduction and targeted delivery of fusion proteins into mitochondria of breast cancer cells. *DNA Repair (Amst)* 4:511–518.

Skulachev, V.P., Anisimov, V.N., Antonenko, Y.N. et al. 2009. An attempt to prevent senescence: A mitochondrial approach. *Biochim Biophys Acta—Bioenerget* 1787:437–461.

Smith, R.A., Porteous, C.M., Gane, A.M., and Murphy, M.P. 2003. Delivery of bioactive molecules to mitochondria in vivo. *Proc Natl Acad Sci USA* 100:5407–5412.

Smith, R.A.J., Porteous, C.M., Coulter, C.V. et al. 1999. Selective targeting of an antioxidant to mitochondria. *Eur J Biochem* 263:709–716.

Vaidya, B., Mishra, N., Dube, D., Tiwari, S., and Vyas, S.P. 2009a. Targeted nucleic acid delivery to mitochondria. *Curr Gene Ther* 9:475–486.

Vaidya, B., Paliwal, R., Rai, S., Khatri, K., Goyal, A.K., Mishra, N., and Vyas, S.P. 2009b. Cell-selective mitochondrial targeting: A new approach for cancer therapy. *Cancer Ther* 7:141–148.

Vaidya, B. and Vyas, S.P. 2012. Transferrin coupled vesicular system for intracellular drug delivery for the treatment of cancer: Development and characterization. *J Drug Target* 20:372–380.

Verner, K. and Lemire, B.D. 1989. Tight folding of a passenger protein can interfere with the targeting function of a mitochondrial presequence. *EMBO J* 8:1491–1495.

Vestweber, D. and Schatz, G. 1989. DNA–protein conjugates can enter mitochondria via the protein import pathway. *Nature* 338:170–172.

Wang, X.X., Li, Y.B., Yao, H.J., Ju, R.J., Zhang, Y., Li, R.J., Yu, Y., Zhang, L., and Lu, W.L. 2011. The use of mitochondrial targeting resveratrol liposomes modified with a dequalinium polyethylene glycol-distearoylphosphatidyl ethanolamine conjugate to induce apoptosis in resistant lung cancer cells. *Biomaterials* 32:5673–5687.

Weissig, V., Boddapati, S.V., Cheng, S.M., and D'Souza, G.G.D. 2006. Liposomes and liposome-like vesicles for drug and DNA delivery to mitochondria. *J Liposome Res* 16:249–264.

Weissig, V., Bodapati, S.V., D'Souza, G.G.M. et al. 2004. Targeting of low molecular weight drugs to mammalian mitochondria. *Drug Des Rev* 1:15–28.

Weissig, V., D'Souza, G.G., and Torchilin, V.P. 2001. DQAsome/DNA complexes release DNA upon contact with isolated mouse liver mitochondria. *J Control Rel* 75:401–408.

Weissig, V., Lasch, J., Erdos, G. et al. 1998. DQAsomes: A novel potential drug and gene delivery system made from Dequalinium. *Pharm Res* 15:334–337.

Weissig, V., Lizano, C., and Torchilin, V.P. 2000. Selective DNA release from DQAsome/DNA complexes at mitochondria-like membranes. *Drug Deliv* 7:1–5.

Weissig, V. and Torchilin, V.P. 2001. Cationic bolasomes with delocalized charge centers as mitochondria-specific DNA delivery systems. *Adv Drug Deliv Rev* 49:127–149.

Whiteman, M., Spencer, J.P., Szeto, H.H., and Armstrong, J.S. 2008. Do mitochondriotropic antioxidants prevent chlorinative stress-induced mitochondrial and cellular injury? *Antioxid Redox Signal* 10:641–650.

Wiedemann, N., Frazier, A.E., and Pfanner, N. 2004. The protein import machinery of mitochondria. *J Biol Chem* 279:14473–14476.

Wongrakpanich, A., Geary, S.M., Joiner, M.L., Anderson, M.E., and Salem, A.K. 2014. Mitochondria-targeting particles. *Nanomedicine (London)* 9:2531–2543.

Yamada, Y., Akita, H., Kamiya, H., Kogure, K., Yamamoto, T., Shinohara, Y., Yamashita, K., Kobayashi, H., Kikuchi, H., and Harashima, H. 2008. MITO-Porter: A liposome-based carrier system for delivery of macromolecules into mitochondria via membrane fusion. *Biochim Biophys Acta* 1778:423–432.

Yamada, Y. and Harashima, H. 2012. Delivery of bioactive molecules to the mitochondrial genome using a membrane-fusing, liposome-based carrier, DF-MITO-Porter. *Biomaterials* 33:1589–1595.

Yamada, Y., Shinohara, Y., Kakudo, T., Chaki, S., Futaki, S., Kamiya, H., and Harashima, H. 2005. Mitochondrial delivery of mastoparan with transferring liposomes equipped with a pH-sensitive fusogenic peptide for selective cancer therapy. *Int J Pharm* 303(1–2):1.

Zhang, L., Yao, H.J., Yu, Y., Zhang, Y., Li, R.J., Ju, R.J., Wang, X.X., Sun, M.G., Shi, J.F., and Lu, W.L. 2012. Mitochondrial targeting liposomes incorporating daunorubicin and quinacrine for treatment of relapsed breast cancer arising from cancer stem cells. *Biomaterials* 33:565–582.

Zhou, J., Zhao, W.Y., Ma, X., Ju, R.J., Li, X.Y., Li, N., Sun, M.G., Shi, J.F., Zhang, C.X., and Lu, W.L. 2013. The anticancer efficacy of paclitaxel liposomes modified with mitochondrial targeting conjugate in resistant lung cancer. *Biomaterials* 34:3626–3638.

Section II

The Role of Mitochondria
in Liver Diseases

8 The Central Role of Mitochondria in Drug-Induced Liver Injury

Lily Dara, Heather Johnson, and Neil Kaplowitz

CONTENTS

ABSTRACT

Liver cell death is a prominent feature of drug-induced liver injury (DILI). The major hepatocyte cell death subroutines, apoptosis and necrosis, occur through mitochondrial pathways (MOMP and MPT). DILI is broadly categorized into direct hepatotoxicity and idiosyncratic hepatotoxicity. Mitochondria are central players in liver pathophysiology. In many instances of direct hepatotoxicity, drugs directly cause mitochondrial damage by interfering with mitochondrial homeostatic processes such as respiration, DNA synthesis, or β-oxidation among others. In idiosyncratic DILI, the involvement of mitochondria is more indirect as the cell death mode in this instance is through adaptive immunity (HLA-linked) and largely apoptotic. Since adaptive immunity seems to play a prominent role in idiosyncratic DILI, host factors such as genetics and mitochondrial fitness may contribute to the susceptibility for injury. In this chapter, we examine the role of mitochondria in hepatotoxicity both as direct targets of toxins and in idiosyncratic DILI.

Keywords: hepatotoxicity, cell death, apoptosis, necrosis, idiosyncratic

INTRODUCTION

Mitochondria are key organelles in aerobic cells with a high energy demand such as liver cells. Hepatocytes contain 1000–2000 mitochondria and have many metabolic functions such as gluconeogenesis, lipid and drug metabolism, bile salt and protein synthesis, and so on, processes that require large amounts of ATP. Liver cells are also the main detoxification system of the body and since most forms of liver cell death occur through the mitochondria, it is no wonder that mitochondria play such a central role in liver physiology and pathology. In this chapter, we will review the role of mitochondria in drug-induced liver injury, either as primary targets of drugs or as secondary targets and obligate participants in liver cell death pathways.

OVERVIEW OF LIVER CELL DEATH

Liver cell death is an essential feature of hepatotoxicity. The liver plays a central role in metabolizing ingested compounds, foods, drugs, and toxins. Hepatocytes are uniquely positioned in a cord-like fashion with their basolateral surface exposed to the space of Disse, where they are in direct contact with elements in the blood stream which pass through the fenestrae of the sinusoidal endothelial cells. This microcirculation facilitates the secretion of synthesized proteins and picks up of drugs and nutrients. It also renders hepatocytes accessible and vulnerable to pathogens and toxins. Hepatocyte death occurs via the main cell death subroutines: necrosis and apoptosis. While these two forms of cell death are distinct biochemically and morphologically, in liver cells both are executed with the participation of mitochondria [1,2].

Broadly, DILI occurs in two contexts: direct toxicity, which is predictable and dose dependent; and idiosyncratic toxicity, which is unpredictable and dose independent beyond a threshold dose [1,3]. The mitochondria are integral to signaling pathways in hepatocytes, both as a platform on which signaling molecules coalesce as active participants in signaling, especially in the cell death pathways. In most instances of direct hepatotoxicity, drugs or their reactive metabolites kill hepatocytes through mitochondrial damage, resulting in mitochondrial permeability transition (MPT). This may occur by directly damaging mitochondrial integrity, interference with the electron transport chain (ETC), blockage of mitochondrial DNA (mtDNA) synthesis, or through inhibition of β-oxidation. The phenotypes of cell death vary in the following instances: interference with the ETC results in accumulation of reactive oxygen species (ROS) and the activation of cellular stress pathways often resulting in MPT, cell rupture, and coagulative necrosis, while interference with β-oxidation results in accumulation of fat droplets and microvesicular steatosis. Long-term impaired mitochondrial respiration, fatty acid oxidation, and regular energy generation eventually lead to lactic acidosis [4]. The role of programmed necrosis in hepatocytes as well as dependence on mitochondria is controversial at present [5,6].

Apoptotic cell death, a feature of immune system–mediated DILI is largely mediated by death receptors (TNFR, FAS, Trail, DR4, DR5, etc.) binding to ligands expressed on or secreted by activated immune cells. Binding of the receptor to its ligand results in the formation of the death-inducing signaling complex (DISC) and activation of initiator caspases, such as caspase 8. In type I cells, caspase 8 can directly activate executioner caspases [3,7] resulting in apoptosis. Type II cells, such as hepatocytes, require the participation of mitochondria for the amplification of caspase activation and execution of cell death. In these cells, caspase 8 cleaves the BCL2 family member bid, resulting in t-bid translocation to mitochondria and leading to mitochondrial outer membrane permeabilization (MOMP) [7,8].

Although classic examples of drugs, such as acetaminophen and valproate, causing direct mitochondrial toxicity exist, most DILI is idiosyncratic, and most drugs affect mitochondria only secondarily and as mediators of liver cell death. Nevertheless, the role of mitochondria in hepatotoxicity is pivotal and mitochondrial fitness is relevant to a compound's potential for toxicity [9]. Thus, even in idiosyncratic DILI (iDILI) the direct effects of drugs on mitochondrial fitness may impact sensitivity to the immune-mediated killing. In this chapter, we examine the role of mitochondria in hepatotoxicity both in iDILI and as the primary targets of direct mitochondrial toxins.

IDIOSYNCRATIC DILI AND MITOCHONDRIA

Most instances of DILI are idiosyncratic. In these cases, the drug toxicity is not dose dependent or reproducible and is largely thought to be immune mediated and related to factors, including the host's genetic makeup and acquired and environmental factors, some of which may impact mitochondrial fitness. Since most compounds are metabolized by the liver, it is hypothesized that the generation of reactive intermediaries may result in the formation of highly reactive molecules, which bind intracellular proteins and activate the host's immune response. While the "hapten hypothesis" does not fully explain iDILI, it is an interesting theory which may explain certain aspects of the process that appear to be clearly immune mediated; for example, certain idiosyncratic toxicities are accompanied by a rash and eosinophilia. The immune-mediated nature of iDILI is also suggested by the formation of circulating antibodies to liver/kidney microsomes (LKM) in halothane and tienilic acid hepatotoxicity [10]. Evidence for humoral immune activation has been shown with diclofenac, where antibodies to metabolite-modified liver proteins were detected in the serum of patients with diclofenac hepatotoxicity [11]. Although the formation of adducts is clearly a red flag in these instances, it is critical to note that many healthy controls receiving diclofenac (12 out of 20) were also positive for the antibodies without evidence of liver disease. Underlying genetic factors, such as cytokine polymorphisms (IL-10 and IL-4), have been suggested to play a role in determining which subjects with adducts exhibit toxicity [11]. The cellular immune response is also an important contributor to iDILI; for example, the lymphocyte proliferation test is frequently positive, when ex vivo samples of patients with known hepatotoxicity are used [9,12]. Once activated, T cells kill hepatocytes, which presumably harbor the reactive metabolite, or "hapten," via death ligands and through the engagement of death receptors with the participation of mitochondria. Further support for the importance of genetic susceptibility has been provided by findings that certain human leukocyte antigen (HLA) polymorphisms confer increased risk of iDILI. Clear HLA associations have been implicated through large genome wide association studies for numerous drugs, examples of which include HLA-DRB1*1501 allele with amoxicillin–clavulanate and lumiracoxib DILI, HLA-B*5701 with flucloxacillin DILI, and HLA-A*A3303 association with ticlopidine [13–18].

While drugs that cause iDILI are not traditional mitochondrial toxins, many clues point toward the contribution of underlying mitochondrial fitness in the potential for toxicity by these agents, many examples of which come from the Spanish DILI registry. Mutations in mitochondrial superoxide dismutase (SOD2 Ala allele) and glutathione peroxidase (GPX1 Leu allele) confer a higher risk of developing cholestatic DILI [19]. Mutations in glutathione S-transferase (GST) have also been shown to contribute to toxicity. In the Spanish DILI registry carriers of double GSTT1 and GSTM1 null genotypes had an increased chance of iDILI regardless of type of drug (OR 3.52 for antibacterials, 5.6 for NSAID, and 2.81 for amoxicillin–clavulanate hepatotoxicity) [20]. In vitro studies also demonstrate that certain drugs that cause idiosyncratic toxicity, such as troglitazone or ciglitazone, may promote MPT in isolated mitochondria [21]. Cell screens on drug effects on isolated mitochondria have suggested that many drugs implicated in iDILI impair mitochondrial function while drugs with no record of causing iDILI do not [22]. It is however important to reconcile these direct effects on mitochondria with the low incidence of overt liver injury. One might speculate that the uniform mitochondrial effects are a predisposing factor by generating a danger signal for or sensitizing hepatocytes (by affecting mitochondrial fitness) to be more susceptible to killing by the immune system, a type of "2-hit" phenomenon.

DIRECT MITOCHONDRIAL TOXICITY

Drugs that are direct mitochondrial toxins damage mitochondria by one or more of the following mechanisms: damage of mtDNA, interference with protein synthesis, impairment of β-oxidation and/or respiration, and inducing mitochondrial membrane permeabilization. Many drugs cause toxicity by multiple mechanisms (Table 8.1).

DRUGS TARGETING MtDNA

Mitochondria are ancient aerobic bacteria that colonized a prokaryotic cell during evolution; they contain their own DNA (mtDNA), RNA, and protein synthesizing machinery [23]. Unlike nuclear DNA, mtDNA does not wrap around histones leaving mtDNA vulnerable to free radicals. This is amplified by the fact that mtDNA resides in proximity to free radicals and generated ROS. Several drugs cause mitochondrial toxicity by affecting the mtDNA. Alcohol, an important generator of hydroxyl radicals and ROS, causes mitochondrial injury partly by depleting mtDNA, which will be discussed in detail elsewhere in this book [24]. Acetaminophen is another potent mitochondrial toxin that also damages mitochondria partly by depleting mtDNA. After an acetaminophen overdose in mice, hepatic mtDNA is markedly depleted, possibly due to the mtDNA damage caused by peroxynitrite and other ROS [25].

TABLE 8.1
List of Common Drugs Known to Cause Drug-Induced Liver Injury and Mitochondrial Toxicity

Drug	mtDNA Damage	Protein Synthesis Interference	β-Oxidation/ Respiration Interference	Membrane Permeabilization/ Rupture	Comments
Abacavir	1°	2°	2°		
Acetaminophen	2°		2°	1°	
Alcohol	1°	2°	2°		
Amineptine					Prevention of VLDL secretion
Amiodorone			1°		Prevention of VLDL secretion
Asprin			1°		Reye's syndrome
Ciprofloxacin				1°	Lysosomal membrane permeabilization results in Bax activation and translocation to mitochondria
Didanosine	1°	2°	2°		
Erythromycins		1°	2°		
Fialuridine	1°	2°	2°		Causes lactic acidosis, myopathy, neuropathy, and microvesicular steatosis
Ganciclovir	1°	2°	2°		
Indinavir				1°	ER stress
Hydroxychloroquin				1°	Lysosomal membrane permeabilization results in Bax activation and translocation to mitochondria
Lamivudine	1°	2°	2°		

(Continued)

TABLE 8.1 (Continued)
List of Common Drugs Known to Cause Drug-Induced Liver Injury and Mitochondrial Toxicity

Drug	mtDNA Damage	Protein Synthesis Interference	β-Oxidation/ Respiration Interference	Membrane Permeabilization/ Rupture	Comments
Linezolide		1°	2°		Lactic acidosis and neuropathy. Decreases in ETC complexes
Methotrexate			1°		Long-term use associated with steatosis, fibrosis, and cirrhosis
Paclitaxel				1°	Lysosomal membrane permeabilization results in Bax activation and translocation to mitochondria
Pirprofen					Prevention of VLDL secretion
Ritonavir				1°	ER stress
Stavudine	1°		2°		
Tamoxifen	2°	2°	1°		Steatohepatitis
Tetracycline					Prevention of VLDL secretion
Rianeptine					Prevention of VLDL secretion
Troglitazone			2°	1°	Triggers MOMP
Valproate	1°		1°		Reactive metabolites: 2,4-diene-valproyl-CoA, 4-ene-valproate. POLG mut
Zalcitabine	1°	2°	2°		
Zidovudine	1°	2°	2°		Also depletes thymidine triphosphate by competing for the same enzyme

1° denotes primary mode of toxicity and 2° denotes secondary modes of toxicity.

Nucleoside analog reverse transcriptase inhibitors (NRTIs), such as zidovudine (AZT), zalcitabine (ddC), didanosine (ddI), stavudine (d4T), lamivudine (3TC), and abacavir (ABC), impair mtDNA replication by incorporating into a growing chain of DNA though their 5′-hydroxyl group, resulting in chain termination [26]. Interestingly, nuclear DNA polymerases do not incorporate these analogues into their genome. Although all NRTIs can cause mitotoxicity by incorporating into the mtDNA, zalcitabine, didanosine, and stavudine are more likely to deplete mtDNA, resulting in decreased respiratory chain complex expression and subsequent interference with the ETC function [27,28]. AZT, a thymidine analogue, can also mitigate mtDNA replication by decreasing thymidine triphosphate, since it competes with thymidine for phosphorylation by the same kinase (thymidine kinase) [29]. NRTIs can also promote point mutations in mtDNA due to nucleoside imbalances, or secondary to oxidative and free radical damage [30,31].

Other nucleoside analogues have also been associated with hepatotoxicity. Fialuridine, a compound that was toxic in clinical trials due to severe pancreatitis, myopathy, neuropathy, lactic acidosis, and microvesicular hepatic steatosis is an example of a nucleoside analogue that is readily incorporated into the mitochondrial genome [32]. Ganciclovir, a commonly used CMV antiviral, can also be incorporated into mtDNA thereby triggering mtDNA loss [33]. Regardless of the drug, impairment of mtDNA synthesis results in decreases in respiratory chain proteins and increased the formation of superoxide anion radicals and ROS, furthering mtDNA deletions [34].

DRUGS THAT INTERFERE WITH MITOCHONDRIAL PROTEIN SYNTHESIS

Since the mitochondrial evolutionary origins reside with bacteria, drugs and especially antibiotics that decrease protein synthesis in bacteria can interfere with mitochondria. Clear similarities between mitochondrial and bacterial ribosomes exist [35,36].

Linezolid, which is an antibiotic for gram-positive bacteria, has side effects of lactic acidosis and neuropathy in patients. A selective decrease in the activity of complexes I, III, and IV in mitochondria from experimentally treated animals and patient biopsy samples have been reported with this antibiotic [37]. The erythromycins, potent 50S ribosomal inhibitors, can inhibit mitochondrial protein synthesis [38]. Although this was postulated to result in sensory neural hearing loss, hepatitis from these drugs seems to occur due to reactive metabolite formation and immune activation with allergic features [39,40], but effects on mitochondrial fitness may contribute.

DRUGS THAT INTERFERE WITH β-OXIDATION AND RESPIRATION

A common form of hepatotoxicity is the inhibition of β-oxidation, which leads to the accumulation of triglycerides (TGs) in the form of lipid droplets in hepatocytes. Characteristically, these droplets are many and small in size and the resultant pathology is described as microvesicular steatosis. This process has been associated with liver failure, encephalopathy, and lactic acidosis [40]. Drugs can directly inhibit fatty acid oxidation by inhibiting the β-oxidation process enzymes, or by sequestering cofactors, such as L-carnitine and coenzyme A (CoA). It is important to note that β-oxidation also may be impaired secondarily due to interferences with the ETC [41,42]. Long-term and chronic inhibition of β-oxidation can lead to macrovesicular steatosis and eventually steatohepatitis, fibrosis, and cirrhosis. This is the case of methotrexate where chronic use over years, especially in patients with underlying metabolic syndrome and obesity, results in cirrhosis [43,44].

Tamoxifen, a breast cancer chemotherapeutic agent and estrogen receptor antagonist, notoriously causes drug-induced steatohepatitis. The drug has multiple effects on mitochondria, including interference with the ETC, depleting mtDNA, impairment of β-oxidation, cumulatively causing oxidative stress and steatohepatitis [45,46] etc. Underlying NAFLD may enhance the steatohepatitis and fibrogenic effects of methotrexate and tamoxifen.

Valproic acid is a branched chain fatty acid used as an anticonvulsant and mood stabilizer, which leads to microvesicular steatosis and rarely a Reye's-like syndrome [47]. Valproate affects the mitochondria in multiple ways, the most important of which are outlined here as follows: being a fatty acid, valproate easily crosses the mitochondrial membrane and enters the matrix where it can uncouple respiration by interfering with the proton gradient [48]. Valproate also sequesters CoA by forming valproyl-CoA moieties, thus interfering with natural fatty acyl-CoA oxidation, as well as pyruvate dehydrogenase, which requires CoA [49,50]. The cyp system metabolizes valproate to form highly reactive intermediaries, such as 2,4-diene-valproyl-CoA, which can inactivate fatty acid oxidation enzymes [51,52]. This reactive metabolite formation explains the increased toxicity associated with poly pharmacy, since concomitant administration of phenytoin or carbamazepine, which are potent cyp inducers, increases the formation of the highly reactive valproate metabolites leading to increased toxicity [53]. In addition to mitochondrial effects, cytoplasmic valproyl-CoA and 4-ene-valproyl-CoA competitively inhibit carnitine-palmitoyl-transferase I activity further interfering with β-oxidation [54]. The 4-ene-valproate is generated in mitochondria via β-oxidation enzymes and is a highly reactive electrophile which targets mitochondrial proteins for covalent binding. Another important determinant of valproate toxicity is amino acid substitutions that decrease the activity of mtDNA polymerase γ (POLG) [55]. This came to light when homozygous or compound heterozygous POLG mutations were discovered to be the underlying cause of Alpers–Huttenlocher syndrome, a rare childhood disorder characterized by epilepsy, developmental delay, and liver disease [55,56]. A striking one-third of patients with Alpers–Huttenlocher disease develop severe hepatotoxicity and liver failure within 3 months of exposure to valproic acid [57,58]. Review of valproate toxicity cases in the Drug-Induced Liver Injury Network (DILIN) revealed that even heterozygous mutations in the *POLG* gene were associated with an increased risk of liver toxicity (OR 23.6) [58].

Aspirin hepatotoxicity is well documented in children where its administration has been linked to a constellation of signs and symptoms including: the development of liver failure, severe steatosis, hypoglycemia, hyperammonemia, coma and death, and so on, termed as *Reye's syndrome*. This occurs usually in the context of viral infections where the mitochondria are stressed and circulating cytokines are released affecting mitochondrial respiration [59,60]. Aspirin is hydrolyzed into salicylic acid, which sequesters CoA. The relative decrease in the availability of CoA prevents long chain fatty acid entrance to the mitochondria for β-oxidation. Furthermore, aspirin acts as an uncoupler, increasing its toxic effects on hepatocytes [41].

Amiodarone is a lipophilic drug that easily permeates mitochondria, where it is protonated in the intermembrane space. Amiodarone then electrophoretically enters the mitochondrial matrix due to the high electrochemical potential across the inner membrane where it concentrates [9]. A high mitochondrial concentration leads to the inhibition of β-oxidation, interference with the ETC, generation of ROS and eventually steatohepatitis and fibrosis indistinguishable from nonalcoholic steatohepatitis (NASH) or alcoholic steatohepatitis (ASH) [61–63].

It is important to note that interference with fatty acid oxidation is not the sole mechanism of drug-induced liver steatosis. Prevention of very low density lipoprotein (VLDL) secretion through interference with apo B lipidation is another mechanism of fat accumulation in the liver. Several steatogenic drugs inhibit not only mitochondrial β-oxidation but also microsomal TG transfer protein activity, thus decreasing apo B lipidation into TG-rich VLDL particles and interfering with hepatic VLDL secretion. Amiodarone is an example of such a drug and others include the following: amineptine, pirprofen, tetracycline, and tianeptine [64].

Drugs That Result in Mitochondrial Membrane Permeabilization

A recent example of a hepatotoxic drug withdrawn from the market is the case of troglitazone. A peroxisome proliferator–associated receptor γ agonist, troglitazone, activated c-Jun terminal kinase (JNK) resulting in tBID translocation to mitochondria, increased ROS, MOMP, and apoptosis [65,66]. Reconciliation of this in vitro finding with the idiosyncratic nature of the injury in patients remains problematic.

Other examples include anticancer drugs and compounds that activate Bax to secondarily target mitochondria and cause apoptosis, such as ciprofloxacin, hydroxychloroquine, and paclitaxel, which can permeabilize lysosomal membranes releasing cathepsins, which activate Bax that then translocates and targets mitochondria [67–69]. Other categories include drugs that permeabilize mitochondria secondarily to endoplasmic reticulum (ER) stress, such as the HIV protease inhibitors ritonavir and indinavir. These drugs cause ER stress and subsequent Bax and JNK activation through CHOP, leading to MOMP or increasing cytosolic calcium levels, thereby promoting MPT [70].

Acetaminophen (APAP), a commonly used analgesic, remains the leading cause of acute liver failure in the United States. It is the most extensively studied hepatotoxic drug as the proximity of animal models to human pathology and the highly reproducible toxicity make it apt for scientific investigation. Acetaminophen is metabolized by the cyp450 system (mainly cyp2E1, 1A2) to a highly reactive metabolite N-acetyl-p-benzoquinoneimine (NAPQI), which covalently binds proteins if not detoxified by glutathione (GSH) [71]. At toxic doses, NAPQI overwhelms the GSH reserve, which is depleted by 80–90%. The metabolite then forms covalent bonds with intracellular proteins, the amount of which correlates to the degree of hepatotoxicity [72–74]. Repletion of GSH via infusion or oral administration of *N*-acetylcysteine (a GSH precursor), abrogates APAP toxicity [75,76]. Once covalent binding occurs and glutathione is depleted, oxidative stress ensues as evidenced by decreased ability of hepatocytes to detoxify peroxide and peroxynitrite and increased reactive nitrogen species accumulation leading to 3-nitrotyrosine formation in the centrilobular zone [77,78]. It has been known for a long time that mitochondria were an essential target of APAP; microscopic examination of mitochondria 2–4 h after a lethal dose of APAP shows clear morphological changes, namely small fragmented mitochondria that have been broken or undergone fission (unpublished data). Furthermore, APAP affects the ETC and alters mitochondrial respiration both in vitro and in isolated hepatocytes from in vivo treated animals [79–81]. It has also been shown that cell death from APAP is necrotic and through mitochondrial membrane rupture and MPT. MPT inhibitors, such as cyclosporin A, inhibited toxicity from APAP in freshly isolated hepatocytes [82]. This has been confirmed in in vivo experiments where cyclophilin D−/− mice were shown to be less sensitive to APAP [83]. During APAP liver toxicity, between covalent binding and the induction of MPT, a series of signalling events occur inside the cell, many of which pertain to the hepatocyte mitochondria. As described above, APAP induces oxidative stress, which then activates the cellular stress machinery including the mitogen-activated protein kinases (MAPKs), MLK3, and ASK1, which ultimately activate and phosphorylate JNK [84,85]. Mice treated with JNK1 and 2 antisense, as well as mice treated with the JNK inhibitor SP600125, are protected from lethal doses of APAP [86,87]. Furthermore, upon activation, JNK translocates to mitochondria where it localizes to the outer mitochondrial membrane [81,87]. The addition of JNK to isolated liver mitochondria directly leads to worsening mitochondrial function [87]. SH3 domain–binding protein 5 (Sab) is the mitochondrial binding partner for p-JNK. Sab is a scaffold protein containing a kinase interaction motif, which is important in the binding of JNK (Figure 8.1). In liver cells, it is exclusively found in the mitochondria and spans the outer mitochondrial membrane. When Sab is knocked down, p-JNK cannot translocate to mitochondria and sustained JNK activation is prevented. The translocation of JNK to the outer mitochondrial membrane during the course of APAP toxicity appears to be a pivotal event, as inhibition of this event by using a Sab-binding peptide (KIM1) or knock down of Sab markedly prevents toxicity. While early JNK activation is through the initial burst of oxidative stress from NAPQI, Sab is required for sustained JNK activation. Data from our lab demonstrate that the binding of p-JNK to Sab in isolated liver mitochondria results in decreased mitochondrial respiration and increased ROS production [88]. The continued ROS production sustains JNK activation in a self-sustaining process which ultimately leads to MPT and necrosis in APAP toxicity (Figure 8.1). In other contexts, this process promotes apoptosis by JNK-mediated activation of MOMP, promoting pro-apoptotic bcl2 family members while inhibiting anti-apoptotic ones. MOMP leads to cytochrome c release and caspase activation, a process which is unable to proceed in APAP toxicity given profound GSH and ATP depletion and severe oxidative stress which promote necrotic cell death.

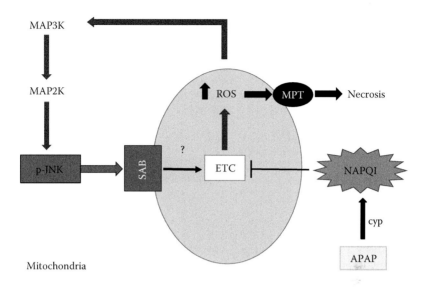

FIGURE 8.1 (See color insert.) Schematic overview of acetaminophen toxicity. Acetaminophen is metabolized to the reactive compound NAPQI via the cyp system. NAPQI is highly reactive and covalently binds proteins both in the cytoplasm and mitochondria. This induces oxidative stress and free radicals which activate the mitogen-activated protein kinase (MAPK) cascade ultimately leading to JNK activation (p-JNK). p-JNK then translocates and binds to Sab, which leads to further interference with the ETC and more ROS formation. This leads to a vicious cycle, which sustains ROS production and JNK activation ultimately resulting in hepatocyte death.

CONCLUSION

Drug-induced hepatotoxicity results in liver cell death, which requires the participation of mitochondria. Most idiosyncratic liver toxins kill hepatocytes through mitochondrial pathways. Cell death in these instances is mediated by the immune system and through death receptor signaling. However, the effects of these drugs on mitochondrial fitness may sensitize hepatocytes to the immune-mediated injury. Direct and reproducible mitotoxins include drugs that interfere with mitochondria directly, such as nucleoside analogues, which interfere with mtDNA; certain antibiotics that prevent mitochondrial protein synthesis; drugs that are steatogenic and interfere with fatty acid oxidation, such as valproate and amiodarone; and finally drugs that cause mitochondrial membrane ruptures, such as acetaminophen, the quintessential mitochondrial toxin.

REFERENCES

1. Han, D., Dara, L., Win, S., Than, T. A., Yuan, L., Abbasi, S. Q., Liu, Z. X. et al. (2013) Regulation of drug-induced liver injury by signal transduction pathways: Critical role of mitochondria. *Trends Pharmacol Sci* 34, 243–253.

2. Jones, D. P., Lemasters, J. J., Han, D., Boelsterli, U. A., and Kaplowitz, N. (2010) Mechanisms of pathogenesis in drug hepatotoxicity putting the stress on mitochondria. *Mol Interv* 10, 98–111.

3. Kaplowitz, N. (2013) Avoiding idiosyncratic DILI: Two is better than one. *Hepatology* 58, 15–17.

4. Pessayre, D., Fromenty, B., Berson, A., Robin, M. A., Letteron, P., Moreau, R., and Mansouri, A. (2012) Central role of mitochondria in drug-induced liver injury. *Drug Metab Rev* 44, 34–87.

5. Tait, S. W., Oberst, A., Quarato, G., Milasta, S., Haller, M., Wang, R., Karvela, M. et al. (2013) Widespread mitochondrial depletion via mitophagy does not compromise necroptosis. *Cell Rep* 5, 878–885.

6. Wang, Z., Jiang, H., Chen, S., Du, F., and Wang, X. (2012) The mitochondrial phosphatase PGAM5 functions at the convergence point of multiple necrotic death pathways. *Cell* 148, 228–243.

7. Belizario, J. E., Alves, J., Occhiucci, J. M., Garay-Malpartida, M., and Sesso, A. (2007) A mechanistic view of mitochondrial death decision pores. *Brazil J Med Biol Res (Revista brasileira de pesquisas medicas e biologicas/Sociedade Brasileira de Biofisica...[et al.])* 40, 1011–1024.

8. Green, D. R. (2005) Apoptotic pathways: Ten minutes to dead. *Cell* 121, 671–674.

9. Pessayre, D., Mansouri, A., Berson, A., and Fromenty, B. (2010) Mitochondrial involvement in drug-induced liver injury. *Handb Exp Pharmacol*, 196, 311–365.

10. Tujios, S. and Fontana, R. J. (2011) Mechanisms of drug-induced liver injury: From bedside to bench. *Nat Rev Gastroenterol Hepatol* 8, 202–211.

11. Aithal, G. P., Ramsay, L., Daly, A. K., Sonchit, N., Leathart, J. B., Alexander, G., Kenna, J. G., Caldwell, J. et al. (2004) Hepatic adducts, circulating antibodies, and cytokine polymorphisms in patients with diclofenac hepatotoxicity. *Hepatology* 39, 1430–1440.

12. Maria, V. A., Pinto, L., and Victorino, R. M. (1994) Lymphocyte reactivity to ex-vivo drug antigens in drug-induced hepatitis. *J Hepatol* 21, 151–158.

13. Hautekeete, M. L., Horsmans, Y., Van Waeyenberge, C., Demanet, C., Henrion, J., Verbist, L., Brenard, R. et al. (1999) HLA association of amoxicillin-clavulanate-induced hepatitis. *Gastroenterology* 117, 1181–1186.

14. Lucena, M. I., Molokhia, M., Shen, Y., Urban, T. J., Aithal, G. P., Andrade, R. J., Day, C. P. et al. (2011) Susceptibility to amoxicillin-clavulanate-induced liver injury is influenced by multiple HLA class I and II alleles. *Gastroenterology* 141, 338–347.

15. O'Donohue, J., Oien, K. A., Donaldson, P., Underhill, J., Clare, M., MacSween, R. N., and Mills, P. R. (2000) Co-amoxiclav jaundice: Clinical and histological features and HLA class II association. *Gut* 47, 717–720.

16. Singer, J. B., Lewitzky, S., Leroy, E., Yang, F., Zhao, X., Klickstein, L., Wright, T. M. et al. (2010) A genome-wide study identifies HLA alleles associated with lumiracoxib-related liver injury. *Nat Genet* 42, 711–714.

17. Alfirevic, A. and Pirmohamed, M. (2012) Predictive genetic testing for drug-induced liver injury: Considerations of clinical utility. *Clin Pharmacol Ther* 92, 376–380.

18. Daly, A. K., Donaldson, P. T., Bhatnagar, P., Shen, Y., Pe'er, I., Floratos, A., Daly, M. J. et al. (2009) HLA-B*5701 genotype is a major determinant of drug-induced liver injury due to flucloxacillin. *Nat Genet* 41, 816–819.

19. Lucena, M. I., Garcia-Martin, E., Andrade, R. J., Martinez, C., Stephens, C., Ruiz, J. D., Ulzurrun, E. et al. (2010) Mitochondrial superoxide dismutase and glutathione peroxidase in idiosyncratic drug-induced liver injury. *Hepatology* 52, 303–312.

20. Lucena, M. I., Andrade, R. J., Martinez, C., Ulzurrun, E., Garcia-Martin, E., Borraz, Y., Fernandez, M. C. et al., and Spanish Group for the Study of Drug-Induced Liver, D. (2008) Glutathione S-transferase m1 and t1 null genotypes increase susceptibility to idiosyncratic drug-induced liver injury. *Hepatology* 48, 588–596.

21. Masubuchi, Y., Kano, S., and Horie, T. (2006) Mitochondrial permeability transition as a potential determinant of hepatotoxicity of antidiabetic thiazolidinediones. *Toxicology* 222, 233–239.

22. Porceddu, M., Buron, N., Roussel, C., Labbe, G., Fromenty, B., and Borgne-Sanchez, A. (2012) Prediction of liver injury induced by chemicals in human with a multiparametric assay on isolated mouse liver mitochondria. *Toxicol Sci* 129, 332–345.

23. Wallace, D. C. (2005) A mitochondrial paradigm of metabolic and degenerative diseases, aging, and cancer: A dawn for evolutionary medicine. *Annu Rev Genet* 39, 359–407.

24. Mansouri, A., Gaou, I., De Kerguenec, C., Amsellem, S., Haouzi, D., Berson, A., Moreau, A. et al. (1999) An alcoholic binge causes massive degradation of hepatic mitochondrial DNA in mice. *Gastroenterology* 117, 181–190.

25. Cover, C., Mansouri, A., Knight, T. R., Bajt, M. L., Lemasters, J. J., Pessayre, D., and Jaeschke, H. (2005) Peroxynitrite-induced mitochondrial and endonuclease-mediated nuclear DNA damage in acetaminophen hepatotoxicity. *J Pharmacol Exp Ther* 315, 879–887.

26. Yarchoan, R., Mitsuya, H., Myers, C. E., and Broder, S. (1989) Clinical pharmacology of 3'-azido-2',3'-dideoxythymidine (zidovudine) and related dideoxynucleosides. *N Engl J Med* 321, 726–738.

27. Walker, U. A., Bauerle, J., Laguno, M., Murillas, J., Mauss, S., Schmutz, G., Setzer, B. et al. (2004) Depletion of mitochondrial DNA in liver under antiretroviral therapy with didanosine, stavudine, or zalcitabine. *Hepatology* 39, 311–317.

28. Brivet, F. G., Nion, I., Megarbane, B., Slama, A., Brivet, M., Rustin, P., and Munnich, A. (2000) Fatal lactic acidosis and liver steatosis associated with didanosine and stavudine treatment: A respiratory chain dysfunction? *J Hepatol* 32, 364–365.

29. Lynx, M. D. and McKee, E. E. (2006) 3'-Azido-3'-deoxythymidine (AZT) is a competitive inhibitor of thymidine phosphorylation in isolated rat heart and liver mitochondria. *Biochem Pharmacol* 72, 239–243.

30. Martin, A. M., Hammond, E., Nolan, D., Pace, C., Den Boer, M., Taylor, L., Moore, H. et al. (2003) Accumulation of mitochondrial DNA mutations in human immunodeficiency virus-infected patients treated with nucleoside-analogue reverse-transcriptase inhibitors. *Am J Human Genet* 72, 549–560.

31. Bartley, P. B., Westacott, L., Boots, R. J., Lawson, M., Potter, J. M., Hyland, V. J., and Woods, M. L., 2nd (2001) Large hepatic mitochondrial DNA deletions associated with L-lactic acidosis and highly active antiretroviral therapy. *Aids* 15, 419–420.

32. Lewis, W., Levine, E. S., Griniuviene, B., Tankersley, K. O., Colacino, J. M., Sommadossi, J. P., Watanabe, K. A. et al. (1996) Fialuridine and its metabolites inhibit DNA polymerase γ at sites of multiple adjacent analog incorporation, decrease mtDNA abundance, and cause mitochondrial structural defects in cultured hepatoblasts. *Proc Natl Acad Sci USA* 93, 3592–3597.

33. Herraiz, M., Beraza, N., Solano, A., Sangro, B., Montoya, J., Qian, C., Prieto, J. et al. (2003) Liver failure caused by herpes simplex virus thymidine kinase plus ganciclovir therapy is associated with mitochondrial dysfunction and mitochondrial DNA depletion. *Human Gene Therapy* 14, 463–472.

34. Lewis, W., Copeland, W. C., and Day, B. J. (2001) Mitochondrial dna depletion, oxidative stress, and mutation: Mechanisms of dysfunction from nucleoside reverse transcriptase inhibitors. *Lab Invest* 81, 777–790.

35. Sharma, M. R., Booth, T. M., Simpson, L., Maslov, D. A., and Agrawal, R. K. (2009) Structure of a mitochondrial ribosome with minimal RNA. *Proc Natl Acad Sci USA* 106, 9637–9642.

36. Denslow, N. D. and O'Brien, T. W. (1978) Antibiotic susceptibility of the peptidyl transferase locus of bovine mitochondrial ribosomes. *Eur J Biochem/FEBS* 91, 441–448.

37. De Vriese, A. S., Coster, R. V., Smet, J., Seneca, S., Lovering, A., Van Haute, L. L., Vanopdenbosch, L. J. et al. (2006) Linezolid-induced inhibition of mitochondrial protein synthesis. *Clin Infect Dis: An Official Publ Infect Dis Soc Am* 42, 1111–1117.

38. Anandatheerthavarada, H. K., Vijayasarathy, C., Bhagwat, S. V., Biswas, G., Mullick, J., and Avadhani, N. G. (1999) Physiological role of the N-terminal processed P4501A1 targeted to mitochondria in erythromycin metabolism and reversal of erythromycin-mediated inhibition of mitochondrial protein synthesis. *J Biol Chem* 274, 6617–6625.

39. Danan, G., Descatoire, V., and Pessayre, D. (1981) Self-induction by erythromycin of its own transformation into a metabolite forming an inactive complex with reduced cytochrome P-450. *J Pharmacol Exp Ther* 218, 509–514.

40. Pessayre, D., Larrey, D., Funck-Brentano, C., and Benhamou, J. P. (1985) Drug interactions and hepatitis produced by some macrolide antibiotics. *J Antimicrob Chemother* 16(Suppl. A), 181–194.

41. Deschamps, D., Fisch, C., Fromenty, B., Berson, A., Degott, C., and Pessayre, D. (1991) Inhibition by salicylic acid of the activation and thus oxidation of long chain fatty acids. Possible role in the development of Reye's syndrome. *J Pharmacol Exp Ther* 259, 894–904.

42. Labbe, G., Pessayre, D., and Fromenty, B. (2008) Drug-induced liver injury through mitochondrial dysfunction: Mechanisms and detection during preclinical safety studies. *Fundam Clin Pharmacol* 22, 335–353.

43. Dahl, M. G. (1969) Methotrexate and the liver. *Brit J Dermatol* 81, 465–467.

44. Barker, J., Horn, E. J., Lebwohl, M., Warren, R. B., Nast, A., Rosenberg, W., Smith, C., International Psoriasis Council (2011) Assessment and management of methotrexate hepatotoxicity in psoriasis patients: Report from a consensus conference to evaluate current practice and identify key questions toward optimizing methotrexate use in the clinic. *J Eur Acad Dermatol Venereol* 25, 758–764.

45. Larosche, I., Letteron, P., Fromenty, B., Vadrot, N., Abbey-Toby, A., Feldmann, G., Pessayre, D. et al. (2007) Tamoxifen inhibits topoisomerases, depletes mitochondrial DNA, and triggers steatosis in mouse liver. *J Pharmacol Exp Ther* 321, 526–535.

46. Moreira, P. I., Custodio, J., Moreno, A., Oliveira, C. R., and Santos, M. S. (2006) Tamoxifen and estradiol interact with the flavin mononucleotide site of complex I leading to mitochondrial failure. *J Biol Chem* 281, 10143–10152.

47. Zimmerman, H. J. and Ishak, K. G. (1982) Valproate-induced hepatic injury: Analyses of 23 fatal cases. *Hepatology* 2, 591–597.

48. Trost, L. C. and Lemasters, J. J. (1996) The mitochondrial permeability transition: A new pathophysiological mechanism for Reye's syndrome and toxic liver injury. *J Pharmacol Exp Ther* 278, 1000–1005.

49. Silva, M. F., Ruiter, J. P., Illst, L., Jakobs, C., Duran, M., de Almeida, I. T., and Wanders, R. J. (1997) Valproate inhibits the mitochondrial pyruvate-driven oxidative phosphorylation in vitro. *J Inherit Metab Dis* 20, 397–400.

50. Ponchaut, S., van Hoof, F., and Veitch, K. (1992) In vitro effects of valproate and valproate metabolites on mitochondrial oxidations. Relevance of CoA sequestration to the observed inhibitions. *Biochem Pharmacol* 43, 2435–2442.

51. Kassahun, K., Farrell, K., and Abbott, F. (1991) Identification and characterization of the glutathione and N-acetylcysteine conjugates of (E)-2-propyl-2,4-pentadienoic acid, a toxic metabolite of valproic acid, in rats and humans. *Drug Metab Dispos* 19, 525–535.

52. Kassahun, K., Hu, P., Grillo, M. P., Davis, M. R., Jin, L., and Baillie, T. A. (1994) Metabolic activation of unsaturated derivatives of valproic acid. Identification of novel glutathione adducts formed through coenzyme A-dependent and -independent processes. *Chemico-Biol Interact* 90, 253–275.

53. Levy, R. H., Rettenmeier, A. W., Anderson, G. D., Wilensky, A. J., Friel, P. N., Baillie, T. A., Acheampong, A. et al. (1990) Effects of polytherapy with phenytoin, carbamazepine, and stiripentol on formation of 4-ene-valproate, a hepatotoxic metabolite of valproic acid. *Clin Pharmacol Ther* 48, 225–235.

54. Aires, C. C., Ijlst, L., Stet, F., Prip-Buus, C., de Almeida, I. T., Duran, M., Wanders, R. J. et al. (2010) Inhibition of hepatic carnitine palmitoyl-transferase I (CPT IA) by valproyl-CoA as a possible mechanism of valproate-induced steatosis. *Biochem Pharmacol* 79, 792–799.

55. Naviaux, R. K. and Nguyen, K. V. (2004) POLG mutations associated with Alpers' syndrome and mitochondrial DNA depletion. *Ann Neurol* 55, 706–712.

56. Huttenlocher, P. R., Solitare, G. B., and Adams, G. (1976) Infantile diffuse cerebral degeneration with hepatic cirrhosis. *Arch Neurol* 33, 186–192.

57. Gopaul, S., Farrell, K., and Abbott, F. (2003) Effects of age and polytherapy, risk factors of valproic acid (VPA) hepatotoxicity, on the excretion of thiol conjugates of (E)-2,4-diene VPA in people with epilepsy taking VPA. *Epilepsia* 44, 322–328.

58. Stewart, J. D., Horvath, R., Baruffini, E., Ferrero, I., Bulst, S., Watkins, P. B., Fontana, R. J. et al. (2010) Polymerase γ gene POLG determines the risk of sodium valproate-induced liver toxicity. *Hepatology* 52, 1791–1796.

59. Partin, J. S., Daugherty, C. C., McAdams, A. J., Partin, J. C., and Schubert, W. K. (1984) A comparison of liver ultrastructure in salicylate intoxication and Reye's syndrome. *Hepatology* 4, 687–690.

60. Hurwitz, E. S., Barrett, M. J., Bregman, D., Gunn, W. J., Schonberger, L. B., Fairweather, W. R., Drage, J. S. et al. (1985) Public Health Service study on Reye's syndrome and medications. Report of the pilot phase. *N Engl J Med* 313, 849–857.

61. Fromenty, B., Fisch, C., Labbe, G., Degott, C., Deschamps, D., Berson, A., Letteron, P. et al. (1990) Amiodarone inhibits the mitochondrial beta-oxidation of fatty acids and produces microvesicular steatosis of the liver in mice. *J Pharmacol Exp Ther* 255, 1371–1376.

62. Fromenty, B., Fisch, C., Berson, A., Letteron, P., Larrey, D., and Pessayre, D. (1990) Dual effect of amiodarone on mitochondrial respiration. Initial protonophoric uncoupling effect followed by inhibition of the respiratory chain at the levels of complex I and complex II. *J Pharmacol Exp Ther* 255, 1377–1384.

63. Berson, A., De Beco, V., Letteron, P., Robin, M. A., Moreau, C., El Kahwaji, J., Verthier, N. et al. (1998) Steatohepatitis-inducing drugs cause mitochondrial dysfunction and lipid peroxidation in rat hepatocytes. *Gastroenterology* 114, 764–774.

64. Letteron, P., Sutton, A., Mansouri, A., Fromenty, B., and Pessayre, D. (2003) Inhibition of microsomal triglyceride transfer protein: Another mechanism for drug-induced steatosis in mice. *Hepatology* 38, 133–140.

65. Bae, M. A. and Song, B. J. (2003) Critical role of c-Jun N-terminal protein kinase activation in troglitazone-induced apoptosis of human HepG2 hepatoma cells. *Mol Pharmacol* 63, 401–408.

66. Shishido, S., Koga, H., Harada, M., Kumemura, H., Hanada, S., Taniguchi, E., Kumashiro, R. et al. (2003) Hydrogen peroxide overproduction in megamitochondria of troglitazone-treated human hepatocytes. *Hepatology* 37, 136–147.

67. Boya, P., Andreau, K., Poncet, D., Zamzami, N., Perfettini, J. L., Metivier, D., Ojcius, D. M. et al. (2003) Lysosomal membrane permeabilization induces cell death in a mitochondrion-dependent fashion. *J Exp Med* 197, 1323–1334.

68. Boya, P., Gonzalez-Polo, R. A., Poncet, D., Andreau, K., Vieira, H. L., Roumier, T., Perfettini, J. L. et al. (2003) Mitochondrial membrane permeabilization is a critical step of lysosome-initiated apoptosis induced by hydroxychloroquine. *Oncogene* 22, 3927–3936.

69. Zhao, H., Cai, Y., Santi, S., Lafrenie, R., and Lee, H. (2005) Chloroquine-mediated radiosensitization is due to the destabilization of the lysosomal membrane and subsequent induction of cell death by necrosis. *Radiat Res* 164, 250–257.

70. Zhou, H., Gurley, E. C., Jarujaron, S., Ding, H., Fang, Y., Xu, Z., Pandak, W. M. et al. (2006) HIV protease inhibitors activate the unfolded protein response and disrupt lipid metabolism in primary hepatocytes. *Am J Physiol Gastrointest Liver Physiol* 291, G1071–G1080.

71. Gillette, J. R., Nelson, S. D., Mulder, G. J., Jollow, D. J., Mitchell, J. R., Pohl, L. R., and Hinson, J. A. (1981) Formation of chemically reactive metabolites of phenacetin and acetaminophen. *Adv Exp Med Biol* 136(Pt B), 931–950.

72. Mitchell, J. R., Jollow, D. J., Potter, W. Z., Gillette, J. R., and Brodie, B. B. (1973) Acetaminophen-induced hepatic necrosis. IV. Protective role of glutathione. *J Pharmacol Exp Ther* 187, 211–217.

73. Jollow, D. J., Thorgeirsson, S. S., Potter, W. Z., Hashimoto, M., and Mitchell, J. R. (1974) Acetaminophen-induced hepatic necrosis. VI. Metabolic disposition of toxic and nontoxic doses of acetaminophen. *Pharmacology* 12, 251–271.

74. Jollow, D. J., Mitchell, J. R., Potter, W. Z., Davis, D. C., Gillette, J. R., and Brodie, B. B. (1973) Acetaminophen-induced hepatic necrosis. II. Role of covalent binding in vivo. *J Pharmacol Exp Ther* 187, 195–202.

75. Smilkstein, M. J., Knapp, G. L., Kulig, K. W., and Rumack, B. H. (1988) Efficacy of oral N-acetylcysteine in the treatment of acetaminophen overdose. Analysis of the national multicenter study (1976 to 1985). *N Engl J Med* 319, 1557–1562.

76. Heard, K. J. (2008) Acetylcysteine for acetaminophen poisoning. *N Engl J Med* 359, 285–292.

77. Hinson, J. A., Pike, S. L., Pumford, N. R., and Mayeux, P. R. (1998) Nitrotyrosine-protein adducts in hepatic centrilobular areas following toxic doses of acetaminophen in mice. *Chem Res Toxicol* 11, 604–607.

78. Sies, H., Sharov, V. S., Klotz, L. O., and Briviba, K. (1997) Glutathione peroxidase protects against peroxynitrite-mediated oxidations. A new function for selenoproteins as peroxynitrite reductase. *J Biol Chem* 272, 27812–27817.

79. Burcham, P. C. and Harman, A. W. (1991) Acetaminophen toxicity results in site-specific mitochondrial damage in isolated mouse hepatocytes. *J Biol Chem* 266, 5049–5054.

80. Donnelly, P. J., Walker, R. M., and Racz, W. J. (1994) Inhibition of mitochondrial respiration in vivo is an early event in acetaminophen-induced hepatotoxicity. *Arch Toxicol* 68, 110–118.

81. Win, S., Than, T. A., Han, D., Petrovic, L. M., and Kaplowitz, N. (2011) c-Jun N-terminal kinase (JNK)-dependent acute liver injury from acetaminophen or tumor necrosis factor (TNF) requires mitochondrial Sab protein expression in mice. *J Biol Chem* 286, 35071–35078.

82. Reid, A. B., Kurten, R. C., McCullough, S. S., Brock, R. W., and Hinson, J. A. (2005) Mechanisms of acetaminophen-induced hepatotoxicity: Role of oxidative stress and mitochondrial permeability transition in freshly isolated mouse hepatocytes. *J Pharmacol Exp Ther* 312, 509–516.

83. Ramachandran, A., Lebofsky, M., Baines, C. P., Lemasters, J. J., and Jaeschke, H. (2011) Cyclophilin D deficiency protects against acetaminophen-induced oxidant stress and liver injury. *Free Radic Res* 45, 156–164.

84. Nakagawa, H., Maeda, S., Hikiba, Y., Ohmae, T., Shibata, W., Yanai, A., Sakamoto, K. et al. (2008) Deletion of apoptosis signal-regulating kinase 1 attenuates acetaminophen-induced liver injury by inhibiting c-Jun N-terminal kinase activation. *Gastroenterology* 135, 1311–1321.

85. Sharma, M., Gadang, V., and Jaeschke, A. (2012) Critical role for mixed-lineage kinase 3 in acetaminophen-induced hepatotoxicity. *Mol Pharmacol* 82, 1001–1007.
86. Gunawan, B. K., Liu, Z. X., Han, D., Hanawa, N., Gaarde, W. A., and Kaplowitz, N. (2006) c-Jun N-terminal kinase plays a major role in murine acetaminophen hepatotoxicity. *Gastroenterology* 131, 165–178.
87. Hanawa, N., Shinohara, M., Saberi, B., Gaarde, W. A., Han, D., and Kaplowitz, N. (2008) Role of JNK translocation to mitochondria leading to inhibition of mitochondria bioenergetics in acetaminophen-induced liver injury. *J Biol Chem* 283, 13565–13577.
88. Win, S., Than, T. A., Fernandez-Checa, J. C., Kaplowitz, N. (2014). JNK interaction with Sab mediates ER stress induced inhibition of mitochondrial respiration and cell death. *Cell Death Dis* 5–989.

9 The Role of Autophagy and Mitophagy in Liver Diseases

Bilon Khambu, Hao Zhang, and Xiao-Ming Yin

CONTENTS

ABSTRACT

Autophagy is an evolutionarily conserved cellular process that involves lysosome-mediated intracellular degradation and recycling of intracellular organelles and proteins to maintain homeostasis during cellular stress. It also serves to selectively remove damaged mitochondria by a process known as *mitophagy*. Both autophagy and mitophagy are vital to the metabolic function and the homeostasis of the liver.

Altered autophagy or mitophagy can be associated with liver pathogenesis in alcoholic liver disease, nonalcoholic fatty liver disease, and ischemia/reperfusion-mediated liver injury. Understanding the molecular mechanisms underlying the induction of autophagy or mitophagy and their regulation and modulation in the disease context will help with developing new therapeutic strategies for treating liver diseases.

Keywords: autophagy, mitophagy, autophagosome, lysosome, ethanol, fatty liver, liver injury

INTRODUCTION

Autophagy (from Greek meaning *self-eating*) is an evolutionarily conserved cellular process where cellular components are transported to the lysosome for degradation. It takes place in eukaryotic cells in normal status. However, cellular stress such as nutrient deprivation, DNA damage, protein aggregation, infection, and hypoxia can markedly enhance autophagy [1–4].

Even though the term *autophagy* was first proposed by Christiana de Duve in 1963, detailed studies and molecular understanding of the autophagy process did not occur until the milestone discoveries of the autophagy-related genes *(ATGs)* through genetic screening in yeast in the 1990s. At least, 32 *ATG* genes have been identified [5], many of which have known orthologous in higher eukaryotes [6]. In addition, autophagy is shown to be connected to a wide range of cellular signaling pathways such as the ubiquitin proteasome system [7] and the membrane trafficking machinery [8]. Most notably, in the past two decades, research in the autophagy field has resulted in a tremendous progress in the understanding of the basic molecular mechanism, which opens the door to the translational research of the role of autophagy in human pathologies. Too little or too much autophagy could be detrimental to cellular physiology. Dysfunction of autophagy has been linked with many human diseases including neurodegenerative, cancer, infection, metabolic diseases, and aging [1,9–15].

Types of Autophagy

Based on morphological and mechanistic characteristics, there are three distinct classes of autophagy, namely, chaperonin-mediated autophagy (CMA), microautophagy, and macroautophagy (Figure 9.1). In CMA, the chaperone protein HSC70 and its cochaperones recognize and unfold protein substrates that carry the KFERQ motif. The protein substrates are then unfolded and bind to LAMP2A on the surface of the lysosome, followed by the translocation into the lysosome for degradation [16]. This process is not involved in the elimination of organelles. During microautophagy, the cytosol or organelles are directly engulfed into the lysosome lumen by invagination of a portion of the lysosomal membrane [17]. Limited knowledge is available on the molecular details about this process and it is best characterized for the degradation of peroxisomes and selected portions of the nucleus. Microautophagy can be both selective and nonselective. Macroautophagy (hereafter referred as autophagy) is the most prevalent and best characterized form of autophagy. It is highly conserved from the yeast to the mammal, both morphologically and with regard to the molecular

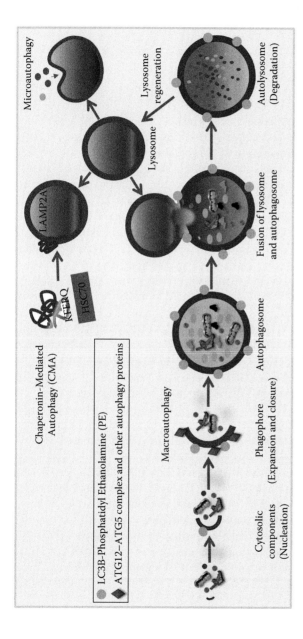

FIGURE 9.1 **(See color insert.)** The definition of autophagy. There are three types of autophagy, which differ in the way that cellular cargoes are delivered into the lysosome. In the CMA, protein cargoes with the signature KFERQ motif are transported to the lysosome with the binding partner, HSC70. The cargo proteins then bind to LAMP2A on the surface of the lysosome, which translocate the cargo proteins into the lysosome for degradation. In microautophagy, the lysosome membrane invaginates to form "phagocytic vesciles," which enwrap molecules on the surface of the lysosomes. The invaginated materials are degraded in the lysosomes. In macroautophagy, the cellular contents (cytosolic proteins or organelles) are enwrapped by autophagosomal membranes in a step known as nucleation. The initial membrane is named as the phagophore, or isolation membrane, which then expands and forms a double-membraned completely enclosed vesicle, known as an autophagosome. The autophagosome containing the cytosolic content is subsequently fused with the lysosome to form the autolysosome, and the content is degraded by the lysosomal enzymes. The molecule LC3 in the mammalian cells is conjugated to PE on the autophagosomal membrane and is a commonly used marker for autophagosomes and autolysosomes, whereas ATG12 and its conjugation partner ATG5 are only presented in the phagophore, but not in the completely circled autophagosomes or autolysosomes.

components that make up the core autophagy machinery. The process basically consists of the formation of a double membranous structure called autophagosome and the fusion of this organelle with the lysosome to form autolysosome, in which the degradation of the sequestered components occurs. The process of autophagy can be broken down to three main steps where distinct groups of ATG proteins contribute to each of these steps.

MOLECULAR STEPS OF MACROAUTOPHAGY

Initiation and Nucleation

Autophagy begins with the formation and expansion of a cup-shaped precursor membrane called a *phagophore* (also called an *isolation membrane*) (Figure 9.1). At present, the precise nature and the origin of phagophore membrane is still unclear and debatable. But the possible membranous origin includes the endoplasmic reticulum (ER) membrane (called the *omegasome*) [18], the ER–Golgi intermediate compartment [19], the mitochondria [20], and the plasma membrane [21]. Accordingly, the exact location where phagophores form differs depending on the location of the cellular cargo. In yeast, the site where phagophore assembles is termed as phagophore assembly site. This is the location where the assembly of ATG proteins occurs in a hierarchical fashion.

In mammalian cells, initiation and regulation of autophagy requires ULK1/2 complex and autophagy-specific Beclin-1 complex. Mammalian ULK1/2 complex consists of ULK1/2 (mammalian orthologous of yeast ATG1), ATG13 and FIP200/RB1cc1 (putative ATG17 homologue), and ATG101/C12orf44 [22–26]. ULK1 is a kinase enzyme whose activity can be directly regulated by AMP-activated protein kinase (AMPK) (in response to glucose starvation) and Mammalian Target of Rapamycin Complex 1 (mTORC1) (in response to amino acid starvation) [27]. AMPK promotes autophagy, while mTORC1 suppresses autophagy (Figure 9.2).

In nutrient-replete condition, ULK1/2–ATG13–FIP200 forms a complex. mTOR1 phosphorylates ULK1 at the residue Ser^{757}, disrupting its interaction with AMPK. Under autophagy-permissive conditions (e.g., starvation), AMPK can directly activate ULK1 by phosphorylation of the residues Ser^{317} and Ser^{777}. Simultaneously, mTOR1 complex is released from the ULK1/2–ATG13–FIP200 complex, resulting in ULK1/2 activation of kinase activity [27]. ULK1/2 then phosphorylates and, presumably, activates ATG13 and FIP200. The activated ULK1 also phosphorylates Beclin-1 on Ser^{14}, which is required for full autophagy induction through the enhancement of the activity of the autophagy-specific Beclin-1 complex [28]. The latter is composed of Beclin-1 (the mammalian ortholog of yeast ATG6), ATG14, and class III phosphatidylinositol-3-kinase (PI3KC3) subunits Vps34 and Vps15 (PIK3R4). This complex generates an abundant amount of phosphatidylinositol 3-phosphate (PI3P) locally, which recruit PI3P-binding proteins such as WIPI-1, WIPI-2, and DFCP1 [29].

Thus, this coordinated event of phosphorylation is important for nucleation of the isolation membrane and for the full autophagic induction. These events occur at the candidate autophagosomal membranes following the recruitment of the molecular complexes, resulting in an increase of PI3P on these membranes.

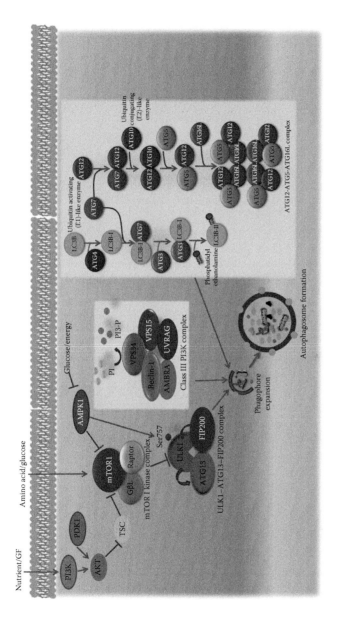

FIGURE 9.2 (**See color insert.**) Signaling pathways in macroautophagy. Macroautophagy is physiologically regulated by nutrients and energy levels. Growth factors, amino acids, and glucose can activate the mTOR complex 1 (mTORC1). mTORC1 suppresses autophagy by inhibiting the ULK1 complex, whereas AMPK1, in response to a lower glucose and/or energy level promotes autophagy by activating the ULK1 complex. The ULK1 complex and the autophagy-specific Beclin-1 complex are responsible for the initial membrane nucleation that leads to the formation of phagophore. The further expansion of phagophore into completely encircled autophagosomes is promoted by the two conjugation systems, where ATG12 is conjugated to ATG5, and LC3 is conjugated to PE on the phagophore membrane. The conjugation reactions are facilitated by ATG7, an E1-like enzyme, and ATG10 or ATG3, an E2-like enzymes. ATG12–ATG5 is further bound to ATG16 to form a large complex, which is also important for LC3 conjugation to PE. Notably, LC3 will need to be proteolytically processed by a cysteine protease, ATG4, before it can be activated by ATG7.

Elongation, Closure of the Isolation Membrane, and Autophagosome Formation

After the formation of the isolation membrane, the membrane expands and envelops the cargo. This process requires ATG9 and two ubiquitin-like conjugation systems. How ATG9 promotes elongation is unclear. The mammalian ATG9 homologue ATG9A is a membrane-spanning protein that localizes to the trans-Golgi network (TGN) and late endosomes in nutrient-rich condition. Following amino acid starvation, this protein translocates to the LC3-labeled autophagosomes in an ULK1- and PI3KC3-dependent manner [30]. WIPI2 (homologue of yeast ATG18) seems to be required as well. Cycling of ATG9 between TGN and later endosome is negatively regulated by p38α MAPK, which competes with ATG9 for binding to p38IP [31]. In the yeast, ATG9 also cycles between the initial autophagosome membrane and peripheral cytoplasmic membranes [32]. ATG9 thus appears to have a conserved role in coordinating membrane transport from donor sources to the phagophore.

There are two ubiquitin protein conjugation systems that participate in autophagy, conjugation of ATG5 to ATG12 to generate ATG12–ATG5, and the conjugation of phosphatidylethanolamine (PE) to ATG8 to generate ATG8–PE (Figure 9.2). The ubiquitin protein conjugation systems are highly conserved from the yeast to the mammals and participate in phagophore expansion. Conjugation of ATG12, an ubiquitin-like molecule, to ATG5 requires ATG7, a molecule similar to the ubiquitin-activating enzyme (E1), and ATG12, an E2-like molecule. The ATG12–ATG5 complex then binds to ATG16 to form a multimer complex in which two units of ATG12–ATG5–ATG16 are represented [33]. The conjugation of ATG8 or its mammalian homologues, such as LC3, to PE on autophagosomal membranes also occurs through the effect of ATG7, and another E2-like molecule, ATG3 [33]. ATG12–ATG5 can facilitate the lipidation of ATG8 (ligase-like activity). Conjugated ATG8 (ATG8–PE) then directly recruits ATG12–ATG5 to membranes by recognizing a noncanonical ATG8-interacting motif (AIM) in ATG12. ATG16 links ATG8–PE/ATG12–ATG5 complexes into a membranous scaffold [34]. The ATG5–ATG12–ATG16 complex disengages the membranes after ATG8 is conjugated. These molecules are thus associated only with the early autophagosomal membrane and are considered as early stage markers. ATG8 remains on the phagophore membrane through the formation of autophagosome and hence is considered as the general marker of autophagosome from the early to the later stage [35]. Both the conjugation steps are required for the autophagosomal membrane to expand and to complete the formation of autophagosome.

There are at least six ATG8 homologues in mammalian cells, which include three LC3 subfamily members (LC3A, LC3B, and LC3C), and three GABARAP subfamily members (GABARAP, GABARAP-L1/ATG8L, and GABRAP-L2/GATE16). LC3 functions at the stage of phagophore elongation, whereas GABARAPs proteins act at a later stage of maturation [36]. Before the activation by ATG7, ATG8 molecules have to first be processed by a cysteine protease, ATG4, at a conserved C-terminal glycine site to form cytosolic LC3-I (Figure 9.2). The exposed glycine is required for these molecules to be conjugated to PE to generate membrane-associated LC3-II [35,37]. There are four mammalian ATG4 homologues including ATG4A, ATG4B,

ATG4C, and ATG4D. Their cleavage activity is different. ATG4A mainly targets to the GABARAP subfamily proteins, whereas ATG4B can effectively cleave all the ATG8 homologues but more so toward the LC3 subfamily members [38]. ATG8–PE is located both in convex face and concave face of the autophagosome. ATG8–PE in concave face acts as an adaptor function and is degraded along with the sequestered substrate. Most ATG8–PE in the convex face of the phagophore is recycled by ATG4-mediated proteolytic cleavage from PE upon autophagosome formation.

Autophagosome–Lysosome Fusion, Cargo Delivery, and Lysosomal Hydrolytic Degradation

Once autophagosomes are formed, they migrate to where lysosomes are located and engage in the fusion with the latter to form the autolysosome. Degradation of the substrates as well as the inner autophagosome membrane occurs in the autolysosome. Studies indicate that in mammalian cells autophagosomes may fuse with the endolysosomal vesicles first to become amphisomes and then with the lysosomes to form autolysosomes. The cytoskeleton seems to be involved in the movement of autophagosomes, which tend to be distributed closer to the cell surface, and toward the perinuclear region, where the lysosomes are enriched. Agents such as nocodazole, which is a microtubule poison, can block the fusion of the autophagosome with the lysosome, perhaps by preventing the movement of autophagosomes. One of the important vexing questions about the formation of autolysosome is how the fusion event is achieved. Numerous studies now indicate that many molecules participate in the fusion process, including Rab7, Sec18, the SNARE proteins Vam3, and the class C Vps/HOPS complex proteins. Recent findings have identified Syntaxin 17 (Stx17) as the autophagosomal SNARE required for fusion with the endosome/lysosome [39]. Stx17 localizes to the outer membrane of a completed autophagosome and interacts with SNAP-29 and the endosomal/lysosomal SNARE family molecule, VAMP8. Stx17 has a unique C-terminal hairpin structure composed of two tandem transmembrane domains that contain glycine zipper-like motifs, which are essential for its association with the autophagosomal membrane.

Once fused, the inner membrane of the autophagosome and the cytoplasm-derived materials contained in the autophagosome are degraded by lysosomal acid hydrolases. Monomeric units of the digested macromolecules, such as amino acids, are exported to the cytosol by lysosomal permeases for cellular reuse.

REGULATION OF AUTOPHAGY

Autophagy is regulated at transcriptional or posttranscriptional level. For example, FOXO3 has been shown to transcriptionally upregulate the several key autophagy-related genes such as LC3, ATG12, GABARAP1, ATG4B, VPS34, Beclin-1, and BNIP3 by directly binding to their promoter regions. Moreover, posttranscriptional regulation of *ATG* gene by mRNA stabilization of ATG8/LC3, ATG12, and ULK1 can occur during amino acid starvation [40].

Autophagy is also posttranscriptionally regulated by mTORC1 and AMPK (Figure 9.2). mTORC1 is well established as a conserved and critical suppressor of autophagy. AMPK, being an established negative regulator of mTORC1, can promote autophagy. Reduction of cellular ATP level, caused by glucose deprivation or mitochondrial dysfunction, activates AMPK, which then phosphorylates tuberous sclerosis complex (TSC), and the TSC1/TSC2 heterodimer, which is a negative regulator of mTOCR1 activation. Growth factor–mediated mTOC1 activation is caused by the desuppression of TSC1/TSC2 by AKT. On the other hand, AMPK can also phosphorylate and inactivate the Raptor subunit of the mTOCR1 complex, inhibiting mTORC1 activity [41].

mTORC1 is a highly conserved serine/threonine kinase capable of integrating signaling from many stimuli including amino acids and growth factors. The Rag GTPase complex tethers mTORC1 to the lysosome. Mammals have four Rag proteins, RagA, RagB, RagC, and RagD, which form heterodimers consisting of RagA or RagB with RagC or RagD. The two members of the heterodimer appear to have opposite nucleotide loading states, so that when RagA/B is bound to GTP, RagC/D is bound to GDP and vice versa. The GATOR1 complex has been shown to possess the GTPase-activating protein (GAP) activity toward RagA and RagB [42]. Cancer cells with inactivating mutations of GATOR1 have hyperactivated mTORC1 and are insensitive to amino acid starvation. These cells are hypersensitive to rapamycin, an mTORC1 inhibitor.

Rag complex are tethered to the lysosome via the Ragulator/Lamtor complex, which acts like a guanine nucleotide exchange factor (GEF) to stimulate Rag activation. Amino acid stimulation promotes Rag activation where RagA/B is GTP bound and RagC/D is GDP bound. Loading of RagA/B with GTP enables the heterodimer to interact with Raptor, a component of mTORC1 [43]. This interaction results in the translocation of mTORC1 from the cytoplasm to the lysosomal surface, where the Rag GTPases dock on the Ragulator/Lamtor complex [44]. The Ragulator/Lamtor complex serves as GEF for RagA and RagB, which allows the latter to bind to mTORC1. The recruitment of mTORC1 to the lysosome brings it into the close proximity to Rheb, another small GTPase, which is absolutely required for mTORC1 activation [45–47].

Recent works have indicated that amino acids accumulate in the lysosomal lumen and initiate a signaling through a mechanism that requires vacuolar H^+-ATPase (v-ATPase) [48]. The depletion of v-ATPase subunits blocks amino acid–induced recruitment of mTORC1 to the lysosomal surface and downstream signaling. The v-ATPase directly interacts with the Ragulator/Lamtor and also with the Rag GTPase, providing a physical link between mTORC1 and the surface of lysosomes following amino acids stimulation, where mTORC1 is further activated by Rheb.

mTORC1, being a kinase complex, negatively regulates the autophagy by phosphorylation of three major components (as discussed earlier): the ULK1/2 kinase complex (discussed earlier), the autophagy-specific Beclin-1 complex through ATG14 subunit [49], and transcription factor EB(TFEB) [50,51]. TFEB controls the transcriptional expression of several autophagy-related genes and lysosome biogenesis genes. Under normal condition, mTORC1 phosphorylates TFEB, promoting the binding to 14-3-3 proteins and hence sequestering the TFEB in the cytoplasm. Under starvation condition, amino acid withdrawal or inactivation of amino acid secretion from the lysosome, mTORC1 is inactivated and the nonphosphorylated TFEB has less affinity binding to the 14-3-3 protein, thus translocating to the nucleus to initiate its action.

SELECTIVE VERSUS NONSELECTIVE AUTOPHAGY

Autophagy can selectively remove the superfluous or damaged organelles such as dysfunctional mitochondria. The concept of selective autophagy for cellular component was raised in 1966 by de Duve and Wattiaux [52] when autophagy was first noted. Later in 1973, Bolender and Weibel [53] reported the removal of expanded ER membrane in the liver following phenobarbital withdrawal. In addition, the turnover of peroxisomes by autophagy was noted in *Hansenula polymorpha* to be induced by selective inactivation of peroxisomal enzymes [54]. These cargo-specific autophagy processes are named after the organelles being engulfed—including mitophagy (mitochondria), pexophagy (peroxisome), erphagy (endoplasmic reticulum), ribophagy (ribosome), and lipophagy (lipid droplets). Although microbes such as bacteria are not endogenous cellular components, autophagic removal of these agents demonstrates a level of selectivity, and hence the term, xenophagy.

MITOPHAGY

Mitochondria are essential for cellular metabolism, signaling, and homeostasis. Numerous metabolic processes, including amino acid biosynthesis, lipid catabolism, and heme and iron–sulfur cluster synthesis are directly or indirectly associated with this organelle. At the same time, however, dysfunctional mitochondria can generate an excessive amount of reactive oxygen species (ROS) that may oxidize lipids, proteins, and DNA, causing cell death. The concept of mitophagy was originated on the basis that cells have to remove altered mitochondria (diseased pool) to maintain a pool of optimally functional mitochondria (healthy pool) [55,56]. General autophagy in starvation could, however, induce mitophagy as a means for nutrients reutilization.

In mammalian cells, autophagic removal of mitochondria has also been shown to be triggered by the loss of mitochondrial transmembrane potentials, for example, following the treatment with an uncoupler, CCCP, or with a potassium ionophore, valinomycin. Mitochondria can also be removed by autophagy during differentiation as in the case of maturation of reticulocytes into red cells [6].

MOLECULAR BASIS OF MITOPHAGY

The current molecular concept of mitophagy is that mitochondrial dysfunction and/or mitochondrial structural alterations generate an "eat-me" signal, which is also called "priming." The mobilized autophagy core machinery can then recognize and sequester the primed mitochondria. Works on different experimental models—mammalian, yeast, and insects—have identified several proteins, such as ATG32, ATG11, Uth1p, and Aup1 in yeast, and NIX, BNIP3, and FUNDC1 in mammalian cells that are key to mitophagy [57].

ATG32 and Yeast Mitophagy

In yeast, mitophagy appears to occur primarily in response to changing metabolic demands. A shift from nutrient-rich medium containing a nonfermentable carbon source to nitrogen starvation medium with a fermentable carbon source can induce mitophagy. ATG32 was the first mitochondrial protein in the yeast identified through genomic screening. It serves as an adaptor for ATG8. When mitophagy is induced, casein kinase 2 phosphorylates ATG32 [58], which allows Atg32 to recruit ATG11 to the mitochondria. Yme1 (mitochondrial-AAA protease) can proteolytically process ATG32 [59], which also allows ATG32 to interact with ATG11. ATG11 is a scaffolding protein and its presence on the surface of mitochondria links the cargo to other ATG proteins and promotes the formation of mitochondria-specific autophagosomes. Finally, ATG11 has been shown to interact with Dnm1, a dynamin-related GTPase, which, together with other fission complex proteins Fis1 and Mdv1-Caf4, is required for mitochondrial fission in yeast [60], suggesting that fragmented mitochondria are efficiently selected for the sequestration—mitochondrial selection and mitochondrial fission are coordinated through these protein complex.

Permeability Transition and Mitochondrial Depolarization

Mitochondrial permeability transition (MPT) is often considered to be a crucial event regulating the initial steps of apoptosis, and has been proposed as a molecular signal initiating mitophagy [61]. Mitochondrial depolarization and compromised mitochondrial outer membrane may potentiate each other in a viscous cycle. Loss of the mitochondrial membrane potentials, increases in cellular ROS (cytosolic derived or mitochondria-deprived), alterations in cellular bioenergetics status, changes in oxygen tension, disturbance of calcium signaling, defects in mitochondrial protein transport, mtDNA damages, and accumulation of protein aggregates in the mitochondria all can precipitate mitophagy.

NIX and BNIP3-Mediated Mitophagy Signaling

Similar to the mitochondrial priming by ATG32 in the yeast model of mitophagy, in mammalian cells BNIP3, NIX [62,63], FUNDC1 [64], and p62/SQSTM1 can prime the target mitochondria for recognition. BNIP3 (Bcl22 and adenovirus E1B 19 kDa interacting protein 3) and NIX (BNIP3-like, or BNIP3L) belong to the BCL-2 family and have a conserved BH3 domain. They can induce cell death and autophagy. BNIP3 and NIX are localized to the outer mitochondrial membranes and can trigger mitochondrial depolarization [61,65], which in turn causes mitophagy (Figure 9.3). The effect of NIX and BNIP3 has been well studied in mitophagy occurring during reticulocyte differentiation into red blood cells. In addition, metabolic defects in the liver-specific *Bnip3*-null mice were observed with increased mitochondrial mass. Increased proportion of mitochondria exhibited loss of mitochondrial membrane potentials, abnormal structures, and reduced oxygen consumption. Elevated ROS, inflammation, and features of steatohepatitis were observed in the livers of *Bnip3*-null mice [66].

These mitophagy adaptors are generally localized at the outer membrane of the mitochondria and have classic tetrapeptide sequence W/F/YxxL/I that mediates

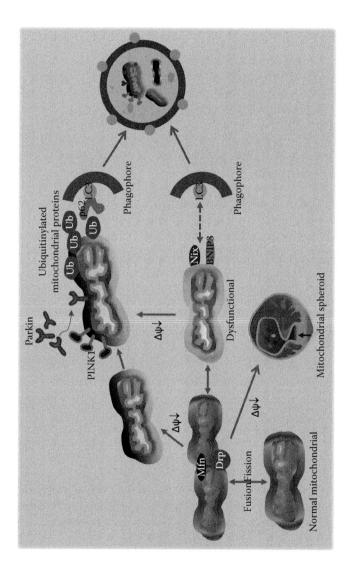

FIGURE 9.3 Regulation of mitophagy and mitochondrial spheroid formation. Normal mitochondria undergo fission and fusion, which are regulated by molecules such as Mfn1/Mfn2 and Drp1. Dysfunctional mitochondria may depolarize and become fragmented. If this is not reversed in time, they may be subjected to removal by autophagy *via* a Nix/BNIP3-medicated mechanism, or by a Parkin-mediated mechanism or by other mechanisms (see the text). However, it is also possible that the dysfunctional mitochondria may transform into a spheroid structure called mitochondrial spheroids, which can trap cytosolic contents. Although the significance of this new form of mitochondria is not quite clear at this moment, its regulation is coupled with Parkin-mediated mitophagy. Therefore, mitochondrial spheroids form only in the absence of Parkin and are dependent on mitofusins. Parkin will cause mitofusin degradation, suppress mitochondrial spheroid formation, but promote mitophagy.

the interaction with LC3. These sequences, known as the LC3-interacting region, thus bridge the autophagosome to the primed mitochondria. Expression of NIX and BNIP3 are also induced at the late stage of erythrocyte maturation and in hypoxia-induced mitophagy. These adaptors are transcriptionally induced by hypoxia-related factors, such as HIF or FOXO3 [67,68].

Interestingly, enforced expression of BNIP3 or NIX induces autophagy in different conditions [69], and that NIX recruits autophagy components independently of its ability to trigger mitochondrial depolarization. It is highly possible that NIX functions as an adaptor protein and recruits components of the autophagy machineries to the mitochondria. An alternative mechanism by which NIX mediates mitophagy is that NIX, being a BH3 domain-containing protein, can compete with Beclin-1 for binding to BCL2 or BCL-X_L, which releases Beclin-1 from the Beclin-1-BCL-2 complex and activates autophagy.

Overexpression of FUNDC1 induces massive mitochondrial fragmentation and knockdown of FUNDC1 enhances mitochondrial fusion, suggesting that mitochondrial fission is required for mitophagy [64]. BCL2L1, but not BCL2, interacts with and inhibits PGAM5, a mitochondrially localized phosphatase, to prevent the dephosphroylation of FUNDC1 at Serine 13 (Ser[13]), which then activates hypoxia-induced mitophagy in mammalian cell [70].

Other BCL2 family members (e.g., BCL-X_L and MCL1) can suppress mitophagy through the inhibition of Parkin translocation to depolarized mitochondria. Consistent with this, Parkin translocation to mitochondria is enhanced by BH3-only proteins or a BH3-only mimetic [71].

PINK1–Parkin–Ubiquitin Signaling Pathway

PINK1 (PTEN-induced putative protein kinase 1) is a mitochondrial kinase located at the outer mitochondrial membrane and Parkin is an E3 ubiquitin ligase located in the cytosol. Mutations of either PINK1 or Parkin have been related to the autosomal recessive forms of Parkinson's disease. Even though PINK1 is well expressed in hepatocytes, no studies on PINK1-dependent mitophagy in the liver have been reported.

Endogenous PINK1 is rapidly degraded in healthy mitochondria. Loss of mitochondrial membrane potentials stabilizes PINK1 at the outer mitochondrial membrane surface, which recruits Parkin to the surface of mitochondria (Figure 9.3) [72,73]. Parkin is known to be important for mitophagy stimulated by certain stimuli, such as carbonyl cyanide m-chlorophenylhydrazone (CCCP), an uncoupler. Recent studies have shown that there are two separate pool of PINK1, full length PINK1, approximately 63 kDa (mitochondrial PINK1 63 or full length PINK1 63) and the short PINK1, approximately 52 kDa (cytosolic PINK1 52 or matured PINK1 52) generated by the cleavage of PINK1 63 by an inner mitochondrial membrane protease PARL. PINK1 52 can spontaneously translocate from the mitochondria to the cytosol. Once in the cytosol, PINK52 represses Parkin translocation to the mitochondria and prevents mitophagy by physically binding to its RING domain in the cytosol [74–76].

Dissipation of mitochondrial membrane potentials seems to be mandatory for mitochondrial fragmentation, and Parkin-dependent mitochondrial turnover does not occur in fragmented mitochondria with intact mitochondrial membrane potential [77]. Once the mitochondria are decorated with Parkin, they cluster and migrate toward perinuclear areas of the cell where they colocalize with autophagy and the lysosomal markers [73,78]. Parkin-labeled mitochondria are shown to be engulfed by autophagic membranes [79]. Parkin ubiquitinates two key mitochondrial fusion mediators, mitofusin 1 and 2 (MFN1 and MFN2), and a number of other mitochondrial outer membrane proteins [80]. Ubiquitination of several mitochondrial substrates by Parkin requires several E2 ubiquitin-conjugating enzymes (UBE2) such as UBE2N, UBE2L3, and UBE2D2/3 [81]. Parkin-mediated ubiquitination of proteins is essential for mitophagy. Consistently, USP30 and USP15, both are mitochondrial deubiquitinases, can antagonize mitophagy when overexpressed [82,83].

There are several possibilities as to how Parkin-mediated ubiquitination of mitochondrial outer membrane proteins can promote mitophagy. One possibility is that the ubiquitin moiety provides the docking site for autophagosome to recognize. The same concept has been demonstrated in selective removal of other types of cargoes, such as protein aggregates, ribosomes, and damaged endosome [84]. An autophagy adaptor p62/SQSTM1 can simultaneously bind to ubiquitin and ATG8/LC3 with two separate domains. Indeed, SQSTM1 can play a role in mitophagy [62], but this may not be essential [85,86]. One reason may be that other adaptor molecules are present, such as NBR1 [87], which can also bind to ubiquitin and LC3 simultaneously. Alternatively, the ubiquitination alone is not sufficient for preparing the mitochondria for mitophagy. Indeed, the proteasome activity is also required for Parkin to promote mitophagy. The proteasome-dependent degradation of ubiquitinated mitofusins and other mitochondrial proteins may contribute to mitochondrial fragmentation and segregation of fragmented mitochondria from the mitochondrial network, rendering the fragmented mitochondria susceptible to mitophagy. Consistently, overexpression of the dominant negative form of Drp1 or wild-type OPA1 promotes fusion and retards mitophagy in mammalian cells. Yet, another explanation for the importance of mitofusin degradation in mitophagy is the process that prevents the formation of mitochondrial spheroids (see following section). Thus, overall how Parkin promotes mitophagy is still quite unclear.

Parkin can also directly interact with the autophagic machinery. Following mitochondrial depolarization, Parkin has been found to interact with AMBRA1, a component of the core autophagic machinery [88]. Several other molecules can regulate Parkin. NIX can promote Parkin translocation to the mitochondria by facilitating mitochondrial depolarization [62]. Another mitochondrial protein, choline dehydrogenase (CHDH), has also been shown to be required for Parkin-mediated mitophagy in the CCCP model. CHDH is found on both the outer and inner membranes of the mitochondria in resting cells. Upon CCCP treatment, CHDH accumulates on the outer membrane in a mitochondrial potential-dependent manner, interacts with SQSTM1, recruits LC3, and stimulate mitophagy [89]. Knockdown of CHDH expression impairs CCCP-induced mitophagy and Parkin-mediated clearance of mitochondria in mammalian cells. Conversely, overexpression of CHDH accelerates Parkin-mediated mitophagy. Other proteins shown to play important roles in Parkin-mediated mitophagy include transglutaminase 2 (TG2).

TG2 generally interacts with the dynamic proteins Drp1 and Fis1. CCCP treatment reduces the interaction of TG2 with Drp1 and Fis1. Knockdown of TG2 in cells increases the number of fragmented mitochondria with decreased membrane potential [90].

Besides CCCP, ER stress can promote Parkin-dependent mitophagy and transiently protect the ischemic brain injury. Both tunicamycin and thapsigargin can prevent Parkin loss, promote its recruitment to the mitochondria, and activate mitophagy during ischemia/reperfusion-induced neuronal injury [91].

MITOPHAGY AND MITOCHONDRIAL SPHEROID FORMATION

As discussed earlier, mitophagy triggered by treatment with CCCP requires Parkin. Surprisingly, in the absence of Parkin, while mitophagy was inhibited, CCCP treatment induces the formation of a special mitochondrial structure, known as mitochondrial spheroids [92] (Figure 9.3). Serial section electron microscopy and electron tomography 3D reconstruction indicate that the mitochondria transform into a phagosome-like spherical structure. It seems that a portion of the mitochondria is compressed to allow the mitochondria to bend around the cytoplasm to form this ball-like structure with an internal lumen, which is surrounded by the mitochondrial membranes. The lumen is not completely delimited, but connects to the cytoplasm through a small orifice. Mitochondrial spheroids can be defined in vivo in the liver following acetaminophen treatment, a condition causing severe oxidative mitochondrial damages and liver injury [93]. Early studies had also described the cup-shaped spheroid mitochondria in normal rat livers [94] and in ethanol-fed rat livers [95]. It is speculated that mitochondria assume this spherical structure to prevent them from being recognized by autophagosomal membranes and hence escapes the mitophagy.

The formation of mitochondrial spheroid depends on ROS, which causes mitochondrial fragmentation, and mitofusins (Mfn1 and Mfn2), which allows the fragmented mitochondria to transform into the spherical structure. Antioxidants or the deletion of Mfn1 or Mfn2 inhibits the spheroid formation. A physiological regulation of mitochondrial spheroids is mediated by Parkin. Following CCCP treatment, Parkin can cause ubiquitination of Mfn1 or Mfn2, which is subsequently degraded. Thus, Parkin inhibits mitochondrial spheroid formation. Furthermore, the knockdown of Mfn1 or Mfn2 will allow CCCP-induced mitophagy to occur even in the absence of Parkin, suggesting that Parkin's ability to degrade the mitofusins is all that needs for mitophagy to occur [96]. One may further speculate that prevention of the spheroid formation, which can be the default pathway under CCCP, is required for mitophagy to occur. Thus, Parkin and mitofusins provide the switch between the pathway to mitochondrial spheroid formation and the pathway to mitophagy.

AUTOPHAGY AND MITOPHAGY IN LIVER DISEASES

Autophagy is involved in many aspects of liver biology. With a high biosynthetic activity and storage capacity, the liver seems to possess the highest autophagic protein turnover rate among the body organs. This protein degradative function is accredited by a high abundance of lysosomes and lysosomal enzymes in the hepatocyte. Liver is one of the best studied organs for autophagy function and regulation. Many of the early observations on autophagic function and structures were made using the liver as the model. Mitochondria-containing autophagosome was first found in hepatocytes [52].

Autophagy plays a critical role both in a normal liver physiology and in liver diseases. Autophagy has been recognized as a prosurvival mechanism, maintaining cell function and survival through context-dependent nonselective or selective degradation of excessive or damaged organelles and proteins. In rodent models of nutrient starvation, autophagy was responsible for the degradation of 35% of total liver protein within 24 hours [97], illustrating the role of autophagy in liver homeostasis and energy conservation. Conversely, the inhibition of autophagy by genetic deletion of autophagy genes, such as ATG7, ATG5, or Vps34, all lead to dramatic increase in liver mass [98–100]. Autophagy deficiency in liver results in the accumulation of protein aggregates, damaged mitochondria, steatosis, and chronic liver injury. The potential roles of autophagy and mitophagy in diseases has been increasingly recognized and investigated. The role of autophagy and mitophagy in alcoholic fatty liver disease, nonalcoholic fatty liver disease (NAFLD), and ischemic/reperfusion liver injury is discussed later.

ALCOHOLIC LIVER DISEASE

Ethanol is mainly metabolized by the liver and liver mitochondria can be a primary target for alcohol toxicity. A major catabolic pathway of ethanol begins with alcohol dehydrogenase (ADH), which generates acetaldehyde. The latter is oxidized predominately by the mitochondrial aldehyde dehydrogenase. In both steps, NADH was generated and is oxidized indirectly by mitochondrial electron transport system. The excessive amount of NADH and thus the reducing capacity in the mitochondrial electron transport system is thought to cause an increased leakage of mitochondrial ROS, causing alcohol-induced oxidative stress. This has been shown to alter oxidative phosphorylation, mitochondrial proteome, and mitochondrial dynamics. Mitochondrial DNA depletion was observed in livers of ethanol-fed mouse [101,102]. Chronic ethanol consumption causes enhanced oxidative damage to mtDNA along with increased strand breakage, and other alterations in the structural integrity of mitochondrial DNA [103], which has been thought to cause steatohepatitis, fibrosis, cirrhosis, and hepatocellular carcinoma.

Mitophagy may be important to clear these dysfunctional and potentially deleterious mitochondria. In support to this concept, ethanol treatment of hepatocytes induces autophagy by ROS-mediated inhibition of mTORC1 [104], and activation of AMPK [105]. Similar induction of autophagy was also observed in hepatoma cell lines expressing ADH and cytochrome p450 2E1 (Cyp2E1) [106]. Ethanol-induced autophagy can selectively target to the damaged mitochondria in addition to the lipid droplets as observed in both acute [104] and chronic [107] ethanol consumption. What is the mechanism of this selectivity has yet to be determined. It is possible that the PINK1–Parkin signaling may be involved. Immunoelectron microscopic studies indicated that the expression of PINK1 was increased in mitochondria from ethanol-treated rats [107]. In addition, ethanol-mediated oxidative stress can also cause the translocation of the inducible form of heme oxygenase-1 to the mitochondria, which in turn increases the recruitment of LC3 to the mitochondria [108]. Furthermore, mitochondrial cyclophilin D has been implicated in autophagy induced in chronically ethanol-fed mice [109].

Autophagic removal of damaged mitochondria by mitophagy eliminates an important source of ROS, inhibits hepatocyte death and liver injury. This is accompanied with mitochondrial biogenesis and expression of antioxidant genes, both promoting recovery. Therapeutics can be designed to rescue mitochondrial function in alcoholic liver disease (ALD). Antioxidants, such as N-*acetylcysteine*, S-adenosylmethionine, betaine, TPP+ (triphenylphosphonium), and cannabidiol could possibly protect against ethanol-induced mitochondrial dysfunction, prevent JNK/MAPK activation and increase autophagy [110].

On the other hand, chronic ethanol consumption may lead to severe mitochondrial phenotypes, and impaired mitophagy. This could be due to the decline in lysosome function and microtubule-mediated transportation of autophagosomes [111,112]. These may be the primary cause for some of the well-defined ethanol-related pathology, such as hepatomegaly and the appearance of Malory–Denk bodies.

NONALCOHOLIC FATTY LIVER DISEASE

NAFLD includes a spectrum of pathologies that include hepatic steatosis, nonalcoholic steatohepatitis, fibrosis, and cirrhosis. A significant level of understanding toward the NAFLD pathogenesis has been achieved, but a lot more about the molecular mechanism has yet to be learnt. The accumulation of triglyceride in hepatocytes has been considered to be the first distinct phase that predisposes the cells to a "second hit," that is, oxidative stress and inflammation, leading to steatohepatitis and liver injury [113,114].

Oxidative stress and mitochondrial dysfunction can be involved in the development of NAFLD [115–118]. Mitochondria from high fat diet (HFD)-fed mice present a more condensed form with increased respiratory capacity and higher ATP production level, compared to the mitochondria from normal diet–fed mice [119]. Quantitative proteomic analysis of these mitochondria reveals abnormalities in the level of many mitochondrial proteins, including mitofilin, Sam50, and Tim proteins. Thus, hepatic mitochondria may play a critical role in the development and pathogenesis of steatosis and NAFLD. Although there are few studies on the role of mitophagy in NAFLD, it is speculated that mitophagy can play protective roles during NAFLD because removing damaged mitochondria would be beneficial.

It is more definite that autophagy plays a significant protective role in NAFLD in the form of lipophagy [120,121]. Cellular lipid droplets can be sequestered by autophagosomes and delivered to lysosomes for degradation. In vivo studies of mouse liver have demonstrated a clear association of LC3 with lipid droplets and the presence of lipid in autophagosome and lysosomes. On the other hand, there is a marked increase in lipids in autophagy-deficient livers. The increase in hepatocellular lipid stores was due to decreased lipolysis but not due to decreased delivery of lipid cargo into the lysosomes, neither to the increased hepatocellular TG synthesis, nor to a reduction in secretion in the form of Very-low-density lipoprotein (VLDL). Electron microscopic studies demonstrated that the inhibition of autophagy in hepatocytes leads to marked increase in the number and size of lipid droplets. Hence, autophagy regulates hepatic lipid stores through its function in removing lipid droplets (lipophagy). Autophagy may also regulate the cellular lipid content through additional mechanisms, including the degradation of factors that mediate lipid metabolism, such as apolipoprotein B [122].

Stimulation of autophagy has shown to be a protective function in NAFLD, for example, by caffeine, by reducing intrahepatic lipid content and stimulating β-oxidation [123,124]. Therapeutic enhancement of autophagy alleviated high fat diet–induced steatosis. Liver injury was also noticeably reduced and insulin resistance was improved as well [125]. Similarly, enforced expression of an autophagy gene, ATG7, in the liver of mice fed with high fat diet, or of ob/ob mice, improved the fatty liver condition and insulin resistance [121].

ISCHEMIA/REPERFUSION-MEDIATED LIVER INJURY

Impaired or insufficient autophagy has also been shown to be the underlying cause of liver injury in ischemia reperfusion (I/R) [126]. During I/R, hepatocytes are exposed to nutrient deprivation, ATP depletion, and increased production of reactive oxygen or nitrogen species (ROS/RNS), all causing mitochondrial dysfunction [127]. Even though the mechanism underlying I/R injury seems to be multifactorial, including calcium overloading, the formation of ROS, and calpain activation, MPT seems to play a major causative role in I/R injury. MPT leads to the uncoupling of oxidative phosphorylation, mitochondrial release of propaoptotic molecules, and cell death [128]. Hence, clearance of these dysfunctional mitochondria by autophagy can be vital for the protection of hepatocytes against death. Induction of autophagy in hepatocytes under I/R by overexpression of ATG7 or Beclin-1, or by nutrient deprivation, promoted mitophagy and prevented MPT, leading to improved cell survival [128,129]. Consistently, the stimulation of autophagy in vivo by carbamazepine or by overexpressing *ATG* genes also increased hepatocytes viability after reperfusion [130].

The evidence of ongoing mitophagy and its importance was first provided in the study by Kim et al. [128]. In the ischemic livers, polarized mitochondria were found to be selectively sequestered in autophagosomes, suggesting the ongoing process of mitophagy. Genetic knockdown of autophagy genes such as ATG7 and Beclin-1 suppressed mitophagy [128]. The inhibition of autophagy by chloroquine or wortmannin aggravated I/R-induced liver injury through the inhibition of autophagy at the late phase. Interestingly, chloroquine could also protect against hepatic I/R by inhibiting inflammatory responses in the early phase [131].

CONCLUSION

Autophagy is an important mechanism for maintaining cellular homeostasis. Many physiological and pathological factors can stimulate autophagy. Autophagy can modulate many processes in the liver, which in turn protect the liver from injury. Mitophagy is a process by which damaged mitochondria are engulfed and removed by the autophagosome. Mitophagy may be particularly active in ethanol- and ischemia/reperfusion-induced liver injury, where mitochondrial damage is frequent. Mitochondrial damage, however, has been noted in other liver diseases such as NAFLD. Mitophagy in drug-induced liver, such as that induced by acetaminophen, is discussed elsewhere in this book. While the basic mechanisms for autophagy and mitophagy have seen great progress in the past several years, those processes in the pathological conditions are far from clear and are the subjects for future studies.

REFERENCES

1. Mathew, R. et al., Autophagy suppresses tumor progression by limiting chromosomal instability. *Genes Dev*, 2007. 21(11): 1367–1381.
2. Fortun, J. et al., Emerging role for autophagy in the removal of aggresomes in Schwann cells. *J Neurosci*, 2003. 23(33): 10672–10680.
3. Singh, S.B. et al., Human IRGM induces autophagy to eliminate intracellular mycobacteria. *Science*, 2006. 313(5792): 1438–1441.
4. Bellot, G. et al., Hypoxia-induced autophagy is mediated through hypoxia-inducible factor induction of BNIP3 and BNIP3L via their BH3 domains. *Mol Cell Biol*, 2009. 29(10): 2570–2581.
5. Klionsky, D.J. et al., A unified nomenclature for yeast autophagy-related genes. *Dev Cell*, 2003. 5(4): 539–545.
6. Kundu, M. and C.B. Thompson, Autophagy: Basic principles and relevance to disease. *Annu Rev Pathol*, 2008. 3: 427–455.
7. Ding, W.X. et al., Linking of autophagy to ubiquitin-proteasome system is important for the regulation of endoplasmic reticulum stress and cell viability. *Am J Pathol*, 2007. 171(2): 513–524.
8. Yi, J. and X.M. Tang, The convergent point of the endocytic and autophagic pathways in leydig cells. *Cell Res*, 1999. 9(4): 243–253.
9. Komatsu, M. et al., Loss of autophagy in the central nervous system causes neurodegeneration in mice. *Nature*, 2006. 441(7095): 880–884.
10. Hara, T. et al., Suppression of basal autophagy in neural cells causes neurodegenerative disease in mice. *Nature*, 2006. 441(7095): 885–889.
11. Schmid, D. and C. Munz, Immune surveillance via self digestion. *Autophagy*, 2007. 3(2): 133–135.
12. Deretic, V. and B. Levine, Autophagy, immunity, and microbial adaptations. *Cell Host Microbe*, 2009. 5(6): 527–549.
13. Kirshenbaum, L.A., Regulation of autophagy in the heart in health and disease. *J Cardiovasc Pharmacol*, 2012. 60(2): 109.
14. Rubinsztein, D.C., P. Codogno, and B. Levine, Autophagy modulation as a potential therapeutic target for diverse diseases. *Nat Rev Drug Discov*, 2012. 11(9): 709–730.
15. Rubinsztein, D.C., G. Marino, and G. Kroemer, Autophagy and aging. *Cell*, 2011. 146(5): 682–695.
16. Kaushik, S. and A.M. Cuervo, Chaperone-mediated autophagy: A unique way to enter the lysosome world. *Trends Cell Biol*, 2012. 22(8): 407–417.

17. Mijaljica, D., M. Prescott, and R.J. Devenish, Microautophagy in mammalian cells: Revisiting a 40-year-old conundrum. *Autophagy*, 2011. 7(7): 673–682.

18. Hamasaki, M. et al., Autophagosomes form at ER-mitochondria contact sites. *Nature*, 2013. 495(7441): 389–393.

19. Hayashi-Nishino, M. et al., A subdomain of the endoplasmic reticulum forms a cradle for autophagosome formation. *Nat Cell Biol*, 2009. 11(12): 1433–1437.

20. Hailey, D.W. et al., Mitochondria supply membranes for autophagosome biogenesis during starvation. *Cell*, 2010. 141(4): 656–667.

21. Ravikumar, B. et al., Plasma membrane contributes to the formation of pre-autophagosomal structures. *Nat Cell Biol*, 2010. 12(8): 747–757.

22. Hosokawa, N. et al., Nutrient-dependent mTORC1 association with the ULK1-ATG13-FIP200 complex required for autophagy. *Mol Biol Cell*, 2009. 20(7): 1981–1991.

23. Mizushima, N., The role of the ATG1/ULK1 complex in autophagy regulation. *Curr Opin Cell Biol*, 2010. 22(2): 132–139.

24. Hosokawa, N. et al., ATG101, a novel mammalian autophagy protein interacting with ATG13. *Autophagy*, 2009. 5(7): 973–979.

25. Mercer, C.A., A. Kaliappan, and P.B. Dennis, A novel, human ATG13 binding protein, ATG101, interacts with ULK1 and is essential for macroautophagy. *Autophagy*, 2009. 5(5): 649–662.

26. Hara, T. et al., FIP200, a ULK-interacting protein, is required for autophagosome formation in mammalian cells. *J Cell Biol*, 2008. 181(3): 497–510.

27. Kim, J. et al., AMPK and mTOR regulate autophagy through direct phosphorylation of Ulk1. *Nat Cell Biol*, 2011. 13(2): 132–141.

28. Russell, R.C. et al., ULK1 induces autophagy by phosphorylating Beclin-1 and activating VPS34 lipid kinase. *Nat Cell Biol*, 2013. 15(7): 741–750.

29. Polson, H.E. et al., Mammalian ATG18 (WIPI2) localizes to omegasome-anchored phagophores and positively regulates LC3 lipidation. *Autophagy*, 2010. 6(4): 506–522.

30. Young, A.R. et al., Starvation and ULK1-dependent cycling of mammalian ATG9 between the TGN and endosomes. *J Cell Sci*, 2006. 119(Pt 18): 3888–3900.

31. Webber, J.L. and S.A. Tooze, Coordinated regulation of autophagy by p38α MAPK through mATG9 and p38IP. *EMBO J*, 2010. 29(1): 27–40.

32. Reggiori, F. et al., ATG9 cycles between mitochondria and the pre-autophagosomal structure in yeasts. *Autophagy*, 2005. 1(2): 101–109.

33. Ohsumi, Y. and N. Mizushima, Two ubiquitin-like conjugation systems essential for autophagy. *Semin Cell Dev Biol*, 2004. 15(2): 231–236.

34. Kaufmann, A. et al., Molecular mechanism of autophagic membrane-scaffold assembly and disassembly. *Cell*, 2014. 156(3): 469–481.

35. Ichimura, Y. et al., In vivo and in vitro reconstitution of ATG8 conjugation essential for autophagy. *J Biol Chem*, 2004. 279(39): 40584–40592.

36. Weidberg, H. et al., LC3 and GATE-16/GABARAP subfamilies are both essential yet act differently in autophagosome biogenesis. *EMBO J*, 2010. 29(11): 1792–1802.

37. Fujita, N. et al., The ATG16L complex specifies the site of LC3 lipidation for membrane biogenesis in autophagy. *Mol Biol Cell*, 2008. 19(5): 2092–2100.

38. Li, M. et al., Kinetics comparisons of mammalian ATG4 homologues indicate selective preferences toward diverse ATG8 substrates. *J Biol Chem*, 2011. 286(9): 7327–7338.

39. Itakura, E., C. Kishi-Itakura, and N. Mizushima, The hairpin-type tail-anchored SNARE syntaxin 17 targets to autophagosomes for fusion with endosomes/lysosomes. *Cell*, 2012. 151(6): 1256–1269.

40. Khambu, B., M. Uesugi, and Y. Kawazoe, Translational repression stabilizes messenger RNA of autophagy-related genes. *Genes Cells*, 2011. 16(8): 857–867.

41. Gwinn, D.M. et al., AMPK phosphorylation of raptor mediates a metabolic checkpoint. *Mol Cell*, 2008. 30(2): 214–226.
42. Bar-Peled, L. et al., A tumor suppressor complex with GAP activity for the Rag GTPases that signal amino acid sufficiency to mTORC1. *Science*, 2013. 340(6136): 1100–1106.
43. Sancak, Y. et al., The Rag GTPases bind raptor and mediate amino acid signaling to mTORC1. *Science*, 2008. 320(5882): 1496–1501.
44. Sancak, Y. et al., Ragulator-Rag complex targets mTORC1 to the lysosomal surface and is necessary for its activation by amino acids. *Cell*, 2010. 141(2): 290–303.
45. Garami, A. et al., Insulin activation of Rheb, a mediator of mTOR/S6K/4E-BP signaling, is inhibited by TSC1 and 2. *Mol Cell*, 2003. 11(6): 1457–1466.
46. Inoki, K. et al., Rheb GTPase is a direct target of TSC2 GAP activity and regulates mTOR signaling. *Genes Dev*, 2003. 17(15): 1829–1834.
47. Tee, A.R. et al., Tuberous sclerosis complex gene products, Tuberin and Hamartin, control mTOR signaling by acting as a GTPase-activating protein complex toward Rheb. *Curr Biol*, 2003. 13(15): 1259–1268.
48. Zoncu, R. et al., mTORC1 senses lysosomal amino acids through an inside-out mechanism that requires the vacuolar H(+)-ATPase. *Science*, 2011. 334(6056): 678–683.
49. Yuan, H.X., R.C. Russell, and K.L. Guan, Regulation of PIK3C3/VPS34 complexes by MTOR in nutrient stress-induced autophagy. *Autophagy*, 2013. 9(12): 1983–1995.
50. Martina, J.A. et al., MTORC1 functions as a transcriptional regulator of autophagy by preventing nuclear transport of TFEB. *Autophagy*, 2012. 8(6): 903–914.
51. Settembre, C. et al., A lysosome-to-nucleus signalling mechanism senses and regulates the lysosome via mTOR and TFEB. *EMBO J*, 2012. 31(5): 1095–1108.
52. de Duve, C. and R. Wattiaux, Functions of lysosomes. *Annu Rev Physiol*, 1966. 28: 435–492.
53. Bolender, R.P. and E.R. Weibel, A morphometric study of the removal of phenobarbital-induced membranes from hepatocytes after cessation of threatment. *J Cell Biol*, 1973. 56(3): 746–761.
54. Veenhuis, M. et al., Degradation and turnover of peroxisomes in the yeast *Hansenula polymorpha* induced by selective inactivation of peroxisomal enzymes. *Arch Microbiol*, 1983. 134(3): 193–203.
55. Lemasters, J.J., Selective mitochondrial autophagy, or mitophagy, as a targeted defense against oxidative stress, mitochondrial dysfunction, and aging. *Rejuven Res*, 2005. 8(1): 3–5.
56. Takeshige, K. et al., Autophagy in yeast demonstrated with proteinase-deficient mutants and conditions for its induction. *J Cell Biol*, 1992. 119(2): 301–311.
57. Ding, W.X. and X.M. Yin, Mitophagy: Mechanisms, pathophysiological roles, and analysis. *Biol Chem*, 2012. 393(7): 547–564.
58. Kanki, T. et al., Casein kinase 2 is essential for mitophagy. *EMBO Rep*, 2013. 14(9): 788–794.
59. Wang, K. et al., Proteolytic processing of ATG32 by the mitochondrial i-AAA protease Yme1 regulates mitophagy. *Autophagy*, 2013. 9(11): 1828–1836.
60. Mao, K. et al., The scaffold protein ATG11 recruits fission machinery to drive selective mitochondria degradation by autophagy. *Dev Cell*, 2013. 26(1): 9–18.
61. Elmore, S.P. et al., The mitochondrial permeability transition initiates autophagy in rat hepatocytes. *FASEB J*, 2001. 15(12): 2286–2287.
62. Ding, W.X. et al., Nix is critical to two distinct phases of mitophagy, reactive oxygen species-mediated autophagy induction and Parkin-ubiquitin-p62-mediated mitochondrial priming. *J Biol Chem*, 2010. 285(36): 27879–27890.
63. Novak, I. et al., Nix is a selective autophagy receptor for mitochondrial clearance. *EMBO Rep*, 2010. 11(1): 45–51.

64. Liu, L. et al., Mitochondrial outer-membrane protein FUNDC1 mediates hypoxia-induced mitophagy in mammalian cells. *Nat Cell Biol*, 2012. 14(2): 177–185.
65. Twig, G. et al., Fission and selective fusion govern mitochondrial segregation and elimination by autophagy. *EMBO J*, 2008. 27(2): 433–446.
66. Glick, D. et al., BNIP3 regulates mitochondrial function and lipid metabolism in the liver. *Mol Cell Biol*, 2012. 32(13): 2570–2584.
67. Sowter, H.M. et al., HIF-1-dependent regulation of hypoxic induction of the cell death factors BNIP3 and NIX in human tumors. *Cancer Res*, 2001. 61(18): 6669–6673.
68. Chinnadurai, G., S. Vijayalingam, and S.B. Gibson, BNIP3 subfamily BH3-only proteins: Mitochondrial stress sensors in normal and pathological functions. *Oncogene*, 2008. 27(Suppl 1): S114–S127.
69. Zhang, J. and P.A. Ney, Role of BNIP3 and NIX in cell death, autophagy, and mitophagy. *Cell Death Differ*, 2009. 16(7): 939–946.
70. Wu, H. et al., The BCL2L1 and PGAM5 axis defines hypoxia-induced receptor-mediated mitophagy. *Autophagy*, 2014. 10(10): 1712–1725.
71. Hollville, E. et al., Bcl-2 family proteins participate in mitochondrial quality control by regulating Parkin/PINK1-dependent mitophagy. *Mol Cell*, 2014. 55(3): 451–466.
72. Jin, S.M. et al., Mitochondrial membrane potential regulates PINK1 import and proteolytic destabilization by PARL. *J Cell Biol*, 2010. 191(5): 933–942.
73. Narendra, D.P. et al., PINK1 is selectively stabilized on impaired mitochondria to activate Parkin. *PLoS Biol*, 2010. 8(1): e1000298.
74. Fedorowicz, M.A. et al., Cytosolic cleaved PINK1 represses Parkin translocation to mitochondria and mitophagy. *EMBO Rep*, 2014. 15(1): 86–93.
75. Becker, D. et al., Pink1 kinase and its membrane potential (Deltapsi)-dependent cleavage product both localize to outer mitochondrial membrane by unique targeting mode. *J Biol Chem*, 2012. 287(27): 22969–22987.
76. Takatori, S., G. Ito, and T. Iwatsubo, Cytoplasmic localization and proteasomal degradation of N-terminally cleaved form of PINK1. *Neurosci Lett*, 2008. 430(1): 13–17.
77. Narendra, D. et al., Parkin is recruited selectively to impaired mitochondria and promotes their autophagy. *J Cell Biol*, 2008. 183(5): 795–803.
78. Vives-Bauza, C. et al., PINK1-dependent recruitment of Parkin to mitochondria in mitophagy. *Proc Natl Acad Sci USA*, 2010. 107(1): 378–383.
79. Yang, J.Y. and W.Y. Yang, Bit-by-bit autophagic removal of parkin-labelled mitochondria. *Nat Commun*, 2013. 4: 2428.
80. Gegg, M.E. et al., Mitofusin 1 and mitofusin 2 are ubiquitinated in a PINK1/parkin-dependent manner upon induction of mitophagy. *Hum Mol Genet*, 2010. 19(24): 4861–4870.
81. Geisler, S. et al., The ubiquitin-conjugating enzymes UBE2N, UBE2L3 and UBE2D2/3 are essential for Parkin-dependent mitophagy. *J Cell Sci*, 2014. 127(Pt 15): 3280–3293.
82. Bingol, B. et al., The mitochondrial deubiquitinase USP30 opposes parkin-mediated mitophagy. *Nature*, 2014. 510(7505): 370–375.
83. Cornelissen, T. et al., The deubiquitinase USP15 antagonizes Parkin-mediated mitochondrial ubiquitination and mitophagy. *Hum Mol Genet*, 2014. 23(19): 5227–5242.
84. Chen, X. et al., Autophagy induced by calcium phosphate precipitates targets damaged endosomes. *J Biol Chem*, 2014. 289(16): 11162–11174.
85. Narendra, D. et al., p62/SQSTM1 is required for Parkin-induced mitochondrial clustering but not mitophagy; VDAC1 is dispensable for both. *Autophagy*, 2010. 6(8): 1090–1106.
86. Okatsu, K. et al., p62/SQSTM1 cooperates with Parkin for perinuclear clustering of depolarized mitochondria. *Genes Cells*, 2010. 15(8): 887–900.

87. Kirkin, V. et al., NBR1 cooperates with p62 in selective autophagy of ubiquitinated targets. *Autophagy*, 2009. 5(5): 732–733.

88. Fimia, G.M. et al., Ambra1 regulates autophagy and development of the nervous system. *Nature*, 2007. 447(7148): 1121–1125.

89. Park, S. et al., Choline dehydrogenase interacts with SQSTM1/p62 to recruit LC3 and stimulate mitophagy. *Autophagy*, 2014. 10(11): 1906–1920.

90. Rossin, F. et al., Transglutaminase 2 ablation leads to mitophagy impairment associated with a metabolic shift towards aerobic glycolysis. *Cell Death Differ*, 2014. 22(3):408–418.

91. Zhang, X. et al., Endoplasmic reticulum stress induced by tunicamycin and thapsigargin protects against transient ischemic brain injury: Involvement of PARK2-dependent mitophagy. *Autophagy*, 2014. 10(10): 1801–1813.

92. Ding, W.X. et al., Electron microscopic analysis of a spherical mitochondrial structure. *J Biol Chem*, 2012. 287(50): 42373–42378.

93. Ni, H.M. et al., Zonated induction of autophagy and mitochondrial spheroids limits acetaminophen-induced necrosis in the liver. *Redox Biol*, 2013. 1(1): 427–432.

94. Stephens, R.J. and R.F. Bils, An atypical mitochondrial form in normal rat liver. *J Cell Biol*, 1965. 24: 500–504.

95. Kiessling, K.H. and U. Tobe, Degeneration of liver mitochondria in rats after prolonged alcohol consumption. *Exp Cell Res*, 1964. 33: 350–354.

96. Ding, W.X. et al., Parkin and mitofusins reciprocally regulate mitophagy and mitochondrial spheroid formation. *J Biol Chem*, 2012. 287(50): 42379–42388.

97. Cuervo, A.M. et al., Activation of a selective pathway of lysosomal proteolysis in rat liver by prolonged starvation. *Am J Physiol*, 1995. 269(5 Pt 1): C1200–C1208.

98. Komatsu, M. et al., Impairment of starvation-induced and constitutive autophagy in ATG7-deficient mice. *J Cell Biol*, 2005. 169(3): 425–434.

99. Takamura, A. et al., Autophagy-deficient mice develop multiple liver tumors. *Genes Dev*, 2011. 25(8): 795–800.

100. Jaber, N. et al., Class III PI3K Vps34 plays an essential role in autophagy and in heart and liver function. *Proc Natl Acad Sci USA*, 2012. 109(6): 2003–2008.

101. Mansouri, A. et al., Acute ethanol administration oxidatively damages and depletes mitochondrial DNA in mouse liver, brain, heart, and skeletal muscles: Protective effects of antioxidants. *J Pharmacol Exp Ther*, 2001. 298(2): 737–743.

102. Mansouri, A. et al., An alcoholic binge causes massive degradation of hepatic mitochondrial DNA in mice. *Gastroenterology*, 1999. 117(1): 181–190.

103. Cahill, A. et al., Chronic ethanol consumption causes alterations in the structural integrity of mitochondrial DNA in aged rats. *Hepatology*, 1999. 30(4): 881–888.

104. Ding, W.X., M. Li, and X.M. Yin, Selective taste of ethanol-induced autophagy for mitochondria and lipid droplets. *Autophagy*, 2011. 7(2): 248–249.

105. Sid, B., J. Verrax, and P.B. Calderon, Role of AMPK activation in oxidative cell damage: Implications for alcohol-induced liver disease. *Biochem Pharmacol*, 2013. 86(2): 200–209.

106. Wu, D. et al., Alcohol steatosis and cytotoxicity: The role of cytochrome P4502E1 and autophagy. *Free Radic Biol Med*, 2012. 53(6): 1346–1357.

107. Eid, N. et al., Elevated autophagic sequestration of mitochondria and lipid droplets in steatotic hepatocytes of chronic ethanol-treated rats: An immunohistochemical and electron microscopic study. *J Mol Histol*, 2013. 44(3): 311–326.

108. Bansal, S., G. Biswas, and N.G. Avadhani, Mitochondria-targeted heme oxygenase-1 induces oxidative stress and mitochondrial dysfunction in macrophages, kidney fibroblasts and in chronic alcohol hepatotoxicity. *Redox Biol*, 2014. 2: 273–283.

109. King, A.L. et al., Involvement of the mitochondrial permeability transition pore in chronic ethanol-mediated liver injury in mice. *Am J Physiol Gastrointest Liver Physiol*, 2014. 306(4): G265–G277.

110. Yang, L. et al., Cannabidiol protects liver from binge alcohol-induced steatosis by mechanisms including inhibition of oxidative stress and increase in autophagy. *Free Radic Biol Med*, 2014. 68: 260–267.

111. Donohue, T.M., Jr., R.K. Zetterman, and D.J. Tuma, Effect of chronic ethanol administration on protein catabolism in rat liver. *Alcohol Clin Exp Res*, 1989. 13(1): 49–57.

112. Donohue, T.M., Jr. et al., Ethanol administration alters the proteolytic activity of hepatic lysosomes. *Alcohol Clin Exp Res*, 1994. 18(3): 536–541.

113. Day, C.P. and O.F. James, Steatohepatitis: A tale of two "hits"? *Gastroenterology*, 1998. 114(4): 842–845.

114. Browning, J.D. and J.D. Horton, Molecular mediators of hepatic steatosis and liver injury. *J Clin Invest*, 2004. 114(2): 147–152.

115. Ibdah, J.A. et al., Mice heterozygous for a defect in mitochondrial trifunctional protein develop hepatic steatosis and insulin resistance. *Gastroenterology*, 2005. 128(5): 1381–1390.

116. Zhou, M. et al., Mitochondrial dysfunction contributes to the increased vulnerabilities of adiponectin knockout mice to liver injury. *Hepatology*, 2008. 48(4): 1087–1096.

117. Rector, R.S. et al., Mitochondrial dysfunction precedes insulin resistance and hepatic steatosis and contributes to the natural history of non-alcoholic fatty liver disease in an obese rodent model. *J Hepatol*, 2010. 52(5): 727–736.

118. Caldwell, S.H. et al., Mitochondrial abnormalities in non-alcoholic steatohepatitis. *J Hepatol*, 1999. 31(3): 430–434.

119. Guo, Y. et al., Quantitative proteomic and functional analysis of liver mitochondria from high fat diet (HFD) diabetic mice. *Mol Cell Proteomics*, 2013. 12(12): 3744–3758.

120. Singh, R. et al., Autophagy regulates lipid metabolism. *Nature*, 2009. 458(7242): 1131–1135.

121. Yang, L. et al., Defective hepatic autophagy in obesity promotes ER stress and causes insulin resistance. *Cell Metab*, 2010. 11(6): 467–478.

122. Pan, M. et al., Presecretory oxidation, aggregation, and autophagic destruction of apoprotein-B: A pathway for late-stage quality control. *Proc Natl Acad Sci USA*, 2008. 105(15): 5862–5867.

123. Sinha, R.A. et al., Caffeine stimulates hepatic lipid metabolism by the autophagy-lysosomal pathway in mice. *Hepatology*, 2014. 59(4): 1366–1380.

124. Salomone, F. et al., Coffee enhances the expression of chaperones and antioxidant proteins in rats with nonalcoholic fatty liver disease. *Transl Res*, 2014. 163(6): 593–602.

125. Lin, C.W. et al., Pharmacological promotion of autophagy alleviates steatosis and injury in alcoholic and non-alcoholic fatty liver conditions in mice. *J Hepatol*, 2013. 58(5): 993–999.

126. Sun, K. et al., Autophagy lessens ischemic liver injury by reducing oxidative damage. *Cell Biosci*, 2013. 3(1): 26.

127. Gujral, J.S. et al., Mechanism of cell death during warm hepatic ischemia-reperfusion in rats: Apoptosis or necrosis? *Hepatology*, 2001. 33(2): 397–405.

128. Kim, J.S. et al., Impaired autophagy: A mechanism of mitochondrial dysfunction in anoxic rat hepatocytes. *Hepatology*, 2008. 47(5): 1725–1736.

129. Rickenbacher, A. et al., Fasting protects liver from ischemic injury through Sirt1-mediated downregulation of circulating HMGB1 in mice. *J Hepatol*, 2014. 61(2): 301–308.

130. Kim, J.S. et al., Carbamazepine suppresses calpain-mediated autophagy impairment after ischemia/reperfusion in mouse livers. *Toxicol Appl Pharmacol*, 2013. 273(3): 600–610.

131. Fang, H. et al., Dual role of chloroquine in liver ischemia reperfusion injury: Reduction of liver damage in early phase, but aggravation in late phase. *Cell Death Dis*, 2013. 4: e694.

10 Mitochondrial Dynamics, Mitophagy and Mitochondrial Spheroids in Drug-Induced Liver Injury

Jessica A. Williams, Hong-Min Ni, and Wen-Xing Ding

CONTENTS

ABSTRACT

Mitochondria are dynamic organelles that play many important roles in maintaining cellular homeostasis and survival, which are important in regulating liver physiology and pathogenesis. The regulation of mitochondrial homeostasis is critical for cellular homeostasis and function. One of the key mechanisms for regulating mitochondrial homeostasis is mitophagy. In this chapter, we discuss the most recent progress on the molecular mechanisms and pathophysiological roles of mitophagy. Specifically, we discuss the role of mitochondrial dynamics on mitophagy and the regulation of mitophagy, dependent on and independent of Parkin. We also discuss the formation of mitochondrial spheroids and mitochondria-derived vesicles as possible alternative mechanisms for the regulation of mitochondrial homeostasis independent of autophagy. Finally, we discuss the zonated changes of mitochondria in acetaminophen-induced liver injury. This chapter provides a comprehensive overview of the regulation of hepatic mitochondrial homeostasis and its role in liver injury and pathogenesis with a focus on drug-induced liver injury.

Keywords: autophagy, mitophagy, Parkin, acetaminophen, necrosis, liver injury

INTRODUCTION

Mitochondria are considered the "power house" of the cell due to being the major site of ATP production. Mitochondria also have other functions including heme synthesis, β-oxidation of fatty acids, and maintenance of calcium homeostasis (Duchen 2004). It is well known that mitochondria act as central executioners of cell death including apoptotic and necrotic cell death. As a result, mitochondrial dysfunction is a key event in progression of liver injury and disease due to accumulation of damaging reactive oxygen species (ROS), decreased oxidative phosphorylation, impaired ATP synthesis, and release of proapoptotic proteins and initiation of cell death. Mitochondrial damage is reflected by decreased respiratory parameters, decreased enzyme activity, accumulation of mitochondrial DNA mutations, and increased oxidative stress (Hill et al. 2012). Mitochondrial dysfunction has been shown to play a role in progression of many different types of liver disease such as alcoholic liver disease, nonalcoholic liver disease, hepatitis C, and cholestasis (Grattagliano et al. 2011).

Mitochondria can maintain homeostasis through many different mechanisms. They have their own proteolytic system, allowing them to degrade misfolded proteins that could potentially disrupt mitochondrial function (Baker and Haynes 2011; Matsushima and Kaguni 2012). In addition, outer mitochondrial membrane proteins that are damaged can be degraded by the proteasome (Karbowski and Youle 2011). Mitochondria can also repair damaged components via constant fission and fusion, which allows for segregation of damaged mitochondria and exchange of material between healthy mitochondria (Twig et al. 2008; van der Bliek et al. 2013). Furthermore, mitochondria-derived vesicles (MDVs) can bud off of mitochondria and fuse with lysosomes to degrade their cargo (Soubannier et al. 2012), or damaged mitochondria can form mitochondrial spheroids and acquire lysosomal markers to possibly serve as an alternative pathway for the removal of damaged mitochondria (Ding et al. 2012a,b; Yin and Ding 2013). Finally, mitophagy initiates engulfment of damaged mitochondria by autophagosomes to trigger their degradation in the lysosome (Youle and Narendra 2011; Ding and Yin 2012; Lemasters 2014). The ability of mitophagy to protect against liver injury is the main focus of this chapter.

MITOPHAGY PROTECTS AGAINST LIVER INJURY

Macroautophagy (hereafter referred to as autophagy) is an evolutionarily conserved process that results in a cell's "self-eating" to degrade cellular proteins and organelles. Autophagy is a protective process that both provides the cell with nutrients in response to starvation and also removes damaged organelles and misfolded or aggregated proteins to prevent injury. Double-membrane autophagosomes engulf these proteins and organelles in the cytoplasm and then fuse with lysosomes to degrade their components via lysosome proteases (Parzych and Klionsky 2014).

Autophagy can be both a selective and nonselective process. Nonselective autophagy breaks down proteins and organelles during starvation in order to provide the cell with necessary nutrients. Selective autophagy removes damaged organelles and protein aggregates using specific receptors and can occur in both nutrient-rich and poor conditions (Reggiori et al. 2012). Mitophagy is a form of selective autophagy that is specific for removing damaged mitochondria, and it is regulated by several different pathways that are further discussed below. Mitophagy can be activated by many different cellular stress conditions including the loss of mitochondrial membrane potential, changes in mitochondrial bioenergetics, accumulation of cellular ROS, mitochondrial DNA damage, or accumulation of protein aggregates in the mitochondria (Liu et al. 2014). Mitophagy is a protective mechanism against liver injury by removing the damaged mitochondria. Removing damaged mitochondria by mitophagy likely prevents the spread of oxidative stress and prevents respiratory chain damage and mitochondrial DNA mutation, which helps to prevent cell death, maintain mitochondrial bioenergetics, and uphold fatty acid oxidation by preserving a healthy population of mitochondria.

Mitochondrial Fission and Fusion and Mitophagy

Mitochondria are dynamic organelles that continuously undergo fission and fusion, and the length of mitochondria is determined by their fission and fusion rates. Mitochondria can be either a large network of fused organelles or they can be divided into many smaller fragments depending on the needs of the cell. These fission and fusion events are necessary for cell survival because they allow the cell to adapt to changing conditions needed for cell growth, division, and distribution of mitochondria during differentiation (van der Bliek et al. 2013). Mitochondrial fission and fusion also allow the cell to adapt to injury because damaged mitochondria can be segregated from healthy mitochondria, leading to the degradation of damaged mitochondria by mitophagy and fusion of healthy mitochondria (Twig et al. 2008).

Mitochondrial fusion in mammals is mediated by the pro fusion genes mitofusin 1 and 2 (MFN1/2) and optic atrophy 1 (OPA1). MFN1/2 are GTPases that are responsible for the fusion of outer mitochondrial membranes, and OPA1 is a dynamin-related GTPase responsible for the fusion of inner mitochondrial membranes. Outer and inner mitochondrial membrane fusions mostly occur simultaneously (Westermann 2010; van der Bliek et al. 2013). OPA1 has eight isoforms that are generated by alternative splicing and alternative processing, which has been suggested to be mediated by presenilin-associated rhomboid-like (PARL) (Cipolat et al. 2006) or by paraplegin, which is an AAA protease present in the mitochondrial matrix (Ishihara et al. 2006). However, the regulation of OPA1 processing by these two proteins is controversial because OPA1 processing still occurred in MEF cells lacking PARL and paraplegin (Duvezin-Caubet et al. 2007). OPA1 is further cleaved under normal conditions to short and long forms (S-OPA1 and L-OPA1) by Yme (Griparic et al. 2007). After treatment with the mitochondrial uncoupler carbonyl cyanide m-chloro phenyl hydrazine (CCCP), which depolarizes mitochondria, L-OPA1 was further cleaved by the inducible protease OMA1, resulting in mitochondrial fragmentation by preventing mitochondrial fusion (Ehses et al. 2009; Head et al. 2009).

Mitochondrial fission in mammals is mediated by dynamin-related protein 1 (Drp1), which is a cytosolic protein that can be recruited to the surface of mitochondria where it interacts with the mitochondrial fission 1 (Fis1) to initiate mitochondrial fission (Westermann 2010). It has been suggested that mitochondria fission factor, which is an outer mitochondrial membrane protein that interacts with Drp1, is an essential factor for Drp1-mediated mitochondrial fission (Gandre-Babbe and van der Bliek 2008; Otera et al. 2010). In addition, mitochondria and endoplasmic reticulum are often tightly associated and in physical contact. It was reported that mitochondrial fission proteins, such as Drp1, are localized at the endoplasmic reticulum-mitochondria contact site and that the endoplasmic reticulum may play a role in the process of mitochondrial fission (Friedman et al. 2011).

It has been suggested that fragmented mitochondria are more easily engulfed by autophagosomes during mitophagy. In addition, elongated mitochondria are spared from autophagosome sequestration (Ding and Yin 2012). After photo-labeling, mitochondria underwent continuous cycles of fission and fusion, and fission events resulted in two sets of daughter mitochondria with either increased or decreased membrane potential. Daughter mitochondria with higher membrane potential proceeded to fusion while depolarized daughter mitochondria were degraded by mitophagy (Twig et al. 2008). Therefore, mitochondrial fission is a necessary step for mitophagy induction. However, mitochondrial fission alone is not enough to induce mitophagy. Mitochondria must be both dysfunctional/depolarized and fragmented and may also possibly need to recruit other autophagy receptor proteins for mitophagy induction to occur (Ding and Yin 2012).

There is a very intriguing puzzle in the mitophagy field because it appears that not all mitochondrial proteins are degraded to the same extent during mitophagy. It has been suggested that outer mitochondrial membrane proteins may be degraded via the proteasomal system whereas the mitochondrial matrix proteins may be removed by autophagy. Although, it is still not clear how autophagy selectively removes some matrix proteins but not others. A recent work proposed a novel mechanism that involves mitochondrial dynamics during mitophagy, which could help to explain this puzzle. Using a proteomic approach in yeast, Abeliovich et al. (2013) elegantly showed that different mitochondrial matrix proteins were degraded at distinct different rates during yeast mitophagy. They found that mitochondria underwent an active matrix protein segregation process, which was regulated by the yeast fission molecule DNM1. The rates of mitophagic degradation of matrix proteins correlated with the degree of physical segregation of specific matrix proteins. These novel findings suggest that mitochondrial dynamics are important for mitochondrial matrix remodeling/segregation and may help to explain why all mitochondrial proteins are not degraded to the same degree by mitophagy.

MITOPHAGY PREVENTS APOPTOSIS AND NECROSIS

Hepatocyte cell death via either apoptosis or necrosis worsens liver disease and injury progression. Cell death by apoptosis in the liver plays major roles in progression of cholestasis, alcoholic liver disease, hepatitis C, and fibrosis (Luedde et al. 2014). Apoptosis is a method of programmed cell death that is characterized by chromatin condensation, DNA fragmentation, cell shrinkage, and cellular fragmentation into membrane-enclosed and organelle-containing apoptotic bodies. Apoptosis occurs by either the extrinsic or intrinsic pathway. The extrinsic pathway is mediated by death receptors Fas, tumor necrosis factor-α (TNF-α) receptor 1 (TNF-R1), and death receptors 4 and 5 (DR4/5) and their corresponding ligands FasL (Fas ligand), TNF-α, and TNF-related apoptosis-inducing ligand, which are all expressed in the liver. After a ligand binds its death receptor, a proteolytic cascade results in activation of effector caspases and leads to cell death. The intrinsic pathway is triggered by DNA damage, oxidative stress, and toxins among other intracellular stress inducers, which causes mitochondrial damage and release of mitochondrial apoptotic factors such as cytochrome c and SMAC/DIABLO to trigger apoptosis. This pathway is regulated by proteins of the B-cell lymphoma-2 (Bcl-2) family. However, in some cell types such as hepatocytes, activation of the extrinsic apoptotic pathway is not sufficient to induce apoptosis. In hepatocytes, upon activation of the extrinsic pathway, caspase 8 cleaves the proapoptotic protein Bid into tBid (truncated bid), which translocates to the mitochondria and activates Bax or Bak. Bax or Bak is then inserted into the mitochondrial membrane, which results in permeabilization of the mitochondrial outer membrane and release of proapoptotic proteins from the mitochondrial intramembrane space into the cytosol and thus activates the intrinsic apoptotic pathway. Proteins released include proteins that promote further caspase activation (cytochrome c and SMAC/DIABLO) and proteins that translocate to the nucleus to degrade DNA (apoptosis-inducing factor and endonuclease G) (Guicciardi et al. 2013). Necrosis in the liver is caused by drug-induced liver injury, such as acetaminophen (APAP) overdose (further discussed below) (Luedde et al. 2014) and ischemia/reperfusion-induced liver injury (Guicciardi et al. 2013). Necrosis results from permeabilization of the cell membrane, loss of membrane potential, and cell membrane rupture, which leads to the release of cellular contents followed by an inflammatory response (Guicciardi et al. 2013). Interestingly, mitochondrial damage also plays a critical role in both APAP overdose and ischemia/reperfusion-induced necrosis. The lack of apoptosis in these conditions is likely due to severe mitochondrial damage resulting in the rapid depletion of cellular ATP levels because ATP is required for caspase activation.

Mitochondria that are damaged produce ROS and release proapoptotic proteins or result in inflammation when necrotic, which amplifies cell death and injury. Therefore, it is important to remove these damaged mitochondria to avoid hepato-cellular death and injury in the liver (Mizushima et al. 2008). Mitophagy has been shown to be protective against both apoptotic and necrotic cell death in the liver by removing damaged mitochondria. For example, mitophagy has been shown to be protective against apoptosis and injury induced by alcoholic liver disease (Ding et al. 2010a; Lin et al. 2013) and also against necrosis in APAP-induced liver injury (Ni et al. 2012, 2013; Lin et al. 2014; Saberi et al. 2014).

MITOPHAGY MAINTAINS MITOCHONDRIAL BIOENERGETICS

Mitochondrial bioenergetics refers to a mitochondrion's capacity and efficiency for ATP production (Liesa and Shirihai 2013), and are regulated by substrate availability, cell energy requirements, and the overall quality and abundance of the mitochondria population (Hill et al. 2012). Interestingly, mitochondrial bioenergetics and mitochondrial dynamics (fission and fusion) have been recently linked. Cells in an environment rich with nutrients have fragmented mitochondria, while cells in a starvation environment have more elongated and fused mitochondria. Furthermore, these elongated mitochondria avoid degradation by mitophagy, once again emphasizing the importance of mitochondrial fragmentation for mitophagy induction as previously discussed (Molina et al. 2009; Gomes et al. 2011).

Removal of the damaged mitochondria via mitophagy is crucial for preventing cell death and injury caused by mitochondrial uncoupling, which is described as a dysfunctional proton gradient in mitochondria that leads to uncoupling of oxidative phosphorylation and ATP production. Uncoupling causes damaged mitochondria and cell death by apoptosis (Hill et al. 2012). Alcoholic liver disease represents an example of where degradation of uncoupled mitochondria by mitophagy is an important protector against hepatocellular death and liver injury. It was shown that the reserve capacity of mitochondria after chronic alcohol consumption is decreased, which contributes to mitochondrial damage. The reserve capacity of mitochondria describes cell energetic status or the maximal energy capacity of mitochondria within a cell (Hill et al. 2012). Acute alcohol treatment was shown to induce autophagy and mitochondrial removal by mitophagy in mice and primary cultured mouse hepatocytes, and inhibition of autophagy exacerbated ethanol-induced liver injury and cell death (Ding et al. 2010a). In addition, activation of autophagy was shown to be protective against chronic alcohol-induced liver injury in mice fed a Lieber DeCarli ethanol diet for 4 weeks, and it was suggested that the removal of damaged mitochondria by mitophagy would also be protective in this model (Lin et al. 2013).

In addition to faulty mitochondrial energetics contributing to mitophagy activation, functional mitochondrial bioenergetics has also been shown to have a role in activating autophagy during starvation conditions. Drugs that modulate oxidative phosphorylation have been shown to inhibit autophagy during starvation. For example, treatment with the ATP synthase inhibitor oligomycin or the electron transport chain complex III inhibitor antimycin A both inhibited autophagy during starvation conditions, suggesting that proper mitochondrial bioenergetics are required for autophagy activation to provide nutrients during starvation (Graef and Nunnari 2011).

Mitophagy is important for removing uncoupled mitochondria to prevent the spread of oxidative stress, cell death, and liver injury to maintain a healthy population of mitochondria that can produce necessary energy efficiently for the remaining cells to survive and proliferate to repair/replace damaged cells. In addition, the role of mitophagy in maintaining a healthy population of mitochondria with properly functioning bioenergetics is important for inducing autophagy during starvation in order to provide the cell with nutrients in starvation conditions.

Mitophagy Maintains FA Oxidation

Mitochondrial fatty acid oxidation (FAO) plays many important functions in the liver including providing substrates for gluconeogenesis, contribution of electrons to the respiratory chain for oxidative phosphorylation and ATP production, generation of acetyl-CoA for gluconeogenesis and ketogenesis (Kompare and Rizzo 2008), and degradation of lipids to prevent hepatocellular steatosis (Gao and Bataller 2011).

Fatty acids are a common source of energy and are often stored as nontoxic triaglycerols. They originate from several sources including de novo lipogenesis, triacylglycerol stores, and plasma nonesterified fatty acids released from adipose tissue (Nguyen et al. 2008). Long-chain fatty acids used for FAO are stored mainly in the adipose tissue as components of triglycerides or phospholipids. Lipases cause these triglycerides to release their fatty acids, which are then transported to the liver and muscle via the bloodstream. For use in FAO, these long-chain fatty acids must first be converted into acylcarnitine esters so they can be transported across the inner mitochondrial membrane into the mitochondrial matrix. The rate-limiting enzyme for shuttling these carnitines across the mitochondrial membrane is carnitine palmitoyltransferase I (CPTI), which is inhibited by malonyl-CoA in fed conditions. However, in fasting conditions, malonyl-CoA concentrations drop and CPTI is activated to begin FAO. Once in the mitochondria, fatty acids are processed by β-oxidation resulting in the release of acetyl-CoA and production of FADH2 and NADH, which donate their electrons to the electron transport chain for ATP production (Kompare and Rizzo 2008; Nguyen et al. 2008).

Several liver diseases are exacerbated due to dysfunctional mitochondria and FAO such as fatty liver caused by alcoholic liver disease (Beyoglu and Idle 2013) and drug-induced liver injury (Begriche et al. 2011). For example, β-oxidation has been shown to be inhibited while fatty acid synthesis is upregulated in alcoholic liver disease, which results in an accumulation of lipids in the liver and development of liver steatosis (Begriche et al. 2011; Gao and Bataller 2011). In addition, liver steatosis can be induced by several different pharmaceuticals such as amiodarone, ibuprofen, APAP, and tamoxifen, which inhibit enzymes needed for effective β-oxidation (Begriche et al. 2011). Consequences of reduced or inhibited β-oxidation include accumulation of fatty acids that are esterified and stored as triglycerides or an accumulation of fatty acids that remain in an unesterified form. If accumulated in their unesterified form, fatty acids can cause further mitochondrial damage and injury. In addition, faulty β-oxidation can impair energy output due to reduced ketone body production and ATP production (Begriche et al. 2011). Because mitophagy acts as an important regulator of mitochondrial homeostasis and helps to maintain a healthy population of mitochondria, it is likely that mitophagy helps defend against lipid accumulation in the liver by removing damaged mitochondria incapable of performing efficient β-oxidation.

Mitophagy and Mitochondrial Biogenesis Must Be Balanced

Mitochondria have developed several methods to adapt to cellular stress induced by drugs or disease including mitophagy activation to degrade damaged mitochondria and stimulation of mitochondrial biogenesis to replace damaged mitochondria removed by mitophagy. The number of mitochondria in cells is balanced by mitochondrial biogenesis and degradation by mitophagy. Changes in demand for mitochondrial function can shift the balance between these, but failing to restore the balance leads to mitochondrial dysfunction, cell death, and disease. For example, a decrease in mitochondria has been associated with cell death, aging, Parkinson's disease, and liver disease. However, failure to remove mitochondria that are damaged can also lead to apoptosis and disease due to an accumulation of damaged mitochondria and overall mitochondrial dysfunction (Hill et al. 2012; Zhu et al. 2013a).

Mitochondrial biogenesis consists of de novo synthesis of mitochondrial components from other cellular precursors, formation of mitochondrial membranes, and division of currently existing mitochondria (Michel et al. 2012). Mitochondrial biogenesis requires mitochondrial DNA transcription and translation, transcription and translation of nuclear DNA, and the assembly of complexes needed for oxidative phosphorylation. In addition, mitochondrial biogenesis requires protein import because most mitochondrial proteins are encoded by nuclear DNA, synthesized in the cytosol, and later transported into the mitochondria (Zhu et al. 2013a; Palikaras and Tavernarakis 2014). Regulators of mitochondrial biogenesis include peroxisome proliferator-activated receptor γ, coactivator 1 α (PGC-1α), nuclear respiratory factors 1 and 2 (Nrf1/2), and mitochondrial transcription factor A (TFAM), and these are activated by nutrient availability, hormones, and growth factors (Palikaras and Tavernarakis 2014).

PGC-1α is the master regulator of mitochondrial biogenesis and is responsible for the transcription of the electron transport chain and fatty acid oxidation genes and plays a role in increasing mitochondrial mass (Jones et al. 2012; Palikaras and Tavernarakis 2014). PGC-1α is located mainly in the cytosol, but it can translocate to the nucleus upon phosphorylation or deacetylation. Once in the nucleus, it can initiate transcription of genes necessary for mitochondrial biogenesis (Wu et al. 1999; Anderson et al. 2008; Chang et al. 2010; Scarpulla et al. 2012). Upregulation of PGC-1α activity is regulated by AMP-activated protein kinase (AMPK), Sirtuin 1 (SIRT1), and cAMP-response element-binding protein (CREB). AMPK and CREB increase PGC-1α activity (Reznick et al. 2007; Schulz et al. 2008;

De Rasmo et al. 2010), and SIRT1 deacetylates PGC-1α to induce its nuclear translocation to initiate mitochondrial biogenesis (Anderson et al. 2008). Glycogen synthase kinase-3β negatively regulates PGC-1α activity by inducing its degradation by the proteasome (Anderson et al. 2008). Nrf1/2 genes are transcriptionally activated by PGC-1α (Wu et al. 1999) and CREB (De Rasmo et al. 2010), and they are responsible for initiating expression of genes encoding respiratory subunits, mitochondrial protein import machinery, and nuclear DNA-encoded cytochrome oxidase subunits. Nrf1/2 also regulate the expression of TFAM (Virbasius et al. 1993; Ongwijitwat and Wong-Riley 2005; Scarpulla et al. 2012; Palikaras and Tavernarakis 2014). TFAM is localized to the mitochondria and regulates the replication of mitochondrial DNA (Ekstrand et al. 2004), and its expression is upregulated by PGC-1α in addition to Nrf1/2 (Scarpulla et al. 2012).

Removing damaged mitochondria limits cellular apoptosis, but mitophagy without compensatory mitochondrial biogenesis may also cause mitochondrial dysfunction and cell death due to putting too much stress on the remaining mitochondria (Palikaras and Tavernarakis 2014). However, accumulation of too many mitochondria due to overactive mitochondrial biogenesis or lack of mitochondrial degradation via mitophagy can also lead to cell death and disease. Therefore, there must be cross talk between mitochondrial biogenesis and mitophagy pathways in order to provide a balanced population of healthy, functioning mitochondria. There are several pathways that simultaneously regulate both mitochondrial biogenesis and autophagy in order to maintain their balance including Parkin and AMPK. Parkin and AMPK promote both mitophagy and mitochondrial biogenesis simultaneously. Parkin, an E3 ubiquitin ligase, ubiquitinates PARIS for its degradation, which results in increased expression of PGC-1α and Nrf1. Because PARIS normally represses expression of these biogenesis genes, degradation of PARIS by Parkin causes up regulation of mitochondrial biogenesis (Shin et al. 2011). In addition, Parkin associates with TFAM and enhances its transcription activity (Kuroda et al. 2006). Parkin has also been shown to interact with mitochondrial DNA to allow for mitochondrial DNA repair during oxidative stress (Rothfuss et al. 2009). At the same time, Parkin induces mitophagy via the Parkin-PTEN-induced putative kinase 1 (PINK1) mitophagy pathway (further discussed below). AMPK, a protein kinase, is activated in response to environmental stress or nutrient deprivation, and it activates mitochondrial biogenesis by either enhancing SIRT1 deacetylation of PGC-1α resulting in its nuclear translocation (Canto et al. 2009), or by enhancing PGC-1α activity through phosphorylation (Jager et al. 2007; Scarpulla et al. 2012). While activating mitochondrial biogenesis, AMPK simultaneously upregulates mitophagy by either suppressing mTOR, which is a negative regulator of autophagy (Mihaylova and Shaw 2011), or by phosphorylating and activating the autophagy-initiating kinase Unc-51 like autophagy activating kinase (ULK1/2) (Kim et al. 2011).

Even though mitophagy is a very important process for managing liver disease by ridding of damaged mitochondria, it is important to remember that proper balance between mitophagy and mitochondrial biogenesis is necessary to maintain an overall healthy population of mitochondria. Removal of damaged mitochondria will prevent cell death and injury, but failure to replace these mitochondria will also lead to mitochondrial dysfunction, cell death, and injury due to an accumulation of dysfunctional mitochondria.

CURRENT MECHANISMS OF MITOPHAGY

Mechanisms of mitophagy have been well studied in yeast and in mammalian cells. However, most have not been studied in the liver. There are several mechanisms for mitophagy induction in mammalian systems, which include both Parkin-dependent and Parkin-independent mechanisms.

PARKIN-DEPENDENT MITOPHAGY

Parkin is an E3 ubiquitin ligase that has been shown to be required for mitophagy induction in mammalian cell models during mitochondrial depolarization. Parkin is well known for its protective role in the brain because loss of Parkin has been shown to play a role in development of autosomal recessive Parkinsonism in humans (Kitada et al. 1998). We recently found that Parkin is not only expressed in the brain, but is also highly expressed in several other tissues including the liver in mice (Ding and Yin 2012).

Parkin has been shown to translocate from the cytosol to depolarized mitochondria after treatment with the mitochondrial uncoupler CCCP, which damages mitochondria by decreasing their membrane potential and initiates their degradation via mitophagy. In addition, Parkin was shown to selectively translocate to damaged mitochondria while avoiding healthy mitochondria (Narendra et al. 2008). The Parkin-dependent mitophagy pathway has been shown to require both Parkin and PINK1, and PINK1 is known to be the upstream of Parkin in this pathway because overexpression of Parkin in PINK1-deficient *Drosophila* partially rescued the PINK1 mutant phenotype while overexpression of PINK1 failed to do so in Parkin-deficient *Drosophila* (Clark et al. 2006; Park et al. 2006; Yang et al. 2006). In addition, overexpression of PINK1 alone can initiate translocation of Parkin to mitochondria without mitochondrial damage (Kawajiri et al. 2010).

PINK1 is normally cleaved in the mitochondria by PARL and then degraded by mitochondrial peptidases (Jin et al. 2010). However, when mitochondria are depolarized, PINK1 is no longer cleaved and degraded and becomes stabilized on the outer mitochondrial membrane (Jin et al. 2010). PINK1 promotes Parkin-mediated mitophagy as an upstream regulator in at least two ways: recruiting Parkin to mitochondria and promoting Parkin E3 ligase activity. The exact mechanism for how PINK1 recruits Parkin is still not completely understood, but it has been shown that Parkin directly interacts with PINK1 (Um et al. 2009). It has been suggested that PINK1 recruits Parkin via phosphorylation because PINK1 was found to phosphorylate Thr175 and Thr217 within Parkin's linker region, which promoted Parkin mitochondrial translocation (Kim et al. 2008). In addition, phosphorylation of Parkin by PINK1 is thought to activate Parkin's E3 ubiquitin ligase activity, enabling it to ubiquitinate mitochondrial proteins (Sha et al. 2010). More recently, it was reported that PINK1 phosphorylated ubiquitin at serine 65 and the phospho-ubiquitin activated Parkin E3 ubiquitin ligase activity through a feed-forward mechanism

(Kane et al. 2014; Koyano et al. 2014). Moreover, four of Parkin's different cognate E2 coenzymes (UBE2D, UBE2L3, UBE2N, and UBE2R1) have also recently been discovered to either positively or negatively regulate Parkin's activation, translocation, and enzymatic functions during mitophagy (Fiesel et al. 2014; Geisler et al. 2014). Neuregulin receptor degradation protein 1 (Nrdp1) is an E3 ubiquitin ligase that has been shown to regulate the proteasomal degradation of Parkin. It was reported recently that C-type lectin domain family 16, member A (Clec16a), a protein that is associated with type 1 diabetes mellitus, interacts with Nrdp1. Loss of Clec16a led to a reduction of Nrdp1 resulting in an increase of Parkin expression. However, pancreas-specific Clec16a knockout mice had abnormal mitochondria with reduced oxygen consumption and ATP concentration in pancreatic β cells due to the defects of mitophagy despite an increased expression of Parkin. Further studies revealed that Clec16a-deficient fibroblasts had impaired fusion of autophagosomes with lysosomes (Soleimanpour et al. 2014). The relevance of Clec16a-and Nrdp1-induced regulation of Parkin-mediated mitophagy in the liver remains to be studied. It is now known that stabilization of PINK1 is important to allow the recruitment of Parkin to depolarized mitochondria (Matsuda et al. 2010; Narendra, Jin et al. 2010; Vives-Bauza et al. 2010). Using a genome-wide small interfering RNA (siRNA) screening for genes that regulate Parkin mitochondrial translocation, it was discovered that translocase of outer mitochondrial membrane 7 (TOMM7) stabilizes PINK1 on the outer mitochondrial membrane whereas SIAH3, a mitochondrial resident protein, destabilizes PINK1 on mitochondria. Furthermore, two other proteins, HSPA1L (an HSP70 family protein) and Bcl2-associated athanogene 4 (BAG4, a nucleotide exchange factor for HSP70), were found to positively and negatively regulate Parkin mitochondrial translocation, respectively, from the same study (Hasson et al. 2013).

Once recruited to the mitochondria, Parkin ubiquitinates several mitochondrial outer membrane proteins including the mitochondrial fusion proteins MFN1/2 in addition to Miro, Translocase of outer mitochondrial membrane 20 (TOMM20), and voltage-dependent anion channel (VDAC) to initiate mitophagy. Ubiquitination of MFN1/2 initiates their subsequent degradation by the proteasome resulting in mitochondrial fission and fragmentation (Gegg et al. 2010; Geisler et al. 2010; Poole et al. 2010; Chan et al. 2011). Mitochondria are under continuous states of fission and fusion, and mitochondrial fission has been shown to be important for mitophagy induction in liver hepatocytes among other cell types (Twig et al. 2008; Kim and Lemasters 2011). Fragmented mitochondria can fuse together if they have normal membrane potential, but loss of membrane potential leads to their segregation and signals them for degradation by mitophagy (Twig et al. 2008). In addition, excessive fusion of mitochondria has been shown to inhibit the mitophagy process (Twig and Shirihai 2011), further suggesting an importance of mitochondrial fragmentation as a perquisite to mitophagy induction. Mitochondrial fission or fragmentation caused by Parkin-induced ubiquitination of MFN1/2 has been suggested to make engulfment of mitochondria easier for autophagosomes during mitophagy (Ding and Yin 2012). Interestingly, Parkin-induced ubiquitination of Miro has been shown to initiate mitochondrial arrest, which can segregate damaged mitochondria from healthy mitochondria prior to mitophagy (Wang et al. 2011). Recently, it was reported that Ubiquitin-specific peptidase 30 (USP30), a deubiquitinase localized to mitochondria, antagonizes Parkin-mediated mitophagy by removing ubiquitin from damaged mito-chondria previously attached by Parkin. Knockdown of USP30 rescues the defective mitophagy caused by pathogenic mutations in Parkin and improves mitochondrial integrity in Parkin- or PINK1-deficient flies and protects flies against paraquat toxic-ity in vivo by ameliorating defects in dopamine levels, motor function, and organ-ismal survival (Bingol et al. 2014). The role of USP30 in mitophagy regulation in mammals and in the liver remains to be studied.

The selective autophagy adaptor protein p62/SQSTM1 (Sequestosome 1) has also been shown to have a role in mitophagy because it recognizes ubiquitinated proteins through its ubiquitin-associated domain and also contains a microtubule-associated protein 1 light chain 3 (LC3)-interacting region (LIR), allowing for recruitment and binding of autophagosomes, which contain LC3 on their membrane. This interaction creates a link between ubiquitinated mitochondria and autophagosomes, resulting in mitophagy and degradation of the mitochondria in the lysosome (Ding et al. 2010b; Geisler et al. 2010; Huang et al. 2011; Manley et al. 2013). In addition, it was shown that Parkin-induced ubiquitination of VDAC does not induce its proteasomal degra-dation like MFN1/2 but instead leads to the recruitment of p62 to the mitochondria (Geisler et al. 2010). However, the necessity of p62 and VDAC for successful mitoph-agy is currently controversial because some have shown that neither is essential for mitophagy induction (Narendra et al. 2010; Okatsu et al. 2010). In addition, Ambra1 has been shown to be recruited to depolarized mitochondria by Parkin where it ini-tiates engulfment of damaged mitochondria by autophagosomes (Van Humbeeck et al. 2011). Therefore, the process of autophagosome recruitment to ubiquitinated mitochondria in this mitophagy pathway is still not completely understood, but mito-chondrial Parkin has been shown to colocalize with the autophagosome protein LC3 (Narendra et al. 2008), indicating that Parkin is important in autophagosome recruit-ment to depolarized mitochondria.

Because of its critical role in regulating mitochondria turn over, Parkin is impor-tant for maintaining mitochondrial function by preserving a healthy population of mitochondria. There was decreased mitochondrial complex I activity in fibroblasts isolated from human patients with Parkin mutations and also in the substantia nigra from patients with Parkinson's disease (Schapira et al. 1990; Mortiboys et al. 2008). In addition, Parkin knockout mice had decreased complex I activity in addition to reduced mitochondrial respiration (Palacino et al. 2004).

PARKIN-INDEPENDENT MITOPHAGY

Mitophagy induction has also been shown to occur via several mechanisms that are independent of Parkin. Parkin-independent mitophagy induction mediators include NIX/BNIP3L, Bcl2/adenovirus E1B 19 kDa protein-interacting protein 3 (BNIP3), Fun14 domain containing 1 (FUNDC1), Smad-specific E3 ubiquitin protein ligase 1 (SMURF1), high-mobility group box 1 (HMGB1), and cardiolipin.

BNIP3

BNIP3 is a proapoptotic mitochondrial protein that contains a Bcl-2 homology 3 (BH3) domain (Yasuda et al. 1998), and it has been shown to have a role in mitophagy induction in hypoxic conditions. BNIP3 is expressed in many tissues including the liver, but it is not highly expressed under normal conditions (Galvez et al. 2006). BNIP3 gene expression is induced during hypoxia by hypoxia-inducing factor-1 (HIF-1), which undergoes enhanced binding to its HIF-1-responsive element on the BNIP3 gene promoter during oxygen deprivation (Bruick 2000). Mitophagy induction in MEF cells has been shown to be regulated by BNIP3 during hypoxia, which allows for the removal of damaged mitochondria and prevention of cell death via protection against ROS accumulation (Zhang et al. 2008). Furthermore, phosphorylation of Ser17 and Ser24 in the LIR of BNIP3 was shown to initiate cell survival via induction of mitophagy by encouraging colocalization with LC3, suggesting that the phosphorylation status of BNIP3 decides whether it will behave as a proapoptotic protein or as a pro-survival protein by initiating mitophagy (Zhu et al. 2013b).

NIX/BNIP3L

NIX, also known as BNIP3L, is a homolog of BNIP3 (Chen et al. 1999). NIX is known to activate autophagy by binding to Bcl-2, which dissociates the complex of Bcl-2 and Beclin-1, a protein necessary for initiation of autophagosome formation (Bellot et al. 2009). In addition, NIX has been shown to interact with the autophagosome membrane protein LC3 (Novak et al. 2010), and may therefore have a role in recruiting autophagosomes to damaged mitochondria.

Like BNIP3, NIX has been shown to play a role in induction of mitophagy during hypoxia due to induction of its expression by HIF-1 (Bruick 2000). NIX and BNIP3 seem to have complementary roles in the mitophagy response during hypoxia because depletion of either NIX or BNIP3 did not affect autophagy, but depletion of both NIX and BNIP3 inhibited the autophagic response to hypoxic conditions. In addition, overexpression of both NIX and BNIP3 induced autophagy under normal oxygen concentrations (Bellot et al. 2009).

In addition to its role in mitophagy induction during hypoxia, NIX is also well known to induce mitophagy during red blood cell maturation. Red blood cells must eliminate their mitochondria during maturation in order to be able to better carry and provide oxygen, and NIX has been shown to be required for these cells to eliminate their mitochondria because mitochondrial entry into autophagosomes for degradation was blocked in NIX-deficient mice, which led to life-span reduction of red blood cells and development of anemia (Schweers et al. 2007; Sandoval et al. 2008). This mechanism of mitophagy induction was also shown to require mitochondrial depolarization (Sandoval et al. 2008).

NIX has also been shown to be involved in mitophagy induction in cells that are undergoing high rates of oxidative phosphorylation, which could lead to increased ROS production, mitochondrial dysfunction, cell death, and injury. When cells are undergoing high rates of oxidative phosphorylation, the GTPase Rheb is recruited to the outer mitochondrial membrane where it interacts with NIX and LC3 to induce mitophagy. This Rheb-dependent mitophagy pathway was shown to require NIX because depletion of NIX inhibited Rheb-induced mitochondrial degradation. It was suggested that ridding of mitochondria undergoing high rates of oxidative phosphorylation via mitophagy helps maintain a healthy population of mitochondria, which promotes efficiency of cellular bioenergetics and mitochondrial energy production (Melser et al. 2013).

Finally, NIX was also shown to promote the recruitment of Parkin to depolarized mitochondria after CCCP treatment in MEF cells. CCCP treatment caused mitochondrial depolarization, ROS accumulation, and translocation of Parkin to mitochondria in Parkin-expressing MEF cells, while NIX-deficient MEF cells failed to induce mitochondrial depolarization and Parkin recruitment, suggesting that NIX may also be required for the Parkin-dependent mitophagy induction pathway (Ding et al. 2010b).

FUNDC1

FUNDC1 is an outer mitochondrial membrane protein that has also been shown to have a role in mitophagy induction under hypoxic conditions. Under oxygenated conditions, FUNDC1 is phosphorylated by the Src kinase on Tyr18 (Liu et al. 2012) and by casein kinase II (CK2). During hypoxia, the mitochondrial phosphatase phosphoglycerate mutase family member 5 (PGAM5) dephosphorylates FUNDC1 on Ser13 (Chen et al. 2014). Hypoxia induces FUNDC1-dependent mitophagy using a different mechanism than NIX or BNIP3. While mRNA expression of NIX and BNIP3 are increased by HIF-1 during hypoxia, mRNA expression of FUNDC1 is actually decreased. When oxygen concentrations are low, phosphorylation of FUNDC1 is inactivated leading to an abundance of it in its dephosphorylated form. FUNDC1 contains an LIR, which allows for it to interact with LC3 on autophagosome membranes, and the binding affinity of FUNDC1 for LC3 increases when FUNDC1 is dephosphorylated (Liu et al. 2012). Upon hypoxia, CK2 is inhibited whereas PGAM5 interacts with FUNDC1 resulting in increased dephosphorylation of FUNDC1 to promote mitophagy (Chen et al. 2014). Moreover, Bcl2-like 1 (BCL2L1), but not BCL2, interacts with and inhibits PGAM5, which prevents the dephosphorylation of FUNDC1 and mitophagy (Wu et al. 2014). Furthermore, the knockdown of FUNDC1 in HeLa cells prevented hypoxia-induced mitophagy (Liu et al. 2012), suggesting that FUNDC1 is required for mitophagy induction in hypoxic conditions. It is currently unknown which has a more important role in mitophagy induction during hypoxia, NIX or FUNDC1, but dephosphorylated FUNDC1 has greater binding affinity for LC3 than NIX (Novak et al. 2010).

SMURF1

SMURF1 is an E3 ubiquitin ligase similar to Parkin that was found to have a role in CCCP-induced mitophagy mediated by Parkin. Unlike Parkin, the ubiquitin ligase function of SMURF1 did not play an important role in mitophagy induction. Instead, the C2 domain of SMURF1 was required for the engulfment of damaged mitochondria by autophagosomes. In addition, SMURF1-deficient mice had accumulated mitochondria that were damaged in their heart, brain, and liver, suggesting that SMURF1 may be another important E3 ligase required for the mitophagy pathway in addition to Parkin (Orvedahl et al. 2011). In the future, it will be interesting to determine how SMURF1-deficient mice respond to drug-induced liver injury since these mice already have some liver phenotype.

HMGB1

HMGB1 is a chromatin-associated nuclear protein that also acts as a damaged-associated molecular pattern molecule when released from necrotic hepatocytes. HMGB1 competes with Bcl-2 for interaction with Beclin1 to induce autophagosome formation (Tang et al. 2010). In addition, HMGB1 regulates trafficking of mitochondria to autophagosomes by inducing expression of heat shock protein β-1 (HSPB1), which is a regulator of the cytoskeleton (Tang et al. 2011). This mechanism of mitophagy is still thought to require the Parkin pathway because the knockdown of Parkin or PINK1 reduced mitochondrial fragmentation and ATP production in HSPB1 restored HMGB1 knockout cells (Tang et al. 2011). It should be noted that most of these conclusions regarding the role of HMGB1 in autophagy/mitophagy were derived from cell culture studies. However, a recent study using tissue-specific deletion of HMGB1 in the mouse liver and heart revealed that deletion of HMGB1 in these two cell types did not alter mitochondrial structure or function, organ function, or long-term survival (Huebener et al. 2014). Therefore, it seems that HMGB1 is dispensable for autophagy, mitochondrial quality control, and organ function in mice.

Cardiolipin

Cardiolipin is a phospholipid dimer synthesized in the inner mitochondrial membrane that is able to translocate to the outer mitochondrial membrane upon mitochondrial damage and depolarization (Ren et al. 2014). Cardiolipin has been shown to initiate the mitophagy pathway upon externalization to the outer mitochondrial membrane in SH-SY5Y cells and primary cortical neurons after treatment with rotenone, which is an electron transport chain complex I inhibitor. Rotenone treatment increased GFP-labeled LC3 puncta, which is a marker of autophagosome formation. Rotenone treatment also increased LC3 colocalization with mitochondria, which is an indicative of mitophagy induction. Prevention of cardiolipin translocation to the outer mitochondrial surface inhibited GFP-LC3 colocalization with mitochondria and mitophagy after rotenone treatment, (Chu et al. 2013), and inhibiting the colocalization of LC3 with mitochondria after rotenone treatment prevented autophagosome engulfment of mitochondria and degradation of the mitochondria in the lysosome (Chu et al. 2014). Interestingly, cardiolipin also functions to induce cellular apoptosis. If cardiolipin is present on the outer mitochondrial membrane in its peroxidized form, it will initiate cell death through apoptosis. However, if it is not peroxidized, it initiates mitophagy to protect the cell from apoptotic cell death (Ren et al. 2014). Chu et al. proposed that cardiolipin will initiate apoptotic cell death if the mitophagy pathway fails to degrade damaged mitochondria (Chu et al. 2013). Whether or not this pathway is important in the liver for initiating mitophagy remains to be determined.

MITOCHONDRIAL QUALITY CONTROL REGULATION BY VESICULAR TRAFFICKING TO THE LYSOSOME

It was recently found that Parkin and PINK1 can stimulate the biosynthesis of MDVs under conditions of mitochondrial oxidative stress. Even though this pathway requires Parkin and PINK1, it is different from canonical mitophagy because it is stimulated by ROS production instead of mitochondrial depolarization. In addition, vesicles bud off of damaged mitochondria and are degraded in the lysosome along with their contents without the involvement of the autophagy pathway (McLelland et al. 2014). It was shown that these vesicles contain oxidized proteins (Soubannier et al. 2012), and it was suggested that these vesicles may regulate mitochondrial quality control using a mechanism faster than mitophagy to prevent complete mitochondrial depolarization while preserving mitochondrial function by selectively degrading damaged mitochondrial contents (Soubannier et al. 2012; McLelland et al. 2014). It is possible that this novel pathway may serve as an alternative pathway to regulate mitochondrial homeostasis in the absence of canonical autophagy such as in Atg5 or Atg7 knockout MEFs, which appear to grow well like wild-type MEFs under normal culture conditions.

PARKIN AND MITOPHAGY IN ACETAMINOPHEN-INDUCED LIVER INJURY

APAP overdose is the most common cause of acute liver failure in the United States. APAP is safe at a therapeutic dose, but can be deadly in overdose situations (Larson 2007). APAP is metabolized to the reactive metabolite NAPQI (*N*-acetyl-*p*-benzoquinone imine) by cytochrome P450s (Cyp), mainly by CYP2E1. NAPQI is normally bound and detoxified by glutathione (GSH) after therapeutic doses of APAP, but upon overdose, GSH levels are depleted and NAPQI is left free to form protein adducts (McGill et al. 2012). When NAQPI forms protein adducts with mitochondrial proteins, mitochondrial dysfunction occurs, including mitochondrial respiration inhibition, oxidative stress, damage of mitochondrial DNA, release of mitochondrial proteins, opening of the membrane permeability transition pore, and depletion of mitochondrial membrane potential and ATP. Mitochondrial dysfunction then leads to hepatocellular necrosis and liver injury (Hinson et al. 2004; Jaeschke et al. 2012; McGill et al. 2012).

Pharmacological induction of autophagy was shown to protect against APAP-induced liver injury by removing damaged mitochondria (Ni et al. 2012). However, the mechanism for removal of these mitochondria is unknown. We investigated the role of Parkin-induced mitophagy in protection against APAP-induced liver injury. We found that Parkin translocated from the cytosol to the mitochondria after APAP treatment in mice followed by increased ubiquitination of mitochondrial proteins and mitophagy induction, suggesting that Parkin-induced mitophagy is likely a protective mechanism against APAP-induced necrosis and liver injury. We are currently investigating the role of Parkin in APAP-induced liver injury using Parkin knockout mice. To our surprise, we found that APAP could induce both Parkin-dependent and Parkin-independent mitophagy. Parkin knockout mice were more resistant to APAP-induced liver injury via multiple mechanisms including inhibition of JNK activation and Mcl-1 degradation, as well as increasing the liver repair process by promoting hepatocyte proliferation independent of its role in mitophagy (Williams et al., 2015).

ZONATION AND MITOCHONDRIAL SPHEROID FORMATION AFTER APAP-INDUCED LIVER INJURY

The liver is divided into periportal, intermediate, and centrilobular zones. These zones have differences in oxygen concentration, CYP450 enzyme expression, and GSH levels with centrilobular hepatocytes having the highest oxygen concentration and expression of CYP450 enzymes and the lowest GSH levels (Lindros 1997; Jungermann and Kietzmann 2000). APAP is mainly metabolized in the centrilobular zone where expression of CYP450s is highest and GSH is the lowest, which causes the majority of hepatocellular necrosis to occur near the central vein (Jaeschke et al. 2002). Treatment of GFP-LC3 transgenic mice with APAP increased the formation of GFP-LC3-positive autophagosomes, which were also localized mainly to the centrilobular zone in the liver (Ni et al. 2012). It is likely that autophagosomes localized to this particular area of the liver in order to degrade mitochondria damaged by APAP adduct formation via mitophagy.

Interestingly, APAP treatment divided the liver into four additional zones based on the distinctive pathogenesis, which was likely due to differences in expression of CYP450s and GSH throughout the three main zones in the liver, as previously discussed. Zone 1, located within the center of the centrilobular zone closest to the central vein, contained necrotic hepatocytes with swollen mitochondria, excessive lipid droplets, and disrupted cellular plasma membranes. This zone contains the highest expression of CYP450s for APAP metabolism and also the lowest concentration of GSH, leading to extensive hepatocellular necrosis and tissue damage. Zone 2, located adjacent to the necrotic areas, contained mitochondrial spheroids. Mitochondrial spheroids are structurally unique mitochondria with a ring or cup-like morphology. They look similar to autophagosomes with the interior lumen being surrounded by mitochondrial membranes, but they have a small opening that connects the spheroid lumen to the cytosol, unlike autophagosomes (Figure 10.1, arrows). The significance of this opening is currently unclear. These spheroids can enwrap contents of the cytosol such as endoplasmic reticulum, lipid droplets, or other mitochondria. They are also positive for lysosome proteins and may have some amount of degradation capacity, but it remains to be proven if they actually can degrade contents within their lumen. Zone 3, which is adjacent to Zone 2, contains normally structured autophagosomes with many of them containing mitochondria. This zone is termed the "autophagy active area." This zone has minimally damaged mitochondria after APAP treatment due to containing less CYP450s and more GSH, so the hepatocytes can adapt and use mitophagy to selectively degrade damaged mitochondria to prevent the spread of cellular necrosis. Zone 4, which is the outermost area surrounding Zone 3, contains proliferating hepatocytes. This area contains very little damage from APAP overdose and therefore does not have a need for mitophagy induction. In contrast, mitochondria in this area must proliferate in order to provide energy needed for hepatocellular proliferation and liver regeneration to repair/replace damaged hepatocytes induced by APAP (Figure 10.2) (Ni et al. 2013).

The formation of mitochondrial spheroids requires the presence of ROS in addition to either MFN1 or MFN2, which allow for mitochondrial fusion. Deletion of

APAP   Ethanol

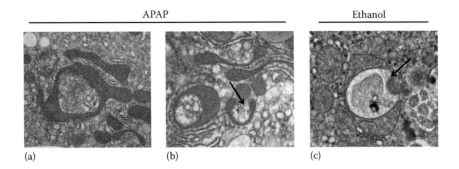

(a)   (b)   (c)

FIGURE 10.1 Induction of mitochondrial spheroids by APAP and acute ethanol treatment in mouse livers. (a,b) Male C57BL/6 wild-type mice were treated with APAP (500 mg/kg, i.p.) for 6 h or (c) male C57BL/6 FXR knockout mice were treated with ethanol (4.5 g/kg, gavage) for 16 h. The mouse liver tissues were fixed and processed for EM studies. Representative EM images of mitochondrial spheroids from APAP-treated (a,b) or ethanol-treated FXR KO mouse livers (c). Arrows denote the openings of mitochondrial spheroids.

either MFN1 or MFN2 inhibits spheroid formation, indicating that both fusion proteins are important for the formation of mitochondrial spheroids (Ding et al. 2012a). Parkin can therefore inhibit the formation of mitochondrial spheroids by inducing proteasomal degradation of MFN1/2 by ubiquitination, which was previously discussed. Indeed, MEF cells, which have an undetectable Parkin expression level, formed mitochondrial spheroids after CCCP treatment. When Parkin was overexpressed in these cells, MFN1 and MFN2 were degraded and mitochondria underwent typical mitophagy instead of forming mitochondrial spheroids (Ding et al. 2012b). These data suggest that Parkin prevents mitochondrial spheroid formation in order for mitophagy to occur.

Interestingly, mitochondrial spheroid formation was found in mice expressing Parkin after APAP treatment in Zone 2. It is important to note that even though APAP induced Parkin translocation from the cytosol to the mitochondria as previously discussed, MFN1 and MFN2 were not degraded by Parkin (Ding et al., unpublished observations). It is possible that degradation of MFN1/2 could not be seen due to the overwhelming amount of mitochondrial proliferation seen in Zone 4 of the liver after APAP treatment. It is also possible that Parkin is inactivated by APAP overdose in Zone 2 of the liver, which would allow for continuous expression of MFN1 and MFN2 and for mitochondrial spheroids to form. Parkin has been shown to be inactivated by posttranslational modifications including ubiquitination, phosphorylation, and nitrosylation (Chung et al. 2004; Walden and Martinez-Torres 2012), but whether or not Parkin is inactivated in the liver after APAP treatment remains to be determined. In addition to APAP, we recently found that acute ethanol treatment also induced the formation of mitochondrial spheroids in farnesoid X receptor (FXR) knockout mouse livers (Manley et al., 2014). We have demonstrated that FXR knockout mice have impaired hepatic autophagy. Thus, the increased mitochondrial spheroids could serve as an alternative mechanism to regulate mitochondrial homeostasis after acute ethanol treatment.

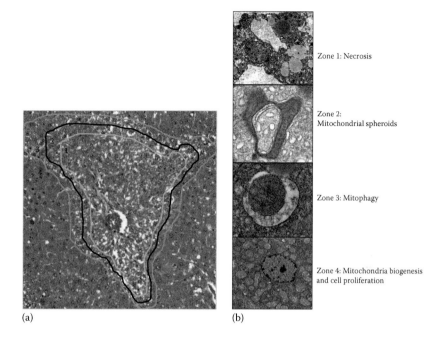

(a)   (b)

FIGURE 10.2 (**See color insert.**) Zonated changes for necrosis, mitochondrial spheroids, mitophagy/autophagy, and mitochondrial biogenesis/hepatocyte proliferation in APAP-induced liver injury. A representative image of the typical histological changes of APAP-induced liver pathogenesis. (a) Male C57BL/6 mice were treated with APAP (500 mg/kg, i.p.) for 6 h and liver tissues were processed for H&E staining. Necrotic areas were mainly detected around the central vein (light gray innermost line circled area, Zone 1). Black line circled areas represent the zone areas that are enriched with mitochondrial spheroids (Zone 2). Light gray outermost line circled areas represent the autophagy activation areas (Zone 3). Outside of the light gray outermost line are the areas with mitochondrial biogenesis and hepatocyte proliferation (Zone 4). (b) Representative EM images illustrate individual zone area changes.

Mitochondrial spheroids are an interesting structure formed in the liver that may provide some protective response against APAP-induced liver injury by degrading damaged mitochondria similar to mitophagy or by blocking the spread of hepatocellular necrosis and oxidative damage by inactivating damaged mitochondria. Interestingly, MEFs depleted of both MFN1 and MFN2 could not form mitochondrial spheroids after CCCP treatment and experienced increased apoptotic cell death (Ding et al., unpublished observations). However, whether or not these spheroids represent a protective mechanism in the liver in addition to mitophagy requires further investigation (Ding et al. 2012a,b; Ni et al. 2013; Yin and Ding 2013).

CONCLUSION

Mitochondria are dynamic organelles that play many important roles in maintaining cellular homeostasis and survival such as ATP production and breakdown of fat by β-oxidation. There are several mechanisms in place to help maintain mitochondrial function including the mitochondrial proteolytic system, proteasomal degradation of damaged outer mitochondrial membrane proteins, mitochondrial fission and fusion, and mitophagy. Mitophagy is an important pathway that helps maintain a population of healthy mitochondria by removing damaged mitochondria to prevent cell death and tissue injury. The most studied mechanism of mitophagy induction has been Parkin-induced mitophagy. However, several other Parkin-independent mechanisms for the induction of mitophagy are surfacing including mitophagy mediated by BNIP3, NIX, FUNDC1, and cardiolipin, or by other E3 ligases, such as SMURF1. In addition, damaged mitochondria may also undergo direct remodeling to form mitochondrial spheroids or form MDVs

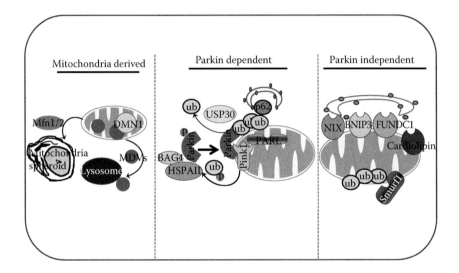

FIGURE 10.3   (See color insert.) A proposed model of three distinctive mechanisms that regulate mitochondrial homeostasis. Following mitochondria damage by various stresses, mitochondria are depolarized and Parkin is translocated to the outer membrane of the mitochondria. This process is regulated by PINK1, which is stabilized on depolarized mitochondria. PINK1 either directly phosphorylates Parkin or ubiquitin to promote Parkin translocation and its ligase activity. Cytosolic HSPA1L also promotes, whereas BAG4 inhibits, Parkin mitochondrial translocation. Once Parkin is translocated to the mitochondria, it promotes canonical-selective mitophagy through mitochondrial ubiquitination, p62 mitochondrial targeting, and LC3-positive autophagosome recruitment (Parkin-dependent mitophagy). For Parkin-independent mitophagy, BNIP3, NIX, FUNDC1, or Cardiolipin can also directly interact with LC3 and recruit autophagosomes to damaged mitochondria. Moreover, other E3 ubiquitin ligases such as SMURF1 can also promote mitochondrial ubiquitination, p62 mitochondrial targeting, and mitophagy. Finally, under certain conditions, mitochondria can undergo direct remodeling to form mitochondrial spheroids that are regulated by Mfn1/2. Moreover, small vesicles that contain a subset group of mitochondrial proteins can be directly generated from damaged mitochondria to form MDVs. The segregation of mitochondria to form MDVs is regulated by the mitochondrial fission molecule DNM1. MDVs are then delivered to lysosomes and are eventually degraded within lysosomes.

that target to lysosomes, which may degrade a specific subset of mitochondrial proteins (Figure 10.3). Finally, the roles of Parkin-induced mitophagy in in vivo versus in vitro models along with the roles of Parkin in pathways independent of mitophagy are beginning to emerge.

ACKNOWLEDGMENTS

The research work in Wen-Xing Ding's lab was supported in part by NIAAA funds (R01 AA020518 and R01 DK102142), the National Center for Research Resources (5P20RR021940), the National Institute of General Medical Sciences (8P20GM103549 and T32 ES007079), and an Institutional Development Award (IDeA) from the National Institute of General Medical Sciences of the National Institutes of Health (P20 GM103418). The authors thank Barbara Fegley from the electron microscopy core facility at the University of Kansas Medical Center for her excellent assistance with the EM studies.

REFERENCES

Abeliovich, H., M. Zarei et al. (2013). Involvement of mitochondrial dynamics in the segregation of mitochondrial matrix proteins during stationary phase mitophagy. *Nat Commun* 4: 2789.

Anderson, R. M., J. L. Barger et al. (2008). Dynamic regulation of PGC-1α localization and turnover implicates mitochondrial adaptation in calorie restriction and the stress response. *Aging Cell* 7(1): 101–111.

Baker, B. M. and C. M. Haynes (2011). Mitochondrial protein quality control during biogenesis and aging. *Trends Biochem Sci* 36(5): 254–261.

Begriche, K., J. Massart et al. (2011). Drug-induced toxicity on mitochondria and lipid metabolism: Mechanistic diversity and deleterious consequences for the liver. *J Hepatol* 54(4): 773–794.

Bellot, G., R. Garcia-Medina et al. (2009). Hypoxia-induced autophagy is mediated through hypoxia-inducible factor induction of BNIP3 and BNIP3L via their BH3 domains. *Mol Cell Biol* 29(10): 2570–2581.

Beyoglu, D. and J. R. Idle (2013). The metabolomic window into hepatobiliary disease. *J Hepatol* 59(4): 842–858.

Bingol, B., J. S. Tea et al. (2014). The mitochondrial deubiquitinase USP30 opposes parkin-mediated mitophagy. *Nature* 510(7505): 370–375.

Bruick, R. K. (2000). Expression of the gene encoding the proapoptotic Nip3 protein is induced by hypoxia. *Proc Natl Acad Sci USA* 97(16): 9082–9087.

Canto, C., Z. Gerhart-Hines et al. (2009). AMPK regulates energy expenditure by modulating NAD$^+$ metabolism and SIRT1 activity. *Nature* 458(7241): 1056–1060.

Chan, N. C., A. M. Salazar et al. (2011). Broad activation of the ubiquitin-proteasome system by Parkin is critical for mitophagy. *Hum Mol Genet* 20(9): 1726–1737.

Chang, J. S., P. Huypens et al. (2010). Regulation of NT-PGC-1α subcellular localization and function by protein kinase A-dependent modulation of nuclear export by CRM1. *J Biol Chem* 285(23): 18039–18050.

Chen, G., J. Cizeau et al. (1999). Nix and Nip3 form a subfamily of pro-apoptotic mitochondrial proteins. *J Biol Chem* 274(1): 7–10.

Chen, G., Z. Han et al. (2014). A regulatory signaling loop comprising the PGAM5 phosphatase and CK2 controls receptor-mediated mitophagy. *Mol Cell* 54(3): 362–377.

Chu, C. T., H. Bayir et al. (2014). LC3 binds externalized cardiolipin on injured mitochondria to signal mitophagy in neurons: Implications for Parkinson disease. *Autophagy* 10(2): 376–378.

Chu, C. T., J. Ji et al. (2013). Cardiolipin externalization to the outer mitochondrial membrane acts as an elimination signal for mitophagy in neuronal cells. *Nat Cell Biol* 15(10): 1197–1205.

Chung, K. K., B. Thomas et al. (2004). S-nitrosylation of parkin regulates ubiquitination and compromises parkin's protective function. *Science* 304(5675): 1328–1331.

Cipolat, S., T. Rudka et al. (2006). Mitochondrial rhomboid PARL regulates cytochrome c release during apoptosis via OPA1-dependent cristae remodeling. *Cell* 126(1): 163–175.

Clark, I. E., M. W. Dodson et al. (2006). Drosophila pink1 is required for mitochondrial function and interacts genetically with parkin. *Nature* 441(7097): 1162–1166.

De Rasmo, D., A. Signorile et al. (2010). cAMP/Ca^{2+} response element-binding protein plays a central role in the biogenesis of respiratory chain proteins in mammalian cells. *IUBMB Life* 62(6): 447–452.

Ding, W. X., F. Guo et al. (2012a). Parkin and mitofusins reciprocally regulate mitophagy and mitochondrial spheroid formation. *J Biol Chem* 287(50): 42379–42388.

Ding, W. X., M. Li et al. (2010a). Autophagy reduces acute ethanol-induced hepatotoxicity and steatosis in mice. *Gastroenterology* 139(5): 1740–1752.

Ding, W. X., M. Li et al. (2012b). Electron microscopic analysis of a spherical mitochondrial structure. *J Biol Chem* 287(50): 42373–42378.

Ding, W. X., H. M. Ni et al. (2010b). Nix is critical to two distinct phases of mitophagy, reactive oxygen species-mediated autophagy induction and Parkin-ubiquitin-p62-mediated mitochondrial priming. *J Biol Chem* 285(36): 27879–27890.

Ding, W. X. and X. M. Yin (2012). Mitophagy: Mechanisms, pathophysiological roles, and analysis. *Biol Chem* 393(7): 547–564.

Duchen, M. R. (2004). Mitochondria in health and disease: Perspectives on a new mitochondrial biology. *Mol Aspects Med* 25(4): 365–451.

Duvezin-Caubet, S., M. Koppen et al. (2007). OPA1 processing reconstituted in yeast depends on the subunit composition of the m-AAA protease in mitochondria. *Mol Biol Cell* 18(9): 3582–3590.

Ehses, S., I. Raschke et al. (2009). Regulation of OPA1 processing and mitochondrial fusion by m-AAA protease isoenzymes and OMA1. *J Cell Biol* 187(7): 1023–1036.

Ekstrand, M. I., M. Falkenberg et al. (2004). Mitochondrial transcription factor A regulates mtDNA copy number in mammals. *Hum Mol Genet* 13(9): 935–944.

Fiesel, F. C., E. L. Moussaud-Lamodiere et al. (2014). A specific subset of E2 ubiquitin-conjugating enzymes regulate Parkin activation and mitophagy differently. *J Cell Sci.* 127(Pt 16): 3488–3504.

Friedman, J. R., L. L. Lackner et al. (2011). ER tubules mark sites of mitochondrial division. *Science* 334(6054): 358–362.

Galvez, A. S., E. W. Brunskill et al. (2006). Distinct pathways regulate proapoptotic Nix and BNIP3 in cardiac stress. *J Biol Chem* 281(3): 1442–1448.

Gandre-Babbe, S. and A. M. van der Bliek (2008). The novel tail-anchored membrane protein Mff controls mitochondrial and peroxisomal fission in mammalian cells. *Mol Biol Cell* 19(6): 2402–2412.

Gao, B. and R. Bataller (2011). Alcoholic liver disease: Pathogenesis and new therapeutic targets. *Gastroenterology* 141(5): 1572–1585.

Gegg, M. E., J. M. Cooper et al. (2010). Mitofusin 1 and mitofusin 2 are ubiquitinated in a PINK1/parkin-dependent manner upon induction of mitophagy. *Hum Mol Genet* 19(24): 4861–4870.

Geisler, S., K. M. Holmstrom et al. (2010). PINK1/Parkin-mediated mitophagy is dependent on VDAC1 and p62/SQSTM1. *Nat Cell Biol* 12(2): 119–131.

Geisler, S., S. Vollmer et al. (2014). The ubiquitin-conjugating enzymes UBE2N, UBE2L3 and UBE2D2/3 are essential for Parkin-dependent mitophagy. *J Cell Sci* 127(Pt 15): 3280–3293.

Gomes, L. C., G. Di Benedetto et al. (2011). During autophagy mitochondria elongate, are spared from degradation and sustain cell viability. *Nat Cell Biol* 13(5): 589–598.

Graef, M. and J. Nunnari (2011). Mitochondria regulate autophagy by conserved signalling pathways. *EMBO J* 30(11): 2101–2114.

Grattagliano, I., S. Russmann et al. (2011). Mitochondria in chronic liver disease. *Curr Drug Targets* 12(6): 879–893.

Griparic, L., T. Kanazawa et al. (2007). Regulation of the mitochondrial dynamin-like protein Opa1 by proteolytic cleavage. *J Cell Biol* 178(5): 757–764.

Guicciardi, M. E., H. Malhi et al. (2013). Apoptosis and necrosis in the liver. *Compr Physiol* 3(2): 977–1010.

Hasson, S. A., L. A. Kane et al. (2013). High-content genome-wide RNAi screens identify regulators of parkin upstream of mitophagy. *Nature* 504(7479): 291–295.

Head, B., L. Griparic et al. (2009). Inducible proteolytic inactivation of OPA1 mediated by the OMA1 protease in mammalian cells. *J Cell Biol* 187(7): 959–966.

Hill, B. G., G. A. Benavides et al. (2012). Integration of cellular bioenergetics with mitochondrial quality control and autophagy. *Biol Chem* 393(12): 1485–1512.

Hinson, J. A., A. B. Reid et al. (2004). Acetaminophen-induced hepatotoxicity: Role of metabolic activation, reactive oxygen/nitrogen species, and mitochondrial permeability transition. *Drug Metab Rev* 36(3–4): 805–822.

Huang, C., A. M. Andres et al. (2011). Preconditioning involves selective mitophagy mediated by Parkin and p62/SQSTM1. *PLoS One* 6(6): e20975.

Huebener, P., G. Y. Gwak et al. (2014). High-mobility group box 1 is dispensable for autophagy, mitochondrial quality control, and organ function in vivo. *Cell Metab* 19(3): 539–547.

Ishihara, N., Y. Fujita et al. (2006). Regulation of mitochondrial morphology through proteolytic cleavage of OPA1. *EMBO J* 25(13): 2966–2977.

Jaeschke, H., G. J. Gores et al. (2002). Mechanisms of hepatotoxicity. *Toxicol Sci* 65(2): 166–176.

Jaeschke, H., M. R. McGill et al. (2012). Oxidant stress, mitochondria, and cell death mechanisms in drug-induced liver injury: Lessons learned from acetaminophen hepatotoxicity. *Drug Metab Rev* 44(1): 88–106.

Jager, S., C. Handschin et al. (2007). AMP-activated protein kinase (AMPK) action in skeletal muscle via direct phosphorylation of PGC-1α. *Proc Natl Acad Sci USA* 104(29): 12017–12022.

Jin, S. M., M. Lazarou et al. (2010). Mitochondrial membrane potential regulates PINK1 import and proteolytic destabilization by PARL. *J Cell Biol* 191(5): 933–942.

Jones, A. W., Z. Yao et al. (2012). PGC-1 family coactivators and cell fate: Roles in cancer, neurodegeneration, cardiovascular disease and retrograde mitochondria-nucleus signalling. *Mitochondrion* 12(1): 86–99.

Jungermann, K. and T. Kietzmann (2000). Oxygen: Modulator of metabolic zonation and disease of the liver. *Hepatology* 31(2): 255–260.

Kane, L. A., M. Lazarou et al. (2014). PINK1 phosphorylates ubiquitin to activate Parkin E3 ubiquitin ligase activity. *J Cell Biol* 205(2): 143–153.

Karbowski, M. and R. J. Youle (2011). Regulating mitochondrial outer membrane proteins by ubiquitination and proteasomal degradation. *Curr Opin Cell Biol* 23(4): 476–482.

Kawajiri, S., S. Saiki et al. (2010). PINK1 is recruited to mitochondria with parkin and associates with LC3 in mitophagy. *FEBS Lett* 584(6): 1073–1079.

Kim, I. and J. J. Lemasters (2011). Mitochondrial degradation by autophagy (mitophagy) in GFP-LC3 transgenic hepatocytes during nutrient deprivation. *Am J Physiol Cell Physiol* 300(2): C308–C317.

Kim, J., M. Kundu et al. (2011). AMPK and mTOR regulate autophagy through direct phosphorylation of Ulk1. *Nat Cell Biol* 13(2): 132–141.

Kim, Y., J. Park et al. (2008). PINK1 controls mitochondrial localization of Parkin through direct phosphorylation. *Biochem Biophys Res Commun* 377(3): 975–980.

Kitada, T., S. Asakawa et al. (1998). Mutations in the parkin gene cause autosomal recessive juvenile parkinsonism. *Nature* 392(6676): 605–608.

Kompare, M. and W. B. Rizzo (2008). Mitochondrial fatty-acid oxidation disorders. *Semin Pediatr Neurol* 15(3): 140–149.

Koyano, F., K. Okatsu et al. (2014). Ubiquitin is phosphorylated by PINK1 to activate parkin. *Nature* 510(7503): 162–166.

Kuroda, Y., T. Mitsui et al. (2006). Parkin enhances mitochondrial biogenesis in proliferating cells. *Hum Mol Genet* 15(6): 883–895.

Larson, A. M. (2007). Acetaminophen hepatotoxicity. *Clin Liver Dis* 11(3): 525–548, vi.

Lemasters, J. J. (2014). Variants of mitochondrial autophagy: Types 1 and 2 mitophagy and micromitophagy (Type 3). *Redox Biol* 2: 749–754.

Liesa, M. and O. S. Shirihai (2013). Mitochondrial dynamics in the regulation of nutrient utilization and energy expenditure. *Cell Metab* 17(4): 491–506.

Lin, C. W., H. Zhang et al. (2013). Pharmacological promotion of autophagy alleviates steatosis and injury in alcoholic and non-alcoholic fatty liver conditions in mice. *J Hepatol* 58(5): 993–999.

Lin, Z., F. Wu et al. (2014). Adiponectin protects against acetaminophen-induced mitochondrial dysfunction and acute liver injury by promoting autophagy in mice. *J Hepatol* 61(4): 825–831.

Lindros, K. O. (1997). Zonation of cytochrome P450 expression, drug metabolism and toxicity in liver. *Gen Pharmacol* 28(2): 191–196.

Liu, L., D. Feng et al. (2012). Mitochondrial outer-membrane protein FUNDC1 mediates hypoxia-induced mitophagy in mammalian cells. *Nat Cell Biol* 14(2): 177–185.

Liu, L., K. Sakakibara et al. (2014). Receptor-mediated mitophagy in yeast and mammalian systems. *Cell Res* 24(7): 787–795.

Luedde, T., N. Kaplowitz et al. (2014). Cell death and cell death responses in liver disease: Mechanisms and clinical relevance. *Gastroenterology* 147(4): 765.e4–783.e4.

Manley, S., H.M. Ni et al. (2014). Farnesoid X receptor regulates forkhead Box O3a activation in ethanol-induced autophagy and hepatotoxicity. *Redox Biol* 2C: 991–1002.

Manley, S., J. A. Williams et al. (2013). Role of p62/SQSTM1 in liver physiology and pathogenesis. *Exp Biol Med (Maywood)* 238(5): 525–538.

Matsuda, N., S. Sato et al. (2010). PINK1 stablized by mitochondrial depolarization recruits Parkin to damaged mitochondria and activates latent Parkin for mitophagy. *J Cell Biol* 189(2):211–21.

Matsushima, Y. and L. S. Kaguni (2012). Matrix proteases in mitochondrial DNA function. *Biochim Biophys Acta* 1819(9–10): 1080–1087.

McGill, M. R., M. R. Sharpe et al. (2012). The mechanism underlying acetaminophen-induced hepatotoxicity in humans and mice involves mitochondrial damage and nuclear DNA fragmentation. *J Clin Invest* 122(4): 1574–1583.

McLelland, G. L., V. Soubannier et al. (2014). Parkin and PINK1 function in a vesicular trafficking pathway regulating mitochondrial quality control. *EMBO J* 33(4): 282–295.

Melser, S., E. H. Chatelain et al. (2013). Rheb regulates mitophagy induced by mitochondrial energetic status. *Cell Metab* 17(5): 719–730.

Michel, S., A. Wanet et al. (2012). Crosstalk between mitochondrial (dys)function and mitochondrial abundance. *J Cell Physiol* 227(6): 2297–2310.

Mihaylova, M. M. and R. J. Shaw (2011). The AMPK signalling pathway coordinates cell growth, autophagy and metabolism. *Nat Cell Biol* 13(9): 1016–1023.

Mizushima, N., B. Levine et al. (2008). Autophagy fights disease through cellular self-digestion. *Nature* 451(7182): 1069–1075.

Molina, A. J., J. D. Wikstrom et al. (2009). Mitochondrial networking protects β-cells from nutrient-induced apoptosis. *Diabetes* 58(10): 2303–2315.

Mortiboys, H., K. J. Thomas et al. (2008). Mitochondrial function and morphology are impaired in parkin-mutant fibroblasts. *Ann Neurol* 64(5): 555–565.

Narendra, D., L. A. Kane et al. (2010). p62/SQSTM1 is required for Parkin-induced mitochondrial clustering but not mitophagy; VDAC1 is dispensable for both. *Autophagy* 6(8): 1090–1106.

Narendra, D., A. Tanaka et al. (2008). Parkin is recruited selectively to impaired mitochondria and promotes their autophagy. *J Cell Biol* 183(5): 795–803.

Narendra D., S.M. Jin et al. (2010). PINK1 is selectively stabilized on impaired mitochondria to activate Parkin. *PLoS Biol* 8(1):e1000298.

Nguyen, P., V. Leray et al. (2008). Liver lipid metabolism. *J Anim Physiol Anim Nutr (Berl)* 92(3): 272–283.

Ni, H. M., A. Bockus et al. (2012). Activation of autophagy protects against acetaminophen-induced hepatotoxicity. *Hepatology* 55(1): 222–232.

Ni, H. M., J. A. Williams et al. (2013). Zonated induction of autophagy and mitochondrial spheroids limits acetaminophen-induced necrosis in the liver. *Redox Biol* 1(1): 427–432.

Novak, I., V. Kirkin et al. (2010). Nix is a selective autophagy receptor for mitochondrial clearance. *EMBO Rep* 11(1): 45–51.

Okatsu, K., K. Saisho et al. (2010). p62/SQSTM1 cooperates with Parkin for perinuclear clustering of depolarized mitochondria. *Genes Cells* 15(8): 887–900.

Ongwijitwat, S. and M. T. Wong-Riley (2005). Is nuclear respiratory factor 2 a master transcriptional coordinator for all ten nuclear-encoded cytochrome c oxidase subunits in neurons? *Gene* 360(1): 65–77.

Orvedahl, A., R. Sumpter, Jr. et al. (2011). Image-based genome-wide siRNA screen identifies selective autophagy factors. *Nature* 480(7375): 113–117.

Otera, H., C. Wang et al. (2010). Mff is an essential factor for mitochondrial recruitment of Drp1 during mitochondrial fission in mammalian cells. *J Cell Biol* 191(6): 1141–1158.

Palacino, J. J., D. Sagi et al. (2004). Mitochondrial dysfunction and oxidative damage in parkin-deficient mice. *J Biol Chem* 279(18): 18614–18622.

Palikaras, K. and N. Tavernarakis (2014). Mitochondrial homeostasis: The interplay between mitophagy and mitochondrial biogenesis. *Exp Gerontol* 56: 182–188.

Park, J., S. B. Lee et al. (2006). Mitochondrial dysfunction in *Drosophila* PINK1 mutants is complemented by Parkin. *Nature* 441(7097): 1157–1161.

Parzych, K. R. and D. J. Klionsky (2014). An overview of autophagy: Morphology, mechanism, and regulation. *Antioxid Redox Signal* 20(3): 460–473.

Poole, A. C., R. E. Thomas et al. (2010). The mitochondrial fusion-promoting factor mitofusin is a substrate of the PINK1/Parkin pathway. *PLoS One* 5(4): e10054.

Reggiori, F., M. Komatsu et al. (2012). Selective types of autophagy. *Int J Cell Biol* 2012: 156272.

Ren, M., C. K. Phoon et al. (2014). Metabolism and function of mitochondrial cardiolipin. *Prog Lipid Res* 55C: 1–16.

Reznick, R. M., H. Zong et al. (2007). Aging-associated reductions in AMP-activated protein kinase activity and mitochondrial biogenesis. *Cell Metab* 5(2): 151–156.

Rothfuss, O., H. Fischer et al. (2009). Parkin protects mitochondrial genome integrity and supports mitochondrial DNA repair. *Hum Mol Genet* 18(20): 3832–3850.

Saberi, B., M. D. Ybanez et al. (2014). Protein kinase C (PKC) participates in acetaminophen hepatotoxicity through c-jun-*N*-terminal kinase (JNK)-dependent and -independent signaling pathways. *Hepatology* 59(4): 1543–1554.

Sandoval, H., P. Thiagarajan et al. (2008). Essential role for Nix in autophagic maturation of erythroid cells. *Nature* 454(7201): 232–235.

Scarpulla, R. C., R. B. Vega et al. (2012). Transcriptional integration of mitochondrial biogenesis. *Trends Endocrinol Metab* 23(9): 459–466.

Schapira, A. H., J. M. Cooper et al. (1990). Mitochondrial complex I deficiency in Parkinson's disease. *J Neurochem* 54(3): 823–827.

Schulz, E., J. Dopheide et al. (2008). Suppression of the JNK pathway by induction of a metabolic stress response prevents vascular injury and dysfunction. *Circulation* 118(13): 1347–1357.

Schweers, R. L., J. Zhang et al. (2007). NIX is required for programmed mitochondrial clearance during reticulocyte maturation. *Proc Natl Acad Sci USA* 104(49): 19500–19505.

Sha, D., L. S. Chin et al. (2010). Phosphorylation of parkin by Parkinson disease-linked kinase PINK1 activates parkin E3 ligase function and NF-κB signaling. *Hum Mol Genet* 19(2): 352–363.

Shin, J. H., H. S. Ko et al. (2011). PARIS (ZNF746) repression of PGC-1α contributes to neurodegeneration in Parkinson's disease. *Cell* 144(5): 689–702.

Soleimanpour, S. A., A. Gupta et al. (2014). The diabetes susceptibility gene Clec16a regulates mitophagy. *Cell* 157(7): 1577–1590.

Soubannier, V., P. Rippstein et al. (2012). Reconstitution of mitochondria derived vesicle formation demonstrates selective enrichment of oxidized cargo. *PLoS One* 7(12): e52830.

Tang, D., R. Kang et al. (2010). Endogenous HMGB1 regulates autophagy. *J Cell Biol* 190(5): 881–892.

Tang, D., R. Kang et al. (2011). High-mobility group box 1 is essential for mitochondrial quality control. *Cell Metab* 13(6): 701–711.

Twig, G., A. Elorza et al. (2008). Fission and selective fusion govern mitochondrial segregation and elimination by autophagy. *EMBO J* 27(2): 433–446.

Twig, G. and O. S. Shirihai (2011). The interplay between mitochondrial dynamics and mitophagy. *Antioxid Redox Signal* 14(10): 1939–1951.

Um, J. W., C. Stichel-Gunkel et al. (2009). Molecular interaction between parkin and PINK1 in mammalian neuronal cells. *Mol Cell Neurosci* 40(4): 421–432.

van der Bliek, A. M., Q. Shen et al. (2013). Mechanisms of mitochondrial fission and fusion. *Cold Spring Harb Perspect Biol* 5(6): pii:a011072.

Van Humbeeck, C., T. Cornelissen et al. (2011). Parkin interacts with Ambra1 to induce mitophagy. *J Neurosci* 31(28): 10249–10261.

Virbasius, C. A., J. V. Virbasius et al. (1993). NRF-1, an activator involved in nuclear-mitochondrial interactions, utilizes a new DNA-binding domain conserved in a family of developmental regulators. *Genes Dev* 7(12A): 2431–2445.

Vives-Bauza, C., C. Zhou et al (2010). PINK1-dependent recruitment of Parkin to mitochondria in mitophagy. *Proc Natl Acad Sci* 107(1): 378–383.

Walden, H. and R. J. Martinez-Torres (2012). Regulation of Parkin E3 ubiquitin ligase activity. *Cell Mol Life Sci* 69(18): 3053–3067.

Wang, X., D. Winter et al. (2011). PINK1 and Parkin target Miro for phosphorylation and degradation to arrest mitochondrial motility. *Cell* 147(4): 893–906.

Williams, J.A., H.M. Ni et al (2015). Chronic deletion and acute knockdown of Parkin have differential responses to acetaminophen-induced mitophagy and liver injury in mice. 290(17): 10934–10946.

Westermann, B. (2010). Mitochondrial fusion and fission in cell life and death. *Nat Rev Mol Cell Biol* 11(12): 872–884.

Wu, H., D. Xue et al. (2014). The BCL2L1 and PGAM5 axis defines hypoxia-induced receptor-mediated mitophagy. *Autophagy* 10(10): 1712–1725.

Wu, Z., P. Puigserver et al. (1999). Mechanisms controlling mitochondrial biogenesis and respiration through the thermogenic coactivator PGC-1. *Cell* 98(1): 115–124.

Yang, Y., S. Gehrke et al. (2006). Mitochondrial pathology and muscle and dopaminergic neuron degeneration caused by inactivation of Drosophila Pink1 is rescued by Parkin. *Proc Natl Acad Sci USA* 103(28): 10793–10798.

Yasuda, M., P. Theodorakis et al. (1998). Adenovirus E1B-19K/BCL-2 interacting protein BNIP3 contains a BH3 domain and a mitochondrial targeting sequence. *J Biol Chem* 273(20): 12415–12421.

Yin, X. M. and W. X. Ding (2013). The reciprocal roles of PARK2 and mitofusins in mitophagy and mitochondrial spheroid formation. *Autophagy* 9(11): 1687–1692.

Youle, R. J. and D. P. Narendra (2011). Mechanisms of mitophagy. *Nat Rev Mol Cell Biol* 12(1): 9–14.

Zhang, H., M. Bosch-Marce et al. (2008). Mitochondrial autophagy is an HIF-1-dependent adaptive metabolic response to hypoxia. *J Biol Chem* 283(16): 10892–10903.

Zhu, J., K. Z. Wang et al. (2013a). After the banquet: Mitochondrial biogenesis, mitophagy, and cell survival. *Autophagy* 9(11): 1663–1676.

Zhu, Y., S. Massen et al. (2013b). Modulation of serines 17 and 24 in the LC3-interacting region of BNIP3 determines pro-survival mitophagy versus apoptosis. *J Biol Chem* 288(2): 1099–1113.

11 Biomarkers of Mitochondrial Damage in the Liver

Mitchell R. McGill and Hartmut Jaeschke

CONTENTS

ABSTRACT

Mitochondrial damage or dysfunction is known to be an important mechanism of injury in many liver diseases. A number of methods for the measurement of mitochondrial damage are available for basic research. While respirometry is the preferred approach, imaging techniques such as the assessment of mitochondrial morphology by electron microscopy and the visualization of mitochondrial membrane potential using dyes that accumulate within functional mitochondria, and various biochemical assays are also available. Importantly, several noninvasive serum and urine biomarkers of mitochondrial damage and oxidative stress have recently been described for use in translational research. Although additional work is needed to validate these markers, they are already allowing the investigation of liver injury mechanisms in humans at the molecular level. The purpose of this chapter is to present an overview of the major methods available to measure mitochondrial damage using tissues, cells, isolated mitochondria, and even the blood of animals and humans.

INTRODUCTION

Mitochondrial damage and dysfunction have been shown to have major roles in many experimental models of liver injury or disease. For example, transgenic expression

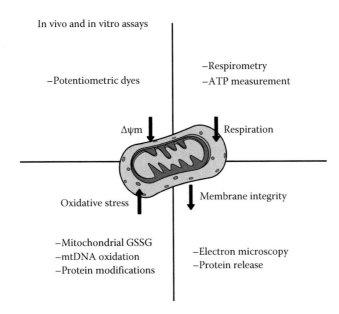

FIGURE 11.1 Major methods to assess mitochondrial function or damage. Loss of membrane potential can be measured using potentiometric dyes. Mitochondrial function can be determined by measuring respiration. Membrane integrity can be assessed by immunoblotting for mitochondrial proteins in the cytosolic space, or by electron microscopy. Finally, oxidative stress in mitochondria can be assessed by measuring oxidized glutathione (GSSG), mtDNA oxidation, or protein oxidation.

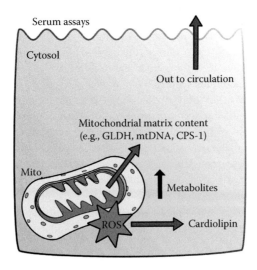

FIGURE 11.2 Release of mitochondrial content into circulation. Mitochondrial content can be released into the cytosol when mitochondria are damaged, and later become detectable in serum. Metabolic intermediates may also accumulate when mitochondria are damaged.

of the hepatitis C virus core protein in various cell lines and in mice leads to the inhibition of mitochondrial respiration and causes oxidative stress (Okuda et al., 2002; Korenaga et al., 2005). Multiple animal models of Wilson disease, caused by hepatic copper overload, display marked abnormalities in mitochondrial morphology and function and have signs of mitochondrial oxidative stress (Sokol et al., 1990; Sternlieb et al., 1995; Zischka et al., 2011). Similar experimental evidence is available for mitochondrial damage in alcoholic liver disease models (García-Ruiz et al., 2013). Many other forms of liver injury, including acetaminophen (APAP) hepatotoxicity (Jaeschke et al., 2012), various other drug-induced liver injuries (Pessayre et al., 2012), and ischemia-reperfusion graft injury in liver transplantation (Theruvath et al., 2008) appear to involve some form of mitochondrial dysfunction as well. In many cases, there is also evidence of mitochondrial damage in the same diseases in humans. For example, changes in mitochondrial morphology have been reported for many years in hepatocytes obtained from Wilson disease patients (Sternlieb, 1968). In viral hepatitis, at least one study has shown that the mitochondria-targeted antioxidant mitoquinone can decrease signs of liver injury, suggesting that mitochondrial oxidative stress is critical in the mechanism of injury (Gane et al., 2010). Finally, recent data suggest that mitochondrial damage occurs during APAP hepatotoxicity in patients, and that this damage could play a major role in the toxicity (McGill et al., 2012, 2014a).

A number of biomarkers are available to study mitochondrial damage in liver injury using both tissue (Figure 11.1) and extracellular fluid (Figure 11.2). These include altered mitochondrial number or morphology in tissues or cultured cells, biochemical markers like reduced mitochondrial respiration and mitochondrial DNA damage, and serum and urine biomarkers of mitochondrial damage. The focus of this chapter is on the history and recent progress in the field of biomarkers of mitochondrial damage. The chapter begins with an overview of methods to measure mitochondrial dysfunction or damage in tissues and cells, and continues with a discussion of recent data on serum and urine biomarkers of mitochondrial damage.

METHODS TO MEASURE MITOCHONDRIAL DAMAGE IN TISSUE AND CELLS

IN SITU APPROACHES

A very crude way to assess mitochondrial stress is to observe the number and morphology of mitochondria in tissue sections. Mitochondria were first observed in histological sections in the 1800s using simple light microscopy (Ernster and Schatz, 1981). The first stain to specifically label this organelle was introduced shortly thereafter by the German pathologist Richard Altmann, although it was not known as the mitochondrion at the time (Ernster and Schatz, 1981; O'Rourke, 2010). Subsequent methods to visualize mitochondria in tissue have included the Gomori trichrome stain, which can be used clinically to diagnose mitochondrial myopathies. However, these methods have largely been supplanted by enzymatic staining techniques (e.g., the cytochrome oxidase reaction in the electron transport chain) (Tanji and Bonilla, 2008), immunohistochemistry (Tanji and Bonilla, 2008), and electron microscopy.

While the former two can provide some indication of mitochondrial function, the latter enables high-resolution views of individual mitochondria.

Electron microscopy is the preferred tool to study mitochondrial damage in liver tissue sections because it allows direct visualization of changes in mitochondrial morphology. For example, mitochondrial swelling and membrane damage have been observed in rodent models of APAP hepatotoxicity (Placke et al., 1987; Ni et al., 2013) and Wilson disease (Zischka et al., 2011), and have even been noted in humans with APAP-induced liver injury (Petersen and Vilstrup, 1979). However, it is important to accurately quantify the observed changes. It is possible that some damaged mitochondria can occasionally be seen even in healthy liver tissue as a result of normal cell processes or organelle turnover. Mitochondria number and morphology can also be determined in cultured cells. This is usually accomplished using either electron microscopy or by labeling the organelle with antibodies in fixed cells or with fluorescent dyes that accumulate within mitochondria in live cells. Although this provides only a gross view of the mitochondria, it can be useful to observe mitochondrial fission or fusion, as has been done in several experimental models of hepatocyte injury (Das et al., 2012; Ramachandran et al., 2013; Galloway et al., 2014; Yu et al., 2014). Although changes in mitochondrial dynamics can occur for other reasons, they may indicate mitochondrial stress.

A better method to assess the general health of mitochondria in cultured cells is to use a potentiometric dye. These cationic compounds accumulate within mitochondria because they are attracted to the electronegative interior of the organelle. Once inside, they become fluorescent. Examples of such dyes include tetramethylrhodamine methyl ester (TMRM) and JC-1 (5,5′-tetrachloro-1,1′,3,3′-tetraethylben zimidazolylcarbocyanine iodide), although it has been noted that the latter can be problematic in some cases (Nicholls, 2012). When the mitochondria lose their membrane potential before or during cell death, the dyes diffuse out of the mitochondria and either lose fluorescence or emit at a different wavelength. These fluorophores can be measured either by microscopy or with a fluorometer. Both TMRM and JC-1 are frequently used to study mitochondrial dysfunction in cultured liver cells (Kweon et al., 2003; Kon et al., 2004; McGill et al., 2011; Pernelle et al., 2011; Zhang and Lemasters, 2013; Xie et al., 2014). One potential problem with these dyes is that strong inhibition of mitochondrial respiration and proton flow may also result in lower fluorescence. Data should be interpreted cautiously. Another way to test mitochondrial permeability is to measure mitochondrial volume or Ca^{2+}-induced swelling (Halestrap, 1989; Zoratti and Szabó, 1995). This approach is based on changes in light scattering caused by differently sized mitochondria. Although this is relatively easy to measure, mitochondrial volume can be affected by several factors, including isolation conditions.

Likely the best way to assess mitochondrial function in vitro is to measure respiration (Brand and Nicholls, 2011) (Figure 11.3). This can be done either in intact cultured cells or in isolated mitochondria from liver after in vivo treatment. The first widely used assay to measure mitochondrial respiration was introduced in the 1950s and was demonstrated using isolated liver mitochondria (Chance and Williams, 1955). In these experiments, the isolated organelle is added to buffer containing phosphates and a partially submerged platinum electrode for the

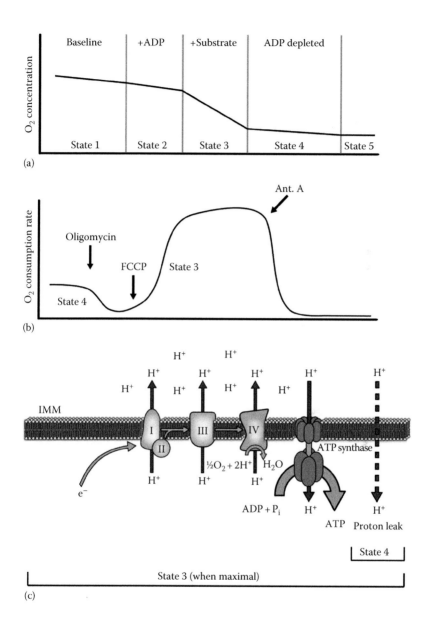

FIGURE 11.3 (See color insert.) Measurement of mitochondrial respiration. (a) Diagram of mitochondrial respiration states determined experimentally. (b) Measurement of states 3 and 4 respiration using a pharmacological approach. (c) Diagram of the electron transport chain showing which components contribute to either state 3 or state 4 respiration. Light gray arrows represent the flow of electrons, while dark gray arrows show the flow of protons in the electron transport chain.

measurement of oxygen (O_2) in the reaction buffer as an indicator of respiration. In normal mitochondrial respiration, electrons from NADH or succinate (products of the Krebs cycle and fatty acid oxidation) are transferred to the electron transport chain complexes in the inner mitochondrial membrane (Figure 11.3). Electron carriers pass the electrons from one complex to the next. At complexes I, III, and IV, the transfer of electrons provides energy to pump protons (H^+) from the mitochondrial matrix out to the intermembrane space. The electron transport chain terminates at complex IV with the transfer of electrons to O_2 to form H_2O. The flow of protons back across the inner membrane to re-enter the matrix drives the synthesis of adenosine triphosphate (ATP) by ATP synthase. Also, the experiments are often carried out in a cuvette, allowing simultaneous detection of oxidation and reduction of NAD^+/NADH by fluorescence.

Several mitochondrial respiration states have been defined using this methodology (Figure 11.3). The initial respiratory rate upon addition of the isolated mitochondria to the buffer is known as state 1 respiration. During this time, endogenous ADP in the sample is converted to ATP. The depletion of ADP prevents further flow of protons back into the matrix. Although the detailed mechanisms of proton pumping in the different complexes are only beginning to be understood, some data suggest that the respiratory rate is controlled largely by the balance of the free energy change in pumping the protons out and the free energy change of the flow of protons back into the matrix. Thus, when the proton concentration is very high in the intermembrane space, the energy from electron transport is insufficient to pump more protons out and respiration ceases. Addition of excess ADP then increases phosphorylation and depletes endogenous substrate (e.g., pyruvate, succinate, etc.) in the samples. This second period is referred to as state 2 respiration. Next, the addition of excess substrate in the presence of the remaining ADP (which is still in excess) leads to rapid respiration, state 3, with phosphorylation again producing ATP. State 3 is the maximal respiratory rate of the mitochondria. As the ADP is consumed, ATP synthesis activity decreases and what respiration remains is largely the result of "accidental" proton leak through the inner membrane. This is known as state 4, and is an indication of inner membrane permeability or leakiness. Finally, depletion of oxygen in the buffer from the continued low rate of respiration creates an anaerobic state called state 5. Respiration rates in states 3 and 4 provide a strong measure of mitochondrial function. The ratio of state 3 to state 4 is called the respiratory control ratio (RCR) and is widely considered the best indicator of mitochondrial function available. A high ratio suggests healthy mitochondria with tight coupling of respiration with phosphorylation, while a low ratio indicates increased mitochondrial permeability. Changes in mitochondrial respiration and the RCR have been observed in some models of liver injury, including APAP toxicity (Meyers et al., 1988).

One caveat in the measurement of respiration in state 4 is that mitochondria preparations are often contaminated with ATPase activity from mitochondria that were damaged during isolation. The ATPase activity is often the result of ATP synthase working in reverse in the absence of a proton gradient in the damaged organelle. It recycles ATP back to ADP, thereby preventing ADP depletion. One way to overcome this is to treat the samples with a compound-like oligomycin that inhibits ATP synthase so that true state 4 respiration (state 4_O) can be measured (Figure 11.3).

A similar approach can be used to measure maximal respiration similar to state 3: drugs that facilitate the movement of protons across the inner membrane without the involvement of ATP synthase effectively uncouple respiration from ADP phosphorylation, allowing respiration to occur at the maximum rate, unaffected by ATP/ADP demands. Carbonyl cyanide *p*-trifluoromethoxyphenylhydrazone (FCCP) is one popular example of an uncoupler. Antimycin A, an inhibitor of complex II, can also be added at the end to stop all mitochondrial respiration so that O_2 consumption from other sources can be determined. Although respiration in intact cells is more complex (Brand and Nicholls, 2011), these compounds can also be used to measure mitochondrial respiration in cultured cells in which the concentrations of ADP and substrate immediately surrounding the mitochondria cannot be controlled (Brand and Nicholls, 2011). When measurement of respiration is not feasible, it may be useful to measure activity of the electron transport chain complexes themselves (Birch-Machin and Turnbull, 2001). Alternatively, immunoblots for electron transport chain subunits can be done. However, it is important to remember that the latter two approaches could be influenced by other factors, such as mitophagy or mitochondrial dynamics.

BIOCHEMICAL METHODS

Various biochemical methods are available for the assessment of mitochondrial health or the measurement of mitochondrial damage in frozen tissue or cell lysates. One approach is to measure ATP levels in the samples. This has been used to show mitochondrial dysfunction caused by APAP (Jaeschke, 1990; Williams et al., 2011; Lee et al., 2014), alcohol (Cunningham and Bailey, 2001), cocaine (Boyer and Petersen, 1991), and other hepatotoxicants. However, it should be remembered that changes in hepatic ATP may reflect altered adenine nucleotide availability in addition to changes in mitochondrial respiration (Brand and Nicholls, 2011). Another option is to measure the release of mitochondrial proteins into the cytosol. Release of the intermembrane protein cytochrome c is commonly measured to assess mitochondrial damage because of the role it plays in apoptosis (Liu et al., 1996; Yin et al., 1999; Feng and Kaplowitz, 2002; Knight and Jaeschke, 2002; Miller et al., 2008; McGill et al., 2015), though release of cytochrome c can also occur during non-apoptotic injury when there is mitochondrial damage (Knight and Jaeschke, 2002; McGill et al., 2015). Release of other mitochondrial proteins can occur as well. In APAP hepatotoxicity, the endonucleases apoptosis-inducing factor and endonuclease G are released and translocate into the nucleus as a result of damage to mitochondria (Bajt et al., 2006).

Oxidative stress can also be measured in mitochondria. This can be achieved in several different ways. Glutathione disulfide (GSSG), a product of GSH oxidation, can be measured by isolating mitochondria from the liver and quickly re-suspending the mitochondrial pellets in a cold acid solution in order to prevent spontaneous oxidation of reduced glutathione (GSH) in the samples before freezing. The samples can then be stored at $-80°C$ and GSSG can later be measured using either a commercial kit or a modified Tietze assay (Jaeschke and Mitchell, 1990; Jaeschke, 1990). This approach has been used to demonstrate mitochondrial oxidative stress during APAP-induced liver injury (Jaeschke, 1990; Knight et al., 2001, 2002). Oxidative modifications of mitochondrial DNA (mtDNA) can also be measured as markers of mitochondrial damage. One commonly used biomarker of oxidative DNA damage is 8-hydroxy-2′-deoxyguanosine (8-OHdG). This marker has been shown to increase in mitochondria isolated from the livers of rodents treated either acutely or chronically with alcohol (Wieland and Lauterburg, 1995; Cahill et al., 1997, 1999), in allograft recipient mice (Nagakawa et al., 2005), and even in humans with nonalcoholic fatty liver disease (Nomoto et al., 2008) and extrahepatic cholestasis (Xu et al., 2012). Chronic or particularly severe oxidative mtDNA damage can also manifest as mtDNA mutations and deletions (Shigenaga et al., 1994), and in some acute cases oxidative mtDNA lesions may interfere with the detection of mtDNA (Mansouri et al., 1999; Cover et al., 2005). Such mtDNA deletions and depletion have been observed in mice with liver stress or injury caused by alcohol (Mansouri et al., 1999; Larosche et al., 2010), APAP (Cover et al., 2005), and endotoxin (Choumar et al., 2011). These kinds of defects have also been observed in samples from patients with alcoholic liver disease (Mansouri et al., 1997a) and Wilson disease (Mansouri et al., 1997b).

Another tissue marker that can sometimes serve as a measure of mitochondrial damage is protein nitration. Protein nitration occurs when peroxynitrite (ONOO$^-$) and its breakdown products react with amino acid residues like tyrosine. Importantly, peroxynitrite forms when superoxide (O_2^-) reacts with nitric oxide (NO). Because mitochondria are the major source of superoxide in some forms of liver injury, protein nitration can sometimes serve as evidence of mitochondrial injury (Knight et al., 2001, 2002; Cover et al., 2005).

CIRCULATING BIOMARKERS

Until recently, little effort had been made to study mitochondrial damage in humans with liver injury or to determine whether or not these mechanisms learned from rodents actually translate to humans, despite the large number of liver diseases believed to involve or affect mitochondria. One reason for this was the lack of convenient, noninvasive biomarkers that can provide information regarding the pathophysiological mechanisms of disease. The first useful tests for liver damage were developed in the early twentieth century. These assays, such as the phthalein clearance tests (Rosenthal, 1922; Rosenthal and White, 1925), the cephalin-cholesterol flocculation test (Hanger, 1938), coagulation time, and serum bilirubin levels were really the measures of hepatic function. They were also laborious and time consuming. The first true serum biomarkers of liver injury were the transaminases (alanine aminotransferase [ALT] and aspartate aminotransferase [AST]), which were developed in the 1950s (Karmen et al., 1955; LaDue and Wroblewski, 1956) and have become the most common clinical biomarkers of liver injury. However, serum levels of these enzymes often cannot accurately predict the outcome for liver injury patients (Antoine et al., 2012; McGill et al., 2014). Also, although they are generally considered markers of oncotic necrosis because the primary mechanism of release of these aminotransferases into circulation is presumed to be leakage through the compromised membranes of dying cells, these aminotransferases do not provide any information regarding the pathophysiology of disease. Newer biomarkers are currently being developed for the latter purpose, including biomarkers of mitochondrial damage (Figure 11.2).

MITOCHONDRIAL MEMBRANE PERMEABILITY OR DAMAGE

Similar to leakage of aminotransferases from dying cells, there is some evidence that mitochondrial content can leak out of damaged mitochondria into the cytosol, and then become detectable in circulation upon cell death. Permeabilization of the outer mitochondrial membrane can occur by translocation of the pro-apoptotic B-cell lymphoma 2 (Bcl-2) family proteins, such as Bax and the truncated form of Bid (García-Sáez et al., 2010). As previously discussed, intermembrane proteins like cytochrome c and endonucleases can exit through these pores and signal in other parts of the cell (García-Sáez et al., 2010; Norberg et al., 2010). In fact, diffusion of cytochrome c into the cytosol and later into extracellular space has been directly observed using immunohistochemical techniques (Miller et al., 2008). Importantly, however, these Bcl-2 proteins are not necessarily required for the leakage of intermembrane proteins. The mitochondrial membrane permeability transition pore (MPTP) is generally thought to span both the inner and outer membranes of mitochondria (Baines, 2009). The composition of the pore is controversial, but it is generally agreed that cyclophilin D is an important component (Baines, 2009). Some release of mitochondrial matrix metabolites can occur through this pore, but it is too narrow to allow passage of most proteins. However, the influx of water into the mitochondria that results from the MPTP opening causes swelling that distends the mitochondrial membranes and eventually causes the outer membrane to burst (Baines, 2009), releasing cytochrome c and other intermembrane proteins. In cases of very severe damage, there is some evidence that both mitochondrial membranes can rupture. In fact, disruption of both the inner and outer membranes has been directly observed by electron microscopy in mouse liver mitochondria during APAP hepatotoxicity (Placke et al., 1987). Thus, apoptotic signaling is not required for the release of mitochondrial content.

Two isoforms of AST exist, one of which is predominantly found within liver mitochondria and is sometimes measured as a biomarker of mitochondrial damage (Panteghini et al., 1984). There are some experimental data showing that mitochondrial AST is only released into the extracellular space after mitochondrial damage and permeability have developed (Shimizu et al., 1994; Zhou et al., 2008). However, possibly due to the difficulty of measuring the subcellular isoforms and the lack of commercially available kits for this purpose, the measurement of mitochondrial AST was never widely adopted for mechanistic studies in humans. More recent data show two other mitochondrial matrix components can be easily measured in serum as circulating biomarkers of severe mitochondrial damage. It was demonstrated in mice that mtDNA and glutamate dehydrogenase (GLDH) increase in serum during APAP hepatotoxicity, but not during the liver injury caused by furosemide that does not involve mitochondria (McGill et al., 2012). Moreover, these mitochondrial damage biomarkers were found to be higher in serum from non survivors of APAP-induced acute liver failure than in serum from survivors (McGill et al., 2014a). Interestingly, preliminary data have recently been presented showing that urine mtDNA may be a useful biomarker of renal mitochondrial damage (Stallons et al., 2014). Importantly, the idea that mitochondrial damage occurs during APAP hepatotoxicity in humans has recently been confirmed by studies using primary human hepatocytes and HepaRG cells (McGill et al., 2011; Xie et al., 2014). Nuclear DNA fragments have also been measured in serum of APAP overdose patients as an indication of mitochondrial damage (McGill et al., 2012, 2014a) because nuclear DNA fragmentation during APAP hepatotoxicity is known to require mitochondrial permeability leading to the release of mitochondrial endonucleases (Bajt et al., 2006, 2008). However, the latter cannot be recommended as a biomarker of mitochondrial damage for general use because it will depend upon the exact mechanism of DNA damage. In fact, one group has shown that different fragmentation patterns result from different mechanisms of cell death, and that these patterns can be detected in serum (Jahr et al., 2001). Apoptosis results in much smaller DNA fragments than APAP-induced necrosis (Jahr et al., 2001). It may be the case that large DNA pieces in serum indicate mitochondrial damage and mitochondrial endonuclease-mediated fragmentation. However, this needs to be further explored. More recently, the matrix enzyme carbamoyl phosphate synthetase 1 (CPS-1) has been shown to be elevated in serum of liver injury patients and it was suggested that it may also be specific for mitochondrial damage (Weerasinghe et al., 2014). Although the latter has not been directly tested, another study did show that CPS-1 leaks out of damaged mitochondria into the cytosol (Brown et al., 2014).

Recent studies have shown that numerous miRNAs are elevated in serum during liver injury in both rodents and humans (Wang et al., 2009; Starkey Lewis et al., 2011; Ward et al., 2014; Krauskopf et al., 2015). Interestingly, this may actually aid diagnosis in the clinic, as serum miRNA profiles can distinguish between liver injuries due to different etiologies (Ward et al., 2014). In addition, it has been shown that some miRNAs are enriched specifically within liver mitochondria (Kren et al., 2009). Thus, it is possible that mitochondrial miRNAs could also be measured in the circulation as specific biomarkers of mitochondrial damage, though this idea has not yet been tested.

One caveat of measuring biomarkers of mitochondrial damage that depend upon membrane damage is that any intact mitochondria present in the serum could affect the results. One way to deal with this is to pellet intact mitochondria in freshly acquired serum by high-speed centrifugation (Jaeschke and McGill, 2013). It has been argued, however, that intact mitochondria could not remain polarized in serum due to the very high Ca^{2+} concentrations (~10,000-fold greater level of free Ca^{2+} than intracellular levels) (Kholmukhamedov et al., 2013). However, this has not been thoroughly tested in the laboratory. While it is likely that mitochondria do depolarize in serum, the MPTP only allows the passage of molecules with molecular weight ≤1.5 kDa, and it is not clear whether the mitochondrial membranes would eventually burst in serum conditions. Thus, it is possible that depolarized mitochondria are still capable of sequestering large macromolecules like mtDNA. This needs to be investigated further.

MITOCHONDRIAL METABOLITES

Proteins and nucleic acids are not the only types of biomarkers of mitochondrial damage that can be measured in circulation during liver injury. When mitochondria are damaged or otherwise compromised, various low molecular weight metabolites that are normally processed by mitochondria can accumulate in the cells and sometimes become detectable in serum. For example, acylcarnitine levels are altered in the circulation of patients with mitochondrial fatty acid oxidation (mFAO) disorders (Rinaldo et al., 2002). Normally, most fatty acids in the cytosol are conjugated with carnitine and transported into mitochondria where they undergo β-oxidation. However, in mFAO disorders, either fatty acid transport or oxidation is compromised and the acylcarnitines accumulate. Similarly, there is evidence that acylcarnitine levels in serum increase during APAP hepatotoxicity in mice as a result of mitochondrial damage (Chen et al., 2009; McGill et al., 2014b). Although an increase was not seen in APAP overdose patients, mouse experiments revealed that this was likely due to the standard-of-care treatment N-acetylcysteine (NAC) (McGill et al., 2014b). Another study found that pediatric patients who received NAC at later time points had higher serum acylcarnitines than patients who received NAC earlier after overdose (Bhattarcharyya et al., 2014). These findings support the use of acylcarnitines as serum biomarkers of mitochondrial dysfunction during liver injury. While little else has been done to develop metabolomic biomarkers specifically for mitochondrial damage in the liver, attempts have been made in other tissues. For example, one study found that creatine is elevated in both the medium of cultured muscle cells during mitochondrial dysfunction and in plasma from patients with respiratory chain disorders (Shaham et al., 2010), and changes in urinary mitochondrial metabolite profiles have been observed in patients with diabetic kidney disease (Sharma et al., 2013). Considerable work is still needed in this area to determine whether or not other mitochondrial metabolites are useful as circulating biomarkers of mitochondrial dysfunction in the liver.

MITOCHONDRIAL OXIDATIVE STRESS

Several biomarkers of oxidative stress and lipid peroxidation are available. One example that is sometimes considered to be specific for oxidative stress in mitochondria is oxidized cardiolipin. Cardiolipin is a phospholipid with a unique structure that is primarily localized to the mitochondrial inner membrane (Paradies et al., 2014). Oxidation of the acyl chains of cardiolipin is thought to be caused by hydroxyl radical (•OH) which can form from the breakdown of hydrogen peroxide H_2O_2, itself a product of O_2^- dismutation. Oxidized cardiolipin is considered by many to be a sensitive indicator of mitochondrial dysfunction and oxidative stress because of the tight associations between the lipid and the electron transport chain complexes and because it usually includes unsaturated fatty acids that readily react (Paradies et al., 2014). Oxidized cardiolipin can be directly measured using mass spectrometry (Martens et al., 2015). However, oxidation of samples during processing and storage can cause inaccurate results. As a result, a better approach may be to measure serum levels of antibodies against oxidized cardiolipin that are produced by the host. The latter has been used to demonstrate the possibility of oxidative stress in patients with alcoholic liver disease (Rolla et al., 2001). It has also been shown that oxidized cardiolipin is elevated by alcohol consumption in hepatitis C patients (Rigamonti et al., 2003).

CONCLUSION

Mitochondrial damage is thought to play a role in numerous liver diseases, as well as acute liver injury of various etiologies. A number of different methods are available for the assessment of mitochondrial damage or dysfunction. The method of choice will depend upon both the mechanism of injury being investigated and the type of sample that is available. When possible, the direct measurement of mitochondrial respiration is preferable to almost any other method. Unfortunately, the latter cannot be done in translational research studies. Instead, recent advances in the development of mechanistic biomarkers have made it possible to test for mitochondrial damage using patient serum samples. Serum mitochondrial damage biomarkers will continue to be developed and refined in the future.

REFERENCES

Antoine, D.J., Jenkins, R.E., Dear, J.W., Williams, D.P., McGill, M.R., Sharpe, M.R., Craig, D.G. et al. 2012. Molecular forms of HMGB1 and keratin-18 as mechanistic biomarkers for mode of cell death and prognosis during clinical acetaminophen hepatotoxicity. *J Hepatol*. 56(5):1070–1079.

Baines, C.P. 2009. The molecular composition of the mitochondrial permeability transition pore. *J Mol Cell Cardiol*. 46(6):850–857.

Bajt, M.L., Cover, C., Lemasters, J.J., Jaeschke, H. 2006. Nuclear translocation of endonuclease G and apoptosis-inducing factor during acetaminophen-induced liver cell injury. *Toxicol Sci*. 94(1):217–225.

Bajt, M.L., Farhood, A., Lemasters, J.J., Jaeschke, H. 2008. Mitochondrial Bax translocation accelerates DNA fragmentation and cell necrosis in a murine model of acetaminophen hepatotoxicity. *J Pharmacol Exp Ther*. 324(1):8–14.

Bhattacharyya, S., Yan, K., Pence, L., Simpson, P.M., Gill, P., Letzig, L.G., Beger, R.D. et al. 2014. Targeted liquid chromatography-mass spectrometry analysis of serum acylcarnitines in acetaminophen toxicity in children. *Biomark Med*. 8(2):147–159.

Birch-Machin, M.A., Turnbull, D.M. 2001. Assaying mitochondrial respiratory complex activity in mitochondria isolated from human cells and tissues. *Methods Cell Biol*. 65:97–117.

Boyer, C.S., Petersen, D.R. 1991. Hepatic biochemical changes as a result of acute cocaine administration in the mouse. *Hepatology*. 14(6):1209–1216.

Brand, M.D., Nicholls, D.G. 2011. Assessing mitochondrial dysfunction in cells. *Biochem J*. 435(2):297–312.

Brown, J.M., Kuhlman, C., Terneus, M.V., Labenski, M.T., Lamyaithong, A.B., Ball, J.G., Valentovic, M.A. 2014. *S*-adenosyl-L-methionine protection of acetaminophen mediated oxidative stress and identification of hepatic 4-hydroxynonenal protein adducts by mass spectrometry. *Toxicol Appl Pharmacol*. 281(2):174–184.

Cahill, A., Stabley, G.J., Wang, X., Hoek, J.B. 1999. Chronic ethanol consumption causes alterations in the structural integrity of mitochondrial DNA in aged rats. *Hepatology*. 30(4):881–888.

Cahill, A., Wang, X., Hoek, J.B. 1997. Increased oxidative damage to mitochondrial DNA following chronic ethanol consumption. *Biochem Biophys Res Commun*. 235(2):286–290.

Chance, B., Williams, G.R. 1955. A simple and rapid assay of oxidative phosphorylation. *Nature*. 175(4469):1120–1121.

Chen, C., Krausz, K.W., Shah, Y.M., Idle, J.R., Gonzalez, F.J. 2009. Serum metabolomics reveals irreversible inhibition of fatty acid beta-oxidation through the suppression of PPARα activation as a contributing mechanism of acetaminophen-induced hepatotoxicity. *Chem Res Toxicol*. 22(4):699–707.

Choumar, A., Tarhuni, A., Lettéron, P., Reyl-Desmars, F., Dauhoo, N., Damasse, J., Vadrot, N. et al. 2011. Lipopolysaccharide-induced mitochondrial DNA depletion. *Antioxid Redox Signal*. 15(11):2837–2854.

Cover, C., Mansouri, A., Knight, T.R., Bajt, M.L., Lemasters, J.J., Pessayre, D., Jaeschke, H. 2005. Peroxynitrite-induced mitochondrial and endonuclease-mediated nuclear DNA damage in acetaminophen hepatotoxicity. *J Pharmacol Exp Ther*. 315(2): 879–887.

Cunningham, C.C., Bailey, S.M. 2001. Ethanol consumption and liver mitochondria function. *Biol Signals Recept*. 10(3–4):271–282.

Das, S., Hajnóczky, N., Antony, A.N., Csordás, G., Gaspers, L.D., Clemens, D.L., Hoek, J.B., Hajnóczky, G. 2012. Mitochondrial morphology and dynamics in hepatocytes from normal and ethanol-fed rats. *Pflugers Arch*. 464(1):101–109.

Ernster, L., Schatz, G. 1981. Mitochondria: A historical review. *J Cell Biol*. 91(3 Pt 2):227s–255s.

Feng, G., Kaplowitz, N. 2002. Mechanism of staurosporine-induced apoptosis in murine hepatocytes. *Am J Physiol Gastrointest Liver Physiol*. 282(5):G825–G834.

Galloway, C.A., Lee, H., Brookes, P.S., Yoon, Y. 2014. Decreasing mitochondrial fission alleviates hepatic steatosis in a murine model of nonalcoholic fatty liver disease. *Am J Physiol Gastrointest Liver Physiol*. 307(6):G632–G641.

Gane, E.J., Weilert, F., Orr, D.W., Keogh, G.F., Gibson, M., Lockhart, M.M., Frampton, C.M., Taylor, K.M., Smith, R.A., Murphy, M.P. 2010. The mitochondria-targeted anti-oxidant mitoquinone decreases liver damage in a phase II study of hepatitis C patients. *Liver Int*. 30(7):19–26.

García-Ruiz, C., Kaplowitz, N., Fernandez-Checa, J.C. 2013. Role of mitochondria in alcoholic liver disease. *Curr Pathobiol Rep*. 1(3):159–168.

García-Sáez, A.J., Fuertes, G., Suckale, J., Salgado, J. 2010. Permeabilization of the outer mitochondrial membrane by Bcl-2 proteins. *Adv Exp Med Biol*. 677:91–105.

Halestrap, A.P. 1989. The regulation of the matrix volume of mammalian mitochondria in vivo and in vitro and its role in the control of mitochondrial metabolism. *Biochim Biophys Acta*. 973(3):355–382.

Hanger, F.M. 1938. The flocculation of cephalin-cholesterol emulsions by pathological sera. *Trans Assoc Am Phys*. 53:148.

Jaeschke, H. 1990. Glutathione disulfide formation and oxidant stress during acetaminophen-induced hepatotoxicity in mice in vivo: The protective effect of allopurinol. *J Pharmacol Exp Ther*. 255(3):935–941.

Jaeschke, H., McGill, M.R. 2013. Serum glutamate dehydrogenase—Biomarkers for liver cell death or mitochondrial dysfunction? *Toxicol Sci*. 134(1):221–222.

Jaeschke, H., Mitchell, J.R. 1990. Use of isolated perfused organs in hypoxia and ischemia/reperfusion oxidant stress. *Methods Enzymol*. 186:752–759.

Jaeschke, H., McGill, M.R., Ramachandran, A. 2012. Oxidant stress, mitochondria, and cell death mechanisms in drug-induced liver injury: Lessons learned from acetaminophen hepatotoxicity. *Drug Metab Rev*. 44(1):88–106.

Jahr, S., Hentze, H., Englisch, S., Hardt, D., Fackelmayer, F.O., Hesch, R.D., Knippers, R. 2001. DNA fragments in the blood plasma of cancer patients: Quantitations and evidence for their origin from apoptotic and necrotic cells. *Cancer Res*. 61(4):1659–1665.

Karmen, A., Wroblewski, F., LaDue, J.S. 1955. Transaminase activity in human blood. *J Clin Invest*. 34(1):126–131.

Kholmukhamedov, A., Schwartz, J.M., Lemasters, J.J. 2013. Isolated mitochondria infusion mitigates ischemia-reperfusion injury of the liver in rats: Mitotracker probes and mitochondrial membrane potential. *Shock*. 39(6):543.

Knight, T.R., Jaeschke, H. 2002. Acetaminophen-induced inhibition of Fas receptor-mediated liver cell apoptosis: Mitochondrial dysfunction versus glutathione depletion. *Toxicol Appl Pharmacol*. 181(2):133–141.

Knight, T.R., Ho, Y.S., Farhood, A., Jaeschke, H. 2002. Peroxynitrite is a critical mediator of acetaminophen hepatotoxicity in murine livers: Protection by glutathione. *J Pharmacol Exp Ther*. 303(2):468–475.

Knight, T.R., Kurtz, A., Bajt, M.L., Hinson, J.A., Jaeschke, H. 2001. Vascular and hepatocellular peroxynitrite formation during acetaminophen toxicity: Role of mitochondrial oxidant stress. *Toxicol Sci*. 62(2):212–220.

Kon, K., Kim, J.S., Jaeschke, H., Lemasters, J.J. 2004. Mitochondrial permeability transition in acetaminophen-induced necrosis and apoptosis of cultured mouse hepatocytes. *Hepatology*. 40(5):1170–1179.

Korenaga, M., Wang, T., Li, Y., Showalter, L.A., Chan, T., Sun, J., Weinman, S.A. 2005. Hepatitis C virus core protein inhibits mitochondrial electron transport and increases reactive oxygen species (ROS) production. *J Biol Chem*. 280(45):37841–37848.

Krauskopf, J., Caiment, F., Claessen, S.M., Johnson, K.J., Warner, R.L., Schomaker, S.J., Burt, D.A., Aubrecht, J., Kleinjans, J.C. 2015. Application of high-throughput sequencing to circulating microRNAs reveals novel biomarkers for drug-induced liver injury. *Toxicol Sci*. 143(2):268–276.

Kren, B.T., Wong, P.Y., Sarver, A., Zhang, X., Zeng, Y., Steer, C.J. 2009. MicroRNAs identified in highly purified liver-derived mitochondria may play a role in apoptosis. *RNA Biol*. 6(1):65–72.

Kweon, Y.O., Paik, Y.H., Schnabl, B., Qian, T., Lemasters, J.J., Brenner, D.A. 2003. Gliotoxin-mediated apoptosis of activated human hepatic stellate cells. *J Hepatol*. 39(1):38–46.

LaDue, J.S., Wroblewski, F. 1956. Serum glutamic pyruvic transaminase SGP-T in hepatic disease: A preliminary report. *Ann Intern Med*. 45(5):801–811.

Larosche, I., Lettéron, P., Berson, A., Fromenty, B., Huang, T.T., Moreau, R., Pessayre, D., Mansouri, A. 2010. Hepatic mitochondrial DNA depletion after an alcohol binge in mice: Probable role of peroxynitrite and modulation by manganese superoxide dismutase. *J Pharmacol Exp Ther*. 332(2):886–897.

Lee, K.K., Imaizumi, N., Chamberland, S.R., Alder, N.N., Boelsterli, U.A. 2014. Targeting mitochondria with methylene blue protects mice against acetaminophen-induced liver injury. *Hepatology*. 61(1):326–336.

Liu, X., Kim, C.N., Yang, J., Jemmerson, R., Wang, X. 1996. Induction of apoptotic program in cell-free extracts: Requirement for dATP and cytochrome c. *Cell*. 86(1):147–157.

Mansouri, A., Fromenty, B., Berson, A., Robin, M.A., Grimbert, S., Beaugrand, M., Erlinger, S., Pessayre, D. 1997a. Multiple hepatic mitochondrial DNA deletions suggest premature oxidative aging in alcoholic patients. *J Hepatol*. 27(1):96–102.

Mansouri, A., Gaou, I., De Kergeunec, C., Amsellem, S., Haouzi, D., Berson, A., Moreau, A. et al. 1999. An alcoholic binge causes massive degradation of hepatic mitochondrial DNA in mice. *Gastroenterology*. 117(1):181–190.

Mansouri, A., Gaou, I., Fromenty, B., Berson, A., Lettéron, P., Degott, C., Erlinger, S., Pessayre, D. 1997b. Premature oxidative aging of hepatic mitochondrial DNA in Wilson's disease. *Gastroenterology*. 113(2):599–605.

Martens, J.C., Keilhoff, G., Halangk, W., Wartmann, T., Gardemann, A., Päge, I., Schild, L. 2015. Lipidomic analysis of molecular cardiolipin species in livers exposed to ischemia/reperfusion. *Mol Cell Biochem*. 400(1–2):253–263.

McGill, M.R., Du, K., Xie, Y., Bajt, M.L., Ding, W.X., Jaeschke, H. 2015. The role of the c-Jun N-terminal kinases 1/2 and receptor-interacting protein kinase 3 in furosemide-induced liver injury. *Xenobiotica*. 45(5):442–449.

McGill, M.R., Li, F., Sharpe, M.R., Williams, C.D., Curry, S.C., Ma, X., Jaeschke, H. 2014b. Circulating acylcarnitines as biomarkers of mitochondrial dysfunction after acetaminophen overdose in mice and humans. *Arch Toxicol*. 88(2):391–401.

McGill, M.R., Sharpe, M.R., Williams, C.D., Taha, M., Curry, S.C., Jaeschke, H. 2012. The mechanism underlying acetaminophen-induced hepatotoxicity in humans and mice involves mitochondrial damage and nuclear DNA fragmentation. *J Clin Invest*. 122(4):1574–1583.

McGill, M.R., Staggs, V.S., Sharpe, M.R., Lee, W.M., Jaeschke, H.; Acute Liver Failure Study Group. 2014a. Serum mitochondrial biomarkers and damage-associated molecular patterns are higher in acetaminophen overdose patients with poor outcome. *Hepatology*. 60(4):1336–1345.

McGill, M.R., Yan, H.M., Ramachandran, A., Murray, G.J., Rollins, D.E., Jaeschke, H. 2011. HepaRG cells: A human model to study mechanisms of acetaminophen hepatotoxicity. *Hepatology*. 53(3):974–982.

Meyers, L.L., Beierschmitt, W.P., Khairallah, E.A., Cohen, S.D. 1988. Acetaminophen-induced inhibition of hepatic mitochondrial respiration in mice. *Toxicol Appl Pharmacol*. 93(3):378–387.

Miller, T.J., Knapton, A., Adeyemo, O., Noory, L., Weaver, J., Hanig, J.P. 2008. Cytochrome c: A non-invasive biomarkers of drug-induced liver injury. *J Appl Toxicol*. 28(7):815–828.

Nagakawa, Y., Williams, G.M., Zheng, Q., Tsuchida, A., Aoki, T., Montgomery, R.A., Klein, A.S., Sun, Z. 2005. Oxidative mitochondrial DNA damage and deletion in hepatocytes of rejecting liver allografts in rats: Role of TNF-α. *Hepatology*. 42(2):208–215.

Ni, H.M., Williams, J.A., Jaeschke, H., Ding, W.X. 2013. Zonated induction of autophagy and mitochondrial spheroids limits acetaminophen-induced necrosis in the liver. *Redox Biol*. 1(1):427–432.

Nicholls, D.G. 2012. Fluorescence measurement of mitochondrial membrane potential changes in cultured cells. *Methods Mol Biol*. 810:119–133.

Nomoto, K., Tsuneyama, K., Takahashi, H., Murai, Y., Takano, Y. 2008. Cytoplasmic fine granular expression of 8-hydroxydeoxyguanosine reflects early mitochondrial oxidative DNA damage in nonalcoholic fatty liver disease. *Appl Immunohisochem Mol Morphol*. 16(1):71–75.

Norberg, E., Orrenius, S., Zhivotovsky, B. 2010. Mitochondrial regulation of cell death: Processing of apoptosis-induced factor (AIF). *Biochem Biophys Res Commun*. 396(1):95–100.

Okuda, M., Li, K., Beard, M.R., Showalter, L.A., Scholle, F., Lemon, S.M., Weinman, S.A. 2002. Mitochondrial injury, oxidative stress, and antioxidant gene expression are induced by hepatitis C virus core protein. *Gastroenterology*. 130(7):2087–2098.

O'Rourke, B. 2010. From bioblasts to mitochondria: Ever expanding roles of mitochondria in cell physiology. *Front Physiol*. 1:7.

Panteghini, M., Malchiodi, A., Calarco, M., Bonora, R. 1984. Clinical and diagnostic significant of aspartate aminotransferase isoenzymes in sera of patients with liver diseases. *J Clin Chem Clin Biochem*. 22(2):153–158.

Paradies, G., Paradies, V., De Benedicts, V., Ruggiero, F.M., Petrosillo, G. 2014. Functional role of cardiolipin in mitochondrial bioenergetics. *Biochim Biophys Acta*. 1837(4):408–417.

Pernelle, K., Le Guevel, R., Glaise, D., Stasio, C.G., Le Charpentier, T., Bouaita, B., Corlu, A., Guguen-Guillouzo, C. 2011. Automated detection of hepatotoxic compounds in human hepatocuytes using HepaRG cells and image-based analysis of mitochondrial dysfunction with JC-1 dye. *Toxicol Appl Pharmacol*. 254(3):256–266.

Pessayre, D., Fromenty, B., Berson, A., Robin, M.A., Lettéron, P., Moreau, R., Mansouri, A. 2012. Central role of mitochondria in drug-induced liver injury. *Drug Metab Rev.* 44(1):34–87.

Petersen, P., Vilstrup, H. 1979. Relation between liver function and hepatocyte ultrastructure in a case of paracetamol intoxication. *Digestion.* 19(6):415–419.

Placke, M.E., Ginsberg, G.L., Wyand, D.S., Cohen, S.D. 1987. Ultrastructural changes during acute acetaminophen-induced hepatotoxicity in the mouse: A time and dose study. *Toxicol Pathol.* 15(4):431–438.

Ramachandran, A., McGill, M.R., Xie, Y., Ni, H.M., Ding, W.X., Jaeschke, H. 2013. Receptor interacting protein kinase 3 is a critical early mediator of acetaminophen-induced hepatocyte necrosis in mice. *Hepatology.* 58(6):2099–2108.

Rigamonti, C., Mottaran, E., Reale, E., Rolla, R., Cipriani, V., Capelli, F., Boldorini, R., Vidali, M., Sartotori, M., Albano, E. 2003. Moderate alcohol consumption increases oxidative stress in patients with chronic hepatitis C. *Hepatology.* 38(1):42–49.

Rinaldo, P., Matern, D., Bennett, M.J. 2002. Fatty acid oxidation disorders. *Annu Rev Physiol.* 64:477–502.

Rolla, R., Vay, D., Mottaran, E., Parodi, M., Vidali, M., Sartori, M., Rigamonti, C., Bellomo, G., Albano, E. 2001. Antiphospholipid antibodies associated with alcoholic liver disease specifically recognize oxidized phospholipids. *Gut.* 49(6):852–859.

Rosenthal, S.M. 1922. An improved method for using phenoltetrachlorphthalein as a liver function test. *J Pharmacol Exp Ther.* 19:385–391.

Rosenthal, S.M., White, E.C. 1925. Clinical application of the bromosulphalein test for hepatic function. *JAMA.* 84(15):1112–1114.

Shaham, O., Slate, N.G., Goldberger, O., Xu, Q., Ramanathan, A., Souza, A.L., Clish, C.B., Sims, K.B., Mootha, V.K. 2010. A plasma signature of human mitochondrial disease revealed through metabolic profiling of spent media from cultured muscle cells. *Proc Natl Acad Sci USA.* 107(4):1571–1575.

Sharma, K., Karl, B., Mathew, A.V., Gangoiti, J.A., Wassel, C.L., Saito, R., Pu, M. et al. 2013. Metabolomics reveals signature of mitochondrial dysfunction in diabetic kidney disease. *J Am Soc Nephrol.* 24(11):1901–1912.

Shigenaga, M.K., Hagen, T.M., Ames, B.N. 1994. Oxidative damage and mitochondrial decay in aging. *Proc Natl Acad Sci USA.* 91(23):10771–10778.

Shimizu, S., Kamiike, W., Hatanaka, N., Nishimura, M., Miyata, M., Inoue, T., Yoshia, Y., Tagawa, K., Matsuda, H. 1994. Enzyme release from mitochondria during reoxygenation of rat liver. *Transplantation.* 57(1):144–148.

Sokol, R.J., Devereaux, M., Mierau, G.W., Hambridge, K.M., Shikes, R.H. 1990. Oxidant injury to hepatic mitochondrial lipids in rats with dietary copper overload. Modification by vitamin E deficiency. *Gastroenterology.* 99:1061–1071.

Stallons, L.J., Arthur, J., Beeson, C., Schnellmann, R. 2014. Urinary mitochondrial DNA as a novel biomarker of mitochondrial dysfunction in human acute kidney injury. *Toxicologist (Toxicol Sci Suppl).* 12:24.

Starkey Lewis, P.J., Dear, J., Platt, V., Simpson, K.J., Craig, D.G., Antoine, D.J., French, N.S. et al. 2011. Circulating microRNAs as potential markers of human drug-induced liver injury. *Hepatology.* 54(5):1767–1776.

Sternlieb, I. 1968. Mitochondrial and fatty changes in hepatocytes of patients with Wilson's disease. *Gastroenterology.* 55(3):354–367.

Sternlieb, I., Quintana, N., Volenberg, I., Schilsky, M.L. 1995. An array of mitochondrial alteration in the hepatocytes of Long-Evans Cinnamon rats. *Hepatology.* 22(6):1782–1787.

Tanji, K., Bonilla, E. 2008. Light microscopic methods to visualize mitochondria on tissue sections. *Methods.* 46(4):274–280.

Theruvath, T.P., Zhong, Z., Pediaditakis, P., Ramshesh, V.K., Currin, R.T., Tikunov, A., Holmuhamedov, E., Lemasters, J.J. 2008. Minocycline and *N*-methyl-4-isoleucine cyclosporing (NIM811) mitigate storage/reperfusion injury after rat liver transplantation through suppression of the mitochondrial permeability transition. *Hepatology*. 47(1):236–246.

Wang, K., Zhang, S., Marzolf, B., Troisch, P., Brightman, A., Hu, Z., Hood, L.E., Galas, D.J. 2009. Circulating microRNAs, potential biomarkers for drug-induced liver injury. *Proc Natl Acad Sci USA*. 106(11):4402–4407.

Ward, J., Kanchagar, C., Veksler-Lublinsky, I., Lee, R.C., McGill, M.R., Jaeschke, H., Curry, S.C., Ambros, V.R. 2014. Circulating microRNA profiles in human patients with acetaminophen hepatotoxicity or ischemic hepatitis. *Proc Natl Acad Sci USA*. 111(33):12169–12174.

Weerasinghe, S.V., Jang, Y.J., Fontana, R.J., Omary, M.B. 2014. Carbamoyl phosphate synthetase-1 is a rapid turnover biomarker in mouse and human acute liver injury. *Am J Gastrointest Liver Physiol*. 307(3):G355–G364.

Wieland, P., Lauterburg, B.H. 1995. Oxidation of mitochondrial proteins and DNA following administration of ethanol. *Biochem Biophys Res Commun*. 213(3):815–819.

Williams, C.D., Koerner, M.R., Lampe, J.N., Farhood, A., Jaeschke, H. 2011. Mouse strain-dependent caspase activation during acetaminophen hepatotoxicity does not result in apoptosis or modulation of inflammation. *Toxicol Appl Pharmacol*. 257(3):449–458.

Xie, Y., McGill, M.R., Dorko, K., Kumer, S.C., Schmitt, T.M., Forster, J., Jaeschke, H. 2014. Mechanisms of acetaminophen-induced cell death in primary human hepatocytes. *Toxicol Appl Pharmacol*. 279(3):266–274.

Xu, S.C., Chen, Y.B., Lin, H., Pi, H.F., Zhang, N.X., Zhao, C.C., Shuai, L. et al. 2012. Damage to mtDNA in liver injury of patients with extrahepatic cholestasis: The protective effects of mitochondrial transcription factor A. *Free Radic Biol Med*. 52(9):1543–1551.

Yin, X.M., Wang, K., Gross, A., Zhao, Y., Zinkel, S., Klocke, B., Roth, K.A., Korsmeyer, S.J. 1999. Bid-deficient mice are resistant to Fas-induced hepatocellular apoptosis. *Nature*. 400(6747):886–891.

Yu, T., Wang, L., Lee, H., O'Brien, D.K., Bronl, S.F., Gores, G.J., Yoon, Y. 2014. Decreasing mitochondrial fission prevents cholestatic liver injury. *J Biol Chem*. 289(49):34074–34088.

Zhang, X., Lemasters, J.J. 2013. Translocation of iron from lysosomes to mitochondria during ischemia predisposes to injury after reperfusion in rat hepatocytes. *Free Radic Biol Med*. 63:243–253.

Zhou, Y.H., Shi, D., Yuan, B., Sun, Q.J., Jiao, B.H., Sun, J.J., Miao, M.Y. 2008. Mitochondrial ultrastructure and release of proteins during liver regeneration. *Indian J Med Res*. 128(2):157–164.

Zischka, H., Lichtmannegger, J., Schmitt, S., Jägemann, N., Schulz, S., Wartini, D., Jennen, L. et al. 2011. Liver mitochondrial membrane crosslinking and destruction in a rat model of Wilson disease. *J Clin Invest*. 121(4):1508–1518.

Zoratti, M., Szabó, I. 1995. The mitochondrial permeability transition. *Biochim Biophys Acta*. 1241(2):139–176.

12 Alterations of Mitochondrial DNA in Liver Diseases

Eric A. Schon and Bernard Fromenty

CONTENTS

ABSTRACT

Mitochondrial DNA (mtDNA) is a tiny double-stranded circular DNA located within the mitochondria of most tissues in mammals. Numerous copies of mtDNA exist in a single cell and its replication occurs continuously, even in cells that do not divide. Another key feature of mtDNA is its high sensitivity to oxidative damage, in comparison to the nuclear genome. Notably, this genome encodes for 13 polypeptides of the respiratory chain, which is mandatory for ATP synthesis and the oxidation of the cofactors NADH and $FADH_2$ generated by numerous cellular metabolic pathways. Hence, any significant qualitative and/or quantitative alterations in mtDNA can have dire consequences on tissue homeostasis and the life of affected individuals. The first part of this chapter summarizes basic concepts associated with mtDNA structure and organization; its replication, transcription, and translation; and the genetics of mitochondrial diseases. The second part focuses on the different mtDNA alterations that can be observed in liver diseases, whether they are inherited or acquired. Indeed, hepatic mtDNA point mutations, deletions, and depletion can be observed in different mitochondrial cytopathies (e.g., mtDNA depletion syndrome, Pearson's syndrome), Wilson disease, alcoholic liver diseases, nonalcoholic fatty liver disease, hepatocellular carcinoma, and liver injuries induced by drugs such as the antiretroviral nucleotide reverse transcriptase inhibitors, fialuridine, linezolid, and acetaminophen. In some of these diseases, hepatic mtDNA alterations may not have direct pathological consequences, but rather reflect the presence of a strong oxidative stress in the liver. On the other hand, mtDNA alterations in other diseases can lead to steatosis (i.e., fatty liver), hepatic cytolysis, cirrhosis, and liver failure.

MITOCHONDRIAL GENETICS, BIOENERGETICS, AND DISEASES

MITOCHONDRIAL DNA AND OXIDATIVE PHOSPHORYLATION

It is commonly thought that all of the information present in each of our cells resides in the DNA contained within the nucleus. In fact, DNA is located in a second organelle as well, namely the mitochondrion, or more precisely, in the hundreds or thousands of mitochondria present in the cytoplasm of each cell. This dual localization of DNA results in two types of genetics: Mendelian genetics for nuclear DNA (nDNA) and population genetics for mitochondrial DNA (mtDNA). Notably, mutations in genes in either compartment can cause mitochondrial disease, as described following.

Mitochondria are approximately the size of bacteria (i.e., ~1 μm). Unique among subcellular organelles, they have two membranes: an outer membrane (MOM) and an inner membrane (MIM). The region between the MOM and MIM is called the *intermembrane space* (IMS), while the central region within the MIM (the "cytoplasm" of the organelle) is called the *matrix*. The tiny size and double-membraned structure of mitochondria are due to the fact that the precursors of mitochondria were prokaryotes that became endiosymbiotic residents of "proto-eukaryotic" cells early in our evolutionary history (Martin, 2010). It is most likely that mitochondria and mtDNA were retained in eukaryotic cells because of their role in oxidative energy metabolism, via the mitochondrial respiratory chain (MRC)/oxidative phosphorylation (OxPhos) system.

The OxPhos system consists of five linked complexes embedded within the MIM, all designed to produce ATP—the energy currency of the cell—from the food we eat (Figure 12.1) (Schon et al., 2012). The chemiosmotic hypothesis, proposed by Peter Mitchell in the 1960s (Mitchell and Moyle, 1967) posits that protons (derived from NADH and $FADH_2$, both produced mainly in the tricarboxylic acid [TCA] cycle and the β-oxidation pathway) are pumped across the MIM (i.e.,

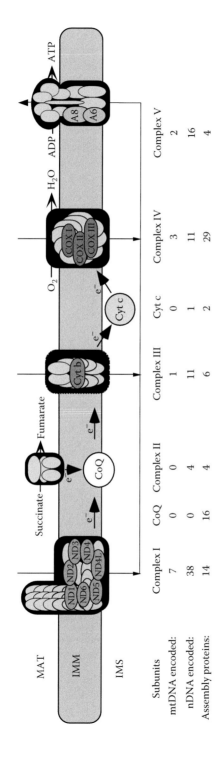

FIGURE 12.1 (See color insert.) The mitochondrial respiratory chain. Blue subunits are encoded by nuclear DNA (nDNA) whereas colored subunits are encoded by mitochondrial DNA (mtDNA). "Assembly proteins" are ancillary polypeptides which are required for the assembly and/or stability of the holocomplex; they are all nDNA encoded. See the text for further description. Other abbreviations used in this figure: ADP, adenosine diphosphate; ATP, adenosine triphosphate; CoQ, coenzyme Q; Cyt c, cytochrome c; e⁻, electron; H⁺, proton.

Subunits	Complex I	CoQ	Complex II	Complex III	Cyt c	Complex IV	Complex V
mtDNA encoded:	7	0	0	1	0	3	2
nDNA encoded:	38	0	4	11	1	11	16
Assembly proteins:	14	16	4	6	2	29	4

from the matrix to the IMS), thereby generating a proton gradient across the MIM. That proton gradient is then dissipated in the opposite direction (i.e., flowing back across the MIM from the IMS to the matrix) to synthesize ATP from ADP and free phosphate. The proton pumping from the matrix to the IMS is accomplished by complexes I, III, and IV, while ATP is generated as protons flow through complex V (ATP synthetase). This "vertical" flow of protons across the MIM is accompanied by the "horizontal" transfer of an equal number of electrons, first from complexes I and II to ubiquinone (an electron carrier, also called coenzyme Q [CoQ]), then to complex III, then to cytochrome c (a second electron carrier), then to complex IV (also called cytochrome c oxidase [COX]), and finally to molecular oxygen to produce water. Complexes I, III, IV, and V contain subunits encoded by both nDNA and mtDNA. Notably, the *only* polypeptides encoded by mtDNA are subunits of these four complexes (Figure 12.2). Complex II (which, besides its electron transfer

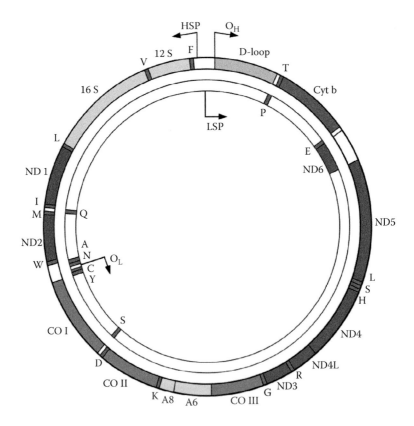

FIGURE 12.2 Human mitochondrial DNA. The genes for the 12S and 16S ribosomal RNAs, the subunits of complexes I (NADH-coenzyme Q oxidoreductase [ND]), III (cytochrome b [Cyt b]), IV (cytochrome c oxidase [CO]), and V (ATP synthase [A]), and 22 tRNAs (1-letter nomenclature) are shown as the origins of mtDNA replication from the so-called "heavy" (O_H) and "light" (O_L) strands, and the corresponding origins of transcription (HSP, LSP).

activity, is the succinate dehydrogenase component of the TCA cycle) contains only nDNA-encoded subunits.

More than 1700 nDNA-encoded gene products are targeted to human mitochondria (i.e., they are synthesized in the cytosol and are then imported into the organelle) (Area-Gomez and Schon, 2014). Of these, about half are required for the well-being and maintenance of the organelle itself (e.g., mitochondrial structure and integrity; replication, transcription, and translation of mtDNA; importation of proteins; transport of small molecules), and are present in the mitochondria of essentially every cell. The other half, on the other hand, are associated with specialized functions such as amino acid metabolism, fatty acid β-oxidation, OxPhos, and apoptosis; many of these are present in the mitochondria of some tissues but not of others. While mutations in any of these gene products could potentially cause pathology, mitochondrial diseases are typically defined as those associated only with defects in OxPhos, either directly (e.g., mutation in a subunit of the respiratory chain) or indirectly (e.g., mutation in a MRC "assembly" protein). Thus, mutations in both nDNA- and mtDNA-encoded genes can give rise to a mitochondrial disorder.

Replication of mtDNA

Replication of any circular DNA requires an "origin" of replication that allows for the synthesis of daughter strand DNA from both the "top" and "bottom" parental strands (called "heavy" and "light" in human mtDNA). As shown in Figure 12.2, the mtDNA origin has apparently been physically separated into two "halves," with an *origin of heavy-strand replication* (O_H) located at "12-o'clock" on the circle and an *origin of light-strand replication* (O_L) located at "8-o'clock," each controlling synthesis of one of the two daughter DNA strands. This unusual architecture has given rise to two competing models of human mtDNA replication, using the mitochondrial-specific DNA polymerase, called polymerase γ (to distinguish it from polymerases α, β, δ, and ε, which control replication and repair of nDNA).

In the "asynchronous, strand displacement" model (Brown et al., 2005), synthesis of one strand begins at O_H, located at the top of the circle, and proceeds continuously in a clockwise direction. Synthesis of the other strand begins at O_L and proceeds continuously in a counterclockwise direction. In the "synchronous, strand-coupled" model (Holt and Reyes, 2012), replication initiates from O_H and proceeds symmetrically in both a clockwise and counterclockwise direction, similar to the classic "theta-replication" mechanism by which circular bacterial genomes divide.

In both models, replication terminates with the creation of a pair of catenated circles (each circle is a double helix containing one "old" parental strand and one "new" daughter strand). Mitochondrial DNA topoisomerase then decatenates the circles, releasing the two daughter molecules.

As discussed in the following section, there are a number of Mendelian mitochondrial disorders that are due to errors in the components of the replication machinery or in the regulation of the organellar nucleotide pools that are required for replication.

TRANSCRIPTION AND TRANSLATION OF mtDNA

Transcription of human mtDNA is "prokaryotic-like". All 37 genes encoded by human mtDNA (see Figure 12.2) are synthesized initially as two large polycistronic precursor transcripts, one (specifying 2 rRNAs, 14 tRNAs, and 12 polypeptides) encoded by the H-strand and starting at the heavy strand promoter (HSP), and the other (specifying 8 tRNAs and 1 polypeptide) by the L-strand and starting at the light strand promoter (LSP). Because the precursor RNAs are single stranded, the tRNA genes that "punctuate" the circle at strategic positions adopt their typical "cloverleaf" conformation and are cleaved at their 5' and 3' ends by two ribonucleases, thereby releasing the tRNAs and the flanking rRNAs and mRNAs (Ojala et al., 1981).

Following cleavage of the precursor polycistronic RNAs, the 3' termini of the mRNAs are polyadenylated, the trinucleotide CCA is added to the 3' termini of the tRNAs, and specific tRNA bases are chemically modified (e.g., conversion of uridine to pseudouridine). These are all accomplished by nDNA-encoded enzymes that are imported into the mitochondria. Translation of the mature mRNAs takes place on mitochondrial ribosomes, which consist of mtDNA-encoded 12S and 16S rRNAs together with imported ribosomal proteins (Christian and Spremulli, 2012). Unlike the translation of nDNA-encoded mRNAs in the cytoplasm, the upstream end of the message contains no added recognition structure ("cap") for translation initiation. Moreover, there is another conceptual problem for translation initiation, as the initiation codon is located at the very beginning of the mature message, with little or no 5'-untranslated region; thus it is not clear how the ribosome finds and binds to the message.

Translation termination at the downstream (3') end of the message is equally baroque. The mRNAs messages often end with a U or a UA; addition of the poly(A) tail to the mRNA converts these terminal nucleotides to UAA, which is a translational stop codon. Because the human mitochondrial genetic code differs from the "universal" code at four positions (AUA encodes Met instead of Ile; UGA encodes Trp instead of Stop; and AGA and AGG encode Stop instead of Arg), termination at mRNAs ending with AGA and AGG is accomplished by an unusual "ribosomal frameshifting" mechanism (Temperley et al., 2010).

As with replication, mutations in proteins involved in transcription and translation have been found to cause mitochondrial diseases (DiMauro et al., 2013).

GENETICS OF MITOCHONDRIAL DISEASES

Mutations in nDNA-encoded genes cause classical Mendelian-inherited diseases (DiMauro et al., 2013). Most prominent among these is Leigh syndrome (LS), a fatal neurodegenerative disorder of infancy. Children with LS typically have a general failure to thrive, psychomotor regression, hypotonia, seizures, and respiratory abnormalities. Liver dysfunction and failure can also be observed in some children (Lee et al., 2009; Quinonez et al., 2013). Neuropathologically, LS is characterized by bilateral symmetrical lesions in the central nervous system, neuronal loss, reactive astrocytosis, and proliferation of cerebral microvessels. LS due to nDNA mutations follows standard Mendelian inheritance patterns, most frequently autosomal-recessive inheritance.

The example of LS as a mitochondrial disease is an interesting one, because it can also be caused by mutations in mtDNA, especially in those genes encoding structural subunits of complex I, IV, and V (DiMauro et al., 2013). In this case, however, we are dealing with population genetics, for the simple reason that mtDNAs are present in multiple copies in each cell—hundreds or even thousands—depending on that cell's bioenergetic requirements. In the normal situation, all the mtDNAs in a cell or tissue are identical, a situation known as homoplasmy. However, in the case of a disease due to a mutation in mtDNA, both normal and mutated mtDNAs may coexist within the same cell or tissue, a situation known as heteroplasmy.

Most mtDNA-based diseases are functionally "recessive," in the sense that the mutant load must typically be very high (e.g., ≥80%) before clinical pathology ensues. In most situations, those mutated mtDNAs came from the mother, because human mtDNAs (and mitochondria) are maternally inherited. But quite often a heteroplasmic mother does not have the disease, because the mutation load in her tissues is below the threshold for dysfunction (typically, it will be ~50%–70%).

The threshold is the level of mutation above which pathology occurs, mainly because there are not enough normal mtDNAs to compensate for one or more missing or functionally deficient MRC subunits derived from the mutated mtDNAs. This threshold applies whether the mutation is in a protein-coding gene (e.g., COX II) or in a protein synthesis gene (e.g., a tRNA). Note, however, that in this example, a mutation in COX II compromises only complex IV function, whereas a mutation in a tRNA compromises the translation of all 13 mtDNA-encoded polypeptides (Schon et al., 2012).

The fact is that the mutant load in a heteroplasmic mother is below the threshold (~50%–70%), whereas that in her affected baby is above the threshold (~80%–90%) highlights another aspect of mitochondrial population genetics: the bottleneck effect. Only a subset of the mtDNAs in the mother's germline repopulate the cells of the child. Of the ~1000 mtDNAs in the mother's primordial germ cells, only about 20–200 "segregating units" eventually repopulate the baby (the term "segregating unit" is used instead of "mtDNA" because each mitochondrion contains ~5 mtDNAs; thus multiple mtDNAs can be inherited together as a "unit"). Of course, because of this random process, the child of a mother with 70% mutation may inherit less than 70% mutation, in which case the child likely will be normal.

There is a second mitochondrial bottleneck in postnatal development. Following oogenesis, the oocyte contains ~100,000 mtDNAs (all amplified from those 20–200 segregating units in the oogonia). At that point, and following fertilization, mtDNA replication and mitochondrial division cease do not recommence until after the blastocyst implantation stage (64–128 cells). Since only a few cells will actually become the fetus (the vast majority of the blastocyst gives rise to the extraembryonic tissues), those ~100,000 mtDNAs are reduced ~100-fold. Thus, any mutated mtDNAs that passed through the germline bottleneck must then pass through the postimplantation developmental bottleneck. The issues of where in development these bottlenecks operate are still a subject of much debate (Carling et al., 2011; Shoubridge and Wai, 2007). From a developmental and evolutionary point of view, these bottlenecks make sense, as they (together with maternal inheritance) help ensure that mtDNA mutations do not spread through the population. However, the price that is paid for this evolutionary "protection" is that mutations are confined to individual pedigrees in which potentially fatal mtDNA mutations are transmitted maternally. Note, however, that because they are confined to maternal lineages, these mutations cannot spread "horizontally" through the population.

Even if a fetus inherits high levels of a pathogenic mtDNA mutation, it is still not guaranteed that the child will be affected postnatally, due to yet another unusual aspect of mitochondrial population genetics: mitotic segregation. After cell division mitochondria are parsed randomly into the two daughter cells, in roughly equal amounts (i.e., ~50:50), the total number of organelles is halved. Following cell division, the genomes replicate and "double-up" to their pre-division copy number (Tang et al., 2000). This partitioning of mitochondria and mtDNAs during cell division is essentially random. In normal individuals who are homoplasmic (i.e., all mtDNAs are identical), the randomness of the partitioning is "invisible" because all the daughter cells contain the same mtDNA genotype. However, in heteroplasmic individuals, the distribution of the mutation in daughter cells is stochastic: most cells will have approximately the same mutation load as the parental cell, but some daughter cells will have more, and others fewer, mutated mtDNAs. This "genetic drift" can occur not only in cells that divide relatively frequently (e.g., blood, liver), but also in cells that divide rarely (e.g., muscle, brain), with clinical consequences.

An excellent example of mitotic segregation is the connection between Pearson's marrow-pancreas syndrome (also referred as Pearson's syndrome) and Kearns–Sayre syndrome (KSS). Pearson's syndrome, a pancytopenia with liver involvement, is due to a spontaneous large-scale (~2–10 kb) partial deletion of mtDNA arising during oogenesis or early embryonic development, with the mtDNA deletion present mainly in hematopoietic tissues. As a result of mitotic segregation, genetic drift, and differential rates of cell division in different tissues, there is a reduction of the mutation load in bone marrow and blood, but an inexorable increase in more slowly dividing tissues, such as muscle and brain (in which mtDNAs still replicate within their nondividing host cells), and perhaps in liver as well (Bianchi et al., 2011), giving rise to KSS, a systemic and fatal multisystem disorder (Fromenty et al., 1997; Obara-Moszynska et al., 2013).

These population genetic principles—mitotic segregation, the threshold effect, maternal inheritance, and the bottleneck—help to explain the remarkably dynamic nature of heteroplasmy at the cellular, tissue, and organismic levels, and give rise to a wide range of disorders.

ALTERATIONS OF mtDNA IN LIVER DISEASES

Genetic Diseases with Associated Hepatopathies

mtDNA Depletion Syndrome

MDS (mtDNA depletion syndrome) is defined as a decrease in the mtDNA copy number in different tissues leading to severe MRC dysfunction. Among the main clinical phenotypes of MDS, a hepatocerebral form can be induced by mutations in nuclear genes encoding deoxyguanosine kinase (DGUOK), DNA polymerase γ (POLG) and a mitochondrial inner membrane protein (MPV17) involved in mtDNA maintenance

(El-Hattab and Scaglia, 2013; Lee and Sokol, 2007). Affected individuals with the hepatocerebral form of MDS have early progressive liver failure and neurologic manifestations such as hypotonia and peripheral neuropathy. Liver histopathology usually shows the presence of several lesions such as extensive microvesicular and macrovacuolar steatosis, ballooning degeneration of hepatocytes, cholestasis, and severe fibrosis, which can progress to cirrhosis (El-Hattab and Scaglia, 2013; Labarthe et al., 2005; Lee and Sokol, 2007; Spinazzola et al., 2008). Liver failure is usually the cause of death after several weeks or months, unless affected patients can benefit from liver transplantation (Ferrari et al., 2005; Grabhorn et al., 2014). Plasma biochemical alterations include hypoglycemia, lactic acidosis, hyperammonemia, increased transaminases, hyperbilirubinemia, and low prothrombine time test (El-Hattab and Scaglia, 2013; Labarthe et al., 2005; Lee and Sokol, 2007).

Hepatic mtDNA depletion in MDS is severe with residual mtDNA levels usually below 20% of the normal values, especially in patients with DGUOK and MPV17 mutations. Reduced mtDNA levels can also be observed in other tissues such as muscle, skin, and blood, although the mtDNA depletion is generally less striking (El-Hattab et al., 2010; Labarthe et al., 2005; Sarzi et al., 2007; Uusimaa et al., 2014). Severe mtDNA depletion in liver is the cause of the MRC deficiency, with a strong reduction of the activity of complexes I, III, IV, and V, whereas complex II activity is not (or moderately) affected (El-Hattab et al., 2010; Labarthe et al., 2005; Sarzi et al., 2007). Importantly, severe MRC deficiency is responsible for secondary alteration of the TCA cycle, thus leading to lactic acidosis, whereas the secondary alteration of the mitochondrial β-oxidation pathway plays a key role in hypoglycemia and microvesicular steatosis (Fromenty and Pessayre, 1995; Mandel et al., 2001).

It is noteworthy that liver failure in patients carrying mutations in the POLG and DGUOK can be triggered by different exogenous factors such as viral infections (Lutz et al., 2009; Shieh et al., 2009) and some drugs, especially the antiepileptic drug sodium valproate (Ferrari et al., 2005; Helbling et al., 2013; Stewart et al., 2010). Importantly, this drug is known to have deleterious effects on mitochondrial function, in particular by impairing the mitochondrial fatty acid oxidation and OxPhos (Fromenty and Pessayre, 1995; Haas et al., 1981; Porceddu et al., 2012). Hence, valproate administration and preexistent mutations in genes regulating mtDNA levels can present additive effects on mitochondrial dysfunction and trigger liver failure. This situation is reminiscent of Reye's syndrome, a severe disease occurring mostly in children and characterized by acute liver failure with microvesicular steatosis and noninflammatory encephalopathy. Indeed, at the time when aspirin was still used in children, Reye's syndrome could occur in young patients receiving this antipyretic agent for viral infections and with latent genetic defects in mitochondrial enzymes involved in the β-oxidation pathway, or in ureagenesis (Fromenty and Pessayre, 1995; Gosalakkal and Kamoji, 2008).

Genetic Hepatopathies with Large mtDNA Deletions

Although most genetic mitochondrial disorders characterized by a severe liver disease are associated with reduced amounts of mtDNA, some reported cases are related to large-scale mtDNA deletions. In this context, liver disease can be observed in children suffering from Pearson's syndrome, as already mentioned in section "Genetics of Mitochondrial Diseases." This syndrome is classically characterized by exocrine pancreatic dysfunction and pancytopenia with severe anemia, neutropenia, and thrombocytopenia (Rötig et al., 1990). However, some affected children also present a hepatic disease, which can lead to liver failure and death (Bianchi et al., 2011; Rötig et al., 1990, 1995). Liver histopathology can reveal the presence of different lesions such as microvesicular and macrovacuolar steatosis, ballooning degeneration of hepatocytes, cholestasis, and fibrosis (Bianchi et al., 2011; McShane et al., 1991; Morris et al., 1997; Rötig et al., 1990). Interestingly, the same hepatic lesions can also be observed in the hepatocerebral form of MDS, thus suggesting that mtDNA depletion and large-scale mtDNA deletions can induce similar pathological abnormalities. Importantly, large mtDNA deletions are deemed to be deleterious because they induce low levels of tRNAs required for adequate protein synthesis in mitochondria (Damas et al., 2014; Manfredi et al., 1997; Schon et al., 2012).

Children with Pearson's syndrome present a single large mtDNA deletion, most frequently the so-called "common 4977 bp deletion," although other single large-scale mtDNA deletions can also be observed (Bianchi et al., 2011; Rötig et al., 1990, 1995). These mtDNA deletions are most frequently sporadic as these rearrangements are usually not found in their parents and siblings (Bianchi et al., 2011; Rötig et al., 1990). The percentage of deleted mtDNA molecules in liver can be very high (>80%) in children with Pearson's syndrome presenting a hepatic disease (Bianchi et al., 2011). In addition to mtDNA deletions, mtDNA duplications with the same breakpoints can also be observed in different tissues including the liver (Muraki et al., 2001; Rötig et al., 1995). However, the direct pathogenic role of mtDNA duplications is unclear because these mtDNA molecules contain all the tRNAs required for translation. In contrast, it has been hypothesized that mtDNA duplication could be pathogenic in an indirect way by generating the corresponding mtDNA deletions via recombination events (Fromenty et al., 1997; Manfredi et al., 1997).

WILSON DISEASE

Wilson disease is an autosomal recessive illness induced by mutations in the *ATP7B* gene, which encodes a metal-transporting P-type adenosine triphosphatase (ATPase) involved in the transmembrane transport of copper within hepatocytes. These mutations thus result in reduced excretion of copper into the bile, leading to the accumulation of this metal within the liver and other tissues such as in the brain and kidneys. Hepatic copper accretion usually induces liver diseases such as chronic hepatitis and cirrhosis and occasionally acute liver failure (Roberts and Schilsky, 2008). Liver histology can show different types of nonspecific lesions including microvesicular and macrovacuolar steatosis, hepatocyte necrosis, inflammatory infiltrates, fibrosis, and macronodular cirrhosis (Davies et al., 1989; Roberts and Schilsky, 2008).

Several studies have shown marked ultrastructural abnormalities of liver mitochondria in patients with Wilson disease (Sternlieb and Feldmann, 1976; Zischka and Lichtmannegger, 2014), which could be a direct consequence of copper accumulation within these organelles (Zischka et al., 2011). Investigations in patients also revealed reduced activity of different MRC complexes in liver (Gu et al., 2000;

Murayama and Ohtake, 2009). A single, or occasionally two, large mtDNA deletions were detected in liver of young patients with Wilson disease, including the common 4977 bp deletion (Mansouri et al., 1997a). However, these mtDNA deletions were present only at very low levels, representing less than 0.01% of the total mtDNA copies (Mansouri et al., 1997a). Interestingly, this is reminiscent to what is observed in liver of old (but healthy) individuals (Lee et al., 1994; Simonetti et al., 1992). In addition to mtDNA deletions, a study also reported reduced hepatic mtDNA content in some patients with Wilson disease (Helbling et al., 2013). Although the exact mechanisms of mitochondrial dysfunction and mtDNA alterations in Wilson disease are still unclear, some clinical and experimental investigations suggested a possible role of copper-induced oxidative stress and lipid peroxidation (Nair et al., 2005; Sauer et al., 2011; Sokol et al., 1994; Zischka et al., 2011; Zischka and Lichtmannegger, 2014). Indeed, similarly to iron, copper is a potent prooxidant metal that can promote the generation of highly deleterious reactive oxygen species (ROS) such as hydroxyl radicals (OH•), thus leading to the oxidative damage of cellular lipids, proteins, and DNA (Alexandrova et al., 2007; Toyokuni and Sagripanti, 1996; Vidyashankar and Patki, 2010). In addition, reduced hepatic activities of major antioxidant enzymes (e.g., superoxide dismutases and catalase) were reported in patients suffering from Wilson disease (Nagasaka et al., 2006). The role of oxidative stress in the occurrence of large mtDNA deletions and other mtDNA alterations will be discussed in more details in the next section.

ALCOHOLIC LIVER DISEASES

Excessive alcohol consumption is leading to different types of liver lesions including microvesicular and macrovacuolar steatosis, necroinflammation, extensive fibrosis, cirrhosis, and hepatocellular carcinoma. A large body of evidence indicates that oxidative stress is a key mechanism whereby ethanol induces acute and chronic liver injury (Leung and Nieto, 2013). Indeed, ethanol abuse is associated with ROS overproduction due to the induction of cytochrome P450 2E1 (CYP2E1) and mitochondrial dysfunction (Cahill et al., 2002; Leung and Nieto, 2013; Robin et al., 2005). In addition, alcohol intoxication reduces ROS detoxication, in particular, by reducing glutathione levels within the cytosol and mitochondria (Garcia-Ruiz and Fernandez-Checa, 2006).

Alcohol-induced ROS formation can damage different cell constituents such as proteins, lipids, and mtDNA (Fromenty and Pessayre, 1995). Several studies have detected multiple mtDNA deletions and point mutations in the liver of alcoholic patients (Caldwell et al., 1999; Fromenty et al., 1995; Kawahara et al., 2007; Mansouri et al., 1997b). Interestingly, the mtDNA deletions were especially detected in patients with microvesicular steatosis (Fromenty et al., 1995; Mansouri et al., 1997b), a liver lesion commonly induced by severe mitochondrial dysfunction (Fromenty and Pessayre, 1995; Massart et al., 2013). However, mtDNA point mutations and deletions were present in liver only at low heteroplasmy levels (<0.1%), similarly to what is observed in old individuals in this tissue (Lee et al., 1994; Liu et al., 1997; Simonetti et al., 1992). Thus, these mtDNA alterations are most probably not pathogenic as such but may rather reflect premature oxidative damage of the mitochondrial genome. Indeed, whereas point mutations could be caused by misreading of oxidized bases (e.g., 8-hydroxydeoxyguanosine [8-OH-dG]) during replication, large mtDNA deletions are thought to be generated during the repair of double-strand breaks (Chen et al., 2011; Richter, 1995). Finally, in addition to these mtDNA alterations, it is noteworthy that significant hepatic mtDNA depletion has also been reported in patients with a history of long-term alcohol drinking (Yin et al., 2004).

Different experimental investigations have been carried out in rodents to determine the detrimental effect of acute, repeated, and chronic ethanol intoxication on the liver mitochondrial genome.

Alcoholic Binge

Hepatic mtDNA was severely depleted in mice 2 h after the intragastric administration of a single binge of ethanol (5 g/kg) (Mansouri et al., 1999). This mtDNA depletion was prevented by manganese superoxide dismutase (MnSOD) overexpression in transgenic mice or by the administration to wild-type mice of the superoxide scavenger Tempol, the peroxynitrite scavenger uric acid or the nitric oxide (NO) synthase inhibitors 1400W and L-NAME (Larosche et al., 2010; Mansouri et al., 2010). These protective effects strongly suggest a role of NO reacting with the superoxide anion (O_2^-) to form the mtDNA-damaging peroxynitrite (Larosche et al., 2010; Mansouri et al., 2010). In addition to mtDNA depletion, some remaining mtDNA molecules harbored oxidative damages blocking the DNA polymerase such as abasic sites and double-strand breaks (Larosche et al., 2010; Mansouri et al., 1999, 2010). mtDNA was also strongly depleted in skeletal muscle, heart, and brain after the alcoholic binge (Mansouri et al., 2001). Notably, several studies suggested that damaged mtDNA molecules harboring numerous strand breaks could be rapidly degraded by mitochondrial endonucleases (Demeilliers et al., 2002; Ikeda and Ozaki, 1997; Kazak et al., 2012). However, mtDNA depletion in liver and other tissues was transient and normal mtDNA levels were restored after a few hours after the binge. The rapid de novo synthesis of mtDNA molecules from intact mtDNA templates could explain this normalization (Mansouri et al., 1999, 2001).

Repeated Alcohol Intoxication

Subsequent experiments in mice have shown that hepatic mtDNA depletion was prolonged, however, after 4 days of a once-a-day alcohol binge (Demeilliers et al., 2002). Indeed, in these repeatedly intoxicated animals, mtDNA depletion persisted for several days after alcohol deprivation (Demeilliers et al., 2002). Further investigations suggested that this prolonged mtDNA depletion was due to the accumulation of nonrepaired, bulky lesions impairing the progress of DNA polymerase γ on the remaining mtDNA templates and that these mtDNA lesions could be induced by lipid peroxidation products (Demeilliers et al., 2002). Prolonged mtDNA depletion was also associated to mitochondrial dysfunction and ultrastructural abnormalities, as well as microvesicular steatosis (Demeilliers et al., 2002).

Chronic Alcohol Intoxication

Excessive ethanol intake for several weeks with different intoxication protocols was also able to damage hepatic mtDNA in rodents with accumulation of oxidized bases (e.g., 8-OH-dG) and reduced mtDNA levels (Cahill et al., 1997, 1999; Zhang et al., 2010). Investigations with another protocol of chronic ethanol intoxication have shown that MnSOD overexpression in mice was associated with severe hepatic mtDNA depletion and damage, whereas these deleterious effects were not observed in wild-type animals (Larosche et al., 2009; Mansouri et al., 2010). Interestingly, ethanol intoxication was associated to higher hepatic iron levels only in the setting of MnSOD overexpression. The concomitant presence of high levels of hydrogen peroxide and iron could thus favor the production of the highly toxic hydroxyl radical, which is particularly deleterious for mtDNA (Larosche et al., 2009; Mansouri et al., 2010). Hence, the effect of MnSOD overexpression on ethanol-induced mtDNA damage is either protective or deleterious, respectively, in the context of acute or chronic intoxication (Larosche et al., 2009, 2010; Mansouri et al., 2010).

In summary, a number of clinical and experimental investigations have shown that acute, repeated, and chronic ethanol intoxication could be deleterious for the hepatic mitochondrial genome via the generation of ROS and reactive nitrogen species (RNS) (Figure 12.3). Among the different mtDNA lesions, mtDNA deletions

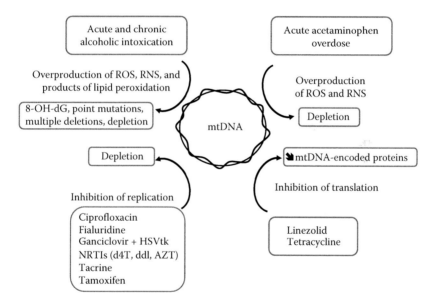

FIGURE 12.3 Alcohol- and drug-induced alterations of mtDNA homeostasis. Alcoholic intoxication and acute acetaminophen overdose can damage mtDNA by way of the overproduction of reactive oxygen species (ROS) and reactive nitrogen species (RNS). Toxic lipid peroxidation products could also be involved during alcoholic intoxication. All these deleterious compounds can directly or indirectly induce different types of mtDNA alterations such as 8-hydroxydeoxyguanosine (8-OH-dG), point mutations, multiple deletions, and depletion. Drugs belonging to different pharmacological classes can also alter mtDNA homeostasis by impairing mtDNA replication or traduction. Other abbreviations used in this figure: NRTIs, nucleoside reverse transcriptase inhibitors; d4T, stavudine; ddI, didanosine; AZT, zidovudine; HSVtk, Herpes simplex virus thymidine kinase.

and point mutations may not be pathogenic since only a small proportion of mtDNA molecules seems affected. In contrast, sustained mtDNA depletion could have a detrimental effect on mitochondrial function, in particular on substrate oxidation (Demeilliers et al., 2002). Although there is evidence that ethanol-induced mtDNA depletion is the consequence of mtDNA degradation by mitochondrial nuclease(s), other mechanisms could be involved. For instance, ethanol intoxication is able to reduce the expression of peroxisome proliferator-activated receptor-γ coactivator-1α (PGC-1α) (Lieber et al., 2008; Yin et al., 2012), a master regulator of mitochondrial biogenesis, function, and mtDNA levels in different tissues including the liver (Aharoni-Simon et al., 2011; Morris et al., 2012). Finally, it is noteworthy that ethanol intoxication is able to oxidatively alter not only the mtDNA but also other mitochondrial constituents such as proteins and membrane lipids (Kamimura et al., 1992; Mantena et al., 2007). All these oxidative damages could contribute to the impairment of mitochondrial function in the context of alcoholic intoxication (Fromenty and Pessayre, 1995; Mantena et al., 2007).

HEPATOCELLULAR CARCINOMA

Hepatocellular carcinoma (HCC), one of the most common cancers worldwide, presents a poor prognosis and high mortality rate. The main factors promoting HCC are hepatitis C virus (HCV), hepatitis B virus (HBV), chronic ethanol intoxication as well as obesity and nonalcoholic fatty liver disease (McGlynn and London, 2011). Several studies have found homoplasmic or heteroplasmic mtDNA point mutations, multiple deletions, and mtDNA depletion in tumor tissues of patients with HCC (Hsu et al., 2013; Lee et al., 2004; Tamori et al., 2004; Yamada et al., 2006; Yin et al., 2004, 2010). According to their localization within the RNA- and protein-coding regions of the mitochondrial genome, some of the detected mtDNA point mutations could be potentially pathogenic, and different mutations in the D-loop region are suspected to impair mtDNA replication (Hsu et al., 2013; Lee et al., 2004; Yin et al., 2010). However, it is still unclear whether these mtDNA alterations have a deleterious impact on protein synthesis as proteome analyses did not revealed reduced levels of mtDNA-encoded OxPhos proteins (Sun et al., 2007; Ye et al., 2013). It is also unknown whether the mtDNA alterations detected in HCC are impairing mitochondrial function and could play a role in the metabolic reprogramming (i.e., the so-called Warburg effect) observed in some cancers including HCC (Beyoglu et al., 2013; Huang et al., 2013).

A primary question which also remains unanswered is to know whether the detected mtDNA alterations are involved in the initiation and progression of HCC (Hsu et al., 2013), although some correlations were reported with several clinicopathologic parameters (Tamori et al., 2004; Yamada et al., 2006). An important findings is that some of the mtDNA alterations observed in HCC can also be observed in noncancerous liver tissues of alcoholic patients (as already mentioned), but also of patients with chronic HBV or HCV infection (Bäuerle et al., 2005; Nishikawa et al., 2001, 2005; Tamori et al., 2004; Wheelhouse et al., 2005). Thus, ethanol intoxication and chronic HBV/HCV infection seem to induce mtDNA alterations via oxidative stress independently of neoplastic transformation. Finally, an interesting observation is that mtDNA mutations were significantly more frequent in HCC with p53 mutations than in HCC with wild-type p53 (Tamori et al., 2004). This suggests that loss of mtDNA integrity could be a secondary consequence of mutations in nuclear-encoded tumor suppressors or oncogenes. Clearly, more investigations are required to determine the causes and the consequences of mtDNA alterations in HCC.

NONALCOHOLIC FATTY LIVER DISEASE

Nonalcoholic fatty liver disease (NAFLD) is currently one of the most frequent liver diseases in western countries, paralleling the continuous increase in the prevalence of obesity (Fabrini et al., 2010). In most cases, NAFLD follows a benign course with the presence of isolated fatty liver. In a few patients, however, nonalcoholic steatohepatitis (NASH) develops and can progress to cirrhosis and HCC (Bhala et al., 2013). There is extensive evidence that several metabolic and mitochondrial adaptations take place in fatty liver, which are deemed to limit fat accretion in the context of insulin resistance and higher de novo lipogenesis (Begriche et al., 2008, 2013). In contrast, mitochondrial dysfunction can be observed in the context of NASH, in particular, at the level of the different MRC complexes (Begriche et al., 2013; Pérez-Carreras et al., 2003). In addition, studies in patients with NASH showed the presence of mtDNA point mutations and large deletions, as well as increased 8-OH-dG levels (Caldwell et al., 1999; Kawahara et al., 2007; Nomoto et al., 2008). However, studies reporting mutations in NASH did not indicate the percentage of heteroplasmy. Thus, similarly to alcoholic intoxication, one cannot exclude the possibility that mtDNA mutations in NASH merely reflect premature oxidative damages secondary to higher mitochondrial ROS production (Begriche et al., 2013). More recently, investigations in individuals with NASH revealed epigenetic methylation of liver mtDNA, in particular in the ND6 gene, and this was associated with a strong reduction in ND6 mRNA levels (Pirola et al., 2013).

Several studies investigated hepatic mtDNA in rodent models of NAFLD. Higher levels of oxidative mtDNA damages including 8-OH-dG have been reported but mtDNA levels were found reduced, unchanged, or increased (Aubert et al., 2012; Begriche et al., 2013; Carabelli et al., 2011; Garcia-Ruiz et al., 2014; Valdecantos et al., 2012; Yuzefovych et al., 2013). Although the reasons of this discrepancy are still unclear, one hypothesis could be the differences in hepatic oxidative stress and mitochondrial adaptations between these models (Begriche et al., 2013). Importantly, these experimental studies did not investigate several time points during the progression of NAFLD. Hence, it is still unknown whether mtDNA alterations are occurring before mitochondrial dysfunction, if any. Further investigations would thus be needed to determine whether mtDNA alterations could directly participate in the pathophysiology of NAFLD, in particular regarding NASH during which oxidative stress and mitochondrial dysfunctions are more prominent compared to a simple fatty liver (Begriche et al., 2013).

DRUG-INDUCED LIVER INJURY

Different hepatotoxic drugs can induce mitochondrial damage and dysfunction (Begriche et al., 2011; Fromenty and Pessayre, 1995; Massart et al., 2013). Among these drugs, some of them are able to impair mtDNA homeostasis by different mechanisms (Figure 12.3). These drugs belong to different pharmacological classes.

Antiviral Drugs
Nucleotide Reverse Transcriptase Inhibitors
Nucleoside reverse transcriptase inhibitors (NRTIs) belong to a major family of drugs that have been developed for the treatment of AIDS. The main NRTIs include stavudine (d4T), zidovudine (AZT), didanosine (ddI), zalcitabine (ddC), lamivudine (3TC), and abacavir (ABC) (Fromenty and Pessayre, 1995; Vella et al., 2012). These drugs are 2′,3′-dideoxynucleoside analogues in which the hydroxyl group (–OH) in the 3′ position on the sugar ring is replaced by either an hydrogen atom or another group unable to form a phosphodiester linkage (Figure 12.4). The outstanding benefit of these antiretroviral drugs is unfortunately overshadowed by the occurrence of severe side effects in a significant number of HIV-infected patients. These adverse effects can involve different tissues such as the liver, heart, skeletal muscle, pancreas, and adipose tissue (Arnaudo et al., 1991; Brown and Glesby, 2011; Fromenty and Pessayre, 1995; Igoudjil et al., 2006). These side effects can be associated to hyperlactatemia, or lactic acidosis, reflecting a major impairment of the TCA cycle (Igoudjil et al., 2006; Matthews et al., 2011). NRTI-induced hepatotoxicity is particularly observed with d4T, ddI, and AZT. Indeed, these drugs can be responsible for acute forms of toxicity including cytolysis and cholestasis as well as chronic liver diseases such as steatosis, steatohepatitis, and cirrhosis (Fromenty and Pessayre, 1995; Nunez, 2010; Wang et al., 2013; Xiao et al., 2013).

It is now acknowledged that most of the side effects induced by NRTIs are the consequence of mitochondrial dysfunction (Fromenty and Pessayre, 1995; Gardner et al., 2013; Igoudjil et al., 2006). In particular, these drugs are able to inhibit mitochondrial DNA (mtDNA) replication, thus inducing mtDNA depletion and OxPhos impairment (Arnaudo et al., 1991; Igoudjil et al., 2006). Indeed, NRTIs are acting as chain terminators through their 5′OH group, which is incorporated into the growing chain of DNA, thanks to the DNA polymerase γ. If the NRTI molecule is not removed by the proofreading activity of this DNA polymerase, no other nucleotides can be incorporated because the DNA chain now lacks a 3′OH end (Figure 12.4). Actually, the ability of NRTIs to inhibit the DNA polymerase γ greatly varies among the analogues. Indeed, in vitro studies on purified DNA polymerase γ indicated that the potency of the inhibition is as follows: ddC > ddI > d4T >> 3TC > AZT > ABC (Jonhson et al., 2001; Kakuda, 2000). Interestingly, this order of potency is also observed in human liver as the so-called "D-drugs" (ddC, ddI, d4T) seem to inhibit DNA polymerase γ and reduce mtDNA levels more strongly than other NRTIs (Walker et al., 2004). Finally, it is noteworthy that some NRTIs such as AZT and d4T could have mitochondrial and metabolic effects through mechanisms unrelated to the inhibition of DNA polymerase γ and mtDNA depletion (Igoudjil et al., 2006, 2007, 2008; Lund et al., 2007).

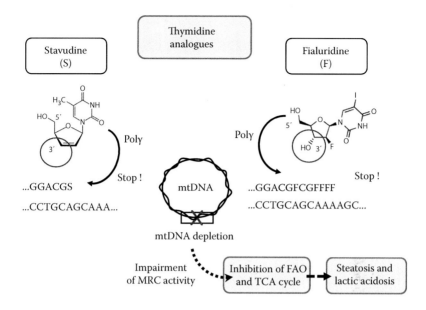

FIGURE 12.4 **(See color insert.)** Alteration of mtDNA synthesis by the thymidine analogues stavudine and fialuridine. The antiretroviral nucleoside reverse transcriptase inhibitor stavudine is a thymidine analogue in which the hydroxyl group (OH) in the 3′ position on the sugar ring is replaced by a hydrogen atom. Stavudine (S) can be incorporated into the growing chain of mtDNA by the DNA polymerase γ (Polγ). However, if stavudine is not removed by the proofreading activity of this DNA polymerase, mtDNA replication is blocked. Indeed, no other nucleotides can be incorporated after stavudine because the DNA chain now lacks a 3′OH end. The anti-HBV fialuridine is also a thymidine analogue but it was never marketed because of severe side effects and death of several patients during clinical trials. Unlike stavudine, fialuridine (F) can be incorporated into mtDNA without immediately stopping mtDNA replication since this drug carries a 3′OH group on the sugar moiety. However, when several adjacent molecules of fialuridine are successively incorporated into a growing chain of mtDNA, DNA Polγ is strongly inhibited. Stavudine and fialuridine-induced blockage of mtDNA replication can lead to mtDNA depletion, impairment of the mitochondrial respiratory chain (MRC) activity, and subsequent inhibition of fatty acid oxidation (FAO) and tricarboxylic acid (TCA) cycle. Alteration of mitochondrial FAO and TCA cycle can secondarily induce severe hepatic steatosis and lactic acidosis, respectively.

Fialuridine

Fialuridine was developed in the early 1990s for the treatment of chronic hepatitis B. However, clinical trials were prematurely interrupted because this drug induced in several patients unmanageable lactic acidosis, microvesicular steatosis, and severe liver failure requiring liver transplantation, or leading to death (Fromenty and Pessayre, 1995; Honkoop et al., 1997; McKenzie et al., 1995). After interruption of these trials, new in vivo and in vitro investigations showed that fialuridine strongly inhibits DNA polymerase γ and triggers mtDNA depletion by an unusual mechanism (Lewis et al., 1996, 1997; Semino-Mora et al., 1997). Indeed, unlike NRTIs such as d4T and AZT whose incorporation into a growing chain of mtDNA interrupts mtDNA replication, fialuridine can be incorporated into mtDNA without immediately terminating mtDNA replication. This is because fialuridine carries a 3′-hydroxyl group on the sugar moiety, thus allowing the subsequent incorporation of other nucleotides. However, when several adjacent molecules of fialuridine are successively incorporated into a growing chain of mtDNA, DNA polymerase γ is strongly inhibited (Figure 12.4). This subsequently blocks further mtDNA replication and leads to mtDNA depletion (Lewis et al., 1996).

Ganciclovir

The nucleoside analogue ganciclovir exhibits broad-spectrum activity against DNA viruses such as cytomegaloviruses, herpes simplex viruses, varicella-zoster virus, and Epstein–Barr virus. This drug can induce liver injury in some patients including cytolysis, cholestasis, and diffuse fibrosis (Biour et al., 2004; Shea et al., 1987; Wang et al., 2013). Ganciclovir can also be administered for the treatment of different cancers in association with an adenoviral vector harboring the gene encoding the Herpes simplex virus thymidine kinase (HSVtk) (Shalev et al., 2000; Sterman et al., 1998; Wirth et al., 2013). The principle of this "suicide gene therapy" is the HSVtk-mediated conversion of ganciclovir into ganciclovir-monophosphate, which is subsequently activated by cellular kinases into ganciclovir triphosphate, thus leading to the termination of DNA synthesis and cell death of the tumor cells (Wirth et al., 2013). However, this therapy can lead to different side effects such as skin rash, thrombocytopenia, anemia, and hepatotoxicity (Immonen et al., 2004; Shalev et al., 2000; Sterman et al., 1998). Experimental in vitro and in vivo studies have shown that HSVtk/ganciclovir-induced liver toxicity was associated with mitochondrial dysfunction and mtDNA depletion (Herraiz et al., 2003; van der Eb et al., 2003). Moreover, some of these investigations showed that ganciclovir is incorporated into both nuclear and mtDNA (Herraiz et al., 2003). Indeed, the acyclic, pseudo-sugar analogue of ganciclovir has two hydroxyl groups, so that incorporation of this drug into DNA does not terminate DNA replication. However, the incorporated ganciclovir molecules may distort the DNA helix and block the next round of DNA replication when the ganciclovir-modified DNA strand serves as a replication template (Thust et al., 2000).

Antibacterial Drugs

Linezolid

The oxazolidinone linezolid is an antibiotic that inhibits bacterial protein synthesis used against drug-resistant, gram-positive pathogens. This drug can induce after several weeks of treatment moderate to severe liver injury in some patients, with the occurrence of increased plasma transaminases and several types of hepatic lesions such as macrovacuolar and microvesicular steatosis as well as bile duct damage (De Bus et al., 2010; De Vriese et al., 2006; Garazzino et al., 2011). In addition, severe lactic acidosis has been observed in some of these patients and in others (Carson et al., 2007; De Bus et al., 2010; Velez and Janech, 2010). Linezolid can also induce neuropathy, skeletal myopathy, and renal failure after prolonged administration (Bressler et al., 2004; De Vriese et al., 2006). Different experimental and clinical investigations have shown that this drug is a potent inhibitor of mitochondrial protein synthesis, and consequently can decrease the activity of MRC complexes that contain mtDNA-encoded proteins (De Vriese et al., 2006; Garrabou et al., 2007; McKee et al., 2006; Nagiec et al., 2005; Soriano et al., 2005). Interestingly, the mtDNA A2706G polymorphism is suspected to favor linezolid-induced lactic acidosis (Carson et al., 2007; Del Pozo et al., 2014; Velez and Janech, 2010). However, further investigations will be needed in order to determine whether this mtDNA polymorphism also increases the risk of linezolid-induced liver injury.

Tetracycline

Tetracycline is a broad-spectrum antibiotic which prevents bacterial growth by inhibiting bacterial protein synthesis. This drug can induce different types of liver injury including cholestasis and microvesicular steatosis, in particular when it was administered at very high doses (Biour et al., 2004; Fromenty and Pessayre, 1995; Wang et al., 2013). Synthetic tetracyclines such as doxycycline and minocycline are now preferred to tetracycline because of easier dose schedules and faster gastrointestinal absorption when taken with food. Several experimental investigations have suggested that tetracycline-induced steatosis could be secondary to two different mechanisms: (1) reduced egress of triglycerides from the liver via an inhibition of the microsomal triglyceride transfer protein (MTP) and (2) inhibition of mitochondrial FAO (Fréneaux et al., 1988; Fromenty and Pessayre, 1995; Labbe et al., 1991; Lettéron et al., 2003). However, the exact mechanism whereby tetracycline inhibits mitochondrial FAO is still unknown. Although a direct inhibition of mitochondrial FAO enzymes is conceivable, it is noteworthy that tetracycline has been shown to inhibit the MRC activity (Yu et al., 2009) and mitochondrial protein synthesis (McKee et al., 2006).

Ciprofloxacin and Nalidixic Acid

Ciprofloxacin and nalidixic acid are 4-quinolone antibiotics that inhibit the bacterial DNA gyrases, which are type II topoisomerases. Ciprofloxacin can induce extensive hepatocellular cytolysis (with possible liver failure), cholestasis, and steatosis (Biour et al., 2004; Orman et al., 2011; Wang et al., 2013), whereas nalidixic acid seems to be less hepatotoxic (Biour et al., 2004). In cultured cells, both drugs progressively decrease mtDNA and impair mitochondrial respiration (Lawrence et al., 1993), while further investigations suggested that ciprofloxacin could impair mtDNA homeostasis via an inhibition of the type II topoisomerase present in mitochondria (Lawrence et al., 1996). Ciprofloxacin could also cause liver toxicity by way of the production of free radicals and lipid peroxidation, independently of mtDNA depletion (Gürbay et al., 2001; Weyers et al., 2002).

Other Drugs

Tamoxifen

Tamoxifen is an antiestrogenic drug used in the treatment of advanced breast cancer. This drug can cause acute and chronic liver injury, in particular, steatosis, steatohepatitis, and cirrhosis (Bruno et al., 2005; Massart et al., 2013; Wang et al., 2013).

Different studies have shown that this drug is able to impair mitochondrial function including the MRC and fatty acid oxidation (Cardoso et al., 2001, 2003; Tuquet et al., 2000). In addition to these effects, other investigations have shown that tamoxifen inhibited topoisomerases in vitro and decreased mtDNA synthesis leading to progressive hepatic mtDNA depletion after chronic treatment in mice (Larosche et al., 2007). Interestingly, tamoxifen is a cationic amphiphilic molecule that can be protonated within the mitochondrial IMS, similarly to other drugs such as amiodarone and perhexiline (Deschamps et al., 1994; Fromenty and Pessayre, 1995). All these drugs can be electrophoretically transported into the mitochondrial matrix by using the membrane potential $\Delta\Psi$, thus achieving high intramitochondrial concentrations (Deschamps et al., 1994; Fromenty et al., 1990; Larosche et al., 2007). However, it is still unknown whether perhexiline and amiodarone could alter mtDNA homeostasis, although some investigations showed that the latter drug induced nDNA damage after chronic treatment in rats (Almeida et al., 2008; Sakr et al., 2013).

Tacrine

Tacrine is a reversible cholinesterase inhibitor prescribed in patients with Alzheimer's disease. However, this drug increases plasma ALT activity in 50% of the recipients and can induce severe hepatocellular injury in some individuals (Watkins et al., 1994; Wang et al., 2013). Hepatic steatosis has also been reported (Biour et al., 2004). Several studies have shown that tacrine is able to uncouple OxPhos and inhibit MRC activity (Berson et al., 1996; Robertson et al., 1998; Zenger et al., 2013). Tacrine is also able to inhibit mtDNA synthesis in vitro and in vivo and induce severe mtDNA depletion after chronic treatment in mice (Mansouri et al., 2003; Robertson et al., 1998). Some investigations suggest that tacrine could impede mtDNA synthesis through topoisomerase inhibition (Mansouri et al., 2003).

Acetaminophen

Acetaminophen (N-acetyl-p-aminophenol, or paracetamol) is one of the most widely prescribed drugs for the management of pain and hyperthermia. Although acetaminophen is usually considered to be safe, the inadvertent or intentional ingestion of an excessive dose of this drug can cause massive hepatocellular necrosis and acute liver failure, which can be fatal (Antoine et al., 2013; Larson, 2007). Hepatic steatosis can also be observed after acetaminophen intoxication (Biour et al., 2004; Ramachandran and Kakar, 2009). Acetaminophen-induced liver injury is caused by cytochrome P450-mediated generation of the highly reactive metabolite N-acetyl-p-benzoquinone imine (NAPQI). Indeed, this electrophilic metabolite depletes hepatic glutathione, increases cell calcium, damages mitochondria, increases ROS formation, induces the expression of inducible NO synthase (thus increasing peroxynitrite formation), and activates c-Jun N-terminal kinase, to finally trigger liver cell necrosis (Fromenty, 2013; Jaeschke et al., 2012). Interestingly, hepatic mtDNA is rapidly depleted after an acetaminophen overdose in mice, possibly due to the mtDNA damage caused by peroxynitrite and other ROS (Figure 12.3) (Aubert et al., 2012; Cover et al., 2005). It is still unknown whether such mtDNA depletion plays a role in acetaminophen-induced steatosis. Although one cannot exclude this latter mechanism, experimental investigations suggested that acetaminophen overdosage could alter lipid homeostasis by impairing the expression and activity of peroxisome proliferator-activated receptor-α (PPARα) (Aubert et al., 2012). Last, it is noteworthy that nDNA and mtDNA fragments can be detected in the circulation of patients with acetaminophen-induced liver injury (McGill et al., 2012, 2014).

ACKNOWLEDGMENTS

This work was supported by grants from the U.S. National Institutes of Health (HD32062), the Department of Defense (W911NF-12-1-0159), the Muscular Dystrophy Association, the Ellison Medical Foundation, and the J. Willard and Alice S. Marriott Foundation. B. Fromenty would like to thank the Institut National de la Santé et de la Recherche Médicale (INSERM) for its constant support. B. Fromenty reports personal funds from the Medicines for Malaria Venture, Medicen Paris Region and Sigma-Tau, and a grant from Société Francophone du Diabète, outside the submitted work.

REFERENCES

Aharoni-Simon M, Hann-Obercyger M, Pen S, Madar Z, and Tirosh O. 2011. Fatty liver is associated with impaired activity of PPARγ-coactivator 1α (PGC1α) and mitochondrial biogenesis in mice. *Lab Invest* 91:1018–1028.

Alexandrova A, Kebis A, Mislanová C, and Kukan M. 2007. Copper impairs biliary epithelial cells and induces protein oxidation and oxidative DNA damage in the isolated perfused rat liver. *Exp Toxicol Pathol* 58:255–261.

Almeida MR, de Oliveira Lima E, da Silva VJ, Campos MG, Antunes LM, Salman AK, and Dias FL. 2008. Genotoxic studies in hypertensive and normotensive rats treated with amiodarone. *Mutat Res* 657:155–159.

Antoine DJ, Dear JW, Lewis PS, Platt V, Coyle J, Masson M, Thanacoody RH et al. 2013. Mechanistic biomarkers provide early and sensitive detection of acetaminophen-induced acute liver injury at first presentation to hospital. *Hepatology* 58:777–787.

Area-Gomez E and Schon EA. 2014. Mitochondrial genetics and disease. *J Child Neurol* 29:1208–1215.

Arnaudo E, Dalakas M, Shanske S, Moraes CT, DiMauro S, and Schon EA. 1991. Depletion of muscle mitochondrial DNA in AIDS patients with zidovudine-induced myopathy. *Lancet* 337:508–510.

Aubert J, Begriche K, Delannoy M, Morel I, Pajaud J, Ribault C, Lepage S et al. 2012. Differences in early acetaminophen hepatotoxicity between obese ob/ob and db/db mice. *J Pharmacol Exp Ther* 342:676–687.

Bäuerle J, Laguno M, Mauss S, Mallolas J, Murillas J, Miquel R, Schmutz G, Setzer B, Gatell JM, and Walker UA. 2005. Mitochondrial DNA depletion in liver tissue of patients infected with hepatitis C virus: Contributing effect of HIV infection? *HIV Med* 6:135–139.

Begriche K, Lettéron P, Abbey-Toby A, Vadrot N, Robin MA, Bado A, Pessayre D, and Fromenty B. 2008. Partial leptin deficiency favors diet-induced obesity and related metabolic disorders in mice. *Am J Physiol Endocrinol Metab* 294:E939–E951.

Begriche K, Massart J, Robin MA, Bonnet F, and Fromenty B. 2013. Mitochondrial adaptations and dysfunctions in nonalcoholic fatty liver disease. *Hepatology* 58:1497–1507.

Begriche K, Massart J, Robin MA, Borgne-Sanchez A, and Fromenty B. 2011. Drug-induced toxicity on mitochondria and lipid metabolism: Mechanistic diversity and deleterious consequences for the liver. *J Hepatol* 54:773–794.

Berson A, Renault S, Lettéron P, Robin MA, Fromenty B, Fau D, Le Bot MA et al. 1996. Uncoupling of rat and human mitochondria: A possible explanation for tacrine-induced liver dysfunction. *Gastroenterology* 110:1878–1890.

Beyoglu D, Imbeaud S, Maurhofer O, Bioulac-Sage P, Zucman-Rossi J, Dufour JF, and Idle JR. 2013. Tissue metabolomics of hepatocellular carcinoma: Tumor energy metabolism and the role of transcriptomic classification. *Hepatology* 58:229–238.

Bhala N, Jouness RI, and Bugianesi E. 2013. Epidemiology and natural history of patients with NAFLD. *Curr Pharm Des* 19:5169–5176.

Bianchi M, Rizza T, Verrigni D, Martinelli D, Tozzi G, Torraco A, Piemonte F et al. 2011. Novel large-range mitochondrial DNA deletions and fatal multisystemic disorder with prominent hepatopathy. *Biochem Biophys Res Commun* 415:300–304.

Biour M, Ben Salem C, Chazouillères O, Grangé JD, Serfaty L, and Poupon R. 2004. Drug-induced liver injury; fourteenth updated edition of the bibliographic database of liver injuries and related drugs. *Gastroenterol Clin Biol* 28:720–759.

Bressler AM, Zimmer SM, Gilmore JL, and Somani J. 2004. Peripheral neuropathy associated with prolonged use of linezolid. *Lancet Infect Dis* 4:528–531.

Brown TA, Cecconi C, Tkachuk AN, Bustamante C, and Clayton DA. 2005. Replication of mitochondrial DNA occurs by strand displacement with alternative light-strand origins, not via a strand-coupled mechanism. *Genes Dev* 19:2466–2476.

Brown TT and Glesby MJ. 2011. Management of the metabolic effects of HIV and HIV drugs. *Nat Rev Endocrinol* 8:11–21.

Bruno S, Maisonneuve P, Castellana P, Rotmensz N, Rossi S, Maggioni M, Persico M et al. 2005. Incidence and risk factors for non-alcoholic steatohepatitis: Prospective study of 5408 women enrolled in Italian tamoxifen chemoprevention trial. *BMJ* 330:932.

Cahill A, Cunningham CC, Adachi M, Ishii H, Bailey SM, Fromenty B, and Davies A. 2002. Effects of alcohol and oxidative stress on liver pathology: The role of the mitochondrion. *Alcohol Clin Exp Res* 26:907–915.

Cahill A, Stabley GJ, Wang X, and Hoek JB. 1999. Chronic ethanol consumption causes alterations in the structural integrity of mitochondrial DNA in aged rats. *Hepatology* 30:881–888.

Cahill A, Wang X, and Hoek JB. 1997. Increased oxidative damage to mitochondrial DNA following chronic ethanol consumption. *Biochem Biophys Res Commun* 235:286–290.

Caldwell SH, Swerdlow RH, Khan EM, Iezzoni JC, Hespenheide EE, Parks JK, and Parker WD. 1999. Mitochondrial abnormalities in non-alcoholic steatohepatitis. *J Hepatol* 31:430–434.

Carabelli J, Burgueño AL, Rosselli MS, Gianotti TF, Lago NR, Pirola CJ, and Sookoian S. 2011. High fat diet-induced liver steatosis promotes an increase in liver mitochondrial biogenesis in response to hypoxia. *J Cell Mol Med* 15:1329–1338.

Cardoso CM, Custodio JB, Almeida LM, and Moreno AJ. 2001. Mechanisms of the deleterious effects of tamoxifen on mitochondrial respiration rate and phosphorylation efficiency. *Toxicol Appl Pharmacol* 176:145–152.

Cardoso CM, Moreno AJ, Almeida LM, and Custodio JB. 2003. Comparison of the changes in adenine nucleotides of rat liver mitochondria induced by tamoxifen and 4-hydroxytamoxifen. *Toxicol In Vitro* 17:663–670.

Carling PJ, Cree LM, and Chinnery PF. 2011. The implications of mitochondrial DNA copy number regulation during embryogenesis. *Mitochondrion* 11:686–692.

Carson J, Cerda J, Chae JH, Hirano M, and Maggiore P. 2007. Severe lactic acidosis associated with linezolid use in a patient with the mitochondrial DNA A2706G polymorphism. *Pharmacotherapy* 27:771–774.

Chen T, He J, Huang Y, and Zhao W. 2011. The generation of mitochondrial DNA large-scale deletions in human cells. *J Hum Genet* 56:689–694.

Christian BE and Spremulli LL. 2012. Mechanism of protein biosynthesis in mammalian mitochondria. *Biochim Biophys Acta* 1819:1035–1054.

Cover C, Mansouri A, Knight TR, Bajt ML, Lemasters JJ, Pessayre D, and Jaeschke H. 2005. Peroxynitrite-induced mitochondrial and endonuclease-mediated nuclear DNA damage in acetaminophen hepatotoxicity. *J Pharmacol Exp Ther* 315:879–887.

Damas J, Samuels DC, Carneiro J, Amorim A, and Pereira F. 2014. Mitochondrial DNA rearrangements in health and disease: A comprehensive study. *Hum Mutat* 35:1–14.

Davies SE, Williams R, and Portmann B. 1989. Hepatic morphology and histochemistry of Wilson's disease presenting as fulminant hepatic failure: A study of 11 cases. *Histopathology* 15:385–394.

De Bus L, Depuydt P, Libbrecht L, Vandekerckhove L, Nollet J, Benoit D, Vogelaers D, and Van Vlierberghe H. 2010. Severe drug-induced liver injury associated with prolonged use of linezolid. *J Med Toxicol* 6:322–326.

Del Pozo JL, Fernández-Ros N, Sáez E, Herrero JI, Yuste JR, and Banales JM. 2014. Linezolid-induced lactic acidosis in two liver transplant patients with the mitochondrial DNA A2706G polymorphism. *Antimicrob Agents Chemother* 58:4227–4229.

Demeilliers C, Maisonneuve C, Grodet A, Mansouri A, Nguyen R, Tinel M, Lettéron P et al. 2002. Impaired adaptive resynthesis and prolonged depletion of hepatic mitochondrial DNA after repeated alcohol binges in mice. *Gastroenterology* 123:1278–1290.

Deschamps D, DeBeco V, Fisch C, Fromenty B, Guillouzo A, and Pessayre D. 1994. Inhibition by perhexiline of oxidative phosphorylation and the beta-oxidation of fatty acids: Possible role in pseudoalcoholic liver lesions. *Hepatology* 19:948–961.

De Vriese AS, Coster RV, Smet J, Seneca S, Lovering A, Van Haute LL, Vanopdenbosch LJ et al. 2006. Linezolid-induced inhibition of mitochondrial protein synthesis. *Clin Infect Dis* 42:1111–1117.

DiMauro S, Schon EA, Carelli V, and Hirano M. 2013. The clinical maze of mitochondrial neurology. *Nat Rev Neurol* 9:429–444.

El-Hattab AW, Li FY, Schmitt E, Zhang S, Craigen WJ, and Wong LJ. 2010. MPV17-associated hepatocerebral mitochondrial DNA depletion syndrome: New patients and novel mutations. *Mol Genet Metab* 99:300–308.

El-Hattab AW and Scaglia F. 2013. Mitochondrial DNA depletion syndromes: Review and updates of genetic basis, manifestations, and therapeutic options. *Neurotherapeutics* 10:186–198.

Fabbrini E, Sullivan S, and Klein S. 2010. Obesity and nonalcoholic fatty liver disease: Biochemical, metabolic, and clinical implications. *Hepatology* 51:679–689.

Ferrari G, Lamantea E, Donati A, Filosto M, Briem E, Carrara F, Parini R, Simonati A, Santer R, and Zeviani M. 2005. Infantile hepatocerebral syndromes associated with mutations in the mitochondrial DNA polymerase-γA. *Brain* 128:723–731.

Fréneaux E, Labbe G, Letteron P, The Le Dinh, Degott C, Genève J, Larrey D, and Pessayre D. 1988. Inhibition of the mitochondrial oxidation of fatty acids by tetracycline in mice and in man: Possible role in microvesicular steatosis induced by this antibiotic. *Hepatology* 8:1056–1062.

Fromenty B. 2013. Bridging the gap between old and new concepts in drug-induced liver injury. *Clin Res Hepatol Gastroenterol* 37:6–9.

Fromenty B, Carrozzo R, Shanske S, and Schon EA. 1997. High proportions of mtDNA duplications in patients with Kearns-Sayre syndrome occur in the heart. *Am J Med Genet* 71:443–452.

Fromenty B, Fisch C, Berson A, Letteron P, Larrey D, and Pessayre D. 1990. Dual effect of amiodarone on mitochondrial respiration. Initial protonophoric uncoupling effect followed by inhibition of the respiratory chain at the levels of complex I and complex II. *J Pharmacol Exp Ther* 255:1377–1384.

Fromenty B, Grimbert S, Mansouri A, Beaugrand M, Erlinger S, Rötig A, and Pessayre D. 1995. Hepatic mitochondrial DNA deletion in alcoholics: Association with microvesicular steatosis. *Gastroenterology* 108:193–200.

Fromenty B and Pessayre D. 1995. Inhibition of mitochondrial beta-oxidation as a mechanism of hepatotoxicity. *Pharmacol Ther* 67:101–154.

Garazzino S, Krzysztofiak A, Esposito S, Castagnola E, Plebani A, Galli L, Cellini M et al. 2011. Use of linezolid in infants and children: A retrospective multicentre study of the Italian Society for Paediatric Infectious Diseases. *J Antimicrob Chemother* 66:2393–2397.

Garcia-Ruiz C and Fernandez-Checa JC. 2006. Mitochondrial glutathione: Hepatocellular survival-death switch. *J Gastroenterol Hepatol* 21(Suppl 3):S3–S6.

García-Ruiz I, Solís-Muñoz P, Fernández-Moreira D, Grau M, Colina F, Muñoz-Yagüe T, and Solís-Herruzo JA. 2014. High-fat diet decreases activity of the oxidative phosphorylation complexes and causes nonalcoholic steatohepatitis in mice. *Dis Model Mech* 7:1287–1296.

Gardner K, Hall PA, Chinnery PF, and Payne BA. 2013. HIV Treatment and associated mitochondrial pathology: Review of 25 years of in vitro, animal, and human studies. *Toxicol Pathol* 42:811–822.

Garrabou G, Soriano A, López S, Guallar JP, Giralt M, Villarroya F, Martínez JA et al. 2007. Reversible inhibition of mitochondrial protein synthesis during linezolid-related hyperlactatemia. *Antimicrob Agents Chemother* 51:962–967.

Gosalakkal JA and Kamoji V. 2008. Reye syndrome and Reye-like syndrome. *Pediatr Neurol* 39:198–200.

Grabhorn E, Tsiakas K, Herden U, Fischer L, Freisinger P, Marquardt T, Ganschow R, Briem-Richter A, and Santer R. 2014. Long-term outcome after liver transplantation for deoxyguanosine kinase (DGUOK) deficiency. A single center experience and a review of the literature. *Liver Transpl* 20:464–472.

Gu M, Cooper JM, Butler P, Walker AP, Mistry PK, Dooley JS, and Schapira AH. 2000. Oxidative-phosphorylation defects in liver of patients with Wilson's disease. *Lancet* 356:469–474.

Gürbay A, Gonthier B, Daveloose D, Favier A, and Hincal F. 2001. Microsomal metabolism of ciprofloxacin generates free radicals. *Free Radic Biol Med* 30:1118–1121.

Haas R, Stumpf DA, Parks JK, and Eguren L. 1981. Inhibitory effects of sodium valproate on oxidative phosphorylation. *Neurology* 31:1473–1476.

Helbling D, Buchaklian A, Wang J, Wong LJ, and Dimmock D. 2013. Reduced mitochondrial DNA content and heterozygous nuclear gene mutations in patients with acute liver failure. *J Pediatr Gastroenterol Nutr* 57:438–443.

Herraiz M, Beraza N, Solano A, Sangro B, Montoya J, Qian C, Prieto J, and Bustos M. 2003. Liver failure caused by herpes simplex virus thymidine kinase plus ganciclovir therapy is associated with mitochondrial dysfunction and mitochondrial DNA depletion. *Hum Gene Ther* 14:463–472.

Holt IJ and Reyes A. 2012. Human mitochondrial DNA replication. *Cold Spring Harb Perspect Biol* 4:a012971.

Honkoop P, Scholte HR, de Man RA, and Schalm SW. 1997. Mitochondrial injury. Lessons from the fialuridine trial. *Drug Saf* 17:1–7.

Hsu CC, Lee HC, and Wei YH. 2013. Mitochondrial DNA alterations and mitochondrial dysfunction in the progression of hepatocellular carcinoma. *World J Gastroenterol* 19:8880–8886.

Huang Q, Tan Y, Yin P, Ye G, Gao P, Lu X, Wang H, and Xu G. 2013. Metabolic characterization of hepatocellular carcinoma using nontargeted tissue metabolomics. *Cancer Res* 73:4992–5002.

Igoudjil A, Abbey-Toby A, Begriche K, Grodet A, Chataigner K, Peytavin G, Maachi M et al. 2007. High doses of stavudine induce fat wasting and mild liver damage without impairing mitochondrial respiration in mice. *Antivir Ther* 12:389–400.

Igoudjil A, Begriche K, Pessayre D, and Fromenty B. 2006. Mitochondrial, metabolic and genotoxic effects of antiretroviral nucleoside reverse-transcriptase inhibitors. *Anti Infect Agents Med Chem* 5: 273–292.

Igoudjil A, Massart J, Begriche K, Descatoire V, Robin MA, and Fromenty B. 2008. High concentrations of stavudine impair fatty acid oxidation without depleting mitochondrial DNA in cultured rat hepatocytes. *Toxicol In Vitro* 22:887–898.

Ikeda S and Ozaki K. 1997. Action of mitochondrial endonuclease G on DNA damaged by L-ascorbic acid, peplomycin, and cis-diamminedichloroplatinum (II). *Biochem Biophys Res Commun* 235:291–294.

Immonen A, Vapalahti M, Tyynelä K, Hurskainen H, Sandmair A, Vanninen R, Langford G, Murray N, and Ylä-Herttuala S. 2004. AdvHSV-tk gene therapy with intravenous ganciclovir improves survival in human malignant glioma: A randomised, controlled study. *Mol Ther* 10:967–972.

Jaeschke H, McGill MR, and Ramachandran A. 2012. Oxidant stress, mitochondria, and cell death mechanisms in drug-induced liver injury: Lessons learned from acetaminophen hepatotoxicity. *Drug Metab Rev* 44:88–106.

Johnson AA, Ray AS, Hanes J, Suo Z, Colacino JM, Anderson KS, and Johnson KA. 2001. Toxicity of antiviral nucleoside analogs and the human mitochondrial DNA polymerase. *J Biol Chem* 276:40847–40857.

Kakuda TN. 2000. Pharmacology of nucleoside and nucleotide reverse transcriptase inhibitor-induced mitochondrial toxicity. *Clin Ther* 22:685–708.

Kamimura S, Gaal K, Britton RS, Bacon BR, Triadafilopoulos G, and Tsukamoto H. 1992. Increased 4-hydroxynonenal levels in experimental alcoholic liver disease: Association of lipid peroxidation with liver fibrogenesis. *Hepatology* 16:448–453.

Kawahara H, Fukura M, Tsuchishima M, and Takase S. 2007. Mutation of mitochondrial DNA in livers from patients with alcoholic hepatitis and nonalcoholic steatohepatitis. *Alcohol Clin Exp Res* 31(1 Suppl.):S54–S60.

Kazak L, Reyes A, and Holt IJ. 2012. Minimizing the damage: Repair pathways keep mitochondrial DNA intact. *Nat Rev Mol Cell Biol* 13:659–671.

Labarthe F, Dobbelaere D, Devisme L, De Muret A, Jardel C, Taanman JW, Gottrand F, and Lombès A. 2005. Clinical, biochemical and morphological features of hepatocerebral syndrome with mitochondrial DNA depletion due to deoxyguanosine kinase deficiency. *J Hepatol* 43:333–341.

Labbe G, Fromenty B, Freneaux E, Morzelle V, Letteron P, Berson A, and Pessayre D. 1991. Effects of various tetracycline derivatives on in vitro and in vivo beta-oxidation of fatty acids, egress of triglycerides from the liver, accumulation of hepatic triglycerides, and mortality in mice. *Biochem Pharmacol* 41:638–641.

Larosche I, Choumar A, Fromenty B, Lettéron P, Abbey-Toby A, Van Remmen H, Epstein CJ et al. 2009. Prolonged ethanol administration depletes mitochondrial DNA in MnSOD-overexpressing transgenic mice, but not in their wild type littermates. *Toxicol Appl Pharmacol* 234:326–338.

Larosche I, Lettéron P, Berson A, Fromenty B, Huang TT, Moreau R, Pessayre D, and Mansouri A. 2010. Hepatic mitochondrial DNA depletion after an alcohol binge in mice: Probable role of peroxynitrite and modulation by manganese superoxide dismutase. *J Pharmacol Exp Ther* 332:886–897.

Larosche I, Lettéron P, Fromenty B, Vadrot N, Abbey-Toby A, Feldmann G, Pessayre D, and Mansouri A. 2007. Tamoxifen inhibits topoisomerases, depletes mitochondrial DNA, and triggers steatosis in mouse liver. *J Pharmacol Exp Ther* 321:526–535.

Larson AM. 2007. Acetaminophen hepatotoxicity. *Clin Liver Dis* 11:525–548.

Lawrence JW, Claire DC, Weissig V, and Rowe TC. 1996. Delayed cytotoxicity and cleavage of mitochondrial DNA in ciprofloxacin-treated mammalian cells. *Mol Pharmacol* 50:1178–1188.

Lawrence JW, Darkin-Rattray S, Xie F, Neims AH, and Rowe TC. 1993. 4-Quinolones cause a selective loss of mitochondrial DNA from mouse L1210 leukemia cells. *J Cell Biochem* 51:165–174.

Lee HC, Li SH, Lin JC, Wu CC, Yeh DC, and Wei YH. 2004. Somatic mutations in the D-loop and decrease in the copy number of mitochondrial DNA in human hepatocellular carcinoma. *Mutat Res* 547:71–78.

Lee HC, Pang CY, Hsu HS, and Wei YH. 1994. Differential accumulations of 4,977 bp deletion in mitochondrial DNA of various tissues in human ageing. *Biochim Biophys Acta* 1226:37–43.

Lee HF, Tsai CR, Chi CS, Lee HJ, and Chen CC. 2009. Leigh syndrome: Clinical and neuroimaging follow-up. *Pediatr Neurol* 40:88–93.

Lee WS and Sokol RJ. 2007. Mitochondrial hepatopathies: Advances in genetics and pathogenesis. *Hepatology* 45:1555–1565.

Lettéron P, Sutton A, Mansouri A, Fromenty B, and Pessayre D. 2003. Inhibition of microsomal triglyceride transfer protein: another mechanism for drug-induced steatosis in mice. *Hepatology* 38:133–140.

Leung TM and Nieto N. 2013. CYP2E1 and oxidant stress in alcoholic and non-alcoholic fatty liver disease. *J Hepatol* 58:395–398.

Lewis W, Griniuviene B, Tankersley KO, Levine ES, Montione R, Engelman L, de Courten-Myers G et al. 1997. Depletion of mitochondrial DNA, destruction of mitochondria, and accumulation of lipid droplets result from fialuridine treatment in woodchucks (*Marmota monax*). *Lab Invest* 76:77–87.

Lewis W, Levine ES, Griniuviene B, Tankersley KO, Colacino JM, Sommadossi JP, Watanabe KA, and Perrino FW. 1996. Fialuridine and its metabolites inhibit DNA polymerase γ at sites of multiple adjacent analog incorporation, decrease mtDNA abundance, and cause mitochondrial structural defects in cultured hepatoblasts. *Proc Natl Acad Sci USA* 93:3592–3597.

Lieber CS, Leo MA, Wang X, and Decarli LM. 2008. Effect of chronic alcohol consumption on Hepatic SIRT1 and PGC-1α in rats. *Biochem Biophys Res Commun* 370:44–48.

Liu VW, Zhang C, Linnane AW, and Nagley P. 1997. Quantitative allele-specific PCR: demonstration of age-associated accumulation in human tissues of the A → G mutation at nucleotide 3243 in mitochondrial DNA. *Hum Mutat* 9:265–271.

Lund KC, Peterson LL, and Wallace KB. 2007. Absence of a universal mechanism of mitochondrial toxicity by nucleoside analogs. *Antimicrob Agents Chemother* 51:2531–2539.

Lutz RE, Dimmock D, Schmitt ES, Zhang Q, Tang LY, Reyes C, Truemper E et al. 2009. De novo mutations in POLG presenting with acute liver failure or encephalopathy. *J Pediatr Gastroenterol Nutr* 49:126–129.

Mandel H, Hartman C, Berkowitz D, Elpeleg ON, Manov I, and Iancu TC. 2001. The hepatic mitochondrial DNA depletion syndrome: Ultrastructural changes in liver biopsies. *Hepatology* 34:776–784.

Manfredi G, Vu T, Bonilla E, Schon EA, DiMauro S, Arnaudo E, Zhang L, Rowland LP, and Hirano M. 1997. Association of myopathy with large-scale mitochondrial DNA duplications and deletions: Which is pathogenic? *Ann Neurol* 42:180–188.

Mansouri A, Demeilliers C, Amsellem S, Pessayre D, and Fromenty B. 2001. Acute ethanol administration oxidatively damages and depletes mitochondrial DNA in mouse liver, brain, heart, and skeletal muscles: Protective effects of antioxidants. *J Pharmacol Exp Ther* 298:737–743.

Mansouri A, Fromenty B, Berson A, Robin MA, Grimbert S, Beaugrand M, Erlinger S, and Pessayre D. 1997b. Multiple hepatic mitochondrial DNA deletions suggest premature oxidative aging in alcoholic patients. *J Hepatol* 27:96–102.

Mansouri A, Gaou I, De Kerguenec C, Amsellem S, Haouzi D, Berson A, Moreau A et al. 1999. An alcoholic binge causes massive degradation of hepatic mitochondrial DNA in mice. *Gastroenterology* 117:181–190.

Mansouri A, Gaou I, Fromenty B, Berson A, Lettéron P, Degott C, Erlinger S, and Pessayre D. 1997a. Premature oxidative aging of hepatic mitochondrial DNA in Wilson's disease. *Gastroenterology* 113:599–605.

Mansouri A, Haouzi D, Descatoire V, Demeilliers C, Sutton A, Vadrot N, Fromenty B, Feldmann G, Pessayre D, and Berson A. 2003. Tacrine inhibits topoisomerases and DNA synthesis to cause mitochondrial DNA depletion and apoptosis in mouse liver. *Hepatology* 38:715–725.

Mansouri A, Tarhuni A, Larosche I, Reyl-Desmars F, Demeilliers C, Degoul F, Nahon P et al. 2010. MnSOD overexpression prevents liver mitochondrial DNA depletion after an alcohol binge but worsens this effect after prolonged alcohol consumption in mice. *Dig Dis* 28:756–775.

Mantena SK, King AL, Andringa KK, Landar A, Darley-Usmar V, and Bailey SM. 2007. Novel interactions of mitochondria and reactive oxygen/nitrogen species in alcohol mediated liver disease. *World J Gastroenterol* 13:4967–4973.

Martin W. 2010. Evolutionary origins of metabolic compartmentalization in eukaryotes. *Philos Trans R Soc Lond B, Biol Sci* 365:847–855.

Massart J, Begriche K, Buron N, Porceddu M, Borgne-Sanchez A, and Fromenty B. 2013. Drug-induced inhibition of mitochondrial fatty acid oxidation and steatosis. *Curr Pathobiol Rep* 1:147–157.

Matthews LT, Giddy J, Ghebremichael M, Hampton J, Guarino AJ, Ewusi A, Carver E et al. 2011. A risk-factor guided approach to reducing lactic acidosis and hyperlactatemia in patients on antiretroviral therapy. *PLoS One* 64:e18736.

McGill MR, Sharpe MR, Williams CD, Taha M, Curry SC, and Jaeschke H. 2012. The mechanism underlying acetaminophen-induced hepatotoxicity in humans and mice involves mitochondrial damage and nuclear DNA fragmentation. *J Clin Invest* 122:1574–1583.

McGill MR, Staggs VS, Sharpe MR, Lee WM, and Jaeschke H. 2014. Serum mitochondrial biomarkers and damage-associated molecular patterns are higher in acetaminophen overdose patients with poor outcome. *Hepatology* 60:1336–1345.

McGlynn KA and London WT. 2011. The global epidemiology of hepatocellular carcinoma: Present and future. *Clin Liver Dis* 15:223–243.

McKee EE, Ferguson M, Bentley AT, and Marks TA. 2006. Inhibition of mammalian mitochondrial protein synthesis by oxazolidinones. *Antimicrob Agents Chemother* 50:2042–2049.

McKenzie R, Fried MW, Sallie R, Conjeevaram H, Di Bisceglie AM, Park Y, Savarese B, Kleiner D, Tsokos M, and Luciano C. 1995. Hepatic failure and lactic acidosis due to fialuridine (FIAU), an investigational nucleoside analogue for chronic hepatitis B. *N Engl J Med* 333:1099–1105.

McShane MA, Hammans SR, Sweeney M, Holt IJ, Beattie TJ, Brett EM, and Harding AE. 1991. Pearson syndrome and mitochondrial encephalomyopathy in a patient with a deletion of mtDNA. *Am J Hum Genet* 48:39–42.

Mitchell P and Moyle J. 1967. Chemiosmotic hypothesis of oxidative phosphorylation. *Nature* 213:137–139.

Morris AA, Lamont PJ, and Clayton PT. 1997. Pearson's syndrome without marrow involvement. *Arch Dis Child* 77:56–57.

Morris EM, Meers GM, Booth FW, Fritsche KL, Hardin CD, Thyfault JP, and Ibdah JA. 2012. PGC-1α overexpression results in increased hepatic fatty acid oxidation with reduced triacylglycerol accumulation and secretion. *Am J Physiol Gastrointest Liver Physiol* 303:G979–G992.

Muraki K, Sakura N, Ueda H, Kihara H, and Goto Y. 2001. Clinical implications of duplicated mtDNA in Pearson syndrome. *Am J Med Genet* 98:205–209.

Murayama K and Ohtake A. 2009. Children's toxicology from bench to bed. Liver Injury (4): Mitochondrial respiratory chain disorder and liver disease in children. *J Toxicol Sci* 34:SP237–SP243.

Nagasaka H, Inoue I, Inui A, Komatsu H, Sogo T, Murayama K, Murakami T et al. 2006. Relationship between oxidative stress and antioxidant systems in the liver of patients with Wilson disease: Hepatic manifestation in Wilson disease as a consequence of augmented oxidative stress. *Pediatr Res* 60:472–477.

Nagiec EE, Wu L, Swaney SM, Chosay JG, Ross DE, Brieland JK, and Leach KL. 2005. Oxazolidinones inhibit cellular proliferation via inhibition of mitochondrial protein synthesis. *Antimicrob Agents Chemother* 49:3896–3902.

Nair J, Strand S, Frank N, Knauft J, Wesch H, Galle PR, and Bartsch H. 2005. Apoptosis and age-dependant induction of nuclear and mitochondrial etheno-DNA adducts in Long-Evans Cinnamon (LEC) rats: Enhanced DNA damage by dietary curcumin upon copper accumulation. *Carcinogenesis* 26:1307–1315.

Nishikawa M, Nishiguchi S, Kioka K, Tamori A, and Inoue M. 2005. Interferon reduces somatic mutation of mitochondrial DNA in liver tissues from chronic viral hepatitis patients. *J Viral Hepat* 12:494–498.

Nishikawa M, Nishiguchi S, Shiomi S, Tamori A, Koh N, Takeda T, Kubo S et al. 2001. Somatic mutation of mitochondrial DNA in cancerous and noncancerous liver tissue in individuals with hepatocellular carcinoma. *Cancer Res* 61:1843–1845.

Nomoto K, Tsuneyama K, Takahashi H, Murai Y, and Takano Y. 2008. Cytoplasmic fine granular expression of 8-hydroxydeoxyguanosine reflects early mitochondrial oxidative DNA damage in nonalcoholic fatty liver disease. *Appl Immunohistochem Mol Morphol* 16:71–75.

Nunez M. 2010. Clinical syndromes and consequences of antiretroviral-related hepatotoxicity. *Hepatology* 52:1143–1155.

Obara-Moszynska M, Maceluch J, Bobkowski W, Baszko A, Jaremba O, Krawczynski MR, and Niedziela M. 2013. A novel mitochondrial DNA deletion in a patient with Kearns-Sayre syndrome: A late-onset of the fatal cardiac conduction deficit and cardiomyopathy accompanying long-term rGH treatment. *BMC Pediatr* 13:27.

Ojala D, Montoya J, and Attardi G. 1981. tRNA punctuation model of RNA processing in human mitochondria. *Nature* 290:470–474.

Orman ES, Conjeevaram HS, Vuppalanchi R, Freston JW, Rochon J, Kleiner DE, and Hayashi PH. 2011. Clinical and histopathologic features of fluoroquinolone-induced liver injury. *Clin Gastroenterol Hepatol* 9:517–523.

Pérez-Carreras M, Del Hoyo P, Martín MA, Rubio JC, Martín A, Castellano G, Colina F, Arenas J, and Solis-Herruzo JA. 2003. Defective hepatic mitochondrial respiratory chain in patients with nonalcoholic steatohepatitis. *Hepatology* 38:999–1007.

Pirola CJ, Gianotti TF, Burgueño AL, Rey-Funes M, Loidl CF, Mallardi P, Martino JS, Castaño GO, and Sookoian S. 2013. Epigenetic modification of liver mitochondrial DNA is associated with histological severity of nonalcoholic fatty liver disease. *Gut* 62:1356–1363.

Porceddu M, Buron N, Roussel C, Labbe G, Fromenty B, and Borgne-Sanchez A. 2012. Prediction of liver injury induced by chemicals in human with a multiparametric assay on isolated mouse liver mitochondria. *Toxicol Sci* 129:332–345.

Quinonez SC, Leber SM, Martin DM, Thoene JG, and Bedoyan JK. 2013. Leigh syndrome in a girl with a novel DLD mutation causing E3 deficiency. *Pediatr Neurol* 48:67–72.

Ramachandran R and Kakar S. 2009. Histological patterns in drug-induced liver disease. *J Clin Pathol* 62:481–492.

Richter C. 1995. Oxidative damage to mitochondrial DNA and its relationship to ageing. *Int J Biochem Cell Biol* 27:647–653.

Roberts EA and Schilsky ML. 2008. Diagnosis and treatment of Wilson disease: An update. *Hepatology* 47:2089–2111.

Robertson DG, Braden TK, Urda ER, Lalwani ND, and de la Iglesia FA. 1998. Elucidation of mitochondrial effects by tetrahydroaminoacridine (tacrine) in rat, dog, monkey and human hepatic parenchymal cells. *Arch Toxicol* 72:362–371.

Robin MA, Sauvage I, Grandperret T, Descatoire V, Pessayre D, and Fromenty B. 2005. Ethanol increases mitochondrial cytochrome P450 2E1 in mouse liver and rat hepatocytes. *FEBS Lett* 579:6895–6902.

Rötig A, Bourgeron T, Chretien D, Rustin P, and Munnich A. 1995. Spectrum of mitochondrial DNA rearrangements in the Pearson marrow-pancreas syndrome. *Hum Mol Genet* 4:1327–1330.

Rötig A, Cormier V, Blanche S, Bonnefont JP, Ledeist F, Romero N, Schmitz J et al. 1990. Pearson's marrow-pancreas syndrome. A multisystem mitochondrial disorder in infancy. *J Clin Invest* 86:1601–1608.

Sakr SA, Zoil Mel-S, and El-Shafey SS. 2013. Ameliorative effect of grapefruit juice on amiodarone-induced cytogenetic and testicular damage in albino rats. *Asian Pac J Trop Biomed* 3:573–579.

Sarzi E, Bourdon A, Chrétien D, Zarhrate M, Corcos J, Slama A, Cormier-Daire V et al. 2007. Mitochondrial DNA depletion is a prevalent cause of multiple respiratory chain deficiency in childhood. *J Pediatr* 150:531–534.

Sauer SW, Merle U, Opp S, Haas D, Hoffmann GF, Stremmel W, and Okun JG. 2011. Severe dysfunction of respiratory chain and cholesterol metabolism in Atp7b(−/−) mice as a model for Wilson disease. *Biochim Biophys Acta* 1812:1607–1615.

Schon EA, DiMauro S, and Hirano M. 2012. Human mitochondrial DNA: Roles of inherited and somatic mutations. *Nat Rev Genet* 13:878–890.

Semino-Mora C, Leon-Monzon M, and Dalakas MC. 1997. Mitochondrial and cellular toxicity induced by fialuridine in human muscle in vitro. *Lab Invest* 76:487–495.

Shalev M, Kadmon D, Teh BS, Butler EB, Aguilar-Cordova E, Thompson TC, Herman JR, Adler HL, Scardino PT, and Miles BJ. 2000. Suicide gene therapy toxicity after multiple and repeat injections in patients with localized prostate cancer. *J Urol* 163:1747–1750.

Shea BF, Hoffman S, Sesin GP, and Hammer SM. 1987. Ganciclovir hepatotoxicity. *Pharmacotherapy* 7:223–236.

Shieh JT, Berquist WE, Zhang Q, Chou PC, Wong LJ, and Enns GM. 2009. Novel deoxyguanosine kinase gene mutations and viral infection predispose apparently healthy children to fulminant liver failure. *J Pediatr Gastroenterol Nutr* 49:130–132.

Shoubridge EA and Wai T. 2007. Mitochondrial DNA and the mammalian oocyte. *Curr Top Dev Biol* 77:87–111.

Simonetti S, Chen X, DiMauro S, and Schon EA. 1992. Accumulation of deletions in human mitochondrial DNA during normal aging: Analysis by quantitative PCR. *Biochim Biophys Acta* 1180:113–122.

Sokol RJ, Twedt D, McKim JM Jr, Devereaux MW, Karrer FM, Kam I, von Steigman G, Narkewicz MR, Bacon BR, and Britton RS. 1994. Oxidant injury to hepatic mitochondria in patients with Wilson's disease and Bedlington terriers with copper toxicosis. *Gastroenterology* 107:1788–1798.

Soriano A, Miró O, and Mensa J. 2005. Mitochondrial toxicity associated with linezolid. *N Engl J Med* 353:2305–2306.

Spinazzola A, Santer R, Akman OH, Tsiakas K, Schaefer H, Ding X, Karadimas CL et al. 2008. Hepatocerebral form of mitochondrial DNA depletion syndrome: Novel MPV17 mutations. *Arch Neurol* 65:1108–1113.

Sterman DH, Treat J, Litzky LA, Amin KM, Coonrod L, Molnar-Kimber K, Recio A et al. 1998. Adenovirus-mediated herpes simplex virus thymidine kinase/ganciclovir gene therapy in patients with localized malignancy: Results of a phase I clinical trial in malignant mesothelioma. *Hum Gene Ther* 9:1083–1092.

Sternlieb I and Feldmann G. 1976. Effects of anticopper therapy on hepatocellular mitochondria in patients with Wilson's disease: An ultrastructural and stereological study. *Gastroenterology* 71:457–461.

Stewart JD, Horvath R, Baruffini E, Ferrero I, Bulst S, Watkins PB, Fontana RJ, Day CP, and Chinnery PF. 2010. Polymerase γ gene POLG determines the risk of sodium valproate-induced liver toxicity. *Hepatology* 52:1791–1796.

Sun W, Xing B, Sun Y, Du X, Lu M, Hao C, Lu Z et al. 2007. Proteome analysis of hepatocellular carcinoma by two-dimensional difference gel electrophoresis: Novel protein markers in hepatocellular carcinoma tissues. *Mol Cell Proteomics* 6:1798–1808.

Tamori A, Nishiguchi S, Nishikawa M, Kubo S, Koh N, Hirohashi K, Shiomi S, and Inoue M. 2004. Correlation between clinical characteristics and mitochondrial D-loop DNA mutations in hepatocellular carcinoma. *J Gastroenterol* 39:1063–1068.

Tang Y, Schon EA, Wilichowski E, Vazquez-Memije ME, Davidson E, and King MP. 2000. Rearrangements of human mitochondrial DNA (mtDNA): New insights into the regulation of mtDNA copy number and gene expression. *Mol Biol Cell* 11:1471–1485.

Temperley R, Richter R, Dennerlein S, Lightowlers RN, and Chrzanowska-Lightowlers ZM. 2010. Hungry codons promote frameshifting in human mitochondrial ribosomes. *Science* 327:301.

Thust R, Tomicic M, Klöcking R, Voutilainen N, Wutzler P, and Kaina B. 2000. Comparison of the genotoxic and apoptosis–inducing properties of ganciclovir and penciclovir in Chinese hamster ovary cells transfected with the thymidine kinase gene of herpes simplex virus-1: Implications for gene therapeutic approaches. *Cancer Gene Ther* 7:107–117.

Toyokuni S and Sagripanti JL. 1996. Association between 8-hydroxy-2'-deoxyguanosine formation and DNA strand breaks mediated by copper and iron. *Free Radic Biol Med* 20:859–864.

Tuquet C, Dupont J, Mesneau A, and Roussaux J. 2000. Effects of tamoxifen on the electron transport chain of isolated rat liver mitochondria. *Cell Biol Toxicol* 16:207–219.

Uusimaa J, Evans J, Smith C, Butterworth A, Craig K, Ashley N, Liao C et al. 2014. Clinical, biochemical, cellular and molecular characterization of mitochondrial DNA depletion syndrome due to novel mutations in the MPV17 gene. *Eur J Hum Genet* 22:184–191.

Valdecantos MP, Pérez-Matute P, González-Muniesa P, Prieto-Hontoria PL, Moreno-Aliaga MJ, and Martínez JA. 2012. Lipoic acid improves mitochondrial function in nonalcoholic steatosis through the stimulation of sirtuin 1 and sirtuin 3. *Obesity* 20:1974–1983.

van der Eb MM, Geutskens SB, van Kuilenburg AB, van Lenthe H, van Dierendonck JH, Kuppen PJ, van Ormondt H et al. 2003. Ganciclovir nucleotides accumulate in mitochondria of rat liver cells expressing the herpes simplex virus thymidine kinase gene. *J Gene Med* 5:1018–1027.

Velez JC and Janech MG. 2010. A case of lactic acidosis induced by linezolid. *Nat Rev Nephrol* 6:236–242.

Vella S, Schwartländer B, Sow SP, Eholie SP, and Murphy RL. 2012. The history of antiretroviral therapy and its implementation in resource-limited areas of the world. *AIDS* 26:1231–1241.

Vidyashankar S and Patki PS. 2010. Liv.52 attenuate copper induced toxicity by inhibiting glutathione depletion and increased antioxidant enzyme activity in HepG2 cells. *Food Chem Toxicol* 48:1863–1868.

Walker UA, Bäuerle J, Laguno M, Murillas J, Mauss S, Schmutz G, Setzer B, Miquel R, Gatell JM, and Mallolas J. 2004. Depletion of mitochondrial DNA in liver under antiretroviral therapy with didanosine, stavudine, or zalcitabine. *Hepatology* 39:311–317.

Wang Y, Lin Z, Liu Z, Harris S, Kelly R, Zhang J, Ge W, Chen M, Borlak J, and Tong W. 2013. A unifying ontology to integrate histological and clinical observations for drug-induced liver injury. *Am J Pathol* 182:1180–1187.

Watkins PB, Zimmerman HJ, Knapp MJ, Gracon SI, and Lewis KW. 1994. Hepatotoxic effects of tacrine administration in patients with Alzheimer's disease. *JAMA* 271:992–998.

Weyers AI, Ugnia LI, García Ovando H, and Gorla NB. 2002. Ciprofloxacin increases hepatic and renal lipid hydroperoxides levels in mice. *Biocell* 26:225–228.

Wheelhouse NM, Lai PB, Wigmore SJ, Ross JA, and Harrison DJ. 2005. Mitochondrial D-loop mutations and deletion profiles of cancerous and noncancerous liver tissue in hepatitis B virus-infected liver. *Br J Cancer* 92:1268–1272.

Wirth T, Parker N, and Ylä-Herttuala S. 2013. History of gene therapy. *Gene* 525:162–169.

Xiao J, Han N, Yang D, and Zhao H. 2013. Liver steatosis in Chinese HIV-infected patients with hypertriglyceridemia: characteristics and independent risk factors. *Virol J* 10:261.

Yamada S, Nomoto S, Fujii T, Kaneko T, Takeda S, Inoue S, Kanazumi N, and Nakao A. 2006. Correlation between copy number of mitochondrial DNA and clinico-pathologic parameters of hepatocellular carcinoma. *Eur J Surg Oncol* 32:303–307.

Ye Y, Huang A, Huang C, Liu J, Wang B, Lin K, Chen Q et al. 2013. Comparative mitochondrial proteomic analysis of hepatocellular carcinoma from patients. *Proteomics Clin Appl* 7:403–415.

Yin H, Hu M, Zhang R, Shen Z, Flatow L, and You M. 2012. MicroRNA-217 promotes ethanol-induced fat accumulation in hepatocytes by down-regulating SIRT1. *J Biol Chem* 287:9817–9826.

Yin PH, Lee HC, Chau GY, Wu YT, Li SH, Lui WY, Wei YH, Liu TY, and Chi CW. 2004. Alteration of the copy number and deletion of mitochondrial DNA in human hepatocellular carcinoma. *Br J Cancer* 90:2390–2396.

Yin PH, Wu CC, Lin JC, Chi CW, Wei YH, and Lee HC. 2010. Somatic mutations of mitochondrial genome in hepatocellular carcinoma. *Mitochondrion* 10:174–182.

Yu HY, Wang BL, Zhao J, Yao XM, Gu Y, and Li Y. 2009. Protective effect of bicyclol on tetracycline-induced fatty liver in mice. *Toxicology* 261:112–118.

Yuzefovych LV, LeDoux SP, Wilson GL, and Rachek LI. 2013. Mitochondrial DNA damage via augmented oxidative stress regulates endoplasmic reticulum stress and autophagy: Crosstalk, links and signaling. *PLoS One* 8:e83349.

Zenger K, Chen X, Decker M, and Kraus B. 2013. In-vitro stability and metabolism of a tacrine-silibinin codrug. *J Pharm Pharmacol* 65:1765–1772.

Zhang X, Tachibana S, Wang H, Hisada M, Williams GM, Gao B, and Sun Z. 2010. Interleukin-6 is an important mediator for mitochondrial DNA repair after alcoholic liver injury in mice. *Hepatology* 52:2137–2147.

Zischka H and Lichtmannegger J. 2014. Pathological mitochondrial copper overload in livers of Wilson's disease patients and related animal models. *Ann N Y Acad Sci* 1315:6–15.

Zischka H, Lichtmannegger J, Schmitt S, Jägemann N, Schulz S, Wartini D, Jennen L et al. 2011. Liver mitochondrial membrane crosslinking and destruction in a rat model of Wilson disease. *J Clin Invest* 121:1508–1518.

13 Epigenetic Modifications of Mitochondrial DNA in Liver Disease
Focus on Nonalcoholic Fatty Liver Disease

Silvia Sookoian and Carlos J. Pirola

CONTENTS

ABSTRACT

It is well known that defects in the mitochondrial genome are responsible for mitochondrial dysfunction as mitochondrial DNA (mtDNA) critically controls the mitochondrial gene expression machinery. Until very recently, mutations and deletions of mtDNA were the only mechanisms to explain changes in the transcriptional mitochondrial profile. An isoform of DNA methyltransferase 1 (DNMT1) is targeted to mtDNA by a conserved N-terminal domain and appears to be responsible for mtDNA methylation of cytosine at the carbon-5 position. Following this, a new concept emerged in the regulation of the human mitochondrial function called *mitochondrial epigenetics.* In this chapter, we introduce the concept of mitochondrial epigenetics in the pathogenesis of nonalcoholic fatty liver disease.

INTRODUCTION OF MITOCHONDRIAL GENOME ORGANIZATION

Mitochondria have their own deoxyribonucleic acid (DNA), and this genetic material is known as mitochondrial DNA or mtDNA. In humans and most multicellular organisms in general, mtDNA is a circular DNA that spans about 16,500 base pairs and contains 37 genes, 13 of which are involved in oxidative phosphorylation [1]. A schematic of human mtDNA structure and coding genes is shown in Figure 13.1. The names of the mtDNA genes encoding proteins of the core subunit of the mitochondrial membrane respiratory chain nicotinamide adenine dinucleotide (NADH) dehydrogenase complex I are as follows: MT-ND1 (mitochondrially encoded NADH dehydrogenase 1), MT-ND2 (mitochondrially encoded NADH dehydrogenase 2), MT-ND3 (mitochondrially encoded NADH dehydrogenase 3), MT-ND4 (mitochondrially encoded NADH dehydrogenase 4), MT-ND4L (mitochondrially encoded

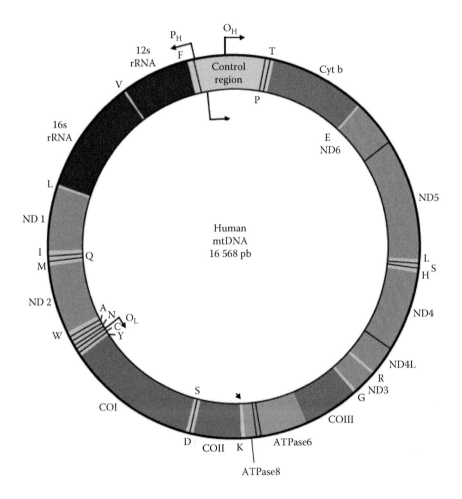

FIGURE 13.1   (See color insert.) Map of the human mitochondrial DNA. (From MITOWEB, MITOMAP: A human mitochondrial genome database, 2013, http://www.mitomap.org.)

NADH dehydrogenase 4L), MT-ND5 (mitochondrially encoded NADH dehydrogenase 5), and MT-ND6 (mitochondrially encoded NADH dehydrogenase 6).

Complex III is constituted by coenzyme Q—cytochrome c reductase/cytochrome b and encoded by MT-CYB.

Members of the cytochrome c oxidase (complex IV, the component of the respiratory chain that catalyzes the reduction of oxygen to water) are as follows: MT-CO1 (mitochondrially encoded cytochrome c oxidase I), MT-CO2 (mitochondrially encoded cytochrome c oxidase II), and MT-CO3 (mitochondrially encoded cytochrome c oxidase III).

In addition, complex V has ATP (adenosine triphosphate) synthase activity and is encoded by MT-ATP6 and MT-ATP8. A representation of the oxidative phosphorylation that takes place in mitochondria and involves protein complexes located in the mitochondrial inner membrane is shown in Figure 13.2; components of the NADH dehydrogenase complex I are highlighted, as they will be a subject of relevant information in the context of this chapter.

The remaining few genes belong to the transfer RNA (ribonucleic acid) family (a tRNA for each amino acid) and two ribosomal RNAs (rRNA) encoded by MT-RNR1 (12S) and MT-RNR2 (16S). Recently, humanin, a biologically active protein in apoptosis, was added to the list of 13 genes; it is also encoded by MT-RNR2 [2].

It has been noted that the mitochondrial genome essentially has bacterial features, for instance, a stretch of roughly 1100 nucleotides is gene free and has been called the *D-loop* (MT-DLOOP: displacement loop and control region). The D-loop contains two hypervariable regions, namely, HVR1 and HVR2, within which mutations accumulate more frequently than anywhere else in the mitochondrial genome.

EPIGENETICS: A BRIEF INTRODUCTION OF EPIGENETIC REGULATION

Epigenetic mechanisms consist in covalent chemical modifications of DNA and histone proteins that ultimately lead to the regulation of gene expression and chromatine organization. Remarkably, epigenetic modifications are highly regulated by environmental stimuli, including nutritional status, are highly dynamic, may operate in a tissue-specific manner, and while epigenetic changes are heritable across cell division, they can also occur de novo. In summary, epigenetic changes are able to change the phenotype of a cell/tissue, thus these mechanisms are extremely important during the development or in response to external stimuli.

DNA methylation occurs particularly but not exclusively in cytosine (C) (5-methylcytosine), where a C is adjacent to a guanine (G) nucleotide (CpG dinucleotides). Paradoxically in normal conditions, it is almost absent in CpG islands (regions rich in CpG dinucleotides as defined below, and in which methylation is mostly associated with silencing of gene expression). In contrast, non-CpG methylation is weakly associated with gene silencing [3].

Enzymes involved in this process are named as DNA methyltransferases (DNMTs) DNMT1, DNMT3A, DNMT3B, DNMT2, and accessory proteins such as Dnmt3L.

A brief description of DNMT's function is given as follows: DNMT1 (DNA (cytosine-5-)-methyltransferase 1) has a role in the establishment and regulation of

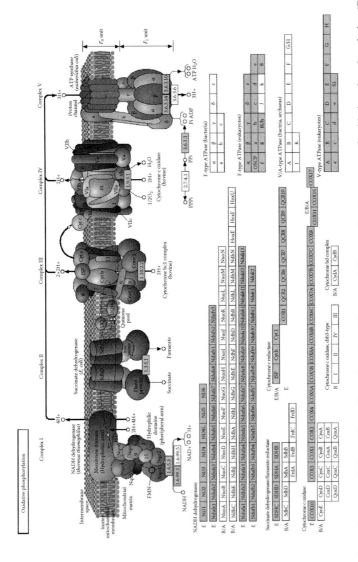

FIGURE 13.2 (**See color insert.**) Kegg (*Kyoto Encyclopedia of Genes and Genomes*) pathway map of oxidative phosphorylation (OXPHOS) in *Homo sapiens*. (Freely available at: http://www.kegg.jp/kegg-bin/highlight_pathway?scale=1.0&map=hsa00190.) Main components of the OXPHOS are the following: NADH-coenzyme Q oxidoreductase (complex I), succinate-Q oxidoreductase (complex II), electron transfer flavoprotein-Q oxidoreductase, Q-cytochrome c oxidoreductase (complex III), cytochrome c oxidase (complex IV), and alternative reductases and oxidases. Components of the OXPHOS encoded by the mtDNA: as depicted on the left side of the figure, complex I is composed of many polypeptides, including seven (ND1, 2, 3, 4, 4L, 5, and 6) encoded by the mtDNA. Complex III is composed of 11 polypeptides, including cytochrome b, *cyt b*, encoded by the mtDNA; complex IV has three components (COI, II, III) from the mtDNA, and complex V has two (ATP6 & 8) components from the mtDNA out of 16 polypeptides.

tissue-specific patterns of methylated cytosine residues, and preferentially methyl-ates hemimethylated DNA CpG island; DNMT2 has been renamed tRNA aspartic acid methyltransferase 1 (TRDMT1) because the gene encodes a protein responsible for the methylation of aspartic acid transfer RNA, specifically at the cytosine-38 residue in the anticodon loop; this enzyme also possesses residual DNA-(cytosine-C5) methyltransferase activity and DNMT3 is responsible for unmethylated CpG island methylation [4].

Around 5% of cytosines are methylated and 50%–60% of genes have CpG-rich islands (regions are typically 300–3000 bp in length with a high content of CpG and C/G% in their five untranslated regions). During fetal development and in adult life, normal somatic cells (and also cancer cells) exhibit alterations in DNA methylation induced by environmental stimuli. As CpGs are paired by CpGs in the opposite strand, methylation in one strand is mirrored by methylation in the other. During replication, methylation in the parent strands directs methylation in the newly repli-cated DNA by recruiting DNMTs. Subsequently, stable transfer of gene methylation patterns to progeny lines is accomplished. CpG methylation is thought to constrain expansive regions of the genome by silencing repetitive sequences or repressing pro-moters by recruiting methyl-CpG-binding proteins (MBD protein family).

Histone posttranslational modifications are complex and diverse and have been considered a second genetic code (acetylation, methylation, ubiquitination, and sumoylation of lysine (K) residues, phosphorylation of serine residues, methylation of arginines (R)). This issue has been extensively reviewed elsewhere [5].

Although DNA methylation is associated with repressed promoters, transcriptional repression via histone methylation and deacetylation precedes DNA methylation.

REGULATION OF MITOCHONDRIAL FUNCTION: THE ROLE OF EPIGENETIC CHANGES

Recently, a new concept has emerged in the regulation of the human mitochondrial function that refers to "mitochondrial epigenetics." This concept was reformulated from the germinal experiments of Shock and coworkers who demonstrated that human mtDNA may be subject to methylation by a novel isoform of DNA methyl-transferase 1 [6].

Indeed, in 1984, Pollack et al. demonstrated the presence of methylation in the mouse mitochondrial DNA that occurred exclusively at the dinucleotide sequence CpG at a frequency of 3%–5% [7]. Nevertheless, Shock et al. introduced new evi-dence about methylation of mtDNA as they reported a mitochondrial isoform of DNA methyltransferase 1 (mtDNMT1), which is the only member of the catalyti-cally active mammalian DNA methyltransferase family targeted to and found in mitochondria [6]. In addition, the authors showed that mtDNMT1 binds to the mito-chondrial genome in a manner proportional to the density of CpG dinucleotides [6]. Furthermore, Shock et al. demonstrated that mtDNMT1 may have an active role in regulating mitochondrial biogenesis, and they also showed that mtDNMT1 expres-sion is regulated in response to hypoxia-induced oxidative stress [6], suggesting a critical role of mtDNMT1 in mitochondrial dysfunction. Moreover, the authors showed that mtDNMT1 is responsive to endogenous factors regulating mitochondrial function, such as nuclear respiratory factor 1 and the peroxisome proliferator-activated γ coactivator-1α (*PPARGC1A*).

Likewise, it was shown that depletion of mtDNA results in changes in the methylation pattern of nuclear genes and in redox status [8].

LIVER EPIGENOME OF HUMAN NAFLD: EPIGENETIC MARKS IN NUCLEAR GENES THAT REGULATE MITOCHONDRIAL FUNCTION AND METABOLISM

Nonalcoholic fatty liver disease (NAFLD) is a common chronic liver disease whose prevalence has reached global epidemic proportions, both in adults and in children [9]. NAFLD develops in the absence of significant alcohol consumption and other causes of secondary hepatic steatosis, and refers to the abnormal accumulation of triglycerides in the liver cells, which may progress from a benign histological disease stage characterized by plain fat accumulation (called *simple steatosis* or *NAFL*) to a more severe histological form characterized by liver cell injury, a mixed inflammatory lobular infiltrate, and variable fibrosis named nonalcoholic steatohepatitis (NASH) [10].

Evidence from diet-induced NAFLD animal models have shown that metabolic insults modify the DNA methylation of candidate gene promoters, for example, high-fat diet (HFD)-induced NAFLD is associated with markedly promoter hypermethylation of glycolytic genes, such as glucokinase (*Gck*) and L-type pyruvate kinase (*LPK*), which is significantly correlated with the downregulation of their transcription and a profound impact on insulin sensitivity [11]. Interestingly, maternal high-fat intake during pregnancy programs, metabolic syndrome–related phenotypes decrease liver mitochondrial DNA copy number and transcriptional activity of liver *Ppargc1a* [12].

Conversely, a high-sucrose diet was not associated with changes in liver DNA *Gck* methylation [13], suggesting that epigenetic changes in the liver might be influenced by some, but not all, nutritional factors. Furthermore, mice fed a lipogenic methyl-deficient diet, which causes liver injury similar to that observed in human NASH, showed aberrant histone modifications and alterations in the expression of Dnmt1 and Dnmt3a proteins in the liver [14]. Finally, recent data have shown that under physiological conditions, protein acetylation is crucial in the regulation of liver gluconeogenesis; for example, the two key enzymes that catalyze the last and first step of glycolysis and gluconeogenesis, pyruvate kinase (PK), and phosphoenolpyruvate carboxy kinase (PEPCK or PCK1), respectively, are both regulated by lysine acetylation [15,16].

The evidence from human studies has illustrated remarkably the concept of the liver epigenome in NAFLD, with particular focus on master regulators of cell metabolism, including *PPARGC1A* that coordinates the regulation of genes involved in energy metabolism by controlling transcriptional programs of mitochondrial biogenesis, adaptive thermogenesis, and fatty acid beta-oxidation. For instance, our group showed the effect of epigenetic changes occurring in the fatty liver on the modulation of insulin resistance (IR) [17]; we observed that the methylation status of *PPARGC1A* promoter is significantly associated with plasma fasting insulin levels and the homeostasis model assessment of IR (HOMA-IR) [17]. In addition, the methylation status of the *PPARGC1A* promoter was inversely correlated with the liver expression of its mRNA, suggesting that methylation of the explored CpG sites in the gene promoter efficiently repressed its transcriptional activity [17]. We also observed an inverse association between the transcriptional activity of *PPARGC1A* and the liver mitochondrial DNA copy number, which also had a direct impact on the status of IR [17].

In addition, an interesting study explored the pre- and post-bariatric changes in the methylation profile of nine genes coding for enzymes that regulate intermediate metabolism and insulin signaling in the liver of morbidly obese patients with NAFLD [18]. The most remarkable finding of this study is that NAFLD-associated methylation changes were partially reversible by therapeutic intervention; for instance, the gene encoding protein-tyrosine phosphatase epsilon (PTPRE) showed both differential expression and differential methylation before and after bariatric surgery [18].

Moreover, the authors observed that the insulin-like growth factor binding protein 2 (IGFBP2) locus was hypermethylated and its mRNA downregulated in NASH [18].

Murphy and colleagues, who recently did global methylation profiling of liver samples of patients with NAFLD at different stages of disease severity by using the Illumina HumanMethylation450 BeadChip platform, observed that patients with advanced NAFLD had a signature of differentially methylated CpG sites that allowed discrimination between advanced versus mild disease [19]. Indeed, the authors showed that advanced NAFLD has a relative hypomethylation state (11% of 52,830 CpG sites) compared with mild NAFLD, specifically in genes associated with tissue repair, for instance, FGFR2 (a fibroblast growth factor receptor family member), genes of the collagen (COL1A1, COL1A2, COL4A1, and COL4A2) and laminin families, and many chemokines [19]. Of note, genes involved in pathways that generate methyl groups, including methylenetetrahydrofolate dehydrogenase 2 (MTHFD2), were significantly hypomethylated in advanced NAFLD [19].

METHYLATION OF HUMAN MITOCHONDRIAL DNA AND ITS ROLE IN NAFLD DISEASE SEVERITY AND PROGRESSION

Mitochondrial dysfunction is largely recognized as critically involved in the development of IR and NAFLD [17,20,21]. In fact, normal activity of the mitochondria determines critically fatty acid beta-oxidation, OXPHOS, and insulin signaling [22,23]. Furthermore, there is a close relation among metabolic stressors, mitochondrial biogenesis, and mtDNA copy number [12,17,20,22,24]. For example, we observed that mitochondrial biogenesis is reduced in the liver of patients with NAFLD, and this reduction is associated with peripheral IR and *PPARGC1A* promoter methylation status [17]; the ratio of mtDNA to nDNA was used as an estimate for the number of mitochondrial genomes per cell, or mtDNA copy number. Nevertheless, recent evidence from our group showed that the modulation of the NAFLD-associated liver epigenome is more complex than expected.

It is well known that defects of the mitochondrial genome are responsible for mitochondrial dysfunction as mtDNA critically controls the mitochondrial gene expression machinery. Until very recently, mutations and deletions of mtDNA were the only mechanisms to explain changes in the transcriptional mitochondrial profile. As introduced previously, an isoform of DNMT1 is targeted to mtDNA by a conserved N-terminal domain and appears to be responsible for mtDNA methylation of cytosine at the carbon-5 position [6]; interaction of mtDNMT1 with the human mitochondrial genome seems to be more evident in the D-loop control region, a noncoding control region of approximately 1.1 kb, which carries the mitochondrial origin of replication and promoters, but mtDNA methylation is observed in some other protein coding regions as well [6].

Notably, the status of 5mC of human mtDNA in the liver tissue was never explored before. In a recent human study, we evaluated the status of liver mtDNA methylation of D-loop (sequence name: gi|251831106:c576-1, c16569-16024) and two other mitochondrial encoded genes, such as MT-ND6 (gene sequence name: gi|251831106:14149-14673) and MT-CO1 (gene sequence name: gi|251831106:5904-7445) [25] in order to explore whether changes in the methylation status in the selected genes were associated with NAFLD biology and pathogenesis. Specifically, we compared the levels of 5mC of the DNA of the mitochondrial genes in the liver of patients with NAFL versus patients with NASH, and we observed that liver MT-ND6-methylated DNA/MT-ND6-unmethylated DNA ratio was significantly associated with NAFLD progression [25], showing that 28.4% of alleles were methylated in NASH versus 20.6% in plain steatosis. Interestingly, we observed that the methylation levels of liver MT-ND6 were significantly associated with the NAFLD activity score (NAS) and also with liver fibrosis, suggesting that patients with more advanced fibrosis had higher levels of liver MT-ND6 methylation [25]. Thus, as a consequence of the interaction of mtD-NMT1 with the mtDNA, the expression pattern of specific mitochondrial genes, such as MT-ND6, was significantly decreased in human NASH [25]. Conversely, the histological disease severity was not associated either with the ratio of D-loop or MT-CO1-methylated DNA/MT-CO1-unmethylated DNA levels [25].

To explain whether these changes in the methylation status of mtDNA genes have an impact on the transcriptional machinery, we measured the liver MT-ND6 mRNA-relative abundance in the liver of patients with NASH and NAFL. Interestingly, we observed that the liver of patients with NASH showed significantly reduced MT-ND6 mRNA levels in comparison with NAFL, and the NAS score was significantly and negatively correlated (Spearman R: -0.50) with MT-ND6 mRNA levels in the whole population of patients with NAFLD [25]. In addition, liver abundance of MT-ND6 mRNA was significantly reduced in patients with a more severe histological disease (fibrosis moderate or severe, graded as F2–3), suggesting that the transcriptional activity of the MT-ND6 is strongly associated with the histological severity of NAFLD [25]. In addition, we observed that changes in liver MT-ND6 mRNA were associated with changes in the protein liver expression as we observed lower levels of liver MT-ND6 expression in patients with NASH in comparison with those showing NAFL [25]; overall, these findings suggest that DNA methylation of MT-ND6 is accompanied by a decrease in its protein level.

As mentioned earlier, DNMT1 is the only member of the DNA methyltransferase family found to have an isoform targeted to the mitochondria, thus we examined the liver transcriptional activity of the DNMT1 isoform among patients with NAFLD in order to answer the question of whether its expression is higher in patients suffering from NASH. Notably, we were able to demonstrate that the liver mRNA DNMT1 levels were significantly higher in patients with NASH compared with NAFL [25].

Finally, to understand the role of mtDNA methylation in target genes and the phenotypes of the metabolic syndrome, we further evaluated whether the status of hepatic 5mC of the above-mentioned mitochondrial genes was associated with metabolic stressors such as peripheral insulin resistance as measured by HOMA-IR, hyperglycemia, hypertriglyceridemia, and hypercholesterolemia. We observed a significant inverse correlation between the hepatic MT-CO1-methylated DNA/ MT-CO1-unmethylated DNA ratio and high-density lipoprotein (HDL) cholesterol levels, and a significant association between MT-CO1-methylated DNA/MT-CO1-unmethylated DNA and body mass index.

To contrast the hypothesis that epigenetic modifications might be reversible by intervention, we also explored whether the observed changes were associated with interventional programs; we observed that physical activity may modulate the methylation status of MT-ND6 [25]. In summary, epigenetics emerged as an interesting target of therapeutic intervention in chronic human diseases because it offers a unique framework of reversible mechanisms that modulate cellular transcriptional machinery.

CONCLUSION

1. Epigenetic modifications are consistently observed in the target tissues that modulate metabolic syndrome–related phenotypes, including the liver; in general, local tissue DNA methylation status is significantly associated with repressed gene transcription.
2. Epigenetics has emerged as an important field in the therapeutic intervention of chronic human diseases because it offers a unique framework of potentially heritable, but reversible, mechanisms that modulate cellular transcriptional machinery and then command signatures of tissue gene expression.
3. The concept of "mitochondrial epigenetics" emerged from recent findings that showed that human mtDNA may be subject to methylation by a novel mitochondrial isoform of DNMT1, which is upregulated by *PPARGC1A*.
4. Mitochondrial epigenetics illustrates that epigenetic changes occurring in the mtDNA are relevant to the biology of complex diseases, including NAFLD.

ACKNOWLEDGMENTS

This study was supported partially by Grants PICT 2010-0441 and PICT 2012-0159 (Agencia Nacional de Promoción Científica y Tecnológica), and UBACYT CM04 (Universidad de Buenos Aires).

REFERENCES

1. Taanman JW. The mitochondrial genome: Structure, transcription, translation and replication. *Biochim Biophys Acta* 1999;1410(2):103–123.
2. Yen K, Lee C, Mehta H et al. The emerging role of the mitochondrial-derived peptide humanin in stress resistance. *J Mol Endocrinol* 2013;50(1):R11–R19.
3. Dodge JE, Ramsahoye BH, Wo ZG et al. De novo methylation of MMLV provirus in embryonic stem cells: CpG versus non-CpG methylation. *Gene* 2002;289(1–2):41–48.
4. Robertson KD, Uzvolgyi E, Liang G et al. The human DNA methyltransferases (DNMTs) 1, 3a and 3b: Coordinate mRNA expression in normal tissues and over-expression in tumors. *Nucleic Acids Res* 1999;27(11):2291–2298.
5. Zhou VW, Goren A, Bernstein BE. Charting histone modifications and the functional organization of mammalian genomes. *Nat Rev Genet* 2011;12(1):7–18.
6. Shock LS, Thakkar PV, Peterson EJ et al. DNA methyltransferase 1, cytosine methylation, and cytosine hydroxymethylation in mammalian mitochondria. *Proc Natl Acad Sci USA* 2011;108(9):3630–3635.
7. Pollack Y, Kasir J, Shemer R et al. Methylation pattern of mouse mitochondrial DNA. *Nucleic Acids Res* 1984;12(12):4811–4824.
8. Smiraglia DJ, Kulawiec M, Bistulfi GL et al. A novel role for mitochondria in regulating epigenetic modification in the nucleus. *Cancer Biol Ther* 2008;7(8):1182–1190.
9. Loomba R, Sanyal AJ. The global NAFLD epidemic. *Nat Rev Gastroenterol Hepatol* 2013;10(11):686–690.
10. Chalasani N, Younossi Z, Lavine JE et al. The diagnosis and management of non-alcoholic fatty liver disease: Practice guideline by the American Gastroenterological Association, American Association for the Study of Liver Diseases, and American College of Gastroenterology. *Gastroenterology* 2012;142(7):1592–1609.
11. Jiang M, Zhang Y, Liu M et al. Hypermethylation of hepatic glucokinase and L-type pyruvate kinase promoters in high-fat diet-induced obese rats. *Endocrinology* 2011;152(4):1284–1289.
12. Burgueno AL, Cabrerizo R, Gonzales MN et al. Maternal high-fat intake during pregnancy programs metabolic-syndrome-related phenotypes through liver mitochondrial DNA copy number and transcriptional activity of liver PPARGC1A. *J Nutr Biochem* 2013;24:6–13.
13. Lomba A, Milagro FI, Garcia-Diaz DF et al. Obesity induced by a pair-fed high fat sucrose diet: Methylation and expression pattern of genes related to energy homeostasis. *Lipids Health Dis* 2010;9:60.
14. Pogribny IP, Tryndyak VP, Bagnyukova TV et al. Hepatic epigenetic phenotype predetermines individual susceptibility to hepatic steatosis in mice fed a lipogenic methyl-deficient diet. *J Hepatol* 2009;51(1):176–186.
15. Li J, Huang J, Li JS et al. Accumulation of endoplasmic reticulum stress and lipogenesis in the liver through generational effects of high fat diets. *J Hepatol* 2012;56:900–907.
16. Xiong Y, Lei QY, Zhao S et al. Regulation of glycolysis and gluconeogenesis by acetylation of PKM and PEPCK. *Cold Spring Harb Symp Quant Biol* 2011;76:285–289.
17. Sookoian S, Rosselli MS, Gemma C et al. Epigenetic regulation of insulin resistance in nonalcoholic fatty liver disease: Impact of liver methylation of the peroxisome proliferator-activated receptor γ coactivator 1α promoter. *Hepatology* 2010;52(6):1992–2000.
18. Ahrens M, Ammerpohl O, von SW et al. DNA methylation analysis in nonalcoholic fatty liver disease suggests distinct disease-specific and remodeling signatures after bariatric surgery. *Cell Metab* 2013;18(2):296–302.

19. Murphy SK, Yang H, Moylan CA et al. Relationship between methylome and transcriptome in patients with nonalcoholic fatty liver disease. *Gastroenterology* 2013;145(5):1076–1087.

20. Carabelli J, Burgueno AL, Rosselli MS et al. High fat diet-induced liver steatosis promotes an increase in liver mitochondrial biogenesis in response to hypoxia. *J Cell Mol Med* 2011;15(6):1329–1338.

21. Sanyal AJ, Campbell-Sargent C, Mirshahi F et al. Nonalcoholic steatohepatitis: Association of insulin resistance and mitochondrial abnormalities. *Gastroenterology* 2001;120(5):1183–1192.

22. Kelley DE, He J, Menshikova EV et al. Dysfunction of mitochondria in human skeletal muscle in type 2 diabetes. *Diabetes* 2002;51(10):2944–2950.

23. Petersen KF, Befroy D, Dufour S et al. Mitochondrial dysfunction in the elderly: Possible role in insulin resistance. *Science* 2003;300(5622):1140–1142.

24. Gianotti TF, Sookoian S, Dieuzeide G et al. A decreased mitochondrial DNA content is related to insulin resistance in adolescents. *Obesity (Silver Spring)* 2008;16(7):1591–1595.

25. Pirola CJ, Gianotti TF, Burgueno AL et al. Epigenetic modification of liver mitochondrial DNA is associated with histological severity of nonalcoholic fatty liver disease. *Gut* 2013;62(9):1356–1363.

26. MITOWEB. MITOMAP: A human mitochondrial genome database, 2013. http://www.mitomap.org (accessed on August 4, 2014).

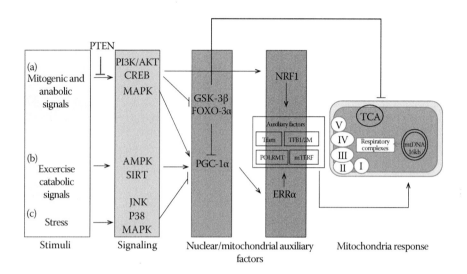

FIGURE 3.2 Cell signaling networks that control mitochondrial gene transcription. (a) In cells exposed to mitogenic or anabolic signals, such as insulin or IGF-1, PI3K/AKT and mitogen-activated protein kinase signaling directly phosphorylates nuclear respiration factor 1 (NRF-1) and estrogen-related receptor α (ERRα) or increased levels of ERRα through non-PKA-dependent phosphorylation of the cAMP response element-binding protein. Both NRF and ERRα signals induce auxiliary factors that act on the mitochondrial genome. In addition, glycogen synthase kinase-3β and FOXO-3α, two substrates of AKT, may also function directly in the mitochondria. A phosphatase and tensin homolog, a tumor suppressor, inhibits this process via its inhibitory effect on the PI3K/AKT signaling pathway. (b) During an endurance exercise or under catabolic conditions, the need to supply ATP induces adenosine monophosphate–activated protein kinase (AMPK) and sirtuin 1 (SIRT1). Acting through deacetylation and activation of peroxisomal proliferating activating factor γ, coactivator-1α, and promoting glucose utilization, AMPK and SIRT1 induce mitochondrial functions to supply ATP demands. (c) Under stress conditions, c-Jun N-terminal kinase and p38MAPK inhibit mitochondrial respiration to slow down the production of reactive oxygen species.

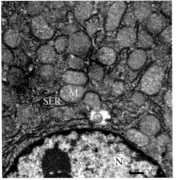
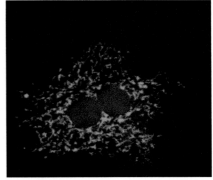

(a) In situ mouse liver mitochodria (EM) (b) Hep G cell mitochondria (mitosis)

FIGURE 4.1 Mitochondrial structure in the liver. (a) Electron micrograph illustrating mitochondria in a normal mouse hepatocyte. Notice that the hepatocyte exhibits electron-dense mitochondria (M) surrounded by an abundant smooth endoplasmic reticulum. N = cell nucleus. Scale bars = 0.2 μm. (b) Fluorescence micrograph of human HepG cells in a culture stained with Mito Tracker Green to target the intracellular mitochondrial network. DAPI stains the cell nucleus, which is in mitosis. 300x.

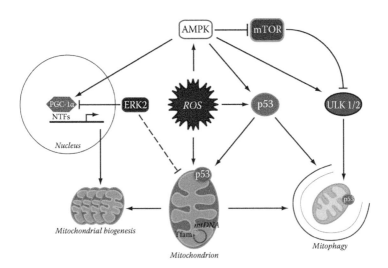

FIGURE 4.5 Regulation of mitochondrial quality control via mitochondrial biogenesis and mitophagy. Mitophagy, in conjunction with mitochondrial biogenesis, regulates changes in mitochondrial number required to meet metabolic demand. AMP-activated protein kinase (AMPK) acutely triggers unc-51-like kinases 1 (ULK1)-dependent mitophagy and simultaneously triggers the biogenesis of new mitochondria via effects on peroxisome proliferator-activated receptor γ coactivator (PGC-1α)-dependent transcription. Conversely, a mammalian target of rapamycin (mTOR) represses mitochondrial biogenesis and ULK1-dependent mitophagy when nutrients are plentiful. The dual processes controlled by AMPK and mTOR determine the net effect of replacing defective mitochondria with new functional mitochondria. AMPK, AMP-activated protein kinase; mTOR, mammalian target of rapamycin; PGC-1α, PPARγ coactivator 1-α; ULK1, the mammalian Atg1 homologs, uncoordinated family member (unc)-51, like kinase 1; ERK2, the extracellular signal-regulated protein kinase 2.

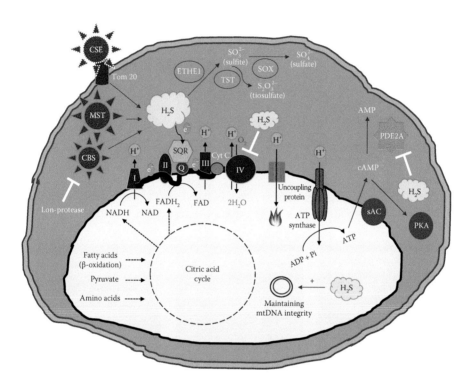

FIGURE 5.3 Schematic representation of the mechanisms by which H$_2$S modulates mitochondrial functions. H$_2$S can be produced in the mitochondrion constitutively by MST and CBS. Moreover, CSE is capable of translocating to the outer mitochondrial membrane under certain stress conditions, resulting in an increment of the intramitochondrial H$_2$S level. The regulatory roles of H$_2$S in the mitochondria are diverse, based on its stimulatory or inhibitory actions. (1) H$_2$S donates electrons to mitochondrial electron transport chain (stimulatory effect). (2) H$_2$S oxidization results in sulfate and thiosulfate end products. (3) H$_2$S inhibits cytochrome c oxidase (complex IV), shutting down the respiration (inhibitory effect). (4) H$_2$S is responsible for the inhibition of mitochondrial PDE2A enzyme. This mode of action increases the intramitochondrial cAMP level that activates the PKA enzyme (stimulatory effect).

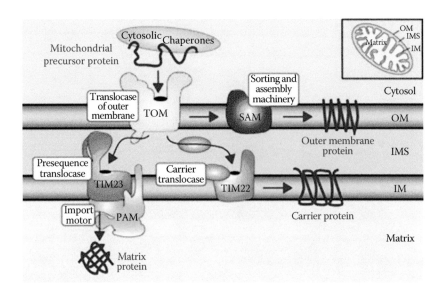

FIGURE 7.2 Protein import machinery into mitochondria. (From Wiedemann, N. et al., *J. Biol. Chem.*, 279, 14473, 2004.)

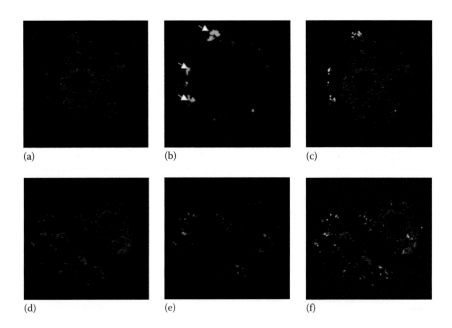

(a)

(b)

(c)

(d)

(e)

(f)

FIGURE 7.6 Representative confocal fluorescence micrographs of BT20 cells stained with Mitotracker® Red CMXRos (red) after exposure for 10 h to fluorescein labeled MLS-pDNA conjugate (green) complexed with cyclohexyl-DQAsomes; (a–c) circular pDNA conjugate, (d–f) linearized pDNA conjugate. (a, d) Green channel, (b, e) red channel, (c, f) overlay of red and green channels with white indicating co-localization of red and green fluorescence. (From D'Souza, G.G. et al., *Mitochondrion*, 5, 352, 2005. With permission.)

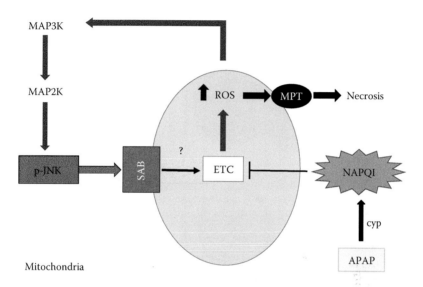

FIGURE 8.1 Schematic overview of acetaminophen toxicity. Acetaminophen is metabolized to the reactive compound NAPQI via the cyp system. NAPQI is highly reactive and covalently binds proteins both in the cytoplasm and mitochondria. This induces oxidative stress and free radicals which activate the mitogen-activated protein kinase (MAPK) cascade ultimately leading to JNK activation (p-JNK). p-JNK then translocates and binds to Sab, which leads to further interference with the ETC and more ROS formation. This leads to a vicious cycle, which sustains ROS production and JNK activation ultimately resulting in hepatocyte death.

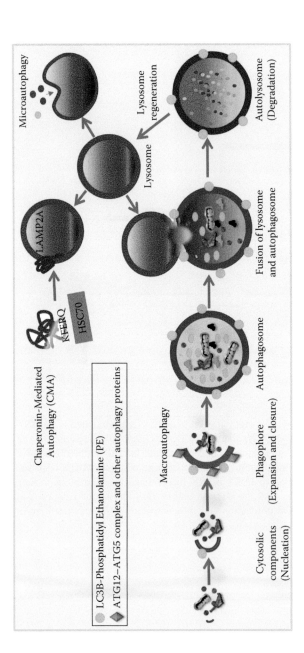

FIGURE 9.1 The definition of autophagy. There are three types of autophagy, which differ in the way that cellular cargoes are delivered into the lysosome. In the CMA, protein cargoes with the signature KFERQ motif are transported to the lysosome with the binding partner, HSC70. The cargo proteins then bind to LAMP2A on the surface of the lysosome, which translocate the cargo proteins into the lysosome for degradation. In microautophagy, the lysosome membrane invaginates to form "phagocytic vesicles," which enwrap molecules on the surface of the lysosomes. The invaginated materials are degraded in the lysosomes. In macroautophagy, the cellular contents (cytosolic proteins or organelles) are enwrapped by autophagosomal membranes in a step known as nucleation. The initial membrane is named as the phagophore, or isolation membrane, which then expands and forms a double-membraned completely enclosed vesicle, known as an autophagosome. The autophagosome containing the cytosolic content is subsequently fused with the lysosome to form the autolysosome, and the content is degraded by the lysosomal enzymes. The molecule LC3 in the mammalian cells is conjugated to PE on the autophagosomal membrane and is a commonly used marker for autophagosomes and autolysosomes, whereas ATG12 and its conjugation partner ATG5 are only presented in the phagophore, but not in the completely circled autophagosomes or autolysosomes.

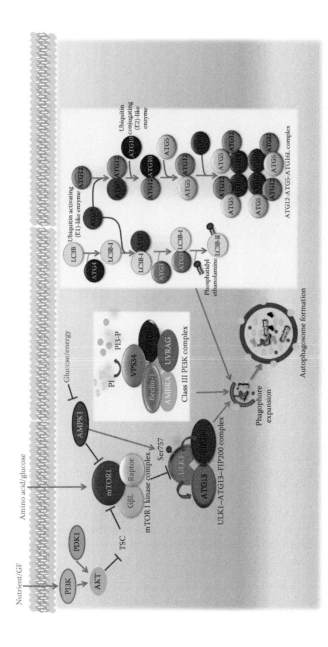

FIGURE 9.2 Signaling pathways in macroautophagy. Macroautophagy is physiologically regulated by nutrients and energy levels. Growth factors, amino acids, and glucose can activate the mTOR complex 1 (mTORC1). mTORC1 suppresses autophagy by inhibiting the ULK1 complex, whereas AMPK1, in response to a lower glucose and/or energy level promotes autophagy by activating the ULK1 complex. The ULK1 complex and the autophagy-specific Beclin-1 complex are responsible for the initial membrane nucleation that leads to the formation of phagophore. The further expansion of phagophore into completely encircled autophagosomes is promoted by the two conjugation systems, where ATG12 is conjugated to ATG5, and LC3 is conjugated to PE on the phagophore membrane. The conjugation reactions are facilitated by ATG7, an E1-like enzyme, and ATG10 or ATG3, an E2-like enzymes. ATG12–ATG5 is further bound to ATG16 to form a large complex, which is also important for LC3 conjugation to PE. Notably, LC3 will need to be proteolytically processed by a cysteine protease, ATG4, before it can be activated by ATG7.

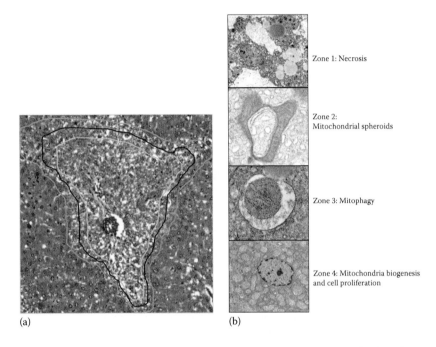

Zone 1: Necrosis

Zone 2:
Mitochondrial spheroids

Zone 3: Mitophagy

Zone 4: Mitochondria biogenesis
and cell proliferation

(a)   (b)

FIGURE 10.2 Zonated changes for necrosis, mitochondrial spheroids, mitophagy/autophagy, and mitochondrial biogenesis/hepatocyte proliferation in APAP-induced liver injury. A representative image of the typical histological changes of APAP-induced liver pathogenesis. (a) Male C57BL/6 mice were treated with APAP (500 mg/kg, i.p.) for 6 h and liver tissues were processed for H&E staining. Necrotic areas were mainly detected around the central vein (light gray innermost line circled area, Zone 1). Black line circled areas represent the zone areas that are enriched with mitochondrial spheroids (Zone 2). Light gray outermost line circled areas represent the autophagy activation areas (Zone 3). Outside of the light gray outermost line are the areas with mitochondrial biogenesis and hepatocyte proliferation (Zone 4). (b) Representative EM images illustrate individual zone area changes.

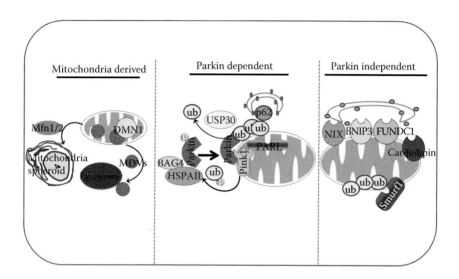

FIGURE 10.3 A proposed model of three distinctive mechanisms that regulate mitochondrial homeostasis. Following mitochondria damage by various stresses, mitochondria are depolarized and Parkin is translocated to the outer membrane of the mitochondria. This process is regulated by PINK1, which is stabilized on depolarized mitochondria. PINK1 either directly phosphorylates Parkin or ubiquitin to promote Parkin translocation and its ligase activity. Cytosolic HSPA1L also promotes, whereas BAG4 inhibits, Parkin mitochondrial translocation. Once Parkin is translocated to the mitochondria, it promotes canonical-selective mitophagy through mitochondrial ubiquitination, p62 mitochondrial targeting, and LC3-positive autophagosome recruitment (Parkin-dependent mitophagy). For Parkin-independent mitophagy, BNIP3, NIX, FUNDC1, or Cardiolipin can also directly interact with LC3 and recruit autophagosomes to damaged mitochondria. Moreover, other E3 ubiquitin ligases such as SMURF1 can also promote mitochondrial ubiquitination, p62 mitochondrial targeting, and mitophagy. Finally, under certain conditions, mitochondria can undergo direct remodeling to form mitochondrial spheroids that are regulated by Mfn1/2. Moreover, small vesicles that contain a subset group of mitochondrial proteins can be directly generated from damaged mitochondria to form MDVs. The segregation of mitochondria to form MDVs is regulated by the mitochondrial fission molecule DNM1. MDVs are then delivered to lysosomes and are eventually degraded within lysosomes.

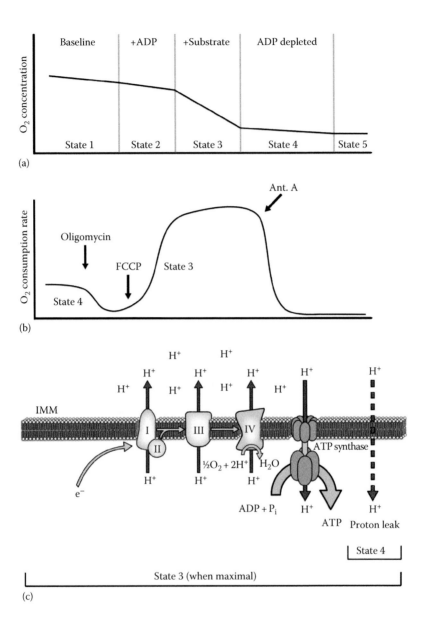

FIGURE 11.3 Measurement of mitochondrial respiration. (a) Diagram of mitochondrial respiration states determined experimentally. (b) Measurement of states 3 and 4 respiration using a pharmacological approach. (c) Diagram of the electron transport chain showing which components contribute to either state 3 or state 4 respiration. Light gray arrows represent the flow of electrons, while dark gray arrows show the flow of protons in the electron transport chain.

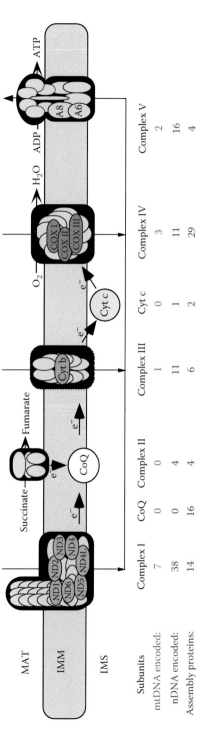

FIGURE 12.1 The mitochondrial respiratory chain. Blue subunits are encoded by nuclear DNA (nDNA) whereas colored subunits are encoded by mitochondrial DNA (mtDNA). "Assembly proteins" are ancillary polypeptides which are required for the assembly and/or stability of the holocomplex; they are all nDNA encoded. See the text for further description. Other abbreviations used in this figure: ADP, adenosine diphosphate; ATP, adenosine triphosphate; CoQ, coenzyme Q; Cyt c, cytochrome c; e⁻, electron; H⁺, proton.

Subunits	Complex I	CoQ	Complex II	Complex III	Cyt c	Complex IV	Complex V
mtDNA encoded:	7	0	0	1	0	3	2
nDNA encoded:	38	0	4	11	1	11	16
Assembly proteins:	14	16	4	6	2	29	4

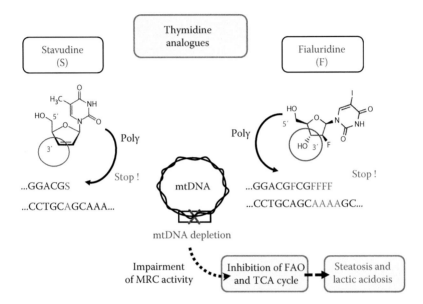

FIGURE 12.4 Alteration of mtDNA synthesis by the thymidine analogues stavudine and fialuridine. The antiretroviral nucleoside reverse transcriptase inhibitor stavudine is a thymidine analogue in which the hydroxyl group (OH) in the 3′ position on the sugar ring is replaced by a hydrogen atom. Stavudine (S) can be incorporated into the growing chain of mtDNA by the DNA polymerase γ (Polγ). However, if stavudine is not removed by the proofreading activity of this DNA polymerase, mtDNA replication is blocked. Indeed, no other nucleotides can be incorporated after stavudine because the DNA chain now lacks a 3′OH end. The anti-HBV fialuridine is also a thymidine analogue but it was never marketed because of severe side effects and death of several patients during clinical trials. Unlike stavudine, fialuridine (F) can be incorporated into mtDNA without immediately stopping mtDNA replication since this drug carries a 3′OH group on the sugar moiety. However, when several adjacent molecules of fialuridine are successively incorporated into a growing chain of mtDNA, DNA Polγ is strongly inhibited. Stavudine and fialuridine-induced blockage of mtDNA replication can lead to mtDNA depletion, impairment of the mitochondrial respiratory chain (MRC) activity, and subsequent inhibition of fatty acid oxidation (FAO) and tricarboxylic acid (TCA) cycle. Alteration of mitochondrial FAO and TCA cycle can secondarily induce severe hepatic steatosis and lactic acidosis, respectively.

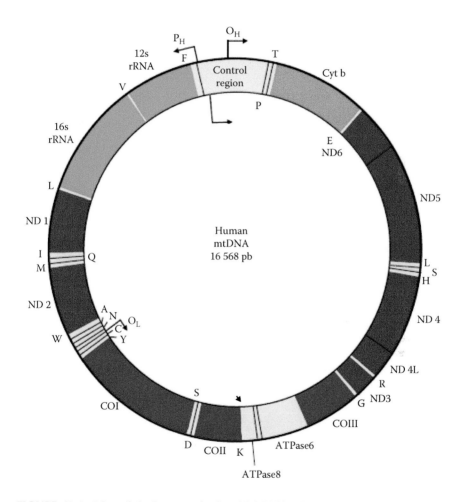

FIGURE 13.1 Map of the human mitochondrial DNA. (From MITOWEB, MITOMAP: A human mitochondrial genome database, 2013, http://www.mitomap.org.)

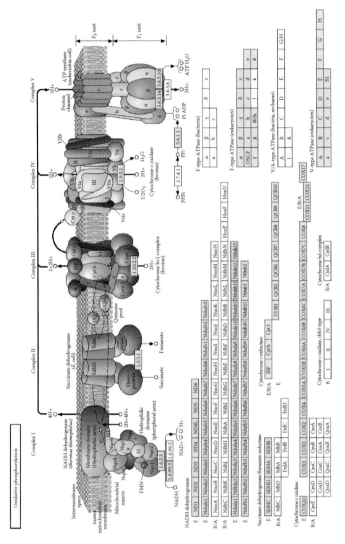

FIGURE 13.2 Kegg (*Kyoto Encyclopedia of Genes and Genomes*) pathway map of oxidative phosphorylation (OXPHOS) in *Homo sapiens*. (Freely available at: http://www.kegg.jp/kegg-bin/highlight_pathway?scale=1.0&map=hsa00190). Main components of the OXPHOS are the following: NADH-coenzyme Q oxidoreductase (complex I), succinate-Q oxidoreductase (complex II), electron transfer flavoprotein-Q oxidoreductase, Q-cytochrome c oxidoreductase (complex III), cytochrome c oxidase (complex IV), and alternative reductases and oxidases. Components of the OXPHOS encoded by the mtDNA: as depicted on the left side of the figure, complex I is composed of many polypeptides, including seven (ND1, 2, 3, 4, 4L, 5, and 6) encoded by the mtDNA. Complex III is composed of 11 polypeptides, including cytochrome b, *cyt b*, encoded by the mtDNA; complex IV has three components (COI, II, III) from the mtDNA, and complex V has two (ATP6 & 8) components from the mtDNA out of 16 polypeptides.

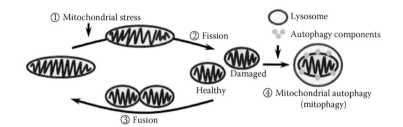

	② Fission	③ Fusion	④ Mitophagy
Role	• Facilitates the segregation of damaged compartment	• Generates new tubular networks	• Removes damaged and superfluous mitochondria
Regulators	• Drp1: Ser616 phosphorylation by CDK1 • Drp1: Ser637 dephosphorylation by clacinurin • Mff, Fis1, Mid49/51: Mitochondrial fission factors	• Mfn 1/2: Modulators for outer mitochondrial membrane fusion • OPA1: Modulator of inner mitochondrial membrane fusion	• Parkin: Mitochondrial translocation • PINK1: Parkin receptor • P62/SQSTM1: Mitophagy adaptor protein • HSPA1L, BAG4, SIAH3: Modulators for Parkin mitochondrial translocation • TOMM7: Modulator for PINK1 stability • Autophagy-related genes: LC3, Atg5, Atg12, etc.

FIGURE 14.1 A schematic representation of the mitochondrial dynamics. Mitochondrial dynamics constitute fission, fusion, and mitophagy as illustrated. Mitochondrial dynamics respond to various environmental cues and stresses (i.e., ROS, genetic mutations, radiation, toxic chemicals, virus/bacterial infections, etc.) that impair mitochondrial function and physiology (①). Under stress conditions, mitochondrial fission allows the segregation of damaged mitochondria from the healthy mitochondrial network initiated by Drp1 recruitment to mitochondria (②) and subsequently facilitates mitophagy that allows the degradation of damaged and superfluous mitochondria (④). In contrast, the mitochondrial fusion process is essential for organizing the fusion of healthy mitochondria which form a functional tubular network (③).

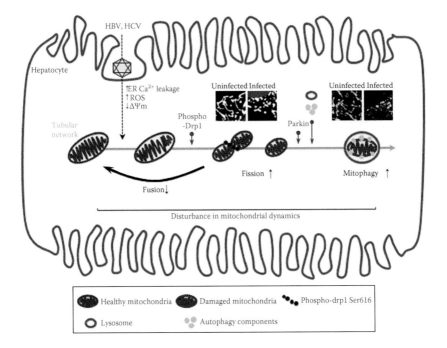

FIGURE 14.2 A schematic representation of mitochondrial dynamics perturbed by HBV and HCV infections. Both HBV and HCV infections induce ER stress and trigger ER calcium leakage followed by the uptake by mitochondria, resulting in mitochondrial oxidative stress and altered membrane potential, thereby inducing mitochondrial depolarization. This scenario leads to mitochondrial fission initiated by mitochondrial translocation of Drp1 which is phosphorylated at Ser616 residue. PINK1 recruits Parkin to the outer membrane of the damaged mitochondria that activates the engulfment of damaged mitochondria by mitophagosome formation initiated by autophagy components. Mitophagosome is further delivered to lysosomes (mitophagy). Mitophagy can be monitored using a novel Mito-mRFP-GFP reporter as shown. (From Kim, S.J. et al., *PLoS Pathog.*, 9(3), e1003285, 2013; Kim, S.J. et al., *PLoS Pathog.*, 9(12), e1003722, 2013; Kim, S.J. et al., *Proc. Natl. Acad. Sci. USA*, 111(17), 6413-8, 2014.)

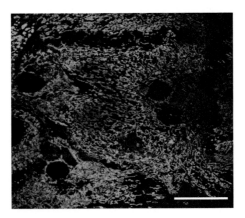

FIGURE 15.1 Mitochondrial morphology of hepatocytes. Primary cultures of mouse hepatocytes were in vivo stained with MitoTracker Green for 15 min and then visualized under the confocal microscope. Scale bar, 10 μm.

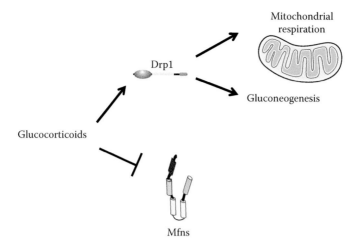

FIGURE 15.2 Model of the hepatic effects of glucocorticoids on mitochondrial dynamics and mitochondrial metabolism. Glucocorticoids (dexamethasone) induce Drp1, repress mitofusins, which leads to rounded-shape mitochondria. Under these conditions, cells show enhanced mitochondrial respiration and gluconeogenic activity.

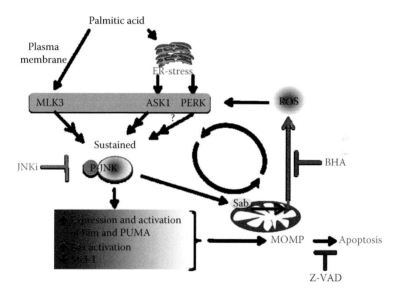

FIGURE 16.1 The role of Sab in lipotoxicity. PA causes decreasing membrane fluidity, expands the lipid raft compartment plasma membrane fluidity, and MLK3 and JNK activation through the clustering of c-Src. Another mechanism for activation of JNK is ER stress. ER stress activates ASK1 through IRE-1α and PERK activation, which leads to JNK activation. Activated JNK translocates to Sab on the mitochondria that causes ROS production. Then it causes sustained JNK activation, which increases expression and activation of Bim and PUMA, increased Bax activation and Mcl-1 degradation leads to outer membrane permeabilization and apoptotic cell death. Inhibition of JNK activation by JNK inhibitor (SP600125) significantly decreased PA-induced cell death. Cell death was also inhibited by the antioxidant BHA and the pancaspase inhibitor Z-VAD-fmk.

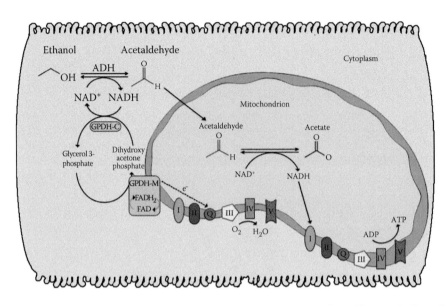

FIGURE 18.1 Role of mitochondria in alcohol metabolism in the liver. The metabolism of alcohol in the liver is dependent on three pathways: (1) The conversion of alcohol to acetaldehyde by alcohol dehydrogenase, which occurs in the cytoplasm and is rate limited by NAD⁺; (2) the conversion of acetaldehyde to acetate by aldehyde dehydrogenase 2, which occurs in the mitochondrial matrix and is rate limited by NAD⁺; and (3) mitochondrial respiration, which oxidizes NADH to regenerate NAD⁺ for alcohol metabolism, which occurs in the inner mitochondrial membrane. Regeneration of NAD⁺ via mitochondrial respiration is an important rate-determining step in alcohol metabolism. NADH generated in the cytoplasm from alcohol metabolism cannot directly cross the inner mitochondrial membrane. Therefore, to regenerate NAD⁺ in the cytoplasm, the glycerol phosphate shuttle transports electrons from cytoplasmic NADH into the mitochondria. Cytoplasmic glycerol phosphate dehydrogenase (GPDH-C) catalyzes the transfer of electrons from NADH to dihydroxyacetone phosphate to form glycerol 3-phosphate. The electrons from glycerol 3-phosphate are then transported to FAD in mitochondria via the catalytic action of mitochondrial GPDH (GPDH-m, GPDH2).

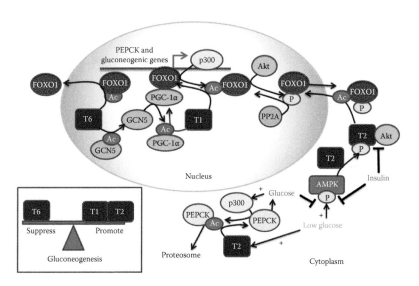

FIGURE 20.1 *Sirtuin regulation of gluconeogenesis.* As shown in the inset, SIRT6 suppresses gluconeogenesis, while SIRT1 and SIRT2 promote it. At the heart of the mechanism is forkhead box protein O1 (FOXO1), which is a master regulator of gluconeogenic gene expression, in concert with PGC-1α. SIRT1 deacetylates both PGC-1α and FOXO1. Deacetylation activates PGC-1α and deacetylation of FOXO1 by SIRT1 promotes nuclear retention and activity. Conversely, when acetylation sites on FOXO1 that are recognized by SIRT6 are deacetylated, FOXO1 is excluded from the nucleus. SIRT6 also deacetylates and activates GCN5, an acetyltransferase that targets PGC-1α. Thus, SIRT6 and SIRT1 directly oppose each other. In the cytoplasm, SIRT2 is activated under nutrient deprivation by AMPK. SIRT2 promotes gluconeogenesis by binding to Akt, which is the kinase that inactivates FOXO1, and by deacetylating FOXO1 to promote its return to the nucleus. SIRT2 also directly deacetylates and stabilizes phosphoenolpyruvate carboxykinase (PEPCK), which catalyzes the rate-limiting step in gluconeogenesis. In a feedback loop, rising glucose levels inactivate AMPK while activating p300, which is the acetyltransferase that acetylates PEPCK, thereby promoting its degradation.

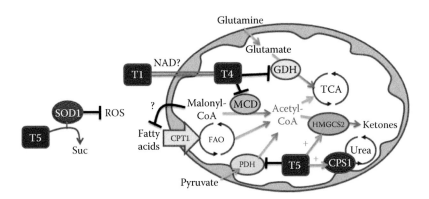

FIGURE 20.2 *Known roles of SIRT4 and SIRT5.* SIRT4, which is strictly mitochondrial, represses glutamine utilization and malonyl-CoA decarboxylase (MCD). Repression of MCD causes malonyl-CoA levels to rise, which may possibly exit the mitochondria and block fatty acid oxidation (FAO) by binding to and inhibiting CPT1. SIRT4 may also cross talk with SIRT1 via NAD^+ levels, as evidenced by increased NAD and SIRT1 activation when SIRT4 is ablated. Ablation of SIRT4 causes FAO to increase by mechanisms that require SIRT1 and PPARα. SIRT5 is both mitochondrial and cytosolic. In the cytosol it reduces reactive oxygen species by the activation of superoxide dismutase 1. In mitochondria it desuccinylates key enzymes in the ketogenic pathway and the urea cycle and inactivates pyruvate dehydrogenase.

14 The Emerging Role of Mitochondrial Dynamics in Viral Hepatitis

Seong-Jun Kim, Gulam Syed, Sohail Mohammad, Mohsin Khan, and Aleem Siddiqui

CONTENTS

ABSTRACT

Mitochondrion is a multifunctional organelle, which plays a central role in vital cellular signaling events including cellular homeostasis. Many viruses target mitochondria to facilitate viral proliferation and mitochondrial aberrations incurred during viral infections, which form the basis for the onset of disease pathogenesis. Alterations to the ultrastructure and function of mitochondria are due to a typical phenotype commonly observed in chronic viral hepatitis caused by hepatitis B and C viruses (HBV and HCV). Both HBV and HCV induce endoplasmic reticulum and oxidative stress that perturb cellular calcium homeostasis and trigger mitochondrial damage and injury. Recent studies demonstrate that HBV and HCV disrupt host cell mitochondrial dynamics and upregulate mitochondrial quality control pathways to eliminate mitochondria damaged during the course of infection. Both HBV and HCV induce mitochondrial fission by triggering the mitochondrial recruitment of fission protein dynamin-related protein 1 (Drp1), which allows the segregation of damaged mitochondria that are subsequently eliminated by Parkin-dependent, mito-chondria-selective autophagy or mitophagy. HBV- and HCV-triggered alterations of mitochondrial dynamics functionally contribute to viral persistence by attenuating mitochondria-mediated apoptosis and innate immune signaling. In this chapter, we discuss the latest observations made in the field of mitochondrial dynamics dur-ing HBV and HCV infection and demonstrate the emerging role of mitochondrial dynamics in chronic viral hepatitis and liver disease pathogenesis.

Keywords: mitochondrial dynamics, fission, fusion, autophagy, mitophagy, Drp1, Parkin, PINK1, viral persistence, apoptosis, innate immunity

INTRODUCTION

Hepatitis is defined as inflammation of the liver and may occur due to various infec-tious and noninfectious causes. It is commonly caused by the infection of hepatotropic viruses, which include hepatitis A, B, C, D, and E viruses [1]. Based on the severity and persistence, hepatitis is acute when it lasts less than 6 months and chronic when it persists longer. Acute liver failure or fulminant hepatitis is a lethal form of acute viral hepatitis. Hepatitis A virus (HAV) is a common cause of acute viral hepatitis [1]. HAV is a common cause of acute viral hepatitis, while hepatitis B (HBV) and hepa-titis C viral (HCV) infections are usually asymptomatic [1]. Some of HBV- or HCV-infected individuals spontaneously clear infection; however, 10–15% or 60–80% of the HBV- or HCV-infected individuals, respectively, become chronically infected [2,3]. Chronic hepatitis can progress to fibrosis (scarring), cirrhosis, and hepatocel-lular carcinoma (HCC) [3,4]. Other virus, host, and environmental factors can also contribute to the rapid progression from chronic hepatitis to cirrhosis, HCC, and liver failure. HBV and HCV infections account for about 80% of HCC cases worldwide, which is the third most lethal cancer responsible for nearly 700,000 deaths per year [5,6]. Hepatitis E virus (HEV) is the most common cause of acute viral hepatitis and jaundice in developing countries, where sporadic infections and epidemics of HEV occur periodically [7]. Occasionally, nonhepatic viruses such as adenovirus, Epstein–Barr virus, herpes simplex virus, and cytomegalovirus can also cause hepatitis [8].

Chronic hepatitis B and C affect approximately 400 million people worldwide and therefore are a looming public health burden. Similar to many other viruses, HBV and HCV promote drastic alterations in cellular physiology, membranes, and organelles to benefit the viral proliferation. Ultrastructural alterations such as endo-plasmic reticulum (ER) ballooning, swollen/round mitochondria, and mitochondria devoid of cristae are common histopathological features observed in the liver biop-sies of chronic HBV and HCV patients [9–11]. Golgi fragmentation has also been noted in HCV-infected cell cultures [12]. Mitochondria play a central role in bioener-getics, metabolism, cellular homeostasis, and innate immune signaling [13–15]. Due to their multitasking capacity, the mitochondria are major targets of many viruses. Many viruses exploit or alter the functions of mitochondria to promote viral infec-tious processes and to escape the surveillance of host defense [16].

Mitochondrial functions are affected directly by viral proteins or indirectly by virus-induced perturbations in cellular physiology. Both HBV and HCV promote

ER stress and consequentially cellular calcium homeostasis is perturbed, leading to mitochondrial dysfunction and damage. Both HBV and HCV enhance reactive oxygen species (ROS) generation and induce oxidative stress which promotes mitochondrial injury [17–19]. Damaged mitochondria are also a major source of ROS and can set off a relay of mitochondrial damage. Some of the HBV and HCV proteins also localize to mitochondria and have been shown to promote mitochondrial dysfunction, but the mechanism of their action remains to be characterized [20–22]. Mitochondria are extremely sensitive to perturbations in cellular environment and mitochondrial homeostasis/quality control is central to sustain cellular homeostasis [23]. Mitochondria are dynamic organelles and constantly undergo fission, fusion, and mitochondria-selective autophagy, termed "mitophagy" [23]. Mitochondrial dynamics (fusion and fission) and mitophagy constitute important arms of the mitochondrial quality control [23]. A tight regulation of these processes, in response to cellular stresses, drives mitochondrial and cellular homeostasis against detrimental physiological or environmental cues. In addition to its role in cellular stress response, this dynamic process has recently emerged as an integral part of other cellular functions, mainly metabolism and innate immunity [24,25].

Mitochondrial liver injury is quite prominent in HBV and HCV infections, yet these viruses significantly overcome cytopathic effects and promote chronic infections in a stealthily manner [10,11,26]. This suggests that mitochondrial dynamics and quality control play a central role in HBV and HCV infections and liver disease pathogenesis. Recent works have shown that both HBV and HCV infections promote mitochondrial fission and shift the balance of mitochondrial dynamics toward mitochondrial fission and mitophagy [27–29]. This shift in mitochondrial dynamics facilitates viability of infected cells and maintenance of persistent infection [27–29]. In this chapter, we will focus on how HBV and HCV infections alter mitochondrial dynamics and how these alterations facilitate chronic viral persistence.

HEPATITIS B VIRUS

HBV is a noncytopathic enveloped virus that belongs to the *Hepadnaviridae* family [30]. It contains a partially double-stranded, relaxed circular DNA (rcDNA) genome that encodes four proteins: surface antigen (HBsAg), core antigen, P (polymerase), and X (HBx), whose synthesis is controlled by at least four promoters and two enhancers [2,30]. Hepatocytes represent the major site of HBV infection. Infection with HBV can result in either acute or chronic infection. Many adults spontaneously clear the virus with only 5–15% rate of chronicity, while in young infants infected at birth, the rate of chronicity is about 80–90% [31]. The relaxed circular genome is converted into covalently closed circular DNA (cccDNA) molecule, which subsequently serves as a template for viral mRNA synthesis in the nuclei of infected hepatocytes. In addition to other viral mRNAs, the HBV DNA encodes a 3.5 kb transcript slightly larger than viral genome-length RNA, termed "pregenomic RNA" (pgRNA). pgRNA encodes viral core/e proteins and polymerase, and serves as template for reverse transcription in viral replication cycle. In the cytoplasm, the pgRNA is encapsidated by the core protein along with the viral polymerase to produce immature cores.

Within these immature cores, the viral polymerase reverse transcribes the pgRNA into genomic RC DNA. After synthesis of 50% or more of the (+) strand, the immature capsids are packaged with HBsAg envelope at the ER and subsequently secreted into the extracellular milieus [2,30]. Hepatocytes represent the major site of HBV infection. HBx is a viral regulatory protein essential for viral life cycle with pleiotropic functions including transcriptional regulation, signal transduction, calcium regulation, cell cycle, apoptosis, and cancer [4,32]. HBx is predominantly cytoplasmic with a minor infection distributed in the nucleus and the mitochondria [21,33–35]. In mitochondria, HBx directly interacts with the voltage-dependent anion channel 3 (VDAC3) on the outer membrane of mitochondria [21]. The precise role of HBx in HBV replication remains to be fully understood. HBx is not directly oncogenic but appears to participate in the processes leading to liver neoplasia [4].

HEPATITIS C VIRUS

HCV is an enveloped, positive sense single-stranded RNA virus that belongs to the *Flaviviridae* family [36]. Following the internalization of the virus by receptor-mediated endocytosis, the 9.6 kb genome is released into the cytoplasm and directly acts as translational template to encode ~3000 amino acids precursor polyprotein [37,38]. The 5′ untranslated region (UTR) contains an internal ribosome entry site required for the initiation of translation and the 3′ UTR is involved in RNA replication [37,38]. The polyprotein is cleaved co- and posttranslationally by viral and host proteases to yield three structural proteins (Core, E1, and E2) and seven nonstructural proteins (p7, NS2, NS3, NS4A, NS4B, NS5A, and NS5B) [37,38]. p7 is a small viral ion-channel protein and NS2 a cysteine protease. NS3/4A contains serine protease and RNA helicase activities. NS4B is an ER-associated highly hydrophobic trans-membrane protein and NS5A is a regulatory protein with many roles in HCV-RNA replication and virion assembly. NS5B is the RNA-dependent RNA polymerase required for the replication of HCV genome [37,38]. The nonstructural proteins associate with the ER and facilitate the formation of ER-derived membranous structures termed "membranous web" that serve as platforms for viral replication [37]. HCV replication is driven by the minus-strand intermediates leading to the generation of several copies of plus-strand RNA genomes which are subsequently encapsidated within the core particles referred to as nucleocapsids. Cytosolic lipid droplets serve as platforms for the assembly of viral nucleocapsids [37–39]. Although the subsequent steps of viral particle morphogenesis and egress are not completely elucidated, it is widely accepted that HCV hijacks the VLDL secretion machinery for its egress via the golgi-secretory route [38,40].

HCV genome displays heterogeneity [41]. There are about six genotypes (20–30% sequence difference) and more than 50 subtypes [41]. Within an infected individual, HCV circulates as a group of different but genetically closely related variants termed as quasispecies (less than 10% sequence difference) due to the high error-prone nature of the HCV RNA polymerase [42]. HCV also primarily infects hepatocytes, and after initial infection, about 20% spontaneously clear infection, whereas the rest develop lifelong chronic infection [3]. Many HCV proteins, particularly core, NS5A, and NS3, are multifunctional and target several host proteins and modulate several cellular pathways related to host-cell metabolism, calcium regulation, apoptosis, cell cycle, and innate immune response [37,38,42]. HCV relies on host lipid pathways in multiple aspects of its life cycle. In an effort to support it proliferation, HCV perturbs host lipid metabolism to promote intracellular enrichment of lipids, which is evident by intracellular accumulation of lipid droplets in infected hepatocytes (hepatosteatosis) [38,40].

PHYSIOLOGICAL PERTURBATION ASSOCIATED WITH HBV AND HCV INFECTIONS

Viruses rely on host-cell energy and exploit cellular machinery for making progeny viral particles. In their pursuit to convert the intracellular environment hospitable for viral proliferation, viruses alter cellular metabolism and physiology. The complex interplay of virus–host interactions determines the outcome of infection and paves the way for the onset of disease pathogenesis. Both HBV and HCV trigger ER stress leading to the unfolded protein response (UPR) [18,43]. Primarily, it is the increased load of viral protein synthesis and folding in the ER that triggers the UPR. In addition, physiological perturbations, such as oxidative stress, deregulated lipid, and calcium homeostasis during HBV/HCV infections, can also trigger ER stress response [18,44]. Many viruses strategically manipulate the UPR signaling to facilitate viral processes and attenuate host antiviral response.

Mutations in the pre-S region of HBV large surface antigen have been shown to promote ER retention of the viral protein leading to ER stress and ER hypertrophy, giving the affected cells a "ground-glass" appearance. The expression of the multifunctional HBx and small surface antigen also leads to ER stress [43,45]. HCV gene expression or expression of some individual HCV proteins like NS4B also triggers ER stress and activates UPR [46–48]. There is no consistent data on the ER stress in chronic hepatitis, which may be due the small proportions of infected liver (7–20% of liver). However, accumulating evidence clearly indicates that UPR plays a critical role in HCV life cycle [46,47]. In vitro infection of human hepatoma cells with cell culture–derived HCV triggers acute ER stress with concomitant activation of all arms of the UPR [49]. Both HBV and HCV also exploit UPR to induce autophagy, which plays a crucial role in their viral life cycle [49–51].

ER stress is implicated in mitochondrial dysfunctions and damage [52,53]. Several findings suggest intimate structural and functional association between ER and mitochondria [52,53]. As a consequence of ER stress, depletion of ER Ca^{2+} pools via inositol 1,4,5-trisphosphate receptors results in the subsequent uptake of Ca^{2+} by the mitochondria that lay in close proximity to the ER via the mitochondrial calcium uniporter (MCU) [54]. This disturbs mitochondrial Ca^{2+} homeostasis and leads to mitochondrial depolarization, dysfunction, inhibition of electron transport, and excess production of mitochondrial ROS [54,55]. This ER/oxidative stress axis may be the underlying basis for mitochondrial dysfunction associated with HBV and HCV infections.

Also, some HBV and HCV proteins have been shown to directly bind to mitochondria and influence mitochondrial functions and trigger ROS production, but the precise mechanism(s) involved is not completely understood [17,18,20,56].

The expression of HBx protein enhances cytosolic and mitochondrial matrix Ca^{2+} levels, promotes loss of mitochondrial transmembrane potential ($\Delta\Psi_m$) and elevates ROS production [32,57]. HBx is primarily localized in cytoplasm but also associates with mitochondria via its interaction with VDAC3 [21]. HBx also elevates cellular calcium levels by promoting calcium influx via store-operated calcium entry [57,58]. Oxidative stress and ROS production induced by HBx alone or in the context of whole genome promotes NF-κB and STAT-3 activation [32,57]. Similar activation is also observed in HBV/HBx transgenic mice [59]. Calcium signaling orchestrates the entire pathway of HBx-mediated mitochondria/nucleus signal transduction and calcium chelators or antioxidants abrogate HBx-mediated NF-κB and STAT-3 activation, respectively [32,57]. The HBV surface antigen (HBsAg) expressing transgenic mice also exhibit enhanced oxidative stress and inflammatory response [60,61]. Similarly, HCV-induced ER stress also leads to oxidative stress via depletion of ER Ca^{2+} stores and subsequent uptake by mitochondria [18,62,63]. A decrease in mitochondrial complex I activity in HCV infection has been shown to be a direct consequence of high ROS levels and oxidation of mitochondrial glutathione pools and glutathionylation of mitochondrial complex I subunits [64,65]. HCV core, NS3, and NS5A protein, when individually expressed, also promote the loss of $\Delta\Psi_m$ and mitochondrial ROS production leading to oxidative stress [20,66]. In isolated mitochondria, HCV core protein is shown to directly stimulate MCU activity and facilitate mitochondrial Ca^{2+} uptake [67]. All the mitochondrial effects of core protein are Ca^{2+} dependent and can be inhibited by blocking Ca^{2+} entry into mitochondria [68]. Nonmitochondrial ROS generation by NADPH oxidases within the infected hepatocytes also contributes to the oxidative stress during HCV infection [69,70].

Oxidative DNA damage and lipid peroxidation, the two common markers of oxidative stress, are of frequent occurrence in chronic hepatitis B or C patients [17,20,71]. Interestingly, HCV infection is linked to higher levels of oxidative stress in comparison to HBV infection [72,73]. Although the direct role of oxidative stress in hepatocellular carcinogenesis is not established, studies suggest that oxidative damage to genomic DNA and inflammation may result in activation of oncogenic signaling contributing to liver disease pathogenesis and the development of HCC [71,72]. The perinuclear clustering of mitochondria commonly observed during HBV/HCV infection may facilitate the direct relay of stress signals and ROS from mitochondria to nucleus cause oxidative base modification of genomic DNA further exacerbating carcinogenesis [74]. Proinflammatory cytokine production mediated by mitochondrial ROS and oxidative stress may also worsen viral hepatitis and hasten progression to end stage liver diseases. Cyclooxygenase 2 (COX2), a key mediator of inflammation, is induced in both HBV- and HCV-infected cells in an ER stress-, ROS-, and NF-κB-dependent manner [75–77]. COX2-mediated inflammation has been associated with hepatic fibrosis, cirrhosis, portal hypertension, and hepatobiliary carcinogenesis [78].

HBV and HCV gene expression does not lead to any cytopathic effect as observed in many asymptomatic HBV and HCV carriers with minimal liver injury [10,26]. However, HBV- and HCV-infected cells are more likely susceptible to exogenous oxidative stress/ER stress induction. In general, hepatocyte damage during viral hepatitis is primarily considered to be immune mediated [26].

MITOCHONDRIAL DYNAMICS: FISSION, FUSION, AND MITOPHAGY

Mitochondria are multifunctional organelles and regulate cellular bioenergetics, metabolism, innate immune signaling, and cell survival [79]. Mitochondria as hubs of various cellular signaling pathways can play a central role in signaling pathways that influence pathogenesis of various diseases [80]. Depending on the cell type, mitochondria occur as a heterogeneous population of many small individual mito-chondria and large interconnected network of tubular mitochondria. Mitochondria are dynamic organelles that undergo fission, fusion, and mitophagy [23] (Figure 14.1). Mitochondrial dynamics is sensitive to subtle physiological perturbations and adjusted according to cellular functional needs [81]. The fine balance between fission and fusion determines the cellular mitochondrial morphology [82]. Mitochondrial dynamics can serve as efficient viral strategy to promote interference of various

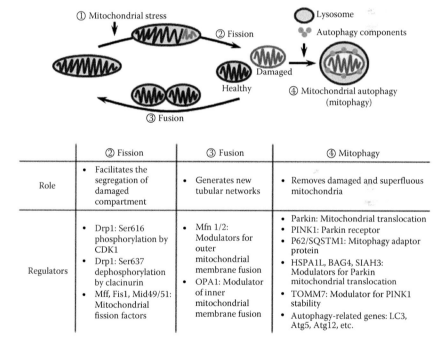

	② Fission	③ Fusion	④ Mitophagy
Role	• Facilitates the segregation of damaged compartment	• Generates new tubular networks	• Removes damaged and superfluous mitochondria
Regulators	• Drp1: Ser616 phosphorylation by CDK1 • Drp1: Ser637 dephosphorylation by clacinurin • Mff, Fis1, Mid49/51: Mitochondrial fission factors	• Mfn 1/2: Modulators for outer mitochondrial membrane fusion • OPA1: Modulator of inner mitochondrial membrane fusion	• Parkin: Mitochondrial translocation • PINK1: Parkin receptor • P62/SQSTM1: Mitophagy adaptor protein • HSPA1L, BAG4, SIAH3: Modulators for Parkin mitochondrial translocation • TOMM7: Modulator for PINK1 stability • Autophagy-related genes: LC3, Atg5, Atg12, etc.

FIGURE 14.1 (See color insert.) A schematic representation of the mitochondrial dynamics. Mitochondrial dynamics constitute fission, fusion, and mitophagy as illustrated. Mitochondrial dynamics respond to various environmental cues and stresses (i.e., ROS, genetic mutations, radiation, toxic chemicals, virus/bacterial infections, etc.) that impair mitochondrial function and physiology (①). Under stress conditions, mitochondrial fission allows the segregation of damaged mitochondria from the healthy mitochondrial network initiated by Drp1 recruitment to mitochondria (②) and subsequently facilitates mitophagy that allows the degradation of damaged and superfluous mitochondria (④). In contrast, the mitochondrial fusion process is essential for organizing the fusion of healthy mitochondria which form a functional tubular network (③).

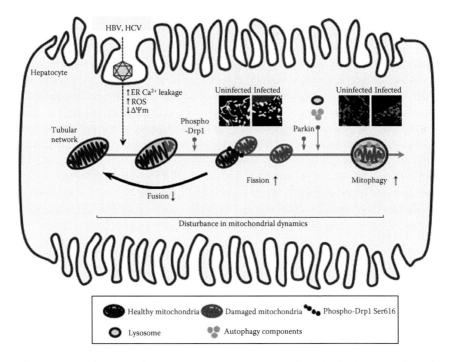

FIGURE 14.2   (See color insert.) A schematic representation of mitochondrial dynamics perturbed by HBV and HCV infections. Both HBV and HCV infections induce ER stress and trigger ER calcium leakage followed by the uptake by mitochondria, resulting in mito-chondrial oxidative stress and altered membrane potential, thereby inducing mitochondrial depolarization. This scenario leads to mitochondrial fission initiated by mitochondrial trans-location of Drp1 which is phosphorylated at Ser616 residue. PINK1 recruits Parkin to the outer membrane of the damaged mitochondria that activates the engulfment of damaged mito-chondria by mitophagosome formation initiated by autophagy components. Mitophagosome is further delivered to lysosomes (mitophagy). Mitophagy can be monitored using a novel Mito-mRFP-GFP reporter as shown. (From Kim, S.J. et al., *PLoS Pathog.*, 9(3), e1003285, 2013; Kim, S.J. et al., *PLoS Pathog.*, 9(12), e1003722, 2013; Kim, S.J. et al., *Proc. Natl. Acad. Sci. USA*, 111(17), 6413-8, 2014.)

cellular signaling pathways [13,83]. Mitochondrial dynamics is involved in the pathogenesis of many human diseases and may also determine the outcome of viral infections and associated diseases [84].

It is safe to envisage that noncytopathic chronic viral infections associated with mitochondrial damage and oxidative stress completely rely on the mitochondrial quality control pathways to sustain cellular homeostasis in infected cells and pro-mote viral persistence [85]. We have recently investigated the effect of HBV and HCV infection on host-cell mitochondrial dynamics and established a direct rela-tionship between viral infections, mitochondrial dynamics, mitochondrial homeo-stasis, and viral persistence [27–29] (Figure 14.2).

The members of dynamin family mediate mitochondrial fusion and fission [82,83] (Figure 14.1). Fission is mediated by the cytosolic dynamin-related protein 1

(Drp1) [82,83]. Mitochondrial fission involves the mitochondrial recruitment of Drp1 from the cytosol which then oligomerizes forming spirals around mitochondria to subsequently constrict and sever the inner and outer mitochondrial membranes [82,83]. Mitochondrial recruitment of Drp1 is mediated by mitochondrial outer membrane proteins, mitochondrial fission factor (MFF), and mitochondrial division 49 and 51 (Mid49 and Mid51), which act as receptors for Drp1 [86] (Figure 14.1). Sites of Drp1 recruitment are often marked by ER protrusion that wraps the mitochondria around [87]. Drp1 recruitment to mitochondria and Drp1 activity is tightly modulated by several posttranslational modification of Drp1, including phosphorylation, nitrosylation, and summoylation [88]. Mitochondrial fusion involves two distinct events: outer mitochondrial membrane (OMM) fusion and inner mitochondrial membrane (IMM) fusion [82]. OMM fusion is mediated by the mitochondrial outer membrane dynamin family member's mitochondrial fusion 1 and 2 (Mfn1 and Mfn2) [82] (Figure 14.1). IMM fusion is mediated by another dynamin family member, optical atrophy 1 (OPA1) [82] (Figure 14.1). Mitochondrial fusion proteins are also regulated by posttranslational modifications and proteolysis [82].

Autophagy involves the engulfment of the target cargo by phagophore, which can be both selective and nonselective, followed by the delivery of the sequestered cargo for degradation in the lysosomes [89]. Nonselective or bulk autophagy has been shown to occur in both HBV and HCV expressing cells [49,50,90]. Selective autophagy involves engulfment of organelles, such as peroxisomes, mitotochondria, lipid droplets etc., and accordingly referred to as pexophagy, mitophagy, or lipophagy [91]. Mitochondria-selective autophagy or mitophagy can be broadly classified as Parkin dependent and Parkin independent [89]. The two major players of Parkin-dependent mitophagy are Parkin, a cytosolic E3-ubiquitin ligase and PINK1, a mitochondrial serine/threonine protein kinase [23] (Figure 14.1). In the healthy mitochondria, PINK1 is rapidly imported to the IMM by concerted action of the translocase of outer mitochondrial membrane (TOM) and inner mitochondrial membrane (TIM) [92]. At the IMM, PINK1 is sequentially processed by mitochondrial processing protease (MPP) and presenilin-associated rhomboid-like protease (PARL), generating a 52 kDa protein, which is finally degraded [92]. In damaged or dysfunctional mitochondria, loss of $\Delta\Psi_m$ activates PINK1 and compromises TOM and TIM activity leading to the tagging of damaged mitochondria by PINK1 [92]. Active PINK1 on the OMM of damaged mitochondria then recruits Parkin subsequently followed by Parkin-dependent elimination of impaired mitochondria by mitophagy [92]. How PINK1 recruits Parkin to the mitochondria is still not very clear, however, direct phosphorylation of Parkin by PINK1 is required to stimulate Parkin's E3 ligase activity. Recent studies also suggest that the phosphorylation of ubiquitin by PINK1 fully activates Parkin's E3-ligase activity and tightens the interaction between Parkin and ubiquitinated OMM proteins [93,94]. The activity of the ubiquitin-proteasome system is critical for Parkin-mediated mitophagy. Recent studies also implicate Parkin-independent modes of mitophagy involving the direct recruitment of LC3 to the mitochondrial outer membrane by proteins containing LC3-interacting regions (LIR) such as AMBRA-1 (activating molecule in Beclin-1-regulated autophagy), Nix, and FUNDC [95]. A recent study also suggests that the IMM-localized mitochondria-specific phospholipid cardiolipin when externalized onto OMM interacts with LC3 to mark mitochondria for degradation [96].

Mitochondrial dynamics is highly vulnerable to physiological perturbations such as oxidative stress, hypoxia, and disturbances in Ca^{2+} homeostasis [97]. Low concentrations of hydrogen peroxide promote mitochondrial fission, reversible mitochondrial swelling, and alterations in mitochondrial morphology [97]. Persistent cytosolic Ca^{2+} influx and depletion of ER Ca^{2+} stores trigger mitochondrial depolarization and subsequent mitochondrial fission [97]. On the contrary, excessive mitochondrial fission is also associated with enhanced ROS production and mutations in mitochondria fusion protein OPA1 is linked to ROS accumulation and downregulation of antioxidant genes and antioxidant capacity. These studies directly implicate the oxidative stress in the modulation of mitochondrial dynamics. Mitochondrial fission helps in the segregation of damaged mitochondria and downregulation of fusion prevents their reintegration into the functional mitochondrial network. The segregated pool of damaged mitochondria is then subsequently eliminated by mitophagy (mitochondria-selective autophagy). Thus, mitochondrial dynamics and mitophagy constitute two important arms of mitochondrial quality control and work in concert to maintain mitochondrial and cellular homeostasis. Mitochondrial fusion and fission also facilitates the exchange of inter-mitochondrial contents and helps in maintenance of mitochondrial integrity.

HBV AND HCV INFECTIONS AND MITOCHONDRIAL DYNAMICS

Both HBV and HCV tightly modulate the mitochondrial dynamics and favor fission over fusion, resulting in prominent mitochondrial fragmentation or fission in HCV-/HBV-infected cells [27–29] (Figure 14.2). We observed that both HBV- and HCV-infected cells displayed perinuclear clustering of the mitochondria, a phenotype quite common in cells subjected to oxidative stress [27,29]. Both HBV and HCV promote mitochondrial translocation of the fission protein Drp1 [27,28] (Figure 14.2). Drp1 expression is also enhanced in HBV- and HCV-infected cells [27,28]. However, overexpression of Drp1 does not usually lead to fission, but stimulation of Drp1 translocation to the mitochondria and induction of its GTPase activity and assembly promote mitochondrial fission. In both HBV and HCV infection, Drp1 translocation to mitochondria is triggered by Drp1 phosphorylation at its serine 616 (S616) residue [27,28]. Drp1 S616 phosphorylation is primarily mediated by the cyclin-dependent kinase 1 (CDK1) [98]. A recent report suggests that high glucose stimulation of liver-derived cells leads to Ca^{2+}-mediated MAP kinase signaling and ERK1/2-dependent Drp1 S616 phosphorylation [99]. In HCV infection, CDK1 is involved as silencing CDK1 resulted in abrogation of HCV-stimulated Drp1 Ser616 phosphorylation [28]. Drp1 translocation to mitochondria is also mediated by calcineurin-dependent dephosphorylation of S637 residue [98]. HCV also stimulated Drp1 S637 dephosphorylation, suggesting that both S616 phosphorylation and S637 dephosphorylation are operative in HCV infection [28]. Since HBV infection is associated with the rise in cytosolic Ca^{2+} levels, it is likely that HBV also stimulates Drp1 S637 dephosphorylation. Disruption of mitochondrial dynamics was observed in cell expression full-length HBV genome or only HBxm but HBx-defective (HBV-ΔX) genome had no effect on mitochondrial dynamics supporting the view that HBx is required and sufficient to disrupt the mitochondrial dynamics [27]. HBx protein is localized to mitochondria via its association with VDAC3 [21]. This association alters the $\Delta\Psi_m$, elevates the levels of ROS, and ultimately causes damage to mitochondria [4,21,58]. It has to be noted that HBV- and HCV-triggered mitochondrial fission peaks with HBV/HCV replication and the oxidative stress associated with it. It remains to be determined if the alteration to mitochondrial dynamics is a transient event and balances out later during the course of infection. Inhibiting mitochondrial fission by silencing Drp1 expression affected the HCV secretion and inhibited aerobic glycolysis implicating the mitochondrial fission directly or indirectly in HCV life cycle and virus-mediated metabolic reprograming, which may likely contribute to the onset of liver disease pathogenesis.

HBV AND HCV INFECTIONS AND MITOPHAGY

Mitochondrial fission/fragmentation facilitates the segregation of damaged mitochondria from the healthy pool [23,82]. Mitochondrial segregation is followed by mitophagy to eliminate the damaged mitochondria and facilitate their rapid turnover [23,82]. Expression of Parkin and PINK, the two major players of Parkin-dependent mitophagy, was enhanced in both HBV- and HCV-infected cells [27,29]. In agreement with the enhanced fission, both HBV- and HCV-infected cells also displayed mitochondrial recruitment of Parkin with concomitant increase in mitophagy of the damaged mitochondria [27,29]. Biochemical analysis of Parkin in the subcellular fractions revealed that Parkin in the mitochondrial fraction was ubiquitinated. Upon recruitment of cytosolic Parkin to the mitochondria, its E3 ubiquitin ligase activity is stimulated several fold [27,29]. On the OMM, Parkin self-ubiquitinates and then subsequently ubquitinates and facilitates proteasomal degradation of its target proteins on the OMM [27,29]. We observed ubiquitination and degradation of several of Parkin's mitochondrial target proteins like VDAC and Mfn2 in HBV- and HCV-infected cells [27,29]. Using the conventional immunofluorescent techniques, we demonstrated Parkin-dependent mitophagosome formation and their subsequent fusion with the lysosomes [27,29]. Alternatively, using the novel mitophagy reporter (tandem-tagged mRFP-EGFP chimeric fluorescence reporter in frame with mitochondrial localization sequence) which exploits the advantage of differential stability of the RFP versus GFP in the acidic environment of the lysosomes, Kim et al. showed complete mitophagy in both HBV- and HCV-infected cells [27,28] (Figure 14.2). Hindering HBV- and HCV-mediated Parkin-dependent mitophagy by silencing Parkin expression affected the viral replication and also partially rescued the mitochondrial oxidative phosphorylation and complex 1 activity [27,29]. This implicates the direct role of Parkin in some of the mitochondrial dysfunctions associated with HBV and HCV infections. In correlation, Parkin silencing rescued the levels of mitochondrial complex proteins in HBV- and HCV-infected cells. Inhibition of mitochondrial fission by silencing Drp1 expression completely halted the HBV- and HCV-mediated Parkin-dependent mitophagy. In addition to the segregation of the damaged mitochondria, asymmetric mitochondrial fission may also generate individual and smaller mitochondrial units more suitable for the elimination by mitophagy compared to the elongated mitochondrial network.

MITOCHONDRIAL DYNAMICS AND MITOPHAGY: HBV/HCV VIRAL PERSISTENCE

HBV and HCV stealthily establish the chronic infection suggesting that these viruses avoid general apoptosis of infected hepatocytes to establish viral persistence [4,26]. In agreement with this notion, hepatocyte death observed in chronic hepatitis B and C is primarily mediated by immune response to infection. However, it is tempting to contemplate that mitochondrial injury incurred by the infected hepatocytes may also promote cell death. Many studies implicate mitochondrial damage in the apoptosis of HBV- and HCV-infected cells [4]. However, these observations contradict the chronicity associated with HBV and HCV infections, suggesting that probably only a minor fraction of the infected cells undergo cell death. Several reports show the activation of both proapoptotic and antiapoptotic signaling in vitro cell cultures and transgenic animal models using individual viral proteins or in the context of infection further deepening the controversy.

Mitochondrial dynamics (fusion and fission) and mitophagy work in conjunction to sustain mitochondrial and cellular homeostasis by facilitating mitochondrial quality control. Recent work strongly correlates with this notion and suggests that both HBV and HCV promote mitochondrial fission followed by mitophagy to facilitate the rapid elimination and turnover of the damaged mitochondria [27–29] (Figure 14.2). Mitochondrial fission is implicated in apoptosis and Drp1-mediated mitochondrial scission is shown to facilitate cytochrome C leakage from the mitochondria [27,28]. Although mitochondrial fission facilitates apoptosis, all fission events may not necessarily lead to apoptosis. The pivotal role of mitochondrial fission in driving mitochondrial quality control and cellular homeostasis further supports this argument. In HBV- and HCV-infected cells, depletion of Drp-1 and Parkin inhibited mitochondrial fission and mitophagy, respectively, which and led to a dramatic increase in mitochondrial apoptotic signaling cascade marked by cytochrome C release into the cytosol and increase in caspase-3 activity and poly (ADP-ribose) polymerase (PARP) cleavage [27,28]. These observations suggest that HBV- and HCV-triggered mitochondrial quality control mechanisms sustain mitochondrial homeostasis and abrogate apoptosis as a consequence of mitochondrial injury accrued during the infection. These studies establish that altered mitochondrial dynamics and mitophagy during HBV/HCV infection are among the major determinants of viral persistence (Figure 14.2).

MITOCHONDRIAL DYNAMICS: INNATE IMMUNE SIGNALING AND INFLAMMASOME

Both HBV and HCV cripple the innate immune signaling by different mechanisms to evade the innate immune response [100]. Upon infection, the HCV RNA genome is sensed by RIG-I, which becomes activated and in turn activates the antiviral mitochondrial signaling protein (MAVS) [101]. HCV NS3/4A protease then cleaves

MAVS that is localized on mitochondria-associated membranes (MAM). MAVS is localized to peroxisome, MAM, and outer mitochondrial membranes (OMM) [102–104]. This cleavage inactivates MAVS and leads to abrogation of downstream signaling for the production of interferons, the antiviral molecules which blunt viral replication [105]. During HCV-induced mitophagy, the mitochondrially localized Parkin-ubiquitinated MAVS associated with OMM via K-48-linked ubiquitination [106]. This ubiquitination of MAVS may cause its inactivation and cessation of downstream signaling for IFN synthesis. These observations suggest that Parkin translocation to mitochondria in HCV-infected cells, in addition to mitophagy, also serves to counteract OMM-localized MAVS signaling and IFN synthesis. The concerted effort of HCV NS3/4A protease and Parkin completely cripples MAVS signaling at both MAMs and OMM. In this manner, HCV completely blocks the IFN production during chronic hepatitis C. Recent study showed that inhibition of mitochondrial fission by silencing Drp1 led to an increase IFN production [28]. This result suggests that mitochondrial fission also somehow contributes to the dampening of innate immunity by HCV.

HBV gene expression also cripples innate immune signaling involving RIG-I and MAVS [107–109]. Our recent observations suggest that Parkin serves as a negative regulator of MAVS signaling and IFN synthesis during HBV infection [110]. In HBV-infected cells, Parkin interacts with MAVS and this interaction is potentiated by HBx, with all three proteins forming a ternary complex [110]. Silencing Parkin or HBx led to an increase in IFN synthesis. Unlike Parkin-mediated K-48-linked MAVS ubiquitination and degradation in HCV-infected cells, in HBV-infected cells, Parkin promoted MAVS linear ubiquitination [110]. MAVS linear ubiquitination leads to the attenuation of its downstream signaling and IFN production [110]. Parkin facilitated the mitochondrial recruitment of the LUBAC complex leading to the linear ubiquitination of MAVS. Silencing LUBAC's catalytic subunit abrogated Parkin-mediated MAVS linear ubiquitination [110]. Collectively, these studies suggest that Parkin serves as the negative regulator of MAVS signaling, in both HBV and HCV infection. Virus-specific variations in Parkin's mechanism of action on MAVS suggest intricate regulation of Parkin's activities by viral-specific factors.

Recent studies have demonstrated that inflammasome is activated in viral hepatitis, but with reduced intensity. Activation of inflammasome in HCV-infected hepatocytes and macrophages has been reported [111–113]. Genetic deletion of genes important for autophagy such as LC3B and Beclin 1 and defective clearance of damaged mitochondria or defective mitophagy also lead to NLRP3 inflammasome activation [114,115]. A recent study suggests that inflammasome-mediated caspase-1 activation results in Parkin cleavage and inhibition of mitophagy [116]. But this may represent a more robust activation of inflammasome and in chronic hepatitis B and C, it is very likely that inflammasome activation occurs at lower intensity and Parkin-mediated mitophagy still continues. During viral hepatitis, enhanced mitochondrial quality control and mitophagy rapidly eliminates damaged mitochondria and restricts the availability of mtROS and mtDNA that can serve as stimuli for NLRP3 inflammasome activation [117,118]. Rarely damaged mitochondria, which escape mitophagic clearance, still serve as stimuli for NLRP3 inflammasome [117].

This is an important area of research and requires further studies to delineate the link between viral hepatitis and intrinsic activation of inflammasome.

MITOCHONDRIAL DYSFUNCTIONS: VIRAL HEPATITIS AND NONALCOHOLIC STEATOHEPATITITS (NASH)

As a consequence of mitochondrial damage, mitochondrial functions such as the β-oxidation of fatty acids is compromised during HCV infection [119]. De novo lipogenesis and lipid uptake is also upregulated during HCV infection [38,119]. As a result of these alterations of the host lipid metabolism, intracellular accumulation of lipid droplets or "hepatosteatosis" is quite prominent in patients with chronic hepatitis C [120,121]. In obese rodent model, mitochondrial dysfunction seems to precede progression to hepatasteatosis and NASH. Mice defective in mitochondrial trifunctional protein that plays a crucial role in β-oxidation of fatty acids develop hepatosteatosis and insulin resistance. Viral hepatitis–associated hepatosteatosis has been implicated in rapid progression into severe end-stage liver diseases such as hepatic fibrosis, cirrhosis, and HCC. Emerging evidence suggests that chronic hepatitis C can complicate nonalcoholic steatohepatitis (NASH) and vice versa. HCV genotype 3, in particular, has been shown to induce NASH via direct cytopathic effects [122]. HCV genotype 3 core protein is also believed to be more potent than the core protein of other genotypes in the induction of intracellular lipid droplet accumulation and alteration of lipid droplet morphology [123]. NASH is associated with dramatic upregulation of various biomarkers of oxidative stress such as mtDNA oxidation, lipid peroxidation, and enhanced ROS production, suggestive of mitochondrial damage and dysfunction [124]. Impaired mitochondrial bioenergetics and functions are also linked with mitochondrial injury during NASH [124]. NASH has been implicated in alteration of mitochondrial dynamics [125]. Our recent observations substantiate that mitochondrial fission can be induced by the treatment with saturated fatty acid, such as palmitic acid, in in vitro cell culture model of NASH (hepatocyte treatment with palmitic acid) via upregulating Drp1 gene expression and activity (unpublished). However, the enhanced mitochondrial fission did not lead to mitophagy as evidenced by the lack of Parkin translocation to the mitochondria and lysosomal delivery of the damaged mitochondria (unpublished). This observation suggests that mitochondrial quality control is compromised in NASH and strongly correlates with the enhanced hepatocyte death and inflammatory signaling observed in NASH. The underlying mechanisms associated with compromised mitochondrial quality control still remain to be elucidated. We envisage PINK1 cleavage in NASH hinders Parkin translocation to mitochondria and subsequent mitophagy. In addition, defective autophagy associated with intracellular accumulation of free fatty acids (obesity) may also contribute to the perturbation of mitophagy in NASH.

CONCLUSION

Human hepatitis B and C viruses cause chronic hepatitis, which can progress to serious liver diseases including cirrhosis and hepatocellular carcinoma. Although dissimilar in their molecular and genome organization, their infection results in similar pathologies of the liver. Mitochondrial liver injury is a prominent histopathological feature observed in the livers of chronic hepatitis B or C patients. Both viruses alter mitochondrial dynamics, physiology, and functions. Recent work has highlighted their ability to upregulate mitochondrial homeostasis by triggering mitochondrial fission, followed by mitophagy. Consequence of these effects manifests in the maintenance of persistent viral infection by attenuating cytopathic effects associated with mitochondrial injury. Moreover, the virus-induced mitochondrial dynamics also helps in subverting the cellular innate immune response to infection. These findings enhance our current understanding of liver disease pathogenesis and open new avenues for investigation and design of potential therapeutic strategies against viral infection and liver disease. Mitochondrial dynamics play a key role in cell survival. Damaged mitochondria are rapidly cleared via mitophagy and mitochondrial quality control system maintains cell homeostasis. HBV or HCV infection usurps this quality control to attenuate apoptosis of infected cells. These changes maintain a delicate balance where viral replication persists without inducing rapid cell death or viral clearance from the host. The role of mitochondria in liver disease pathogenesis is an emerging field and holds the promise in elucidating many metabolic and physiological perturbations that occur during viral hepatitis. Their functional role in chronic hepatitis and other relevant syndromes, including development of HCC, will certainly be highlighted in future endeavors of research. Finally, the potential of newer targets may be revealed to specifically blunt mitochondrial liver injury.

ACKNOWLEDGMENTS

Research in the laboratory (A.S.) is supported by grants from the U.S. National Institutes of Health (DK077704, DK08379, and AI085087) and the Michael J. Fox Foundation. S.M. was supported by Grant T32 DK07202.

REFERENCES

1. Ou JHJ (2002) *Hepatitis Viruses.* Boston, MA: Kluwer Academic, p. xviii, 281pp.
2. Koziel MJ, Siddiqui A (2005) Hepatitis B virus and hepatitis delta virus. In Mandell GL, Bennett JE, Dolin R, editors. *Principles and Practice of Infectious Diseases,* 6: 1864–1890. Churchill Livingstone, UK: Elsevier.
3. Shepard CW, Finelli L, Alter MJ (2005) Global epidemiology of hepatitis C virus infection. *Lancet Infect Dis* 5: 558–567.
4. Bouchard MJ, Navas-Martin S (2011) Hepatitis B and C virus hepatocarcinogenesis: Lessons learned and future challenges. *Cancer Lett* 305: 123–143.
5. Ferlay J, Shin HR, Bray F, Forman D, Mathers C et al. (2010) Estimates of worldwide burden of cancer in 2008: GLOBOCAN 2008. *Int J Cancer* 127: 2893–2917.
6. Wasley A, Alter MJ (2000) Epidemiology of hepatitis C: Geographic differences and temporal trends. *Semin Liver Dis* 20: 1–16.
7. Jameel S (1999) Molecular biology and pathogenesis of hepatitis E virus. *Expert Rev Mol Med* 1999: 1–16.
8. Gallegos-Orozco JF, Rakela-Brodner J (2010) Hepatitis viruses: Not always what it seems to be. *Rev Med Chil* 138: 1302–1311.

9. Kumar V, Cotran RS, Robbins SL (2003) *Robbins Basic Pathology*. Philadelphia, PA: Saunders, p. xii, 873pp.

10. Yamane D, McGivern DR, Masaki T, Lemon SM (2013) Liver injury and disease pathogenesis in chronic hepatitis C. *Curr Top Microbiol Immunol* 369: 263–288.

11. Barbaro G, Di Lorenzo G, Asti A, Ribersani M, Belloni G et al. (1999) Hepatocellular mitochondrial alterations in patients with chronic hepatitis C: Ultrastructural and biochemical findings. *Am J Gastroenterol* 94: 2198–2205.

12. Amako Y, Syed GH, Siddiqui A (2011) Protein kinase D negatively regulates hepatitis C virus secretion through phosphorylation of oxysterol-binding protein and ceramide transfer protein. *J Biol Chem* 286: 11265–11274.

13. Raimundo N (2014) Mitochondrial pathology: Stress signals from the energy factory. *Trends Mol Med* 20: 282–292.

14. Pourcelot M, Arnoult D (2014) Mitochondrial dynamics and the innate antiviral immune response. *FEBS J* 281: 3791–3802.

15. Lazarou M (2015) Keeping the immune system in check: A role for mitophagy. *Immunol Cell Biol* 93: 3–10.

16. Ohta A, Nishiyama Y (2011) Mitochondria and viruses. *Mitochondrion* 11: 1–12.

17. Waris G, Siddiqui A (2003) Regulatory mechanisms of viral hepatitis B and C. *J Biosci* 28: 311–321.

18. Tardif KD, Waris G, Siddiqui A (2005) Hepatitis C virus, ER stress, and oxidative stress. *Trends Microbiol* 13: 159–163.

19. Rawat S, Clippinger AJ, Bouchard MJ (2012) Modulation of apoptotic signaling by the hepatitis B virus X protein. *Viruses* 4: 2945–2972.

20. Wang T, Weinman SA (2013) Interactions between hepatitis C virus and mitochondria: Impact on pathogenesis and innate immunity. *Curr Pathobiol Rep* 1: 179–187.

21. Rahmani Z, Huh KW, Lasher R, Siddiqui A (2000) Hepatitis B virus X protein colocalizes to mitochondria with a human voltage-dependent anion channel, HVDAC3, and alters its transmembrane potential. *J Virol* 74: 2840–2846.

22. Tsutsumi T, Matsuda M, Aizaki H, Moriya K, Miyoshi H et al. (2009) Proteomics analysis of mitochondrial proteins reveals overexpression of a mitochondrial protein chaperon, prohibitin, in cells expressing hepatitis C virus core protein. *Hepatology* 50: 378–386.

23. Youle RJ, Narendra DP (2011) Mechanisms of mitophagy. *Nat Rev Mol Cell Biol* 12: 9–14.

24. Liesa M, Shirihai OS (2013) Mitochondrial dynamics in the regulation of nutrient utilization and energy expenditure. *Cell Metab* 17: 491–506.

25. West AP, Shadel GS, Ghosh S (2011) Mitochondria in innate immune responses. *Nat Rev Immunol* 11: 389–402.

26. Guidotti LG, Chisari FV (2006) Immunobiology and pathogenesis of viral hepatitis. *Annu Rev Pathol* 1: 23–61.

27. Kim SJ, Khan M, Quan J, Till A, Subramani S et al. (2013) Hepatitis B virus disrupts mitochondrial dynamics: Induces fission and mitophagy to attenuate apoptosis. *PLoS Pathog* 9: e1003722.

28. Kim SJ, Syed GH, Khan M, Chiu WW, Sohail MA et al. (2014) Hepatitis C virus triggers mitochondrial fission and attenuates apoptosis to promote viral persistence. *Proc Natl Acad Sci USA* 111: 6413–6418.

29. Kim SJ, Syed GH, Siddiqui A (2013) Hepatitis C virus induces the mitochondrial translocation of Parkin and subsequent mitophagy. *PLoS Pathog* 9: e1003285.

30. Seeger C, Mason WS (2000) Hepatitis B virus biology. *Microbiol Mol Biol Rev* 64(1): 51–68.

31. Hyams KC (1995) Risks of chronicity following acute hepatitis B virus infection: A review. *Clin Infect Dis* 20: 992–1000.

32. Bouchard MJ, Schneider RJ (2004) The enigmatic X gene of hepatitis B virus. *J Virol* 78: 12725–12734.

33. Clippinger AJ, Bouchard MJ (2008) Hepatitis B virus HBx protein localizes to mitochondria in primary rat hepatocytes and modulates mitochondrial membrane potential. *J Virol* 82: 6798–6811.

34. Henkler F, Hoare J, Waseem N, Goldin RD, McGarvey MJ et al. (2001) Intracellular localization of the hepatitis B virus HBx protein. *J Gen Virol* 82: 871–882.

35. Siddiqui A, Jameel S, Mapoles J (1987) Expression of the hepatitis B virus X gene in mammalian cells. *Proc Natl Acad Sci USA* 84: 2513–2517.

36. Pawlotsky JM (2004) Pathophysiology of hepatitis C virus infection and related liver disease. *Trends Microbiol* 12: 96–102.

37. Moradpour D, Penin F, Rice CM (2007) Replication of hepatitis C virus. *Nat Rev Microbiol* 5: 453–463.

38. Bartenschlager R, Penin F, Lohmann V, Andre P (2011) Assembly of infectious hepatitis C virus particles. *Trends Microbiol* 19: 95–103.

39. Miyanari Y, Atsuzawa K, Usuda N, Watashi K, Hishiki T et al. (2007) The lipid droplet is an important organelle for hepatitis C virus production. *Nat Cell Biol* 9: 1089–1097.

40. Syed GH, Amako Y, Siddiqui A (2010) Hepatitis C virus hijacks host lipid metabolism. *Trends Endocrinol Metab* 21: 33–40.

41. Simmonds P (1999) Viral heterogeneity of the hepatitis C virus. *J Hepatol* 31(Suppl. 1): 54–60.

42. Tellinghuisen TL, Evans MJ, von Hahn T, You S, Rice CM (2007) Studying hepatitis C virus: Making the best of a bad virus. *J Virol* 81: 8853–8867.

43. Xu Z, Jensen G, Yen TS (1997) Activation of hepatitis B virus S promoter by the viral large surface protein via induction of stress in the endoplasmic reticulum. *J Virol* 71: 7387–7392.

44. Kim I, Xu W, Reed JC (2008) Cell death and endoplasmic reticulum stress: Disease relevance and therapeutic opportunities. *Nat Rev Drug Discov* 7: 1013–1030.

45. Wang HC, Wu HC, Chen CF, Fausto N, Lei HY et al. (2003) Different types of ground glass hepatocytes in chronic hepatitis B virus infection contain specific pre-S mutants that may induce endoplasmic reticulum stress. *Am J Pathol* 163: 2441–2449.

46. Tardif KD, Mori K, Kaufman RJ, Siddiqui A (2004) Hepatitis C virus suppresses the IRE1-XBP1 pathway of the unfolded protein response. *J Biol Chem* 279: 17158–17164.

47. Saeed M, Suzuki R, Watanabe N, Masaki T, Tomonaga M et al. (2011) Role of the endoplasmic reticulum-associated degradation (ERAD) pathway in degradation of hepatitis C virus envelope proteins and production of virus particles. *J Biol Chem* 286: 37264–37273.

48. Li S, Ye L, Yu X, Xu B, Li K et al. (2009) Hepatitis C virus NS4B induces unfolded protein response and endoplasmic reticulum overload response-dependent NF-κB activation. *Virology* 391: 257–264.

49. Sir D, Chen WL, Choi J, Wakita T, Yen TS et al. (2008) Induction of incomplete autophagic response by hepatitis C virus via the unfolded protein response. *Hepatology* 48: 1054–1061.

50. Sir D, Tian Y, Chen WL, Ann DK, Yen TS et al. (2010) The early autophagic pathway is activated by hepatitis B virus and required for viral DNA replication. *Proc Natl Acad Sci USA* 107: 4383–4388.

51. Dreux M, Chisari FV (2010) Viruses and the autophagy machinery. *Cell Cycle* 9: 1295–1307.

52. Brenner C, Galluzzi L, Kepp O, Kroemer G (2013) Decoding cell death signals in liver inflammation. *J Hepatol* 59: 583–594.

53. Giorgi C, De Stefani D, Bononi A, Rizzuto R, Pinton P (2009) Structural and functional link between the mitochondrial network and the endoplasmic reticulum. *Int J Biochem Cell Biol* 41: 1817–1827.

54. Rizzuto R, Marchi S, Bonora M, Aguiari P, Bononi A et al. (2009) Ca(2+) transfer from the ER to mitochondria: When, how and why. *Biochim Biophys Acta* 1787: 1342–1351.

55. Bravo R, Gutierrez T, Paredes F, Gatica D, Rodriguez AE et al. (2012) Endoplasmic reticulum: ER stress regulates mitochondrial bioenergetics. *Int J Biochem Cell Biol* 44: 16–20.

56. Quarato G, Scrima R, Agriesti F, Moradpour D, Capitanio N et al. (2013) Targeting mitochondria in the infection strategy of the hepatitis C virus. *Int J Biochem Cell Biol* 45: 156–166.

57. Waris G, Huh KW, Siddiqui A (2001) Mitochondrially associated hepatitis B virus X protein constitutively activates transcription factors STAT-3 and NF-κB via oxidative stress. *Mol Cell Biol* 21: 7721–7730.

58. Bouchard MJ, Wang LH, Schneider RJ (2001) Calcium signaling by HBx protein in hepatitis B virus DNA replication. *Science* 294: 2376–2378.

59. Wang C, Yang W, Yan HX, Luo T, Zhang J et al. (2012) Hepatitis B virus X (HBx) induces tumorigenicity of hepatic progenitor cells in 3,5-diethoxycarbonyl-1,4-dihy-drocollidine-treated HBx transgenic mice. *Hepatology* 55: 108–120.

60. Hsieh YH, Su IJ, Wang HC, Chang WW, Lei HY et al. (2004) Pre-S mutant surface antigens in chronic hepatitis B virus infection induce oxidative stress and DNA damage. *Carcinogenesis* 25: 2023–2032.

61. Wang HC, Chang WT, Chang WW, Wu HC, Huang W et al. (2005) Hepatitis B virus pre-S2 mutant upregulates cyclin A expression and induces nodular proliferation of hepatocytes. *Hepatology* 41: 761–770.

62. Benali-Furet NL, Chami M, Houel L, De Giorgi F, Vernejoul F et al. (2005) Hepatitis C virus core triggers apoptosis in liver cells by inducing ER stress and ER calcium depletion. *Oncogene* 24: 4921–4933.

63. Qadri I, Iwahashi M, Capasso JM, Hopken MW, Flores S et al. (2004) Induced oxidative stress and activated expression of manganese superoxide dismutase during hepatitis C virus replication: Role of JNK, p38 MAPK and AP-1. *Biochem J* 378: 919–928.

64. Kim KY, Stevens MV, Akter MH, Rusk SE, Huang RJ et al. (2011) Parkin is a lipid-responsive regulator of fat uptake in mice and mutant human cells. *J Clin Invest* 121: 3701–3712.

65. Korenaga M, Wang T, Li Y, Showalter LA, Chan T et al. (2005) Hepatitis C virus core protein inhibits mitochondrial electron transport and increases reactive oxygen species (ROS) production. *J Biol Chem* 280: 37481–37488.

66. Brault C, Levy PL, Bartosch B (2013) Hepatitis C virus-induced mitochondrial dysfunctions. *Viruses* 5: 954–980.

67. Li Y, Boehning DF, Qian T, Popov VL, Weinman SA (2007) Hepatitis C virus core protein increases mitochondrial ROS production by stimulation of Ca²⁺ uniporter activity. *FASEB J* 21: 2474–2485.

68. Wang T, Campbell RV, Yi MK, Lemon SM, Weinman SA (2010) Role of hepatitis C virus core protein in viral-induced mitochondrial dysfunction. *J Viral Hepat* 17: 784–793.

69. Boudreau HE, Emerson SU, Korzeniowska A, Jendrysik MA, Leto TL (2009) Hepatitis C virus (HCV) proteins induce NADPH oxidase 4 expression in a transforming growth factor β-dependent manner: A new contributor to HCV-induced oxidative stress. *J Virol* 83: 12934–12946.

70. de Mochel NS, Seronello S, Wang SH, Ito C, Zheng JX et al. (2010) Hepatocyte NAD(P) H oxidases as an endogenous source of reactive oxygen species during hepatitis C virus infection. *Hepatology* 52: 47–59.

71. Higgs MR, Chouteau P, Lerat H (2014) 'Liver let die': Oxidative DNA damage and hepatotropic viruses. *J Gen Virol* 95: 991–1004.

72. Fujita N, Sugimoto R, Ma N, Tanaka H, Iwasa M et al. (2008) Comparison of hepatic oxidative DNA damage in patients with chronic hepatitis B and C. *J Viral Hepat* 15: 498–507.

73. Farinati F, Cardin R, Bortolami M, Burra P, Russo FP et al. (2007) Hepatitis C virus: From oxygen free radicals to hepatocellular carcinoma. *J Viral Hepat* 14: 821–829.

74. Al-Mehdi AB, Pastukh VM, Swiger BM, Reed DJ, Patel MR et al. (2012) Perinuclear mitochondrial clustering creates an oxidant-rich nuclear domain required for hypoxia-induced transcription. *Sci Signal* 5: ra47.

75. Cho HK, Cheong KJ, Kim HY, Cheong J (2011) Endoplasmic reticulum stress induced by hepatitis B virus X protein enhances cyclo-oxygenase 2 expression via activating transcription factor 4. *Biochem J* 435: 431–439.

76. Waris G, Siddiqui A (2005) Hepatitis C virus stimulates the expression of cyclooxygenase-2 via oxidative stress: Role of prostaglandin E2 in RNA replication. *J Virol* 79: 9725–9734.

77. Yu Y, Gong R, Mu Y, Chen Y, Zhu C et al. (2011) Hepatitis B virus induces a novel inflammation network involving three inflammatory factors, IL-29, IL-8, and cyclooxygenase-2. *J Immunol* 187: 4844–4860.

78. Hu KQ (2003) Cyclooxygenase 2 (COX2)-prostanoid pathway and liver diseases. *Prostaglandins Leukot Essent Fatty Acids* 69: 329–337.

79. Green DR, Galluzzi L, Kroemer G (2014) Cell biology. Metabolic control of cell death. *Science* 345: 1250256.

80. Galluzzi L, Kepp O, Kroemer G (2012) Mitochondria: Master regulators of danger signalling. *Nat Rev Mol Cell Biol* 13: 780–788.

81. Jezek P, Plecita-Hlavata L (2009) Mitochondrial reticulum network dynamics in relation to oxidative stress, redox regulation, and hypoxia. *Int J Biochem Cell Biol* 41: 1790–1804.

82. Chan DC (2006) Mitochondria: Dynamic organelles in disease, aging, and development. *Cell* 125: 1241–1252.

83. Youle RJ, Karbowski M (2005) Mitochondrial fission in apoptosis. *Nat Rev Mol Cell Biol* 6: 657–663.

84. Archer SL (2013) Mitochondrial dynamics—Mitochondrial fission and fusion in human diseases. *N Engl J Med* 369: 2236–2251.

85. Vescovo T, Refolo G, Romagnoli A, Ciccosanti F, Corazzari M et al. (2014) Autophagy in HCV infection: Keeping fat and inflammation at bay. *Biomed Res Int* 2014: 265353.

86. Loson OC, Song Z, Chen H, Chan DC (2013) Fis1, Mff, MiD49, and MiD51 mediate Drp1 recruitment in mitochondrial fission. *Mol Biol Cell* 24: 659–667.

87. Friedman JR, Lackner LL, West M, DiBenedetto JR, Nunnari J et al. (2011) ER tubules mark sites of mitochondrial division. *Science* 334: 358–362.

88. Haun F, Nakamura T, Lipton SA (2013) Dysfunctional mitochondrial dynamics in the pathophysiology of neurodegenerative diseases. *J Cell Death* 6: 27–35.

89. Chan NC, Salazar AM, Pham AH, Sweredoski MJ, Kolawa NJ et al. (2011) Broad activation of the ubiquitin-proteasome system by Parkin is critical for mitophagy. *Hum Mol Genet* 20: 1726–1737.

90. Ke PY, Chen SS (2011) Activation of the unfolded protein response and autophagy after hepatitis C virus infection suppresses innate antiviral immunity in vitro. *J Clin Invest* 121: 37–56.

91. Levine B, Kroemer G (2008) Autophagy in the pathogenesis of disease. *Cell* 132: 27–42.

92. Jin SM, Youle RJ (2012) PINK1- and Parkin-mediated mitophagy at a glance. *J Cell Sci* 125: 795–799.

93. Kondapalli C, Kazlauskaite A, Zhang N, Woodroof HI, Campbell DG et al. (2012) PINK1 is activated by mitochondrial membrane potential depolarization and stimulates Parkin E3 ligase activity by phosphorylating Serine 65. *Open Biol* 2: 120080.

94. Randow F, Youle RJ (2014) Self and nonself: How autophagy targets mitochondria and bacteria. *Cell Host Microbe* 15: 403–411.

95. Ding WX, Yin XM (2012) Mitophagy: Mechanisms, pathophysiological roles, and analysis. *Biol Chem* 393: 547–564.

96. Chu CT, Bayir H, Kagan VE (2014) LC3 binds externalized cardiolipin on injured mitochondria to signal mitophagy in neurons: Implications for Parkinson disease. *Autophagy* 10: 376–378.

97. Mattson MP, Gleichmann M, Cheng A (2008) Mitochondria in neuroplasticity and neurological disorders. *Neuron* 60: 748–766.

98. Knott AB, Perkins G, Schwarzenbacher R, Bossy-Wetzel E (2008) Mitochondrial fragmentation in neurodegeneration. *Nat Rev Neurosci* 9: 505–518.

99. Yu T, Jhun BS, Yoon Y (2011) High-glucose stimulation increases reactive oxygen species production through the calcium and mitogen-activated protein kinase-mediated activation of mitochondrial fission. *Antioxid Redox Signal* 14: 425–437.

100. Pichlmair A, Reis e Sousa C (2007) Innate recognition of viruses. *Immunity* 27: 370–383.

101. Horner SM, Gale M, Jr. (2013) Regulation of hepatic innate immunity by hepatitis C virus. *Nat Med* 19: 879–888.

102. Dixit E, Boulant S, Zhang Y, Lee AS, Odendall C et al. (2010) Peroxisomes are signaling platforms for antiviral innate immunity. *Cell* 141: 668–681.

103. Horner SM, Liu HM, Park HS, Briley J, Gale M, Jr. (2011) Mitochondrial-associated endoplasmic reticulum membranes (MAM) form innate immune synapses and are targeted by hepatitis C virus. *Proc Natl Acad Sci USA* 108: 14590–14595.

104. Seth RB, Sun L, Ea CK, Chen ZJ (2005) Identification and characterization of MAVS, a mitochondrial antiviral signaling protein that activates NF-κB and IRF 3. *Cell* 122: 669–682.

105. Li XD, Sun L, Seth RB, Pineda G, Chen ZJ (2005) Hepatitis C virus protease NS3/4A cleaves mitochondrial antiviral signaling protein off the mitochondria to evade innate immunity. *Proc Natl Acad Sci USA* 102: 17717–17722.

106. Kim SJ, Siddiqui A (2014) Hepatitis C virus-induced altered mitochondrial dynamics modulates innate antiviral immunity. In *21st International Symposium on Hepatitis C and Related Viruses*, Banff, Alberta, Canada, p. 72.

107. Kumar M, Jung SY, Hodgson AJ, Madden CR, Qin J et al. (2011) Hepatitis B virus regulatory HBx protein binds to adaptor protein IPS-1 and inhibits the activation of β interferon. *J Virol* 85: 987–995.

108. Wei C, Ni C, Song T, Liu Y, Yang X et al. (2010) The hepatitis B virus X protein disrupts innate immunity by downregulating mitochondrial antiviral signaling protein. *J Immunol* 185: 1158–1168.

109. Wang X, Li Y, Mao A, Li C, Tien P (2010) Hepatitis B virus X protein suppresses virus-triggered IRF3 activation and IFN-β induction by disrupting the VISA-associated complex. *Cell Mol Immunol* 7: 341–348.

110. Khan M, Kim SJ, Syed GH, Siddiqui A (2014) Hepatitis B virus-mediated altered mitochondrial dynamics cripples innate-immune response. In *2014 International Meeting on Molecular Biology of Hepatitis B Viruses*, Los Angeles, CA, p. O-70.

111. Burdette D, Haskett A, Presser L, McRae S, Iqbal J et al. (2012) Hepatitis C virus activates interleukin-1β via caspase-1-inflammasome complex. *J Gen Virol* 93: 235–246.

112. Negash AA, Ramos HJ, Crochet N, Lau DT, Doehle B et al. (2013) IL-1β production through the NLRP3 inflammasome by hepatic macrophages links hepatitis C virus infection with liver inflammation and disease. *PLoS Pathog* 9: e1003330.

113. Shrivastava S, Mukherjee A, Ray R, Ray RB (2013) Hepatitis C virus induces interleukin-1β (IL-1β)/IL-18 in circulatory and resident liver macrophages. *J Virol* 87: 12284–12290.

114. Nakahira K, Haspel JA, Rathinam VA, Lee SJ, Dolinay T et al. (2011) Autophagy proteins regulate innate immune responses by inhibiting the release of mitochondrial DNA mediated by the NALP3 inflammasome. *Nat Immunol* 12: 222–230.
115. Zhou R, Yazdi AS, Menu P, Tschopp J (2011) A role for mitochondria in NLRP3 inflammasome activation. *Nature* 469: 221–225.
116. Yu J, Nagasu H, Murakami T, Hoang H, Broderick L et al. (2014) Inflammasome activation leads to Caspase-1-dependent mitochondrial damage and block of mitophagy. *Proc Natl Acad Sci USA* 111: 15514–15519.
117. Shimada K, Crother TR, Karlin J, Dagvadorj J, Chiba N et al. (2012) Oxidized mitochondrial DNA activates the NLRP3 inflammasome during apoptosis. *Immunity* 36: 401–414.
118. Kim SR, Kim DI, Kim SH, Lee H, Lee KS et al. (2014) NLRP3 inflammasome activation by mitochondrial ROS in bronchial epithelial cells is required for allergic inflammation. *Cell Death Dis* 5: e1498.
119. Syed GH, Siddiqui A (2011) Effects of hypolipidemic agent nordihydroguaiaretic acid on lipid droplets and hepatitis C virus. *Hepatology* 54: 1936–1946.
120. Negro F (2010) Abnormalities of lipid metabolism in hepatitis C virus infection. *Gut* 59: 1279–1287.
121. Clement S, Pascarella S, Negro F (2009) Hepatitis C virus infection: Molecular pathways to steatosis, insulin resistance and oxidative stress. *Viruses* 1: 126–143.
122. Bedossa P, Moucari R, Chelbi E, Asselah T, Paradis V et al. (2007) Evidence for a role of nonalcoholic steatohepatitis in hepatitis C: A prospective study. *Hepatology* 46: 380–387.
123. Goossens N, Negro F (2014) Is genotype 3 of the hepatitis C virus the new villain? *Hepatology* 59: 2403–2412.
124. Mantena SK, King AL, Andringa KK, Eccleston HB, Bailey SM (2008) Mitochondrial dysfunction and oxidative stress in the pathogenesis of alcohol- and obesity-induced fatty liver diseases. *Free Radic Biol Med* 44: 1259–1272.
125. Babbar M, Sheikh MS (2013) Metabolic stress and disorders related to alterations in mitochondrial fission or fusion. *Mol Cell Pharmacol* 5: 109–133.

15 Mitochondrial Dynamics and Liver Metabolism

*María Isabel Hernández-Alvarez
and Antonio Zorzano*

CONTENTS

ABSTRACT

Mitochondria are very dynamic organelles that undergo rapid changes in shape, apparent number, and cell localization. Some of these alterations are a consequence of specific proteins that participate in mitochondrial fusion or fission processes. In this chapter, we analyze the current information available on how the activity of proteins involved in mitochondrial dynamics impacts liver metabolism or the capacity of liver cells to respond to hormones. Recent data indicate that the mitochondrial fusion protein Mfn2 maintains normal hepatic metabolism and its ablation causes reduced mitochondrial respiration, reduced glucose tolerance, and enhanced hepatic glucose production. Mfn2 deficiency also causes impaired insulin signaling and insulin resistance. Glucocorticoids induce hepatic Drp1 expression and a dominant negative form of Drp1 impairs some of the metabolic functions of glucocorticoids in hepatoma cells. The liver represents a unique organ in which to analyze, in depth, the mechanisms that link mitochondrial dynamics and disease.

Keywords: mitochondrial fusion, Mfn2, mitochondrial fission, Drp1, insulin resistance, glucocorticoid action

MITOCHONDRIAL FUNCTION AND HEPATIC METABOLISM

Hepatocytes are rich in mitochondria, which permits to cope with their high metabolic requirements. In this respect, there is data linking mitochondrial dysfunction with the progression of hepatic diseases (Perez-Carreras et al. 2003). Hepatic manifestations of mitochondrial disorders range from hepatic steatosis, cholestasis, and chronic liver disease with insidious onset to neonatal liver failure, and they are frequently associated with neuromuscular symptoms (Sokol and Treem 1999). Defects in mitochondrial function can lead to impaired oxidative phosphorylation (OXPHOS), increased generation of reactive oxygen species, impairment of other metabolic pathways, and activation of mechanisms of cellular death (Duchen and Szabadkai 2010). Mitochondrial hepatopathies can be classified into primary, in which the mitochondrial defect is the primary cause of the disease, and secondary, in which mitochondrial function is affected by a nonmitochondrial gene. The primary mitochondrial disorder can be caused by mutations affecting mitochondrial DNA (mtDNA) or by nuclear genes that encode mitochondrial proteins or cofactors (Bandyopadhyay and Dutta 2005).

As to primary mitochondrial hepatopathies caused by mutations in mtDNA, the most severe mtDNA rearrangement disease is the Pearson syndrome; the proteins affected by this deletion include respiratory chain enzymes (complex I is the most severely affected), 2 subunits of complex V, 1 subunit of complex IV, and 5 transfer RNA genes (Kapsa et al. 1994). The liver involvement is manifested as marked hepatomegaly, hepatic steatosis, and cirrhosis. Liver failure and death in some cases have been reported before the age of 4 years (Morikawa et al. 1993). mtDNA depletion syndrome is defined as a reduction in the mtDNA copy number in different tissues, leading to insufficient syntheses of respiratory chain complexes I, III, IV, and V (Moraes et al. 1991; DiMauro and Schon 2003). Affected patients generally present with a severe form of liver failure in infancy (Morris et al. 1998; Labarthe et al. 2005). Most patients with mtDNA diseases caused by deletions that cross two or more gene boundaries (intergenic mutation) generally do not reproduce. Therefore, most intergenic mutations arise de novo, resulting in sporadic disease (Holt et al. 1988).

It is now clear that most mitochondrial diseases with primary involvement of the liver are caused by nuclear DNA (nDNA) mutations. A wide array of pathogenic mutations has been identified in nDNA-encoded mitochondrial OXPHOS structural (Procaccio and Wallace 2004) or assembly genes such as the complex IV (COX) or *Cox10* (Antonicka et al. 2003).

Liver-specific inactivation of *Cox10* caused mitochondrial hepatopathy similar to what is found in some children with mitochondrial disease. These mice show reduced body weight and activity, and start dying between 45 and 65 days. They had severe liver dysfunction, with reduced COX activity, increased SDH activity, increased mitochondrial proliferation and lipid accumulation, reduced stored glycogen, and diminished ATP levels (Diaz et al. 2008).

A mutation in *BCS1L* (ubiquinol–cytochrome *c* reductase complex) has been found to be associated with mitochondrial neonatal liver failure (de Lonlay et al.

2001). De Lonlay et al. reported deficient activity of complex III of the respiratory chain in the liver, fibroblasts, or muscle in affected infants with hepatic failure, lactic acidosis, renal tubulopathy, and variable degrees of encephalopathy (de Lonlay et al. 2001).

Alpers-Huttenlocher syndrome (delayed-onset liver disease) presents hepatopathy with or without acute liver failure (Harding 1990). Typically, the onset of symptoms occurs between 2 months and 8 years of life and is characterized by hepatomegaly, jaundice, and progressive coagulopathy and hypoglycemia (Narkewicz et al. 1991). In some patients, NADH oxidoreductase (complex I) deficiency has been found in liver mitochondria (Cormier-Daire et al. 1997).

Reye's syndrome is the best known form of secondary mitochondrial hepatopathy. This disease is caused by the interaction of viral infection (influenza, varicella, enterovirus, and other viruses) and salicylate use with some underlying undefined metabolic (defects in fatty acid oxidation) or genetic (reduced activity of an uncoupling protein) predisposition (Orlowski 1999). Salicylate impairs mitochondrial fatty acid oxidation by reversible inhibition of LCHAD activity (Glasgow et al. 1999). Other examples of secondary mitochondrial hepatopathies include Wilson disease, valproic acid hepatotoxicity, and the effects of nucleoside reverse transcriptase inhibitors. Liver biopsies of Wilson disease show microvesicular steatosis in the absence of hepatic inflammation or necrosis and characteristic swelling and pleomorphism of mitochondria under electron microscopy (Sternlieb 1968).

HEPATIC MITOCHONDRIA–RELATED DISEASES

Fatty liver is another hepatopathy related to deficiencies in enzymes involved in mitochondrial fatty acid oxidation (beta-oxidation), a pathway that activates free fatty acids to acyl-CoA esters and further metabolizes them to acetyl-CoA. Decreased fatty acid oxidation with subsequent fatty infiltration of the liver may contribute to the genesis of steatosis, particularly in the setting of hyperinsulinemia (Farrell and Larter 2006). Hepatic fat accumulation is a well-recognized complication of diabetes with a reported frequency of 40%–70% (Levinthal and Tavill 1999).

Nonalcoholic steatohepatitis (NASH) is a variant of fatty liver in which fat in the hepatocytes is accompanied by lobular inflammation and steatonecrosis. The diagnosis can only be made in the absence of alcohol abuse or other causes of liver disease, particularly hepatitis C (Levinthal and Tavill 1999). In patients with diabetes and steatohepatitis, Mallory bodies such as those seen in alcoholic liver disease may be seen (Hübscher 2006). NASH has been associated most commonly with obese women with diabetes, but the disease is certainly not limited to patients with this clinical profile. There is certainly a higher prevalence in type 2 diabetic patients treated with insulin (Gupte et al. 2004).

Hepatocarcinogenesis is considered a multistep process involving subsequent gene mutations that control proliferation and/or apoptosis in the hepatocytes

(Hsu et al. 2013a). Genes encoding mitochondrial proteins are involved in the mechanisms underlying hepatocellular carcinoma (HCC). For instance, human mitochondrial ATPase subunit 6 (mtATPase 6) has been reported to be differentially expressed in hepatoma cell lines (Oh et al. 1998). In addition, mtDNA mutants in T6787C (cytochrome *c* oxidase subunit I, COI), G7976A (COII), G9267A (COIII), and A11708G (ND4) may result in amino acid substitutions in the highly conserved regions of mitochondrial genes. These mtDNA mutations have the potential to cause mitochondrial dysfunction in HCC (Yin et al. 2010). Furthermore, mitochondrial dysfunction reduces intracellular ATP content and represses the expression of hypoxia-inducible factor-1α (HIF-1α) through the activation of the AMP-activated protein kinase (AMPK)-mTOR pathways in hepatoma cells (Hsu et al. 2013b). The activation of reversing signaling from the mitochondria to the nucleus may play an important role in the malignant progression of HCC.

MITOCHONDRIAL MORPHOLOGY

Morphological alterations of hepatic mitochondria have been observed in nonalcoholic fatty liver disease (NAFLD), suggesting the involvement of alterations in mitochondrial dynamics during the progression of liver disease (Caldwell et al. 1999; Sanyal et al. 2001). Mitochondria were observed as round and swollen with the loss of discernable crista structure; instead, the mitochondria contained paracrystalline inclusion bodies. Sanyal et al. correlated defects in the mitochondrial morphology with an enhanced oxidative stress by demonstrating significant increases in nitrotyrosine in hepatocytes from patients with fatty liver disease and NASH (Sanyal et al. 2001).

MITOCHONDRIAL DYNAMICS

Mitochondria are very dynamic organelles and constantly undergo changes in shape, size, number, and location. Mitochondria form tubular or branched reticular networks, which undergo a regulated balance between fusion and fission processes (Nunnari et al. 1997; Yaffe 2003). Mitochondrial filaments or networks are relevant in multiple functions of mitochondria such as in apoptosis, Ca^{2+} homeostasis, mitochondrial quality control, and oxidative metabolism (Nunnari and Suomalainen 2012). Hepatocytes form mitochondrial networks as shown in Figure 15.1.

Mitochondrial fission in mammals is mediated by Drp1 (dynamin-related protein), Fis1 (fission-1 homologue protein), and Mff (mitochondrial fission factor). Drp1 is a soluble GTPase located in the cytosol and is expressed in all tissues assayed (Smirnova et al. 1998). Mitochondrial fission requires the recruitment of Drp1 to the outer mitochondrial membrane, where it forms dotted structures located on future mitochondrial scission sites (Smirnova et al. 2001). In order to localize on the mitochondrial membrane, Drp1 must interact with Fis1 (Koch et al. 2005) or Mff (Otera et al. 2010), both located in the outer mitochondrial membrane. Fis1 and Mff, in

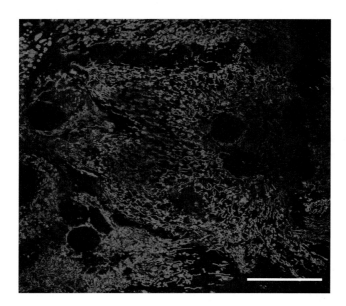

FIGURE 15.1 (See color insert.) Mitochondrial morphology of hepatocytes. Primary cultures of mouse hepatocytes were in vivo stained with MitoTracker Green for 15 min and then visualized under the confocal microscope. Scale bar, 10 μm.

contrast to other fusion and fission proteins, do not show GTPase activity. Recently, MiD49 and MiD51 have been described to have a role in the recruitment of Drp1 to mitochondria, although their role in mitochondrial fission is not completely elucidated (Palmer et al. 2011, 2013; Zhao et al. 2011).

Mitochondrial fusion recruits the exchange of content DNA, and metabolites between neighboring mitochondria (Detmer and Chan 2007). To ensure an appropriate mitochondrial compartmentalization, both inner and outer mitochondrial membranes need to fuse (Eura et al. 2003). Optic atrophy protein-1 (OPA1) mediates fusion of the inner mitochondrial membrane and is involved in maintaining the morphology of mitochondrial cristae and in the control of apoptosis (Frezza et al. 2006; Baricault et al. 2007); its downregulation leads to aberrant cristae remodeling and to the release of cytochrome *c*. Fusion of the outer mitochondrial membrane depends on two GTPase family members: mitofusin-1 (Mfn1) and mitofusin-2 (Mfn2). Mfn1 and Mfn2 proteins are ubiquitously expressed, but differences in expression levels are observed among tissues: Mfn1 predominates in heart, liver, pancreas, adrenal glands, and testis, whereas Mfn2 is more abundant in heart, skeletal muscle, brain, and brown adipose tissue (Bach et al. 2003; Eura et al. 2003).

Mfn2 REGULATES HEPATIC MITOCHONDRIAL METABOLISM

Mfn2 is an integral outer mitochondrial membrane protein with both N-terminal and C-terminal ends exposed to the cytosol (Rojo et al. 2002). Like Mfn1, Mfn2 contains a heptad repeat domain (HR2) localized in the C-terminal region of the protein. The N-terminal GTPase activity is crucial for Mfn2-mediated mitochondrial fusion (Santel and Fuller 2001; Rojo et al. 2002). However, similar to that reported for Mfn1, the GTPase activity of Mfn2 is not necessary for the perinuclear mitochondrial aggregates caused by Mfn2 overexpression (Santel and Fuller 2001; Santel et al. 2003; Huang et al. 2007). Mfn2 is relevant in human disease. Mutations in *Mfn2* gene cause Charcot–Marie–Tooth 2A, an autosomal-dominant axonal neuropathy (Zuchner et al. 2004).

Mfn2 is essential for embryonic development, and global knockout (KO) mice have been reported to die in midgestation due to placental defects. When the placental defect is rescued, Mfn2-deficient mice develop a rapid cerebellar degeneration (Chen et al. 2003, 2007).

Our laboratory has demonstrated that liver-specific Mfn2 deficiency causes mitochondrial dysfunction, which is characterized by reduced state 3 and state 4 respiration, reduced glucose oxidation, and diminished activity of respiratory complexes I + III and II + III, which occurs in the absence of changes in ATP concentrations or in mitochondrial mass (Sebastian et al. 2012). Liver-specific Mfn2 knockout mice show a transient hyperglycemia, and impaired glucose tolerance when subjected to a normal diet. The impaired glucose tolerance of liver-specific Mfn2 knockout mice was characterized by enhanced hepatic glucose production by gluconeogenesis, and by increased expression of genes encoding for gluconeogenic enzymes (pyruvate carboxylase, PEPCK, or glucose-6-phosphatase) and regulators of gluconeogenesis such as PGC-1α. Furthermore, knockout mice showed a high hepatic activity of CRTC2 (a CREB coactivator) in response to fasting, and reduced phosphorylation of FOXO1. Overall, the results suggest that Mfn2 deficiency enhances the activity of CRTC2 and FOXO1, and the induction of target genes and gluconeogenesis (Sebastian et al. 2012). In keeping with those data, transient in vivo repression of liver Mfn2 using a RNA interference technique (shRNA) causes increased fasting glycemia, enhanced hepatic glucose production, and reduced hepatic fatty acid synthesis leading to hypertriglyceridemia (Chen and Xu 2009). In keeping with these data, adenoviral-mediated expression of Mfn2 reduced hepatic steatosis induced by a high-fat diet in rats (Gan et al. 2013).

Another relevant aspect is whether the metabolic effects of Mfn2 are dependent on changes in mitochondrial morphology. This has been documented by expressing a short form of human Mfn2 (ΔMfn2) that is inactive as a mitochondrial fusion protein in muscle and liver cells (Segales et al. 2013). Data obtained in hepatoma FaO cells and in mouse liver indicate that ΔMfn2 stimulates gluconeogenic activity. These data are in keeping with the reduced gluconeogenesis from lactate/pyruvate reported in FaO cells upon Mfn2 silencing (Segales et al. 2013). The effects of ΔMfn2 are not a consequence of changes in the expression of gluconeogenic genes and may be a consequence of enhanced mitochondrial pyruvate uptake or enhanced TCA activity (Segales et al. 2013). In fact, this view is in keeping with evidence indicating that pyruvate carboxylase and its mitochondrial uptake control the gluconeogenic flux in hepatocytes, and the observations that TCA activity is linked to hepatic gluconeogenesis. Of note, changes in liver metabolism associated with the hepatic expression of ΔMfn2 caused enhanced circulating glucose and insulin during fasting (Segales et al. 2013).

The expression of ΔMfn2 caused a pattern of changes in cell respiration that was similar in hepatoma FaO cells and in muscle cells, indicating that the effects of Mfn2 protein are detectable in different cell contexts. ΔMfn2 enhances oxygen consumption under routine and maximally stimulated conditions in both cell types, and this occurs in the absence of alterations in the rate of proton leak (Segales et al. 2013). These data suggest that the higher rates of oxygen consumption are not explained by reduced mitochondrial energy efficiency caused by proton leak. Furthermore, the effects of ΔMfn2 are independent of changes in mitochondrial mass. These results are in keeping with the data obtained in cells or in tissues upon Mfn2's loss of function (Sebastian et al. 2012; Segales et al. 2013). Under these conditions, Mfn2's deficiency caused mitochondrial dysfunction, which was characterized by enhanced respiratory leak and unchanged mitochondrial mass (Segales et al. 2013).

Mfn2 REGULATES HEPATIC INSULIN SIGNALING

The liver plays a crucial role in the development of insulin resistance and in type 2 diabetes (Kotronen et al. 2008). Several lines of evidence indicate that defects in liver mitochondrial oxidative function can induce hepatic insulin resistance (Perez-Carreras et al. 2003; Petersen et al. 2003, 2004). The development of NAFLD is strongly associated with hepatic insulin resistance. This relationship is most apparent when NAFLD is induced in rats by a high fat diet for 3 days (Perry et al. 2014).

Recent evidence indicates that hepatic Mfn2 is subjected to both hormonal and metabolic control. Mfn2 is repressed by glucocorticoids, which has been documented in cultured hepatoma cells and also in mouse liver after in vivo treatment with dexamethasone (Segales et al. 2013). In addition, mice subjected to a high fat diet, mainly enriched in lard, caused a reduced hepatic Mfn2 in parallel to insulin resistance and oxidative stress (Lionetti et al. 2014). Induction of excessive lipid accumulation in hepatoma HepG2 cells by incubation with long-chain fatty acids causes Mfn2 repression, and under these conditions, omega-3 polyunsaturated fatty acids such as eicosapentaenoic acid or docosahexaenoic acid increased the expression of Mfn2 (Zhang et al. 2011). Furthermore, patients with extrahepatic cholestasis show a reduced expression of hepatic Mfn2, and in this regard, glycochenodeoxycholic acid, the main toxic component of bile acid in patients with extrahepatic cholestasis, repressed Mfn2 in human liver cells (Chen et al. 2013).

As mentioned earlier, hepatic Mfn2's loss of function reduces glucose disposal, enhances plasma insulin, and increases insulin resistance, which strongly supports the view that this situation alters the cellular capacity to respond to insulin. Moreover, exposure of these mice to a high fat diet caused a substantial hyperinsulinemia under basal conditions or during a glucose tolerance test, and an impaired glucose tolerance is also documented. In addition, the response to an insulin tolerance test is also markedly impaired in liver-specific Mfn2 knockout mice (Sebastian et al. 2012). Our group has reported that liver-specific Mfn2 ablation reduced insulin signaling in liver. The reduced stimulated insulin signaling was characterized by a lower activation of insulin receptors, IRS1's and IRS2's association to the p85 subunit of PI 3-kinase, or Akt phosphorylation as well as a reduced IRS2 expression (Sebastian et al. 2012).

Moreover, genetic repression of Mfn2 reduced insulin signaling and significantly prevented the inhibitory action of insulin on gluconeogenic genes in liver HuH7 cells. Furthermore, overexpression of Mfn2 increased both insulin signaling and insulin action (Tubbs et al. 2014).

REGULATION OF MITOCHONDRIAL FISSION AND Drp1

In addition to mitochondrial fusion proteins, alterations in mitochondrial fission also contribute to the defects in mitochondrial metabolism, increased generation of reactive oxygen species, a higher susceptibility of cells to undergo apoptosis, a different capacity to generate ATP, and mitochondria with heterogeneous mtDNA distribution (Galloway and Yoon 2013). Mitochondria become rapidly fragmented in high glucose concentrations with a concomitant increase of ROS (Yu et al. 2006).

Given that Drp1 undergoes changes in localization from the cytosol to the mitochondria, many studies have focused on the mitochondrial recruitment and the posttranslational regulation of this protein. It is likely that the detailed understanding of the regulation of Drp1 localization will explain how some cellular processes mediate mitochondrial fragmentation. Drp1 activity can be modulated by apoptosis, neural function, cardiac and muscle differenciation, cell cycle and for the second messenger molecule cAMP through PKA (protein kinase A), but also for calcium through CaMKI and for the phosphatase calcineurin (Wasiak et al. 2007). In addition, a number of studies have shown that Drp1 and its effects on mitochondrial dynamics are regulated by phosphorylation, thereby providing significant clues as to the intracellular signals that induce mitochondrial fission. A first report demonstrated that cyclic-dependent kinase 1 (Cdk1/cyclin B) phosphorylates rat Drp1 in serine 585 (a residue located in the GED/assembly domain) during mitosis. This phosphorylation enhances mitochondrial fission (Taguchi et al. 2007), thereby allowing the proper distribution of mitochondria within daughter nascent cells. On the other hand, ERK1/2-mediated Drp1 phosphorylation increases mitochondrial fission during high-glucose condition and the recovery signal could involve dephosphorylation of Drp1 at S616. In contrast, the Ca^{2+}-dependent phosphatase calcineurin dephosphorylates Drp1 at the cyclic AMP–dependent protein kinase phosphorylation site (S637) (Yu et al. 2011). High-glucose challenging of cells derived from the liver and cardiovascular system evokes a Ca^{2+} transient that activates the mitogen-activated protein (MAP) kinase extracellular signal–regulated kinase 1/2 (ERK1/2) (Galloway and Yoon 2012). Importantly, these results demonstrate that the high-glucose-induced Ca^{2+} transient and ERK1/2 activation are upstream components activating mitochondrial fission by increasing the Drp1 translocation to the mitochondria (Galloway and Yoon 2012).

ROLE OF Drp1 ON GLUCOCORTICOID ACTION

Glucocorticoids, hormones released into the blood by the adrenal glands, are essential for survival. Du et al. (2009) have shown that glucocorticoids directly affect mitochondrial function in neurons, suggesting a mechanism by which these hormones can influence neuronal function and survival. In the liver, glucocorticoids, such as dexamethasone, enhance hepatic gluconeogenesis partly through changes in mitochondrial function as reported in rat liver mitochondria (Roussel et al. 2003, 2004; Arvier et al. 2007a,b) and hepatic cell lines (Desquiret et al. 2008). It has been reported that some of the effects of glucocorticoids on respiratory chain activity are dependent on the transcription of genes encoding enzymes of oxidative phosphorylation (Chen et al. 2005).

Furthermore, we recently reported that dexamethasone caused a marked modification of mitochondrial morphology in hepatoma FaO cells; thus, mitochondria shifted from elongated tubules to donut-like shaped mitochondria accompanied by a marked increase in Drp1 expression (Hernandez-Alvarez et al. 2013). The observation that dexamethasone induces Drp1 and cellular respiration permits to propose that dexamethasone stimulates cell respiration by a mechanism that requires Drp1 activity (Hernandez-Alvarez et al. 2013) (Figure 15.2).

Moreover, the inhibition of mitochondrial fission by the expression of Drp1^{K38A} (a dominant negative form of the protein) almost completely blocked the effects of dexamethasone, enhancing gluconeogenesis from lactate/pyruvate (which requires mitochondrial steps). This indicates that dexamethasone enhances gluconeogenesis, at least in part, by activation of mitochondrial function, which requires Drp1 activity (Hernandez-Alvarez et al. 2013) (Figure 15.2).

Interestingly, the formation of donut-like mitochondria seems to be independent of Drp1 activity. In fact, the effects of dexamethasone or the effects of hypoxia on mitochondrial morphology are not cancelled upon expression of the dominant-negative Drp1^{K38A} (Liu and Hajnoczky 2011). Thus, it does not seem that the changes

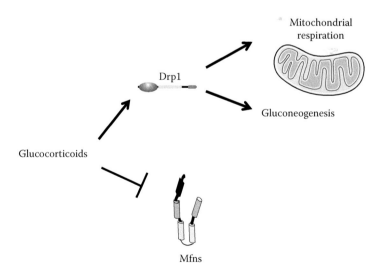

FIGURE 15.2 (**See color insert.**) Model of the hepatic effects of glucocorticoids on mitochondrial dynamics and mitochondrial metabolism. Glucocorticoids (dexamethasone) induce Drp1, repress mitofusins, which leads to rounded-shape mitochondria. Under these conditions, cells show enhanced mitochondrial respiration and gluconeogenic activity.

induced in mitochondrial morphology are relevant to the effects of dexamethasone on mitochondrial respiration or on gluconeogenesis. Rapid fragmentation of mito-chondria concomitantly increases total mitochondrial surface area, and it may also increase the accessibility of metabolic substrates (e.g., pyruvate) to carrier proteins (Lionetti et al. 2014).

IMPLICATIONS OF MITOCHONDRIAL–ENDOPLASMIC RETICULUM CONTACT SITES ON METABOLISM

The physical and functional communication between mitochondria and the ER is modulated by fission and fusion of mitochondria. ER–mitochondria contact sites are close appositions between these two organelles that facilitate both signaling and the passage of ions and lipids from one compartment to another. The ER membranes that bridge with mitochondria are known as mitochondria-associated membranes (MAMs) (Vance 1990). MAMs play an essential role in several cellular func-tions, including lipid transport, Ca^{2+} signaling, and apoptosis (Hayashi et al. 2009). A number of mitochondrial or ER-bound proteins are important for maintaining structural communication between the two organelles at the MAMs (Hayashi et al. 2009; Simmen et al. 2010). In particular, communication between the organelles is modulated by a family of chaperone proteins. The voltage-dependent anion channel (VDAC) is physically linked to the inositol 1,4,5-triphosphate receptor (IP_3R) via the molecular chaperone grp75 (Szabadkai et al. 2006). Overexpression of the cytosolic form of grp75 selectively increases IP_3-induced Ca^{2+} uptake into the mitochondrial matrix, whereas overexpression of the mitochondrial form of grp75 does not have this effect. Another protein that modulates the interaction between mitochondria and the ER is phosphofurin acidic cluster sorting protein-2 (PACS-2), which is known to integrate ER–mitochondrial communication and apoptosis signaling (Simmen et al. 2005). PACS-2 depletion induces mitochondrial fragmentation, dissociates the ER from mitochondria, and blocks apoptosis signaling. More recently, Sigma-1 receptors have been shown to be located at the MAMs, where they form complexes with BiP (Hayashi and Su 2007). Sigma-1 receptors dissociate from BiP and bind to type-3 IP_3Rs under conditions of ER Ca^{2+} depletion. Thus, type-3 IP3Rs are not degraded by proteasomes. Ca^{2+} depletion appears to induce a prolonged Ca^{2+} signal-ing event from the ER to the mitochondria, via IP_3Rs. Together, the data suggest that Sigma-1 receptors are involved in maintaining normal Ca^{2+} signaling from the ER to mitochondria (Hayashi and Su 2007).

Mitofusin-2 tethers the ER to mitochondria via the formation of both homotypic and heterotypic complexes. The tethering effect of mitofusin-2 appears to play a role in the control of Ca^{2+} flow between mitochondria and the ER (de Brito and Scorrano 2008). In addition, Mfn2-mediated ER–mitochondrial contact plays a role in autophagosome biogenesis (Hamasaki et al. 2013). These physical contacts are persistent and maintained under dynamic conditions (Friedman et al. 2010), sug-gesting that the ER–mitochondrial interface is vital for function. In agreement with this view, silencing Mfn2 in liver cells reduced the interaction between ER and mitochondria (Tubbs et al. 2014).

Furthermore, recent evidence linking MAM–mitochondria contact sites to the regulation of glucose metabolism is the finding that the kinase Akt is highly enriched in MAMs (Betz et al. 2013). In addition to Akt being localized to MAMs, mTORC2 (the mammalian target of rapamycin complex), which enhances cell growth and is antiapoptotic, is also enriched in MAMs (Betz et al. 2013). Insulin promoted the association of mTORC2 with MAMs and increased the number of connections between the ER and mitochondria. Furthermore, the association between mTORC2 with MAMs increased the phosphorylation of MAM-associated Akt (Betz et al. 2013). In mice in which rictor (a component required for mTORC activity) was specifically eliminated from the liver, the connections between the MAM and mitochondria were disrupted and the mice developed glucose intolerance, hyperinsulinemia, and increased gluconeogenesis (Hagiwara et al. 2012).

In addition to physical and functional connections to the ER, a recent study demonstrates that mitochondria are physically and functionally tethered to melanosomes, specialized lysosome-related organelles of pigment cells (Daniele et al. 2014). Mitochondria-melanosome tethering is mediated by Mfn2 and plays a role in melanosome biogenesis.

Perhaps through its participation in ER–mitochondrial contact sites, Mfn2 plays a role in the unfolding protein response (UPR) (Sebastian et al. 2012; Munoz et al. 2013). Specifically, we have demonstrated that the loss of function of Mfn2 induces chronic activation of ER stress, in parallel to metabolic dysfunction and insulin resistance. In this regard, amelioration of ER stress with chronic TUDCA treatment normalized the abnormal glucose homeostasis, excessive oxidative stress, and deficient insulin signaling of liver-specific Mfn2 knockout mice (Sebastian et al. 2012). In addition, we have demonstrated that ER stress is upstream of the alterations in mitochondrial function caused by Mfn2 deficiency. Thus, selective knockdown of the PERK UPR branch in Mfn2-deficient cells improved mitochondrial respiration and reduced ROS production, indicating that, at least partially, mitochondrial dysfunction is caused by sustained PERK activation. Our studies demonstrate that PERK is physically and functionally associated with Mfn2, and that the mitochondria–ER contact sites regulate UPR and mitochondrial metabolism (Munoz et al. 2013). Furthermore, chronic treatment with NAC normalized insulin signaling and glucose tolerance in liver-specific KO mice (Sebastian et al. 2012). In all, our data support the view that Mfn2 deficiency causes mitochondrial dysfunction and ER stress, which leads to enhanced ROS production, enhanced JNK activity, and inactivation of IRS1, a key protein in insulin signaling. All these data are coherent with the model shown in Figure 15.3.

In all, the mitochondria–ER contact sites can be considered as potential novel targets to modulate cellular metabolism and other aspects of cell function.

FUTURE PROSPECTS

Mitochondrial dynamics is bound to play a key role in tissues, and alterations in mitochondrial fusion or fission proteins may participate in the development of as-yet unrecognized pathological processes. In the specific context of the liver, the most

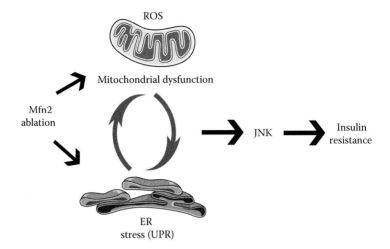

FIGURE 15.3 Model of the mechanisms by which Mfn2 deficiency causes insulin resistance in liver. Mfn2's loss of function causes ER stress (UPR response) and mitochondrial dysfunction (enhanced ROS production). These two alterations, in turn, activate JNK, which causes insulin resistance. Both pathways show a positive cross-talk (gray arrows).

relevant questions to be answered regarding the impact of mitochondrial dynamics relate to (a) the modulation of mitochondrial fusion and fission proteins during the fed to fasting transition in which major metabolic changes occur, and in which the liver goes from the state in which it uses glucose as a major substrate to the state in which it synthesizes and releases glucose to maintain glucose homeostasis and (b) the role of mitochondrial fusion or fission in the mitochondrial dysfunction reported in liver disease ranging from steatosis to NASH, cirrhosis and HCC, or other liver cancers.

ACKNOWLEDGMENTS

This study was supported by research grants from MINECO (SAF2008-03803 and SAF2013-40987R), Grants 2009SGR915 and 2014SGR48 from the "Generalitat de Catalunya," EFSD (2013), CIBERDEM ("Instituto de Salud Carlos III"), INTERREG IV-B-SUDOE-FEDER (DIOMED, SOE1/P1/E178). A.Z. is the recipient of an ICREA Acadèmia (Generalitat de Catalunya).

REFERENCES

Antonicka, H., S. C. Leary et al. (2003). Mutations in COX10 result in a defect in mitochondrial heme A biosynthesis and account for multiple, early-onset clinical phenotypes associated with isolated COX deficiency. *Hum Mol Genet* 12(20): 2693–2702.

Arvier, M., L. Lagoutte et al. (2007a). Adenine nucleotide translocator promotes oxidative phosphorylation and mild uncoupling in mitochondria after dexamethasone treatment. *Am J Physiol Endocrinol Metab* 293(5): E1320–E1324.

Arvier, M., L. Lagoutte et al. (2007b). Adenine nucleotide translocator promotes oxidative phosphorylation and mild uncoupling in mitochondria after dexamethasone treatment. *Am J Physiol Endocrinol Metab* 293(5): 14.

Bach, D., S. Pich et al. (2003). Mitofusin-2 determines mitochondrial network architecture and mitochondrial metabolism. A novel regulatory mechanism altered in obesity. *J Biol Chem* 278(19): 17190–17197.

Bandyopadhyay, S. and A. Dutta (2005). Mitochondrial hepatopathies. *JAPI* 53: 973–978.

Baricault, L., B. Segui et al. (2007). OPA1 cleavage depends on decreased mitochondrial ATP level and bivalent metals. *Exp Cell Res* 313(17): 3800–3808.

Betz, C., D. Stracka et al. (2013). Feature Article: mTOR complex 2-Akt signaling at mitochondria-associated endoplasmic reticulum membranes (MAM) regulates mitochondrial physiology. *Proc Natl Acad Sci USA* 110(31): 12526–12534.

Caldwell, S. H., R. H. Swerdlow et al. (1999). Mitochondrial abnormalities in non-alcoholic steatohepatitis. *J Hepatol* 31(3): 430–434.

Cormier-Daire, V., D. Chretien et al. (1997). Neonatal and delayed-onset liver involvement in disorders of oxidative phosphorylation. *J Pediatr* 130(5): 817–822.

Chen, H., S. A. Detmer et al. (2003). Mitofusins Mfn1 and Mfn2 coordinately regulate mitochondrial fusion and are essential for embryonic development. *J Cell Biol* 160(2): 189–200.

Chen, H., J. M. McCaffery et al. (2007). Mitochondrial fusion protects against neurodegeneration in the cerebellum. *Cell* 130(3): 548–562.

Chen, J.-Q., J. D. Yager et al. (2005). Regulation of mitochondrial respiratory chain structure and function by estrogens/estrogen receptors and potential physiological/pathophysiological implications. *Biochim Biophys Acta: Mol Cell Res* 1746(1): 1–17.

Chen, X. and Y. Xu (2009). Liver-specific reduction of Mfn2 protein by RNAi results in impaired glycometabolism and lipid homeostasis in BALB/c mice. *J Huazhong Univ Sci Technol Med Sci* 29(6): 689–696.

Chen, Y., L. Lv et al. (2013). Mitofusin 2 protects hepatocyte mitochondrial function from damage induced by GCDCA. *PLoS One* 8(6): e65455.

Daniele, T., I. Hurbain et al. (2014). Mitochondria and melanosomes establish physical contacts modulated by Mfn2 and involved in organelle biogenesis. *Curr Biol* 24(4): 393–403.

de Brito, O. M. and L. Scorrano (2008). Mitofusin 2 tethers endoplasmic reticulum to mitochondria. *Nature* 456(7222): 605–610.

de Lonlay, P., I. Valnot et al. (2001). A mutant mitochondrial respiratory chain assembly protein causes complex III deficiency in patients with tubulopathy, encephalopathy and liver failure. *Nat Genet* 29(1): 57–60.

Desquiret, V., N. Gueguen et al. (2008). Mitochondrial effects of dexamethasone imply both membrane and cytosolic-initiated pathways in HepG2 cells. *Int J Biochem Cell Biol* 40(8): 1629–1641.

Detmer, S. A. and D. C. Chan (2007). Functions and dysfunctions of mitochondrial dynamics. *Nat Rev Mol Cell Biol* 8(11): 870–879.

Diaz, F., S. Garcia et al. (2008). Pathophysiology and fate of hepatocytes in a mouse model of mitochondrial hepatopathies. *Gut* 57(2): 232–242.

DiMauro, S. and E. A. Schon (2003). Mitochondrial respiratory-chain diseases. *N Engl J Med* 348(26): 2656–2668.

Du, J., Y. Wang et al. (2009). Dynamic regulation of mitochondrial function by glucocorticoids. *Proc Natl Acad Sci* 106(9): 3543–3548.

Duchen, M. R. and G. Szabadkai (2010). Roles of mitochondria in human disease. *Essays Biochem* 47: 115–137.

Eura, Y., N. Ishihara et al. (2003). Two mitofusin proteins, mammalian homologues of FZO, with distinct functions are both required for mitochondrial fusion. *J Biochem* 134(3): 333–344.

Farrell, G. C. and C. Z. Larter (2006). Nonalcoholic fatty liver disease: From steatosis to cirrhosis. *Hepatology* 43(2 Suppl 1): S99–S112.

Frezza, C., S. Cipolat et al. (2006). OPA1 controls apoptotic cristae remodeling independently from mitochondrial fusion. *Cell* 126(1): 177–189.

Friedman, J. R., B. M. Webster et al. (2010). ER sliding dynamics and ER–mitochondrial contacts occur on acetylated microtubules. *J Cell Biol* 190(3): 363–375.

Galloway, C. A. and Y. Yoon (2012). What comes first, misshape or dysfunction? The view from metabolic excess. *J Gen Physiol* 139(6): 455–463.

Galloway, C. A. and Y. Yoon (2013). Mitochondrial morphology in metabolic diseases. *Antioxid Redox Signal* 19(4): 415–430.

Gan, K. X., C. Wang et al. (2013). Mitofusin-2 ameliorates high-fat diet-induced insulin resistance in liver of rats. *World J Gastroenterol* 19(10): 1572–1581.

Glasgow, J. F., B. Middleton et al. (1999). The mechanism of inhibition of beta-oxidation by aspirin metabolites in skin fibroblasts from Reye's syndrome patients and controls. *Biochim Biophys Acta* 31(1): 115–125.

Gupte, P., D. Amarapurkar et al. (2004). Non-alcoholic steatohepatitis in type 2 diabetes mellitus. *J Gastroenterol Hepatol* 19(8): 854–858.

Hagiwara, A., M. Cornu et al. (2012). Hepatic mTORC2 activates glycolysis and lipogenesis through Akt, glucokinase, and SREBP1c. *Cell Metab* 15(5): 725–738.

Hamasaki, M., N. Furuta et al. (2013). Autophagosomes form at ER-mitochondria contact sites. *Nature* 495(7441): 389–393.

Harding, B. N. (1990). Progressive neuronal degeneration of childhood with liver disease (Alpers-Huttenlocher syndrome): A personal review. *J Child Neurol* 5(4): 273–287.

Hayashi, T., R. Rizzuto et al. (2009). MAM: More than just a housekeeper. *Trends Cell Biol* 19(2): 81–88.

Hayashi, T. and T. P. Su (2007). Sigma-1 receptor chaperones at the ER-mitochondrion interface regulate Ca(2+) signaling and cell survival. *Cell* 131(3): 596–610.

Hernandez-Alvarez, M. I., J. C. Paz et al. (2013). Glucocorticoid modulation of mitochondrial function in hepatoma cells requires the mitochondrial fission protein Drp1. *Antioxid Redox Signal* 19(4): 366–378.

Holt, I. J., A. E. Harding et al. (1988). Deletions of muscle mitochondrial DNA in patients with mitochondrial myopathies. *Nature* 331(6158): 717–719.

Hsu, C. C., H. C. Lee et al. (2013a). Mitochondrial DNA alterations and mitochondrial dysfunction in the progression of hepatocellular carcinoma. *World J Gastroenterol* 19(47): 8880–8886.

Hsu, C. C., C. H. Wang et al. (2013b). Mitochondrial dysfunction represses HIF-1α protein synthesis through AMPK activation in human hepatoma HepG2 cells. *Biochim Biophys Acta* 10(51): 18.

Huang, P., T. Yu et al. (2007). Mitochondrial clustering induced by overexpression of the mitochondrial fusion protein Mfn2 causes mitochondrial dysfunction and cell death. *Eur J Cell Biol* 86(6): 289–302.

Hübscher, S. (2006). Histological assessment of non-alcoholic fatty liver disease. *Histopathology* 49(5): 450–465.

Kapsa, R., G. N. Thompson et al. (1994). A novel mtDNA deletion in an infant with Pearson syndrome. *J Inherit Metab Dis* 17(5): 521–526.

Koch, A., Y. Yoon et al. (2005). A role for Fis1 in both mitochondrial and peroxisomal fission in mammalian cells. *Mol Biol Cell* 16(11): 5077–5086.

Kotronen, A., A. Seppala-Lindroos et al. (2008). Tissue specificity of insulin resistance in humans: Fat in the liver rather than muscle is associated with features of the metabolic syndrome. *Diabetologia* 51(1): 130–138.

Labarthe, F., D. Dobbelaere et al. (2005). Clinical, biochemical and morphological features of hepatocerebral syndrome with mitochondrial DNA depletion due to deoxyguanosine kinase deficiency. *J Hepatol* 43(2): 333–341.

Levinthal, G. N. and A. S. Tavill (1999). Liver disease and diabetes mellitus. *Clin Diabetes* 17(2): 478–535.

Lionetti, L., M. P. Mollica et al. (2014). High-lard and high-fish-oil diets differ in their effects on function and dynamic behaviour of rat hepatic mitochondria. *PLoS One* 9(3): e92753.

Liu, X. and G. Hajnoczky (2011). Altered fusion dynamics underlie unique morphological changes in mitochondria during hypoxia-reoxygenation stress. *Cell Death Differ* 18(10): 1561–1572.

Moraes, C. T., S. Shanske et al. (1991). mtDNA depletion with variable tissue expression: A novel genetic abnormality in mitochondrial diseases. *Am J Hum Genet* 48(3): 492–501.

Morikawa, Y., N. Matsuura et al. (1993). Pearson's marrow/pancreas syndrome: A histological and genetic study. *Virchows Arch A: Pathol Anat Histopathol* 423(3): 227–231.

Morris, A. A., J. W. Taanman et al. (1998). Liver failure associated with mitochondrial DNA depletion. *J Hepatol* 28(4): 556–563.

Munoz, J. P., S. Ivanova et al. (2013). Mfn2 modulates the UPR and mitochondrial function via repression of PERK. *EMBO J* 32(17): 2348–2361.

Narkewicz, M. R., R. J. Sokol et al. (1991). Liver involvement in Alpers disease. *J Pediatr* 119(2): 260–267.

Nunnari, J., W. Marshall et al. (1997). Mitochondrial transmission during mating in *Saccharomyces cerevisiae* is determined by mitochondrial fusion and fission and the intramitochondrial segregation of mitochondrial DNA. *Mol Biol Cell* 8(7): 1233–1242.

Nunnari, J. and A. Suomalainen (2012). Mitochondria: In sickness and in health. *Cell* 148(6): 1145–1159.

Oh, S., N. Kim et al. (1998). Identification of differentially expressed genes in human hepatoblastoma cell line (HepG2) and HBV-X transfected hepatoblastoma cell line (HepG2-4x). *Mol Cells* 8(2): 212–218.

Orlowski, J. P. (1999). Whatever happened to Reye's syndrome? Did it ever really exist? *Crit Care Med* 27(8): 1582–1587.

Otera, H., C. Wang et al. (2010). Mff is an essential factor for mitochondrial recruitment of Drp1 during mitochondrial fission in mammalian cells. *J Cell Biol* 191(6): 1141–1158.

Palmer, C. S., K. D. Elgass et al. (2013). Adaptor proteins MiD49 and MiD51 can act independently of Mff and Fis1 in Drp1 recruitment and are specific for mitochondrial fission. *J Biol Chem* 288(38): 27584–27593.

Palmer, C. S., L. D. Osellame et al. (2011). MiD49 and MiD51, new components of the mitochondrial fission machinery. *EMBO Rep* 12(6): 565–573.

Perez-Carreras, M., P. Del Hoyo et al. (2003). Defective hepatic mitochondrial respiratory chain in patients with nonalcoholic steatohepatitis. *Hepatology* 38(4): 999–1007.

Perry, R. J., V. T. Samuel et al. (2014). The role of hepatic lipids in hepatic insulin resistance and type 2 diabetes. *Nature* 510(7503): 84–91.

Petersen, K. F., D. Befroy et al. (2003). Mitochondrial dysfunction in the elderly: Possible role in insulin resistance. *Science* 300(5622): 1140–1142.

Petersen, K. F., S. Dufour et al. (2004). Impaired mitochondrial activity in the insulin-resistant offspring of patients with type 2 diabetes. *N Engl J Med* 350(7): 664–671.

Procaccio, V. and D. C. Wallace (2004). Late-onset Leigh syndrome in a patient with mitochondrial complex I NDUFS8 mutations. *Neurology* 62(10): 1899–1901.

Rojo, M., F. Legros et al. (2002). Membrane topology and mitochondrial targeting of mitofusins, ubiquitous mammalian homologs of the transmembrane GTPase Fzo. *J Cell Sci* 115(8): 1663–1674.

Roussel, D., J. F. Dumas et al. (2003). Dexamethasone treatment specifically increases the basal proton conductance of rat liver mitochondria. *FEBS Lett* 541(1–3): 75–79.

Roussel, D., J. F. Dumas et al. (2004). Kinetics and control of oxidative phosphorylation in rat liver mitochondria after dexamethasone treatment. *Biochem J* 382(Pt 2): 491–499.

Santel, A., S. Frank et al. (2003). Mitofusin-1 protein is a generally expressed mediator of mitochondrial fusion in mammalian cells. *J Cell Sci* 116(13): 2763–2774.

Santel, A. and M. T. Fuller (2001). Control of mitochondrial morphology by a human mitofusin. *J Cell Sci* 114(5): 867–874.

Sanyal, A. J., C. Campbell-Sargent et al. (2001). Nonalcoholic steatohepatitis: Association of insulin resistance and mitochondrial abnormalities. *Gastroenterology* 120(5): 1183–1192.

Sebastian, D., M. I. Hernandez-Alvarez et al. (2012). Mitofusin 2 (Mfn2) links mitochondrial and endoplasmic reticulum function with insulin signaling and is essential for normal glucose homeostasis. *Proc Natl Acad Sci USA* 109(14): 5523–5528.

Segales, J., J. C. Paz et al. (2013). A form of mitofusin 2 (Mfn2) lacking the transmembrane domains and the COOH-terminal end stimulates metabolism in muscle and liver cells. *Am J Physiol Endocrinol Metab* 305(10): E1208–E1221.

Simmen, T., J. E. Aslan et al. (2005). PACS-2 controls endoplasmic reticulum-mitochondria communication and Bid-mediated apoptosis. *EMBO J* 24(4): 717–729.

Simmen, T., E. M. Lynes et al. (2010). Oxidative protein folding in the endoplasmic reticulum: Tight links to the mitochondria-associated membrane (MAM). *Biochim Biophys Acta* 8(73): 27.

Smirnova, E., L. Griparic et al. (2001). Dynamin-related protein Drp1 is required for mitochondrial division in mammalian cells. *Mol Biol Cell* 12(8): 2245–2256.

Smirnova, E., D. L. Shurland et al. (1998). A human dynamin-related protein controls the distribution of mitochondria. *J Cell Biol* 143(2): 351–358.

Sokol, R. J. and W. R. Treem (1999). Mitochondria and childhood liver diseases. *J Pediatr Gastroenterol Nutr* 28(1): 4–16.

Sternlieb, I. (1968). Mitochondrial and fatty changes in hepatocytes of patients with Wilson's disease. *Gastroenterology* 55(3): 354–367.

Szabadkai, G., K. Bianchi et al. (2006). Chaperone-mediated coupling of endoplasmic reticulum and mitochondrial Ca^{2+} channels. *J Cell Biol* 175(6): 901–911.

Taguchi, N., N. Ishihara et al. (2007). Mitotic phosphorylation of dynamin-related GTPase Drp1 participates in mitochondrial fission. *J Biol Chem* 282(15): 11521–11529.

Tubbs, E., P. Theurey et al. (2014). Mitochondria-associated endoplasmic reticulum membrane (MAM) integrity is required for insulin signaling and is implicated in hepatic insulin resistance. *Diabetes* 63(10): 3279–3294.

Vance, J. E. (1990). Phospholipid synthesis in a membrane fraction associated with mitochondria. *J Biol Chem* 265(13): 7248–7256.

Wasiak, S., R. Zunino et al. (2007). Bax/Bak promote sumoylation of DRP1 and its stable association with mitochondria during apoptotic cell death. *J Cell Biol* 177(3): 439–450.

Yaffe, M. P. (2003). The cutting edge of mitochondrial fusion. *Nat Cell Biol* 5(6): 497–499.

Yin, P. H., C. C. Wu et al. (2010). Somatic mutations of mitochondrial genome in hepatocellular carcinoma. *Mitochondrion* 10(2): 174–182.

Yu, T., B. S. Jhun et al. (2011). High-glucose stimulation increases reactive oxygen species production through the calcium and mitogen-activated protein kinase-mediated activation of mitochondrial fission. *Antioxid Redox Signal* 14(3): 425–437.

Yu, T., J. L. Robotham et al. (2006). Increased production of reactive oxygen species in hyperglycemic conditions requires dynamic change of mitochondrial morphology. *Proc Natl Acad Sci USA* 103(8): 2653–2658.

Zhang, Y., L. Jiang et al. (2011). Mitochondrial dysfunction during in vitro hepatocyte steatosis is reversed by omega-3 fatty acid-induced up-regulation of mitofusin 2. *Metabolism* 60(6): 767–775.

Zhao, J., T. Liu et al. (2011). Human MIEF1 recruits Drp1 to mitochondrial outer membranes and promotes mitochondrial fusion rather than fission. *EMBO J* 30(14): 2762–2778.

Zuchner, S., I. V. Mersiyanova et al. (2004). Mutations in the mitochondrial GTPase mitofusin 2 cause Charcot-Marie-Tooth neuropathy type 2A. *Nat Genet* 36(5): 449–451.

16 The Role of JNK in Lipotoxicity in the Liver

Sanda Win, Tin Aung Than, and Neil Kaplowitz

CONTENTS

ABSTRACT

"Lipotoxicity" refers to the accumulation of lipids which induces endoplasmic reticulum stress and mitochondrial dysfunction, cellular dysfunction, and death resulting in organ dysfunction. c-Jun N-terminal protein kinases (JNK) are a family of serine/threonine kinases that are activated by reactive oxygen species (ROS), pathogens, toxins, drugs, endoplasmic reticulum stress, free fatty acids, metabolic changes, environmental stresses and growth factors, and cytokines. Activated JNK induce multiple biologic events through the transcription factor activator protein-1 and transcription-independent control of effector molecules. JNK regulate cell death and survival, differentiation, proliferation, ROS accumulation, metabolism, insulin signaling, and carcinogenesis in the liver. The biologic functions of JNK are isoform, cell type, and context dependent. Important activators of JNK in many cell lines are ROS. JNK play a major role in the mechanism of lipotoxicity-induced mitochondrial dysfunction. The activation of JNK affects the development of hepatic steatosis and insulin resistance. Sustained JNK activation by saturated fatty acids plays a role in lipotoxicity and the pathogenesis of NASH. Sab is found exclusively in the mitochondrial outer membrane. It contains C-terminal JNK docking sites facing the cytoplasm and is predicted to contain one transmembrane hydrophobic domain with an N-terminal stretch facing the intermembrane space. The translocation of P-JNK to Sab on mitochondria is required in lipotoxicity. The interaction of JNK with the mitochondrial outer membrane protein Sab (SH3BP5), which is a binding target and substrate of JNK, impairs mitochondrial respiration and increases ROS production, cell death, and lipotoxicity.

INTRODUCTION

The fatty accumulation in the liver (hepatic steatosis) is referred to nonalcoholic fatty liver disease (NAFLD). Approximately, 20% of the U.S. population suffers from NAFLD, and the prevalence is increasing [1]. The exact reasons and mechanisms by which the disease progresses from hepatic steatosis to NASH, characterized by hepatic inflammation and fibrosis, are not known. The potential causes of hepatic inflammation in the development of steatohepatitis include oxidative stress, adipose tissue-derived cytokines, or translocation of endotoxin from the intestinal lumen, all of which may induce this "second hit" phenomenon superimposed on NAFLD [2]. The activities of JNK in the liver, accompanied by the adipose tissues and muscles, are believed to be important in the development of NAFLD and NASH. JNK plays a major role in the mechanism of lipotoxicity. The mechanism of lipotoxicity-induced mitochondrial dysfunction has been investigated with saturated fatty acids, such as palmitic acid (PA) [3–11]. Strong activation of JNK has been observed in the liver, fat, and muscle tissues in mice placed on a high-fat diet (HFD) and genetically (ob/ob) obese mice [3,12,13]. We previously demonstrated that sustained JNK activation and cell death in several models of hepatotoxicity, such as acetaminophen overdose, TNF/galactosamine, and tunicamycin-induced ER stress, depends on the expression of Sab, a mitochondrial JNK docking protein [14,15]. We also demonstrated that the addition of P-JNK plus ATP, but not either alone, to isolated hepatic mitochondria induced an inhibition of respiration and enhanced ROS production [15]. JNK activation is also associated with ER stress–mediated insulin resistance. ER stress causes JNK activation through ASK1 association with IRE1 [15,16]. However, the ER stress induced by hepatic XBP-1 deficiency is therefore not associated with JNK-mediated insulin resistance. In mouse hepatocytes, PA increased P-PERK and downstream target CHOP in PMH but failed to activate the IRE-1α arm of the UPR. The specific inhibition of PERK prevented JNK activation and cell death, indicating a major role upstream of JNK activation [9]. JNK activation partially depends on JIP1, a scaffold protein that contributes to insulin resistance and hepatic steatosis [17]. Double-stranded RNA-dependent protein kinase (PKR) is another upstream regulator of JNK that has been associated with metabolic disease. PKR senses high levels of nutrients and obese states to activate JNK, which causes insulin resistance and hepatic steatosis [18]. JNK activity regulates pathophysiologic processes, such as hepatocyte death, steatosis, inflammation, and insulin resistance, which are associated with NASH, fibrosis, and hepatocellular carcinoma (HCC). Preclinical studies in animal models or human cells have indicated that reagents that inhibit JNK might be used to treat patients with liver diseases, including acute liver failure, I/R injury, fibrosis, HCC, and NASH [19–21]. Several JNK inhibitors, such as SP600125, D-JNKI1, and BI-78D3, have been tested in preclinical studies. CC-930 is currently being tested in a phase 2 clinical trial for idiopathic pulmonary fibrosis [22], and dual inhibitors of JIP and JNK have been identified [23]. Further work is required to assess new JNK inhibitors, alone or in combination with other therapeutics in the treatment and prevention of NASH, insulin resistance, and other liver diseases [24].

MITOCHONDRIA AND JNK ACTIVATION IN LIPOTOXICITY

Mitochondria are the key organelles in cell death signaling pathways in different experimental liver disease models including ER stress or metabolic stress–induced cell death or acute liver failure. Liver mitochondria are the cellular powerhouses that generate ATP by using substrates derived from fat and glucose. Mitochondria play an important role in hepatocyte metabolism, being the primary site for the oxidation of fatty acids and oxidative phosphorylation. The mitochondrial function abnormalities associated with NAFLD include ultrastructural lesions, depletion of mtDNA, decreased activity of respiratory chain complexes, and impaired mitochondrial β-oxidation. NAFLD is often found in patients with insulin resistance, obesity, and type 2 diabetes, the same metabolic conditions in which there is decreased oxygen consumption and ATP production, reduced total mtDNA and mtDNA transcription factor A, and reduced content of respiratory proteins in the fat, muscle, and liver [25].

There are three isoforms of JNK in mammals: JNK1, JNK2, and JNK3. At least 14 MAP3Ks have been found to activate JNK. MKK4 activates JNK and p38; MKK7 is specifically associated with cytokine-induced JNK activation through phosphorylation of the Thr residue of JNK [26]. JNK signaling is associated with cell death, survival, differentiation, proliferation, and tumorigenesis in hepatocytes. JNK1 and JNK2 are expressed in almost all cells, including liver parenchymal cells, whereas JNK3 is mainly expressed in brain, heart, and testis [19,27]. At least 10 alternative splicing variants indicate the diversity of JNK proteins, but their functional significance is unclear. The JNK proteins, including splicing variants, range from 46 to 55 kilodaltons in size. The JNKs have a common substrate docking site in their C-terminus and a glutamate-aspartate domain in their N-terminus. JNK has at least 50 proteins substrates, which include c-Jun, JunB, JunD, activating transcription factor-2, p53, c-Myc, serum response factor, Itch, insulin receptor substrate (IRS)-1, JNK interacting protein (JIP)-1,14-3-3, Sab (SH3BP5), Bcl-2, Bcl-xL, Bid, Bim, Bad, Bax, and Mcl-1 [28,29]. The substrate, c-Jun, is a representative target of JNKs. c-Jun dimerizes with JunB, JunD, or Fos to form the transcription factor activator protein (AP)-1, whereas serum response factor controls expression of the Fos proteins that dimerize with the Jun proteins [30]. JNK is involved in inflammation and fibrosis in nonparenchymal liver cells, such as hepatic macrophages (Kupffer cells) and hepatic stellate cells (HSCs).

JNK activity regulates pathophysiologic processes, such as hepatocyte death, steatosis, inflammation, and insulin resistance, which are associated with NASH, fibrosis, and HCC. These proteins control multiple cellular processes, acting either as transcription factors or by controlling protein degradation, localization, and signaling. In adipocytes and fibroblasts, saturated FFAs cause aggregation of the membrane-anchored tyrosine kinase c-Src within lipid rafts, leading to MLK3 activation and subsequent JNK activation [31]. In the context of primary mouse hepatocytes and PA, the Src inhibitor, PP2, inhibited JNK activation at 1 h but not at 4, 6, 8 h and had a modest protective effect on cell death induced by PA. In contrast, a highly specific PERK inhibitor blocked JNK activation at all-time points and protected against PA-induced cell death [9]. Considerable work has focused on the ability of JNK to directly or indirectly regulate the members of the Bcl2 family (e.g., Bax and bcl-xl) and consequently influence apoptosis [32,33]. JNK has been shown to promote apoptosis by promoting Bax translocation to mitochondria by directly phosphorylating Bax [34] or by phosphorylation of 14-3-3, which anchors Bax in cytoplasm [35] In addition, JNK has been shown to translocate to mitochondria and phosphorylate bcl-xl (inactivate) in mitochondrial membranes [32,36,37]. Similarly, JNK translocation to mitochondria can induce cytochrome c and SMAC release from the intermembrane space, leading to apoptosis [38,39].

The distinct functions of the isoforms JNK1 and JNK2 in the pathogenesis of liver diseases have been characterized in Jnk1$^{-/-}$ and Jnk2$^{-/-}$ mice. Studies of genetically engineered mice and human cells and tissues have been used to understand the role of the JNK and/or c-Jun pathways in different types of liver injury (TNF-induced liver injury, fibrosis and carcinogenesis, and steatohepatitis). NAFLD is a hepatic manifestation of the metabolic syndrome, which is characterized by obesity and insulin resistance. NAFLD ranges from simple steatosis to steatosis with hepatic inflammation and fibrosis, known as nonalcoholic steatohepatitis (NASH). NASH has a high risk factor for cirrhosis. JNK1 and JNK2 each participate in hepatic injury, steatosis, insulin resistance, and obesity [40]. The HFD and genetically induced obesity activate JNK in the liver. The saturated FFA, palmitate, upregulates p53-mediated apoptosis (PUMA) and Bax in a JNK1-dependent manner [41]. Puma$^{-/-}$ hepatocytes are resistant to FFA-induced lipoapoptosis, indicating the importance of this death regulation pathway. In addition, FFAs cause degradation of Mcl-1 through JNK1, which contributes to lipoapoptosis [42,43]. Steatotic hepatocytes are more susceptible to TNF-induced apoptosis, which involves the activation of ASK1 and JNK [44,45]. As noted earlier, liver expresses 2 JNK isoforms. Most pathologic processes are associated with JNK1; the severe phenotypes observed following deletion of JNK2 probably result from compensatory JNK1 activation. Furthermore, JNK has a major role in the mechanism of lipotoxicity-induced mitochondrial dysfunction [3,5–8,10,11].

Once P-JNK translocates to mitochondria, it binds to Sab (SH3 domain–binding protein that preferentially associates with Btk), a scaffold protein on the outer membrane of mitochondria that contains a kinase interaction motif [46]. Sab is essential in APAP hepatotoxicity, and we observed that silencing Sab protected mice from liver injury caused by APAP, even in the presence of excessive mitochondrial GSH depletion and covalent binding [14]. The interaction of JNK with mitochondrial outer membrane protein Sab (SH3BP5), which is a binding target and substrate of JNK, impaired mitochondrial respiration, increased ROS production, and enhanced cell death and hepatotoxicity [9,14,15]. We have identified Sab as a crucial binding partner in mitochondria, important in mediating injury. However, the mechanism by which JNK–Sab modulates the electron transport chain (ETC), mitochondrial ROS generation, and MPT remains under investigation. Sustained activation of JNK is an effector signaling molecule strongly associated with cell death [9,14,15] (Figure 16.1).

Lipotoxicity causes decreased mitochondrial respiration (state III) and lower respiratory control ratio (state III/state IV) in isolated liver mitochondria that is due to reduced protein level of mitochondrial complex I (NADH dehydrogenase).

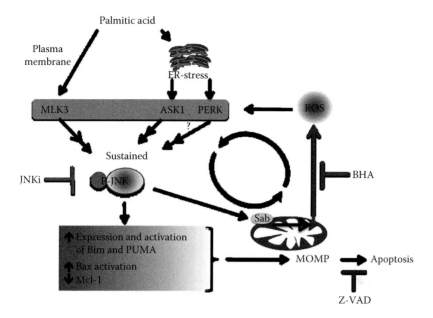

FIGURE 16.1 **(See color insert.)** The role of Sab in lipotoxicity. PA causes decreasing membrane fluidity, expands the lipid raft compartment plasma membrane fluidity, and MLK3 and JNK activation through the clustering of c-Src. Another mechanism for activation of JNK is ER stress. ER stress activates ASK1 through IRE-1α and PERK activation, which leads to JNK activation. Activated JNK translocates to Sab on the mitochondria that causes ROS production. Then it causes sustained JNK activation, which increases expression and activation of Bim and PUMA, increased Bax activation and Mcl-1 degradation leads to outer membrane permeabilization and apoptotic cell death. Inhibition of JNK activation by JNK inhibitor (SP600125) significantly decreased PA-induced cell death. Cell death was also inhibited by the antioxidant BHA and the pancaspase inhibitor Z-VAD-fmk.

Mitochondrial respiratory chain complex I is inactivated by NADPH oxidase NOX4 [47,48], and low activity of complexes I, III, IV, and V has been identified [49]. When mitochondrial respiration is impaired and ROS release is enhanced, JNK activation is sustained as a consequence of its interaction with Sab, promoting cell death and hepatotoxicity.

FATTY ACID OXIDATION AND MITOCHONDRIA RESPIRATION

Fatty acid oxidation occurs in subcellular organelles. β-oxidation occurs in mitochondria and peroxisomes [50,51]. The oxidation of fatty acids in the liver fuels the synthesis of ketone bodies, 3-hydroxybutyrate, and acetoacetate. Fatty acid catabolism involves three stages. The first stage of fatty acid catabolism is beta-oxidation. Mitochondrial β-oxidation is the dominant oxidative pathway for the disposal of fatty acids under normal physiologic conditions. Beta-oxidation is the process by which fatty acid molecules are broken down in the mitochondria to generate acetyl-coA, which enters the citric acid cycle, generating NADH and FADH2, which are used by the ETC. Beta-oxidation can begin if fatty acid is present in mitochondrial matrix. The activity of the enzyme carnitine palmitoyl transferase-I (CPT-I) regulates long-chain FFA's entry into the mitochondria. Short-chain and medium-chain FFAs freely enter the mitochondria. Long-chain fatty acid is dehydrogenated to create a trans double bond between C2 and C3 to produce trans-delta 2-enoyl CoA. Trans-delta2-enoyl CoA is hydrated at the double bond to produce L-B-hydroxyacyl CoA. Then, this releases the first two carbon units as acetyl CoA and a fatty acyl CoA minus two carbons. The process continues until all of the carbons in the fatty acid are turned into acetyl CoA, an intermediate that goes through the citric acid cycle for the production of reducing agents and ATP. If glucose and energy levels are elevated, acetyl-CoA is converted to citrate, which can leave the mitochondrial matrix into the cytosol through the tricarboxylate carrier. Citrate regenerates acetyl-CoA, which is converted to malonyl-CoA by acetyl-CoA carboxylase. Malonyl-CoA plays important roles in both hepatic fatty acid oxidation and lipid synthesis. Malonyl-CoA is the initial component for fatty acid synthesis. High malonyl-CoA levels also inhibit CPT-I enzyme activity, thus robustly decreasing fatty acid oxidation by reducing the rate of fatty acid entry into the mitochondria. Thus, periods of caloric overconsumption and excessive energy supply increase malonyl-CoA levels, which promotes hepatic fatty acid synthesis (storage) and suppresses fatty acid oxidation (catabolism). Conversely, in the fasting state, hepatic malonyl-CoA levels are low, allowing extensive mitochondrial import of long-chain FFAs and high rates of β-oxidation. Increased oxygen consumption rate occurs in response to PA-induced mitochondrial β-oxidation. Cellular oxygen consumption of PMHs in response to PA steadily increases with dose and time. However, higher dose of PA markedly decreased the reserve capacity and mitochondrial respiration compared to low dose. The activation of mitochondrial respiration in response to low concentration is not observed at the higher concentration of PA [9]. This toxic effect of PA on mitochondria is mediated by the activation of JNK and its interaction with Sab.

ENDOPLASMIC RETICULUM STRESS AND
JNK ACTIVATION IN LIPOTOXICITY

The main functions of the ER are to maintain normal protein folding prior to subsequent protein transport to the Golgi apparatus. In the context of the insulin resistance, ER stress causes the misfolding of proteins in the ER lumen, and results in an unfolded protein response (UPR) as exemplified by the pancreatic β-cell, which upregulates insulin biosynthesis to counteract insulin resistance and subsequent hyperglycemia. Although UPR is an adaptive response, sustained ER stress has been shown to eventually contribute to β-cell apoptosis [52]. ER stress has been demonstrated to be a primary signal precipitating inflammation and activation of the stress kinase, JNK [53], that contributes to the development of type 2 diabetes, including liver and adipose tissue [16]. ER stress is elevated in the diabetic heart accompanied by increased ER stress–related inflammation and activation of JNK in cardiac myocytes of db/db mice [54] and ER stress–induced apoptosis, that is possibly via the action of ER signals such as calcium release, JNK, or Bax translocation to mitochondria [55]. GSK-3β is also involved in FFA-induced JNK activation, independent of FFA-induced ER stress responses [11]. ER stress leads to JNK activation in lipoapoptosis [7,56–59]. ER stress is known to activate ASK1 through the IRE-1α pathway [60,61], but this has been shown to be dispensable in PA toxicity, and another mechanism for ER stress activation of JNK via MLK3 activation was suggested [62]. Although we found no evidence of activation of IRE-1α in PMH as reflected by the absence of XBP-1 splicing or protection by chemical chaperone inhibition of ER stress, PERK was activated and a PERK inhibitor blocked JNK activation and cell death, supporting a selective role for PA-induced PERK activation upstream of JNK [9]. Others have suggested that PERK and PKR can activate JNK [18,63–66]. PA activated both kinases in PMH, but the PERK inhibitor, which selectively inhibited PERK, protected against JNK activation, mitochondrial dysfunction, and apoptosis [9,67,68]. The mechanism for PA-induced selective PERK activation and the MAPK signaling pathway, which leads to JNK activation, is not known but likely involves MLK3 [69–71]. Interestingly, ER stress in hematopoietic stem cells has recently been reported to selectively activate PERK and predispose to apoptosis [72]. Thus, selective PERK activation in response to ER stress is possible and is context dependent. PERK activation contributes upstream of JNK- and Sab-mediated lipotoxicity in hepatocyte.

ROS AND JNK ACTIVATION IN LIPOTOXICITY

ROS are formed from the normal metabolism of oxygen and have important roles in cell signaling and homeostasis. ROS are produced from the NADPH oxidase (NOX) complexes in cell membranes, mitochondria, peroxisomes, and endoplasmic reticulum [73,74]. Oxidative phosphorylation is the production of energy for the cell biological function into a usable form, adenosine triphosphate (ATP), by mitochondria via the ETC. The ETC involves the transport of protons (hydrogen ions) across the inner mitochondrial membrane. Oxidation-reduction reactions occur through the ETC, that is, electrons are passed through a series of proteins with each acceptor protein along the chain having a greater reduction potential than the previous. Then in the final step, the ETC reduces an oxygen molecule to water. In normal conditions, the oxygen is reduced to produce water. If oxygen is instead prematurely and incompletely reduced due to a block in the ETC, superoxide radical ($\cdot O_2^-$) is released, most well documented from complex I and complex III [75]. The oxidative stress derived from mitochondria may contribute to neurological diseases (e.g., Parkinson's disease and Alzheimer's disease), cardiovascular diseases (atherosclerosis and myocardial infarction), and cancers. Excessive oxidative stress can cause significant impairment in the effectiveness of antioxidant defenses, such as depletion of glutathione [76]. The effects of oxidative stress depend upon the magnitude of these changes, with the cell being able to overcome small perturbations and regain its original state [77]. Intensity and duration of ROS and JNK activation cause different responses. JNK plays an important role in the stress response and can be activated by various stressors, including ROS and various cytokines. A transient activation of JNK upregulates signaling pathways and cellular response to stress such as adaptive signaling and inflammation. Moderate intensity of JNK activation causes cellular dysfunction like insulin resistance. Chronic and sustained JNK activation causes increased ROS that can activate or inhibit important signal transduction pathways. When a critical threshold of mitochondrial stress or injury occurs, JNK is activated and targets mitochondria, promoting mitochondrial dysfunction and amplifying oxidative stress, which can induce MPT or mitochondrial outer membrane permeabilization (MOMP) (Figure 16.1). Excessive ROS can induce apoptosis through both the extrinsic and intrinsic pathways [78]. In the extrinsic pathway of apoptosis, ROS are generated by Fas ligand as an upstream event for Fas activation. In turn, ROS are required for Fas phosphorylation at the tyrosine residue, which is a signal for subsequent recruitment of Fas-associated protein with death domain and caspase-8 and for apoptosis induction [78–81]. In addition, ROS are required for the ubiquitination and subsequent degradation of the FLICE inhibitory protein to further enhance Fas activation [82,83]. In the intrinsic pathway, ROS causes inactivation of pore-stabilizing proteins (Bcl-2 and Bcl-xL) as well as activation of pore-destabilizing proteins (Bcl-2-associated X protein, Bcl-2 homologous antagonist/killer) to facilitate cytochrome c release. An even higher ROS level can result in both apoptosis and necrosis in cancer cells. ROS can also induce cell death through autophagy, which is a self-catabolic process involving sequestration of cytoplasmic contents for degradation in lysosomes. Increased ROS activate or inhibit important signal transduction pathways. ROS accumulation is an important activator of MAPK pathway, leading to JNK activation (Figure 16.2). Moderate accumulation of oxidative stress can cause apoptosis, while more intense stresses may cause necrosis and cell death and damage to DNA, implicated in aging and cancer.

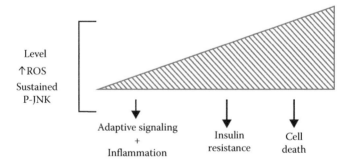

FIGURE 16.2 Intensity and duration of JNK activation cause different responses. JNK plays an important role in the stress response and can be activated by various stressors, including ROS and various cytokines. A transient activation of JNK upregulates adaptive signaling pathways and inflammation. Moderately intense and sustained JNK activation can cause cellular dysfunction like insulin resistance. High levels of sustained JNK activation causes increased ROS that activates or inhibits important signal transduction pathways. When a critical threshold of mitochondrial stress or injury occurs, JNK is activated and targets mitochondria to induce MPT or MOMP that causes cell death by mediating mitochondrial dysfunction.

ROLE OF JNK SIGNALING TO MITOCHONDRIA IN LIPOTOXICITY

ASK-1 is activated by oxidative stress, ER stress, and inflammatory cytokines such as TNFα [84]. Activation of apoptosis signal-regulating kinase 1 (ASK-1), a member of the MAP3K family, can regulate both the MKK4/MKK7-JNK and the MKK3/MKK6-p38 MAPK signaling cascades [85]. In resting cells, ASK-1 forms an inactive complex with reduced thioredoxin (Trx). Under conditions of stress by TNFα or ROS, ASK-1 dissociates from Trx and becomes activated [86]. Oxidation of Trx by ROS causes dissociation of ASK-1 from the oxidized Trx, which switches the inactive form of ASK-1 to the active kinase. Activated ASK-1 then promotes activation (phosphorylation) of the downstream MAPKK; then MKK4/MKK7 can phosphorylate JNK and p38 MAPK and contribute to the liver injury [87]. The pathways for saturated fatty acid–induced JNK activation have been extensively studied and evidence supports a role for Src-dependent activation of the MAP3K, MLK3 [62,88–90]. Recently, autophagy-mediated degradation of KEAP-1 has been demonstrated to be upstream of JNK in PA-induced apoptosis, possibly upstream of MLK3 [91]. The role of ER stress in activating ASK-1 has also been suggested [60], but recent evidence indicates that ER stress is somehow linked to MLK3 activation [5,90,92]. PERK inhibitor blocked JNK activation and cell death, supporting a selective role for PA-induced PERK activation upstream of JNK. MLK3 can phosphorylate and activate the MAP2K isoforms MKK4 and MKK7 [93]. Compound deficiency of MKK4 plus MKK7 prevents stress-induced JNK activation [94]. FFA did not activate JNK in MEF prepared from Mkk4−/− mice or Mkk7−/− mice. This observation indicated that both MKK4 and MKK7 are required for FFA-stimulated JNK activation in MEF. The combined requirement for MKK4 and MKK7 may reflect the co-operative phosphorylation of the JNK Thr-Pro-Tyr dual phosphorylation motif on Tyr by MKK4 and on Thr by MKK7 [94,95].

The JNKs are important signaling molecules in multiple pathways in liver physiology, and they contribute to hepatocellular injury caused by bile acids, fatty acid toxins, and TNF cytotoxicity. In addition, JNK-mediated activation of transcription factors, for example, AP-1, promotes inflammatory cytokine production and cell proliferation and disease pathogenesis. Thus, depending on the context, duration, and level of activation of JNK, JNK can promote cell death or survival.

The activation of the mitogen-activated protein kinase JNK has been identified as a pivotal step in FFA-induced Bax activation and hepatocyte apoptosis [96]. Others have shown that JNK can be activated by saturated FFA-stimulated MAP3K, MLK3 [88], and/or secondary to saturated FFA-induced ER stress [97,98]. In PA hepatotoxicity, we found that sustained JNK activation required the release of ROS from mitochondria, leading to the upstream activation of the MAPK cascade. The activation loop was initiated and amplified by binding of activated JNK (P-JNK) to the mitochondrial adaptor protein Sab (SH3 homology–associated BTK binding protein or SH3BP5), and subsequent inhibition of mitochondrial respiration and cell death (Figure 16.3) [9,14,15,40,99,100]. D-Galactosamine inhibits RNA synthesis, thus inhibiting NF-κB transcription of genes that turn off ROS and/or JNK. It is likely that JNK binding to mitochondrial Sab is key in mediating liver injury

in these models, because silencing Sab was found to protect against TNF-a plus
D-galactosamine-induced liver injury [14]. Most liver injury models in which mitochondrial dysfunction and cytochrome c release are central appear to be mediated
by JNK involvement. JNK also activates the proapoptotic Bax, which translocates
to mitochondria in most JNK-dependent toxicities [101]. JNK has also been suggested to phosphorylate and inactivate bcl-2 and bcl-xl, two antiapoptotic proteins
on the outer membrane of mitochondria [102]. The bcl-2 family of proteins are
important regulators of apoptotic cell death in many model systems, for example,
TNF-α-induced apoptosis [42] and bid in cholestatic liver injury [103]. Mitochondria
are key direct targets of JNK in a self-amplifying ROS cycle and subsequent necrosis by ROS-mediated MPT in the acetaminophen toxicity model or bcl-2 family-
mediated apoptosis in the TNF/galactosamine, ER stress, and lipotoxicity models,
pointing to the role of mitochondria in a final common pathway of hepatocellular
death. JNK-dependent Bax activation has been reported in lipotoxicity [96]. Beside
direct JNK-mediated Bax activation, JNK can mediate PUMA upregulation as a
transcriptional target of the JNK/activated AP-1 transcription factor complex. Both
JNK1 and JNK2 have been implicated in liver injury. However, depending on the
stimulus, these two JNK isoforms differentially contribute to hepatocyte apoptosis [41]. When a critical threshold of mitochondrial stress or injury occurs, JNK is
activated and targets mitochondria to induce MPT (necrosis) or MOMP (apoptosis).
P-JNK1 and 2 translocate to mitochondria and bind to and phosphorylate Sab, which
leads to further increase in mitochondrial ROS generation that is needed to sustain
JNK activity (Figure 16.3). The massive activation of JNK within a short period, and
sustained JNK activation followed the release of ROS from mitochondria leading

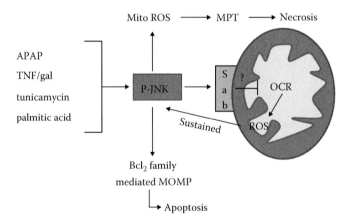

FIGURE 16.3   Model of JNK-Sab-mediated mitochondrial impairment. Different models of
liver injury, necrosis, apoptosis, ER stress, and lipotoxicity cause JNK activation. The translocation of active JNK and interaction with Sab, the mitochondrial outer membrane protein, and
the JNK-binding target cause impaired respiration, leading to ROS release from the mitochondria. Then sustained JNK activation occurs and causes MPT and MOMP mediating mitochondrial dysfunction, leading to cell death. The intramitochondrial mechanism of impaired
respiration oxygen consumption rate (OCR) and ROS production is currently unknown.

to upstream activation of the MAPK cascade and induction of mitochondrial permeability transition (MPT) in damaged mitochondria (e.g., acetaminophen toxicity) that causes increased permeability of the mitochondrial inner membrane and mitochondrial swelling and necrotic death. The interaction of JNK with Sab has been shown to enhance mitochondrial ROS generation by complex I up to 80% [104]. Thus, the increased mitochondrial ROS generation by JNK binding to Sab may be important in sustaining JNK activation through a feed-forward loop [14]. In addition, elevated FFA level is intimately associated with oxidative stress and high level of ROS. The role of oxidative stress in the activation and regulation of JNK and its specific effect on downstream associated transcription factors are important factors in PA-induced apoptosis. The findings of our in vitro experiments have demonstrated that high concentrations of FFAs cause high rates of cell death and the important role of JNK interaction with mitochondrial Sab in promoting ROS release sustained JNK activation and lipoapoptosis. The upstream activation of JNK largely depends upon ER PERK activation and the downstream mitochondrial dysfunction depends on the self-sustaining JNK activation due to ROS production. Thus, PA initiates JNK activation in a mitochondrial-independent stage, which then induces impaired mitochondrial function in a Sab-dependent fashion and this contributes to cell death due to mitochondrial ROS production and sustained JNK activation, leading to the modification of Bcl2 family of regulators of the intrinsic pathway of apoptosis. In conclusion, we have found that the lipotoxicity of PA in PMH involves the translocation of JNK to mitochondria and the interplay of JNK with mitochondrial Sab, which leads to mitochondrial dysfunction and triggers apoptosis.

ACKNOWLEDGMENTS

This work was supported by U.S. National Institutes of Health Grants RO1-DK067215 (NK) and RO1-AA014428 (NK), the USC Research Center for Liver Disease (P30-DK48522) Cellular and Tissue Imaging, Cell Separation and Culture, and Analytical/Metabolic/Instrumentation Cores, and the Southern California Research Center for Alcoholic Liver and Pancreatic Disease (P30-AA11999).

REFERENCES

1. Lazo M, Hernaez R, Bonekamp S, Kamel IR, Brancati FL, Guallar E, Clark JM. Non-alcoholic fatty liver disease and mortality among US adults: Prospective cohort study. *BMJ* 2011; 343:d6891.
2. Adams LA, Angulo P. Treatment of non-alcoholic fatty liver disease. *Postgrad Med J* 2006; 82(967):315–322.
3. Solinas G, Naugler W, Galimi F, Lee MS, Karin M. Saturated fatty acids inhibit induction of insulin gene transcription by JNK-mediated phosphorylation of insulin-receptor substrates. *Proc Natl Acad Sci USA* 2006; 103:16454–16459.
4. Yuan Q, Zhao S, Wang F, Zhang H, Chen ZJ, Wang J, Wang Z, Du Z, Ling EA, Liu Q, Hao A. Palmitic acid increases apoptosis of neural stem cells via activating c-Jun N-terminal kinase. *Stem Cell Res* Mar 2013; 10(2):257–266.
5. Ibrahim SH, Gores GJ. Who pulls the trigger: JNK activation in liver lipotoxicity? *J Hepatol* 2012; 56:17–19.

6. Feldstein AE, Werneburg NW, Li Z, Bronk SF, Gores GJ. Bax inhibition protects against free fatty acid-induced lysosomal permeabilization. *Am J Physiol Gastrointest Liver Physiol* 2006; 290:G1339–G1346.

7. Akazawa Y, Cazanave S, Mott JL et al. Palmitoleate attenuates palmitate-induced Bim and PUMA up-regulation and hepatocyte lipoapoptosis. *J Hepatol* 2010; 52:586–593.

8. Rockenfeller P, Ring J, Muschett V et al. Fatty acids trigger mitochondrion-dependent necrosis. *Cell Cycle* 2010; 9:2836–2842.

9. Win S, Than TA, Le BA, Garcia-Ruiz C, Fernández-Checa JC, Kaplowitz N. Sab (Sh3bp5) dependence of JNK mediated inhibition of mitochondrial respiration in palmitic acid induced hepatocyte lipotoxicity. *J Hepatol* 2015; 62(6):1367–1374. doi: http://dx.doi.org/10.1016/j.jhep.2015.01.032

10. Kakisaka K, Cazanave SC, Fingas CD et al. Mechanisms of lysophosphatidylcholine-induced hepatocyte lipoapoptosis. *Am J Physiol Gastrointest Liver Physiol* 2012; 302:G77–G84.

11. Ibrahim SH, Akazawa Y, Cazanave SC et al. Glycogen synthase kinase-3 (GSK-3) inhibition attenuates hepatocyte lipoapoptosis. *J Hepatol* 2011; 54:765–772.

12. Hirosumi J, Tuncman G, Chang L, Görgün CZ, Uysal KT, Maeda K, Karin M, Hotamisligil GS. A central role for JNK in obesity and insulin resistance. *Nature* 2002; 420:333–336.

13. Vernia S, Cavanagh-Kyros J, Barrett T, Jung DY, Kim JK, Davis RJ. Diet-induced obesity mediated by the JNK/DIO2 signal transduction pathway. *Genes Dev* 2013; 27(21):2345–2355.

14. Win S, Than TA, Han D, Petrovic LM, Kaplowitz N. c-Jun N-terminal kinase (JNK)-dependent acute liver injury from acetaminophen or tumor necrosis factor (TNF) requires mitochondrial Sab protein expression in mice. *J Biol Chem* 2011; 286:35071–35078.

15. Win S, Than TA, Fernandez-Checa JC, Kaplowitz N. JNK interaction with Sab mediates ER stress induced inhibition of mitochondrial respiration and cell death. *Cell Death Dis* 2014; 5:e989.

16. Ozcan U, Cao Q, Yilmaz E, Lee AH, Iwakoshi NN, Ozdelen E, Tuncman G, Görgün C, Glimcher LH, Hotamisligil GS. Endoplasmic reticulum stress links obesity, insulin action, and type 2 diabetes. *Science* 2004; 306:457–461.

17. Morel C, Standen CL, Jung DY, Gray S, Ong H, Flavell RA, Kim JK, Davis RJ. Requirement of JIP1-mediated c-Jun N-terminal kinase activation for obesity-induced insulin resistance. *Mol Cell Biol* 2010; 30:4616–4625.

18. Nakamura T, Furuhashi M, Li P, Cao H, Tuncman G, Sonenberg N, Gorgun CZ, Hotamisligil GS. Double-stranded RNA-dependent protein kinase links pathogen sensing with stress and metabolic homeostasis. *Cell* 2010; 140:338–348.

19. Wagner EF, Nebreda AR. Signal integration by JNK and p38 MAPK pathways in cancer development. *Nat Rev Cancer* 2009; 9:537–549.

20. Hui L, Zatloukal K, Scheuch H, Stepniak E, Wagner EF. Proliferation of human HCC cells and chemically induced mouse liver cancers requires JNK1-dependent p21 downregulation. *J Clin Invest* 2008; 118:3943–3953.

21. Stebbins JL, De SK, Machleidt T et al. Identification of a new JNK inhibitor targeting the JNK-JIP interaction site. *Proc Natl Acad Sci USA* 2008; 105:16809–16813.

22. Plantevin Krenitsky V, Nadolny L, Delgado M et al. Discovery of CC-930, an orally active anti-fibrotic JNK inhibitor. *Bioorg Med Chem Lett* 2012; 22:1433–1438.

23. Chen T, Kablaoui N, Little J, Timofeevski S, Tschantz WR, Chen P, Feng J, Charlton M, Stanton R, Bauer P. Identification of small-molecule inhibitors of the JIP-JNK interaction. *Biochem J* 2009; 420:283–294.

24. Bogoyevitch MA, Ngoei KR, Zhao TT, Yeap YY, Ng DC. c-Jun N-terminal kinase (JNK) signaling: Recent advances and challenges. *Biochim Biophys Acta* 2010; 1804:463–475.

25. Valerio A, Cardile A, Cozzi V et al. TNF-α downregulates eNOS expression and mitochondrial biogenesis in fat and muscle of obese rodents. *J Clin Invest* 2006; 116:2791–2798.
26. Haeusgen W, Herdegen T, Waetzig V. The bottleneck of JNK signaling: Molecular and functional characteristics of MKK4 and MKK7. *Eur J Cell Biol* 2011; 90:536–544.
27. Davis RJ. Signal transduction by the JNK group of MAP kinases. *Cell* 2000; 103:239–252.
28. Kallunki T, Deng T, Hibi M, Karin M. c-Jun can recruit JNK to phosphorylate dimerization partners via specific docking interactions. *Cell* 1996; 87:929–939.
29. Bogoyevitch MA, Kobe B. Uses for JNK: The many and varied substrates of the c-Jun N-terminal kinases. *Microbiol Mol Biol Rev* 2006; 70:1061–1095.
30. Karin M. The regulation of AP-1 activity by mitogen-activated protein kinases. *Philos Trans R Soc Lond B Biol Sci* 1996; 351:127–134.
31. Holzer RG, Park EJ, Li N et al. Saturated fatty acids induce c-Src clustering within membrane subdomains, leading to JNK activation. *Cell* 2011; 147:173–184.
32. Fan M, Goodwin M, Vu T, Brantley-Finle C, Gaarde WA, Chamber TC. Vinblastine-induced phosphorylation of Bcl-2 and Bcl-XL is mediated by JNK and occurs in parallel with inactivation of the Raf-1/MEK/ERK cascade. *J Biol Chem* 2000; 275:29980–29985.
33. Schroeter H, Boyd CS, Ahmed R, Spencer JP, Duncan RF, Rice-Evans C, Cadenas E. c-Jun N-terminal kinase (JNK)-mediated modulation of brain mitochondria function: New target proteins for JNK signalling in mitochondrion-dependent apoptosis. *Biochem J* 2003; 372:359–369.
34. Kim BJ, Ryu SW, Song BJ. JNK- and p38 kinase-mediated phosphorylation of Bax leads to its activation and mitochondrial translocation and to apoptosis of human hepatoma HepG2 cells. *J Biol Chem* 2006; 281:21256–21265.
35. Tsuruta F, Sunayama J, Mori Y, Hattori S, Shimizu S, Tsujimoto Y, Yoshioka K, Masuyama N, Gotoh Y. JNK promotes Bax translocation to mitochondria through phosphorylation of 14–3–3 proteins. *EMBO J* 2004; 23:1889–1899.
36. Yamamoto K, Ichijo H, Korsmeyer SJ. BCL-2 Is phosphorylated and inactivated by an ASK1/Jun N-terminal protein kinase pathway normally activated at G2/M. *Mol Cell Biol* 1999; 19(12):8469–8478.
37. Kharbanda S, Saxena S, Yoshida K et al. Translocation of SAPK/JNK to mitochondria and interaction with Bcl-x(L) in response to DNA damage. *J Biol Chem* 2000; 275:322–327.
38. Aoki H, Kang PM, Hampe J, Yoshimura K, Noma T, Matsuzaki M, Izumo S. Direct activation of mitochondrial apoptosis machinery by c-Jun N-terminal kinase in adult cardiac myocytes. *J Biol Chem* 2002; 277:10244–10250.
39. Chauhan D, Li G, Hideshima T, Podar K, Mitsiades C, Mitsiades N, Munshi N, Kharbanda S, Anderson KC. JNK-dependent release of mitochondrial protein, Smac, during apoptosis in multiple myeloma (MM) cells. *J Biol Chem* 2003; 278:17593–17596.
40. Seki E, Brenner DA, Karin M. A liver full of JNK: Signaling in regulation of cell function and disease pathogenesis, and clinical approaches. *Gastroenterology* 2012; 143:307–320.
41. Cazanave SC, Mott JL, Elmi NA, Bronk SF, Werneburg NW, Akazawa Y, Kahraman A, Garrison SP, Zambetti GP, Charlton MR, Gores GJ. JNK1-dependent PUMA expression contributes to hepatocyte lipoapoptosis. *J Biol Chem* 2009; 284(39):26591–602.
42. Kodama Y, Taura K, Miura K, Schnabl B, Osawa Y, Brenner DA. Antiapoptotic effect of c-Jun N-terminal Kinase-1 through Mcl-1 stabilization in TNF-induced hepatocyte apoptosis. *Gastroenterology* 2009; 136:1423–1434.
43. Masuoka HC, Mott J, Bronk SF, Werneburg NW, Akazawa Y, Kaufmann SH, Gores GJ. Mcl-1 degradation during hepatocyte lipoapoptosis. *J Biol Chem* 2009; 284:30039–30048.

44. Zhang W, Kudo H, Kawai K, Fujisaka S, Usui I, Sugiyama T, Tsukada K, Chen N, Takahara T. Tumor necrosis factor-α accelerates apoptosis of steatotic hepatocytes from a murine model of non-alcoholic fatty liver disease. *Biochem Biophys Res Commun* 2010; 391:1731–1736.

45. Tobiume K, Matsuzawa A, Takahashi T, Nishitoh H, Morita K, Takeda K, Minowa O, Miyazono K, Noda T, Ichijo H. ASK1 is required for sustained activations of JNK/p38 MAP kinases and apoptosis. *EMBO Rep* Mar 15, 2001; 2(3):222–228.

46. Wiltshire C, Matsushita M, Tsukada S, Gillespie DA, May GH. A new c-Jun N-terminal kinase (JNK)-interacting protein, Sab (SH3BP5), associates with mitochondria. *Biochem J* 2002; 367:577–585.

47. Kozieł R, Pircher H, Kratochwil M, Lener B, Hermann M, Dencher NA, Jansen-Dürr P. Mitochondrial respiratory chain complex I is inactivated by NADPH oxidase NOX4. *Biochem J* 2013; 452:231–239.

48. Santamaria E, Avila MA, Latasa MU, Rubio A, Martin-Duce A, Lu SC, Mato JM, Corrales FJ. Functional proteomics of nonalcoholic steatohepatitis: Mitochondrial proteins as targets of *S*-adenosylmethionine. *Proc Natl Acad Sci USA* 2003; 100:3065–3070.

49. Perez-Carreras M, Del Hoyo P, Martin MA, Rubio JC, Martin A, Castellano G, Colina F, Arenas J, Solis-Herruzo JA. Defective hepatic mitochondrial respiratory chain in patients with nonalcoholic steatohepatitis. *Hepatology* 2003; 38:999–1007.

50. Rao MS, Reddy JK. PPARα in the pathogenesis of fatty liver disease. *Hepatology* 2004; 40:783–786.

51. Reddy JK, Hashimoto T. Peroxisomal beta-oxidation and peroxisome proliferator-activated receptor α: An adaptive metabolic system. *Annu Rev Nutr* 2001; 21:193–230.

52. Muoio DM, Newgard CB. Mechanisms of disease: Molecular and metabolic mechanisms of insulin resistance and β-cell failure in type 2 diabetes. *Nat Rev Mol Cell Biol* 2008; 9(3):193–205.

53. Hotamisligil GS. Role of endoplasmic reticulum stress and c-Jun NH2-terminal kinase pathways in inflammation and origin of obesity and diabetes. *Diabetes* 2005; 54(Suppl 2): S73–S78.

54. Dong F, Ren J. Adiponectin improves cardiomyocyte contractile function in db/db diabetic obese mice. *Obesity* 2009; 17(2):262–268.

55. Di Sano F, Ferraro E, Tufi R, Achsel T, Piacentini M, Cecconi F. Endoplasmic reticulum stress induces apoptosis by an apoptosome-dependent but caspase 12-independent mechanism. *J Biol Chem* 2006; 281:2693–2700.

56. Borradaile NM, Han X, Harp JD, Gale SE, Ory DS, Schaffer JE. Disruption of endoplasmic reticulum structure and integrity in lipotoxic cell death. *J Lipid Res* 2006; 47:2726–2737.

57. Kitai Y, Ariyama H, Kono N, Oikawa D, Iwawaki T, Arai H. Membrane lipid saturation activates IRE1α without inducing clustering. *Genes Cells* 2013; 18:798–809.

58. Ben Mosbah I, Alfany-Fernández I, Martel C et al. Endoplasmic reticulum stress inhibition protects steatotic and non-steatotic livers in partial hepatectomy under ischemia-reperfusion. *Cell Death Dis* 2010; 1:e52.

59. Malhi H, Kaufman RJ. Endoplasmic reticulum stress in liver disease. *J Hepatol* 2011; 54:795–809.

60. Nishitoh H, Matsuzawa A, Tobiume K et al. ASK1 is essential for endoplasmic reticulum stress-induced neuronal cell death triggered by expanded polyglutamine repeats. *Genes Dev* 2002; 16:1345–1355.

61. Kaplowitz N, Than TA, Shinohara M, Ji C. Endoplasmic reticulum stress and liver injury. *Semin Liver Dis* 2007; 27:367–377.

62. Sharma M, Urano F, Jaeschke A. Cdc42 and Rac1 are major contributors to the saturated fatty acid-stimulated JNK pathway in hepatocytes. *J Hepatol* 2012; 56:192–198.

63. Zhao Y, Tian T, Huang T et al. Subtilase cytotoxin activates MAP kinases through PERK and IRE1 branches of the unfolded protein response. *Toxicol Sci* 2011; 120:79–86.
64. Verfaillie T, Rubio N, Garg AD et al. PERK is required at the ER-mitochondrial contact sites to convey apoptosis after ROS-based ER stress. *Cell Death Differ* 2012; 19:1880–1891.
65. Peidis P, Papadakis AI, Muaddi H, Richard S, Koromilas AE. Doxorubicin bypasses the cytoprotective effects of eIF2α phosphorylation and promotes PKRmediated cell death. *Cell Death Differ* 2011; 18:145–154.
66. Zhang P, Langland JO, Jacobs BL, Samuel CE. Protein kinase PKR-dependent activation of mitogen-activated protein kinases occurs through mitochondrial adapter IPS-1 and is antagonized by vaccinia virus E3L. *J Virol* 2009; 83:5718–5725.
67. Axten JM, Medina JR, Feng Y et al. Discovery of 7-methyl-5-(1-{[3-(trifluoromethyl)phenyl]acetyl}-2,3-dihydro-1H-indol-5-yl)-7Hpyrrolo[2,3-d]pyrimidin-4-amine (GSK2606414), a potent and selective first-inclass inhibitor of protein kinase R (PKR)-like endoplasmic reticulum kinase (PERK). *J Med Chem* 2012; 55:7193–7207.
68. Atkins C, Liu Q, Minthorn E et al. Characterization of a novel PERK kinase inhibitor with antitumor and antiangiogenic activity. *Cancer Res* 2013; 73:1993–2002.
69. Brancho D, Ventura JJ, Jaeschke A, Doran B, Flavell RA, Davis RJ. Role of MLK3 in the regulation of mitogen-activated protein kinase signaling cascades. *Mol Cell Biol* 2005; 25:3670–3681.
70. Ibrahim SH, Gores GJ, Hirsova P et al. Mixed lineage kinase 3 deficient mice are protected against the high fat high carbohydrate diet-induced steatohepatitis. *Liver Int* 2014; 34:427–437.
71. Gadang V, Kohli R, Myronovych A, Hui DY, Perez-Tilve D, Jaeschke A. MLK3 promotes metabolic dysfunction induced by saturated fatty acid-enriched diet. *Am J Physiol Endocrinol Metab* 2013; 305:E549–E556.
72. Van Galen P, Kreso A, Mbong N et al. The unfolded protein response governs integrity of the haematopoietic stem-cell pool during stress. *Nature* 2014; 510:268–272.
73. Muller F. The nature and mechanism of superoxide production by the electron transport chain: Its relevance to aging. *AGE* 2000; 23(4):227–253.
74. Han D, Williams E, Cadenas E. Mitochondrial respiratory chain-dependent generation of superoxide anion and its release into the intermembrane space. *Biochem J* 2001; 353(Pt 2):411–416.
75. Li X, Fang P, Mai J, Choi ET, Wang H, Yang XF. Targeting mitochondrial reactive oxygen species as novel therapy for inflammatory diseases and cancers. *J Hematol Oncol* 2013; 6:19.
76. Schafer FQ, Buettner GR. Redox environment of the cell as viewed through the redox state of the glutathione disulfide/glutathione couple. *Free Radic Biol Med* 2001; 30(11):1191–1212.
77. Ozben T. Oxidative stress and apoptosis: Impact on cancer therapy. *J Pharm Sci* 2007; 96(9):2181–2196.
78. Denning TL, Takaishi H, Crowe SE, Boldogh I, Jevnikar A, Ernst PB. Oxidative stress induces the expression of Fas and Fas ligand and apoptosis in murine intestinal epithelial cells. *Free Radic Biol Med* 2002; 33:1641–1650.
79. Medan D, Wang L, Toledo D, Lu B, Stehlik C, Jiang BH, Shi X, Rojanasakul Y. Regulation of Fas (CD95)-induced apoptotic and necrotic cell death by reactive oxygen species in macrophages. *J Cell Physiol* 2005; 203:78–84.
80. Reinehr R, Becker S, Eberle A, Grether-Beck S, Haus-Singer D. Involvement of NADPH oxidase isoforms and Src family kinases in CD95-dependent hepatocyte apoptosis. *J Biol Chem* 2005; 280:27179–27194.

81. Uchikura K, Wada T, Hoshino S, Nagakawa Y, Aiko T, Bulkley GB, Klein AS, Sun Z. Lipopolysaccharides induced increases in Fas ligand expression by Kupffer cells via mechanisms dependent on reactive oxygen species. *Am J Physiol Gastrointest Liver Physiol* 2004; 287: G620–G626.

82. Wang X, Zhang J, Xu T. Cyclophosphamide as a potent inhibitor of tumor thioredoxin reductase in vivo. *Toxicol Appl Pharmacol* 2007; 218:88–95.

83. Gupta SC, Hevia D, Patchva S, Park B, Koh W, Aggarwal BB Upsides and downsides of reactive oxygen species for cancer: The roles of reactive oxygen species in tumorigenesis, prevention, and therapy. *Antioxid Redox Signal* 2012; 16(11):1295–322.

84. Kyriakis JM, Avruch J. Mammalian mitogen-activated protein kinase signal transduction pathways activated by stress and inflammation. *Physiol Rev* 2001; 81(2):807–869.

85. Matsuzawa A, Ichijo H. Redox control of cell fate by MAP kinase: Physiological roles of ASK1-MAP kinase pathway in stress signaling. *Biochimica et Biophysica Acta* 2008; 1780(11):1325–1336.

86. Fujino G, Noguchi T, Matsuzawa A, Yamauchi S, Saitoh M, Takeda K, Ichijo H. Thioredoxin and TRAF family proteins regulate reactive oxygen speciesdependent activation of ASK1 through reciprocal modulation of the N-terminal homophilic interaction of ASK1. *Mole Cell Biol* 2007; 27(23):8152–8163.

87. Liu H, Nishitoh H, Ichijo H, Kyriakis JM. Activation of apoptosis signal-regulating kinase 1 (ASK1) by tumor necrosis factor receptor-associated factor 2 requires prior dissociation of the ASK1 inhibitor thioredoxin. *Mole Cell Biol* 2000; 20(6):2198–2208.

88. Jaeschke A, Davis RJ. Metabolic stress signaling mediated by mixed-lineage kinases. *Mol Cell* 2007; 27:498–508.

89. Kant S, Swat W, Zhang S, Zhang ZY, Neel BG, Flavell RA, Davis RJ. TNF-stimulated MAP kinase activation mediated by a Rho family GTPase signaling pathway. *Genes Dev* 2011; 25:2069–2078.

90. Kant S, Barrett T, Vertii A, Noh YH, Jung DY, Kim JK, Davis RJ. Role of the mixed lineage protein kinase pathway in the metabolic stress response to obesity. *Cell Rep* 2013; 4:681–688.

91. Cazanave SC, Wang X, Zhou H, Rahmani M, Grant S, Durrant DE, Klaassen CD, Yamamoto M, Sanyal AJ. Degradation of Keap1 activates BH3-only proteins Bim and PUMA during hepatocyte lipoapoptosis. *Cell Death Differ* 2014; 21:1303–1312.

92. Lizunov V, Chlanda P, Kraft M, Zimmerberg J. Long, saturated chains: Tasty domains for kinases of insulin resistance. *Dev Cell* 2011; 21:604–606.

93. Gallo KA, Johnson GL. Mixed-lineage kinase control of JNK and p38 MAPK pathways. *Nat Rev Mol Cell Biol* 2002; 3:663–672.

94. Tournier C, Dong C, Turner TK, Jones SN, Flavell RA, Davis RJ. MKK7 is an essential component of the JNK signal transduction pathway activated by proinflammatory cytokines. *Genes Dev* 2001; 15:1419–1426.

95. Lawler S, Fleming Y, Goedert M, Cohen P. Synergistic activation of SAPK1/JNK1 by two MAP kinase kinases in vitro. *Curr Biol* 1998; 8:1387–1390.

96. Malhi H, Bronk SF, Werneburg NW, Gores GJ. Free fatty acids induce JNK-dependent hepatocyte lipoapoptosis. *J Biol Chem* 2006; 281:12093–12101.

97. Wei Y, Wang D, Topczewski F, Pagliassotti MJ. Saturated fatty acids induce endoplasmic reticulum stress and apoptosis independently of ceramide in liver cells. *Am J Physiol Endocrinol Metab* 2006; 291:E275–E281.

98. Urano F, Wang X, Bertolotti A, Zhang Y, Chung P, Harding HP, Ron D. Coupling of stress in the ER to activation of JNK protein kinases by transmembrane protein kinase IRE1. *Science* 2000; 287:664–666.

99. Shinohara M, Ybanez MD, Win S, Than TA, Jain S, Gaarde WA, Han D, Kaplowitz N. Silencing glycogen synthase kinase-3β inhibits acetaminophen hepatotoxicity and attenuates JNK activation and loss of glutamate cysteine ligase and myeloid cell leukemia sequence 1. *J Biol Chem* 2010; 285:8244–8255.
100. Schwabe RF. Cell death in the liver—All roads lead to JNK. *Gastroenterology* 2006; 131:314–316.
101. Hanawa N, Shinohara M, Saberi B, Gaarde WA, Han D, Kaplowitz N. Role of JNK translocation to mitochondria leading to inhibition of mitochondria bioenergetics in acetaminophen- induced liver injury. *J Biol Chem* 2008; 283:13565–13577.
102. Latchoumycandane C, Goh CW, Ong MM, Boelsterli UA. Mitochondrial protection by the JNK inhibitor leflunomide rescues mice from acetaminophen-induced liver injury. *Hepatology* 2007; 45:412–421.
103. Higuchi H, Miyoshi H, Bronk SF, Zhang H, Dean N, Gores GJ. Bid antisense attenuates bile acid-induced apoptosis and cholestatic liver injury. *J Pharmacol Exp Ther* 2001; 299:866–873.
104. Chambers JW, Lograsso PV. Mitochondrial c-Jun N-terminal kinase (JNK) signaling initiates physiological changes resulting in amplification of reactive oxygen species generation. *J Biol Chem* 2011; 286:16052–16062.

17 Mitochondrial Function, Dysfunction, and Adaptation in the Liver during the Development of Diabetes

Andras Franko, Martin Hrabê de Angelis, and Rudolf J. Wiesner

CONTENTS

ABSTRACT

Due to its epidemiological dimensions, there are tremendous efforts to understand the ultimate pathways that lead from modern Western lifestyles to the development of insulin resistance and, finally, overt type 2 diabetes (T2DM), which is often accompanied by nonalcoholic fatty liver disease (NAFLD). The insulin-resistant liver is intimately involved in T2DM, since it importantly contributes to high circulating blood glucose levels due to the unsuppressed release of glucose, even in the fasted state. There is a large body of literature on the "involvement" of mitochondrial dysfunction in the liver in the development of T2DM. However, it is unclear if mitochondrial dysfunction causes hepatic insulin resistance, thereby truly contributing to the development of T2DM and NAFLD, or if it is just a consequence. Also, the term *mitochondrial dysfunction* has been used in a very uncritical way. Finally, there seems to be a continuum of mitochondrial changes during the development of NAFLD, from the initial benign steatosis to nonalcoholic steatohepatitis (NASH). In this chapter, we summarize the current knowledge on mitochondrial functions and their failure and critically review the existing literature on these processes in the liver during the development of T2DM and NASH.

Keywords: type 2 diabetes mellitus, type 1 diabetes mellitus, mitochondrial function, insulin resistance, mitochondrial metabolism

INTRODUCTION

Due to its epidemiological dimensions, there are tremendous efforts to understand the ultimate pathways which lead from modern western lifestyle to the development of insulin resistance and, finally, overt type 2 diabetes (T2DM). The insulin-resistant liver is intimately involved in this process, since it importantly contributes to high circulating blood glucose, even in the fasted state, as a consequence of unsuppressed release of glucose derived from gluconeogenesis (DeFronzo et al. 1982). There is a large body of literature on the "involvement" of mitochondrial dysfunction in the liver in the development of T2DM. However, in our opinion, it is still unclear if mitochondrial dysfunction causes hepatic insulin resistance, thereby truly contributing to the development of T2DM, or if it is just a consequence, if it exists at all. Even if the former were the case, it is unclear what would be the initial cause for the mitochondrial problem. As an alternative, it may be the metabolic environment of developing T2DM leading to mitochondrial dysfunction, which in turn may, or may not, be involved in further aggravation of disease progression. Again, it would be unclear which of the circulating factors or altered intracellular pathways would be to blame ultimately for causing the mitochondrial problem.

Also, in our opinion, the term "mitochondrial dysfunction" has been used in a very uncritical way. Finally, there seems to be a continuum of events during the development of nonalcoholic fatty liver disease (NAFLD), which is very often found in patients with T2DM, from the initial benign steatosis to nonalcoholic steatohepatitis (NASH).

In this book chapter, we summarize and critically review the existing literature on mitochondrial function during the development of T2DM and NASH, which may explain why so many controversial data have been reported. An extensive review on the same topic, however with a major focus on the role of reactive oxygen and nitrogen species (ROS and RNS) generated by the mitochondria in NASH, has been published very recently (Begriche et al. 2013). Thus, we discuss in this chapter mostly those studies in which the state of the patient or the experimental animal regarding insulin resistance and diabetic state has been reported.

MITOCHONDRIAL FUNCTIONS

Another important problem is that mitochondrial function is restricted by many authors to the organelle's role in ATP production and production of ROS, while many other and probably even more vital functions are generally overlooked (Figure 17.1).

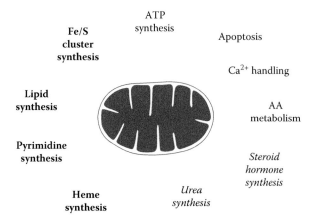

FIGURE 17.1 Essential functions that cannot be replaced by cytosolic activities (bold: left); Special functions typical for a special cell type (italics; lower right).

ATP Synthesis

The best-known function of mitochondria is the production of ATP by converting the redox potential of NADH and $FADH_2$ into an electrochemical potential across the inner mitochondrial membrane, which is then used by the ATP synthase to produce high-energy phosphate (Figure 17.1). In order to perform this task, the inner mitochondrial membrane is organized into membranous compartments called "cristae," which are packed with the four large membrane complexes of the electron transport chain and the ATP synthase, which is located at the tips of these cristae, thus contributing to or even causing their curvature (Davies et al. 2012). Since 13 essential subunits of these complexes are encoded by the small mitochondrial genome, any changes of the mtDNA sequence or its expression may cause a severe disturbance of this whole system of oxidative phosphorylation (OXPHOS). NADH and $FADH_2$ are derived from the oxidative degradation of carbohydrate or amino acid–derived pyruvate or fatty acids to CO_2.

Ca^{2+} Handling

Probably the second most well-known function is the involvement of mitochondria in cellular Ca^{2+} handling. It has long been known that isolated organelles are able to take up this divalent cation, driven by the electrochemical potential (negatively charged inside). However, only recently the Ca^{2+} channel responsible for this process was cloned, and was found to be surprisingly complex, being composed of the true channel, the mitochondrial Ca^{2+} uniporter (MCU) (Baughman et al. 2011, De Stefani et al. 2011), and several associated regulatory proteins (MICUs) (Mallilankaraman et al. 2012, Kovacs-Bogdan et al. 2014, De Stefani and Rizzuto 2014). With the detailed characterization of the properties of the channel, it is now clear that it only operates at very high Ca^{2+} concentrations, which would immediately lead to cell death, if they would be present all over the cytosol. Therefore, it is now well accepted that mitochondria are taking up Ca^{2+} only at cellular microdomains within the cytosol, where Ca^{2+} does indeed reach micromolar concentrations, that is, at contact sites with the endoplasmic reticulum and maybe at the subsarcolemmal region of muscle cells and cardiomyocytes (Pendin et al. 2014). One of the purposes of this mechanism may be the activation of TCA enzymes in order to match oxidative metabolism with high Ca^{2+} turnover, for example, during muscle contraction (Denton and McCormack 1990). Another mechanism may be shaping of Ca^{2+} transients in time and space (Pendin et al. 2014).

However, ATP synthesis and Ca^{2+} handling seem to be dispensable for many cell types, at least for a considerable time (see below), and other functions of the organelle, which are less well known, are probably the true essential mitochondrial functions which are needed to maintain cellular homeostasis.

Synthesis of Membrane Lipids

It is now clear that mitochondria are necessary for the synthesis of all cellular membrane lipids. Precursors are imported from parts of the endoplasmic reticulum called mitochondria-associated membranes (MAMs), converted into products in the inner mitochondrial membrane and either retained or exported again. This is best described for the import of phosphatidylserine from MAMs via the outer to the inner membrane, where some of it is converted to phosphatidylethanolamine, which remains in both mitochondria or is transferred to all other cellular membranes, that is, the endoplasmatic reticulum, the nuclear envelope, and the plasma membrane (Tatsuta et al. 2014). Thus, mitochondria are needed to synthetize the lipid building blocks for all cellular compartments during regular cellular turnover, but even more during enhanced cell proliferation, which may occur in the liver under pathological conditions.

Synthesis of Nucleic Acids

Mitochondria are essential to synthesize uridine, the common precursor for all pyrimidine derivatives during their de novo synthesis from carbamoyl-phosphate and aspartate. They catalyze the conversion of dihydroorotate to orotate by an electron transfer step by dihydroorotate dehydrogenase in the inner mitochondrial membrane (Berg et al. 2002). The final product of this pathway, uridine, is used to produce UTP needed for RNA synthesis, and also for the generation of UTP glucose, which is needed for glycogen synthesis as well as the synthesis of complex carbohydrates (Berg et al. 2002). UTP is converted to other pyrimidine-triphosphates CTP and TTP, needed for RNA and, as deoxynucleotides, for DNA synthesis, respectively. Although the nucleotides needed for RNA synthesis in a nondividing liver cell are mostly recycled from the preexisting pool by salvage pathways, during proliferation or enhanced transcription of genes, for example, during the acute phase response, mitochondria are necessary to provide essential nucleic acid building blocks.

Amino Acid Metabolism

Mitochondria are deeply involved via deamination and transamination reactions in amino acid metabolism (Berg et al. 2002). Therefore, especially organs like the liver, which not only uses amino acids to cover synthesis of its own proteins during regular turnover, but which produces and excretes large amounts of proteins, are highly dependent on this mitochondrial function.

Heme Synthesis

Mitochondria synthetize 5-aminolevulinate from succinyl-CoA and glycine and export it to the cytosol as the precursor for all kinds of heme groups (porphyrines), which are not only essential parts for hemoglobin and mitochondrial RC complexes, but also for the wide variety of cytochromes present in the endoplasmic reticulum, especially in the liver (Berg et al. 2002).

Synthesis of FeS Clusters

While all of the above-mentioned processes can be transferred to other parts of the cell, as can be seen in unicellular organisms without mitochondria, the one and only truly essential mitochondrial function seems to be the synthesis of FeS clusters (Stehling et al. 2014). These organisms, which are unable to respire, live in anaerobic environments, however they contain double-membrane compartments called mitosomes, which produce these highly complex protein machineries (Shiflett and Johnson 2010). FeS clusters are intricate parts of the mitochondrial electron transport chain; however, they are also needed in other cellular compartments like the nucleus, where they are parts of DNA repair complexes, for example.

Apoptosis

Apoptosis is often considered as a mitochondrial dysfunction, since it is a mechanism destroying the cell. However, we rather consider this process being an extremely important *function*, since it is highly controlled and only occurs when the decision has been made that a cell is no longer important or even dangerous for the whole organism. When a cellular defect has been detected by various sophisticated surveillance mechanisms, either caused by intrinsic problems (most importantly irreparable DNA damage) or by extrinsic causes like infection with a pathogen, mitochondria finally execute apoptosis by releasing cytochrome c and other factors from the tightly closed intra-cristae space through highly regulated cristae junctions to activate the death cascade (Martinou and Youle 2011).

In addition to these ubiquitous functions, there are tissue-specific functions of the organelles, with urea synthesis in the liver being the probably best-known example, where the initial steps, the synthesis of carbamoylphosphate and citrulline, take place in the matrix. Also the synthesis of steroid hormones in the adrenal cortex from cholesterol is performed within mitochondria (Berg et al. 2002). However, when their proteome was compared, it was found that, quite surprisingly, only about one-third of the ca. 1.100 reliably annotated mitochondrial proteins encoded in the nucleus are shared by 14 different mouse tissues, while the remaining proteins are cell type specific and tissue pairs share only about 75% of their protein equipment (Pagliarini et al. 2008).

MITOCHONDRIAL DYSFUNCTION

Mitochondrial dysfunction is, in the first place, considered as the inability to produce sufficient ATP to maintain cellular homeostasis. However, low levels of ATP are not necessarily a sign of mitochondrial problems. In the second place, the overproduction of ROS is thought to be one of the most common consequences of mitochondrial dysfunction, leading to cellular damage, and third, the inability to handle cytosolic Ca^{2+}, independent of ATP levels, is also believed to be one of the main problems when mitochondria do not work properly. As already mentioned above, it is the *inability* to properly execute apoptosis which should be considered as mitochondrial dysfunction and not the process of apoptosis itself. This inability is a hallmark of almost any cancer cell thereby escaping destruction although having accumulated the necessary chromosomal rearrangements which lead to tumor formation (Hanahan and Weinberg 2011).

ATP Synthesis

Surprisingly, the synthesis of high-energy phosphate by the OXPHOS system is dispensable in mammalian cells. Several cell lines have been generated lacking completely the mitochondrial genome, therefore containing a nonfunctional RC (called ρ0 cells; King and Attardi 1996, Bayona-Bafaluy et al. 2003). These cells need uridine for nucleic acid synthesis, since they cannot oxidize dihydroorotate (see above). They also need high concentrations of pyruvate in the medium, which simply serves as an electron acceptor, being reduced to lactate, thereby allowing the ATP generating flux from glucose to lactate through glycolysis running at maximum speed. Only few such cell lines have been established from single clones after treating cells with mitochondrial replication inhibitors, indicating that a considerable amount of rearrangements in the nuclear genome has to occur to enable cells to maintain that state forever. Such cell lines as well as cell lines containing severe mutations of mtDNA very often proliferate as fast as control cells, although they contain low levels of ATP. However, also the content of ADP and AMP is low, so that the energy charge, the ratio of high-to-low energy phosphate, remains maintained. We concluded that low ATP levels may be an adaption to enable maximal glycolytic flux under such conditions, which is then the only source of ATP, since glycolysis is inhibited by high ATP concentrations (von Kleist-Retzow et al. 2007). In these cells, ATP produced by glycolysis is even imported into the mitochondria and hydrolyzed to ADP and phosphate by reverse action of the ATP synthase. This process generates a net negative charge inside, thus maintaining the electrochemical gradient needed to import all the enzymes and substrates needed to perform the essential synthesis reactions discussed above (Appleby et al. 1999).

However, completely unexpected, cells can even survive in vivo for a surprisingly long time without a functional RC. This state was achieved by cell-type specific depletion of the mitochondrial transcription factor A (TFAM), which is essential for the replication and maintenance of mtDNA. For example, skeletal muscle–specific TFAM KO mice contain no detectable TFAM at 1 month, followed by no detectable RC subunits at 4 months due to slow mitochondria turnover (Wredenberg et al. 2006). However, animals were still alive and had to be sacrificed only a few weeks later due to general weakness and respiratory failure (Wredenberg, personal communication). An even more impressive gap of at least 1 month between the total absence of RC subunits and death of the animals was observed in the brain cortex after TFAM knockout in cortical neurons (Sorensen et al. 2001). Finally, we have shown previously that epidermal stem cells and their descendants can even hyperproliferate and form a well-functioning epidermis without any mtDNA and consequently without RC (Baris et al. 2011). In conclusion, these results show that cells can survive without ATP-producing mitochondria for a surprisingly long time even in vivo, emphasizing again that ATP generation is obviously not the most essential function of the organelles.

Ca^{2+} Handling

A Ca^{2+} handling defect can be seen in model cells with mitochondrial impairment due to mtDNA mutations (Brini et al. 1999, von Kleist-Retzow et al. 2007). Surprisingly, mice without the mitochondrial Ca^{2+} uniporter MCU have a very moderate skeletal muscle phenotype, showing that also this function is dispensable in vivo, at least in most tissues under basal conditions (Pan et al. 2013). However, in cells which experience high fluctuations of Ca^{2+} like the heart or pacemaking dopaminergic neurons, the inability to remove from the cytosol a sufficient amount of Ca^{2+} from locations of high concentrations may indeed impede cell function under conditions of high Ca^{2+} load.

ROS Production

It is still found in many reviews that "5% of the oxygen consumed by mitochondria" is first converted to potentially damaging superoxide, which is only then detoxified by scavenger systems. These numbers might be true for isolated mitochondrial preparations, but certainly not in vivo. The actual production of superoxide is extremely difficult to measure, however assays using newly developed in vivo indicators have shown that the true rate is probably very low (Cocheme et al. 2011). Only after poisoning with RC blockers, in the absence of proper free radical scavenging systems or in certain situations of high inner membrane potential, but low ATP consumption rate, free radicals like superoxide, H_2O_2, and RNS will truly accumulate and may then damage mitochondrial nucleic acids, proteins, and membrane lipids (Murphy 2009). However, we have shown that such free radicals generated by the RC do not cause mutations in the nuclear genome (Hoffmann et al. 2004), clearly showing that mitochondrial radicals are probably not causally involved in chromosome aberrations, be it in cancer or in the progress of aging.

Also, one of the most popular theories of aging, the free radical theory, postulating a vicious cycle of mutations of mtDNA causing RC dysfunction, leading to the generation of ROS causing more mutations of mtDNA, etc., has been elegantly disproven recently. One of the strongest arguments was provided when it was shown that a mouse model of premature aging, which accumulates a large number of mutations in their mtDNA (mutator mouse), does not show any signs of enhanced ROS production or downstream modifications of proteins or lipids (Trifunovic et al. 2004, 2005). The second argument is that such a vicious cycle should lead to random accumulation of mtDNA mutations during aging, while single cell analysis has shown that clonal expansion of single founder mtDNA mutations occurs during a lifetime (Khrapko et al. 1999, Wiesner et al. 2006).

Finally, it is also a widespread misconception that a nonfunctioning RC would inevitably lead to increased ROS production. On the contrary, it has been shown many times that the absence of the RC results in lowered intracellular ROS levels (Trifunovic et al. 2005, Schauen et al. 2006). Finally, there is some evidence that a certain ROS "tone" coming from the mitochondrial RC is a necessary prerequisite maintaining cellular homeostasis, since these ROS may be important signaling molecules (Schauen et al. 2006, Owusu-Ansah and Banerjee 2009).

DIABETES

FROM OBESITY VIA INSULIN RESISTANCE TO TYPE 2 DIABETES

The prevalence of diabetes is still increasing and 10 years ago the number of people who will suffer from diabetes in 2030 was estimated to be 366 million (Wild et al. 2004). Unfortunately, these days there are already more than 387 million people who live with diabetes and according to the new estimations for 2035, 592 million people are supposed to have diabetes, reaching then a 10% prevalence in adults (updated sixth diabetes atlas 2014 of the international diabetes federation (Cho et al. 2014). What is the ultimate cause for T2DM? At the moment, there is no obvious answer for this question, but insulin resistance and β-cell dysfunction are two major contributors.

The β cells are responsible to sense blood glucose levels and after a meal they secrete insulin, which tells the muscle and fat tissue to take up glucose, while it signals to the liver to stop glucose production from precursors (gluconeogenesis). If these peripheral tissues fail to work properly, insulin resistance evolves. An impairment of insulin secretion and peripherial glucose disposal lead to diabetes. A polygenetic predisposition definitely impacts the development of diabetes and there are about 70 established genetic loci identified, with numbers still increasing (Mahajan et al. 2014).

However, exogenous or environmental factors, like diet and exercise, are also the main contributors to develop T2DM. It was shown that about 90% of the new diabetes cases could be assigned to an adverse lifestyle (Krebs and Roden 2004), with the major lifestyle contributor being food. Excessive and prolonged high-calorie intake, in the form of highly palatable food which contains high fat as well as high sucrose, worsens or even directly leads to insulin resistance and NAFLD (Johnson et al. 2013, Kahle et al. 2013, Vos and Lavine 2013). In animal studies, rodents are usually treated with diets containing high fat and high sucrose to evoke insulin resistance and T2DM (Franko et al. 2014, Verbeek et al. 2014). It is not quite clear why it is hard to stop overeating, in rodents as well as in man, but high fat and sucrose diets were shown to lead to addiction like behaviors in rats and it is discussed whether highly palatable food could also be addictive for humans (Benton 2010, Garber and Lustig 2011, Gearhardt et al. 2011).

A second important factor is the sedentary nature of modern lifestyle which has probably an even higher impact. In the Australian diabetes, obesity, and lifestyle study (AusDiab), more than 4800 individuals without diagnosed diabetes were analyzed for self-reported sitting and TV viewing time. Sitting time was associated with high BMI, fasting insulin, and 2 h post-load plasma glucose (Thorp et al. 2010). A meta-analysis, summarizing data from more than 175,000 individuals reported by four studies, demonstrated that prolonged TV viewing is associated with an increased risk of T2DM. The risk score for T2DM with greater TV viewing remained significant even when obesity-associated parameters were also adjusted. Vice versa, the reduction of TV viewing time was shown to be beneficial in children in terms of BMI and energy intake even when physical activity remained unchanged (Grontved and Hu 2011). Furthermore, weight loss *per se* is also a powerful tool to reduce diabetes incidence and lifestyle interventions were demonstrated to be more effective than antidiabetic drugs (Krebs and Roden 2004).

Obesity

Obesity, which is usually originated from an imbalance between food intake and energy expenditure, is postulated to be a risk factor independent of insulin resistance and is associated with a 70%–80% risk to develop NAFLD and a 15%–20% risk to develop NASH (Choudhury and Sanyal 2005). Excess nutrient intake and obesity *per se* may also impact mitochondrial performance. Obese subjects showed a downregulated expression of energy metabolism genes in the liver compared to lean controls (Pihlajamaki et al. 2009). Morbidly obese, nondiabetic subjects exhibited lower skeletal muscle mitochondrial respiration compared to obese and lean controls and laparoscopic adjustable gastric banding bariatric surgery on these very obese patients led to a pronounced weight loss as well as elevated mitochondrial respiration in muscle (Vijgen et al. 2013). Roux-en-y gastric bypass was also applied to high fat, high sucrose (HF-HSD)-fed mice and resulted in weight loss, but also insulin resistance, high fat mass and low complex I activity as well as ATP content in the liver were normalized (Verbeek et al. 2014). These results suggest that obesity may disturb mitochondrial performance and that pronounced weight loss is able to revert the suppressed mitochondrial capacity. Since most NAFLD/NASH patients are obese, it is not possible at this moment to distinguish between alterations in mitochondrial function due to obesity, NAFLD or, finally, NASH, respectively.

Insulin Resistance

Insulin resistance and diabetes were reported to be associated with steatosis and NASH, however there are NASH patients who show normal glucose metabolism. In human studies, the HOMA (homeostasis model assessment) index is usually used to assess insulin resistance, which is calculated from fasting insulin and fasting glucose values. Most of the patients with NAFLD exhibit insulin resistance or diabetes (Cortez-Pinto et al. 1999), however about 10%–20% of them do not (Larter et al. 2010). Chitturi and colleagues reported that 98% of the NASH patients analyzed exhibited insulin resistance, revealed by high HOMA-IR (Chitturi et al. 2002). Furthermore, 87% and 50% of the NASH patients, respectively, were characterized by impaired glucose metabolism and overt T2DM (Chitturi et al. 2002). Therefore, insulin resistance may play an important role in the development of NAFLD. Prolonged fat and sucrose intake leads to ectopic lipid accumulation in the liver and this was postulated to cause liver insulin resistance by changing the phosphorylation state of insulin receptor machinery (Lowell and Shulman 2005, Szendroedi and Roden 2009). Insulin binds to its tyrosine kinase receptor (IR), which undergoes a conformational change, so that the tyrosine kinase domain is activated via autophosphorylation. This phosphorylation step recruits the insulin receptor substrate proteins (IRS) and the IR phosphorylates IRS on tyrosine residues. The phosphorylated IRS serves as a scaffold recruiting several effector kinases such as phospho-inositol-3-kinase (PI3K). These kinases activate other target proteins (like protein kinase B; PKB/AKT) which modulate transcription factors like FOXO1 and in turn modify gene regulation and control metabolic pathways, as well as cell proliferation and cell survival/death processes (for a current review see Boucher 2014). This highly complex system could be impaired by many different ways. Here we only point out the role of lipid intermediates like diacylglycerols, which can activate classical and novel PKC family members such as α, β and δ, ϵ, and these kinases are able to phosphorylate IR and IRS on serine residues. Serine phosphorylation inhibits tyrosine phosphorylation and in turn the recruitment of downstream substrates, which could result in diminished insulin signaling efficiency (Lowell and Shulman 2005, Szendroedi and Roden 2009, Boucher et al. 2014).

However, insulin resistance itself could also contribute to fat accumulation (Biddinger et al. 2008, Semple et al. 2009), therefore, it is hard to conclude whether steatosis causes insulin resistance or vice versa, since both states can induce the other (Larter et al. 2010). Furthermore, the role of insulin resistance in the development of NASH is not well understood, since liver steatosis further develops to NASH in some patients with insulin resistance, however in others it does not (Choudhury and Sanyal 2005).

Diabetes

Pancreatic β cells are flexible cells, they can expand their mass and number, can increase insulin production and secretion, and they can even form new β cells upon demand (Chang-Chen et al. 2008). Thus, the insulin-resistant state can be preserved for decades due to increased insulin secretion, but eventually β-cell failure and in turn diabetes develop. At the time of diagnosis, about 70% or 25%–50% reduction in β-cell mass were found in T1DM or T2DM patients, respectively (Chang-Chen et al. 2008). There are many reasons why β cells cannot infinitely compensate for insulin resistance. First, the function of β cells is especially dependent on mitochondrial performance, since insulin secretion is tightly coupled to mitochondrial ATP production. If ATP production in the mitochondria is suppressed, ATP-dependent potassium channels are not properly closed, membrane depolarization as well as calcium influx are diminished, resulting in disturbed insulin secretion (Lowell and Shulman 2005). High lipid levels and their intermediates were shown to directly disturb β-cell function, a phenomenon called lipotoxicity. Pancreatic lipid levels were demonstrated to be doubled in patients with T2DM and negatively correlated with β-cell performance (Szendroedi and Roden 2009). A physiological rise of glucose concentration is known to stimulate insulin secretion, however permanently high glucose levels were demonstrated to be detrimental for β-cell function, known as glucotoxicity. Chronic hyperglycemia caused diminished β-cell mass and function and ROS are one of the postulated major mediators of this process (Chang-Chen et al. 2008). Patients with T2DM are usually characterized by high lipid and glucose levels, therefore they are even facing "glucolipotoxicity." The effect of hyperglycemia and hyperlipidemia are assumed to synergize, since the detrimental effects of lipotoxicity on β-cell death is enhanced by high glucose levels. A proposed mechanism of glucolipotoxicity includes a suppressed fatty acid oxidation by high malonyl-CoA levels, which inhibits the mitochondrial fatty acid transporter CPT1 (Chang-Chen et al. 2008). Endoplasmic reticulum stress was also found to play a crucial role in β-cell dysfunction due to obesity and diabetes (Chang-Chen et al. 2008).

POSSIBLE ROLE OF MITOCHONDRIAL ALTERATIONS IN THE DEVELOPMENT OF DIABETES

Since mitochondria play a central role in nutrient metabolism, it is not surprising that mitochondrial alterations in the liver have been "associated" with steatosis, NASH, and diabetes (Pessayre and Fromenty 2005). However, it is still debated whether mitochondrial changes, if they exist at all, are the cause or the consequence of, or alternatively, an adaptation to steatosis or diabetes (Larter et al. 2010, Koliaki and Roden 2014). There are dozens of reports pointing to the "changes in mitochondrial function" when patients with NAFLD or mouse models with steatosis or NASH were analyzed (for an excellent summary, see Begriche et al. 2013). However, the vast majority of these studies unfortunately do not report on circulating insulin or glucose levels or other important measures of glucose metabolism, like tolerance tests (Brady et al. 1985, Chavin et al. 1999, Serviddio et al. 2008a,b), so final conclusions regarding insulin resistance and mitochondrial function are impossible to be drawn from these studies at this point.

In this review, we are mainly summarizing data about liver mitochondrial alterations in NAFLD, but our main focus will be on studies with reported insulin resistance and/or diabetes. We will also shortly discuss mitochondrial alterations in skeletal muscle and heart, since they may also influence the development and progression of disease. One should again note that it is not an easy task to distinguish between hyperlipidemia in the circulation, ectopic lipid accumulation in the liver, liver insulin resistance, diminished whole body glucose tolerance, and manifested diabetes. All these single factors can preexist alone or in combination for years or decades without turning into more serious metabolic problems.

ROS, A Possible Contributor to Mitochondrial Dysfunction

What is the mechanism for mitochondrial impairment in NAFLD? Reactive oxygen species are one of the postulated key molecules, which possibly lead to mitochondrial dysfunction upon steatosis (for detailed information see Chapter 1). According to the "ROS theory," increased supply of energy, especially fatty acids, cannot be compensated completely by increased fatty acid transport and β-oxidation (Pessayre and Fromenty 2005). Fatty acids are accumulating and may additionally disturb the insulin signaling pathways, as discussed above (Lowell and Shulman 2005). Elevated fatty acid oxidation results in higher levels of the reducing equivalents NADH and $FADH_2$, which are donating electrons to the mitochondrial complexes I and II. Electron transport runs at high speed and reactive superoxide molecules are created by complexes I and III. These superoxide molecules are reduced by the mitochondrial Mn-SOD to hydrogen peroxide, which in turn is converted by glutathione peroxidase or catalase to water (Patti and Corvera 2010, Begriche et al. 2013). On the other hand, superoxide and hydrogen peroxide are known to activate the uncoupling proteins (UCPs) (Echtay et al. 2002), which are proton transporters and dissipate the electron flow from creating a proton gradient, which is used to drive the ATP synthase. Thus, UCP activation by ROS may result in diminished ATP production.

What Makes Liver Mitochondria Special?

Liver mitochondria have a smaller size, less cristae and total cellular area covered by mitochondria is much lower in the liver when compared to, for example, the heart (Veltri et al. 1990). Liver mitochondria show less maximal activities for CIII and CIV, however state 3 respiration and respiratory control ratio are similar in liver and muscle mitochondria (Benard et al. 2006). Furthermore, in liver mitochondria, the maximal capacity for the entry of electrons is primarily carried out by CII rather than CI (Holmstrom et al. 2012, Lapuente-Brun et al. 2013, Franko et al. 2014), while muscle shows a similar contribution of both complexes (Holmstrom et al. 2012) (Franko and Wiesner unpublished data). On the other hand, the maximal possible rate of ATP synthesis was higher in liver than in muscle mitochondria (Schmid et al. 2008).

What Makes Liver Mitochondria Special in Diabetic Situations?

Mitochondria in the liver are also different from the muscle in diabetic situations; here we summarize data reported from type 1 diabetic animal models. Heart and skeletal muscle mitochondria were found to have a defect in insulin-deficient diabetes in terms of respiration and proton conductance, whereas liver mitochondria did not show any reduction in these parameters (Herlein et al. 2009). Heart mitochondria isolated from mice with a type 1-like diabetes state displayed less oxygen consumption compared to healthy controls, whereas kidney and liver mitochondria exhibited unchanged respiration (Bugger et al. 2009). In the nonobese diabetic (NOD) mouse, which is a genetic model for T1DM, mitochondrial respiration was reported to be decreased in skeletal muscle in the fasted state, however it remained unaltered in liver mitochondria, while in the fed state, liver mitochondria displayed a higher respiration, and muscle mitochondria showed an unchanged or decreased oxygen consumption, depending on complex I and II substrates (Jelenik et al. 2014). When we analyzed insulin-deficient, type 1 diabetic mice, we also noticed a striking difference in mitochondrial capacity comparing muscle and liver tissues. Mitochondrial performance was severely suppressed in the mitochondria of soleus muscle (Franko et al. 2012), however isolated liver mitochondria displayed normal complex II respiration or even elevated complex I oxygen consumption (Franko et al. 2014) (for details, see the section "Mitochondrial Function in the Liver in Type 1 Diabetes"). These results indicate that changes in mitochondrial capacity in type 1 diabetic states are tissue specific and conclusions drawn from one organ cannot be generalized. Moreover, heart and muscle mitochondrial function seems to be impaired in insulin-resistant and/or diabetic states, while liver rather can compensate for the disturbed metabolic state.

MITOCHONDRIAL FUNCTION IN THE LIVER DURING THE PROGRESSION OF STEATOSIS TO NASH

In the following chapters, we keep the terms steatosis, NAFLD, or NASH like they were used by the authors of the quoted references. To simplify the nomenclature, the term *fatty liver* is used equivalent to steatosis, defined by >5.5% intrahepatic fat content (Koliaki and Roden 2013). For a detailed review about NAFLD, see Chapters 18 and 12 or Begriche et al. (2013). In this section, we summarize the published studies on mitochondrial function in NAFLD, although insulin resistance and the diabetic state were not reported.

HUMAN STUDIES

High levels of oxidized glutathione were found in isolated liver mitochondria of NASH patients, who did not show any other metabolic disorders. Furthermore, mRNA levels of the uncoupling protein UCP-2 were increased (Serviddio et al. 2008a), suggesting mitochondrial oxidative stress and mitochondrial uncoupling as a compensation for it in the liver of these NASH patients.

RODENT STUDIES

The leptin-deficient ob/ob mice serve as a mild model for NASH with hepatocellular necroinflammation and fibrosis (Begriche et al. 2013). In young, 5-week old ob/ob mice, lipid-driven respiration of liver mitochondria as well as activity of CPT-1, the main carrier for fatty acids in the inner mitochondrial membrane, was unaltered compared to lean controls. In contrast, in old 6–9 months old ob/ob mice, liver mitochondrial respiration and CPT-1 activity were found to be increased (Brady et al. 1985). We also analyzed 7 months old ob/ob mice, which were normoglycemic, and liver mitochondria showed elevated respiration via CI or CII substrates as well as elevated CPT-1 activity (Franko and Wiesner, unpublished data). These results suggest that in the liver disease progression or ageing processes could affect mitochondrial function and the elevated mitochondrial capacity reported in older ob/ob mice could be an adaptation process, which is probably triggered by the oversupply of fatty acids. Chavin et al also studied young, 10–12-weeks old ob/ob mice and liver mitochondrial respiration was increased by CII substrate, but not with CI substrates compared to lean controls. Mitochondria exhibited a decreased membrane potential and upregulated UCP-2 levels, while liver ATP levels were reduced compared to lean controls (Chavin et al. 1999). These data suggest that UCP-2 upregulation and the reduced mitochondrial membrane potential could be a possible way to attenuate the redox pressure caused by the energy oversupply. The data also indicate that one should take care by choosing the appropriate substrate for respiration studies, since changes in the inner membrane RC complex stoichiometry may take place as an adaptation. Insulin receptor substrate (IRS) proteins are the key regulators for insulin signaling pathway connecting receptor activation to gene regulation. A diminished mitochondrial respiration was observed in 8–10-weeks old IRS-1/IRS-2 double KO mice as well as in 16–20-weeks old leptin-deficient ob/ob and leptin receptor mutant db/db mice compared to respective controls, in parallel with reduced mitochondrial number, respiratory control ratio, ATP level, and CI activity. In these animals, the level of heme, which is a cofactor needed for CIII and CIV, was also lower, in association with decreased CIII and CIV protein levels. FOXO1, a transcription factor activated via the insulin signaling pathway, as well as PGC-1α, a transcriptional coactivator responsible for mitochondrial biogenesis, were shown to control mitochondrial function in these insulin-resistant states (Cheng et al. 2009).

The methionine-choline-deficient (MCD) diet is proposed to cause more severe liver inflammation and fibrosis compared to ob/ob mice, however without causing insulin resistance. Three-week treatment resulted in normal CI-driven respiration, however CII respiration was increased in these animals, possibly again as an adaptation, but liver ATP level was decreased. After 7-and 11-weeks treatment, CII respiration was normal but liver ATP content still decreased (Serviddio et al. 2008a). Thus, the basal liver ATP level is probably not a reliable marker for mitochondrial performance (see below discussion of Cortez-Pinto et al. 1999), since liver ATP level was decreased at all time points, but mitochondrial respiration was not compromised at all.

MITOCHONDRIAL FUNCTION IN THE LIVER IN INSULIN-RESISTANT STATES

In this section, we summarize some of the published data on mitochondrial function in animals or patients, whose insulin-resistant state was reported. Subjects with fatty liver or NASH are also included in this section, since they are very probably insulin resistant (see above). Our decision criteria to include a study in this paragraph were based on reported mild elevation in blood glucose levels (but not manifested diabetes with high fasted blood glucose), an elevated HOMA-IR, impaired glucose tolerance, and/or whole body or hepatic insulin resistance assessed by hyperinsulinemic–euglycemic clamp studies. Still one should bear in mind that the insulin-resistant state of the subjects determined by different methods will not necessarily lead to a homogenous group. HOMA-IR is an estimate for insulin resistance, calculated from fasting insulin and glucose values, while an euglycemic–hyperinsulinemic clamp study serves as the ultimate in vivo measurement. Also, the severity of insulin resistance probably impacts on the mitochondrial phenotype, as in extreme insulin-resistant states mitochondrial dysfunction was observed, at least in the muscle (Sleigh et al. 2011, Franko et al. 2012). However, according to our knowledge there is no common accepted one and only method to verify insulin resistance. Euglycemic–hyperinsulinemic clamp studies represent the gold standard for testing the efficacy of insulin in vivo, although because of its laborious, costly and time-consuming nature, only few laboratories perform this method routinely, especially in rodent studies (Fuchs et al. 2012). During the clamp procedure, a constant insulin dose is infused i.v., together with an i.v., glucose infusion. Glucose infusion is adjusted to maintain euglycemia in the presence of high insulin levels, so the infused dose will vary depending on the insulin sensitivity of the subject or animal. In an insulin-resistant subject, less glucose is needed to maintain euglycemia compared to a normal insulin-sensitive subject, since insulin is not able to decrease blood glucose levels properly due to hepatic and peripheral tissue insulin resistance. Furthermore, this technique is frequently combined with the use of radioactive isotopes, providing a unique possibility to dissect liver, muscle, or fat insulin sensitivity (Muniyappa et al. 2008).

HUMAN STUDIES

Intravenous fructose injection leads to a transient depletion of hepatic ATP pool due to a rapid conversion of fructose to glucose-6-phosphate, thus it was used to analyze liver mitochondrial function in obese NASH patients with probably mild insulin resistance. Basal and fructose-depleted ATP levels were similar in NASH patients compared to healthy controls, however ATP levels recovered to basal level after 1 h in control subjects, while recovery was impaired in NASH patients, suggesting a severe mitochondrial dysfunction (Cortez-Pinto et al. 1999). These data also clearly show that the basal ATP level is not an appropriate marker for mitochondrial performance, since NASH patient had normal values, and the diminished mitochondrial performance became only apparent after fructose depletion. In obese patients with steatosis or NASH, overall insulin sensitivity was assessed by euglycemic–hyperinsulinemic clamps and showed that the glucose infusion rate was lower compared to healthy controls, indicating whole body insulin resistance. However, these NASH patients exhibited similar hepatic insulin sensitivity compared to patients with steatosis. Transmission microscopy revealed swollen mitochondria with paracrystalline inclusion bodies in the liver biopsies of patients with NASH, but steatosis patients did not show such mitochondrial abnormalities (Sanyal et al. 2001). In accordance with these findings, liver homogenates of overweight NASH patients with an elevated HOMA-IR index, suggested that patients were insulin resistant, exhibited reduced activities for all five RC complexes (Perez-Carreras et al. 2003). Furthermore, liver samples of morbidly obese but nondiabetic subjects with elevated HOMA-IR and increased liver fat content exhibited a decreased expression of OXPHOS genes compared to lean controls (Pihlajamaki et al. 2009). Sunny et al. studied obese NAFLD patients with high hepatic TG levels and showed that impaired hepatic insulin sensitivity compared to body fat, BMI, and age-matched subjects with low hepatic TG. Despite hepatic insulin resistance, oxidative and anaplerotic fluxes determined in liver mitochondria were elevated in NAFLD patients (Sunny et al. 2011). These data indicate that liver steatosis and insulin resistance do not necessarily lead to mitochondrial damage, and only severe forms of NASH are associated with mitochondrial abnormalities.

RODENT STUDIES

High fat diet in the combination with high sucrose (HF-HSD) supplementation is a powerful tool to evoke insulin resistance. Sucrose is a disaccharide composed of one glucose and one fructose molecule. Fructose is known to deplete the mitochondrial ATP pool in the liver (Cortez-Pinto et al. 1999) and is postulated to be a risk factor for the development of NAFLD (Ouyang et al. 2008). HF-HSD treatment of mice for 4 weeks evoked impaired glucose tolerance and a tendency for elevated HOMA-IR, but without diabetes. This short-term HF-HSD feeding neither caused symptoms of NASH nor did it reduce mitochondrial activity (Verbeek et al. 2014) (Verbeek, personal communication). These results support the data from humans, indicating that mitochondrial function remains preserved in most insulin-resistant states.

MITOCHONDRIAL FUNCTION IN THE LIVER IN OVERT TYPE 2 DIABETES

Diabetes is defined by fasting plasma glucose ≥ 7.0 mmol/L (126 mg/dL) for human subjects (WHO 2006, 2014). This definition was used to identify published data in this category. One should note that in rodent studies glucose is usually measured in whole blood taken from the tail vein, since plasma samples cannot conveniently be collected for routine and repeated measurements. Tail vein blood glucose levels are usually lower than plasma samples due to the presence of erythrocytes; however, fasting blood glucose levels in mice are usually higher than in humans.

HUMAN STUDIES

Nuclear genes encoding OXPHOS proteins were shown to be upregulated in the liver of T2DM patients compared to age and BMI-matched nondiabetic controls (Misu et al. 2007). In a second study, this was also shown in the liver with steatosis of obese patients with T2DM compared to nonobese patients with T2DM (Takamura et al. 2008). On the other hand, morbidly obese patients with T2DM and steatosis were reported to exhibit reduced expression of OXPHOS genes compared to lean, healthy controls (Pihlajamaki et al. 2009). Interestingly, this study also analyzed obese, non-diabetic subjects and gene expressions of this group showed very similar downregulation in terms of OXPHOS genes as the obese, diabetic group (Pihlajamaki et al. 2009). The apparent controversy in these studies is certainly due to the fact that they have used (1) different controls for their comparisons (lean healthy versus lean diabetics), (2) the studied subjects originated from different ethnicity (Japanese versus Caucasian Americans), and (3) the degree of obesity was different (morbidly obese vs. mildly obese subjects BMI 52 vs. 27) (Patti and Corvera 2010). Taken together, it remains still controversial whether and how mitochondrial gene expression is influenced by the diabetic situation in the human liver. Szendroedi et al. reported for the first time, using in vivo ^{31}P-NMR, that the liver of overweight patients with T2DM showed less energy-rich γ-ATP, however also less inorganic phosphate (Pi) levels than nondiabetic, age and BMI-matched or lean controls (Szendroedi et al. 2009). One possible explanation of these findings is "the energy deficit" hypothesis: excessive gluconeogenesis and/or lipogenesis, which are not properly suppressed due to insulin resistance, may evoke an ATP synthesis/demand dysbalance, leading to a drop of intracellular ATP if the hepatocytes are not able to compensate this situation. The lowered Pi levels, however, may indicate that the energy charge of the adenylate system is maintained. These T2DM patients were extensively characterized by euglycemic–hyperinsulinemic clamp studies, which revealed the whole body as well as hepatic insulin resistance (Szendroedi et al. 2009). To analyze in vivo ATP turnover in the liver, in contrast to steady-state levels, the ^{31}P-NMR method was further developed using saturation transfer (Petersen et al. 2004) by the same group. Overweight T2DM patients, who exhibited hepatic insulin resistance, were reported to show a reduced liver ATP synthetic flux rate (fATP) compared to nondiabetic age and BMI-matched controls. On the other hand, hepatic lipid content was not altered and the prevalence of steatosis was similar between both groups (Schmid et al. 2011). These results suggest that a lower mitochondrial ATP production rate is not necessarily due to ectopic lipid accumulation.

RODENT STUDIES

Liver mitochondrial function was compared in lean, diabetic Goto-Kakizaki rats to control Wistar rats and complex I- and complex II-driven mitochondrial respiration rate were found to be increased. The elevated mitochondrial respiration was associated with higher respiratory control ratios and less lipid peroxidation was found (Ferreira et al. 1999). In a later study of the same group, the higher mitochondrial respiration rate and increased membrane potential were associated with elevated complex activities in 6 months old, but not in 3-month-old Goto-Kakizaki rats (Ferreira et al. 2003). Buchner et al. compared two congenic mouse strains, called 6C1 and 6C2, derived from chromosome substitutions. The 6C1 strain fed with HF-HSD showed higher body weight, higher blood glucose levels, impaired glucose tolerance, and insulin resistance compared to 6C2. Genes coding for mitochondrial proteins were upregulated in 6C1 liver compared to 6C2 in parallel with elevated CI-driven mitochondrial respiration (Buchner et al. 2008, 2011). In our study, HF-HSD feeding of C57BL/6 mice for 6 months caused increased body weight, high liver triglycerides, increased serum insulin, fasted glucose and triglyceride levels, and thus T2DM. The oxygen consumption of isolated liver mitochondria revealed normal CI- and CII-driven respiration, however an increased capacity to oxidize glycerol-3-phosphate, feeding electrons into the quinone pool before CIII, was found. Elevated mitochondrial respiration was not associated with higher mitochondrial RC protein levels, but TCA enzyme activities were increased in HF-HSD mice. Mitochondrial membrane potential measurements exhibited unchanged proton leak kinetics in HF-HSD mice (Franko et al. 2014). These results strongly support the concept that mitochondrial dysfunction is not a necessary prerequisite for steatosis, since massive lipid accumulation was observed in the presence of normal mitochondrial function. Our results and the former results rather indicate an adaptation process of the liver mitochondria under insulin-resistant and diabetic conditions.

On the other hand, there are other reports showing attenuated mitochondrial performance in rodents with diabetes. Sand rats (*Psammomys obesus*) were fed with a high caloric diet (HFD), which led to an increased body weight and insulin levels as well as manifested diabetes together with elevated liver triglyceride levels. Liver mitochondria isolated from HFD animals exhibited reduced mitochondrial respiration via CI substrates, whereas respiration via CII substrate remained unchanged (Bouderba et al. 2012). Holmstrom et al. compared obese, diabetic leptin receptor mutant db/db mice to lean controls and found reduced CI- and CII-driven respiration in liver homogenates. The protein levels of CI and CIV subunits were reduced in association with lower PGC-1α expression, however the mtDNA copy number was increased in db/db mice (Holmstrom et al. 2012). Very recently, Verbeek et al. reported an excellent longitudinal study on 4-, 12-, and 20-week HF-HSD-treated C57/BL6J mice. Interestingly, all three treatment groups showed elevated citrate synthase activity. The liver of 20-week HF-HSD mice exhibited obvious signs for NASH, reflected by hematoxylin and eosin stainings as well as collagen morphometry. Furthermore, these mice not only exhibit insulin resistance, assessed by glucose tolerance test and HOMA-IR, but also increased fasting blood glucose levels (Verbeek et al. 2014 and Verbeek, personal communication). Complex I and IV enzymes activities as well as ATP content were significantly decreased in the 20-week HF-HSD-treated group, however mtDNA content remained unchanged. Furthermore, ultrastructural analysis of the 20-week treated mice revealed mitochondrial membrane damage, rearranged cristae, and swollen mitochondria. HF-HSD treatment for 12 weeks evoked similar but less severe signs for NASH in the liver compared to 20-week treatment, but mitochondrial morphological and functional impairments were comparable to the 20-week group (Verbeek et al. 2014). Since 4-week HF-HSD treatment caused an impaired glucose tolerance, but neither NASH nor elevated blood glucose levels, we postulate from this and other studies (Szendroedi et al. 2009) that insulin resistance can precede manifested NASH.

MITOCHONDRIAL FUNCTION IN THE LIVER IN TYPE 1 DIABETES

To our knowledge, there are no human studies on liver mitochondrial performance in this state.

RODENT STUDIES

The Akita mice carrying a mutation in the insulin 2 gene are used as a model for T1DM with severe hyperglycemia. In the liver of these mice, an upregulation of mitochondrial genes and proteins were found, however neither mitochondrial respiration, analyzed via CI or CII substrates, nor mitochondrial mass or morphology were changed (Bugger et al. 2009). A chemical called streptozotocin (STZ) can be used to destroy pancreatic β cells in healthy animals and thus evokes an insulin-deficient state resembling T1DM. In the liver of STZ-treated mice, elevated levels of mRNAs for mitochondrial genes and for mitochondrial biogenesis regulating genes, as well as the cognate proteins were observed, together with higher mtDNA level and increased mitochondrial mass (Liu et al. 2009). These and results on insulin-treated hepatocytes (Liu et al. 2009) indicate that insulin may suppress mitochondrial biogenesis in the liver.

Importantly, one needs to take into account the time factor, namely the duration of diabetes, which does also impact the mitochondrial phenotype. Ferreira et al. investigated STZ-treated male Wistar rats 3 or 9 weeks after STZ injection: Although both STZ groups presented blood glucose levels higher than 400 mg/dL and reduced body weights, mitochondrial respiration stimulated by FCCP, respiratory control ratios, and the mitochondrial membrane potential were only elevated in the 9-week group, but were normal in the 3-week group. Both STZ groups showed elevated activities for CII and CIV at 3 and 9 weeks after injection (Ferreira et al. 2003). Furthermore, liver mitochondrial function was also studied in STZ-treated male Sprague–Dawley rats 2 or 8 weeks after injection. Mitochondrial respiration via CI or CII substrates, respiratory control ratios as well as mitochondrial membrane potential remained unaffected in both STZ groups presenting with severe hyperglycemia. Mitochondrial H_2O_2 production was elevated in the 2-week group but was reduced in the 8-week group (Herlein et al. 2009). We have also investigated the long-term effect of type 1 diabetes in male C57BL/6N mice and have not observed any sings for mitochondrial dysfunction. Euglycemic–hyperinsulinemic clamps proved hepatic insulin resistance in the treated mice, which was accompanied by very low insulin and very high blood glucose levels after 2 months. Mitochondrial protein levels were increased in liver mitochondria in parallel with elevated CI-driven respiration and with better coupling (Franko et al. 2014). Finally, the impact of the duration of diabetes (acute vs. chronic) on mitochondrial performance was also investigated in a recent study of Jelenik et al., where female NOD (nonobese diabetic) mice were examined, which represent an excellent genetic model for T1DM with autoimmune characteristics. Three different insulin-resistant NOD mice were investigated and compared to wild-type mice: (1) normoglycemic but insulin-resistant—NOD mice, (2) acute diabetic—NOD mice, and (3) chronic diabetic—NOD mice. The insulin-resistant and acute diabetic NOD mice displayed hepatic insulin resistance, verified by euglycemic–hyperinsulinemic clamps. In the fasted state normal mitochondrial respiration was observed in the liver of all mice (although no data were collected from chronic diabetic NOD mice). On the other hand, in the fed state an elevated mitochondrial respiration was found in the acute diabetic NOD compared to wild-type and insulin-resistant NOD mice. Mitochondria from the fed chronic diabetic

NOD mice also exhibited a higher respiration rate compared to wild-type controls, however it was lower than in acute diabetic mice. The expression of genes regulating mitochondrial biogenesis was upregulated in acute diabetic NOD mice compared to wild-type controls (Jelenik et al. 2014). These results indicate that the insulin-resistant state per se is not able to change mitochondrial function in the liver of T1DM mice, however acute diabetes evokes a prominent elevation of mitochondrial oxygen consumption, which later declines in the chronic diabetic state, but still remains higher than healthy controls. This study also points to the importance of the nutrition state, which clearly influences mitochondrial performance in the liver as well as in the muscle (Jelenik et al. 2014), maybe by influencing the assembly of RC supercomplexes in the inner membrane (Lapuente-Brun et al. 2013).

Taken together, these data suggest that in the T1DM state with high glucose but low insulin levels, in contrast to T2DM with usually elevation of both parameters, liver mitochondrial function is not impaired but rather compensates for the disturbed metabolic state.

DOES LIVER MITOCHONDRIAL DYSFUNCTION PER SE CAUSE LIVER INSULIN RESISTANCE?

An altered mitochondrial capacity could arise as a consequence of insulin resistance or the diabetes state. Thus, in the former sections, we collected data on mitochondrial function from studies which investigated liver ranging from steatosis to NASH. On the contrary, primary mitochondrial dysfunction was also postulated to play an important role in the development of insulin resistance. Therefore, in this section, we summarize some of the reports, in which mitochondrial proteins were deleted or in which drugs were used in vivo to manipulate mitochondrial function, followed by the analysis of the insulin-resistant status of the liver.

In mice, ablation of the mitochondrial long-chain acyl-CoA dehydrogenase, responsible for fatty acid oxidation, led to hepatic insulin resistance, diminished IRS-2 phosphorylation, and PI3K as well as AKT2 activities, which are the key enzymes transmitting the insulin signal to the cell, in association with decreased energy consumption and higher body fat mass (Zhang et al. 2007). These results suggest that primary mitochondrial dysfunction, in this case failure of β-oxidation, is possibly involved in the development of hepatic insulin resistance. Apoptosis-inducing factor (AIF) is probably involved in apoptosis and necessary for normal respiratory chain function. The deletion of this gene in mouse liver (LAIFKO mice) caused a coordinated downregulation of OXPHOS genes, diminished CI and CIV activities, ATP levels and CII-driven respiration, however respiratory control ratio was elevated. On the other hand, LAIFKO mice exhibited a better glucose and insulin tolerance as well as whole body insulin sensitivity and were protected against HFD-induced obesity (Pospisilik et al. 2007). These results indicate that a primary OXPHOS defect alone does not cause insulin resistance but, rather, can result in even higher insulin sensitivity.

Treatment with 2,4-dinitrophenol (DNP), which is a mitochondrial uncoupler, showed beneficial effects on HFD-fed rats in terms of hepatic insulin resistance in association with higher IRS-1 and IRS-2 tyrosin phosphorylation as well as PI3K and AKT2 activities (Samuel et al. 2004). These results suggest that mild mitochondrial uncoupling could be favorable, since it increases fatty acid oxidation to cover increased energy expenditure and thus slows down the development of T2DM.

CONCLUSION

In conclusion, we do not have a clear picture yet about the relation between mitochondrial function in the liver and the development of insulin resistance and T2DM. Even the term mitochondrial dysfunction is not well defined, since there is not a single well accepted method easily used ex vivo representing mitochondrial performance in vivo. According to our experience, respiration measurements of freshly isolated mitochondria are a very reliable method to investigate maximal mitochondrial functional capacity. On the other hand, other methods give also valuable information about mitochondrial function, although from a different angle, and the more methods are used, the clearer the picture becomes. In some studies, upregulated levels of mitochondrial transcripts were observed (Misu et al. 2007, Takamura et al. 2008), and when mitochondrial oxygen consumption was measured, gene expression positively correlated with respiration (Buchner et al. 2011), while other studies did not report a correlation between these two parameters (Bugger et al. 2009, Holmstrom et al. 2012). Mitochondrial RC levels were reported to correlate with mitochondrial respiration in some animal models (Holmstrom et al. 2012, Franko et al. 2014), however the correlation was missing in others (Bugger et al. 2009, Franko et al. 2014). mtDNA copy number is claimed to be a good marker for mitochondrial mass in some (Liu et al. 2009), however other studies disagree (Wiesner 1997, Franko et al. 2008, Kim et al. 2008). In insulin-resistant/diabetic animals, mtDNA levels changed in parallel with mitochondrial respiration (Jelenik et al. 2014), however no correlation was also found (Holmstrom et al. 2012, Jelenik et al. 2014). In order to assess mitochondrial capacity, RC activities are also routinely measured. Mitochondrial oxygen consumption was shown to be only significantly impaired when RC activities are severely suppressed (Begriche et al. 2013) and respiration and activity measurements could change in the same direction, as expected (Serviddio et al. 2008b, Bouderba et al. 2012), but also behave differently (Serviddio et al. 2008b, Bouderba et al. 2012). These results indicate that mitochondrial transcript and protein levels as well as RC complex activities could, but not always do complement mitochondrial respiration measurements and conclusions drawn from the former methods generalized as mitochondrial function should be interpreted carefully. One should also appreciate that oxygen consumption protocols are different, since some laboratories investigate complex I and II respiration separately (Bugger et al. 2009, Herlein et al. 2009, Franko et al. 2014), while others analyze both in combination (Holmstrom et al. 2012, Jelenik et al. 2014). Most importantly, experiments studying liver mitochondrial capacity using complex I or complex II substrates separately could reveal significant differences between the two pathways, however the difference could disappear when both complex I and II substrates are applied together (Lapuente-Brun et al. 2013). Furthermore, the respiration rates for CI and CII substrates in liver mitochondria in insulin-resistant/diabetic states do not necessarily supplement each other: some studies reported higher CII and normal CI-driven respiration (Chavin et al. 1999,

Serviddio et al. 2008b), while others found reduced CI but normal CII-driven oxygen consumption (Bouderba et al. 2012) or both respiration rates were observed to be elevated (Ferreira et al. 1999) or even remained unchanged (Franko et al. 2014). Liver ATP levels in basal states were reported to be reduced (Szendroedi et al. 2009) or shown to be normal (Cortez-Pinto et al. 1999) in insulin-resistant patients. Moreover, ATP recovery after fructose injection and ATP synthesis flux were diminished in insulin-resistant subjects (Cortez-Pinto et al. 1999, Schmid et al. 2011). However, a low ATP level does not necessarily result from disturbed oxygen consumption, since in some obese animals a higher mitochondrial respiration capacity was observed although ATP level was reduced (Chavin et al. 1999, Serviddio et al. 2008b). In conclusion, the chosen protocol for measuring mitochondrial performance may strongly determine the outcome of the experiments and conclusions based on only one or two methods should be drawn carefully.

Another important issue is the diverse background of patients and animals. In this review, we summarized the data from patients and rodents displaying steatosis only, but some developed NASH with inflammation and fibrosis. Moreover, the overweight or obese state could also influence mitochondrial function (Patti and Corvera 2010, Verbeek et al. 2014) and some studies compared the analyzed patients with BMI-matched controls, some did not. When insulin produced by the pancreatic β cells cannot compensate anymore for insulin resistance, the blood glucose level rises and T2DM develops. If this stage is prolonged for a longer period, glucotoxicity will impair the function of organs (Krebs and Roden 2004, Roseman 2005). Furthermore, the results generated by Jelenik et al. importantly point to the nutrition state of the studied rodents, since in the fasting state no significant changes were found for diabetic, insulin-resistant rodents, however in the fed state a pronounced elevation in mitochondrial respiration was reported (Jelenik et al. 2014).

Finally, longitudinal studies provide an excellent opportunity for investigating mitochondrial capacity at different time points and they have called attention to the

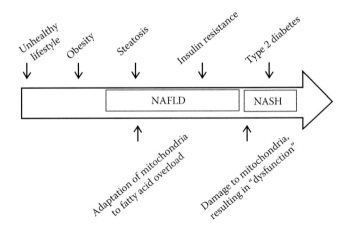

FIGURE 17.2 Possible changes of mitochondrial function during the "typical" development of T2DM.

disease duration, which was shown to significantly influence the mitochondrial performance. Most studies found that short-term duration of the disease evoked a compensatory rise in mitochondrial oxygen consumption, which usually vanished upon disease progression (Ferreira et al. 2003, Serviddio et al. 2008b, Jelenik et al. 2014); however, other authors did not observe significant changes among different time points (Ferreira et al. 2003, Herlein et al. 2009). Therefore, the found discrepancies on mitochondrial performance could probably be explained by the distinct disease states of the studied patients or animals.

Altogether, according to our own results (Franko et al. 2014) and other studies (Schiff et al. 2009, Turner 2013), mitochondria in the liver seem to be capable to adapt to the metabolic environment of NAFLD and insulin-resistant and diabetic states, however, seem to be damaged as soon as NASH develops (Figure 17.2).

REFERENCES

American Diabetes Association. 2014. Diagnosis and classification of diabetes mellitus. *Diabetes Care* 37 (Suppl. 1):S81–S90. doi: 10.2337/dc14-S081.

Appleby, R. D., W. K. Porteous, G. Hughes, A. M. James, D. Shannon, Y. H. Wei, and M. P. Murphy. 1999. Quantitation and origin of the mitochondrial membrane potential in human cells lacking mitochondrial DNA. *Eur J Biochem* 262 (1):108–116.

Baris, O. R., A. Klose, J. E. Kloepper, D. Weiland, J. F. Neuhaus, M. Schauen, A. Wille et al. 2011. The mitochondrial electron transport chain is dispensable for proliferation and differentiation of epidermal progenitor cells. *Stem Cells* 29 (9):1459–1468. doi: 10.1002/stem.695.

Baughman, J. M., F. Perocchi, H. S. Girgis, M. Plovanich, C. A. Belcher-Timme, Y. Sancak, and X. R. Bao. 2011. Integrative genomics identifies MCU as an essential component of the mitochondrial calcium uniporter. *Nature* 476 (7360):341–345. doi: 10.1038/nature10234.

Bayona-Bafaluy, M. P., G. Manfredi, and C. T. Moraes. 2003. A chemical enucleation method for the transfer of mitochondrial DNA to rho(o) cells. *Nucleic Acids Res* 31 (16):e98.

Begriche, K., J. Massart, M. A. Robin, F. Bonnet, and B. Fromenty. 2013. Mitochondrial adaptations and dysfunctions in nonalcoholic fatty liver disease. *Hepatology* 58 (4):1497–1507. doi: 10.1002/hep.26226.

Benard, G., B. Faustin, E. Passerieux, A. Galinier, C. Rocher, N. Bellance, J. P. Delage, L. Casteilla, T. Letellier, and R. Rossignol. 2006. Physiological diversity of mitochondrial oxidative phosphorylation. *Am J Physiol Cell Physiol* 291 (6):C1172–C1182. doi: 10.1152/ajpcell.00195.2006.

Benton, D. 2010. The plausibility of sugar addiction and its role in obesity and eating disorders. *Clin Nutr* 29 (3):288–303. doi: 10.1016/j.clnu.2009.12.001.

Berg, J. M, J. L. Tymoczko, and L. Stryer. 2002. *Biochemistry*. W. H. Freeman, New York.

Biddinger, S. B., A. Hernandez-Ono, C. Rask-Madsen, J. T. Haas, J. O. Aleman, R. Suzuki, E. F. Scapa et al. 2008. Hepatic insulin resistance is sufficient to produce dyslipidemia and susceptibility to atherosclerosis. *Cell Metab* 7 (2):125–134. doi: 10.1016/j.cmet.2007.11.013.

Boucher, J., A. Kleinridders, and C. R. Kahn. 2014. Insulin receptor signaling in normal and insulin-resistant states. *Cold Spring Harb Perspect Biol* 6 (1):a009191. doi: 10.1101/cshperspect.a009191.

Bouderba, S., M. N. Sanz, C. Sanchez-Martin, M. Y. El-Mir, G. R. Villanueva, D. Detaille, and E. A. Koceir. 2012. Hepatic mitochondrial alterations and increased oxidative stress in nutritional diabetes-prone *Psammomys obesus* model. *Exp Diabetes Res* 2012:430176. doi: 10.1155/2012/430176.

Brady, L. J., P. S. Brady, D. R. Romsos, and C. L. Hoppel. 1985. Elevated hepatic mitochondrial and peroxisomal oxidative capacities in fed and starved adult obese (ob/ob) mice. *Biochem J* 231 (2):439–444.

Brini, M., P. Pinton, M. P. King, M. Davidson, E. A. Schon, and R. Rizzuto. 1999. A calcium signaling defect in the pathogenesis of a mitochondrial DNA inherited oxidative phosphorylation deficiency. *Nat Med* 5 (8):951–954. doi: 10.1038/11396.

Buchner, D. A., L. C. Burrage, A. E. Hill, S. N. Yazbek, W. E. O'Brien, C. M. Croniger, and J. H. Nadeau. 2008. Resistance to diet-induced obesity in mice with a single substituted chromosome. *Physiol Genomics* 35 (1):116–122. doi: 10.1152/physiolgenomics.00033.2008.

Buchner, D. A., S. N. Yazbek, P. Solinas, L. C. Burrage, M. G. Morgan, C. L. Hoppel, and J. H. Nadeau. 2011. Increased mitochondrial oxidative phosphorylation in the liver is associated with obesity and insulin resistance. *Obesity* (*Silver Spring*) 19 (5):917–924. doi: 10.1038/oby.2010.214.

Bugger, H., D. Chen, C. Riehle, J. Soto, H. A. Theobald, X. X. Hu, B. Ganesan, B. C. Weimer, and E. D. Abel. 2009. Tissue-specific remodeling of the mitochondrial proteome in type 1 diabetic akita mice. *Diabetes* 58 (9):1986–1997. doi: 10.2337/db09-0259.

Chang-Chen, K. J., R. Mullur, and E. Bernal-Mizrachi. 2008. β-cell failure as a complication of diabetes. *Rev Endocr Metab Disord* 9 (4):329–343. doi: 10.1007/s11154-008-9101-5.

Chavin, K. D., S. Yang, H. Z. Lin, J. Chatham, V. P. Chacko, J. B. Hoek, E. Walajtys-Rode et al. 1999. Obesity induces expression of uncoupling protein-2 in hepatocytes and promotes liver ATP depletion. *J Biol Chem* 274 (9):5692–5700.

Cheng, Z., S. Guo, K. Copps, X. Dong, R. Kollipara, J. T. Rodgers, R. A. Depinho, P. Puigserver, and M. F. White. 2009. FOXO-1 integrates insulin signaling with mitochondrial function in the liver. *Nat Med* 15 (11):1307–1311. doi: 10.1038/nm.2049.

Chitturi, S., S. Abeygunasekera, G. C. Farrell, J. Holmes-Walker, J. M. Hui, C. Fung, R. Karim et al. 2002. NASH and insulin resistance: Insulin hypersecretion and specific association with the insulin resistance syndrome. *Hepatology* 35 (2):373–379. doi: 10.1053/jhep.2002.30692.

Cho, N. H., D. Whiting, L. Guariguata, P. A. Montoya, N. Forouhi, I. Hambleton, R. Li et al. 2014. *IDF Diabetes Atlas update poster*, 6th edn. Brussels, Belgium: International Diabetes Federation.

Choudhury, J. and A. J. Sanyal. 2005. Insulin resistance in NASH. *Front Biosci* 10:1520–1533.

Cocheme, H. M., C. Quin, S. J. McQuaker, F. Cabreiro, A. Logan, T. A. Prime, I. Abakumova et al. 2011. Measurement of H_2O_2 within living Drosophila during aging using a ratiometric mass spectrometry probe targeted to the mitochondrial matrix. *Cell Metab* 13 (3):340–350. doi: 10.1016/j.cmet.2011.02.003.

Cortez-Pinto, H., J. Chatham, V. P. Chacko, C. Arnold, A. Rashid, and A. M. Diehl. 1999. Alterations in liver ATP homeostasis in human nonalcoholic steatohepatitis: A pilot study. *JAMA* 282 (17):1659–1664.

Davies, K. M., C. Anselmi, I. Wittig, J. D. Faraldo-Gomez, and W. Kuhlbrandt. 2012. Structure of the yeast F1Fo-ATP synthase dimer and its role in shaping the mitochondrial cristae. *Proc Natl Acad Sci USA* 109 (34):13602–13607. doi: 10.1073/pnas.1204593109.

De Stefani, D., A. Raffaello, E. Teardo, I. Szabo, and R. Rizzuto. 2011. A forty-kilodalton protein of the inner membrane is the mitochondrial calcium uniporter. *Nature* 476 (7360):336–340. doi: 10.1038/nature10230.

De Stefani, D. and R. Rizzuto. 2014. Molecular control of mitochondrial calcium uptake. *Biochem Biophys Res Commun* 449 (4):373–376. doi: 10.1016/j.bbrc.2014.04.142.

DeFronzo, R. A., D. Simonson, and E. Ferrannini. 1982. Hepatic and peripheral insulin resistance: A common feature of type 2 (non-insulin-dependent) and type 1 (insulin-dependent) diabetes mellitus. *Diabetologia* 23 (4):313–319.

Denton, R. M. and J. G. McCormack. 1990. Ca²⁺ as a second messenger within mitochondria of the heart and other tissues. *Annu Rev Physiol* 52:451–466. doi: 10.1146/annurev. ph.52.030190.002315.

Echtay, K. S., D. Roussel, J. St-Pierre, M. B. Jekabsons, S. Cadenas, J. A. Stuart, J. A. Harper et al. 2002. Superoxide activates mitochondrial uncoupling proteins. *Nature* 415 (6867):96–99. doi: 10.1038/415096a.

Ferreira, F. M., C. M. Palmeira, M. J. Matos, R. Seica, and M. S. Santos. 1999. Decreased susceptibility to lipid peroxidation of Goto-Kakizaki rats: Relationship to mitochondrial antioxidant capacity. *Life Sci* 65 (10):1013–1025.

Ferreira, F. M., C. M. Palmeira, R. Seica, A. J. Moreno, and M. S. Santos. 2003. Diabetes and mitochondrial bioenergetics: Alterations with age. *J Biochem Mol Toxicol* 17 (4):214–222. doi: 10.1002/jbt.10081.

Franko, A., S. Mayer, G. Thiel, L. Mercy, T. Arnould, H. T. Hornig-Do, R. J. Wiesner, and S. Goffart. 2008. CREB-1α is recruited to and mediates upregulation of the cytochrome c promoter during enhanced mitochondrial biogenesis accompanying skeletal muscle differentiation. *Mol Cell Biol* 28 (7):2446–2459. doi: 10.1128/MCB.00980–07.

Franko, A., J. C. von Kleist-Retzow, M. Bose, C. Sanchez-Lasheras, S. Brodesser, O. Krut, W. S. Kunz et al. 2012. Complete failure of insulin-transmitted signaling, but not obesity-induced insulin resistance, impairs respiratory chain function in muscle. *J Mol Med (Berl)* 90 (10):1145–1160. doi: 10.1007/s00109–012–0887-y.

Franko, A., J. C. von Kleist-Retzow, S. Neschen, M. Wu, P. Schommers, M. Bose, A. Kunze et al. 2014. Liver adapts mitochondrial function to insulin resistant and diabetic states in mice. *J Hepatol* 60 (4):816–823. doi: 10.1016/j.jhep.2013.11.020.

Fuchs, H., S. Neschen, J. Rozman, B. Rathkolb, S. Wagner, T. Adler, L. Afonso et al. 2012. Mouse genetics and metabolic mouse phenotyping. In *Genetics Meets Metabolomics: From Experiment to Systems Biology*, K. Suhre (ed.), pp. 85–106. Springer Science+Business Media, LLC, Springer, New York.

Garber, A. K. and R. H. Lustig. 2011. Is fast food addictive? *Curr Drug Abuse Rev* 4 (3):146–162.

Gearhardt, A. N., C. M. Grilo, R. J. DiLeone, K. D. Brownell, and M. N. Potenza. 2011. Can food be addictive? Public health and policy implications. *Addiction* 106 (7):1208–1212. doi: 10.1111/j.1360–0443.2010.03301.x.

Grontved, A. and F. B. Hu. 2011. Television viewing and risk of type 2 diabetes, cardiovascular disease, and all-cause mortality: A meta-analysis. *JAMA* 305 (23):2448–2455. doi: 10.1001/jama.2011.812.

Hanahan, D. and R. A. Weinberg. 2011. Hallmarks of cancer: The next generation. *Cell* 144 (5):646–674. doi: 10.1016/j.cell.2011.02.013.

Herlein, J. A., B. D. Fink, Y. O'Malley, and W. I. Sivitz. 2009. Superoxide and respiratory coupling in mitochondria of insulin-deficient diabetic rats. *Endocrinology* 150 (1):46–55. doi: 10.1210/en.2008-0404.

Hoffmann, S., D. Spitkovsky, J. P. Radicella, B. Epe, and R. J. Wiesner. 2004. Reactive oxygen species derived from the mitochondrial respiratory chain are not responsible for the basal levels of oxidative base modifications observed in nuclear DNA of Mammalian cells. *Free Radic Biol Med* 36 (6):765–773. doi: 10.1016/j. freeradbiomed.2003.12.019.

Holmstrom, M. H., E. Iglesias-Gutierrez, J. R. Zierath, and P. M. Garcia-Roves. 2012. Tissue-specific control of mitochondrial respiration in obesity-related insulin resistance and diabetes. *Am J Physiol Endocrinol Metab* 302 (6):E731–E739. doi: 10.1152/ ajpendo.00159.2011.

Jelenik, T., G. Sequaris, K. Kaul, D. M. Ouwens, E. Phielix, J. Kotzka, B. Knebel et al. 2014. Tissue-specific differences in the development of insulin resistance in a mouse model for type 1 diabetes. *Diabetes*. doi: 10.2337/db13-1794.

Johnson, R. J., T. Nakagawa, L. G. Sanchez-Lozada, M. Shafiu, S. Sundaram, M. Le, T. Ishimoto, Y. Y. Sautin, and M. A. Lanaspa. 2013. Sugar, uric acid, and the etiology of diabetes and obesity. *Diabetes* 62 (10):3307–3315. doi: 10.2337/db12–1814.

Kahle, M., M. Horsch, B. Fridrich, A. Seelig, J. Schultheiss, J. Leonhardt, M. Irmler et al. 2013. Phenotypic comparison of common mouse strains developing high-fat diet-induced hepatosteatosis. *Mol Metab* 2 (4):435–446. doi: 10.1016/j.molmet.2013.07.009.

Khrapko, K., N. Bodyak, W. G. Thilly, N. J. van Orsouw, X. Zhang, H. A. Coller, T. T. Perls, M. Upton, J. Vijg, and J. Y. Wei. 1999. Cell-by-cell scanning of whole mitochondrial genomes in aged human heart reveals a significant fraction of myocytes with clonally expanded deletions. *Nucleic Acids Res* 27 (11):2434–2341.

Kim, M. J., C. Jardel, C. Barthelemy, V. Jan, J. P. Bastard, S. Fillaut-Chapin, S. Houry, J. Capeau, and A. Lombes. 2008. Mitochondrial DNA content, an inaccurate biomarker of mitochondrial alteration in human immunodeficiency virus-related lipodystrophy. *Antimicrob Agents Chemother* 52 (5):1670–1676. doi: 10.1128/aac.01449–07.

King, M. P. and G. Attardi. 1996. Isolation of human cell lines lacking mitochondrial DNA. *Methods Enzymol* 264:304–313.

Koliaki, C. and M. Roden. 2013. Hepatic energy metabolism in human diabetes mellitus, obesity and non-alcoholic fatty liver disease. *Mol Cell Endocrinol* 379 (1–2):35–42. doi: 10.1016/j.mce.2013.06.002.

Koliaki, C. and M. Roden. 2014. Do mitochondria care about insulin resistance? *Mol Metab* 3 (4):351–353. doi: 10.1016/j.molmet.2014.04.004.

Kovacs-Bogdan, E., Y. Sancak, K. J. Kamer, M. Plovanich, A. Jambhekar, R. J. Huber, M. A. Myre, M. D. Blower, and V. K. Mootha. 2014. Reconstitution of the mitochondrial calcium uniporter in yeast. *Proc Natl Acad Sci USA* 111 (24):8985–8990. doi: 10.1073/pnas.1400514111.

Krebs, M. and M. Roden. 2004. Nutrient-induced insulin resistance in human skeletal muscle. *Curr Med Chem* 11 (7):901–908.

Lapuente-Brun, E., R. Moreno-Loshuertos, R. Acin-Perez, A. Latorre-Pellicer, C. Colas, E. Balsa, E. Perales-Clemente et al. 2013. Supercomplex assembly determines electron flux in the mitochondrial electron transport chain. *Science* 340 (6140):1567–1570. doi: 10.1126/science.1230381.

Larter, C. Z., S. Chitturi, D. Heydet, and G. C. Farrell. 2010. A fresh look at NASH pathogenesis. Part 1: The metabolic movers. *J Gastroenterol Hepatol* 25 (4):672–690. doi: 10.1111/j.1440–1746.2010.06253.x.

Liu, H. Y., E. Yehuda-Shnaidman, T. Hong, J. Han, J. Pi, Z. Liu, and W. Cao. 2009. Prolonged exposure to insulin suppresses mitochondrial production in primary hepatocytes. *J Biol Chem* 284 (21):14087–14095. doi: 10.1074/jbc.M807992200.

Lowell, B. B. and G. I. Shulman. 2005. Mitochondrial dysfunction and type 2 diabetes. *Science* 307 (5708):384–387. doi: 10.1126/science.1104343.

Mahajan, A., M. J. Go, W. Zhang, J. E. Below, K. J. Gaulton, T. Ferreira, M. Horikoshi et al. 2014. Genome-wide trans-ancestry meta-analysis provides insight into the genetic architecture of type 2 diabetes susceptibility. *Nat Genet* 46 (3):234–244. doi: 10.1038/ng.2897.

Mallilankaraman, K., C. Cardenas, P. J. Doonan, H. C. Chandramoorthy, K. M. Irrinki, T. Golenar, G. Csordas et al. 2012. MCUR1 is an essential component of mitochondrial Ca2+ uptake that regulates cellular metabolism. *Nat Cell Biol* 14 (12):1336–1343. doi: 10.1038/ncb2622.

Martinou, J. C. and R. J. Youle. 2011. Mitochondria in apoptosis: Bcl-2 family members and mitochondrial dynamics. *Dev Cell* 21 (1):92–101. doi: 10.1016/j.devcel.2011.06.017.

Misu, H., T. Takamura, N. Matsuzawa, A. Shimizu, T. Ota, H. Sakurai, H. Ando et al. 2007. Genes involved in oxidative phosphorylation are coordinately upregulated with fasting hyperglycaemia in livers of patients with type 2 diabetes. *Diabetologia* 50 (2):268–277. doi: 10.1007/s00125-006-0489-8.

Muniyappa, R., S. Lee, H. Chen, and M. J. Quon. 2008. Current approaches for assessing insulin sensitivity and resistance in vivo: Advantages, limitations, and appropriate usage. *Am J Physiol Endocrinol Metab* 294 (1):E15–E26. doi: 10.1152/ajpendo.00645.2007.

Murphy, M. P. 2009. How mitochondria produce reactive oxygen species. *Biochem J* 417 (1):1–13. doi: 10.1042/bj20081386.

Ouyang, X., P. Cirillo, Y. Sautin, S. McCall, J. L. Bruchette, A. M. Diehl, R. J. Johnson, and M. F. Abdelmalek. 2008. Fructose consumption as a risk factor for non-alcoholic fatty liver disease. *J Hepatol* 48 (6):993–999. doi: 10.1016/j.jhep.2008.02.011.

Owusu-Ansah, E. and U. Banerjee. 2009. Reactive oxygen species prime Drosophila haematopoietic progenitors for differentiation. *Nature* 461 (7263):537–541. doi: 10.1038/nature08313.

Pagliarini, D. J., S. E. Calvo, B. Chang, S. A. Sheth, S. B. Vafai, S. E. Ong, G. A. Walford et al. 2008. A mitochondrial protein compendium elucidates complex I disease biology. *Cell* 134 (1):112–123. doi: 10.1016/j.cell.2008.06.016.

Pan, X., J. Liu, T. Nguyen, C. Liu, J. Sun, Y. Teng, M. M. Fergusson et al. 2013. The physiological role of mitochondrial calcium revealed by mice lacking the mitochondrial calcium uniporter. *Nat Cell Biol* 15 (12):1464–1472. doi: 10.1038/ncb2868.

Patti, M. E. and S. Corvera. 2010. The role of mitochondria in the pathogenesis of type 2 diabetes. *Endocr Rev* 31 (3):364–395. doi: 10.1210/er.2009–0027.

Pendin, D., E. Greotti, and T. Pozzan. 2014. The elusive importance of being a mitochondrial Ca(2+) uniporter. *Cell Calcium* 55 (3):139–145. doi: 10.1016/j.ceca.2014.02.008.

Perez-Carreras, M., P. Del Hoyo, M. A. Martin, J. C. Rubio, A. Martin, G. Castellano, F. Colina, J. Arenas, and J. A. Solis-Herruzo. 2003. Defective hepatic mitochondrial respiratory chain in patients with nonalcoholic steatohepatitis. *Hepatology* 38 (4):999–1007. doi: 10.1053/jhep.2003.50398.

Pessayre, D. and B. Fromenty. 2005. NASH: A mitochondrial disease. *J Hepatol* 42 (6):928–940. doi: 10.1016/j.jhep.2005.03.004.

Petersen, K. F., S. Dufour, D. Befroy, R. Garcia, and G. I. Shulman. 2004. Impaired mitochondrial activity in the insulin-resistant offspring of patients with type 2 diabetes. *N Engl J Med* 350 (7):664–671. doi: 10.1056/NEJMoa031314.

Pihlajamaki, J., T. Boes, E. Y. Kim, F. Dearie, B. W. Kim, J. Schroeder, E. Mun et al. 2009. Thyroid hormone-related regulation of gene expression in human fatty liver. *J Clin Endocrinol Metab* 94 (9):3521–3529. doi: 10.1210/jc.2009–0212.

Pospisilik, J. A., C. Knauf, N. Joza, P. Benit, M. Orthofer, P. D. Cani, I. Ebersberger et al. 2007. Targeted deletion of AIF decreases mitochondrial oxidative phosphorylation and protects from obesity and diabetes. *Cell* 131 (3):476–491. doi: 10.1016/j.cell.2007.08.047.

Roseman, H. M. 2005. Progression from obesity to type 2 diabetes: Lipotoxicity, glucotoxicity, and implications for management. *J Manag Care Pharm* 11 (6 Suppl. B):S3–S11.

Samuel, V. T., Z. X. Liu, X. Qu, B. D. Elder, S. Bilz, D. Befroy, A. J. Romanelli, and G. I. Shulman. 2004. Mechanism of hepatic insulin resistance in non-alcoholic fatty liver disease. *J Biol Chem* 279 (31):32345–32353. doi: 10.1074/jbc.M313478200.

Sanyal, A. J., C. Campbell-Sargent, F. Mirshahi, W. B. Rizzo, M. J. Contos, R. K. Sterling, V. A. Luketic, M. L. Shiffman, and J. N. Clore. 2001. Nonalcoholic steatohepatitis: Association of insulin resistance and mitochondrial abnormalities. *Gastroenterology* 120 (5):1183–1192. doi: 10.1053/gast.2001.23256.

Schauen, M., D. Spitkovsky, J. Schubert, J. H. Fischer, J. Hayashi, and R. J. Wiesner. 2006. Respiratory chain deficiency slows down cell-cycle progression via reduced ROS generation and is associated with a reduction of p21CIP1/WAF1. *J Cell Physiol* 209 (1):103–112. doi: 10.1002/jcp.20711.

Schiff, M., S. Loublier, A. Coulibaly, P. Benit, H. O. de Baulny, and P. Rustin. 2009. Mitochondria and diabetes mellitus: Untangling a conflictive relationship? *J Inherit Metab Dis* 32 (6):684–698. doi: 10.1007/s10545-009-1263-0.

Schmid, A. I., M. Chmelik, J. Szendroedi, M. Krssak, A. Brehm, E. Moser, and M. Roden. 2008. Quantitative ATP synthesis in human liver measured by localized 31P spectroscopy using the magnetization transfer experiment. *NMR Biomed* 21 (5):437–443. doi: 10.1002/nbm.1207.

Schmid, A. I., J. Szendroedi, M. Chmelik, M. Krssak, E. Moser, and M. Roden. 2011. Liver ATP synthesis is lower and relates to insulin sensitivity in patients with type 2 diabetes. *Diabetes Care* 34 (2):448–453. doi: 10.2337/dc10-1076.

Semple, R. K., A. Sleigh, P. R. Murgatroyd, C. A. Adams, L. Bluck, S. Jackson, A. Vottero et al. 2009. Postreceptor insulin resistance contributes to human dyslipidemia and hepatic steatosis. *J Clin Invest* 119 (2):315–322. doi: 10.1172/jci37432.

Serviddio, G., F. Bellanti, R. Tamborra, T. Rollo, N. Capitanio, A. D. Romano, J. Sastre, G. Vendemiale, and E. Altomare. 2008a. Uncoupling protein-2 (UCP2) induces mitochondrial proton leak and increases susceptibility of non-alcoholic steatohepatitis (NASH) liver to ischaemia-reperfusion injury. *Gut* 57 (7):957–965. doi: 10.1136/gut.2007.147496.

Serviddio, G., F. Bellanti, R. Tamborra, T. Rollo, A. D. Romano, A. M. Giudetti, N. Capitanio, A. Petrella, G. Vendemiale, and E. Altomare. 2008b. Alterations of hepatic ATP homeostasis and respiratory chain during development of non-alcoholic steatohepatitis in a rodent model. *Eur J Clin Invest* 38 (4):245–252. doi: 10.1111/j.1365-2362.2008.01936.x.

Shiflett, A. M. and P. J. Johnson. 2010. Mitochondrion-related organelles in eukaryotic protists. *Annu Rev Microbiol* 64:409–429. doi: 10.1146/annurev.micro.62.081307.162826.

Sleigh, A., P. Raymond-Barker, K. Thackray, D. Porter, M. Hatunic, A. Vottero, C. Burren et al. 2011. Mitochondrial dysfunction in patients with primary congenital insulin resistance. *J Clin Invest* 121 (6):2457–461. doi: 10.1172/JCI46405.

Sorensen, L., M. Ekstrand, J. P. Silva, E. Lindqvist, B. Xu, P. Rustin, L. Olson, and N. G. Larsson. 2001. Late-onset corticohippocampal neurodepletion attributable to catastrophic failure of oxidative phosphorylation in MILON mice. *J Neurosci* 21 (20):8082–8090.

Stehling, O., C. Wilbrecht, and R. Lill. 2014. Mitochondrial iron-sulfur protein biogenesis and human disease. *Biochimie* 100:61–77. doi: 10.1016/j.biochi.2014.01.010.

Sunny, N. E., E. J. Parks, J. D. Browning, and S. C. Burgess. 2011. Excessive hepatic mitochondrial TCA cycle and gluconeogenesis in humans with nonalcoholic fatty liver disease. *Cell Metab* 14 (6):804–810. doi: 10.1016/j.cmet.2011.11.004.

Szendroedi, J., M. Chmelik, A. I. Schmid, P. Nowotny, A. Brehm, M. Krssak, E. Moser, and M. Roden. 2009. Abnormal hepatic energy homeostasis in type 2 diabetes. *Hepatology* 50 (4):1079–1086. doi: 10.1002/hep.23093.

Szendroedi, J. and M. Roden. 2009. Ectopic lipids and organ function. *Curr Opin Lipidol* 20 (1):50–56.

Takamura, T., H. Misu, N. Matsuzawa-Nagata, M. Sakurai, T. Ota, A. Shimizu, S. Kurita et al. 2008. Obesity upregulates genes involved in oxidative phosphorylation in livers of diabetic patients. *Obesity (Silver Spring)* 16 (12):2601–2069. doi: 10.1038/oby.2008.419.

Tatsuta, T., M. Scharwey, and T. Langer. 2014. Mitochondrial lipid trafficking. *Trends Cell Biol* 24 (1):44–52. doi: 10.1016/j.tcb.2013.07.011.

Thorp, A. A., G. N. Healy, N. Owen, J. Salmon, K. Ball, J. E. Shaw, P. Z. Zimmet, and D. W. Dunstan. 2010. Deleterious associations of sitting time and television viewing time with cardiometabolic risk biomarkers: Australian Diabetes, Obesity and Lifestyle (AusDiab) study 2004–2005. *Diabetes Care* 33 (2):327–334. doi: 10.2337/dc09-0493.

Trifunovic, A., A. Hansson, A. Wredenberg, A. T. Rovio, I. Dufour, I. Khvorostov, J. N. Spelbrink, R. Wibom, H. T. Jacobs, and N. G. Larsson. 2005. Somatic mtDNA mutations cause aging phenotypes without affecting reactive oxygen species production. *Proc Natl Acad Sci USA* 102 (50):17993–17998. doi: 10.1073/pnas.0508886102.

Trifunovic, A., A. Wredenberg, M. Falkenberg, J. N. Spelbrink, A. T. Rovio, C. E. Bruder, Y. M. Bohlooly et al. 2004. Premature ageing in mice expressing defective mitochondrial DNA polymerase. *Nature* 429 (6990):417–423. doi: 10.1038/nature02517.

Turner, N. 2013. *Mitochondrial Metabolism and Insulin Action.* In *Type 2 Diabetes.* InTech.

Veltri, K. L., M. Espiritu, and G. Singh. 1990. Distinct genomic copy number in mitochondria of different mammalian organs. *J Cell Physiol* 143 (1):160–164. doi: 10.1002/jcp.1041430122.

Verbeek, J., M. Lannoo, E. Pirinen, D. Ryu, P. Spincemaille, I. Vander Elst, P. Windmolders et al. 2014. Roux-en-y gastric bypass attenuates hepatic mitochondrial dysfunction in mice with non-alcoholic steatohepatitis. *Gut.* doi: 10.1136/gutjnl-2014-306748.

Vijgen, G. H., N. D. Bouvy, J. Hoeks, S. Wijers, P. Schrauwen, and W. D. van Marken Lichtenbelt. 2013. Impaired skeletal muscle mitochondrial function in morbidly obese patients is normalized one year after bariatric surgery. *Surg Obes Relat Dis* 9 (6):936–941. doi: 10.1016/j.soard.2013.03.009.

von Kleist-Retzow, J. C., H. T. Hornig-Do, M. Schauen, S. Eckertz, T. A. Dinh, F. Stassen, N. Lottmann et al. 2007. Impaired mitochondrial Ca^{2+} homeostasis in respiratory chain-deficient cells but efficient compensation of energetic disadvantage by enhanced anaerobic glycolysis due to low ATP steady state levels. *Exp Cell Res* 313 (14):3076–3089. doi: 10.1016/j.yexcr.2007.04.015.

Vos, M. B. and J. E. Lavine. 2013. Dietary fructose in nonalcoholic fatty liver disease. *Hepatology* 57 (6):2525–2531. doi: 10.1002/hep.26299.

WHO. 2006. *Definition and Diagnosis of Diabetes Mellitus and Intermediate Hyperglycaemia*: World Health Organization, Geneva, Switzerland.

Wiesner, R. J. 1997. Adaptation of mitochondrial gene expression to changing cellular energy demands. *News Physiol Sci* 12:178–183.

Wiesner, R. J., G. Zsurka, and W. S. Kunz. 2006. Mitochondrial DNA damage and the aging process: Facts and imaginations. *Free Radic Res* 40 (12):1284–1294. doi: 10.1080/10715760600913168.

Wild, S., G. Roglic, A. Green, R. Sicree, and H. King. 2004. Global prevalence of diabetes: Estimates for the year 2000 and projections for 2030. *Diabetes Care* 27 (5):1047–1053.

Wredenberg, A., C. Freyer, M. E. Sandstrom, A. Katz, R. Wibom, H. Westerblad, and N. G. Larsson. 2006. Respiratory chain dysfunction in skeletal muscle does not cause insulin resistance. *Biochem Biophys Res Commun* 350 (1):202–207. doi: 10.1016/j.bbrc.2006.09.029.

Zhang, D., Z. X. Liu, C. S. Choi, L. Tian, R. Kibbey, J. Dong, G. W. Cline, P. A. Wood, and G. I. Shulman. 2007. Mitochondrial dysfunction due to long-chain Acyl-CoA dehydrogenase deficiency causes hepatic steatosis and hepatic insulin resistance. *Proc Natl Acad Sci USA* 104 (43):17075–17080. doi: 10.1073/pnas.0707060104.

18 Mitochondrial Dynamics and Adaptation in Alcoholic Liver Disease and Nonalcoholic Fatty Liver Disease

*Jerome Garcia, Ho Leung, Bradley Blackshire,
Carl Decker, Christopher Kyaw, and Derick Han*

CONTENTS

ABSTRACT

Alcoholic liver disease (ALD) and nonalcoholic fatty liver disease (NAFLD) have been traditionally linked with mitochondrial dysfunction in the liver. This may, however, be an incomplete description of mitochondrial dynamics that occur in the liver. Mitochondria can undergo many types of changes in response to stress or metabolic changes to help cells adapt. Metabolic stresses such as alcohol may cause some mitochondrial injury and damage. However, following injury, mitochondria can remodel and alter to help the liver adapt to stress. Dynamic mitochondrial remodeling and adaptation may be important mechanisms for the liver to deal with metabolic stresses that cause steatosis and, likely, the predominant, early mitochondrial alterations that occur in ALD and NAFLD.

INTRODUCTION

Fat accumulation in the liver (steatosis) is associated with alcohol consumption, obesity, and many metabolic disorders [1,2]. The rapid increase in steatosis parallels the dramatic rise of obesity and metabolic syndromes that have been occurring over the past three decades. Steatosis can be caused by alcohol consumption, obesity, and many metabolic disorders. Although steatosis itself is somewhat benign, a subset of patients (~15% to 30%) will dramatically and unpredictably progress to steatohepatitis (fatty liver and inflammation) and, in severe cases, cirrhosis (fibrosis of liver and extensive hepatocyte death) [1,2]. Why only a subset of patients with steatosis develop more severe liver disease remains unknown.

Liver diseases associated with steatosis are divided into two broad categories: alcoholic liver disease (ALD) and nonalcoholic fatty liver disease (NAFLD). ALD, as the name suggests, is associated with chronic alcohol intake, while NAFLD is primarily associated with obesity and metabolic disorders. Although ALD and NAFLD may originate from different causes, they share many biochemical and histological similarities. Mitochondrial dysfunction has been traditionally considered an important and early event in the development of both ALD and NAFLD [3–5]. In various animal models of steatosis (i.e., chronic alcohol feeding, high-fat diet [HFD], methionine-choline-deficient [MCD] diet), changes in mitochondrial morphology, increased mitochondrial reactive oxygen species (ROS) generation, increased mtDNA damage, and declines in mitochondrial respiration have all been observed in the liver [5–8]. Since mitochondria play a central role in metabolism and cell death (mitochondrial-dependent apoptosis), observed mitochondria dysfunction in ALD and NAFLD animal models has lead to the popular notion that mitochondria dysfunction plays an important role in the pathogenesis of these liver diseases. However, newer observations suggest that mitochondrial dysfunction may be an incomplete assessment of the mitochondrial changes that accompany ALD and NAFLD. Our recent work with alcohol-fed mice suggests that mitochondrial adaptation and remodeling may be the predominant mitochondrial alterations that occur during steatosis, at least in the early stages [9]. Mitochondria are very dynamic organelles that adapt and alter (i.e., fusion–fission changes, respiratory complex remodeling, morphological changes) following stress and injury [10,11]. In this chapter, we review the alterations and adaptations that occur in mitochondria during ALD and NAFLD and their importance in the pathogenesis of these liver diseases.

ALCOHOLIC LIVER DISEASE

THE ROLE OF MITOCHONDRIA IN ALCOHOL METABOLISM IN THE LIVER

ALD affects over 2 million people and causes over 15,000 deaths annually in the United States. The liver, as the major site of alcohol metabolism, is prone to alcohol-induced injury. Alcohol is primarily metabolized by two enzymes in the liver: alcohol dehydrogenase (ADH) in the cytoplasm and aldehyde dehydrogenase 2 (ALDH2) in mitochondria (reactions givein in Equations 18.1 and 18.2).

$$\text{Ethanol} + \text{NAD}^+ \leftrightarrow \text{acetaldehyde} + \text{NADH} + \text{H}^+ \quad \text{catalyzed by ADH} \quad (18.1)$$

$$\text{Acetaldehyde} + \text{NAD}^+ + \text{H}_2\text{O} \leftrightarrow \text{acetate} + \text{NADH} + \text{H}^+ \quad \text{catalyzed by ALDH2}$$
$$(18.2)$$

For alcohol metabolism to proceed, NAD^+ needs to be regenerated by NADH oxidation, which occurs primarily in mitochondria. Since both ADH and ALDH2 are kinetically limited by NAD^+ levels, mitochondrial respiration plays an important role in alcohol metabolism [12–14]. Interestingly, alcohol feeding has not been shown to alter ADH and ALDH2 protein levels in rats [14–16]. The importance of mitochondrial respiration in regenerating NAD^+ needed for alcohol metabolism is shown in Figure 18.1. Cytochrome p450 2E1 isoform (CYP2E1) is also upregulated by chronic alcohol feeding; however, this pathway mainly becomes significant when alcohol levels are very high [17].

The lack of enhanced protein levels of ADH and ALDH2 protein levels in rats with alcohol feeding is somewhat surprising since it is well established that animals adapt to alcohol feeding and develop an enhanced ability to metabolize alcohol in the liver [15,16,18]. Even a single dose of alcohol is known to enhance alcohol metabolism in the liver [19]. This acute adaptation of the liver to alcohol is linked with a "hypermetabolic state," which is characterized by a rapid increase in oxygen uptake by the liver following alcohol intake [12,18,20]. The increased oxygen consumption that occurs in the hypermetabolic state may be due to increased mitochondrial respiration needed to regenerate NAD^+ for alcohol metabolism [12]. However, direct evidence that the hypermetabolic state is caused by enhanced mitochondrial respiration in the liver has not been established. Similarly, enhanced oxygen consumption in the liver has been observed with chronic alcohol feeding, but again, the direct role of mitochondria in enhanced oxygen consumption has not been firmly established.

MITOCHONDRIAL DYSFUNCTION IN ALD

Given the importance of mitochondria in alcohol metabolism, it has been surprising that mitochondrial dysfunction in the liver appears to be an early change of chronic alcohol feeding. Work performed in the eighties demonstrated that chronic feeding of alcohol (Lieber–DeCarli oral diet) to rats causes a decline in mitochondrial respiration (state III, respiration in the presence of substrates and ADP) and a decline in

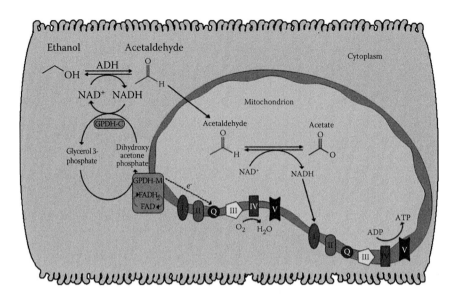

FIGURE 18.1 **(See color insert.)** Role of mitochondria in alcohol metabolism in the liver. The metabolism of alcohol in the liver is dependent on three pathways: (1) The conversion of alcohol to acetaldehyde by alcohol dehydrogenase, which occurs in the cytoplasm and is rate limited by NAD^+; (2) the conversion of acetaldehyde to acetate by aldehyde dehydrogenase 2, which occurs in the mitochondrial matrix and is rate limited by NAD^+; and (3) mitochondrial respiration, which oxidizes NADH to regenerate NAD^+ for alcohol metabolism, which occurs in the inner mitochondrial membrane. Regeneration of NAD^+ via mitochondrial respiration is an important rate-determining step in alcohol metabolism. NADH generated in the cytoplasm from alcohol metabolism cannot directly cross the inner mitochondrial membrane. Therefore, to regenerate NAD^+ in the cytoplasm, the glycerol phosphate shuttle transports electrons from cytoplasmic NADH into the mitochondria. Cytoplasmic glycerol phosphate dehydrogenase (GPDH-C) catalyzes the transfer of electrons from NADH to dihydroxyacetone phosphate to form glycerol 3-phosphate. The electrons from glycerol 3-phosphate are then transported to FAD in mitochondria via the catalytic action of mitochondrial GPDH (GPDH-m, GPDH2).

the respiratory control ratio (RCR; state III/state IV) in isolated liver mitochondria [7,21,22]. A decline in the RCR indicates that mitochondria were uncoupled, which would decrease ATP production by the respiratory chain. Chronic oral alcohol feeding to rats has been shown to also decrease ribosomal activity and synthesis of respiratory complex proteins in the liver [23,24]. These findings suggest that alcohol feeding reduces the bioenergetic activity of mitochondria and its ability to produce ATP due to mitochondrial uncoupling and decreases the expression of respiratory proteins in rat liver.

Various alterations to liver mitochondria have been reported with alcohol treatment. In cultured primary hepatocytes, alcohol exposure has also been suggested to enhance ROS generation by mitochondria. Utilizing the redox dye dichlorofluorescin (DCFH), alcohol treatment to primary hepatocytes was shown to enhance DCFH oxidation to dichlorofluorescein (DCF), suggesting that ROS generation increased with alcohol exposure [25]. Subsequent experiments with mitochondrial inhibitors (i.e., antimycin, rotenone) demonstrated that complex I and III in the respiratory chain were responsible for mitochondrial ROS generation following alcohol treatment to primary cultured hepatocytes [6]. In addition to experiments utilizing DCFH, other studies have demonstrated that alcohol feeding enhances oxidative stress in the liver. Alcohol feeding to rats has been shown to increase lipid peroxidation and mitochondrial DNA (mtDNA) oxidation in the liver [26]. Bolus doses of alcohol have been shown to also cause degradation of mtDNA in the liver, possibly due to mtDNA oxidative damage [27]. Glutathione (GSH), the major antioxidant in cells and mitochondria, has been shown to decrease in liver mitochondria from rats fed alcohol chronically [28]. Subsequently, it was demonstrated that mitochondrial GSH levels decline in the liver following alcohol feeding because alcohol induces accumulation of cholesterol in mitochondrial membranes. Cholesterol has been shown to interfere with GSH transport into mitochondria [29]. Mitochondria import GSH through various transporters to sustain its GSH pool, as mitochondria lack the enzymes necessary to synthesize GSH [30]. Alcohol-fed rats (oral feeding) have increased levels of posttranslational redox modifications to mitochondrial proteins in the liver [31,32]. Posttranslational redox changes such as nitrosylation and glutathionylation can cause conformational changes in proteins that can alter protein activity [33]. Finally, liver mitochondria from alcohol-fed rats have been shown to have greater abnormal morphologies compared to control [34]. Taken together, these findings suggest that mitochondrial dysfunction in the liver occurs with alcohol feeding. Given the central role of mitochondria in regulating metabolism and cell death, mitochondrial dysfunction caused by alcohol will likely play an important pathogenic role in ALD. Indeed, the hypothesis that mitochondrial dysfunction is important in ALD has become widely accepted and an important dogma in the field.

REASSESSMENT OF MITOCHONDRIAL ALTERATIONS IN ALD

Although the evidence is compelling that mitochondrial dysfunction occurs with chronic alcohol intake, mitochondrial dysfunction probably represents an incomplete understanding of mitochondrial changes that occur in ALD. In addition, because of the many problems and discrepancies in animal models, it is unclear if mitochondrial dysfunction is the major mitochondrial alteration that occurs in ALD. Chronic alcohol feeding is a major stress to hepatocytes that will cause oxidative damage and injury to mitochondria. However, mitochondria are very dynamic organelles and can adapt to stress. Thus, mitochondrial dysfunction may not always be a consequence of oxidative stress caused by alcohol and the mitochondrial dysfunction paradigm in ALD needs some reassessment.

MITOCHONDRIAL ALTERATIONS DIFFER BETWEEN ALCOHOL ANIMAL MODELS

Central to the mitochondrial dysfunction hypothesis in ALD has been the observation of that there is a decline in mitochondrial respiration (state III, mitochondrial substrates plus ADP) in the liver following alcohol feeding to rats. These findings have been primarily obtained from studies using the oral Lieber–DeCarli alcohol diet in rats [7,21,22]. The decline in mitochondrial respiration in the liver following alcohol feeding has, however, not been observed or assessed in all animal alcohol models. Alcohol feeding to mice and rats had been traditionally performed using either the Lieber–DeCarli oral diet or using the intragastric infusion model. Oral alcohol feeding to mice and rats using the Lieber–DeCarli diet has been the most studied alcohol model, due to its ease of implementation. However, the Lieber–DeCarli oral alcohol model only produces a mild form of liver injury, primarily fatty liver (steatosis) due to the fact that mice and rats will only consume limited amounts of alcohol [35]. On the other hand, the intragastric alcohol infusion model produces more severe liver injury with some pathophysiological features such as fibrosis and inflammation, which are also observed in alcoholic humans. In this model, alcohol is force-fed to mice and rats using a pump and a surgically implanted catheter [36,37]. Since oral alcohol feeding to rats caused a decline in mitochondrial respiration in the liver, it has been generally assumed that mitochondrial dysfunction was a common phenomenon of alcohol feeding. However, we observed that in the mouse alcohol model, chronic alcohol feeding (both intragastric and oral) enhances state III mitochondrial respiration in the liver [9]. Intragastric alcohol feeding, which produces more liver injury than oral feeding, was particularly effective in enhancing mitochondrial respiration and an approximately twofold increase in state III respiration was observed in the liver following 2 weeks of alcohol feeding. Oral alcohol feeding, on the other hand, only enhanced mitochondria respiration in the liver by ~30%. Other researchers have also observed a slight increase or no change in mitochondrial respiration in the livers of mice fed alcohol orally [38]. The enhanced mitochondrial respiration following alcohol feeding was associated with increased levels of proteins in the respiratory complexes and increased total levels of mitochondrial pyridine nucleotide (NADH, NAD^+) [9].

Decline in mitochondrial respiration does not correlate with liver injury caused by alcohol. Rats fed alcohol generally have the least liver injury (ALT ~40 to 60 following 6 weeks) and the greatest decline in mitochondrial respiration (~30% to 60%). Mice fed alcohol intragastrically have much greater liver injury (ALT ~ >100 U/L following 4 weeks) but have the greatest mitochondrial respiration in the liver (approximately twofold increase). The reason why the liver mitochondria of mice and rats respond so differently to alcohol remains unclear. There is a need for more experiments to better understand the mitochondrial changes observed in the liver with alcohol feeding. Most experiments have relied on one time point, and future temporal studies may shed light on mitochondrial dynamics in the liver. We observed that an increase in mitochondrial state III peaked at 2 weeks following intragastric alcohol feeding to mice and at 4 weeks the increase declined by ~20% [9]. If alcohol feeding continued to or passed 6 weeks, perhaps mitochondria respiration in the liver may decline further. In addition, experiments are needed to measure mitochondrial respiration in rats fed alcohol intragastrically, which would generate more liver injury in rats than the Lieber–DeCarli model. Recently, a new model of alcohol binge drinking has been developed and shows greater liver injury than the Lieber–DeCarli oral feeding model [39]. Mitochondrial alterations in the liver of these mice in the binge model also need to be explored. Overall, our data with mice suggest that mitochondrial alterations do not appear to correlate with liver injury caused by alcohol, but mitochondrial alterations in many other alcohol models still need to be examined.

Liver mitochondria of mice and rats appear to respond very differently to alcohol feeding. Mitochondrial alterations in the liver are not the only difference between mice and rats in response to alcohol feeding. Mice are overall more sensitive to alcohol-induced liver injury possibly due to lower expression of betaine homocysteine methyltransferase (BHMT) [40]. BHMT is important in detoxifying homocysteine, which may play a role in endoplasmic reticulum stress and liver injury caused by alcohol. Moreover, there may be strain differences within mice and rats that have not been extensively explored. The differences in response to alcohol between mice and rats lead to an important question of which animal model is more relevant in alcoholic patients. On the other hand, humans represent a genetically diverse population unlike inbred strains of mice and rats. Therefore, it is possible that some people may respond to alcohol with mitochondrial changes seen in rats, while others may experience changes similarly observed in mice. Lieber showed that baboons given alcohol orally develop mitochondrial alterations including declines in mitochondrial respiration in the liver similarly seen in rats [41]. Mitochondrial respiration and activities of respiratory complexes (i.e., succinate dehydrogenase [complex II], cytochrome oxidase [complex IV]) declined in the liver of baboons. However, in liver biopsies of patients with ALD, no changes in activities of succinate dehydrogenase and cytochrome oxidase were observed [42]. On the contrary, the enzyme activities of many mitochondrial matrix proteins (glutamate dehydrogenase, malate dehydrogenase, and aspartate aminotransferase) were greatly increased in the liver of ALD patients, leading the authors to suggest that some type of mitochondrial adaptation was occurring. Clearly, more studies are needed in patients, particularly at different stages of ALD, to fully understand the extent of mitochondrial changes that occur in the liver of patients with ALD.

Effect of Alcohol Feeding on Mitochondrial ROS Generation in the Liver

Metabolism is associated with the generation of ROS, particularly from mitochondria, which are the major sources of ROS in cells [43,44]. The respiratory chain, mainly complex I and III, generate superoxide both toward the intermembrane space, which can diffuse out of mitochondria through voltage-dependent anion channels [45,46], and toward the matrix, where Mn-SOD converts superoxide to H_2O_2 [47]. As previously discussed, chronic alcohol feeding has been suggested to induce oxidative stress and increase mitochondrial ROS generation in the liver. Important biomarkers of oxidative stress (i.e., lipid peroxidation, DNA oxidation) increase in the liver with alcohol feeding in rats and mice [27,48]. Antioxidant treatment (vitamin E, coenzyme Q, etc) and upregulation of Cu-Zn SOD have been shown to reduce alcohol-induced injury in the liver [26,27,49]. It is clear that there is significant oxidative stress in the liver with alcohol feeding; however, whether enhanced mitochondrial ROS generation is the underlying cause of alcohol-induced oxidative stress remains uncertain. The early and seminal works of Bailey and Cunningham suggest that alcohol treatment to primary hepatocytes enhances mitochondrial ROS generation [6,25]. Unfortunately, these findings are difficult to interpret since the redox dye, DCFH, was used as the primary method to measure ROS in hepatocytes. Later studies have clearly established that DCFH is a very nonspecific marker for ROS measurement [50–53]. DCF fluorescence can be modulated by many factors independent of ROS including GSH, iron, antioxidants, oxidases, and cytochromes. In some cases, DCFH itself can redox cycle and enhance ROS generation in cells. ROS have an extremely short half-life and therefore are difficult to measure in biological systems. Even newer redox dyes such as hydroethidine are nonspecific for superoxide, unless coupled with HPLC or other systems [54,55]. The mitochondrial respiratory chain may be the major source of ROS in hepatocytes following alcohol feeding; however, other candidates such as CYP2E1, which can generate superoxide and is unregulated by alcohol, must also be considered. Further experiments using newer and more specific techniques to measure ROS are needed to determine if mitochondrial ROS generation in the liver is truly increased with alcohol feeding.

JNK Translocation to Mitochondria May Mediate Apoptosis in ALD

Mitochondria play a central regulatory role in apoptosis, as it houses key apoptotic proteins such as cytochrome c and apoptosis-inducing factor. Although it remains in question whether mitochondrial bioenergetic changes and ROS generation is important in ALD, mitochondria may still play an important pathologic role in ALD by mediating c-Jun N-terminal kinase (JNK)-induced apoptosis. Mitochondria have been shown to be regulated by proapoptotic proteins such as JNK during tumor necrosis factor (TNF) or acetaminophen-induced liver injury [56–58]. TNF is a pluripotent cytokine that has been shown to play an important pathological role in various liver diseases including ALD. In ALD, TNF is believed to trigger a low level of hepatocyte apoptosis [50]. TNF knockout mice are more resistant to liver injury caused by chronic alcohol feeding than control mice [59,60]. Hepatocytes are normally resistant to TNF, but stresses, such as ROS and GSH depletion that occur with alcohol feeding, can sensitize hepatocytes to TNF-induced apoptosis [50,61,62]. TNF-induced apoptosis involves JNK activation and translocation to mitochondria. JNK is believed to bind to Sab on the outer membrane of mitochondria, which triggers the mitochondrial permeability transition (MPT), leading to apoptosis or necrosis in hepatocytes [58]. JNK inhibitor treatment has been shown to reduce alcohol-induced liver injury in mice. ALD, therefore, likely involves TNF-induced JNK activation, which translocates to the mitochondria to induce MPT.

NONALCOHOLIC FATTY LIVER DISEASE

Pathology of NAFLD

NAFLD is a fatty liver disease that occurs in patients who consume fewer than 40 grams of alcohol a day. Largely associated with obesity, NAFLD has become one of the many associated health consequences of the high-fat, high-carbohydrate diets of the Westernized world and is a growing epidemic [1,2]. Excessive accumulation of fatty acid droplets in hepatocytes characterizes NAFLD, although steatosis of other organs such as the heart can also occur. Obesity, which can initiate the pathogenesis of metabolic syndrome and type 2 diabetes, is a major risk factor for NAFLD. Insulin resistance is commonly associated with NAFLD. Besides obesity, NAFLD can occur in patients on long-term regimens of certain classes of drugs, toxins, or other factors (excluding alcohol-related liver injury). While steatosis maybe a somewhat benign condition, 10%–20% of NAFLD patients can progress to nonalcoholic steatohepatitis (NASH), which is characterized by necroinflammation and some fibrosis. NASH patients are at risk of developing cirrhosis and requiring liver transplants. It is unclear why a small percent of NAFLD patients progress to NASH. At present, there is no cure for NASH. The best management for patients with NASH is to try to lower obesity and control diabetes, hyperlipidemia, and other complications that arise with obesity. Vitamin E supplements have been shown to lower liver damage associated with NASH and have been used clinically in the treatment of NASH in many patients.

ANIMAL MODELS FOR NAFLD AND NASH

Many animal models of NAFLD exist, particularly mice and rats with obesity phenotypes. These include various mutants with impaired leptin synthesis (ob/ob mice) or impaired leptin receptors (db/db mouse, fa/fa mice, Zucker rats). Due to the relationship of obesity with NAFLD, obese mice and rats are the useful models for studying hepatic steatosis. In an attempt to replicate diets that promote NAFLD, various models of diet-induced hepatic steatosis have also been developed. "Western" diet models use high fat and/or high glucose to replicate the scenario of the majority of humans that have NAFLD. A Western diet model, consisting of 10–50 weeks of 45%–60% calories from saturated fat, induced various liver pathologies including slight inflammation and slight fibrosis. An appropriate animal model that represents the clinical manifestations of human NASH patients has not been fully developed. Obese mice fail to progress to NASH and leave much to be addressed in the underlying mechanism of NASH pathogenesis. An alternative dietary model is the MCD diet. Although the clinical relevance of the model can be questioned, the MCD diet causes lipid accumulation in hepatocytes, and histologically, there are some necroinflammation and fibrosis, with some similar histopathology to humans.

MITOCHONDRIAL ALTERATIONS WITH NAFLD

Mitochondria are the major site of fatty acid metabolism (beta-oxidation), and inhibition of mitochondrial function can directly lead to accumulation of fatty acids in the liver. Drugs that inhibit beta-oxidation and/or mitochondrial respiration can induce steatosis in hepatocytes by decreasing lipid degradation pathways [5,63]. However, steatosis induced by mitochondrial dysfunction may be limited to a handful of drugs and toxins that target mitochondria. It is unlikely that steatosis-associated obesity and metabolic syndrome are caused by the inhibition of beta-oxidation in mitochondria. Steatosis appears to cause a number of mitochondrial changes in the liver as adaptations to excess accumulation of lipids due to enhanced fatty acid synthesis or excess fatty acids in diet. In the next section, we will examine the major mitochondrial changes that have been observed with NAFLD and NASH in animal models and in patients.

Mitochondrial Bioenergetic Changes Associated with NAFLD

There is a great variability in mitochondrial changes that are affected temporally in various NAFLD models. In rats fed an MCD diet, an increase in mitochondrial respiration in the liver is observed, which declines to near normal levels over time [8]. In rats fed an HFD, little change in mitochondrial respiration except a decline in beta-oxidation was observed with 8 weeks feeding [64]. In A/J mice, an HFD for 10 days caused an upregulation of genes involved in oxidative phosphorylation and increased mitochondrial respiration (states 2, 3, and 4) [65]. This study also showed increased mitochondrial uncoupling after 10 days of HFD that corresponded with the increased mitochondrial respiration. This study showed little changes in liver mitochondria in C57BL/6 mice following HFD, only in A/J mice. On the other hand, another study found that feeding C57BL/6 mice a HFD caused a decline in mitochondrial respiration (state III) in the liver, which corresponded with decreased levels of many proteins in the respiratory complexes [3]. Complex I activity was observed to decrease in the liver of ob/ob mice (obesity model), but activities of other complexes were not observed to be altered [66]. In another study, it was observed that injection of leptin in ob/ob mice, which caused weight loss, caused a decline in mitochondrial respiration, mass, and protein levels [67]. This study suggests that mitochondrial respiration and mass is related to obesity, with higher mitochondria number and respiration occurring with greater obesity and steatosis occurs. This is in agreement with our ongoing work, which suggests that weight gain in ob/ob mice increases mitochondrial mass and bioenergetic activity in the liver (unpublished results).

Mitochondrial dysfunction has been suspected in NASH patients because they display decreased ATP resynthesis rates following fructose challenge, which transiently depletes hepatic ATP. These data suggest that ATP resynthesis, which occurs primarily in mitochondria, is altered in NASH patients but only indirectly suggests that mitochondrial bioenergetic activity is altered in the liver. Direct measurements of mitochondrial activity in patients have provided mixed results. One study reported that the activities of respiratory complexes were significantly decreased (complex I—37%, II—41.5%, III—29.4%, IV—37.5%, and V—37.6%) in the liver of NASH patients [68]. Another study measuring mitochondrial respiration in NASH patients using cybrids observed no noticeable changes in liver mitochondrial respiration [69]. In this study, however, abnormal mitochondrial morphology, which has also been seen in some animal models of NAFLD, was observed. Finally, another study observed an enhancement of beta-oxidation in the liver of NASH patients, even in the presence of significant oxidative and nitrosative stress [70]. Clearly, more studies are needed to understand the extent of mitochondrial changes that occur in NASH and their contribution to the pathogenesis of the disease.

Mitochondria as Sources of ROS in NAFLFD

NAFLD is associated with oxidative stress in the liver and many oxidative stress biomarkers are enhanced in the liver. Lipid peroxidation products, mtDNA oxidation, and upregulation of antioxidant enzymes such as superoxide dismutase have been observed with NAFLD [5,71]. Enhanced mitochondrial ROS generation in the liver has been suspected as a source of oxidative stress in the liver but has not been extensively demonstrated. Surprisingly, CYP2E1 has been shown to be upregulated in the liver in some NAFLD animal models. The function of CYP2E1 in NAFLD is unclear, unlike in ALD where CYP2E1 plays an important role in alcohol metabolism. Oxidative stress in NAFLD may therefore originate from many sources including CYP2E1 and the mitochondrial respiratory chain.

Mitochondria as the Final Target of JNK-Induced Apoptosis in NAFLD

Although the extent and importance of mitochondrial bioenergetic changes and mitochondrial ROS generation in NAFLD remains uncertain, mitochondria are still likely to play an important pathologic role in NAFLD by mediating JNK-induced apoptosis. Fatty acids are injurious to hepatocytes and can induce lipotoxicity, mediated by JNK-induced apoptosis [72]. The toxicity of free fatty acids to hepatocytes may be the reason why fatty acids are stored as triacylglycerols in lipid droplets in the liver [73]. Palmitate treatment to primary hepatocytes will induce apoptosis that can be inhibited by treatment with JNK inhibitors [72]. Later studies have demonstrated that JNK binding to Sab, on the outer mitochondrial membrane, is essential in lipotoxicity (Chapter 16). While JNK seems central in mediating lipotoxicity in hepatocytes, its role in animal models appears to be more complicated. JNK1 appears to have an important role in promoting steatosis and insulin resistance induced by HFD, while JNK 2 appears to have a protective role and prevent hepatocyte cell death and mitochondrial-dependent death pathways [74]. The discrepancy of JNK being injurious to mitochondria in cultured hepatocytes, while protecting mitochondria (JNK 2) in an animal steatosis model requires further investigation.

MITOCHONDRIAL DYSFUNCTION OR MITOCHONDRIAL ADAPTATION IN ALD AND NAFLD

It is clear that mitochondria are altered and damaged to some extent in ALD and NAFLD disease models. While mitochondria damage does occur in these diseases, the paradigm of mitochondria dysfunction being central to ALD and, to some extent, NAFLD represents an incomplete picture of mitochondrial dynamics in the liver. In many animal models, mitochondrial function is often enhanced with steatosis, which suggests mitochondrial adaptation is a major response to fatty liver. "Mitochondrial remodeling or adaptation" better characterizes the changes that occur in liver mitochondria during ALD and NAFLD. Clearly, mitochondria injury can promote steatosis as observed with drug-induced steatosis, in which drugs directly inhibit mitochondria function to promote steatosis and liver injury [5,63]. However, in most forms of steatosis, the fat likely comes first to affect mitochondria and cause adaptive changes. In recent years, the dynamic and adaptive nature of mitochondria has begun to be better characterized and appreciated. Mitochondria can undergo many types of changes in response to stress or metabolic changes to adapt to the needs of the cell. In skeletal muscle cells, the energy demand of acute or chronic exercise has been shown to cause mitochondrial biogenesis and remodeling (i.e., increased components of the mitochondria respiratory chain and proteins involved in β-oxidation) as an adaptation [75–77]. Similarly, liver mitochondria can adapt and alter in response to stress and metabolic changes. Dynamic mitochondrial adaptations are likely the predominant early mitochondrial changes observed in most models of ALD and NAFLD.

Signaling Pathways Regulating Mitochondrial Dynamics and Adaptation

Mitochondrial biogenesis in most cells is regulated by peroxisome proliferator-activated receptor γ coactivator-1α (PGC-1α), the master regulator of mitochondria. PGC-1α knockout mice have decreased levels of key mitochondrial proteins (i.e., cytochrome c, ATP synthase, cytochrome oxidase [COX]) [78,79], while overexpression of PGC-1α promotes mitochondrial biogenesis [80]. PGC-1α does not directly bind DNA, but acts as a coactivator that helps promote transcription of mitochondrial genes through protein–protein interactions with key transcription factor. PGC-1α has been shown to bind and activate many transcription factors including nuclear respiratory factor-1 (NRF-1), which transcribes subunits of all five respiratory complexes [81]. Other important transcription factors regulated by PGC-1α include NRF-2, estrogen-related receptor (ERRα), and peroxisome proliferator-activated receptor α and γ (PPARα, PPARγ), which transcribe various mitochondrial proteins. Two other coactivators that share a sequence homology with PGC-1α have also been identified: PGC-1β and PGC-1-related coactivator. Both NRF-1 and PGC-1α are upregulated by exercise to mediate mitochondrial biogenesis and remodeling [75,82]. HFD has also been shown to upregulate PGC-1α in muscle tissue [83]. The half-life of PGC-1α in cells is extremely short (~2 h), allowing cells to dynamically regulate the mitochondrial biogenesis according to the energy needs of cells [80]. In alcohol-fed rats (Lieber–DeCarli diet), PGC-1α proteins levels remained unchanged, although the protein was found to be acetylated [84]. On the other hand, in mice fed alcohol intragastrically, it was observed that PGC-1α levels increased in a time-dependent manner with alcohol feeding [9,85].

While PGC-1α regulates the levels of respiratory complex proteins, a family of conserved GTPases regulates mitochondrial morphology. Mitochondria are dynamic organelles that are undergoing constant fusion–fission rates to exchange mtDNA, proteins and other constituents [86,87]. Stresses such as serum deprivation can dramatically alter mitochondrial fusion–fission rates and mitochondrial morphology [10]. In embryonic fibroblasts, starvation causes a decrease in mitochondrial fission to produce elongated mitochondria that are more resistant to mitophagy (mitochondrial autophagy) and have greater cristae surface area, thereby greater mitochondrial respiration [88]. Mitochondrial fusion and fission are primarily controlled by four highly conserved GTPases: mitofusins (MFN1 and MFN2) in mitochondrial outer membranes promote fusion; optic atrophy (OPA1) in the inner mitochondrial membrane

(IMM) promotes fusion; and dynamin-related protein1 (DRP1) from the cytoplasm regulates fission [86]. Phosphorylation of DRP1 by PKA at serine 637 has been shown to sequester DRP1 in the cytoplasm and inhibit its translocation to mitochondria that promote fission [89]. Mitochondrial fusion–fission has also been shown to regulate mitochondrial fragmentation to mediate cell death [86,89]. Changes in mitochondrial morphology have been observed in various liver steatosis models and are likely due to alterations in mitochondrial fission–fusion that need to be further explored.

MITOCHONDRIAL DYSFUNCTION OR MITOCHONDRIAL ADAPTATION IN ALD

Many mitochondrial alterations observed in various steatosis models are likely adaptations to fatty deposits that build up in the liver. In mice fed alcohol, we observed that mitochondrial respiration was increased in the liver as an adaptation to increase NAD^+ regeneration, the limiting substrate for alcohol metabolism [9]. In support of this idea, we observed that isolated liver mitochondria from alcohol-fed mice have greater acetaldehyde-driven mitochondrial respiration and acetaldehyde metabolism than control. The observation that oral alcohol feeding to rats causes a decline in mitochondrial respiration has been central to the mitochondrial dysfunction hypothesis in ALD. However, a decline in respiratory complex proteins and mitochondrial respiration does not necessarily signal mitochondrial dysfunctional and could be a normal adaptation of the liver to metabolic changes and stress. Several arguments can be made that a decline in mitochondrial respiration is an adaptive response rather than a sign of mitochondria dysfunction following oral alcohol feeding to rats. First, as previously mentioned, proteins in the mitochondrial respiratory chain can increase in liver and muscle when energy or metabolic demand is high. It may similarly be that a decline in mitochondrial respiratory proteins and decline in respiration may be adaptive changes because of low energy demand due to alcohol intake. Alcohol feeding is associated with excess energy intake, and thus energy demand in the liver may be low. Therefore, it is possible that the liver may be downregulating mitochondrial respiration as a response to the excess energy levels from alcohol intake. Rats have been shown to have excess mitochondrial respiratory capacity and thus have excessive mitochondrial proteins [90]. In addition, the decline in mitochondrial respiration in rat liver has always been difficult to reconcile with the observation of increased oxygen consumption (hypermetabolic state) in the liver of alcohol-fed rats. Enhanced oxygen consumption by the liver is seen with acute alcohol feeding and chronic alcohol feeding. In liver slices of chronic alcohol-fed rats, oxygen consumption is greater than control in the presence of alcohol [16,91]. If mitochondrial respiration is suppressed, how is oxygen consumption increased with alcohol feeding? It is also possible that liver mitochondria in rats remodel such that some mitochondrial respiration pathways are downregulated (i.e., succinate-driven respiration), while other mitochondrial respiration pathways may be upregulated (i.e., glycerol phosphate–driven respiration). We are presently investigating this hypothesis that liver mitochondria in rats adapt in a different way such that mitochondria respiration through complex I and II is downregulated, but respiration through other pathways such as glycerol phosphate dehydrogenase II is upregulated following alcohol feeding.

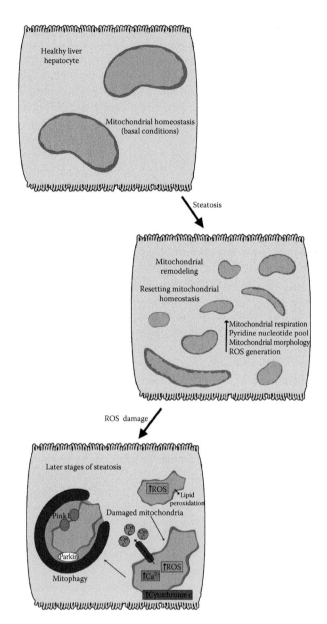

FIGURE 18.2 Model of mitochondrial remodeling and adaptation in the liver following stress. Mitochondrial homeostasis is reset in the liver by metabolic stresses such as alcohol or high-fat diet. The resetting of mitochondrial homeostasis is associated with greater mitochondrial respiration, greater number of mitochondria, and alterations in mitochondrial morphology. While mitochondrial remodeling helps the liver adapt to stresses, long-term mitochondrial alterations can also have negative consequences such as enhanced mitochondrial reactive oxygen species (ROS) generation that may promote liver injury. Increased mitochondrial ROS generation may lead to mitochondrial injury that promotes mitophagy and initiates apoptosis.

Another factor that is commonly cited as evidence of mitochondrial dysfunction in ALD is the decline in RCR, which suggests that there is mitochondrial uncoupling in the liver. In many ALD and NAFLD models, mitochondrial uncoupling has been observed, even in cases where mitochondrial respiration is enhanced. Since mitochondria uncoupling reduces the ATP production capacity of mitochondria, it is often linked to mitochondrial dysfunction. However, mitochondrial uncoupling may not necessarily be a negative phenomenon. Mitochondrial uncoupling can have positive attributes, such as decreased mitochondrial ROS generation due to shorter half-life of ubisemiquinone radical responsible for ROS generation in complex III [92]. ALD and NAFLD are diseases with excess energy in the form of fatty acids, thus likely where ATP levels are not limiting. Under these conditions of energy excess, uncoupled mitochondria may be an adaption to regenerate NAD^+ needed for alcohol metabolism and beta-oxidation, without generating excess ATP and generating less ROS. Uncoupling proteins (UCP) regulate mitochondrial uncoupling, and further research is needed to understand the possible relationship between UCP and steatosis in the liver.

MODEL OF MITOCHONDRIAL ADAPTATION IN ALD AND NAFLD

Our model is based on the concept that mitochondria are dynamic and help the liver adapt to chronic alcohol intake (Figure 18.2). Alcohol intake and/or enhanced steatosis cause mitochondrial remodeling and resetting of mitochondrial homeostasis in the liver to adapt to metabolic changes. Mitochondrial remodeling and adaptation are likely the major mitochondrial alterations observed during the early stages of steatosis. Mitochondrial adaptation and dysfunction are not mutually exclusive; some mitochondrial changes such as enhanced mitochondrial respiration may have long-term negative consequences (i.e., enhanced ROS generation), which promotes liver injury. Although enhanced mitochondrial ROS generation has not been definitively established in most steatosis models, generally enhanced mitochondrial respiration is generally linked to greater superoxide generation by the mitochondrial respiratory chain. Long-term enhanced mitochondrial ROS generation can cause lipid peroxidation and other forms of oxidative damage that promote mitophagy and hepatocyte apoptosis.

REFERENCES

1. Altamirano J and Bataller R. Alcoholic liver disease: Pathogenesis and new targets for therapy. *Nat Rev Gastroenterol Hepatol.* 2011;8(9):491–501.
2. Scaglioni F, Ciccia S, Marino M, Bedogni G, and Bellentani S. ASH and NASH. *Dig Dis.* 2011;29(2):202–210.
3. Mantena SK, King AL, Andringa KK, Eccleston HB, and Bailey SM. Mitochondrial dysfunction and oxidative stress in the pathogenesis of alcohol- and obesity-induced fatty liver diseases. *Free Radic Biol Med.* 2008;44(7):1259–1272.
4. Hoek JB, Cahill A, and Pastorino JG. Alcohol and mitochondria: A dysfunctional relationship. *Gastroenterology.* 2002;122(7):2049–2063.
5. Pessayre D, Berson A, Fromenty B, and Mansouri A. Mitochondria in steatohepatitis. *Semin Liver Dis.* 2001;21(1):57–69.

6. Bailey SM, Pietsch EC, and Cunningham CC. Ethanol stimulates the production of reactive oxygen species at mitochondrial complexes I and III. *Free Radic Biol Med.* 1999;27(7–8):891–900.

7. Spach PI and Cunningham CC. Control of state 3 respiration in liver mitochondria from rats subjected to chronic ethanol consumption. *Biochim Biophys Acta.* 1987;894(3):460–467.

8. Serviddio G, Bellanti F, Tamborra R, Rollo T, Romano AD, Giudetti AM, Capitanio N, Petrella A, Vendemiale G, and Altomare E. Alterations of hepatic ATP homeostasis and respiratory chain during development of non-alcoholic steatohepatitis in a rodent model. *Eur J Clin Invest.* 2008;38(4):245–252.

9. Han D, Ybanez MD, Johnson HS, McDonald JN, Mesropyan L, Sancheti H, Martin G et al. Dynamic adaptation of liver mitochondria to chronic alcohol feeding in mice: Biogenesis, remodeling, and functional alterations. *J Biol Chem.* 2012;287(50):42165–42179.

10. Han D, Dara L, Win S, Than TA, Yuan L, Abbasi SQ, Liu ZX, and Kaplowitz N. Regulation of drug-induced liver injury by signal transduction pathways: Critical role of mitochondria. *Trends Pharmacol Sci.* 2013;34(4):243–253.

11. Liesa M and Shirihai OS. Mitochondrial dynamics in the regulation of nutrient utilization and energy expenditure. *Cell Metabol.* 2013;17(4):491–506.

12. Israel Y and Orrego H. Hypermetabolic state and hypoxic liver damage. *Recent Dev Alcohol.* 1984;2:119–133.

13. Videla L and Israel Y. Factors that modify the metabolism of ethanol in rat liver and adaptive changes produced by its chronic administration. *Biochem J.* 1970;118(2):275–281.

14. Hasumura Y, Teschke R, and Lieber CS. Characteristics of acetaldehyde oxidation in rat liver mitochondria. *J Biol Chem.* 1976;251(16):4908–4913.

15. Tobon F and Mezey E. Effect of ethanol administration on hepatic ethanol and drug-metabolizing enzymes and on rates of ethanol degradation. *J Lab Clin Med.* 1971;77(1):110–121.

16. Videla L, Bernstein J, and Israel Y. Metabolic alterations produced in the liver by chronic ethanol administration. Increased oxidative capacity. *Biochem J.* 1973;134(2):507–514.

17. Teschke R, Hasumura Y, and Lieber CS. Hepatic microsomal alcohol-oxidizing system. Affinity for methanol, ethanol, propanol, and butanol. *J Biol Chem.* 1975;250(18):7397–7404.

18. Israel Y, Kalant H, Orrego H, Khanna JM, Videla L, and Phillips JM. Experimental alcohol-induced hepatic necrosis: Suppression by propylthiouracil. *Proc Natl Acad Sci USA.* 1975;72(3):1137–1141.

19. Thurman RG, Paschal D, Abu-Murad C, Pekkanen L, Bradford BU, Bullock K, and Glassman E. Swift increase in alcohol metabolism (SIAM) in the mouse: Comparison of the effect of short-term ethanol treatment on ethanol elimination in four inbred strains. *J Pharmacol Exp Ther.* 1982;223(1):45–49.

20. Bradford BU and Rusyn I. Swift increase in alcohol metabolism (SIAM): Understanding the phenomenon of hypermetabolism in liver. *Alcohol.* 2005;35(1):13–17.

21. Cederbaum AI, Lieber CS, and Rubin E. Effects of chronic ethanol treatment of mitochondrial functions damage to coupling site I. *Arch Biochem Biophys.* 1974;165(2):560–569.

22. Bernstein JD and Penniall R. Effects of chronic ethanol treatment upon rat liver mitochondria. *Biochem Pharmacol.* 1978;27(19):2337–2342.

23. Cunningham CC, Coleman WB, and Spach PI. The effects of chronic ethanol consumption on hepatic mitochondrial energy metabolism. *Alcohol Alcohol.* 1990;25(2–3):127–136.

24. Venkatraman A, Landar A, Davis AJ, Chamlee L, Sanderson T, Kim H, Page G et al. Modification of the mitochondrial proteome in response to the stress of ethanol-dependent hepatotoxicity. *J Biol Chem.* 2004;279(21):22092–22101.

25. Bailey SM and Cunningham CC. Contribution of mitochondria to oxidative stress associated with alcoholic liver disease. *Free Radic Biol Med.* 2002;32(1):11–16.

26. Arteel GE. Oxidants and antioxidants in alcohol-induced liver disease. *Gastroenterology.* 2003;124(3):778–790.

27. Mansouri A, Demeilliers C, Amsellem S, Pessayre D, and Fromenty B. Acute ethanol administration oxidatively damages and depletes mitochondrial dna in mouse liver, brain, heart, and skeletal muscles: Protective effects of antioxidants. *J Pharmacol Exp Ther.* 2001;298(2):737–743.

28. Colell A, Garcia-Ruiz C, Morales A, Ballesta A, Ookhtens M, Rodes J, Kaplowitz N, and Fernandez-Checa JC. Transport of reduced glutathione in hepatic mitochondria and mitoplasts from ethanol-treated rats: Effect of membrane physical properties and S-adenosyl-L-methionine. *Hepatology.* 1997;26(3):699–708.

29. Fernandez-Checa JC, Kaplowitz N, Garcia-Ruiz C, and Colell A. Mitochondrial glutathione: Importance and transport. *Semin Liver Dis.* 1998;18(4):389–401.

30. Garcia J, Han D, Sancheti H, Yap LP, Kaplowitz N, and Cadenas E. Regulation of mitochondrial glutathione redox status and protein glutathionylation by respiratory substrates. *J Biol Chem.* 2010;285(51):39646–39654.

31. Moon KH, Hood BL, Kim BJ, Hardwick JP, Conrads TP, Veenstra TD, and Song BJ. Inactivation of oxidized and S-nitrosylated mitochondrial proteins in alcoholic fatty liver of rats. *Hepatology.* 2006;44(5):1218–1230.

32. Venkatraman A, Landar A, Davis AJ, Ulasova E, Page G, Murphy MP, Darley-Usmar V, and Bailey SM. Oxidative modification of hepatic mitochondria protein thiols: Effect of chronic alcohol consumption. *Am J Physiol Gastrointest Liver Physiol.* 2004;286(4):G521–G527.

33. Han D, Hanawa N, Saberi B, and Kaplowitz N. Mechanisms of liver injury. III. Role of glutathione redox status in liver injury. *Am J Physiol Gastrointest Liver Physiol.* 2006;291(1):G1–G7.

34. Arai M, Leo MA, Nakano M, Gordon ER, and Lieber CS. Biochemical and morphological alterations of baboon hepatic mitochondria after chronic ethanol consumption. *Hepatology.* 1984;4(2):165–174.

35. Arteel GE. Animal models of alcoholic liver disease. *Dig Dis.* 2010;28(6):729–736.

36. Tsukamoto H, French SW, Benson N, Delgado G, Rao GA, Larkin EC, and Largman C. Severe and progressive steatosis and focal necrosis in rat liver induced by continuous intragastric infusion of ethanol and low fat diet. *Hepatology.* 1985;5(2):224–232.

37. Tsukamoto H, Towner SJ, Ciofalo LM, and French SW. Ethanol-induced liver fibrosis in rats fed high fat diet. *Hepatology.* 1986;6(5):814–822.

38. Venkatraman A, Shiva S, Wigley A, Ulasova E, Chhieng D, Bailey SM, and Darley-Usmar VM. The role of iNOS in alcohol-dependent hepatotoxicity and mitochondrial dysfunction in mice. *Hepatology.* 2004;40(3):565–573.

39. Bertola A, Mathews S, Ki SH, Wang H, and Gao B. Mouse model of chronic and binge ethanol feeding (the NIAAA model). *Nat Protoc.* 2013;8(3):627–637.

40. Shinohara M, Ji C, and Kaplowitz N. Differences in betaine-homocysteine methyltransferase expression, endoplasmic reticulum stress response, and liver injury between alcohol-fed mice and rats. *Hepatology.* 2010;51(3):796–805.

41. Popper H and Lieber CS. Histogenesis of alcoholic fibrosis and cirrhosis in the baboon. *Am J Pathol.* 1980;98(3):695–716.

42. Jenkins WJ and Peters TJ. Mitochondrial enzyme activities in liver biopsies from patients with alcoholic liver disease. *Gut.* 1978;19(5):341–344.

43. Han D, Antunes F, Daneri F, and Cadenas E. Mitochondrial superoxide anion production and release into intermembrane space. *Methods Enzymol.* 2002;349:271–280.

44. Cadenas E and Davies KJ. Mitochondrial free radical generation, oxidative stress, and aging. *Free Radic Biol Med.* 2000;29(3–4):222–230.

45. Han D, Antunes F, Canali R, Rettori D, and Cadenas E. Voltage-dependent anion channels control the release of the superoxide anion from mitochondria to cytosol. *J Biol Chem*. 2003;278(8):5557–5563.

46. Han D, Williams E, and Cadenas E. Mitochondrial respiratory chain-dependent generation of superoxide anion and its release into the intermembrane space. *Biochem J*. 2001;353(Pt 2):411–416.

47. Han D, Canali R, Rettori D, and Kaplowitz N. Effect of glutathione depletion on sites and topology of superoxide and hydrogen peroxide production in mitochondria. *Mol Pharmacol*. 2003;64(5):1136–1144.

48. Rouach H, Fataccioli V, Gentil M, French SW, Morimoto M, and Nordmann R. Effect of chronic ethanol feeding on lipid peroxidation and protein oxidation in relation to liver pathology. *Hepatology*. 1997;25(2):351–355.

49. Wheeler MD, Kono H, Yin M, Rusyn I, Froh M, Connor HD, Mason RP, Samulski RJ, and Thurman RG. Delivery of the Cu/Zn-superoxide dismutase gene with adenovirus reduces early alcohol-induced liver injury in rats. *Gastroenterology*. 2001;120(5):1241–1250.

50. Han D, Ybanez MD, Ahmadi S, Yeh K, and Kaplowitz N. Redox regulation of tumor necrosis factor signaling. *Antioxid Redox Signal*. 2009;11(9):2245–63.

51. Bonini MG, Rota C, Tomasi A, and Mason RP. The oxidation of 2,7,-dichlorofluorescin to reactive oxygen species: A self-fulfilling prophesy? *Free Radic Biol Med*. 2006;40(6):968–975.

52. Corda S, Laplace C, Vicaut E, and Duranteau J. Rapid reactive oxygen species production by mitochondria in endothelial cells exposed to tumor necrosis factor-α is mediated by ceramide. *Am J Respir Cell Mol Biol*. 2001;24(6):762–768.

53. Tampo Y, Kotamraju S, Chitambar CR, Kalivendi SV, Keszler A, Joseph J, and Kalyanaraman B. Oxidative stress-induced iron signaling is responsible for peroxide-dependent oxidation of dichlorodihydrofluorescein in endothelial cells: Role of transferrin receptor-dependent iron uptake in apoptosis. *Circ Res*. 2003;92(1):56–63.

54. Zielonka J, Hardy M, and Kalyanaraman B. HPLC study of oxidation products of hydroethidine in chemical and biological systems: Ramifications in superoxide measurements. *Free Radic Biol Med*. 2009;46(3):329–338.

55. Zielonka J, Srinivasan S, Hardy M, Ouari O, Lopez M, Vasquez-Vivar J, Avadhani NG, and Kalyanaraman B. Cytochrome c-mediated oxidation of hydroethidine and mito-hydroethidine in mitochondria: Identification of homo- and heterodimers. *Free Radic Biol Med*. 2008;44(5):835–846.

56. Zhou Q, Lam PY, Han D, and Cadenas E. c-Jun N-terminal kinase regulates mitochondrial bioenergetics by modulating pyruvate dehydrogenase activity in primary cortical neurons. *J Neurochem*. 2008;104(2):325–335.

57. Hanawa N, Shinohara M, Saberi B, Gaarde WA, Han D, and Kaplowitz N. Role of JNK translocation to mitochondria leading to inhibition of mitochondria bioenergetics in acetaminophen-induced liver injury. *J Biol Chem*. 2008;283(20):13565–13577.

58. Win S, Than TA, Han D, Petrovic LM, and Kaplowitz N. c-Jun N-terminal kinase (JNK)-dependent acute liver injury from acetaminophen or tumor necrosis factor (TNF) requires mitochondrial Sab protein expression in mice. *J Biol Chem*. 2011;286(40):35071–35078.

59. Ji C, Deng Q, and Kaplowitz N. Role of TNF-α in ethanol-induced hyperhomocysteinemia and murine alcoholic liver injury. *Hepatology*. 2004;40(2):442–451.

60. Yin M, Wheeler MD, Kono H, Bradford BU, Gallucci RM, Luster MI, and Thurman RG. Essential role of tumor necrosis factor α in alcohol-induced liver injury in mice. *Gastroenterology*. 1999;117(4):942–952.

61. Han D, Hanawa N, Saberi B, and Kaplowitz N. Hydrogen peroxide and redox modulation sensitize primary mouse hepatocytes to TNF-induced apoptosis. *Free Radic Biol Med*. 2006;41(4):627–639.

62. Matsumaru K, Ji C, and Kaplowitz N. Mechanisms for sensitization to TNF-induced apoptosis by acute glutathione depletion in murine hepatocytes. *Hepatology.* 2003;37(6):1425–1434.

63. Fromenty B and Pessayre D. Inhibition of mitochondrial beta-oxidation as a mechanism of hepatotoxicity. *Pharmacol Ther.* 1995;67(1):101–154.

64. Flamment M, Rieusset J, Vidal H, Simard G, Malthiery Y, Fromenty B, and Ducluzeau PH. Regulation of hepatic mitochondrial metabolism in response to a high fat diet: A longitudinal study in rats. *J Physiol Biochem.* 2012;68(3):335–44.

65. Poussin C, Ibberson M, Hall D, Ding J, Soto J, Abel ED, and Thorens B. Oxidative phosphorylation flexibility in the liver of mice resistant to high-fat diet-induced hepatic steatosis. *Diabetes.* 2011;60(9):2216–2224.

66. Finocchietto PV, Holod S, Barreyro F, Peralta JG, Alippe Y, Giovambattista A, Carreras MC, and Poderoso JJ. Defective leptin-AMP-dependent kinase pathway induces nitric oxide release and contributes to mitochondrial dysfunction and obesity in ob/ob mice. *Antioxid Redox Signal.* 2011;15(9):2395–2406.

67. Singh A, Wirtz M, Parker N, Hogan M, Strahler J, Michailidis G, Schmidt S et al. Leptin-mediated changes in hepatic mitochondrial metabolism, structure, and protein levels. *Proc Natl Acad Sci USA.* 2009;106(31):13100–13105.

68. Perez-Carreras M, Del Hoyo P, Martin MA, Rubio JC, Martin A, Castellano G, Colina F, Arenas J, and Solis-Herruzo JA. Defective hepatic mitochondrial respiratory chain in patients with nonalcoholic steatohepatitis. *Hepatology.* 2003;38(4):999–1007.

69. Caldwell SH, Swerdlow RH, Khan EM, Iezzoni JC, Hespenheide EE, Parks JK, and Parker WD, Jr. Mitochondrial abnormalities in non-alcoholic steatohepatitis. *J Hepatol.* 1999;31(3):430–434.

70. Sanyal AJ, Campbell-Sargent C, Mirshahi F, Rizzo WB, Contos MJ, Sterling RK, Luketic VA, Shiffman ML, and Clore JN. Nonalcoholic steatohepatitis: Association of insulin resistance and mitochondrial abnormalities. *Gastroenterology.* 2001;120(5):1183–1192.

71. Satapati S, Sunny NE, Kucejova B, Fu X, He TT, Mendez-Lucas A, Shelton JM, Perales JC, Browning JD, and Burgess SC. Elevated TCA cycle function in the pathology of diet-induced hepatic insulin resistance and fatty liver. *J Lipid Res.* 2012;53(6):1080–1092.

72. Ibrahim SH and Gores GJ. Who pulls the trigger: JNK activation in liver lipotoxicity? *J Hepatol.* 2012;56(1):17–19.

73. Yamaguchi K, Yang L, McCall S, Huang J, Yu XX, Pandey SK, Bhanot S, Monia BP, Li YX, and Diehl AM. Inhibiting triglyceride synthesis improves hepatic steatosis but exacerbates liver damage and fibrosis in obese mice with nonalcoholic steatohepatitis. *Hepatology.* 2007;45(6):1366–1374.

74. Singh R, Wang Y, Xiang Y, Tanaka KE, Gaarde WA, and Czaja MJ. Differential effects of JNK1 and JNK2 inhibition on murine steatohepatitis and insulin resistance. *Hepatology.* 2009;49(1):87–96.

75. Baar K, Wende AR, Jones TE, Marison M, Nolte LA, Chen M, Kelly DP, and Holloszy JO. Adaptations of skeletal muscle to exercise: Rapid increase in the transcriptional coactivator PGC-1. *FASEB J.* 2002;16(14):1879–1886.

76. Holloszy JO and Booth FW. Biochemical adaptations to endurance exercise in muscle. *Annu Rev Physiol.* 1976;38:273–291.

77. Mole PA, Oscai LB, and Holloszy JO. Adaptation of muscle to exercise. Increase in levels of palmityl Coa synthetase, carnitine palmityltransferase, and palmityl Coa dehydrogenase, and in the capacity to oxidize fatty acids. *J Clin Invest.* 1971;50(11):2323–2330.

78. Leone TC, Lehman JJ, Finck BN, Schaeffer PJ, Wende AR, Boudina S, Courtois M et al. PGC-1α deficiency causes multi-system energy metabolic derangements: Muscle dysfunction, abnormal weight control and hepatic steatosis. *PLoS Biol.* 2005;3(4):e101.

79. Lin J, Wu PH, Tarr PT, Lindenberg KS, St-Pierre J, Zhang CY, Mootha VK et al. Defects in adaptive energy metabolism with CNS-linked hyperactivity in PGC-1α null mice. *Cell*. 2004;119(1):121–135.

80. Scarpulla RC. Transcriptional paradigms in mammalian mitochondrial biogenesis and function. *Physiol Rev*. 2008;88(2):611–638.

81. Scarpulla RC. Metabolic control of mitochondrial biogenesis through the PGC-1 family regulatory network. *Biochim Biophys Acta*. 2011;1813(7):1269–1278.

82. Murakami T, Shimomura Y, Yoshimura A, Sokabe M, and Fujitsuka N. Induction of nuclear respiratory factor-1 expression by an acute bout of exercise in rat muscle. *Biochim Biophys Acta*. 1998;1381(1):113–122.

83. Garcia-Roves P, Huss JM, Han DH, Hancock CR, Iglesias-Gutierrez E, Chen M, and Holloszy JO. Raising plasma fatty acid concentration induces increased biogenesis of mitochondria in skeletal muscle. *Proc Natl Acad Sci USA*. 2007;104(25):10709–10713.

84. Lieber CS, Leo MA, Wang X, and Decarli LM. Alcohol alters hepatic FoxO1, p53, and mitochondrial SIRT5 deacetylation function. *Biochem Biophys Res Commun*. 2008;373(2):246–252.

85. Oliva J, French BA, Li J, Bardag-Gorce F, Fu P, and French SW. Sirt1 is involved in energy metabolism: The role of chronic ethanol feeding and resveratrol. *Exp Mol Pathol*. 2008;85(3):155–159.

86. Otera H and Mihara K. Molecular mechanisms and physiologic functions of mitochondrial dynamics. *J Biochem*. 2011;149(3):241–251.

87. Chan DC. Mitochondrial fusion and fission in mammals. *Annu Rev Cell Dev Biol*. 2006;22:79–99.

88. Gomes LC, Di Benedetto G, and Scorrano L. During autophagy mitochondria elongate, are spared from degradation and sustain cell viability. *Nat Cell Biol*. 2011;13(5):589–598.

89. Reddy PH, Reddy TP, Manczak M, Calkins MJ, Shirendeb U, and Mao P. Dynamin-related protein 1 and mitochondrial fragmentation in neurodegenerative diseases. *Brain Res Rev*. 2011;67(1–2):103–118.

90. Davey GP, Peuchen S, and Clark JB. Energy thresholds in brain mitochondria. Potential involvement in neurodegeneration. *J Biol Chem*. 1998;273(21):12753–12757.

91. Israel Y, Kalant H, Khanna JM, Orrego H, Phillips MJ, and Stewart DJ. Ethanol metabolism, oxygen availability and alcohol induced liver damage. *Adv Exp Med Biol*. 1977;85A:343–358.

92. Cadenas E and Boveris A. Enhancement of hydrogen peroxide formation by protophores and ionophores in antimycin-supplemented mitochondria. *Biochem J*. 1980;188(1):31–37.

19 Acid Sphingomyelinase, Mitochondria, and Liver Diseases

Carmen Garcia-Ruiz and José C. Fernández-Checa

CONTENTS

ABSTRACT

Ceramide is the most studied sphingolipid due to its emerging role as a second messenger that regulates multiple cellular functions. Acid sphingomyelinase (ASMase) functions as a specific mechanism of ceramide generation upon sphingomyelin hydrolysis at acidic compartments. In addition to its established role in apoptosis, recent evidence has shown a novel function of ASMase in liver fibrosis, endoplasmic reticulum (ER) stress, autophagy, and lysosomal membrane permeabilization, which impact the development of liver diseases. Moreover, ASMase regulates mitochondrial function, antioxidant defense, and reactive oxygen species (ROS) generation due to the stimulation of cholesterol trafficking to mitochondria by a mechanism involving ASMase-induced ER stress. In this chapter, we summarize the role of ASMase in Wilson disease, hepatic ischemia/reperfusion injury, and alcoholic and nonalcoholic fatty liver diseases. This evidence suggests that inhibiting ASMase may be of relevance in liver diseases in preventing ceramide generation and preserving lysosomal sphingomyelin content and function.

INTRODUCTION

Ceramide is the prototype sphingolipid that has been the focus of attention of bio-medical research in the last decades due to its role as a critical determinant of the structural properties of membrane bilayers and as a second lipid messenger that regulate many cell processes. In particular, ceramide has been shown to play a key role in cell stress, apoptosis, senescence, differentiation, and metabolism, hence emerging as a critical factor in human diseases (Garcia-Ruiz et al., 2015a). The structural components of ceramide include the sugar backbone sphingosine to which a fatty acyl moiety is attached in an amide bond. The length of the attached fatty acid define a heterogeneous family of ceramides that exhibit singular biological properties, which are synthesized by ceramide synthases (CerS), a family of enzymes that display specificity toward a selective fatty acids. Initially described as a proapoptotic mediator, ceramide can be metabolized to other derivatives such as sphingosine 1-phosphate that plays antiapoptotic functions, whose role was best described in the context of cancer cell biology (Morales et al., 2007).

Cells have several pathways to generate ceramide, predominantly de novo synthesis in the endoplasmic reticulum (ER) or sphingomyelin (SM) hydrolysis by sphingomyelinases (SMases) in specific cell sites. Stimulated de novo synthesis or SMases activation provides sustained or transient ceramide generation, respectively, that mediate the biological effects of specific external stimuli, including stress, death receptors, radiation, chemotherapy, or viral infection. The role of ceramide in linking external stimuli to cell responses is mediated by targeting intracellular organelles such as mitochondria, lysosomes, or ER. In this book chapter, we will briefly describe the role of ceramide and acid SMase (ASMase), a specific mechanism of ceramide generation, in liver diseases. In particular, we provide evidence for ASMase as a critical player in metabolic liver diseases such as Wilson disease, alcoholic and nonalcoholic steatohepatitis (NASH), and in hepatic ischemia/reperfusion (I/R) injury due to the generation of ceramide and the regulation of lysosomal SM homeostasis that impacts mitochondria, autophagy, and lysosomal membrane permeabilization (LMP). Thus, emerging evidence shows that targeting ASMase may be of relevance in liver diseases.

CERAMIDE GENERATION: DE NOVO SYNTHESIS AND SM HYDROLYSIS

DE NOVO SYNTHESIS

Cells generate ceramide by several pathways that differ in kinetics and cellular localization. The de novo pathway of ceramide generation occurs in the ER. In this pathway, the amino acid serine is conjugated with palmitoyl-CoA in a step catalyzed by the rate-limiting enzyme serine palmitoyl transferase (SPT). The product of the reaction sphinganine is acylated by ceramide synthases to dihydroceramide. Subsequent dehydrogenation catalyzed by dihydroceramide desaturase generates ceramide. In addition to acylation of the sphinganine, CerS also catalyze reacylation of sphingosine in the salvage pathway. Interestingly, six different ceramide synthases have been identified (Hannun and Obeid, 2008; Mizutani et al., 2008), which exhibit tissue-specific expression and variable substrate selectivity, thereby providing the basis for the generation of singular ceramide species of variable acyl chain in particular tissues. In this regard, ceramide synthase CerC2 is widely expressed and of major importance in liver and preferentially incorporates long chain C20–C24 acyl residues to generate C20–C24 ceramide. CerS3 is preferentially expressed in skin and catalyzes the acylation of very long acyl chains up to C34:0 to sphinganine. Ceramide synthase CerC5 specifically catalyzes the generation of C16 ceramide, while ceramide synthase CerC6 shows slightly wider substrate selectivity as it is involved in C14, C16, and C18 ceramide synthesis (Grösch et al., 2012; Hannun and Obeid, 2008). It is important to emphasize that ceramides with different acyl chain lengths are generated in specific physiological and pathophysiological contexts in a tissue- and cell-dependent fashion. Despite this defined specific profile of ceramide synthesized by the different CerS, there are compensatory mechanisms that offset the absence of specific ceramide species. Thus, an increase in a particular CerS may regulate a specific ceramide pool that may affect the integrity and function of individual cell compartments, such as lysosomes, ER, or mitochondria. For instance, CerS2 knockout mice exhibit a compensatory increase in the levels of C16 in the liver triggering hepatocyte apoptosis and proliferation that progress to hepatocellular hyperplasia. These changes in ceramide homeostasis translate to increased rates of hepatocyte apoptosis, mitochondrial dysfunction, and mitochondrial reactive oxygen species (ROS) generation, as well as proliferation that progress to the widespread formation of nodules of regenerative hepatocellular hyperplasia in aged mice. Progressive hepatomegaly and noninvasive liver tumors are observed in 10-month-old CerS2 null mice (Pewzner-Jung et al., 2010). An important factor that controls the de novo ceramide synthesis involves the availability of the substrate palmitoyl-CoA, which is required for sphinganine synthesis and whose level increases in obesity and metabolic syndrome and related disorders (i.e., nonalcoholic fatty liver disease [NAFLD]) (Bartke and Hannun, 2009; Breslow and Weisman, 2010).

SM Hydrolysis

While ceramide generation by de novo synthesis is slow but sustained, cells can instantaneously generate ceramide by a mechanism involving SM hydrolysis by the activation of SMases. In response to many deleterious stimuli causing cell stress, or apoptosis, cells activate SMases leading to a rapid and transient release of ceramide in specific sites that engage particular signaling pathways (Canals et al., 2011; Hannun and Obeid, 2008; Morales et al., 2007; Smith and Schuchmann, 2008). Several mammalian SMases have been characterized, which are classified according to their optimal pH (alkaline, neutral, or acid). Neutral sphingomyelinase (NSMase) and ASMase are the most studied enzymes in ceramide generation, which have been involved in important pathophysiological processes. NSMase-induced ceramide generation has been described as a critical lipid mediator in inflammatory diseases and *Xenopus laevis* oocyte maturation (Angulo et al., 2011; Coll et al., 2007). Moreover, in line with pioneering studies showing that ceramide directly targets mitochondrial respiratory chain components and stimulates ROS

(Garcia-Ruiz et al., 1997), NSMase activation has been shown to play a key role in apoptosis. Recent evidence has shown that NSMase cooperates with Bak and Bax to promote the mitochondrial pathway of apoptosis (Chipuk et al., 2012). In addition, bacterial NSMase has been identified as a key component of the therapeutic effects of the probiotic formulation VSL#3 in inflammatory bowel disease due to the selective induction of apoptosis of activated mucosal immune cells. In line with these findings, the beneficial effects of VSL#3 are prevented by NSMase inhibition with GW48469 (Angulo et al., 2011).

Generation of mice deficient in ASMase has been a valuable tool to show the relevance of ASMase in mediating cell death induced by a variety of stress stimuli and death ligands. ASMase has been extensively described as a signaling intermediate in cell death pathways and metabolic liver diseases (Garcia-Ruiz et al., 2003; Lin et al., 2000; Mari and Fernandez-Checa, 2007). ASMase catalyzes the formation of ceramide from SM primarily within the endolysosomal compartment, although it can be secreted extracellularly through Golgi trafficking as a secretory ASMase (S-SMase) (Canals et al., 2011; Smith and Schuchman, 2008). The dependence on Zn^{2+} for proper function is a differential feature between the lysosomal ASMase and the secreted form, with the former being Zn^{2+} independent. Both isoforms derive from a proinactive form whose proteolytic processing within the C terminal leads to the maturation of the endosomal/lysosomal ASMase and the secretory form (Jenkins et al., 2011). Consistent with this process, the mature ASMase counterpart (65 kDa) is sensitive to the lysosomotropic inhibitor desipramine/imipramine unlike the pro-ASMase form. While the role NSMase in liver diseases has been limited to the study of factor associated with NSMase activation, recent evidence demonstrates that ASMase plays a critical role in liver diseases by regulating multiple pathways, including ER stress, autophagy, lysosomal function, and mitochondrial cholesterol trafficking.

ASMase AND MITOCHONDRIA: THE ROLE OF ER STRESS AND CHOLESTEROL REGULATION

Besides the role of ASMase in tumor necrosis factor (TNF)/Fas-mediated apoptosis (Garcia-Ruiz et al., 2003; Mari et al., 2004), recent evidence has shown that ASMase regulates mitochondrial function by inducing the trafficking of cholesterol to mitochondrial membranes, which in turn, controls mitochondrial respiration and ROS generation (Mari et al., 2014). The ER plays a key role in Ca^{2+} homeostasis and is the place where proteins and lipids are synthesized. Altered protein folding or disturbed lipid homeostasis in ER (e.g., increased PC/PE ratio or ceramide synthesis) results in ER stress, which may ultimately lead to cell death (Fu et al., 2011; Malhi and Kaufman 2011). To restore homeostasis, the ER turns on the unfolded protein response (UPR). UPR is controlled by glucose-regulated protein 78 (GRP78), which regulates three specific transducers, inositol-requiring enzyme 1α, protein kinase R–like ER kinase, and activating transcription factor 6α, which act in concert to increase ER content, expand the ER protein-folding capacity, degrade misfolded proteins, and reduce the load of new proteins entering the ER. Moreover, ER stress has been shown to regulate hepatic steatosis, hence playing a major role in metabolic liver diseases, particularly alcoholic liver disease (ALD) (Ji and Kaplowitz, 2006). Mice with liver-specific deletion of GRP78 exhibit impaired global UPR and sensitization to alcohol-induced ER stress, steatosis, and injury (Ji et al., 2011). These effects are thought to be caused by alcohol-induced hyperhomocysteinemia. In line with these observations, nutritional therapy with betaine supplementation to alcohol-fed mice decreased hyperhomocysteinemia preventing alcohol-mediated ER stress, hepatic steatosis, and liver injury (Ji and Kaplowitz, 2003). Interestingly, betaine treatment has also been shown to attenuate alcohol-induced alterations to the mitochondrial respiratory chain proteome (Kharbanda et al., 2012). However, despite this evidence, recent data reported that mice fed a chow diet supplemented with homocysteine (Hcy), which exhibited increased plasma Hcy levels (three to seven fold), displayed no pathophysiologic changes or ER stress (Henkel and Gree, 2009; Henkel et al., 2012). Furthermore, the supplementation of methionine- and choline-deficient (MCD) diet with Hcy attenuated MCD-induced hepatic UPR activation and liver injury. These results suggest that Hcy at pathophysiological concentrations plays a minor role in ER stress and steatosis and that the associations between Hcy and UPR are not causally related.

Recent evidence has suggested a ceramide–ER stress link and the requirement for ASMase activation. For instance, evidence from mice deficient in cystathionine β-synthase has shown that the link between hyperhomocysteinemia and the subsequent glomerular injury is dependent on ASMase-mediated ceramide generation (Boini et al., 2012). These findings are in line with the evidence showing that exogenous ASMase but not NSMase directly induced ER stress in isolated hepatocytes by disrupting ER Ca^{2+} homeostasis (Fernandez et al., 2013), supporting the idea that altered lipid composition in the ER regulates ER Ca^{2+} homeostasis and subsequent ER stress susceptibility possibly through the modulation of Ca^{2+} pump SERCA. Interestingly, specific ASMase activation has been observed in association with hyperhomocysteinemia and decreased SAM/SAH ratio in mice fed MCD (Caballero et al., 2010), a nutritional model that induces ER stress (Henkel et al., 2012). Consistent with this novel function of ASMase in inducing ER stress, recent evidence indicated that ASMase deletion abrogated alcohol-induced hepatic steatosis and ER stress (Fernandez et al., 2013). Intriguingly, however, the level of hyperhomocysteinemia and the decrease in SAM/SAH ratio caused by alcohol intake were similar in ASMase$^{+/+}$ and ASMase$^{-/-}$ mice, suggesting that the resistance of ASMase knockout mice to ER stress caused by alcohol seems to be independent of hyperhomocysteinemia. Moreover, ASMase$^{-/-}$ mice are refractory to alcohol-induced mitochondrial cholesterol accumulation, an important event in hepatocyte apoptosis regulation due to mitochondrial GSH depletion (Josekutty et al., 2013; Mari et al., 2006, 2008). The resistance of ASMase null mice to alcohol-induced increase in mitochondrial cholesterol accumulation and maintenance of mGSH pool parallels the lower susceptibility toward LPS and concanavalin-A-mediated liver injury, which is observed only in ASMase$^{+/+}$ mice. Consistent with the critical role of steroidogenic acute regulatory domain StARD1, the founding member of StARD protein family in the trafficking of cholesterol to mitochondrial inner membrane (Anuka et al., 2013), alcohol increased the StARD1 expression in ASMase$^{+/+}$ mice but not in ASMase$^{-/-}$, indicating a correlation between alcohol-induced ER stress and StARD1 expression. Moreover, these data suggest that StARD1 is an ER stress–regulated gene. Indeed, the addition of tunicamycin to primary hepatocytes induced ER stress and StARD1 upregulation that was prevented by tauroursodeoxycholic acid (Fernandez et al., 2013; Ozcan et al., 2004), further suggesting that StARD1 is an ER stress target gene. As expected for SREBP-2-regulated genes, feeding mice a high-cholesterol diet downregulated hydroxymethylglutaryl CoA reductase but not StARD1, indicating that StARD1 upregulation is ER stress dependent. Therefore, the link between ER stress and mitochondrial cholesterol accumulation and subsequent mGSH depletion due to alcohol intake requires a novel ASMase–ER stress–StARD1 axis. These results point to ASMase as a novel targetable approach to prevent ER stress in the context of ALD and preserve mitochondrial function and antioxidant defenses.

ROLE OF ASMase IN AUTOPHAGY AND LYSOSOMAL MEMBRANE PERMEABILIZATION

In addition to ER stress, autophagy and LMP regulate lipid metabolism and cell death and therefore have emerged as critical players in metabolic liver diseases. Autophagy is a proteolytic pathway that regulates the turnover of organelles and cellular debris (Czaja et al., 2013). In this process, intracellular organelles or protein aggregates are sequestered in a double membrane structure called the phagophore that evolves into the autophagosome, which fuses with lysosomes to generate autolysosomes where the cargo contents are degraded. Autophagy is considered a protective mechanism by channeling cellular components for energy supply when nutrient availability is limited. Although autophagy has been predominantly studied in the context of cell death regulation, where it plays a dual role, recent findings described a new role of autophagy in liver fibrogenesis, lipid metabolism, and hepatic steatosis in a process called lipophagy. For instance, pharmacologic or genetic autophagy inhibition attenuated hepatic stellate cells (HSC) activation and fibrogenesis as mice with a genetic deletion of Atg7 in HSCs have reduced fibrosis following sustained liver injury (Hernandez-Gea and Friedman, 2012). The link between lipophagy and HSC activation implies that autophagy increases the energy production by liberating FA from retinyl esters to serve as an energy source for HSCs. Moreover, mice with liver-specific deletion of Atg7, a key protein that regulates autophagosome formation, exhibited defective autophagy and increased triglyceride and cholesterol storage underlying hepatic steatosis (Singh et al., 2009). Other studies, however, reported conflicting results about the role of autophagy in lipid metabolism and hepatic steatosis. For instance, mice with liver-specific deletion of FIP200, a core subunit of the Atg1 complex, are protected from starvation and high-fat diet (HFD)-induced hepatic steatosis (Ma et al., 2013). Moreover, mice with liver-specific deletion of Atg7 fed HFD have reduced hepatic steatosis (Kim et al., 2013).

Although autophagy is regulated by the expression of a number of specialized proteins that regulate the initiation of autophagosomes and maturation into autolysosomes (Czaja et al., 2013), this process is also controlled by ceramide and ASMase. Ceramide participates in lysosome fusion to cell plasma membranes, endosomes, phagosomes, and other organelles, and it is known to regulate cytoskeleton and microtubule assembly, and ceramide can directly interact with LC3I, which may facilitate the targeting of lysosomes to autophagosomes (Sentelle et al., 2012; Trajkovic et al., 2008; Zeidan et al., 2008). Moreover, hepatocytes from ASMase$^{-/-}$ mice exhibit defective autophagic flux even though this outcome is accompanied by decreased hepatic steatosis induced by HFD (Fucho et al., 2014). In line with these findings, mouse coronary arterial smooth muscle cells from ASMase$^{-/-}$ mice exhibit a defect in the fusion of autophagosomes with lysosomes due to impaired lysosomal function (Li et al., 2014). Knockdown of ASMase suppressed the induction of autophagy in leukemia HL-60 cells induced by amino acid deprivation and ASMase is required for the upregulation of Atg5 expression and autophagy induction in HepG2 cells (Park et al., 2008; Taniguchi et al., 2012). The mechanism of ASMase-mediated regulation of autophagy not only involves ceramide but also lysosomal SM homeostasis. For instance, decreased SM in lysosomes by ASMase can regulate the TRPLM1/lysosomal Ca^{2+}/dynein axis. Dynein is a multi-subunit microtubule motor protein complex, involved in the trafficking of autophagosomes with lysosomes to form autolysosomes that translates to an increased number of autophagosomes (Xu et al., 2013; Yamamoto et al., 2010).

Lysosomes contain specialized hydrolytic enzymes that are used for digestion and removal of protein and intracellular organelles, which ultimately regulate the turnover of macromolecules (Mari and Fernandez-Checa, 2014). Besides this role, lysosomes also promote cell death in a process causing LMP, which allows the release of lysosomal content for cell death initiation. LMP is caused by a wide range of stimuli, including ROS, saturated fatty acids, sphingosine, or cell death effectors such as Bax. LMP releases lysosome contents to the cytosol, including cathepsins (e.g., cathepsin B, CTSB), which in turn target mitochondria, leading to apoptosis. Cathepsin B can cause caspase-independent cell death or recruit mitochondria by BH3-interacting domain death agonist cleavage. Related to the regulation of autophagy at the level of fusion autophagosomes with lysosomes, ASMase has been shown to regulate LMP (Fucho et al., 2014). Although the primary phenotype of ASMase deficiency is increased SM levels in lysosomes, ASMase deficiency leads to lysosomal cholesterol accumulation (LCA). The localization of cholesterol in this compartment reflects the high affinity of SM to bind cholesterol, which decreases the efflux of cholesterol out of lysosomes. The characteristic LCA in ASMase$^{-/-}$ hepatocytes is common to other lysosomal storage diseases, such as Niemann Pick type C (NPC) disease. In fibroblasts treatment with U18666A, a cationic amphiphile disrupts intracellular cholesterol trafficking and reproduces the NPC phenotype, causing LCA and protection against LMP-mediated apoptosis (Appleqvist et al., 2011). Hepatocytes from ASMase$^{-/-}$ mice exhibit LCA, resulting in the prevention against amphiphilic lysosomotropic detergents, which induce cell death and caspase activation following LMP (Fucho et al., 2014). As LMP has been shown to contribute to palmitic acid (PA)-induced apoptosis, ASMase$^{-/-}$ hepatocytes are also resistant to PA-induced lipotoxicity, an effect that was reversed by depleting LCA with the oxysterol 25-hydroxycholesterol (Fucho et al., 2014). Thus, LCA resulting from ASMase deficiency modulates the autophagy–lysosomal degradation pathway and ameliorates LMP and the subsequent lipotoxicity of saturated fatty acids. Overall, these emerging findings indicate that ASMase regulate autophagy and LMP and hence is a druggable target of relevance in liver disease.

ASMase IN WILSON DISEASE

Wilson disease is an autosomal recessive disorder with a prevalence of about 1 in 30,000 births. The disease is caused by inactivating mutations in ATP7B, an enzyme involved in the secretion of Cu^{2+} from the liver. The defect results in the accumulation of Cu^{2+} in hepatocytes and other tissues including neurons, blood, or muscle cells (Cumings, 1948; Gitlin, 2003). An excess of Cu^{2+} ions in cells and tissues induces severe disorders including progressive hepatic cirrhosis, chronic active hepatitis, or even progressive hepatic failure, Fanconi syndrome, neurological and psychiatric symptoms, cardiomyopathy, osteomalacia, and in some individuals, anemia. Although Cu^{2+} is an essential trace element of the human diet and is required as a cofactor for the function of diverse proteins, Cu^{2+} triggers the release of ROS that seem to be crucially involved in the induction of cell death by Cu^{2+} (Krumschnabel et al., 2005; Pourahmad et al., 2001; Sokol et al., 1994). The release of ROS after cellular treatment with Cu^{2+} has been described as occurring predominantly in lysosomes and mitochondria, and lysosomal membrane damage precedes Cu^{2+} cytotoxicity. Furthermore, Cu^{2+} triggers, at least in erythrocytes, the formation of lipid peroxides and inhibits the activity of antioxidant enzymes, finally resulting in oxidative cell damage, denaturation of hemoglobin, and hemolytic anemia.

Recent studies have shown that Cu^{2+} triggers hepatocyte apoptosis through activation of ASMase and the release of ceramide (Figure 19.1) (Lang et al., 2007). Genetic deficiency of ASMase prevented Cu^{2+}-induced hepatocyte apoptosis, while

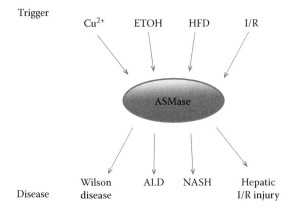

FIGURE 19.1 Role of acid sphingomyelinase in liver diseases. As a stress mediator, acid sphingomyelinase (ASMase) becomes activated in response to various triggers such as Cu^{2+}, alcohol, high-fat diet, or during ischemia/reperfusion (I/R). ASMase activation hence regulates lysosomal ceramide/sphingomyelin homeostasis, which impacts downstream responses including endoplasmic reticulum stress, autophagy, and lysosomal membrane permeabilization that ultimately regulate the onset of liver diseases including Wilson disease, alcoholic liver disease, nonalcoholic steatohepatitis, and hepatic I/R injury.

ASMase inhibition with desipramine in rats with a mutation in the *Atp7b* gene, a genetic model of Wilson disease, protects against Cu^{2+}-induced hepatocyte death and liver failure. In line with these findings, individuals with Wilson disease showed elevated plasma levels of Asm and displayed a constitutive increase of ceramide and phosphatidylserine-positive erythrocytes. The concentrations of free Cu^{2+} (1–3 mM) required to elicit activation of the ASMase, ceramide release, and apoptosis of hepatocytes are in the range of concentrations encountered in plasma of individuals with Wilson disease, thus supporting the clinical significance of the activation of ASMase by Cu^{2+} and its involvement in Wilson disease.

Moreover, a functional link between methionine metabolism and ASMase activation has been established in Wilson disease. As indicated earlier, MCD diet and SAM depletion results in ASMase activation (Caballero et al., 2010) and recent findings linked abnormal methionine metabolism and Wilson disease. In this regard, using tx-j mouse, a model of Wilson disease, it has been recently reported that Cu^{2+} decreases SAH hydrolase expression and activity, resulting in reduced SAM/SAH ratio and global DNA hypomethylation preceding liver disease, and these effects were reduced by Cu^{2+} chelation or betaine administration to provide methyl groups (Medici et al., 2013). These findings indicate that Cu^{2+} accumulation may lead to altered SAM/SAH ration, which in turn activates ASMase mediating Wilson disease. It remains to be established whether betaine treatment indirectly inhibited ASMase by hypermethylation or if ASMase inhibition by desipramine restored SAM/SAH and normalized DNA methylation. Thus, these findings indicate that ASMase-induced ceramide generation plays a novel role in Wilson disease and that ASMase targeting may be of relevance in the treatment of this disease.

ASMase IN HEPATIC I/R INJURY

Liver damage induced by I/R has an important impact in different clinical settings such as liver resection, transplantation, trauma, or hemorrhagic shock. Hepatocyte ischemia during liver transplantation, resection, and shock causes anoxia, depletion of glycolytic substrates, loss of adenosine triphosphate, and acidosis. The mechanisms responsible for hepatic I/R injury are not well understood despite the identification of several intercellular and molecular mechanisms involved in the necrotic and apoptotic death of hepatocytes. Molecular events include nuclear factor κB (NF-κB) activation, TNF generation, JNK activation, mitochondrial permeability transition (MPT), and ROS overproduction (Jaeschke and Lemasters, 2003; Rudiger and Clavien 2002; Uehara et al., 2005). Previous studies using mice deficient in *TNF* gene or *TNF receptor* 1 (TNFR1) have identified TNF as a critical mediator in warm hepatic I/R injury (Colletti et al., 1990; Rudiger et al., 2002). Moreover, increased hepatic ceramide levels and SMases have been observed in cold ischemia and warm reperfusion (Alessenko et al., 2002; Bradham et al., 1997). In this regard, the reperfusion of the ischemic liver stimulates NSMase, while ASMase becomes depressed (Alessenko et al., 2002). Consistent with the key role of TNF in hepatic I/R and of ASMase in TNF signaling, it has been shown that ASMase participates in hepatic I/R injury

(Figure 19.1) (Llacuna et al., 2006). In this context, hepatic ceramide levels transiently increase after the reperfusion phase of the ischemic liver in mice, because of an early activation of ASMase followed by acid ceramidase stimulation. In vivo administration of an ASMase inhibitor, imipramine, or ASMase knockdown by siRNA decreased ceramide generation during I/R and attenuated serum ALT levels, hepatocellular necrosis, cytochrome c release, and caspase-3 activation. ASMase-induced ceramide generation activated JNK resulting in Bim_L phosphorylation and translocation to mitochondria, events that were prevented by the inhibition of ASMase by imipramine. In contrast, blockade of ceramide catabolism by N-oleoylethanolamine, an acid ceramidase inhibitor, enhanced ceramide levels and potentiated I/R injury compared with vehicle-treated mice. In the context of hepatic I/R injury, ASMase-induced ceramide generation targets mitochondria by a JNK-dependent mechanism. In agreement with these findings, the potent activation of JNK by exogenous ASMase has been reported in primary hepatocytes (Gupta et al., 2004). However, the cytotoxic role of JNK may be due not only to its putative interaction with mitochondria but also to the activation of downstream targets, including BH3-only members of the Bcl-2 family such as Bim, which can be regulated by transcriptional and posttranslational mechanisms (Kurinna et al., 2004; Ley et al., 2005). For instance, Bim_L, a Bim splice variant normally associated with microtubules through its interaction with the dynein light chain 1 that prevents its contact with other Bcl-2 family members, can be regulated at the transcriptional level or by its phosphorylation state (Ley et al., 2005). The phosphorylation of Bim_L at threonine56 by JNK unleashes Bim_L from microtubules, allowing its redistribution and binding to Bcl-2. In line with this mechanism, the phosphorylation of Bim_L and its translocation to mitochondria during I/R have been reported, in agreement with previous findings in lung cancer cells, and this step is prevented by JNK inhibition (SP600125) or ASMase inhibition (imipramine). Overall, current evidence indicates that the modulation of ceramide generation by ASMase during I/R regulates the mitochondrial pathway of hepatocellular death through JNK activation and subsequent mitochondrial translocation of Bim_L, suggesting that the modulation of ceramide may be a novel therapeutic approach to prevent postischemic liver injury. Thus, the emerging role of ASMase in hepatic I/R-mediated injury has important clinical implications such as liver transplantation, surgery, and resections implying that targeting ASMase may be beneficial in these conditions.

ASMase IN FATTY LIVER DISEASE AND GLUCOSE HOMEOSTASIS

Fatty liver disease is one of the most prevalent liver diseases and encompasses both ALD and NAFLD. Although both ALD and NAFLD differ in etiology, both share similar biochemical and histological characteristics. Both diseases start with steatosis, which can progress to more advanced stages, such as steatohepatitis, in which the fatty liver is accompanied by inflammation, liver injury, hepatocyte ballooning, and liver fibrosis. Due to the association with obesity, insulin resistance (IR), and diabetes type 2, the prevalence of NAFLD is expected to increase globally and is one of the most important health concerns in the world (Angulo 2002).

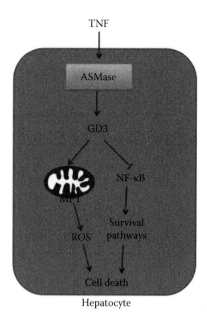

Hepatocyte

FIGURE 19.2 Role of acid sphingomyelinase in tumor necrosis factor–induced hepatocellular apoptosis. Tumor necrosis factor (TNF) overexpression has been shown to mediate hepatocyte apoptosis and participate in prevalent liver diseases. TNF activates acid sphingomyelinase, which in turn induce hepatocyte cell death by a dual mechanism including ganglioside GD3 generation and nuclear factor κB (NF-κB) inactivation. GD3 has been shown to traffic to mitochondria and target mitochondrial respirator chain inducing reactive oxygen species generation and mitochondrial permeability transition. In addition, NF-κB suppression prevents the recruitment of survival pathways that further contribute to hepatocyte demise.

Despite intense research, the molecular mechanisms governing the transition from steatosis to steatohepatitis are still poorly understood. As described earlier, ASMase has emerged as a novel regulator of key players in steatohepatitis such as ER stress, which impacts mitochondrial cholesterol homeostasis, autophagy, and LMP. Moreover, ASMase mediates hepatocellular apoptosis and liver fibrosis and promotes both ALD and NAFLD. For instance, ASMase has been reported to mediate TNF-induced hepatocellular apoptosis and TNF/Gal-mediated fulminant liver failure and Fas-induced lethal hepatitis (Garcia-Ruiz et al., 2003; Lin et al., 2000; Mari et al., 2004). The mechanism of ASMase-induced hepatocyte apoptosis by TNF/Fas involves the recruitment of mitochondria through ganglioside GD3 generation, which elicits apoptosis by a dual mechanism of ROS generation and NF-κB inhibition (Figure 19.2) (Colell et al., 2001; Garcia-Ruiz et al., 2002). In addition to its role in cell death, ASMase regulates HSC activation and liver fibrogenesis. Selective stimulation of ASMase, but not NSMase, occurs during the transdifferentiation of primary mouse HSCs to myofibroblast-like cells, coinciding with the processing of the downstream effectors cathepsin B and D (Moles et al., 2010). ASMase antagonism blunted cathepsin B/D processing and prevented the activation and proliferation of mouse and human HSCs. Moreover, it has been shown that amitriptyline, an ASMase inhibitor, reduces established hepatic fibrosis induced by CCl$_4$ in mice (Quillin et al., 2014).

Consistent with these findings, recent evidence has shown a critical role for ASMase in fatty liver disease. First, ASMase becomes activated in both ALD and NAFLD (Fernandez et al., 2013; Moles et al., 2010), which reflects its role in stress signaling, consistent with the role of ROS and oxidative stress in fatty liver disease due to mitochondrial dysfunction (Garcia-Ruiz et al., 2015a). Alcohol feeding activates ASMase and liver biopsies from patients with acute alcoholic hepatitis exhibit increased ASMase mRNA levels (Deaciuc et al., 2000; Fernandez et al., 2013; Liangpunsakul et al., 2012). Moreover, ASMase$^{-/-}$ mice are resistant to alcohol-induced lipogenesis, macrosteatosis, mitochondrial cholesterol loading, and subsequent LPS sensitization and concanavalin-A-mediated liver injury (Fernandez et al., 2013). The lack of the effect of alcohol to stimulate mitochondrial cholesterol trafficking was due to the repression of StARD1, a protein essential for the regulation of cholesterol movement to the mitochondrial inner membrane that is regulated by ER stress. Moreover, ASMase inhibition with amitriptyline in wild-type mice prevented alcohol-induced steatosis, liver injury, and LPS sensitization, without compromising liver regeneration. ASMase is overexpressed in adipose tissue of ob/ob mice, in mice fed an MCD diet, and in liver and serum samples from patients with NASH (Caballero et al., 2010; Grammatikos et al., 2014; Moles et al., 2010). Moreover, ASMase deletion, superimposed on a genetic background of low-density lipoprotein (LDL) receptor deficiency, prevented diet-induced steatosis (Deevska et al., 2009). ASMase$^{-/-}$ mice, fed an HFD, were resistant to HFD-induced activation of lipogenic enzymes (Fucho et al., 2014). These findings translated into resistance to HFD-mediated steatosis, with similar findings observed when mice were fed an MCD diet. Furthermore, ASMase inhibition with amitriptyline protected wild-type mice against HFD-induced steatosis, liver injury, inflammation, and fibrosis.

As mentioned earlier, ceramide has been identified as a bioactive mediator in various cellular functions, including apoptosis, senescence, cell differentiation, and metabolism. Ceramide accumulation contributes to the development of type 2 diabetes (Summers and Nelson, 2005). Indeed, inhibition of ceramide synthesis by myriocin, a SPT inhibitor, or dihydroceramide desaturase improves IR induced by glucocorticoid or saturated fat (Holland and Summers 2007). Ceramide induces IR by inactivation of AKT through protein phosphatase-2A and PKC-ζ, or by the inhibition of AKT translocation to the plasma membrane (Holland et al., 2008; Stratford et al., 2001; Wymann and Schneiter 2008). In addition, ceramide inhibits AMP-activated protein kinase (AMPK) activation in hepatoma cells (Liangpunsakul et al., 2010). The role of ASMase in glucose homeostasis and IR is less established with conflicting results. ASMase activity is increased in the serum of type 2 diabetic patients, induces GLUT4 translocation to the plasma membrane, and increases glucose uptake by adipocytes (Al-Makdissy et al., 2003; David et al., 1998; Gorska et al., 2003; Liu et al., 2004). However, previous findings in mice with ASMase and LDL receptor deletion prevented diet-induced hyperglycemia. Intriguingly, these findings were accompanied by a paradoxical increase in hepatic ceramide and de novo ceramide synthesis due to SPT activation (Deevska et al., 2009). In contrast,

ASMase overexpression improved glucose metabolism in diabetic db/db mice, and accordingly, ASMase$^{-/-}$ mice exhibited higher blood glucose levels than wild-type mice upon glucose tolerance tests (Osawa et al., 2009). Moreover, inhibition of ASMase with amitriptyline in wild-type mice reduced body weight gain, hepatomegaly, and epididymal white adipose tissue gain upon HFD without the effect of amitriptyline on food intake, thus further confirming the role of ASMase in glucose homeostasis (Fucho et al., 2014). Moreover, amitriptyline normalized hyperglycemia induced by HFD and reduced blood glucose levels during glucose (GTT) and insulin tolerance tests (ITT) following HFD feeding. The beneficial effects of amitriptyline on body weight and glucose tolerance seen in wild-type mice fed HFD were abrogated in ASMase$^{-/-}$ mice, confirming the dependence of amitriptyline on ASMase to regulate body weight and glucose homeostasis. Furthermore, consistent with the previous findings (Osawa et al., 2009), the effects of HFD feeding in blood glucose homeostasis following GTT and ITT were aggravated in ASMase$^{-/-}$ mice compared to wild-type mice. Moreover, the effects of ASMase in increasing glucose uptake and glycogen accumulation are mediated through the activation of AKT and glycogen synthase kinase-3β. In addition, ASM induced upregulation of glucose transporter 2 accompanied by the suppression of AMPK phosphorylation. Overall, these findings provide evidence that ASMase is required for physiological tuning of glucose homeostasis.

CONCLUSION

ASMase functions as a specific means of ceramide generation in defined cellular sites, predominantly in endolysosomes but also in plasma membrane, which supports the engagement of signaling platforms. Besides ceramide generation, the consequence of ASMase activation results in the degradation of SM in lysosomes, which impacts the lipid composition of lysosomal membranes and secondarily modulates the cholesterol content. While ASMase activation has been mainly recognized as a critical player in death receptor–mediated apoptosis, ASMase also regulates lysosomal cathepsins, which in turn, modulate HSC transdifferentiation and liver fibrosis. While the ceramide generation due to ASMase activation is transient and subject to degradation via acid ceramidase activation, which converts ceramide into sphingosine that becomes phosphorylated to sphingosine 1-phosphate by sphingosine kinases, the alteration in lysosomal membrane composition (altered SM/cholesterol) lasts over time and has an important impact in lysosomal function and in the regulation of autophagy at the level of fusion of autophagosomes with lysosomes to generate autolysosomes as well as in LMP, which have emerged as critical mechanisms involved in lipotoxicity of saturated fatty acids, lipid metabolism, and fibrosis. Besides these targets, ASMase is known to modulate methionine metabolism and SAM/SAH ratio, establishing a reciprocal functional link between SAM/SAH and ASMase activation (Garcia-Ruiz et al., 2015a). Accordingly, ASMase acts as a versatile mechanism that regulates the multiple pathways of relevance in diverse liver diseases. Thus, pharmacological inhibition of ASMase may be of clinical relevance in the treatment of prevalent metabolic liver diseases, such as ALD and particularly NAFLD. In this regard, while novel and more effective ASMase inhibitors are being developed, tricyclic antidepressants and other cationic amphiphilic drugs, which are already in the clinic for the treatment of depression (desipramine, nortriptyline, and amitriptyline), malaria (chloroquine), allergies (terfenadine), or hypertension (amlodipine), have been shown to efficiently antagonize ASMase by preventing its proteolytic activation (Petersen et al., 2013). Overall, due to the ability to activate and recruit multiple pathways that control lipid metabolism, autophagy, cell death, and glucose homeostasis, ASMase may be an attractive target for the treatment of liver diseases.

REFERENCES

Al-Makdissy N, Younsi M, Pierre S, Ziegler O, and Donner M. 2003. Sphingomyelin/cholesterol ratio: An important de-terminant of glucose transport mediated by GLUT-1 in 3T3-L1 preadipocytes. *Cell Signal* 15:1019–1030.

Alessenko AV, Galperin EI, Dudnik LB, Korobko VG, Mochalova ES, Platonova LV et al. 2002. Role of tumor necrosis factor α and sphingomyelin cycle activation in the induction of apoptosis by ischemia/reperfusion of the liver. *Biochemistry (Mosc)* 67:1347–1355.

Angulo P. 2002. Nonalcoholic fatty liver disease. *N Engl J Med* 346:1221–1231.

Angulo S, Morales A, Danese S, Llacuna L, Masamunt MC, Pultz N et al. 2011. Probiotic sonicates selectively induce mucosal immune cells apoptosis through ceramide generation via neutral sphingomyelinase. *PLoS One* 6:e16953.

Anuka E, Gal M, Stocco DM, Orly J. 2013. Expression and roles of steroidogenic acute regulatory (StAR) protein in 'non-classical', extra-adrenal and extra-gonadal cells and tissues. *Mol Cell Endocrinol* 371:47–61.

Appelqvist H, Nilsson C, Garner B, Brown AJ, Kågedal K, Ollinger K. 2011. Attenuation of the lysosomal death pathway by lysosomal cholesterol accumulation. *Am J Pathol* 178(2):629–639.

Boini KM, Xia M, Abais JM et al. 2012. Acid sphingomyelinase gene knockout ameliorates hyperhomocysteinemic glomerular injury in mice lacking cystathionine-b-synthase. *PLoS One* 7:e45020.

Bartke N, Hannun YA. 2009. Bioactive sphingolipids: metabolism and function. *J Lipid Res* 50:S91–S96.

Bradham CA, Stachlewitz RF, Gao W, Qian T, Jayadev S, Jenkins G et al. 1997. Reperfusion after liver transplantation in rats differentially activates the mitogen-activated protein kinases. *Hepatology* 25:1128–1135.

Breslow DK, Weissman JS. 2010. Membranes in balance: Mechanisms of sphingolipid homeostasis. *Mol Cell* 40:267–279.

Caballero F, Fernández A, Matías N et al. 2010. Specific contribution of methionine and choline in nutritional nonalcoholic steatohepatitis: Impact on mitochondrial S-adenosyl-L-methio-nine and glutathione. *J Biol Chem* 285:18528–18536.

Canals D, Perry DM, Jenkins RW, Hannun YA. 2011. Drug targeting of sphingolipid metabolism: Sphingomyelinases and ceramidases. *Br J Pharmacol* 163:694–712.

Chipuk JE, Mcstay GP, Bharti A, Kuwana T, Clarke CJ, Siskind LJ, Obeid LM, Green, DR. 2012. Sphingolipid metabolism cooperates with Bak and Bax to promote the mitochondrial pathway of apoptosis. *Cell* 148:988–1000.

Colell A, García-Ruiz C, Roman J, Ballesta A, Fernández-Checa JC. 2001. Ganglioside GD3 enhances apoptosis by suppressing the nuclear factor-κ B-dependent survival pathway. *FASEB J* 15:1068–1670.

Coll O, Morales A, Fernández-Checa JC, Garcia-Ruiz C. 2007. Neutral sphingomyelinase-induced ceramide triggers germinal vesicle breakdown and oxidant-dependent apoptosis in *Xenopus laevis* oocytes. *J Lipid Res* 48:1924–1935.

Colletti LM, Remick DG, Burtch GD, Kunkel SL, Strieter RM, Campbell DA Jr. 1990. Role of tumor necrosis factor-α in the pathophysiologic alterations after hepatic ischemia/ reperfusion injury in the rat. *J Clin Invest* 85:1936–1943.

Cumings JN 1948. Copper and iron content of brain and liver in normal and in hepatolenticular degeneration. *Brain* 71:410–415.

Czaja MJ, Ding WX, Donohue TM Jr, Friedman SL, Kim JS, Komatsu M et al. 2013. Functions of autophagy in normal and diseased liver. *Autophagy* 9(8):1131–1158.

David TS, Ortiz PA, Smith TR, Turinsky J. 1998 Sphingomyelinase has an insulin-like effect on glucose transporter translocation in adipocytes. *Am J Physiol* 274:R1446–R1453.

Deaciuc IV, Nikolova-Karakashian M, Fortunato F, Lee EY, Hill DB, McClain CJ. 2000. Apoptosis and dysregulated ceramide metabolism in a murine model of alcohol-enhanced lipopolysaccharide hepatotoxicity. *Alcohol Clin Exp Res* 24:1557–1565.

Deevska GM, Rozenova KA, Giltiay NV, Chambers MA, White J, Boyanovsky BB et al. 2009. Acid sphingomyelinase deficiency prevents diet-induced hepatic triacylglycerol accumulation and hyperglycemia in mice. *J Biol Chem* 284(13):8359–8368.

Fernandez A, Matias N, Fucho R, Ribas V, Von Montfort C, Nuño N et al. 2013. ASMase is required for chronic alcohol induced hepatic endoplasmic reticulum stress and mitochondrial cholesterol loading. *J Hepatol* 59:805–813.

Fucho R, Martinez L, Baulies A, Torres S, Tarrats N, Fernandez A et al. 2014. Asmase regulates autophagy and lysosomal membrane permeabilization and its inhibition prevents early stage nonalcoholic steatohepatitis. *J Hepatol* 61(5):1126–1134.

Fu S, Yang L, Li P, Hofmann O et al. 2011. Aberrant lipid metabolism disrupts calcium homeostasis causing liver endoplasmic reticulum stress in obesity. *Nature* 473:528–531.

Garcia-Ruiz C, Colell A, Mari M, Morales A, Calvo M, Enrich C, Fernandez-Checa JC. 2003. Defective TNF-α-mediated hepatocellular apoptosis and liver damage in acidic sphingomyelinase knockout mice. *J Clin Invest* 111:197–208.

García-Ruiz C, Colell A, Marí M, Morales A, Fernández-Checa JC. 1997. Direct effect of ceramide on the mitochondrial electron transport chain leads to generation of reactive oxygen species. Role of mitochondrial glutathione. *J Biol Chem* 272:11369–11377.

García-Ruiz C, Colell A, Morales A, Calvo M, Enrich C, Fernández-Checa JC. 2002. Trafficking of ganglioside GD3 to mitochondria by tumor necrosis factor-α. *J Biol Chem* 277: 36443.

Garcia-Ruiz C, Mato JM, Vance D, Kaplowitz N, Fernández-Checa JC. 2015a. Acid sphingomyelinase-ceramide system in steatohepatitis: A novel target regulating multiple pathways. *J Hepatol* 62:219–233.

Garcia-Ruiz C, Morales A, Fernandez-Checa JC. 2015b. Glycosphingolipids and cell death. One aim, many ways. *Apoptosis*, in press.

Gitlin JD. 2003. Wilson disease. *Gastroenterology* 125:1868–1877.

Gorska M, Baranczuk E, Dobrzyn A. 2003. Secretory Zn2-dependent sphingomyelinase activity in the serum of patients with type 2 diabetes is elevated. *Horm Metab Res* 35:506–507.

Grammatikos G, Mühle C, Ferreiros N, Schroeter S, Bogdanou D, Schwalm S et al. 2014. Serum acid sphingomyelinase is upregulated in chronic hepatitis C infection and non alcoholic fatty liver disease. *Biochim Biophys Acta* 1841(7):1012–1120.

Grösch S, Schiffmann S, Geisslinger G. 2012. Chain length-specific properties of ceramides. *Prog Lipid Res* 51:50–62.

Gupta S, Natarajan R, Payne SG, Studer EJ, Spiegel S, Dent P et al. 2004. Deoxycholic acid activates c-Jun N-terminal kinase pathway via FAS receptor activation in primary hepatocytes. *J Biol Chem* 279:5821–5828.

Hannun YA, Obeid LM. 2008. Principles of bioactive lipid signalling: Lessons from sphingolipids. *Nat Rev Mol Cell Biol* 9:139–150.

Henkel AS, Dewey AM, Anderson KA et al. 2012. Reducing endoplasmic reticulum stress does not improve steatohepatitis in mice fed a methionine- and choline-deficient diet. *Am J Physiol Gastrointest Liver Physiol* 303:G54–G59.

Henkel AS, Elias MS, Green RM. 2009. Homocysteins supplementation attenuates the unfolded protein response in a murine nutritional model of steatohepatitis. *J Biol Chem* 284:31807–31816.

Hernández-Gea V, Friedman SL. 2012. Autophagy fuels tissue fibrogenesis. *Autophagy* 8(5):849–850.

Holland WL, Brozinick JT, Wang LP, Hawkins ED, Sargent KM, Liu Y et al. 2007. Inhibition of ceramide synthesis ameliorates glucocorticoid-, saturated-fat-, and obesity-induced insulin resistance. *Cell Metab* 5:167–179.

Holland WL, Summers SA. 2008. Sphingolipids, insulin resistance, and metabolic disease: New insights from in vivo manipulation of sphingolipid metabolism. *Endocr Rev* 29:381–402.

Jaeschke H, Lemasters JJ. 2003. Apoptosis versus oncotic necrosis in hepatic ischemia/reperfusion injury. *Gastroenterology* 125:1246–1257.

Jenkins RW, Idkowiak-Baldys J, Simbari F, Canals D, Roddy P, Riner CD, Clarke CJ, Hannun YA. 2011. A novel mechanism of lysosomal acid sphingomyelinase maturation: Requirement for carboxyl-terminal proteolytic processing. *J Biol Chem* 286:3777–3788.

Ji C, Kaplowitz N. 2006. ER stress: Can the liver cope? *J Hepatol* 45(2):321–333.

Ji C, Kaplowitz N. 2003. Betaine decreases hyperhomocysteinemia, endoplasmic reticulum stress, and liver injury in alcohol-fed mice. *Gastroenterology* 124(5):1488–1499.

Ji C, Kaplowitz N, Lau MY, Kao E, Petrovic LM, Lee AS. 2011. Liver-specific loss of glucose-regulated protein 78 perturbs the unfolded protein response and exacerbates a spectrum of liver diseases in mice. *Hepatology* 54(1):229–239.

Josekutty J, Iqbal J, Iwawaki T et al. 2013. Microsomal triglyceride transfer protein inhibition induces endoplasmic reticulum stress and increases gene transcription via Ire1α/cJun to enhance plasma ALT/AST. *J Biol Chem* 288:14372–14383.

Kharbanda KK, Todero SL, King AL et al. 2012. Betaine treatment attenuates chronic ethanol-induced hepatic steatosis and alterations to the mitochondrial respiratory chain proteome. *Int J Hepatol* 2012: 962183.

Kim KH, Jeong YT, Oh H, Kim SH, Cho JM, Kim YN et al. 2013. Autophagy deficiency leads to protection from obesity and insulin resistance by inducing Fgf21 as a mitokine. *Nat Med* 19(1):83–92.

Krumschnabel G, Manzl C, Berger C, Hofer B. 2005. Oxidative stress, mitochondrial permeability transition, and cell death in Cu-exposed trout hepatocytes. *Toxicol Appl Pharmacol* 209:62–73.

Kurinna SM, Tsao CC, Nica AF, Jiffar T, Ruvolo PP. 2004. Ceramide promotes apoptosis in lung cancer-derived A549 cells by a mechanism involving c-Jun NH2-terminal kinase. *Cancer Res* 64:7852–7856.

Lang PA, Schenck M, Nicolay JP, Becker JU, Kempe DS, Lupescu A et al. 2007. Liver cell death and anemia in Wilson disease involve acid sphingomyelinase and ceramide. *Nat Med* 13(2):164–170.

Ley R, Ewings KE, Hadfield K, Cook SJ. 2005. Regulatory phosphorylation of Bim: Sorting out the ERK from the JNK. *Cell Death Differ* 12:1008–1014.

Li X, Xu M, Pitzer AL, Xia M, Boini KM, Li PL, Zhang Y. 2014. Control of autophagy maturation by acid sphingomyelinase in mouse coronary arterial smooth muscle cells: Protective role in atherosclerosis. *J Mol Med* 92(5):473–485.

Liangpunsakul S, Rahmini Y, Ross RA, Zhao Z, Xu Y, Crabb DW. 2012. Imipramine blocks ethanol-induced ASMase activation, ceramide generation, and PP2A activation, and ameliorates hepatic steatosis in ethanol-fed mice. *Am J Physiol* 302(5):G515–G523.

Liangpunsakul S, Sozio MS, Shin E, Zhao Z, Xu Y, Ross RA, Zeng Y, Crabb DW. 2010. Inhibitory effect of ethanol on AMPK phosphorylation is mediated in part through elevated ceramide levels. *Am J Physiol Gastrointest Liver Physiol* 298:G1004–G1012.

Lin T, Genestier L, Pinkoski MJ, Castro A, Nicholas S, Mogil R et al. 2000. Role of acidic sphingomyelinase in Fas/CD95-mediated cell death. *J Biol Chem* 275(12):8657–8663.

Liu P, Leffler BJ, Weeks LK, Chen G, Bouchard CM, Strawbridge AB, Elmendorf JS. 2004. Sphingomyelinase activates GLUT4 translocation via a cholesterol-dependent mechanism. *Am J Physiol Cell Physiol* 286:C317–C329.

Llacuna L, Marí M, Garcia-Ruiz C, Fernandez-Checa JC, Morales A. 2006. Critical role of acidic sphingomyelinase in murine hepatic ischemia-reperfusion injury. *Hepatology* 44(3):561–572.

Ma D, Molusky MM, Song J, Hu CR, Fang F, Rui C et al. 2013. Autophagy deficiency by hepatic FIP200 deletion uncouples steatosis from liver injury in NAFLD. *Mol Endocrinol* 27(10):1643–1654.

Malhi H, Kaufman RJ. 2011. Endoplasmic reticulum stress in liver disease. *J Hepatol* 54:795–809.

Marí M, Caballero F, Colell A et al. 2006. Mitochondrial free cholesterol loading sensitizes to TNF- and Fas-mediated steatohepatitis. *Cell Metab* 4:185–198.

Mari M, Colell A, Morales A et al. 2008. Mechanism of mitochondrial glutathione-dependent hepatocellular susceptibility to TNF despite NF-κB activation. *Gastroenterology* 134:1507–1520.

Mari M, Colell A, Morales A, Paneda C, Varela-Nieto I, Garcia-Ruiz C, Fernandez-Checa JC. 2004. Acidic sphingomyelinase downregulates the liver-specific methionine adenosyltransferase 1A, contributing to tumor necrosis factor-induced lethal hepatitis. *J Clin Invest* 113:895–904.

Mari M, Fernandez-Checa JC. Lysosomes. In *Pathobiology of Human Diseases: A Dynamic Enciclopedia of Disease Mechanisms*, section ed. P. Bernardi. Elsevier, 97–107, 2014.

Marí M, Fernández-Checa JC. 2007. Sphingolipid signalling and liver diseases. *Liver Int* 27:440–450.

Mari M, Morales A, Colell A, Garcia-Ruiz C, Fernandez-Checa JC. 2014. Mitochondrial cholesterol accumulation in alcoholic liver disease: Role of ASMase and endoplasmic reticulum stress. *Redox Biol* 3:100–108.

Medici V, Shibata NM, Kharbanda KK, LaSalle JM, Woods R, Liu S et al. 2013. Wilson's disease: Changes in methionine metabolism and inflammation affect global DNA methylation in early liver disease. *Hepatology* 57(2):555–565.

Mizutani Y, Kihara A, Chiba H, Tojo H, Igarashi Y. 2008. 2-Hydroxy-ceramide synthesis by ceramide synthase family: enzymatic basis for the preference of FA chain length. *J Lipid Res* 49:2356–2364.

Moles A, Tarrats N, Morales A, Domínguez M, Bataller R, Caballería J, García-Ruiz C, Fernández-Checa JC, Marí M. 2010. Acidic sphingomyelinase controls hepatic stellate cell activation and in vivo liver fibrogenesis. *Am J Pathol* 177:1214–1224.

Morales A, Lee H, Goñi F, Kolesnick R, Fernandez-Checa J. 2007. Sphingolipids and cell death. *Apoptosis* 12:923–939.

Osawa Y, Seki E, Kodama Y, Suetsugu A, Miura K, Adachi M et al. 2009. Acid sphingomyelinase regulates glucose and lipid metabolism in hepatocytes through AKT activation and AMP-activated protein kinase suppression. *FASEB J* 25:1133–1144.

Ozcan U, Cao Q, Yilmaz E, Lee AH, Iwakoshi NN, Ozdelen E et al. 2004. Endoplasmic reticulum stress links obesity, insulin action, and type 2 diabetes. *Science* 306:457–461.

Park MA, Zhang G, Martin AP, Hamed H, Mitchell C, Hylemon PB et al. 2008. Vorinostat and sorafenib increase ER stress, autophagy and apoptosis via ceramide-dependent CD95 and PERK activation. *Cancer Biol Ther* 7(10):1648–1662.

Petersen NH, Olsen OD, Groth-Pedersen L, Ellegaard AM, Bilgin M, Redmer S et al. 2013. Transformation-associated changes in sphingolipid metabolism sensitize cells to lysosomal cell death induced by inhibitors of acid sphingomyelinase. *Cancer Cell* 24(3):379–393.

Pewzner-Jung Y, Brenner O, Braun S, Laviad EL, Ben-Dor S, Feldmesser E et al. 2010. A critical role for ceramide synthase 2 in liver homeostasis: II. Insights into molecular changes leading to hepatopathy. *J Biol Chem* 285:10911–10923.

Pourahmad J, Ross S, O'Brien PJ. 2001. Lysosomal involvement in hepatocyte cytotoxicity induced by Cu(2+) but not Cd(2+). *Free Radic Biol Med* 30:89–97.

Quillin RC 3rd, Wilson GC, Nojima H, Freeman CM, Wang J, Schuster RM et al. 2014. Inhibition of acidic sphingomyelinase reduces established hepatic fibrosis in mice. *Hepatol Res* 45:305–314.

Rudiger HA, Clavien PA. 2002; Tumor necrosis factor α, but not Fas, mediates hepatocellular apoptosis in the murine ischemic liver. *Gastroenterology* 122:202–210.

Sentelle RD, Senkal CE, Jiang W, Ponnusamy S, Gencer S, Selvam SP et al. 2012. Ceramide targets autophagosomes to mitochondria and induces lethal mitophagy. *Nat Chem Biol* 8:831–838.

Singh R, Kaushik S, Wang Y, Xiang Y, Novak I, Komatsu M, Tanaka K, Cuervo AM, Czaja MJ. 2009. Autophagy regulates lipid metabolism. *Nature* 458(7242):1131–1135.

Smith EL, Schuchman EH. 2008. The unexpected role of acid sphingomyelinase in cell death and the pathophysiology of common diseases. *FASEB J* 22:3419–3431.

Sokol RJ et al. 1994. Oxidant injury to hepatic mitochondria in patients with Wilson's disease and Bedlington terriers with copper toxicosis. *Gastroenterology* 107:1788–1798.

Stratford S, DeWald DB, Summers SA. 2001. Ceramide dissociates 3'-phosphoinositide production from pleckstrin homology domain translocation. *Biochem J* 354:359–368.

Summers SA, Nelson DH. 2005. A role for sphingolipids in producing the common features of type 2 diabetes, metabolic syndrome X, and Cushing's syndrome. *Diabetes* 54:591–602.

Taniguchi M, Kitatani K, Kondo T, Hashimoto-Nishimura M, Asano S, Hayashi A et al. 2012. Regulation of autophagy and its associated cell death by "sphingolipid rheostat": Reciprocal role of ceramide and sphingosine 1-phosphate in the mammalian target of rapamycin pathway. *J Biol Chem* 287(47):39898–39910.

Trajkovic K, Hsu C, Chiantia S, Rajendran L, Wenzel D, Wieland F, Schwille P, Brügger B, Simons M. 2008. Ceramide triggers budding of exosome vesicles into multivesicular endosomes. *Science* 319(5867):1244–1247.

Uehara T, Bennett B, Sakata ST, Satoh Y, Bilter GK, Westwick JK et al. 2005. JNK mediates hepatic ischemia reperfusion injury. *J Hepatol* 42:850–859.

Wymann MP, Schneiter R. 2008. Lipid signalling in disease. *Nat Rev Mol Cell Biol* 9:162–176.

Xu M, Li XX, Xiong J, Xia M, Gulbins E, Zhang Y, Li PL. 2013. Regulation of autophagic flux by dynein-mediated autophagosomes trafficking in mouse coronary arterial myocytes. *Biochim Biophys Acta* 1833:3228–3236.

Yamamoto M, Suzuki SO, Himeno M. 2010. The effects of dynein inhibition on the autophagic pathway in glioma cells. *Neuropathology* 30:1–6.

Zeidan YH, Jenkins RW, Hannun YA. 2008. Remodeling of cellular cytoskeleton by the acid sphingomyelinase/ceramide pathway. *J Cell Biol* 181(2):335–350.

20 Regulation of Liver Metabolism by Sirtuins
Emerging Roles

Eric S. Goetzman

CONTENTS

ABSTRACT

Sirtuins are NAD^+-dependent deacylases that are becoming recognized as important regulators of metabolism in the liver. SIRT1 controls metabolism globally through its actions in the nucleus while SIRT3 deacetylates and regulates numerous key proteins involved in mitochondrial energy metabolism. This chapter aggregates recent literature on the lesser studied sirtuins (SIRT2, SIRT4, SIRT5, SIRT6, and SIRT7) and their emerging roles in governing hepatic metabolism. SIRT2, SIRT4, and SIRT5 are nonnuclear sirtuins whose target proteins are just beginning to be elucidated. SIRT2, stimulated by nutrient deprivation, positively regulates gluconeogenesis and suppresses lipogenesis. SIRT6 appears to work in opposition to SIRT2, suppressing gluconeogenesis and glucose metabolism in general. SIRT4 represses glutamine and fatty acid oxidation, while SIRT5 promotes ketogenesis and the urea cycle. SIRT6 and SIRT7, which are nuclear-localized, suppress ribosomal biogenesis and hypoxia inducible factor-1 α (HIF1α) signaling. Understanding how the seven sirtuin enzymes communicate across multiple cellular compartments to coordinate metabolism is expected to lead to new therapies for hepatic disease.

INTRODUCTION

Hepatic energy metabolism is a dynamic process that balances a myriad of incoming signals in order to maintain whole-body homeostasis. In this regard, the liver is unrivaled by any other organ. Sensing of extracellular nutrients and hormones by hepatocytes is transduced in real time into a network of regulatory mechanisms that ultimately decide which nutrients will be taken up in what amounts and what the metabolic fate of these carbon sources will be. Dysregulation of hepatic metabolism lies at the core of many diseases ranging from obesity and the metabolic syndrome to cancer. In the age of gene-targeting in mice and sophisticated molecular genetics, we have learned much about the molecular mechanisms governing metabolism in the liver, and this has produced new pharmacotherapies that have been successful in treating human disease. Yet there is still much more to be discovered. One area of research that has rapidly developed in recent years surrounds a family of seven NAD^+-dependent lysine deacylase enzymes known as the sirtuins, aptly named SIRT1–SIRT7 (Table 20.1). As a family, the sirtuins integrate nutrient sensing in the form of NAD^+, arguably one of the cell's most critical bioenergetic cofactors, with the epigenetic, transcriptional, translational, and posttranslational landscape of the liver. This is accomplished through histone deacetylation, transcription factor deacetylation, regulation of ribosomal genes, and lysine deacylation of a host of cytoplasmic and mitochondrial proteins including components of the cytoskeleton, glycolytic pathway, electron transport chain, and the tricarboxylic acid cycle, among others.

Since the first characterization of the human sirtuin cDNAs just 15 years ago [1], a dearth of knowledge has been published on the mammalian sirtuins. The best studied sirtuins are SIRT1 and SIRT3. Entering SIRT1 as a keyword in PubMed produces over 3100 abstracts that mention this term; for SIRT3 it is just over 400. There are numerous excellent recent reviews regarding the roles of SIRT1 and SIRT3 in the regulation of metabolism [2–6]. Less is known about the remaining five sirtuin family members, and reviews integrating the primary literature are lacking. SIRT2 and SIRT6 as keywords produce 338 and 202 results, respectively, that mention these terms. SIRT4, SIRT5, and SIRT7 are each mentioned less than 100 times in the literature. The emerging role of the understudied sirtuins SIRT2, SIRT4, SIRT5, SIRT6, and SIRT7 in regulating metabolism will be the subject of this chapter.

SIRT2

SIRT2 is best known as a cytosolic protein that interacts with microtubules and deacetylates tubulin [7]. However, SIRT2 may also transiently reside in the nucleus. During mitosis, SIRT2 shuttles from the cytoplasm to the nucleus where it serves as a histone

TABLE 20.1
Summary of Known Sirtuin Functions

Sirtuin	Location	Major Enzymatic Function	Role in Liver	References
SIRT1	Nucleus, cytoplasm	Deacetylase	Promotes FAO, gluconeogenesis, and mitochondrial biogenesis; protects against ER stress	[2,5]
SIRT2	Cytoplasm, nucleus	Deacetylase	Regulates glucose homeostasis and represses lipogenesis	[10,16,21,23]
SIRT3	Mitochondria	Deacetylase	Promotes FAO, ketogenesis, and mitochondrial respiration	[5,6]
SIRT4	Mitochondria	?	Suppresses FAO and glutamine utilization	[29–33]
SIRT5	Mitochondria, cytoplasm	Desuccinylase, demalonylase, deglutarylase	Suppresses pyruvate oxidation, promotes urea cycle and ketogenesis	[37,43,44]
SIRT6	Nucleus	Deacetylase	Suppresses ribosomal biogenesis, glycolysis, gluconeogenesis, lipogenesis, and cholesterol synthesis	[51–57]
SIRT7	Nucleus	Deacetylase	Suppresses ribosomal biogenesis, lipogenesis, and HIF1α signaling; promotes expression of mitochondrial proteins	[64–69]

deacetylase [8]. The recently uncovered links between SIRT2 and liver metabolism, function, and cancer may be partially related to its role as a histone deacetylase and regulator of cell cycle [9], but are also likely to include direct interaction and deacetylation of target proteins in the cytoplasm. An intriguing thread has emerged that implicates SIRT2 as a glucose-sensitive deacetylase that regulates several key players in cellular energy metabolism, which will be described in the following two sections.

SIRT2 AND GLUCOSE HOMEOSTASIS

An intriguing study by Ramakrishnan et al. [10] demonstrated that SIRT2 can interact with and influence the activity of Akt. Activation of the phosphatidylinositol 3-kinase (PI3K)–Akt signaling pathway is crucial to the stimulation of glucose uptake and cellular metabolism by insulin. The PI3K–Akt signaling cascade regulates many downstream pathways as reviewed by others [11]; reduced signaling through this cascade is associated with insulin resistance and metabolic syndrome, while overactivation is associated with many human cancers [12]. In the liver, a critical aspect of glucose signaling involves the transition from fasted to fed states. In the fasted state, the liver converts lactate, glycerol, amino acids, and other nonglucose carbon sources into glucose for export to glucose-dependent organs such as the brain. Upon refeeding, this pathway, known as gluconeogenesis, must be quickly shut down, which is carried out by Akt. Akt phosphorylates thereby inactivates forkhead box protein O1 (FOXO1), a critical transcriptional regulator of gluconeogenic genes. It has been suggested using double knockout mice lacking both Akt and FOXO1 that Akt's major role in the liver may be to counteract FOXO1 [13].

Ramakrishnan et al. [10] found that SIRT2 binds to Akt in insulin-responsive cells and increases its activity. SIRT2 inhibitors diminished Akt activity and the cellular response to insulin, while SIRT2 overexpression enhanced it. However, the authors could not detect any acetylation of Akt and hence any deacetylation by SIRT2, and thus it is not clear whether SIRT2 activation of Akt involves lysine deacetylation or a deacetylation-independent mechanism. Possibly, the most intriguing observation was that SIRT2: Akt binding was promoted by glucose deprivation and required AMP-activated protein kinase (AMPK) phosphorylation of SIRT2. SIRT2 appeared to be required for maximal Akt activation following insulin exposure, but at the same time insulin induced dissociation of the SIRT2–Akt complex.

The studies of Ramakrishnan et al. could lead to the conclusion that SIRT2 positively regulates Akt, which would in turn downregulate gluconeogenesis via nuclear exclusion and inactivation of FOXO1. However, other studies provide evidence that SIRT2 promotes gluconeogenesis rather than repressing it. First, Wang et al. [14] showed that in 3T3L1 cells, SIRT2 can bind to and deacetylate FOXO1 in the cytosol, which in turn leads to a reduction in Akt phosphorylation of FOXO1 and promotes its return to the nucleus. This mechanism parallels what has been shown for SIRT1 inside the nucleus, where SIRT1 deacetylation of FOXO1 appears to make it a better substrate for the phosphatase PP2A, thereby removing phosphorylation and keeping FOXO1 inside the nucleus where it can activate transcription of phosphoenolpyruvate carboxykinase (PEPCK) and other critical gluconeogenic genes [15]. Second, SIRT2 can deacetylate and stabilize the PEPCK protein directly [16]. Acetylation of PEPCK by the acetyltransferase p300 promotes its degradation by increasing binding of PEPCK to an E3 ubiquitin ligase known as UBR5. The action of p300 is positively regulated by glucose, such that when glucose is high, PEPCK acetylation levels are high, and its half-life is short, via interaction with UBR5 and degradation by the proteasome. When glucose levels fall, SIRT2 activity increases and PEPCK becomes deacetylated. Overexpression of SIRT2 increases the steady-state levels of the PEPCK protein, while SIRT2 knockdown results in reduced PEPCK. The importance of this mechanism in vivo was demonstrated by SIRT2 knockdown in mouse liver using adenoviral delivery of short hairpin RNAs. Hepatic knockdown of SIRT2 led to reduced PEPCK protein levels and a 35% drop in blood glucose values.

The studies of Ramakrishnan did not functionally dissect the effects of the SIRT2–Akt interaction. The primary measure was phosphorylation levels of Akt itself in response to supraphysiological levels of insulin in cultured cells. Others have noted that SIRT2 interactions with target proteins are promoted by low glucose (see following section), and thus a model in which nutrient deprivation activates SIRT2 via AMPK as part of a program to restore intracellular ATP levels makes sense. But the actions of AMPK and Akt are at odds with each other, and therefore activation of Akt by SIRT2 is counterintuitive. Rather, it may be that SIRT2 binds to Akt and sequesters it during nutrient deprivation, and this interaction dissipates with insulin administration as noted by Ramakrishnan et al. Sequestration of Akt by

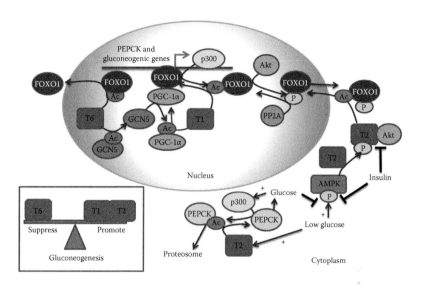

FIGURE 20.1 (See color insert.) *Sirtuin regulation of gluconeogenesis.* As shown in the inset, SIRT6 suppresses gluconeogenesis, while SIRT1 and SIRT2 promote it. At the heart of the mechanism is forkhead box protein O1 (FOXO1), which is a master regulator of gluconeogenic gene expression, in concert with PGC-1α. SIRT1 deacetylates both PGC-1α and FOXO1. Deacetylation activates PGC-1α and deacetylation of FOXO1 by SIRT1 promotes nuclear retention and activity. Conversely, when acetylation sites on FOXO1 that are recognized by SIRT6 are deacetylated, FOXO1 is excluded from the nucleus. SIRT6 also deacetylates and activates GCN5, an acetyltransferase that targets PGC-1α. Thus, SIRT6 and SIRT1 directly oppose each other. In the cytoplasm, SIRT2 is activated under nutrient deprivation by AMPK. SIRT2 promotes gluconeogenesis by binding to Akt, which is the kinase that inactivates FOXO1, and by deacetylating FOXO1 to promote its return to the nucleus. SIRT2 also directly deacetylates and stabilizes phosphoenolpyruvate carboxykinase (PEPCK), which catalyzes the rate-limiting step in gluconeogenesis. In a feedback loop, rising glucose levels inactivate AMPK while activating p300, which is the acetyltransferase that acetylates PEPCK, thereby promoting its degradation.

SIRT2 combined with deacetylation of cytosolic FOXO1 and PEPCK would serve to increase the activity of existing PEPCK protein while also promoting for increased production of PEPCK by FOXO1. This model is illustrated in Figure 20.1.

OTHER GLUCOSE-DEPENDENT INTERACTIONS OF SIRT2 WITH TARGET PROTEINS

Two other targets of SIRT2 relevant to liver function have been characterized, and in both the interactions are glucose sensitive. The first is keratin-8 (K8). The intermediate filament cytoskeleton in hepatocytes is unique among simple epithelia in that it is comprised solely of K8 and keratin-18 (K18), and no other keratins. K8 knockout mice are sensitive to liver injury and K8 gene variants have been linked to several forms of human liver disease [17]. A loss in the integrity of the intermediate filament cytoskeleton renders hepatocytes fragile and vulnerable to apoptosis. K8 also plays a key role in the formation of Mallory–Denk bodies that are large cytoplasmic protein aggregates linked to numerous liver diseases, most notably steatohepatitis (both alcoholic and nonalcoholic) and hepatocellular carcinoma [18]. Mallory–Denk bodies are comprised primarily of misfolded, ubiquitinated, and cross-linked forms of K8/K18. High-fat diet, alcohol, and other environmental stressors modulate the degree of posttranslational modifications to K8/K18 as well as their relative abundance; these modifications and an increase in the K8/K18 ratio promote the formation of Mallory–Denk bodies [19].

Mass spectrometry proteomic surveys identified seven lysine acetylation sites on K8 [20]. Snider et al. later showed that one of these, K207, is the primary acetylation site on human K8 [21]. Substitution of this residue with glutamine, often used to mimic lysine acetylation, led to alterations in filament organization and reduced solubility of K8. SIRT2 was found to coimmunoprecipitate with K8. As with Akt, the interaction of SIRT2 with K8 was strongest under low-glucose conditions and dissipated upon the addition of high-glucose media. Inhibition or knockdown of SIRT2 led to increased K8 acetylation, decreased K8 solubility, and the appearance of perinuclear aggregates of K8/K18. The implication of these studies is that SIRT2 plays a role in the mechanical integrity of hepatocytes and in regulating the reorganization of intermediate filaments, which is seen during mitosis and cell migration and in the cellular stress response.

Under fasting conditions and nutrient deprivation, lipid oxidation by mitochondria increases to supply ATP, while in the fed state, the liver funnels carbon derived from excess glucose into newly synthesized lipids and exports them as very-low-density lipoprotein (VLDL) particles. The building block of these lipids is acetyl-CoA, and the pathway by which acetyl-CoA is sequentially condensed to make fatty acids, is called lipogenesis. The acetyl-CoA needed for lipogenesis is derived from citrate exported from the mitochondrial TCA cycle and requires the action of a key metabolic enzyme known as ATP citrate lyase (ACLY). ACLY catalyzes the formation of acetyl-CoA from citrate at the expense of ATP. Increased activity of ACLY has previously been linked to cancer [22]. As with K8, ACLY was identified as an acetylated cytosolic protein by mass spectrometry proteomics surveys. More recently, Lin et al. elucidated the functional effects of ACLY acetylation and demonstrated ACLY to be a SIRT2 target protein [23]. Unlike most acetylated proteins, ACLY demonstrates a gain of function when acetylated. As with PEPCK, the mechanism was revealed to be a cross talk between lysine ubiquitylation and acetylation, but this time acetylation blocks ubiquitylation rather than promoting it. As with PEPCK, the acetylation becomes increased when glucose is abundant, but contrary to PEPCK, this causes increased stability and half-life of ACLY by preventing ubiquitylation at the same lysine residues. SIRT2, but not the other sirtuins, was coimmunoprecipitated with ACLY. Overexpression of SIRT2 led to a reduction of ACLY acetylation and a simultaneous increase in ubiquitylation. The knockdown of SIRT2 in mouse liver caused an increase in ACLY protein abundance presumably due to an inhibition of degradation. The interaction between ACLY and SIRT2 is strong under low-glucose conditions and dissipates under high glucose. While phosphorylation of SIRT2 was not investigated, it is tempting to speculate that AMPK is once again the link between glucose levels and SIRT2 function.

Further work is necessary to establish the role of AMPK in regulating SIRT2. For instance, both AMPK and SIRT2 demonstrate nucleocytoplasmic shuttling. In the nucleus, AMPK activation leads to suppression of the cell cycle, presumably shutting down proliferation in the face of nutrient deprivation [24]. SIRT2 has been shown to be a stress-regulated mitotic checkpoint protein [25]. While it is speculation at this point, perhaps SIRT2 is a downstream effector of AMPK in both the nucleus and the cytoplasm. Under nutrient replete conditions, SIRT2 would remain largely inactive. During nutrient deprivation, the activation of SIRT2 via AMPK would assist in repressing mitosis and maintaining genomic integrity while simultaneously promoting for the restoration of metabolic balance by its actions on the cytoplasmic targets described earlier. Another major question is how the sirtuins cross talk to regulate glucose homeostasis. As shown in Figure 20.1 and discussed in the section *SIRT6 and Glucose Metabolism*, SIRT6 appears to directly oppose the actions of SIRT1 and SIRT2 on the gluconeogenic pathway.

SIRT4

SIRT4 is the least understood of the three mitochondrial sirtuin enzymes. It was originally reported to be an ADP-ribosyltransferase enzyme that modulated insulin secretion by ADP-ribosylating glutamate dehydrogenase (GDH) [26,27]. ADP ribose is one of the products of the sirtuin deacylation reaction that cleaves NAD$^+$ to nicotinamide and ADP ribose. SIRT4 appeared to modify lysines with the ADP ribose cleavage product. However, further study of the ADP-ribosyltransferase activity of sirtuins has revealed that this is a side reaction that occurs at a very low rate [28]. Thus, SIRT4 remains an orphan sirtuin without a substrate, although there has been one report of SIRT4 deacetylating mitochondrial malonyl-CoA decarboxylase (MCD) [29]. SIRT4 knockout mice display some metabolic abnormalities. GDH activity appears to be reduced, albeit perhaps not due to direct ADP-ribosylation. As a result, SIRT4 KO mice have enhanced glutamine utilization by the TCA cycle, increased amino acid–stimulated insulin secretion, higher plasma insulin, and lower fasting blood glucose concentrations [29,30].

There is mounting evidence that SIRT4 regulates hepatic fatty acid metabolism (Figure 20.2). First, it has been suggested that the deactivation of MCD by SIRT4 through deacetylation could cause malonyl-CoA levels to rise and malonyl-CoA is a known inhibitor of fatty acid oxidation (FAO) [29]. Further, Nasrin et al. [31] observed that adenoviral knockdown of SIRT4 either in primary mouse hepatocytes or in intact animals caused an induction in the expression of FAO genes such as medium-chain acyl-CoA dehydrogenase and carnitine palmitoyltransferase 1 (CPT1), the gatekeeper of long-chain fatty acid entry into mitochondria. Flux through the FAO pathway was significantly increased. Knocking down SIRT4 also increased the mRNAs for peroxisome proliferator-activated receptors α and delta (PPARα and PPARδ), PGC-1α, and SIRT1. These transcription factors are well-known positive regulators of mitochondrial gene expression and may explain the increased expression of FAO genes and FAO flux. Moreover, the amount of phosphorylated AMPK becomes increased in the context of SIRT4 knockdown [32]. AMPK increases mitochondrial FAO by phosphorylating ACC, which in turn leads to an increased activity of CPT1. This can also contribute to the increased FAO observed in the absence of SIRT4.

Laurent et al. [33] further investigated the connection between SIRT4 and PPARα. An evaluation of hepatic gene expression changes by microarray in SIRT4

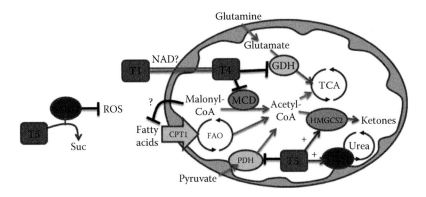

FIGURE 20.2 **(See color insert.)** *Known roles of SIRT4 and SIRT5.* SIRT4, which is strictly mitochondrial, represses glutamine utilization and malonyl-CoA decarboxylase (MCD). Repression of MCD causes malonyl-CoA levels to rise, which may possibly exit the mitochondria and block fatty acid oxidation (FAO) by binding to and inhibiting CPT1. SIRT4 may also cross talk with SIRT1 via NAD$^+$ levels, as evidenced by increased NAD and SIRT1 activation when SIRT4 is ablated. Ablation of SIRT4 causes FAO to increase by mechanisms that require SIRT1 and PPARα. SIRT5 is both mitochondrial and cytosolic. In the cytosol it reduces reactive oxygen species by the activation of superoxide dismutase 1. In mitochondria it desuccinylates key enzymes in the ketogenic pathway and the urea cycle and inactivates pyruvate dehydrogenase.

KO mice revealed a pattern of expression very similar to that of PPARα knockout mice. Again, as found by Nasrin et al., loss of SIRT4 resulted in enhanced FAO, and overexpression of SIRT4 decreased FAO. In vitro experiments with PPARα reporter plasmids indicated that the presence of SIRT4 interfered with PPARα transcriptional activation. This effect was dependent upon SIRT1, which is known to bind to PPAR response elements and influence transcription by deacetylating and activating the critical PPAR cofactor PGC-1α. NAD$^+$ levels were higher in SIRT4 KO cells. The authors speculate that the loss of SIRT4 causes a retrograde signaling to the nucleus via NAD$^+$, such that increased NAD$^+$ levels activate SIRT1 that in turn increases the transcription from PPAR response elements via positive regulation of PGC-1α.

Overall, the literature points to a suppressive effect of SIRT4 on cellular metabolism. However, particularly in regard to mitochondrial FAO, the suppressive effect is inferred from the fact that loss of SIRT4 leads to an increased FAO. The molecular mechanisms appear to be very complex and involve second messenger effects. The actual reaction catalyzed by SIRT4 and the identity of its target proteins remain obscure.

SIRT5

Human SIRT5 localizes primarily to mitochondria but has also been detected in the cytosol [34]. SIRT5 was originally presumed to be a lysine deacetylase and was reported to regulate the urea cycle via deacetylation of carbamoyl phosphate synthetase-1 (CPS1) [35]. More recently, it has been shown that the deacetylase activity of SIRT5 is weak and that its preferred substrates are malonyllysine, glutaryllysine, and succinyllysine [36,37]. The origin of these lysine modifications is unclear but may involve nonenzymatic acylation due to deprotonated lysine residues reacting chemically with malonyl-CoA, glutaryl-CoA, and succinyl-CoA, respectively [38,39]. In keeping with this, glutaryl-CoA dehydrogenase (GCDH) knockout mice, which have high levels of intramitochondrial glutaryl-CoA, demonstrate hyperglutamylation of mitochondrial proteins.

SIRT5 IN THE CYTOSOL

To date, the only cytoplasmic target protein that has been definitively identified for SIRT5 is superoxide dismutase 1 (SOD1) [40]. SOD1 is an abundant protein in the cytosol and also in the intermembrane space of the mitochondria. It plays the important role of removing superoxide by conversion to oxygen and hydrogen peroxide, with the latter subject to subsequent removal by catalase. Lin et al. [40] delineated a mechanism in which succinylation at lysine K123 reduces SOD1 activity, which can be restored by SIRT5. Functionally, overexpression of SIRT5 led to a reduction of reactive oxygen species through increased activation of SIRT5. This mechanism certainly may be important for containing cellular reactive oxygen species, but there may also be a larger context. SOD1 was recently shown to play a second role as a nutrient sensor that integrates the levels of superoxide, oxygen, and glucose in order to regulate the rate of mitochondrial respiration [41]. SIRT5 may impinge upon this regulatory circuit through its actions on SOD1. Through its dual localization in cytosol and mitochondria, SIRT5 may ultimately be established as a determinant of aerobic versus anaerobic glucose metabolism. Its overexpression in HEK293 cells leads to both increased glycolytic rate and increased mitochondrial glucose oxidation [34]. The increase in glycolytic rate disappears when SIRT5 overexpressing cells are switched to low-glucose media. Another study that suggests a potential role for SIRT5 regulating glycolysis is that of Peng et al. [42]. HeLa cells were grown in stable isotope-labeled malonate and mass spectrometry was used to identify malonylated proteins. The list was comprised almost entirely of cytosolic proteins and included the glycolytic enzymes phosphoglycerate kinase-1, glyceraldehyde phosphate dehydrogenase, α enolase, and glucose-6-phosphate isomerase. As SIRT5 is the only known lysine demalonylase, it may be logical to assume that these enzymes may turn out to be SIRT5 targets.

SIRT5 IN THE MITOCHONDRIA

Recent large-scale quantitative proteomics studies comparing mitochondrial succinylation levels between wild-type and SIRT5 knockout mice have generated long lists of potential SIRT5 target proteins awaiting validation and further study [43,44]. Park et al. presented evidence that the pyruvate dehydrogenase (PDH) and succinate dehydrogenase (SDH) complexes are both SIRT5 targets [44]. They reported that lysine succinylation increased the function of both complexes and that SIRT5 suppresses mitochondrial energy metabolism via desuccinylation of PDH and SDH. However, the lysine residues involved in these regulatory mechanisms and the molecular details were lacking. Rardin et al. [43] studied succinylation of the enzyme 3-hydroxy-3-methylglutaryl-CoA synthase 2 (HMGCS2) as a regulator of ketogenesis in the SIRT5 knockout mouse. HMGCS2 catalyzes the rate-limiting step in hepatic ketogenesis, the process by which ketone bodies are synthesized from acetyl-CoA during starvation. Ketone bodies cannot be utilized by the liver itself but are an important fuel source for heart, muscle, and brain. SIRT5 knockout mice were demonstrated to have modestly reduced FAO, which generates the acetyl-CoA required for ketogenesis during starvation and also reduced rates of ketone body formation. The latter was attributable to hypersuccinylation and loss of activity of HMGCS2. The mechanism was further localized to two lysine residues in the substrate-binding pocket, which, when converted to glutamic acid in site-directed mutagenesis experiments, led to a complete loss of HMGCS2 catalytic activity. It was concluded that changing the positively charged lysines in the binding pocket to negative charges, as occurs with succinylation, is sufficient to reduce flux through the ketogenic pathway.

One of the first roles described for SIRT5, prior to its preferred substrates being discovered, was as a regulator of the urea cycle through lysine deacetylation of CPS1 [35]. CPS1 catalyses the ATP-dependent transfer of ammonia from glutamine to bicarbonate. CPS1 is known to be heavily acetylated and succinylated. The importance of SIRT5 as a CPS1 deacetylase could be called into question now that SIRT5 has been shown to have very low deacetylase activity, and it is not yet understood what role succinylation plays in regulating CPS1 or the urea cycle. However, Tan et al. have investigated the effects of lysine glutarylation on CPS1 in the context of the GCDH knockout mouse [37]. GCDH is part of the pathway of lysine and tryptophan degradation, and its deficiency in humans causes a life-threatening metabolic disorder known as glutaric academia. Glutaryl-CoA accumulates in GCDH-deficient mitochondria and leads to hyperglutarylation of mitochondrial proteins, most likely through nonenzymatic reaction with deprotonated lysines. Glutarylation reduces CPS1 enzymatic activity. SIRT5 deglutarylates and restores CPS1 function. The physiological relevance of the mechanism is shown by experiments in which the knockdown of SIRT5 in cultured cells leads to elevated urea levels in the media.

SIRT6

The past 5 years have seen an explosion in knowledge regarding SIRT6, during which time it went from a virtually unexplored member of the sirtuin family to being recognized as an important regulator of metabolism, aging, and cancer [45]. SIRT6 is a nuclear sirtuin whose diverse functions have multiple implications for health and disease in the liver, and these functions are detailed here.

ACTIVATION BY FATTY ACIDS

Early investigations into SIRT6 indicated that it had no lysine deacetylase activity. However, more recent work demonstrates that SIRT6 can deacetylate two histone marks (H3K9, H3K56) as well as a DNA repair enzyme known as C-terminal binding protein interacting protein (CtIP) [46]. The physiological relevance of the deacetylation of histones was called into question, however, by the work of the Denu Laboratory at the University of Wisconsin that performed kinetic analysis on SIRT6 with H3K9 as a substrate and found that the catalytic efficiency is approximately 1000-fold less than that of a well-studied yeast H3K9 deacetylase known as Hst2 [47]. Two groups subsequently published data showing that SIRT6 preferentially deacetylates lysines modified with longer acyl-chains (8–14 carbons) [48,49]. But even more remarkable is the observation that long-chain free fatty acids, particularly myristic acid (C14), can stimulate the H3K9 deacetylase activity of SIRT6 by as much as 35-fold by inducing conformational changes in SIRT6 that greatly lower the K_m for H3K9 substrate [48]. While still ~28-fold lower than the yeast deacetylase Hst2, this at least places SIRT6's deacetylase activity within the physiologically important realm.

SIRT6 AND GLUCOSE METABOLISM

If the phenotype of whole-body null mice is used as the benchmark of physiological importance, then SIRT6 ranks second only behind SIRT1. Most SIRT1 whole-body null mice die within 1 week of birth; in contrast, SIRT6 mice perish around 4 weeks after birth [50]. While SIRT1 null mice show multiple developmental defects, SIRT6 null mice die primarily of metabolic derangements. They appear normal at birth albeit smaller than wild-type littermates. Within 2 weeks, they begin to exhibit neonatal wasting, characterized by severe hypoglycemia and lymphocytic apoptosis that essentially eliminates the lymphocyte population. At the cellular level, the phenotype is primarily one of genomic instability and DNA damage, as observed in multiple cell types.

A second group studied the hypoglycemic phenotype of SIRT6 null mice in greater detail using an independently created knockout strain [51]. When maintained on a mixed strain background, lethality at 4 weeks was 60% compared to 100% in the original study of Mostoslavsky et al. [50] whose SIRT6 deficient mice were on a pure strain background. This suggests that modifier genes play a role. Lethality dropped from 60% to 17% when mice were supplemented with glucose in the drinking water, illustrating that it is indeed glucose deprivation that ultimately leads to early death. The authors further showed that SIRT6 knockout mice have ~40% less circulating insulin, yet maintain modestly higher levels of glucose uptake in several organs, most notably the spleen. In the liver, the increase in glucose uptake was negligible. Upon injection with insulin, glucose uptake is substantially higher in SIRT6 knockout mice in multiple organs, including liver, but again most notably the spleen where it is nearly 10-fold higher than in wild-type controls. The reason is enhanced insulin signaling upstream of AKT and increased abundance of glucose transporter-1 (GLUT1). SIRT6 normally functions to put the brakes on insulin signaling, and in its absence, insulin signaling becomes dysregulated.

The overarching effect of SIRT6 on glucose metabolism appears to be repressed. SIRT6 represses both the formation of glucose via gluconeogenesis in the liver and the utilization of glucose by peripheral organs. Repression of gluconeogenesis is achieved by the inhibition of FOXO1 [52,53], while repression of glucose uptake and glycolysis is achieved by the inhibition of hypoxia-inducible factor 1α (HIF1α) [54]. Zhang et al. [52] demonstrated that the tumor suppressor p53 activates transcription of the SIRT6 gene and that the resulting increase in SIRT6 attenuates gluconeogenesis by deacetylating FOXO1, thereby causing its exclusion from the nucleus. Note that this mechanism is in direct opposition to those mediated by SIRT1 and SIRT2, in which deacetylation of FOXO1 promotes nuclear localization and increased activation of gluconeogenic genes (Figure 20.1) [15]. It is not clear why deacetylation promotes exclusion in one case and retention in the other, but it may involve cross talk with phosphorylation by Akt that is required for efficient export of FOXO1 from the nucleus. SIRT6 targets a different set of lysines on FOXO1 and then SIRT1 that could differentially influence the subsequent interaction of FOXO1 with Akt and the phosphatase PP2A. In addition to directly inactivating FOXO1, SIRT6 also indirectly inactivates the transcription of gluconeogenic genes by deacetylating general control nonrepressed protein 5 (GCN5), a nuclear lysine acetyltransferase (Figure 20.1) [53]. The deacetylation of GCN5 increases its activity resulting in increased acetylation of the critical transcriptional coactivator PGC-1α. PGC-1α coactivates the transcription of gluconeogenic genes with FOXO1, and when acetylated, this activity is reduced.

The interaction of SIRT6 with HIF1α was detailed by Zhong et al. [54]. The major role for HIF1α is to increase glycolysis and suppress mitochondrial respiration in response to low oxygen and low glucose. Overexpression of SIRT6 reduces the transcriptional activation of a reporter gene bearing numerous hypoxia-responsive elements. HIF1α and SIRT6 can be coimmunoprecipitated from both cultured cells and mouse muscle tissue. In the absence of SIRT6, the amount and stability of HIF1α increases. In SIRT6 knockout cells, inhibiting HIF1α attenuates the accelerated glucose uptake, suggesting that the defects in glucose metabolism in SIRT6 knockout mice may be largely HIF1α dependent. Indeed, treating SIRT6 whole-body knockout mice with a small molecule HIF1α inhibitor led to a doubling in blood glucose within 3 h.

SIRT6 and Hepatic Lipid Metabolism

Liver-specific ablation of SIRT6 does not lead to the severe hypoglycemia or early death that were seen in the whole-body knockout, attesting to the fact that the hypoglycemia is likely caused by increased glucose uptake by the periphery and not the liver. However, liver-specific SIRT6 knockout mice do develop a metabolic phenotype, which is fatty liver [55]. Hepatocytes of SIRT6 liver-specific knockout mice incorporate fatty acids into triglycerides at a higher rate than wild-type hepatocytes. SIRT6 appears to regulate the expression of many metabolic genes by binding to their promoters and deacetylating the H3K9 epigenetic mark, resulting in repression. In the absence of SIRT6, liver expression of glycolytic and lipogenic genes becomes enhanced, while the expression of the rate-limiting mitochondrial FAO gene CPT1 is decreased. The result is a 50% drop in the rate of mitochondrial FAO and enhanced uptake of both glucose and fatty acids, which are preferentially shunted into stored lipid.

The overexpression of SIRT6 in mice reduces LDL cholesterol. Conversely, the knockout of SIRT6 in the liver leads to an increased serum cholesterol. On the molecular level, this is mediated by forkhead box protein O3 (FOXO3), which recruits SIRT6 to the promoter of the master regulator of cholesterol synthesis, sterol regulatory binding protein-2 (SREBP2) [56]. SIRT6 deacetylates histones at the SREBP2 promoter region, causing a repressive chromatin state and reduced expression of SREBP2, with downstream reductions in the expression of the SREBP2 target genes in the cholesterol synthesis pathway. Similarly, FOXO3 draws SIRT6 to the promoter of a gene named proprotein convertase subtilisin/kexin type 9 (PCSK9) and represses its expression [57]. PCSK9 binds to and promotes recycling of the LDL receptor, and high-level expression of this gene results in increased circulating cholesterol due to a paucity of the LDL receptor. By its repression, SIRT6 promotes the uptake of LDL. This *targeted silencing* theme, whereby FOXO3 recruits SIRT6 to the promoters of genes in order to epigenetically silence them, may occur with other transcription factors as well. SIRT6 has been shown to interact with Myc on the promoter of ribosomal biosynthesis genes, silencing them and thereby counteracting the proliferative activity of Myc [58]. Since Myc regulates lipogenesis and anabolic processes in general, counteracting Myc may be another mechanism by which SIRT6 suppresses hepatic lipid metabolism.

SIRT6 AND LIVER INFLAMMATION

While adding glucose to the drinking water prolongs the lifespan of SIRT6 knockout mice, most still die within the first year of life, indicating that other factors contribute to premature death. Xiao and colleagues discovered severe liver inflammation and hepatic fibrosis in SIRT6 knockout mice [59]. This inflammation became prominent around 7–8 months of age. It was not present in liver-specific SIRT6 knockout mice but was recapitulated in mice where SIRT6 was knocked out only in cells of myeloid lineage or specifically in T-cells. The conclusion is that SIRT6 plays an important role in immune regulation. Immune cell activation is known to involve changes in cellular metabolism, and the spleen from SIRT6 knockout mice takes up a tremendous amount of glucose [51]. These observations may all be linked. The chronic inflammation in the liver may also be related to the reports of liver tumors in SIRT6 knockout mice.

SIRT7

A keyword search of the PubMed database as of September 2014 reveals a mere 81 publications that contain the term "SIRT7," and only a fraction of these describe studies focused on elucidating SIRT7 function. As such, it competes with SIRT4 for the title of the least understood sirtuin family member. SIRT7 is present in the nucleus and may also localize to the cytoplasm [60]. It has a high degree of overlap with SIRT6 in terms of proteins it interacts with and with some of the pathways it regulates (Figure 20.3) [61]. It appears to have activity as a deacetylase, although some functions of SIRT7 appear to be independent of this activity. Mice show the highest expression of SIRT7 in spleen, liver, and testis and the lowest in the muscle, heart, and brain [62]. Despite low expression in heart, it is inflammatory cardiomyopathy that causes a 59% shortening of lifespan in the SIRT7 knockout mouse [63]. SIRT7 knockout mice also display kyphosis, or exaggerated curvature of the spine, and subcutaneous fat wasting. The molecular mechanisms behind these phenotypes have not been determined.

SIRT7 AND HEPATIC LIPID METABOLISM

Three recent investigations have linked SIRT7 to hepatic steatosis. All three studies employed SIRT7 knockout mice. Two demonstrated hepatic steatosis in the absence

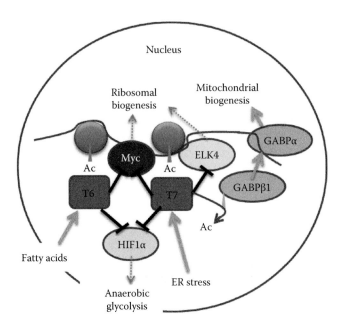

FIGURE 20.3 *Overlapping functions of SIRT6 and SIRT7 in the nucleus.* SIRT6 resides in the nucleus where it deacetylates histones in a fatty-acid stimulated manner. It interacts with Myc, deacetylating histones at Myc-regulated promoters involved in ribosomal biogenesis, thereby repressing these genes and counteracting Myc. SIRT6 also represses HIF1α. SIRT7, also a nuclear-localized histone deacetylase, is activated by ER stress. It too counteracts Myc, and also ELK4, thereby limiting ribosomal biogenesis. SIRT7 deacetylates GA binding protein β1, which increases its interaction with its partner GABPα and stimulates transcription of mitochondrial genes. Finally, SIRT7 interacts with and represses HIF1α.

of SIRT7, while the third study showed the opposite effect, that is, protection against steatosis. The reasons for the completely opposite findings are not yet clear.

The first published study on SIRT7 and steatosis was by Shin et al. [64], who showed that SIRT7 knockout mice have increased fat deposition in the liver as visualized by oil red O staining. The increased lipid was confirmed to be triglycerides. Several inflammatory markers were increased in the liver. The SIRT7 knockout mice were leaner than control, corroborating the original observation of Vakhrusheva et al. [63] regarding subcutaneous fat reduction. Serum triglycerides were low and it was shown to be due to impaired VLDL secretion. Thus, it is possible that the reduction in fat mass was due to an inability to properly traffic lipids from the liver to adipose tissue for storage. Shin et al. proceeded to work out a potential molecular mechanism for the fatty liver phenotype. They showed that SIRT7 plays a role in alleviating endoplasmic reticulum (ER) stress and that in its absence ER stress increases, with the subsequent downstream alterations in hepatic lipid metabolism eventually resulting in the steatosis. Chemicals that induce ER stress increased the T7 expression; overexpression of T7 reduced the expression of genes known to be associated with the cellular response to ER stress.

One defense mechanism against ER stress is for the stressed cell to reduce protein synthesis, which helps prevent the unfolded protein response. The oncoprotein

Myc is a master regulator of ribosome biogenesis; it turns out that during the early response to ER stress, SIRT7 binds to Myc at the promoters of genes required for ribosome biogenesis and transcriptionally silences those genes. SIRT7 deacetylates histone 3 at lysine 18 (H318K). This epigenetic mark has been associated with cancer. SIRT7 does not have its own preferred DNA binding sequence. Rather, in the context of ER stress, SIRT7 is drawn to promoters where Myc is actively promoting the transcription and counteracts Myc via H318K deacetylation. Reintroduction of SIRT7 into the livers of SIRT7 knockout mice using an adenovirus reduced the ER stress, suppressed de novo lipogenesis, and corrected the fatty liver phenotype. Knocking down Myc in SIRT7 knockout mouse liver resulted in the same effect, indicating that it is indeed Myc that drives the liver phenotype in the absence of SIRT7. These studies demonstrate SIRT7 to be an attractive therapeutic target for the treatment of fatty liver disease.

A few months after the aforementioned studies were published, Yoshizawa and colleagues put forth a paper demonstrating the complete opposite effect of SIRT7 on hepatic steatosis [65]. Yoshizawa et al. used male mice of the SIRT7 knockout line in which Vakhrusheva et al. had originally described the inflammatory cardiomyopathy phenotype, which is on a C57BL/6J strain background. In contrast, Shin et al. had used an independently created SIRT7 knockout on a mixed C57Bl/6 and 129Sv background and the gender of the mice was not specified. Perhaps these differences are sufficient to produce opposite phenotypes. Yoshizawa et al. fed their SIRT7 knockout strain a high-fat diet and documented resistance to fatty liver and obesity and increased insulin sensitivity. Mice fed normal chow had normal livers. In the high-fat fed SIRT7 knockout mice, serum leptin was reduced. Resting energy expenditure data indicated an increased reliance on FAO. Whole-body temperature was increased, which was attributed to an increase in brown adipose tissue UCP1. However, no further analysis of brown adipose tissue was made so it is not clear whether the effect on UCP1 is direct or a secondary effect due to alterations in circulating leptin, thyroid hormone, or other serum factors known to impinge upon the nonshivering thermogenesis pathway. With regard to the liver phenotype, the authors suggested a mechanism centered on a nuclear receptor known as testicular receptor 4 (TR4). TR4 knockout mice had previously been shown to be resistant to the effects of high-fat feeding in terms of weight gain, hepatic steatosis, and insulin resistance. SIRT7 knockout mice had reduced abundance of TR4, and replacing TR4 abolished the beneficial phenotype. It was further shown that SIRT7 interacts with, although does not necessarily deacetylate, an E3 ubiquitin ligase complex consisting of DCAF1/DDB1/CUL4b that ubiquitinates and ultimately signals the degradation of TR4.

The third and most recent study of hepatic steatosis in SIRT7 knockout mice comes from Ryu and colleagues [66]. Like the Shin et al. study, they described enhanced steatosis in SIRT7 knockout mice but reported elevated serum lipids in contrast to the reduced triglycerides seen by Shin et al. Shin et al. had focused on the regulation of ribosomal biogenesis via Myc; Ryu et al. extended the role for SIRT7 to include mitochondrial ribosomes. They demonstrated a correlation between the abundance of SIRT7 and mRNA levels for several nuclear-encoded mitochondrial ribosomal binding proteins across multiple strains of mice. This correlation was confirmed in a set of human liver biopsies. The influence of SIRT7 overexpression of these genes was traced back to the transcription factor GA binding protein-β-1 (GABPβ1), also dubbed nuclear respiratory factor 2 for its involvement in regulating the expression of subunits of the mitochondrial respiratory chain. SIRT7 deacetylates GABPβ1 at three key lysines, and deacetylated GABPβ1 shows increased heterodimerization with its obligate partner GABPα. The result is increased transcriptional activation of target genes.

SIRT7 AND HYPOXIA

The hypoxia-inducible factors 1 and 2 (HIF1, HIF2) are largely responsible for transducing the signal of low oxygen tension into compensatory changes in gene expression. Recently, it was shown that SIRT7 physically interacts with HIF1 and HIF2 [67]. SIRT7 overexpression led to decreased HIF1 and HIF2 protein levels, decreased HIF transcriptional activity, and reduced expression of HIF target genes in multiple cell lines including Hep3B, a human hepatoma cell line. Knocking down SIRT7 had the opposite effect. The deacetylase activity of SIRT7 was not required for the effect on the HIF proteins since overexpressing a catalytically inactive SIRT7 mutant produced the same results as overexpressing the wild-type protein. The authors demonstrated that SIRT7 is a negative regulator of the hypoxia response pathway but did not elucidate the mechanism. However, they did rule out changes in ubiquitination or increased targeting of the HIF proteins for degradation, because the proteasome was not required for the observed effects.

SIRT7 AS A TUMOR PROMOTER

Ribosomal biogenesis is critical for cellular growth and proliferation. Shin et al. [64] showed that SIRT7 can counteract the transcriptional stimulation of ribosomal genes by Myc. Similarly, SIRT7 physically binds with the ELK4 and is thereby targeted to ELK4-responsive promoters where it can deacetylate H3K18 and suppress ELK4 targets, which include numerous ribosomal genes and genes involved in RNA processing and translation [68]. In studies by Tsai et al. [69], knocking down SIRT7 led to reduced cellular protein synthesis. Based on its ability to silence genes required for growth and proliferation, combined with its inhibitory effects on the HIF proteins, it would be tempting to conclude that SIRT7 is a tumor suppressor. However, recent studies indicate that it is not that simple. SIRT7 knockdown in HT1080 and U2OS cancer cells appears to greatly reduce two important hallmarks of transformed cells—anchorage-independent cellular growth and proliferation in low serum [68]. In subcutaneous xenografts of U251 cancer cells in mice, SIRT7 knockdown severely limited tumor formation. Kim et al. [70] showed that SIRT7 expression is increased in human hepatocellular carcinoma. Knockdown of SIRT7 reduced proliferation rates in three different liver cancer cell lines. Thus, there is a paradox between evidence showing that SIRT7 counters ER stress, hypoxia signaling, and oncoprotein-mediated activation of ribosomal biogenesis and evidence showing that SIRT7 is itself a tumor promoter. The explanation for the paradox may lie in imbalances in the cross talk between SIRT7-associated processes. The recruitment of SIRT7 to Myc, ELK4, and perhaps other promoter subsets will depend upon the relative amounts of each of these factors, and the end result—H3K18 deacetylation and repression—is also ultimately dependent upon the availability of NAD^+. For example, deregulation of Myc in the context of cancer could result in greater amounts of SIRT7 being drawn to Myc-regulated promoters, and therefore less SIRT7 would be available to target ELK4 regulated promoters, to bind with regulated promoters or to repress HIF proteins. Clearly, more research is needed in this area to discern the role of SIRT7 in cancer.

FUTURE DIRECTIONS

Much progress has been made in the past 15 years with regard to sirtuin functions in metabolic pathways, particularly for SIRT1 and SIRT3. Recent evidence suggests that the other sirtuins also play important roles in regulating cellular metabolism. Liver expresses all seven sirtuins, and one of the major challenges that remains is determining how the sirtuins integrate a common metabolic signal—NAD^+—into a coordinated regulation of numerous metabolic pathways across multiple cellular compartments. SIRT1, SIRT6, and SIRT7 have broad effects on metabolism by deacetylating histones and transcription factors. SIRT2, SIRT3, SIRT4, and SIRT5 have targeted effects by deacylating specific enzymes. Additionally, SIRT4 may be involved in signaling between the mitochondria and other cellular compartments. Investigating such *retrograde signaling* and cross talk between the mitochondrial, cytoplasmic, and nuclear sirtuins is important for understanding the pathogenesis of liver disease as well as for the development of new therapies.

REFERENCES

1. Frye, R. A. (1999) Characterization of five human cDNAs with homology to the yeast SIR2 gene: Sir2-like proteins (sirtuins) metabolize NAD and may have protein ADP-ribosyltransferase activity. *Biochemical and Biophysical Research Communications* 260, 273–279.
2. Kemper, J. K., Choi, S. E., and Kim, D. H. (2013) Sirtuin 1 deacetylase: A key regulator of hepatic lipid metabolism. *Vitamins and Hormones* 91, 385–404.
3. Li, X. (2013) SIRT1 and energy metabolism. *Acta Biochimica et Biophysica Sinica* 45, 51–60.
4. Brenmoehl, J. and Hoeflich, A. (2013) Dual control of mitochondrial biogenesis by sirtuin 1 and sirtuin 3. *Mitochondrion* 13, 755–761.
5. Nogueiras, R., Habegger, K. M., Chaudhary, N., Finan, B., Banks, A. S., Dietrich, M. O., Horvath, T. L., Sinclair, D. A., Pfluger, P. T., and Tschop, M. H. (2012) Sirtuin 1 and sirtuin 3: Physiological modulators of metabolism. *Physiological Reviews* 92, 1479–1514.
6. Green, M. F. and Hirschey, M. D. (2013) SIRT3 weighs heavily in the metabolic balance: A new role for SIRT3 in metabolic syndrome. *The Journals of Gerontology: Series A, Biological Sciences and Medical Sciences* 68, 105–107.
7. North, B. J., Marshall, B. L., Borra, M. T., Denu, J. M., and Verdin, E. (2003) The human Sir2 ortholog, SIRT2, is an NAD^+-dependent tubulin deacetylase. *Molecular Cell* 11, 437–444.
8. Eskandarian, H. A., Impens, F., Nahori, M. A., Soubigou, G., Coppee, J. Y., Cossart, P., and Hamon, M. A. (2013) A role for SIRT2-dependent histone H3K18 deacetylation in bacterial infection. *Science* 341, 1238858.

9. Inoue, T., Hiratsuka, M., Osaki, M., and Oshimura, M. (2007) The molecular biology of mammalian SIRT proteins: SIRT2 in cell cycle regulation. *Cell Cycle* 6, 1011–1018.

10. Ramakrishnan, G., Davaakhuu, G., Kaplun, L., Chung, W. C., Rana, A., Atfi, A., Miele, L., and Tzivion, G. (2014) Sirt2 deacetylase is a novel AKT binding partner critical for AKT activation by insulin. *The Journal of Biological Chemistry* 289, 6054–6066.

11. Schultze, S. M., Jensen, J., Hemmings, B. A., Tschopp, O., and Niessen, M. (2011) Promiscuous affairs of PKB/AKT isoforms in metabolism. *Archives of Physiology and Biochemistry* 117, 70–77.

12. Cohen, M. M., Jr. (2013) The AKT genes and their roles in various disorders. *American Journal of Medical Genetics. Part A* 161A, 2931–2937.

13. Lu, M., Wan, M., Leavens, K. F., Chu, Q., Monks, B. R., Fernandez, S., Ahima, R. S., Ueki, K., Kahn, C. R., and Birnbaum, M. J. (2012) Insulin regulates liver metabolism in vivo in the absence of hepatic Akt and Foxo1. *Nature Medicine* 18, 388–395.

14. Wang, F. and Tong, Q. (2009) SIRT2 suppresses adipocyte differentiation by deacety-lating FOXO1 and enhancing FOXO1's repressive interaction with PPARγ. *Molecular Biology of the Cell* 20, 801–808.

15. Frescas, D., Valenti, L., and Accili, D. (2005) Nuclear trapping of the forkhead tran-scription factor FoxO1 via Sirt-dependent deacetylation promotes expression of gluco-genetic genes. *The Journal of Biological Chemistry* 280, 20589–20595.

16. Jiang, W., Wang, S., Xiao, M., Lin, Y., Zhou, L., Lei, Q., Xiong, Y., Guan, K. L., and Zhao, S. (2011) Acetylation regulates gluconeogenesis by promoting PEPCK1 degrada-tion via recruiting the UBR5 ubiquitin ligase. *Molecular Cell* 43, 33–44.

17. Strnad, P., Paschke, S., Jang, K. H., and Ku, N. O. (2012) Keratins: Markers and modu-lators of liver disease. *Current Opinion in Gastroenterology* 28, 209–216.

18. Strnad, P., Zatloukal, K., Stumptner, C., Kulaksiz, H., and Denk, H. (2008) Mallory-Denk-bodies: Lessons from keratin-containing hepatic inclusion bodies. *Biochimica et Biophysica Acta* 1782, 764–774.

19. Kucukoglu, O., Guldiken, N., Chen, Y., Usachov, V., El-Heliebi, A., Haybaeck, J., Denk, H., Trautwein, C., and Strnad, P. (2014) High-fat diet triggers Mallory-Denk body forma-tion through misfolding and crosslinking of excess keratin 8. *Hepatology* 60, 169–178.

20. Leech, S. H., Evans, C. A., Shaw, L., Wong, C. H., Connolly, J., Griffiths, J. R., Whetton, A. D., and Corfe, B. M. (2008) Proteomic analyses of intermediate filaments reveals cytokeratin8 is highly acetylated--implications for colorectal epithelial homeostasis. *Proteomics* 8, 279–288.

21. Snider, N. T., Leonard, J. M., Kwan, R., Griggs, N. W., Rui, L., and Omary, M. B. (2013) Glucose and SIRT2 reciprocally mediate the regulation of keratin 8 by lysine acetyla-tion. *The Journal of Cell Biology* 200, 241–247.

22. Zaidi, N., Swinnen, J. V., and Smans, K. (2012) ATP-citrate lyase: A key player in can-cer metabolism. *Cancer Research* 72, 3709–3714.

23. Lin, R., Tao, R., Gao, X., Li, T., Zhou, X., Guan, K. L., Xiong, Y., and Lei, Q. Y. (2013) Acetylation stabilizes ATP-citrate lyase to promote lipid biosynthesis and tumor growth. *Molecular Cell* 51, 506–518.

24. Koh, H. and Chung, J. (2007) AMPK links energy status to cell structure and mitosis. *Biochemical and Biophysical Research Communications* 362, 789–792.

25. Inoue, T., Hiratsuka, M., Osaki, M., Yamada, H., Kishimoto, I., Yamaguchi, S., Nakano, S., Katoh, M., Ito, H., and Oshimura, M. (2007) SIRT2, a tubulin deacetylase, acts to block the entry to chromosome condensation in response to mitotic stress. *Oncogene* 26, 945–957.

26. Ahuja, N., Schwer, B., Carobbio, S., Waltregny, D., North, B. J., Castronovo, V., Maechler, P., and Verdin, E. (2007) Regulation of insulin secretion by SIRT4, a mitochondrial ADP-ribosyltransferase. *The Journal of Biological Chemistry* 282, 33583–33592.

27. Haigis, M. C., Mostoslavsky, R., Haigis, K. M., Fahie, K., Christodoulou, D. C., Murphy, A. J., Valenzuela, D. M. et al. (2006) SIRT4 inhibits glutamate dehydrogenase and opposes the effects of calorie restriction in pancreatic β cells. *Cell* 126, 941–954.
28. Du, J., Jiang, H., and Lin, H. (2009) Investigating the ADP-ribosyltransferase activity of sirtuins with NAD analogues and 32P-NAD. *Biochemistry* 48, 2878–2890.
29. Laurent, G., German, N. J., Saha, A. K., de Boer, V. C., Davies, M., Koves, T. R., Dephoure, N. et al. (2013) SIRT4 coordinates the balance between lipid synthesis and catabolism by repressing malonyl CoA decarboxylase. *Molecular Cell* 50, 686–698.
30. Jeong, S. M., Xiao, C., Finley, L. W., Lahusen, T., Souza, A. L., Pierce, K., Li, Y. H. et al. (2013) SIRT4 has tumor-suppressive activity and regulates the cellular metabolic response to DNA damage by inhibiting mitochondrial glutamine metabolism. *Cancer Cell* 23, 450–463.
31. Nasrin, N., Wu, X., Fortier, E., Feng, Y., Bare, O. C., Chen, S., Ren, X., Wu, Z., Streeper, R. S., and Bordone, L. (2010) SIRT4 regulates fatty acid oxidation and mitochondrial gene expression in liver and muscle cells. *The Journal of Biological Chemistry* 285, 31995–32002.
32. Ho, L., Titus, A. S., Banerjee, K. K., George, S., Lin, W., Deota, S., Saha, A. K. et al. (2013) SIRT4 regulates ATP homeostasis and mediates a retrograde signaling via AMPK. *Aging* 5, 835–849.
33. Laurent, G., de Boer, V. C., Finley, L. W., Sweeney, M., Lu, H., Schug, T. T., Cen, Y. et al. (2013) SIRT4 represses peroxisome proliferator-activated receptor α activity to suppress hepatic fat oxidation. *Molecular and Cellular Biology* 33, 4552–4561.
34. Barbi de Moura, M., Uppala, R., Zhang, Y., Van Houten, B., and Goetzman, E. S. (2014) Overexpression of mitochondrial sirtuins alters glycolysis and mitochondrial function in HEK293 cells. *PLoS One* 9, e106028.
35. Nakagawa, T., Lomb, D. J., Haigis, M. C., and Guarente, L. (2009) SIRT5 Deacetylates carbamoyl phosphate synthetase 1 and regulates the urea cycle. *Cell* 137, 560–570.
36. Du, J., Zhou, Y., Su, X., Yu, J. J., Khan, S., Jiang, H., Kim, J. et al. (2011) Sirt5 is a NAD-dependent protein lysine demalonylase and desuccinylase. *Science* 334, 806–809.
37. Tan, M., Peng, C., Anderson, K. A., Chhoy, P., Xie, Z., Dai, L., Park, J. et al. (2014) Lysine glutarylation is a protein posttranslational modification regulated by SIRT5. *Cell Metabolism* 19, 605–617.
38. Wagner, G. R. and Hirschey, M. D. (2014) Nonenzymatic protein acylation as a carbon stress regulated by sirtuin deacylases. *Molecular Cell* 54, 5–16.
39. Wagner, G. R. and Payne, R. M. (2013) Widespread and enzyme-independent Nepsilon-acetylation and Nepsilon-succinylation of proteins in the chemical conditions of the mitochondrial matrix. *The Journal of Biological Chemistry* 288, 29036–29045.
40. Lin, Z. F., Xu, H. B., Wang, J. Y., Lin, Q., Ruan, Z., Liu, F. B., Jin, W., Huang, H. H., and Chen, X. (2013) SIRT5 desuccinylates and activates SOD1 to eliminate ROS. *Biochemical and Biophysical Research Communications* 441, 191–195.
41. Reddi, A. R. and Culotta, V. C. (2013) SOD1 integrates signals from oxygen and glucose to repress respiration. *Cell* 152, 224–235.
42. Peng, C., Lu, Z., Xie, Z., Cheng, Z., Chen, Y., Tan, M., Luo, H. et al. (2011) The first identification of lysine malonylation substrates and its regulatory enzyme. *Molecular & Cellular Proteomics: MCP* 10(M111), 012658.
43. Rardin, M. J., He, W., Nishida, Y., Newman, J. C., Carrico, C., Danielson, S. R., Guo, A. et al. (2013) SIRT5 regulates the mitochondrial lysine succinylome and metabolic networks. *Cell Metabolism* 18, 920–933.
44. Park, J., Chen, Y., Tishkoff, D. X., Peng, C., Tan, M., Dai, L., Xie, Z. et al. (2013) SIRT5-mediated lysine desuccinylation impacts diverse metabolic pathways. *Molecular Cell* 50, 919–930.

45. Kugel, S. and Mostoslavsky, R. (2014) Chromatin and beyond: The multitasking roles for SIRT6. *Trends in Biochemical Sciences* 39, 72–81.

46. Kaidi, A., Weinert, B. T., Choudhary, C., and Jackson, S. P. (2010) Human SIRT6 promotes DNA end resection through CtIP deacetylation. *Science* 329, 1348–1353.

47. Pan, P. W., Feldman, J. L., Devries, M. K., Dong, A., Edwards, A. M., and Denu, J. M. (2011) Structure and biochemical functions of SIRT6. *The Journal of Biological Chemistry* 286, 14575–14587.

48. Feldman, J. L., Baeza, J., and Denu, J. M. (2013) Activation of the protein deacetylase SIRT6 by long-chain fatty acids and widespread deacylation by mammalian sirtuins. *The Journal of Biological Chemistry* 288, 31350–31356.

49. Jiang, H., Khan, S., Wang, Y., Charron, G., He, B., Sebastian, C., Du, J. et al. (2013) SIRT6 regulates TNF-α secretion through hydrolysis of long-chain fatty acyl lysine. *Nature* 496, 110–113.

50. Mostoslavsky, R., Chua, K. F., Lombard, D. B., Pang, W. W., Fischer, M. R., Gellon, L., Liu, P. et al. (2006) Genomic instability and aging-like phenotype in the absence of mammalian SIRT6. *Cell* 124, 315–329.

51. Xiao, C., Kim, H. S., Lahusen, T., Wang, R. H., Xu, X., Gavrilova, O., Jou, W., Gius, D., and Deng, C. X. (2010) SIRT6 deficiency results in severe hypoglycemia by enhancing both basal and insulin-stimulated glucose uptake in mice. *The Journal of Biological Chemistry* 285, 36776–36784.

52. Zhang, P., Tu, B., Wang, H., Cao, Z., Tang, M., Zhang, C., Gu, B. et al. (2014) Tumor suppressor p53 cooperates with SIRT6 to regulate gluconeogenesis by promoting FoxO1 nuclear exclusion. *Proceedings of the National Academy of Sciences of the United States of America* 111, 10684–10689.

53. Dominy, J. E., Jr., Lee, Y., Jedrychowski, M. P., Chim, H., Jurczak, M. J., Camporez, J. P., Ruan, H. B. et al. (2012) The deacetylase SIRT6 activates the acetyltransferase GCN5 and suppresses hepatic gluconeogenesis. *Molecular Cell* 48, 900–913.

54. Zhong, L., D'Urso, A., Toiber, D., Sebastian, C., Henry, R. E., Vadysirisack, D. D., Guimaraes, A. et al. (2010) The histone deacetylase Sirt6 regulates glucose homeostasis via Hif1α. *Cell* 140, 280–293.

55. Kim, H. S., Xiao, C., Wang, R. H., Lahusen, T., Xu, X., Vassilopoulos, A., Vazquez-Ortiz, G. et al. (2010) Hepatic-specific disruption of SIRT6 in mice results in fatty liver formation due to enhanced glycolysis and triglyceride synthesis. *Cell Metabolism* 12, 224–236.

56. Tao, R., Xiong, X., DePinho, R. A., Deng, C. X., and Dong, X. C. (2013) Hepatic SREBP-2 and cholesterol biosynthesis are regulated by FoxO3 and SIRT6. *Journal of Lipid Research* 54, 2745–2753.

57. Tao, R., Xiong, X., DePinho, R. A., Deng, C. X., and Dong, X. C. (2013) FoxO3 transcription factor and Sirt6 deacetylase regulate low density lipoprotein (LDL)-cholesterol homeostasis via control of the proprotein convertase subtilisin/kexin type 9 (Pcsk9) gene expression. *The Journal of Biological Chemistry* 288, 29252–29259.

58. Sebastian, C., Zwaans, B. M., Silberman, D. M., Gymrek, M., Goren, A., Zhong, L., Ram, O. et al. (2012) The histone deacetylase SIRT6 is a tumor suppressor that controls cancer metabolism. *Cell* 151, 1185–1199.

59. Xiao, C., Wang, R. H., Lahusen, T. J., Park, O., Bertola, A., Maruyama, T., Reynolds, D. et al. (2012) Progression of chronic liver inflammation and fibrosis driven by activation of c-JUN signaling in SIRT6 mutant mice. *The Journal of Biological Chemistry* 287, 41903–41913.

60. Kiran, S., Chatterjee, N., Singh, S., Kaul, S. C., Wadhwa, R., and Ramakrishna, G. (2013) Intracellular distribution of human SIRT7 and mapping of the nuclear/nucleolar localization signal. *The FEBS Journal* 280, 3451–3466.

61. Lee, N., Kim, D. K., Kim, E. S., Park, S. J., Kwon, J. H., Shin, J., Park, S. M. et al. (2014) Comparative interactomes of SIRT6 and SIRT7: Implication of functional links to aging. *Proteomics* 14, 1610–1622.

62. Ford, E., Voit, R., Liszt, G., Magin, C., Grummt, I., and Guarente, L. (2006) Mammalian Sir2 homolog SIRT7 is an activator of RNA polymerase I transcription. *Genes & Development* 20, 1075–1080.

63. Vakhrusheva, O., Smolka, C., Gajawada, P., Kostin, S., Boettger, T., Kubin, T., Braun, T., and Bober, E. (2008) SIRT7 increases stress resistance of cardiomyocytes and prevents apoptosis and inflammatory cardiomyopathy in mice. *Circulation Research* 102, 703–710.

64. Shin, J., He, M., Liu, Y., Paredes, S., Villanova, L., Brown, K., Qiu, X. et al. (2013) SIRT7 represses Myc activity to suppress ER stress and prevent fatty liver disease. *Cell Reports* 5, 654–665.

65. Yoshizawa, T., Karim, M. F., Sato, Y., Senokuchi, T., Miyata, K., Fukuda, T., Go, C. et al. (2014) SIRT7 controls hepatic lipid metabolism by regulating the ubiquitin-proteasome pathway. *Cell Metabolism* 19, 712–721.

66. Ryu, D., Jo, Y. S., Lo Sasso, G., Stein, S., Zhang, H., Perino, A., Lee, J. U. et al. (2014) A SIRT7-dependent acetylation switch of GABPβ1 controls mitochondrial function. *Cell Metabolism* 20(5), 856–869.

67. Hubbi, M. E., Hu, H., Kshitiz, G. D. M., and Semenza, G. L. (2013) Sirtuin-7 inhibits the activity of hypoxia-inducible factors. *The Journal of Biological Chemistry* 288, 20768–20775.

68. Barber, M. F., Michishita-Kioi, E., Xi, Y., Tasselli, L., Kioi, M., Moqtaderi, Z., Tennen, R. I. et al. (2012) SIRT7 links H3K18 deacetylation to maintenance of oncogenic transformation. *Nature* 487, 114–118.

69. Tsai, Y. C., Greco, T. M., and Cristea, I. M. (2014) Sirtuin 7 plays a role in ribosome biogenesis and protein synthesis. *Molecular & Cellular Proteomics: MCP* 13, 73–83.

70. Kim, J. K., Noh, J. H., Jung, K. H., Eun, J. W., Bae, H. J., Kim, M. G., Chang, Y. G. et al. (2013) Sirtuin7 oncogenic potential in human hepatocellular carcinoma and its regulation by the tumor suppressors MiR-125a-5p and MiR-125b. *Hepatology* 57, 1055–1067.

Index